Lecture Notes in Computer Science 15434

Founding Editors

The series Lecture Notes in Computer Science (LNCS), including its subseries Lecture Notes in Artificial Intelligence (LNAI) and Lecture Notes in Bioinformatics (LNBI), has established itself as a medium for the publication of new developments in computer science and information technology research, teaching, and education.

LNCS enjoys close cooperation with the computer science R & D community, the series counts many renowned academics among its volume editors and paper authors, and collaborates with prestigious societies. Its mission is to serve this international community by providing an invaluable service, mainly focused on the publication of conference and workshop proceedings and postproceedings. LNCS commenced publication in 1973.

Donglei Du · Lu Han · Dachuan Xu
Editors

Combinatorial Optimization and Applications

17th International Conference, COCOA 2024
Beijing, China, December 6–8, 2024
Proceedings, Part I

Springer

Editors
Donglei Du
University of New Brunswick
New Brunswick, NB, Canada

Lu Han
Beijing University of Posts
and Telecommunications
Beijing, China

Dachuan Xu
Beijing University of Technology
Beijing, China

ISSN 0302-9743 ISSN 1611-3349 (electronic)
Lecture Notes in Computer Science
ISBN 978-981-96-4444-5 ISBN 978-981-96-4445-2 (eBook)
https://doi.org/10.1007/978-981-96-4445-2

This Springer imprint is published by the registered company Springer Nature Singapore Pte Ltd.
The registered company address is: 152 Beach Road, #21-01/04 Gateway East, Singapore 189721, Singapore

Preface

The 17th Annual International Conference on Combinatorial Optimization and Applications (COCOA 2024) took place in Beijing, China, during December 6–8, 2024. COCOA 2024 provided an excellent venue for researchers in the area of combinatorial optimization and its applications, including algorithm design, theoretical and experimental analysis, and applied research of general algorithmic interest. The Program Committee received a total of 124 submissions, among which 53 were accepted for presentation at the conference. Each contributed paper was subject to a rigorous peer review process, receiving three double-blind reviews from reviewers selected from the Program Committee. We would like to express our sincere appreciation to everyone who made COCOA 2024 a success by volunteering their time and effort: the authors, the Program Committee members, and the reviewers. We thank Springer for accepting the proceedings of COCOA 2024 for publication in the Lecture Notes in Computer Science (LNCS) series. Our special thanks also extend to the other chairs and the conference Organizing Committee members for their excellent work.

Donglei Du
Lu Han
Dachuan Xu

Organization

General Chair

Du, Ding-Zhu	University of Texas at Dallas, USA

Program Committee Co-chairs

Du, Donglei	University of New Brunswick, Canada
Han, Lu	Beijing University of Posts and Telecommunications, China
Xu, Dachuan	Beijing University of Technology, China

Web Co-chairs

Sai Ji	Hebei University of Technology, China
Lili Mei	Hangzhou Dianzi University, China

Publication Co-chairs

Chenchen Fu	Southeast University, China
Ruiqi Yang	Beijing University of Technology, China

Finance Chair

Yicheng Xu	Shenzhen Institute of Advanced Technology, Chinese Academy of Sciences, China

Registration Chair

Jian Sun	Nanjing Normal University, China

Local Chairs

Zhongzheng Tang	Beijing University of Posts and Telecommunications, China
Chenchen Wu	Tianjin University of Technology, China

Program Committee Members

Bein, Wolfgang	University of Nevada, USA
Calinescu, Gruia	Illinois Institute of Technology, USA
Chau, Vincent	Southeast University, China
Chen, Xujin	Academy of Mathematics and Systems Science, Chinese Academy of Sciences, China
Chen, Yong	Hangzhou Dianzi University, China
Cheng, Yukun	Jiangnan University, China
DasGupta, Bhaskar	University of Illinois Chicago, USA
De Bonis, Annalisa	Università degli Studi di Salerno, Italy
Diao, Zhuo	Central University of Finance and Economics, China
Fan, Neng	University of Arizona, USA
Feng, Qilong	Central South University, China
Guo, Longkun	Fuzhou University, China
Ji, Sai	Hebei University of Technology, China
Jiang, Shaofeng	Peking University, China
Khachay, Michael	Krasovsky Institute of Mathematics and Mechanics, Russia
Lee, Joong-Lyul	University of North Carolina at Pembroke, USA
Li, Xianyue	Lanzhou University, China
Li, Minming	City University of Hong Kong, China
Li, Jianping	Yunnan University, China
Lin, Guohui	University of Alberta, Canada
Liu, Bin	Ocean University of China, China
Mei, Lili	Hangzhou Dianzi University, China
Nguyen, Viet Hung	Clermont Auvergne University, France
Ogihara, Mitsunori	University of Miami, USA
Satpute, Meghana	University of Texas at Dallas, USA
Tang, Zhongzheng	Beijing University of Posts and Telecommunications, China
Tong, Weitian	Eastern Michigan University, USA
Wu, Chenchen	Tianjin University of Technology, China

Xu, Yicheng	Shenzhen Institute of Advanced Technology, Chinese Academy of Sciences, China
Yang, Ruiqi	Beijing University of Technology, China
Zhang, Ruilong	Technical University of Munich, Germany
Zhang, Peng	Shandong University, China
Zhang, Zhao	Zhejiang Normal University, China
Zhang, Xiaoyan	Nanjing Normal University, China
Zhang, Yong	Shenzhen Institute of Advanced Technology, Chinese Academy of Sciences, China
Ziegler, Martin	Korea Advanced Institute of Science and Technology, South Korea
Zissimopoulos, Vassilis	National and Kapodistrian University of Athens, Greece

Additional Reviewers

Chen, Lin
Chen, Xianrun
Dai, Sijia
Duan, Chengcheng
Filippos, Mavropoulos
Guo, Xinru
Ju, Jiachen
Khachai, Daniil
Lazaropoulos, Nikos
Li, Mengzhen
Lian, Yuefang
Liang, Wei
Lou, Jianing
Luo, Junjie
Polevoy, Gleb
Popa, Alexandru
Sigalas, Ioannis
Wang, Guihao
Wang, Fengjuan
Wang, Changjun
Wang, Chenhao
Xia, Xinlan
Xiao, Hao
Ye, Qingjie
Yu, Wei
Zhang, Yubai
Zhang, An
Zhang, Yubo
Zhao, Lei
Zoros, Dimitris

Contents – Part I

Contents – Part II

An Optimization Strategy for Fresh Agricultural Product Supply Chain with Altruistic Preference Under Blockchain Technology

Yi Zhao[1], Juan Gao[2], and Chunming Xu[2(✉)]

[1] Guangzhou Huali College, Jiangmen 529149, People's Republic of China
zhaoyi@stud.tjut.edu.cn

[2] Institute of Operations Research and Systems Engineering, College of Science, Tianjin University of Technology, Tianjin 300384, People's Republic of China
chunmingxu@tjut.edu.cn

Abstract. The quality and safety issues of product have received widespread attention in the fresh agricultural supply chain (FASC). Blockchain technology (BT) provides a new solution for the FASC. In this situation, altruistic behavior also plays an important role. The study incorporates blockchain technology and altruistic preference of supply chain members into a FASC consisting of a supplier and a powerful retailer. The study analyses the optimal strategies of FASC members under the basic scenarios without and with BT and two scenarios based on blockchain system that is the unilateral altruisitc preference of the retailer and the supplier. The study finds that the adoption of BT is related to the cost of building a blockchain platform. Furthermore, the supplier's altruisitic preference can increase the profit of the retailer and the supplier under certain conditions. The retailer's altruistic preference can increase the profit of the supplier but reduces himself profit and we design a revenue sharing contract to coordinate the FASC.

Keywords: Blockchain technology · Altruistic preference · Fresh agriculture supply chain · Traceability system

1 Introduction

In reality, fresh agricultural products are indispensable in people's daily life, and the consumption of fresh agricultural products by consumers is enormous. However, the characteristics of a short quality guarantee period and difficult storage make it difficult to maintain the freshness and easy to damage during transportation of fresh agricultural products. Also, negative reports on food safety are emerge in an endless stream. In China, the employee of FRESHIPPO changed the packaging date of the product by changing the label. In August 2020, the US Centers for Disease Control and Prevention reported the outbreak

D. Du et al. (Eds.): COCOA 2024, LNCS 15434, pp. 1–17, 2025.
https://doi.org/10.1007/978-981-96-4445-2_1

of Salmonella, which was sourced from onions sold in supermarkets. These not only pose a serious threat to consumers' health, but also bring inestimable losses to the fresh agricultural supply chain (FASC). Therefore, how to control the quality and safety of fresh agricultural products and maintain consumer trust is an urgent problem to be solved in the FASC.

Blockchain technology (BT), as a distributed ledger technology, is widely mentioned in supply chain traceability [1,2]. The characteristics of the BT such as decentralization, open, transparent, and information tamper proofing enable every participant's published information like the entire process of raw materials, cultivation and breeding, processing, transportation to be viewed and cannot be tampered at blockchain nodes, which improves the credibility of information traceability of products in the minds of consumers [3]. In practice, some companies have applied the traceability of BT to their FASC. For example, Walmart cooperates with IBM to deploy the BT to track American mangoes and Chinese pork. At present, many scholars have also studied the application of BT in supply chain traceability. Galimberti et al. [4] studied the application of traditional traceability technologies such as RFID and DNA analysis in the agricultural product industry to prevent adulteration fraud and other issues. George et al. [5] investigated the major existing traceability methods and concluded that the BT can achieve more reliable traceability than traditional traceability technologies. Wu et al. [6] established the models of agricultural product supply chain based on the traceability with BT and provided some theoretical guidance for decision analysis of the agricultural product supply chain. However, the BT does indeed bring high-quality traceability to enterprises and consumers, but it also encounters some bottlenecks in practical implementation. The deployment and adoption of BT is expensive, some supply chain enterprises bear great stress in the BT investment due to their limited economic strength.

Numerous case studies have shown that members of the FASC consider the benefits of other members and provide assistance to them when making decisions. For example, JD helps farmers develop a breeding system and optimize the breeding environment. Walmart provides a technical training on crop screening and agricultural sustainable practices for its agricultural product suppliers. In Shandong, China, hypermarkets have established a deep cooperative relationship with rural cooperatives. The hypermarkets provide free transportation and packaging of agricultural products for the cooperatives, and the cooperatives actively dispatch production personnel to assist supermarkets in transportation. The mutual assistance and cooperation among supply chain members is a favorable manifestation of altruistic preference.

Currently, the research related to the altruism appears more in the field of economics, while the behavior application in management science is relatively rare. Some scholars studied the impacts of social preference and individual behavior on interest relationships through experimental methods and demonstrated the existence of the altruistic preference among decision-makers in economics [7,8]. On the basis of experimental research, Loch and Wu [9] first developed a social altruistic preference model in the context of supply chain transactions with one supplier and one retailer. Wang et al. [10] further demonstrated that the altruistic behavior from an active participant can improve supply chain profits and system efficiency under certain conditions. Unlike existing literature on the impact of the altruistic preference on a general products supply chain, this study mainly focuses on exploring whether the altruistic preference is beneficial for the introduction of the BT.

As mentioned above, high quality traceability of products through BT system is crucial for FASC, but the investment and adoption in BT remains a significant burden for some small and medium-sized enterprises under FASC. Thus, it is very meaningful how to encourage and support more enterprises to participate in BT and achieve mutual prosperity and coexistence in the entire supply chain. Our main purpose is to investigate whether the altruistic behavior of mutual assistance and cooperation in FASC is conducive to the introduction of the BT. More specifically, the study will explore the following questions.

(1) How to determine the optimal decisions of the FASC participants with the altruistic preference under BT?
(2) How the altruistic preference affect the profit of the FASC system ? Is the altruistic preference beneficial for adopting the BT in the FASC?
(3) Does the altruistic preference from one participant in the FASC benefit all participants under BT? If not, is there a more effective contract to achieve a win-win situation?

To answer these questions, the study incorporates the altruistic preference into a two-level FASC consisting of a supplier and a powerful retailer and explore the optimal strategies of supply chain members in the benchmark scenarios without and with BT (scenario N and scenario B). Then the optimal strategies of the supply chain under BT are gived, including scenario with the retailer's unilateral altruistic preference (scenario BR) and scenario with the retailer's unilateral altruistic preference (scenario BS).

2 Notations, Problem Description, and Assumptions

The notations involved in this study are summarized in Table 1. We use $i \in \{n, b, br, bs\}$ to represent different scenarios. Specifically, $i = n$ represents the scenario without BT, and $i = b$ represents the case with BT. For the adoption of BT, the study considers four specific scenarios, $i = br$ represents the scenario with the retailer's unilateral altruistic preference (i.e. scenario BR), $i = bs$ represents the scenario with the supplier's unilateral altruistic preference (i.e. scenario BS).

The study considers a two-echelon supply chain consisting of a supplier and a strong retailer. The structure of the FASC is shown in Fig. 1. Before the sales season, the retailer determines the sales price p_i and the service investment t_i, and initiates order with the supplier. The customer demand D_i is assumed to be deterministic. Considersing the damage caused by loading, unloading and squeezing during transportation of fresh agricultural product, the retailer usually

Table 1. Notations for parameters and variables.

Notation	Definition
q	Initial market size
a	Consumer traceability awareness
k	The cost coefficient of service investment
c_b	The marginal cost with BT
f	The cost of building the blockchain platform
d	The marginal amount of tokens
β	The transportation efficiency factor, $\beta \in (0, 1)$
θ_n	Freshness of the product without BT, $\theta_n \in (0, 1)$
θ_b	Freshness of the product with BT, $\theta_b \in (0, 1)$ and $\theta_b > \theta_n$
η_r	The retailer's altruistic preference
η_s	The supplier's altruistic preference
p_i	The sales price, $i \in \{n, b, br, bs\}$, decision variable
t_i	The service investment of the retailer, decision variable
w_i	The wholesale price, decision variable
D_i	Market demand
Π_{r-i}	The retailer's profit
Π_{s-i}	The supplier's profit
Π_{sc-i}	The entire supply chain's profit

has to order more to offset these transportation losses, thus the actual order quantity is D_i/β, $\beta \in (0, 1)$, where β is the transportation efficiency factor, which refers to the proportion of intact products during transportation to all products. The supplier sells the product to the retailer at wholesale price w_i after receiving the order, and then the retailer sells the product to consumers at p_i.

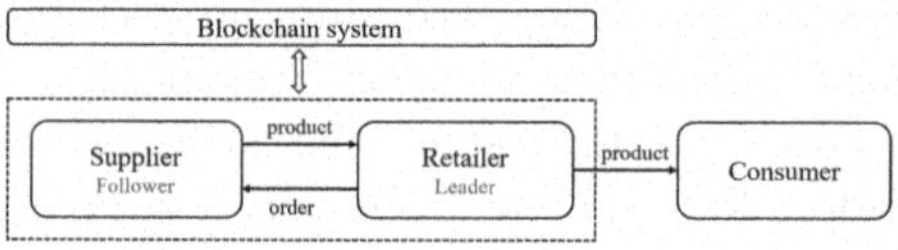

Fig. 1. The structure of the FASC

Considering a perishable nature of fresh agriculture product, the study assumes that freshness directly affects consumers' willingness to purchase. The higher the freshness, the more willing consumers are to purchase. Also, to attract more end consumers, the retailer will invest in retail service. Following Wu et al. and Niu et al., we assume the demand without BT is $D_n = \theta_n q - p_n + t_n$, where q represents the initial market demand, θ_n represents the freshness of fresh agriculture product without BT, p_n represents the sale price without BT, and t_n represents the retailer's service investment without BT.

The adoption of BT in the FASC greatly improves the speed of information transmission and product circulation, which achieving an efficient trading and distribution. Furthermore, the traceability of BT has a positive impact on consumers, and consumers have a high level of recognition for products with BT. For example, the use of the anti-counterfeiting and tracing platform under BT has brought an overall sales growth of 9.97% to JD. Thus, the study here assumes the demand with BT is $D_j = \theta_b q - p_j + t_j + a$, $j \in \{b, br, bs\}$. Where θ_b represents the freshness of fresh agriculture product with BT, $\theta_b > \theta_n$, p_j represents the sales price with BT, t_j represents the retailer's service investment with BT, and a represents consumers' traceability awareness.

Observing in the actual operation of the FASC, supply chain member often gives consideration to the profits of the others when making decisions. To characterize the altruistic preference among members of the FASC. According to Loch and Wu [9], the study introduces the utility function of the retailer's altruistic preference: $U_r = \Pi_r + \eta_r \Pi_s$ and the utility function of the supplier's altruistic preference: $U_s = \Pi_s + \eta_s \Pi_r$, where Π_r and Π_s represent the profits of the retailer and the supplier, respectively, and η_r and η_s denote the altruistic preference coefficients of the retailer and the supplier, respectively. The study assumes that each member will never care more about the interests of the others than about his or her own interests, thus $\eta_r \in [0, 1]$ and $\eta_s \in [0, 1]$.

In addition, some assumptions are given here.

Assumption 1, the service investment of the retailer leads to a cost. The service investment has the characteristic of diminishing marginal utility, the cost can be written as $\frac{1}{2}kt_i^2$, which is a common assumption [11].

Assumption 2, to ensure the market demand is positive, we assume that the initial market size is large enough, it satisfies the condition $q > \frac{(1+\beta)(c_b - t) - \beta a}{\beta \theta_b}$.

3 Basic Model

In this section, the study first explores the optimal strategies of FASC members under the scenario without BT. Then the study analyses the optimal strategies of the FASC members under the scenario with BT.

3.1 Scenario N

In scenario N, BT is not adopted in the FASC. Therefore, the profit functions of the retailer and the supplier are represented as follows:

$$\Pi_{r-n}(p_n, t_n) = p_n D_n - w_n D_n / \beta - \frac{k}{2} t_n^2, \tag{1}$$

$$\Pi_{s-n}(w_n) = w_n D_n / \beta. \tag{2}$$

Solving Eqs. (1) and (2), we can obtain Theorem 1.

Theorem 1. *In scenario N, if* $k > \frac{\beta}{2(1+\beta)}$, *the equilibrium results of the FASC are provided by,*

$$p_n^* = \frac{k(2+\beta)\theta_n q}{2k(1+\beta)-\beta}, \tag{3}$$

$$t_n^* = \frac{\beta\theta_n q}{2k(1+\beta)-\beta}, \tag{4}$$

$$w_n^* = \frac{k\beta\theta_n q}{2k(1+\beta)-\beta}. \tag{5}$$

Proof. Refer to proof in Appendix 1.

Remark. Theorem 1 gives a sufficient condition for the equilibrium results of the FASC and this condition ensures that the profit function is concave. In fact, if $k < \frac{\beta}{2(1+\beta)}$, then we easily get $t_n^* = 0$, which is contrary to our original assumption of the service investment, so we here omit it.

3.2 Scenario B

In scenario B, BT is adopted in the FASC. We consider that, as the leader of the FASC, the retailer initiates the building BT platform and bears the fixed cost of adopting BT. The retailers and the supplier can obtain token incentives through commodity transactions from the BT-based platform [12]. Thus, the profit functions of the retailer and the supplier are represented as follows:

$$\Pi_{r-b}(p_b, t_b) = (p_b - c_b + d)D_b - w_b D_b/\beta - \frac{k}{2}t_b^2 - f, \tag{6}$$

$$\Pi_{s-b}(w_b) = (w_b - c_b + d)D_b/\beta. \tag{7}$$

Solving Eqs. (6) and (7), we can obtain Theorem 2.

Theorem 2. *In scenario B, if* $k > \frac{\beta}{2(1+\beta)}$, *the equilibrium results of the FASC are given by,*

$$p_b^* = \frac{k(2+\beta)(\theta_b q + a) - (1-k)(1+\beta)(c_b - d)}{2k(1+\beta)-\beta}, \tag{8}$$

$$t_b^* = \frac{\beta(\theta_b q + a) - (1+\beta)(c_b - d)}{2k(1+\beta)-\beta}, \tag{9}$$

$$w_b^* = \frac{k\beta(\theta_b q + a) + (k - \beta + k\beta)(c_b - d)}{2k(1+\beta)-\beta}. \tag{10}$$

Proof. Refer to proof in Appendix 2.

Remark. Similar to Theorem 1, a sufficient condition is given for the equilibrium results, $k > \frac{\beta}{2(1+\beta)}$. This condition ensures that the profit function is concave.

Proposition 1. *Let* $\hat{f}_1 = \frac{k[2k(2+\beta)-\beta]}{2\beta[2k(1+\beta)-\beta]^2}\left\{[\beta(\theta_b q + a) - (1+\beta)(c_b - d)]^2 - (\beta\theta_n q)^2\right\}$, *if* $f < \hat{f}_1$, *then* $\Pi_{sc-b} > \Pi_{sc-n}$, *otherwise,* $\Pi_{sc-b} < \Pi_{sc-n}$.

Proof. Refer to proof in Appendix 3.

Proposition 1 indicates that it is conditional for the adoption of BT in the FASC. The profit of the FASC in scenario B is higher then that in scenario N when the cost of building a blockchain platform is below a certain threshold. Specifically, the threshold is related to initial market size, consumers' traceability awareness, the cost of adopting BT, the cost coefficient of service investment, token quantity, the freshness of agricultural products, and the transportation efficiency factor.

4 Models with the Altruistic Preference and BT

In the actual operation of the FASC, members of the supply chain are interdependent frequently. That is, the enterprise also concerns about the profits of other members when considering his/her own profit. Thus, we here consider the altruistic preferences of the FASC members under BT.

4.1 Scenario BR

In scenario BR, we consider the retailer has an unilateral altruistic preference under BT. According to Loch and Wu [9], the utility functions of the retailer and the supplier are provided as follows:

$$U_{r-br} = \Pi_{r-br} + \eta_r \Pi_{s-br}, \tag{11}$$

$$U_{s-br} = \Pi_{s-br}, \tag{12}$$

specifically, that is,

$$U_{r-br}(p_{br}, t_{br}) = (p_{br} - c_b + d)\, D_{br} - w_{br} D_{br}/\beta - \frac{k}{2} t_{br}^2 - f + \eta_r \left[(w_{br} - c_b + d)\, D_{br}/\beta\right], \tag{13}$$

$$U_{s-br}(w_{br}) = (w_{br} - c_b + d)\, D_{br}/\beta. \tag{14}$$

Solving Eqs. (13) and (14), we can obtain Theorem 3.

Theorem 3. *In scenario BR, let* $\hat{k_1} = \frac{\beta}{2(1+\beta-2\eta_r)}$, *if* $k > \hat{k_1}$, *the equilibrium results of the FASC are as given by, +*

$$p_{br}^* = \frac{k(2+\beta-2\eta_r)(\theta_b q + a) - (1-k)(1+\beta)(c_b - d)}{2k(1+\beta-\eta_r) - \beta}, \tag{15}$$

$$t_{br}^{*} = \frac{\beta\left(\theta_b q + a\right) - (1+\beta)\left(c_b - d\right)}{2k\left(1+\beta-\eta_r\right) - \beta}, \tag{16}$$

$$w_{br}^{*} = \frac{k\beta\left(\theta_b q + a\right) + \left[k\left(1+\beta-2\eta_r\right) - \beta\right]\left(c_b - d\right)}{2k\left(1+\beta-\eta_r\right) - \beta}. \tag{17}$$

Proof. Refer to proof in Appendix 4.

Remark. To ensure that the profit function is concave and the nonnegative of profits for the FASC members, so we assume that $k > \frac{\beta}{2(1+\beta-2\eta_r)}$.

4.2 Scenario BS

In scenario BS, we consider the supplier has a unilateral altruistic preference under BT, According to Loch and Wu [9], the utility functions of the retailer and the supplier are represented as follows:

$$U_{r-br} = \Pi_{r-bs}, \tag{18}$$

$$U_{s-bs} = \Pi_{s-bs} + \eta_s \Pi_{r-bs}, \tag{19}$$

specifically, that is,

$$U_{r-bs}(p_{bs}, t_{bs}) = \left(p_{bs} - c_b + d\right) D_{bs} - w_{bs} D_{bs}/\beta - \frac{k}{2} t_{bs}^2 - f, \tag{20}$$

$$U_{s-bs}(w_{bs}) = \left(w_{bs} - c_b + d\right) D_{bs}/\beta + \eta_s \left[\left(p_{bs} - c_b + d\right) D_{bs} - w_{bs} D_{bs}/\beta - \frac{k}{2} t_{bs}^2 - f\right]. \tag{21}$$

Solving Eqs. (20) and (21), we can obtain Theorem 4.

Theorem 4. *In scenario BS, let* $\hat{k_2} = \frac{\beta}{2(1-\eta_s)(1+\beta-\eta_s+\beta\eta_s)}$, *if* $k > \hat{k_2}$ *and* $\eta_s < \frac{3-\sqrt{1+8\beta}}{4(1-\beta)}$, *the equilibrium results of the FASC are as given by*

$$p_{bs}^{*} = \frac{k\left(1-\eta_s\right)\left(2+\beta-2\eta_r+2\beta\eta_s\right)\left(\theta_b q + a\right) - (1+\beta)\left[1-k\left(1-\eta_s\right)\right]\left(c_b - d\right)}{2k\left(1-\eta_s\right)\left(1+\beta-\eta_s+\beta\eta_s\right) - \beta}, \tag{22}$$

$$t_{bs}^{*} = \frac{\beta\left(\theta_b q + a\right) - (1+\beta)\left(c_b - d\right)}{2k\left(1-\eta_s\right)\left(1+\beta-\eta_s+\beta\eta_s\right) - \beta}, \tag{23}$$

$$w_{bs}^{*} = \frac{\left\{\begin{array}{l} k\beta\left(1-3\eta_s+2\eta_s^2-2\beta\eta_s^2\right)\left(\theta_b q + a\right) + \left[k\left(1+\right.\right. \\ \left.\left.\beta-\eta_s+3\beta\eta_s-2\beta\eta_s^2+2\beta^2\eta_s^2\right) - \beta\right]\left(c_b - d\right) \end{array}\right\}}{2k\left(1-\eta_s\right)\left(1+\beta-\eta_s+\beta\eta_s\right) - \beta}. \tag{24}$$

Proof. Refer to proof in Appendix 5.

Remark. To ensure that the profit function is concave and the positivity of profits for the FASC members, we assume that $k > \hat{k_2}$ and $\eta_s < \frac{3-\sqrt{1+8\beta}}{4(1-\beta)}$, respectively.

Proposition 2. *In scenarios with BS and B, for the effect of the supplier's unilateral altruistic preference, Let* $\hat{k_{bs}} = \frac{\beta(2-3\eta_s)}{2(1-\eta_s)[\eta_s(1-\eta_s)+\beta(2-\eta_s+\eta_s^2)]}$, *we have the following properties.*

(1) $D^*_{bs} > D^*_b$, $\Pi^*_{r-bs} > \Pi^*_{r-b}$, *and* $\Pi^*_{sc-bs} > \Pi^*_{sc-b}$;
(2) If $k \in (\hat{k_2}, \hat{k_{bs}})$, *then* $\Pi^*_{s-bs} > \Pi^*_{s-b}$, *and if* $k > \hat{k_{bs}}$, *then* $\Pi^*_{s-bs} < \Pi^*_{s-b}$.

Proof. Refer to proof in Appendix 6.

Proposition 2 shows that the supplier's unilateral altruistic behavior can increase the customer demand and the overall profit of the entire FASC. Interestingly, when the service investment cost coefficient is low ($k \in (\hat{k_2}, \hat{k_{bs}})$), the supplier's unilateral altruistic behavior can coordinate the entire FASC, but when the service investment cost coefficient is high ($k > \hat{k_{bs}}$), the supplier's unilateral altruistic behavior reduces his own profit.

Proposition 3. *In scenarios with BR and B, for the effect of the retailer's unilateral altruistic preference, we have the following properties,* $D^*_{br} > D^*_b$, $\Pi^*_{r-br} < \Pi^*_{r-b}$, $\Pi^*_{s-br} > \Pi^*_{s-b}$, *and* $\Pi^*_{sc-br} > \Pi^*_{sc-b}$.

Proof. Refer to proof in Appendix 7.

Proposition 3 indicates that, the retailer's unilateral altruistic preference can increase the demand and the profits of the supplier and the entire FASC, but it reduces the her own profit. This is because the strong retailer with altruistic preference will give up part of her profit in the face of the weak supplier, this may harm the retailer's profit, but from the perspective of the entire supply chain, the unilateral altruistic preference of the retailer is beneficial to the entire supply chain, which can actively promoting cooperation among supply chain members and maintaining sustainable development of the supply chain.

In the next, to coordinate the supply chain and improve the profit of each supply chain member, we will seek a coordination mechanism in scenario BR.

Coordination Mechanism in Scenario BR, From Proposition 3, it is easy to see that the retailer with unilateral altruistic preference under BT always beneficial to the supplier and the entire FASC, but not to the retailer. We attempt to design a revenue sharing contract that will make each FASC members benefit from the retailer with unilateral altruistic preference under BT. In this contract, the supplier allocate a portion of revenue to the retailer. We assume that λ is the proportion of revenue retained by the supplier, where $\lambda \in [0, 1]$, then $1 - \lambda$ is the proportion of revenue provieded by the supplier to the retailer. Therefore, the profit functions of the retailer and the supplier is represented as follows:

$$\Pi_{r-s}(p_s, t_s) = (p_s - c_s + d)D_s - w_s D_s/\beta - \frac{k}{2}t_s^2 + (1-\lambda)wD_s/\beta - f, \quad (25)$$

$$\Pi_{s-s}(w_s) = (\lambda w_s - c_s + d)D_s/\beta. \tag{26}$$

The retailer with a unilateral altruistic preference under BT, the utilities of the retailer and the supplier are represented as follows:

$$U_{r-s} = \Pi_{r-s} + \eta_r \Pi_{s-s}, \tag{27}$$

$$U_{s-s} = \Pi_{s-s}. \tag{28}$$

Solving Eqs. (27) and (28), we can obtain the following optimal profits of the retailer and the supplier under revenue sharing contract.

$$\Pi^*_{r-s} = \frac{k\left[2k\left(\lambda\left(1-2\eta_r\right)+\beta\right)-\beta\right]\left[\beta\left(\theta_b q+a\right)-\left(1+\beta\right)\left(c_b-d\right)\right]^2}{2\beta\left[2k\left(\lambda\left(1-\eta_r\right)+\beta\right)-\beta\right]^2} - f, \tag{29}$$

$$\Pi^*_{s-s} = \frac{\lambda k^2\left[\beta\left(\theta_b q+a\right)-\left(1+\beta\right)\left(c_b-d\right)\right]^2}{\beta\left[2k\left(\lambda\left(1-\eta_r\right)+\beta\right)-\beta\right]^2}. \tag{30}$$

Proof. Refer to proof in Appendix 8.

Proposition 4. *If* $\frac{(2k-1)\beta}{2k\eta_r+(2k-1)\beta} < \lambda < \frac{(2k-1)\beta}{2k\eta_r^2+(2k-1)\beta}$, *then* $\Pi^*_{r-s} > \Pi^*_{r-b}$ *and* $\Pi^*_{s-s} > \Pi^*_{s-b}$, *that is, the supply chain realizes a Pareto improvement.*

Proof. Refer to proof in Appendix 9.

Proposition 4 indicates that, each of the FASC members can more benefit from the revenue sharing contract when the revenue sharing ratio meets certain condition, which achieves a win-win situation for the FASC members, resulting in a Pareto improvement in the FASC.

5 Numerical Examples

This section conducts several numerical examples. Through numerical examples, this section explores the effects of agricultural product freshness θ_b and the transportation efficiency factor β on the optimal decision-making of the FASC members in scenarios BR and BS.

(1) The impact of agricultural product freshness
Figure 2 was generated by setting $q = 10$, $a = 1.2$, $c_b = 1.0$, $d = 0.4$, $k = 0.9$, $f = 5$, $\beta = 0.7$, $\eta_r = 0.2$, $\eta_s = 0.2$, and $\theta_n = 0.6$. Figure 2 shows that the freshness of fresh agricultural products has a positive impact on the sales price, the service investment, and the wholesale price under scenario BR and scenario BS, respectively. As the freshness of fresh agricultural products increases, it means that the quality of fresh agricultural products is better and consumers are more willing to purchase. This will encourage the retailer to increase the sales price and further improve the service investment. The increase in demand also means an increase in upstream order, and the supplier will correspondingly increase the wholesale price.

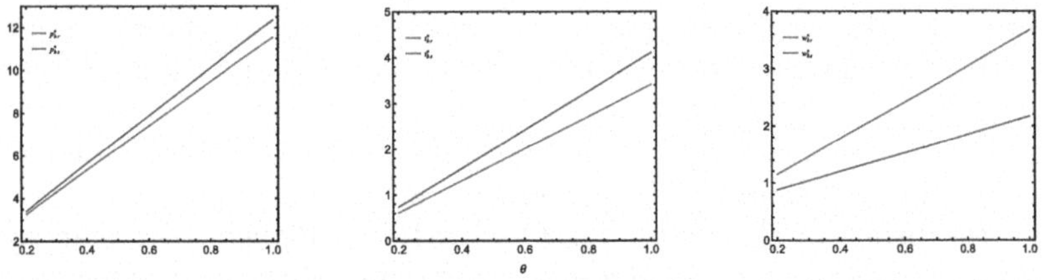

(1)The impact of θ_b on sales price

(2)The impact of θ_b on service investment

(3)The impact of θ_b on wholesale price

Fig. 2. The impact of θ_b

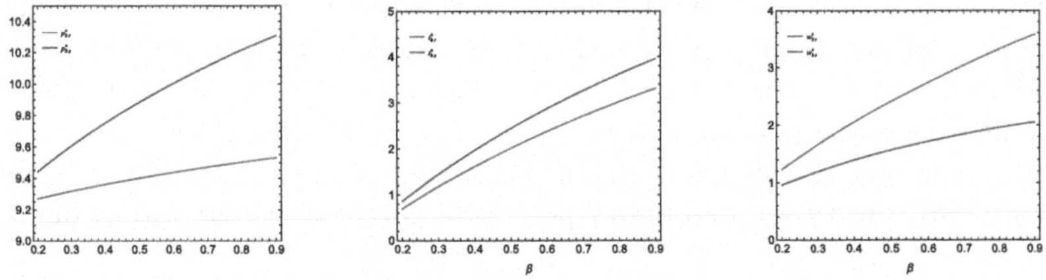

(1)The impact of β on sales price

(2)The impact of β on service investment

(3)The impact of β on wholesale price

Fig. 3. The impact of β

(2) The impact of the transportation efficiency factor
Figure 3 was generated by setting $q = 10$, $a = 1.2$, $c_b = 1.0$, $d = 0.4$, $k = 0.9$, $f = 5$, $\eta_r = 0.2$, $\eta_s = 0.2$, $\theta_b = 0.8$, and $\theta_n = 0.6$. Figure 3 shows that the impacts of the transportation efficiency factor on the sales price, the service investment, and the wholesale price under scenario BR and scenario BS, respectively. We observe that the sales price, the service investment, and the wholesale price all increase as the enhancement of transportation efficiency factor, which implies that transportation efficiency factor has a positive effect on the optimal strategies of the FASC members.

6 Conclusions

In practice, the safety and quality of fresh agricultural products can not be effectively guaranteed. The adoption of BT can control the products quality and establish the trust system. To study the operation of the FASC, this study considers a traceability-based of two-echelon supply chain including a powerful retailer and a supplier. We study the optimal strategies of the FASC members without and with BT, and obtain the conditions for FASC to adopt BT through profit comparisons. By comparing the profits of the FASC with BT in two scenarios, the retailer has an unilateral altruistic preference and the supplier has an unilateral altruistic preference. We obtain the specific conditions for the

FASC to adopt BT and the impact of altruistic preference. Then, we design a revenue sharing contract to coordinate the retailer and the supplier under the FASC with BT and the retailer has an unilateral altruistic preference. Finally, numerical examples are carried out to verify the model and obtain some results.

Some main conclusions can be gained as follows:

First, the profit of the entire FASC with BT is higher then that scenario without BT when the cost of building a blockchain platform is below a certain threshold, that is, the FASC can benefit from the traceability of the blockchain.

Furthermore, the retailer's unilateral altruistic preference can increase the profits of the supplier and the entire FASC, but for the retailer, the altruistic behavior can reduce it own profit. The supplier's unilateral altruistic preference can increase the profits of the retailer and the entire FASC, but for the supplier, the altruistic behavior can increase the profit of himself when the service investment coefficient is low, and it can reduce the profit of himself when the service investment coefficient is high.

Finally, we design a revenue sharing contract, which can achieve a win-win situation for supply chain members when the revenue sharing ratio meets certain condition.

There are some limitations in our study. For example, we only explore the operation of one supply chain, and chain to chain competition will be an interesting topic. In addition, the operation of the supply chain with BT under stochastic demand may also be incorporated in our study.

Acknowledgements. This work was supported by Humanities and Social Science Foundation of the Education of Ministry of China (No. 23YJA630105).

Appendix

Appendix 1: Proof of Theorem 1. Let $p_n = w_n + \phi_n$. By Eq. (2), the first-order and the second-order derivatives of Π_{s-n} with respect to w_n can be taken, that is, $\frac{\partial \Pi_{s-n}}{\partial w_n} = \frac{\theta_n q + t_n - 2w_n - \phi_n}{\beta}$ and $\frac{\partial^2 \Pi_{s-n}}{\partial {w_n}^2} = -\frac{2}{\beta} < 0$. By $\frac{\partial \Pi_{s-n}}{\partial w_n} = 0$, we have

$$w_n = \frac{\theta_n q + t_n - \phi_n}{2}. \tag{31}$$

Substituting w_n into Eq. (1), then the first-order and the second-order derivatives of Π_{r-n} with respect to ϕ_n and t_n can be taken, i.e. $\frac{\partial \Pi_{r-n}}{\partial \phi_n} = \frac{\theta_n q + t_n - (1+\beta)\phi_n}{2\beta}$, $\frac{\partial^2 \Pi_{r-n}}{\partial {\phi_n}^2} = -\frac{1+\beta}{2\beta}$, $\frac{\partial \Pi_{r-n}}{\partial t_n} = \frac{(-1+\beta)\theta_n q - (2k\beta+1-\beta)t_n + \phi_n}{2\beta}$, $\frac{\partial^2 \Pi_{r-n}}{\partial {t_n}^2} = -\frac{2k\beta+1-\beta}{2\beta}$, and $\frac{\partial^2 \Pi_{r-n}}{\partial \phi_n \partial t_n} = \frac{\partial^2 \Pi_{r-n}}{\partial t_n \partial \phi_n} = \frac{1}{2\beta}$. By the Hessian matrix $H(\phi_n, t_n) = \begin{bmatrix} \frac{\partial^2 \Pi_{r-n}}{\partial {\phi_n}^2} & \frac{\partial^2 \Pi_{r-n}}{\partial \phi_n \partial t_n} \\ \frac{\partial^2 \Pi_{r-n}}{\partial t_n \partial \phi_n} & \frac{\partial^2 \Pi_{r-n}}{\partial {t_n}^2} \end{bmatrix}$ is negative definite, $\frac{\partial \Pi_{r-n}}{\partial \phi_n} = 0$, and $\frac{\partial \Pi_{r-n}}{\partial t_n} = 0$, we have if $k > \frac{\beta}{2(1+\beta)}$, then

$$\phi_n^* = \frac{2k\theta_n q}{2k(1+\beta) - \beta}, \tag{32}$$

$$t_n^* = \frac{\beta\theta_n q}{2k(1+\beta)-\beta}. \tag{33}$$

According to $p_n = w_n + \phi_n$ and Eqs. (31), (32), and (33), the equilibrium results can be given, $p_n^* = \frac{k(2+\beta)\theta_n q}{2k(1+\beta)-\beta}$, $t_n^* = \frac{\beta\theta_n q}{2k(1+\beta)-\beta}$, and $w_n^* = \frac{k\beta\theta_n q}{2k(1+\beta)-\beta}$.

Appendix 2: Proof of Theorem 2. Let $p_b = w_b + \phi_b$. By Eq. (7), the first-order and the second-order derivatives of Π_{s-b} with respect to w_b can be taken, that is, $\frac{\partial \Pi_{s-b}}{\partial w_b} = \frac{\theta_b q + t_b - 2w_b - \phi_b + c_b - d + a}{\beta}$ and $\frac{\partial^2 \Pi_{s-b}}{\partial {w_b}^2} = -\frac{2}{\beta} < 0$. By $\frac{\partial \Pi_{s-b}}{\partial w_b} = 0$, we have

$$w_b = \frac{\theta_b q + t_b - \phi_b + c_b - d + a}{2}. \tag{34}$$

Substituting w_b into Eq. (6), then the first-order and the second-order derivatives of Π_{r-b} with respect to ϕ_b and t_b can be taken, i.e. $\frac{\partial \Pi_{r-b}}{\partial \phi_b} = \frac{\theta_b q + t_b - (1+\beta)\phi_b + a}{2\beta}$, $\frac{\partial^2 \Pi_{r-b}}{\partial {\phi_b}^2} = -\frac{1+\beta}{2\beta}$, $\frac{\partial \Pi_{r-b}}{\partial t_b} = \frac{(-1+\beta)\theta_b q - (2k\beta+1-\beta)t_b + \phi_b - \beta(c_b - d) - (1-\beta)a}{2\beta}$, $\frac{\partial^2 \Pi_{r-b}}{\partial {t_b}^2} = -\frac{2k\beta+1-\beta}{2\beta}$, and $\frac{\partial^2 \Pi_{r-b}}{\partial \phi_b \partial t_b} = \frac{\partial^2 \Pi_{r-b}}{\partial t_b \partial \phi_b} = \frac{1}{2\beta}$. By the Hessian matrix $H(\phi_b, t_b) = \begin{bmatrix} \frac{\partial^2 \Pi_{r-b}}{\partial {\phi_b}^2} & \frac{\partial^2 \Pi_{r-b}}{\partial \phi_b \partial t_b} \\ \frac{\partial^2 \Pi_{r-b}}{\partial t_b \partial \phi_b} & \frac{\partial^2 \Pi_{r-b}}{\partial {t_b}^2} \end{bmatrix}$ is negative definite, $\frac{\partial \Pi_{r-b}}{\partial \phi_b} = 0$, and $\frac{\partial \Pi_{r-b}}{\partial t_b} = 0$, we have if $k > \frac{\beta}{2(1+\beta)}$, then

$$\phi_b^* = \frac{2k(\theta_b q + a) - c_b + d}{2k(1+\beta)-\beta}, \tag{35}$$

$$t_b^* = \frac{\beta(\theta_b q + a) - (1+\beta)(c_b - d)}{2k(1+\beta)-\beta}. \tag{36}$$

According to $p_b = w_b + \phi_b$ and Eqs. (34), (35), and (36), the equilibrium results can be given, $p_b^* = \frac{k(2+\beta)(\theta_b q + a) - (1-k)(1+\beta)(c_b - d)}{2k(1+\beta)-\beta}$, $t_b^* = \frac{\beta(\theta_b q + a) - (1+\beta)(c_b - d)}{2k(1+\beta)-\beta}$, and $w_b^* = \frac{k\beta(\theta_b q + a) + (k - \beta + k\beta)(c_b - d)}{2k(1+\beta)-\beta}$.

Appendix 3: Proof of Proposition 1. According to Eqs. (1)–(5) and Eqs. (6)–(10), we can obtain Π_{sc-b} and Π_{sc-n}. Thus, we have $\Pi_{sc-b} - \Pi_{sc-n} = \frac{k[2k(2+\beta)-\beta][\beta(\theta_b q + a) - (1+\beta)(c_b - d)]^2}{2\beta[2k(1+\beta)-\beta]^2} - f - \frac{k\beta[2k(2+\beta)-\beta](\theta_n q)^2}{2[2k(1+\beta)-\beta]^2}$. Thus, if $f < \frac{k[2k(2+\beta)-\beta]}{2\beta[2k(1+\beta)-\beta]^2}\left\{[\beta(\theta_b q + a) - (1+\beta)(c_b - d)]^2 - (\beta\theta_n q)^2\right\}$, then $\Pi_{sc-b} > \Pi_{sc-n}$.

Appendix 4: Proof of Theorem 3. Let $p_{br} = w_{br} + \phi_{br}$. By Eq. (14), the first-order and the second-order derivatives of U_{s-br} with respect to w_{br} can be taken, that is, $\frac{\partial U_{s-br}}{\partial w_{br}} = \frac{\theta_b q + t_{br} - 2w_{br} - \phi_{br} + c_b - d + a}{\beta}$ and $\frac{\partial^2 U_{s-br}}{\partial {w_{br}}^2} = -\frac{2}{\beta} < 0$. By $\frac{\partial U_{s-br}}{\partial w_{br}} = 0$, we have

$$w_{br} = \frac{\theta_b q + t_{br} - \phi_{br} + c_b - d + a}{2}. \tag{37}$$

Substituting w_{br} into Eq. (13), then the first-order and the second-order derivatives of U_{r-br} with respect to ϕ_{br} and t_{br} can be taken, i.e. $\frac{\partial U_{r-br}}{\partial \phi_{br}} = \frac{(1-\eta_r)(\theta_b q+t_{br}+a)-(1+\beta-\eta_r)\phi_{br}+\eta_r(c_b-d)}{2\beta}$, $\frac{\partial^2 U_{r-br}}{\partial {\phi_{br}}^2} = -\frac{1+\beta-\eta_r}{2\beta}$, $\frac{\partial U_{r-br}}{\partial t_{br}} = \frac{(-1+\beta+\eta_r)(\theta_b q+t_{br}+a)+(1-\eta_r)\phi_{br}-(\beta+\eta_r)(c_b-d)-2k\beta t_{br}}{2\beta}$, $\frac{\partial^2 U_{r-br}}{\partial {t_{br}}^2} = -\frac{2k\beta+1-\beta-\eta_r}{2\beta}$, and $\frac{\partial^2 U_{r-br}}{\partial \phi_{br}\partial t_{br}} = \frac{\partial^2 U_{r-br}}{\partial t_{br}\partial \phi_{br}} = \frac{1-\eta_r}{2\beta}$. By the Hessian matrix $H(\phi_{br}, t_{br}) = \begin{bmatrix} \frac{\partial^2 U_{r-br}}{\partial {\phi_{br}}^2} & \frac{\partial^2 U_{r-br}}{\partial \phi_{br}\partial t_{br}} \\ \frac{\partial^2 U_{r-br}}{\partial t_{br}\partial \phi_{br}} & \frac{\partial^2 U_{r-br}}{\partial {t_{br}}^2} \end{bmatrix}$ is negative definite, $\frac{\partial U_{r-br}}{\partial \phi_{br}} = 0$, and $\frac{\partial U_{r-br}}{\partial t_{br}} = 0$, we have if $k > \frac{\beta}{2k(1+\beta-\eta_r)}$, then

$$\phi_{br}^* = \frac{2k(1-\eta_r)(\theta_b q + a) - (1 - 2k\eta_r)(c_b - d)}{2k(1+\beta-\eta_r) - \beta}, \tag{38}$$

$$t_{br}^* = \frac{\beta(\theta_b q + a) - (1+\beta)(c_b - d)}{2k(1+\beta-\eta_r) - \beta}. \tag{39}$$

According to $p_{br} = w_{br} + \phi_{br}$ and Eqs. (37), (38), and (39), the equilibrium results can be given, $p_{br}^* = \frac{k(2+\beta-2\eta_r)(\theta_b q+a)-(1-k)(1+\beta)(c_b-d)}{2k(1+\beta-\eta_r)-\beta}$, $t_{br}^* = \frac{\beta(\theta_b q+a)-(1+\beta)(c_b-d)}{2k(1+\beta-\eta_r)-\beta}$, and $w_{br}^* = \frac{k\beta(\theta_b q+a)+[k(1+\beta-2\eta_r)-\beta](c_b-d)}{2k(1+\beta-\eta_r)-\beta}$.

Appendix 5: Proof of Theorem 4. Let $p_{bs} = w_{bs} + \phi_{bs}$. By Eq. (21), the first-order and the second-order derivatives of U_{s-bs} with respect to w_{bs} can be taken, that is, $\frac{\partial U_{s-bs}}{\partial w_{bs}} = \frac{[1-(1-\beta)\eta_s](theta_b q+t_{bs}-2w_{bs}-\phi_{bs})+(1+\beta\eta_s)(c_b-d)-\beta\eta_s\phi_{bs}}{\beta}$ and $\frac{\partial^2 U_{s-bs}}{\partial {w_{bs}}^2} = -\frac{2[1-(1-\beta)\eta_s]}{\beta} < 0$. By $\frac{\partial U_{s-bs}}{\partial w_{bs}} = 0$, we have

$$w_{bs} = \frac{\left\{\begin{array}{l} (1-\eta_s+\beta\eta_s)(\theta_b q + t_{bs} + a) - (1 \\ -\eta_s + 2\beta\eta_s)\phi_{bs} + (1+\beta\eta_s)(c_b - d) \end{array}\right\}}{2(1-\eta_s+\beta\eta_s)}. \tag{40}$$

Substituting w_{bs} into Eq. (20), then the first-order and the second-order derivatives of U_{r-bs} with respect to ϕ_{bs} and t_{bs} can be taken, i.e. $\frac{\partial U_{r-bs}}{\partial \phi_{bs}} = \frac{(1-\eta_s+\beta\eta_s)^2(\theta_b q+t_{bs}+a)-\left(1+\beta+2\eta_s-\eta_s^2+\beta\eta_s^2\right)\phi_{bs}}{2\beta[1-(1-\beta)\eta_s]^2} - \frac{\beta\eta_s(2-\eta_s+\beta\eta_s)(c_b-d)}{2\beta[1-(1-\beta)\eta_s]^2}$, $\frac{\partial^2 U_{r-bs}}{\partial {\phi_{bs}}^2} = -\frac{(1-\eta_s)(1+\beta-\eta_s+\beta\eta_s)}{2\beta[1-(1-\beta)\eta_s]^2}$, $\frac{\partial U_{r-bs}}{\partial t_{bs}} = \frac{(-1+\beta)(\theta_b q+a)-(2k\beta+1-\beta)t_{bs}+\phi_{bs}-\beta(c_b-d)}{2\beta}$, $\frac{\partial^2 U_{r-bs}}{\partial {t_{bs}}^2} = -\frac{2k\beta+1-\beta}{2\beta}$, and $\frac{\partial^2 U_{r-bs}}{\partial \phi_{bs}\partial t_{bs}} = \frac{\partial^2 U_{r-bs}}{\partial t_{bs}\partial \phi_{bs}} = \frac{1}{2\beta}$. By the Hessian matrix $H(\phi_{bs}, t_{bs}) = \begin{bmatrix} \frac{\partial^2 U_{r-bs}}{\partial {\phi_{bs}}^2} & \frac{\partial^2 U_{r-bs}}{\partial \phi_{bs}\partial t_{bs}} \\ \frac{\partial^2 U_{r-bs}}{\partial t_{bs}\partial \phi_{bs}} & \frac{\partial^2 U_{r-bs}}{\partial {t_{bs}}^2} \end{bmatrix}$ is negative definite, $\frac{\partial U_{r-bs}}{\partial \phi_{bs}} = 0$, and $\frac{\partial U_{r-bs}}{\partial t_{bs}} = 0$, we have if $k > \frac{\beta}{2(1-\eta_s)(1+\beta-\eta_s+\beta\eta_s)}$, then

$$\phi_{bs}^* = \frac{\left\{\begin{array}{l} 2k[1-(1-\beta)\eta_s]^2(\theta_b q + a) - [1 \\ +2k\beta\eta_s(2-(1-\beta)\eta_s)](c_b - d) \end{array}\right\}}{2k(1-\eta_s)(1+\beta-\eta_s+\beta\eta_s) - \beta}, \tag{41}$$

$$t_{bs}^* = \frac{\beta(\theta_b q + a) - (1+\beta)(c_b - d)}{2k(1-\eta_s)(1+\beta-\eta_s+\beta\eta_s) - \beta}. \tag{42}$$

According to $p_{bs} = w_{bs} + \phi_{bs}$ and Eqs. (40), (41), and (42), the equilibrium results can be given, $p_{bs}^* = \frac{k(1-\eta_s)(2+\beta-2\eta_r+2\beta\eta_s)(\theta_b q+a)-(1+\beta)[1-k(1-\eta_s)](c_b-d)}{2k(1-\eta_s)(1+\beta-\eta_s+\beta\eta_s)-\beta}$, $t_{bs}^* = \frac{\beta(\theta_b q+a)-(1+\beta)(c_b-d)}{2k(1-\eta_s)(1+\beta-\eta_s+\beta\eta_s)-\beta}$, and

$$w_{bs}^* = \frac{\left\{\begin{array}{l} k\beta\left(1-3\eta_s+2\eta_s^2-2\beta\eta_s^2\right)(\theta_b q+a)+\left[k\left(1\right.\right. \\ \left.\left.+\beta-\eta_s+3\beta\eta_s-2\beta\eta_s^2+2\beta^2\eta_s^2\right)-\beta\right](c_b-d) \end{array}\right\}}{2k(1-\eta_s)(1+\beta-\eta_s+\beta\eta_s)-\beta}.$$

Appendix 6: Proof of Proposition 2. According to Eqs. (6)–(10) and Eqs. (18)–(24), we can obtain D_b^*, Π_{r-b}, Π_{s-b}, Π_{sc-b}, D_{bs}^*, Π_{r-bs}, Π_{s-bs}, and Π_{sc-bs}. Let $D_1(\eta_s) = \frac{k(1-\eta_s)[\beta(\theta_b q+a)-(1+\beta)(c_b-d)]}{2k(1-\eta_s)(1+\beta-\eta_s+\beta\eta_s)-\beta}$, then $D_{bs}^* = D_1$ and $D_b^* = D_1(0)$, because of $\frac{\partial D_{bs}^*}{\partial\eta_s} = \frac{k\left[2k(1-\beta)(1-\eta_s)^2+\beta\right][\beta(\theta_b q+a)-(1+\beta)(c_b-d)]}{[2k(1-\eta_s)(1+\beta-\eta_s+\beta\eta_s)-\beta]^2} > 0$, thus, we have $D_{bs}^* > D_b^*$. $\Pi_{r-b}^* = \frac{k[\beta(\theta_b q+a)-(1+\beta)(c_b-d)]^2}{2\beta[2k(1+\beta)-\beta]} - f$, $\Pi_{r-bs}^* = \frac{k[\beta(\theta_b q+a)-(1+\beta)(c_b-d)]^2}{2\beta[2k(1-\eta_s)(1+\beta-\eta_s+\beta\eta_s)-\beta]} - f$, because of $2k(1+\beta) - \beta > 2k(1-\eta_s)(1+\beta-\eta_s+\beta\eta_s) - \beta$, thus, $\Pi_{r-bs}^* > \Pi_{r-b}^*$. Let $\Pi_{sc-1}(\eta_s) = \frac{k\left[2k(1-\eta_s)^2(2+\beta-2\eta_s+2\beta\eta_s)-\beta\right][\beta(\theta_b q+a)-(1+\beta)(c_b-d)]^2}{2\beta[2k(1-\eta_s)(1+\beta-\eta_s+\beta\eta_s)-\beta]^2}$, then $\Pi_{sc-bs}^* = \Pi_{sc-1}$ and $\Pi_{sc-b}^* = \Pi_{sc-1}(0)$, because of $\frac{\partial\Pi_{sc-bs}^*}{\partial\eta_s} = \frac{\left\{\begin{array}{l} 2k^2[1-(1-\beta)\eta_s]\left[2k(1-\beta)(1-\eta_s)^3\right. \\ \left.-3\beta\eta_s+\beta\right][\beta(\theta_b q+a)-(1+\beta)(c_b-d)]^2 \end{array}\right\}}{\beta[2k(1-\eta_s)(1+\beta-\eta_s+\beta\eta_s)-\beta]^3}$, $k > \frac{\beta}{2(1-\eta_s)(1+\beta-\eta_s+\beta\eta_s)}$, and $\eta_s < \frac{3-\sqrt{1+8\beta}}{4(1-\beta)}$, we have $1 - 3\eta_s + 2\eta_s^2 - 2\beta\eta_s^2 > 0$ and $2k(1-\beta)(1-\eta_s)^3 - 3\beta\eta_s+\beta > \frac{\beta(1-\beta)(1-\eta_s)^2+\beta(1-3\eta_s)(1+\beta-\eta_s+\beta\eta_s)}{1+\beta-\eta_s+\beta\eta_s} = \frac{\beta}{1+\beta-\eta_s+\beta\eta_s}\{(1-\beta)(1-\eta_s)^2 + (1-3\eta_s)(1+\beta-\eta_s+\beta\eta_s)\} = \frac{\beta}{1+\beta-\eta_s+\beta\eta_s}(1 - 3\eta_s + 2\eta_s^2 - 2\beta\eta_s^2) > 0$, then, $\frac{\partial\Pi_{sc-bs}^*}{\partial\eta_s} > 0$, thus, we have $\Pi_{sc-bs}^* > \Pi_{sc-b}^*$. Let $\Pi_{s-1}(\eta_s) = \frac{K^2(1-\eta_s)\left(1-3\eta_s+2\eta_s^2-2\beta\eta_s^2\right)[\beta(\theta_b q+a)-(1+\beta)(c_b-d)]^2}{2\beta[2k(1-\eta_s)(1+\beta-\eta_s+\beta\eta_s)-\beta]^2}$, then $\Pi_{s-bs}^* = \Pi_{s-1}$ and $\Pi_{s-b}^* = \Pi_{s-1}(0)$,

because of $\frac{\partial\Pi_{s-bs}^*}{\partial\eta_s} = \frac{\left\{\begin{array}{l} 2k^2[1-(1-\beta)\eta_s][\beta(2-3\eta_s)-2k(1-\eta_s) \\ \left((1-\eta_s)\eta_s+\beta\left(2-\eta_s+\eta_s^2\right)\right)][\beta(\theta_b q+a)- \\ (1+\beta)(c_b-d)]^2 \end{array}\right\}}{\beta[2k(1-\eta_s)(1+\beta-\eta_s+\beta\eta_s)-\beta]^3}$, thus, if $k < \frac{\beta(2-3\eta_s)}{2(1-\eta_s)[(1-\eta_s)\eta_s+\beta(2-\eta_s+\eta_s^2)]}$, then, $\frac{\partial\Pi_{s-bs}^*}{\partial\eta_s} > 0$, i.e. $\Pi_{s-bs}^* > \Pi_{s-b}^*$, otherwise, $\Pi_{s-bs}^* < \Pi_{s-b}^*$.

Appendix 7: Proof of Proposition 3. According to Eqs. (6)–(10) and Eqs. (11)–(17), we can obtain D_b^*, Π_{r-b}, Π_{s-b}, Π_{sc-b}, D_{br}^*, Π_{r-br}, Π_{s-br}, and Π_{sc-br}. $D_b^* = \frac{k[\beta(\theta_b q+a)-(1+\beta)(c_b-d)]}{2k(1+\beta)-\beta}$ and $D_{br}^* = \frac{k[\beta(\theta_b q+a)-(1+\beta)(c_b-d)]}{2k(1+\beta-\eta_r)-\beta}$, because of $2k(1+\beta) - \beta > 2k(1+\beta-\eta_r) - \beta$, then $D_{br}^* > D_b^*$. Let

$\Pi_{r-2}(\eta_r) = \frac{k[2k(1-2\eta_r)+(2k-1)\beta][\beta(\theta_b q+a)-(1+\beta)(c_b-d)]^2}{2\beta[2k(1+\beta-\eta_r)-\beta]^2} - f$, then $\Pi^*_{r-br} = \Pi_{r-2}$ and $\Pi^*_{r-b} = \Pi_{r-2}(0)$, because of $\frac{\partial \Pi^*_{r-br}}{\partial \eta_r} = -\frac{4k^3\eta_r[\beta(\theta_b q+a)-(1+\beta)(c_b-d)]^2}{\beta[2k(1+\beta-\eta_r)-\beta]^3} < 0$, thus, we have $\Pi^*_{r-br} < \Pi^*_{r-b}$. $\Pi^*_{s-b} = \frac{k^2[\beta(\theta_b q+a)-(1+\beta)(c_b-d)]^2}{\beta[2k(1+\beta)-\beta]^2}$ and $\Pi^*_{s-br} = \frac{k^2[\beta(\theta_b q+a)-(1+\beta)(c_b-d)]^2}{\beta[2k(1+\beta-\eta_r)-\beta]^2}$, because of $2k(1+\beta)-\beta > 2k(1+\beta-\eta_r)-\beta$, then $\Pi^*_{s-br} > \Pi^*_{s-b}$. Let $\Pi_{sc-2}(\eta_r) = \frac{k[4k(1-\eta_r)+(2k-1)\beta][\beta(\theta_b q+a)-(1+\beta)(c_b-d)]^2}{2\beta[2k(1+\beta-\eta_r)-\beta]^2} - f$, then $\Pi^*_{sc-br} = \Pi_{sc-2}$ and $\Pi^*_{sc-b} = \Pi_{sc-2}(0)$, because of $\frac{\partial \Pi^*_{sc-br}}{\partial \eta_r} = \frac{4k^3(1-\eta_r)[\beta(\theta_b q+a)-(1+\beta)(c_b-d)]^2}{\beta[2k(1+\beta-\eta_r)-\beta]^3} > 0$, thus, we have $\Pi^*_{sc-br} > \Pi^*_{sc-b}$.

Appendix 8: Proof of Coordination mechanism in scenario BR. Let $p_s = w_s + \phi_s$. By Eq. (28), the first-order and the second-order derivatives of U_{s-s} with respect to w_s can be taken, that is, $\frac{\partial U_{s-s}}{\partial w_s} = \frac{\lambda(\theta_b q+t_s-2w_s-\phi_s+a)+c_b-d}{\beta}$ and $\frac{\partial^2 U_{s-s}}{\partial {w_s}^2} = -\frac{2\lambda}{\beta} < 0$. By $\frac{\partial U_{s-s}}{\partial w_s} = 0$, we have

$$w_s = \frac{\lambda(\theta_b q + t_s - \phi_s + a) + c_b - d}{2\lambda}. \tag{43}$$

Substituting w_s into Eq. (27), then the first-order and the second-order derivatives of U_{r-s} with respect to ϕ_s and t_s can be taken, i.e. $\frac{\partial U_{r-s}}{\partial \phi_s} = \frac{\lambda^2(1-\eta_r)(\theta_b q+t_s-\phi_s+a)+[\lambda(\beta+\eta_r)-\beta](c_b-d)-\lambda\beta\phi_s}{2\lambda\beta}$, $\frac{\partial^2 U_{r-s}}{\partial {\phi_s}^2} = -\frac{\lambda(1-\eta_r)+\beta}{2\beta}$, $\frac{\partial U_{r-s}}{\partial t_s} = \frac{[\beta-\lambda(1-\eta_r)](\theta_b q+a)+[\beta-\lambda(1-\eta_r)-2k\beta]t_s-\lambda(1-\eta_r)\phi_s-(\beta-\eta_r)(c_b-d)}{2\beta}$, $\frac{\partial^2 U_{r-s}}{\partial {t_s}^2} = \frac{\beta-\lambda(1-\eta_r)-2k\beta}{2\beta}$, and $\frac{\partial^2 U_{r-s}}{\partial \phi_s \partial t_s} = \frac{\partial^2 U_{r-s}}{\partial t_s \partial \phi_s} = \frac{\lambda(1-\eta_r)}{2\beta}$. By the Hessian matrix $H(\phi_s, t_s) = \begin{bmatrix} \frac{\partial^2 U_{r-s}}{\partial {\phi_s}^2} & \frac{\partial^2 U_{r-s}}{\partial \phi_s \partial t_s} \\ \frac{\partial^2 U_{r-s}}{\partial t_s \partial \phi_s} & \frac{\partial^2 U_{r-s}}{\partial {t_s}^2} \end{bmatrix}$ is negative definite, $\frac{\partial U_{r-s}}{\partial \phi_s} = 0$, and $\frac{\partial U_{r-s}}{\partial t_s} = 0$, we have if $k > \frac{\beta}{2(\lambda-\lambda\eta_r+\beta)}$, then

$$\phi_s^* = \frac{\left\{\begin{array}{l} 2k\lambda^2(1-\eta_r)(\theta_b q + a) - [(2k-1)\beta \\ +\lambda(1+\beta-2k\beta-2k\eta_r)](c_b - d) \end{array}\right\}}{\lambda[2k(\lambda - \lambda\eta_r + \beta) - \beta]}, \tag{44}$$

$$t_s^* = \frac{\beta(\theta_b q + a) - (1+\beta)(c_b - d)}{2k(\lambda - \lambda\eta_r + \beta) - \beta}. \tag{45}$$

According to $p_s = w_s + \phi_s$ and Eqs. (43), (44), and (45), the equilibrium results can be given, $p_s^* = \frac{k[2\lambda(1-\eta_r)+\beta](\theta_b q+a)-(1-k)(1+\beta)(c_b-d)}{2k(\lambda-\lambda\eta_r+\beta)-\beta}$, $t_s^* = \frac{\beta(\theta_b q+a)-(1+\beta)(c_b-d)}{2k(\lambda-\lambda\eta_r+\beta)-\beta}$, $w_s^* = \frac{k\lambda\beta(\theta_b q+a)+[\beta(2k-1)+k\lambda(1-\beta-2\eta_r)](c_b-d)}{\lambda[2k(\lambda-\lambda\eta_r+\beta)-\beta]}$, and $D_s^* = \frac{k[\beta(\theta_b q+a)-(1+\beta)(c_b-d)]}{2k(\lambda-\lambda\eta_r+\beta)-\beta}$. Substituting p_s^*, t_s^*, w_s^*, and D_s^* into Eqs. (25) and (26), we can obtain $\Pi^*_{r-s} = \frac{k[2k(\lambda-2\lambda\eta_r+\beta)-\beta][\beta(\theta_b q+a)-(1+\beta)(c_b-d)]^2}{2\beta[2k(\lambda-\lambda\eta_r+\beta)-\beta]^2} - f$ and $\Pi^*_{s-s} = \frac{\lambda k^2[\beta(\theta_b q+a)-(1+\beta)(c_b-d)]^2}{\beta[2k(\lambda-\lambda\eta_r+\beta)-\beta]^2}$.

Appendix 9: Proof of Proposition 4. According to Eqs. (6), (7), (29), and (30), we have $\Pi^*_{s-s} - \Pi^*_{s-b} = \frac{k^2[\beta(\theta_b q+a)-(1+\beta)(c_b-d)]^2}{\beta}\{\frac{\lambda}{[2k(\lambda-\lambda\eta_r+\beta)-\beta]^2} - \frac{1}{[2k(1+\beta)-\beta]^2}\}$ and $\Pi^*_{r-s} - \Pi^*_{r-b} = \frac{k[\beta(\theta_b q+a)-(1+\beta)(c_b-d)]^2}{2\beta}\left\{\frac{1}{[2k(\lambda-\lambda\eta_r+\beta)-\beta]} - \frac{2k\lambda\eta_r}{[2k(\lambda-\lambda\eta_r+\beta)-\beta]^2} - \frac{1}{[2k(1+\beta)-\beta]}\right\}$. Because of $0 < \lambda < 1$, thus, if $\left[\frac{2k\lambda(1-\eta_r)+(2k-1)\beta}{2k+(2k-1)\beta}\right]^2 < \lambda^2 < \lambda < \frac{(1-\lambda+\lambda\eta_r)[2k\lambda(1-\eta_r)+(2k-1)\beta]}{\eta_r[2k+(2k-1)\beta]}$, then $\Pi^*_{s-s} > \Pi^*_{s-b}$ and $\Pi^*_{r-s} > \Pi^*_{r-b}$. Therefore, if $\lambda > \frac{(2k-1)\beta}{2k\eta_r+(2k-1)\beta}$, then $\lambda > \lambda^2 > \left[\frac{2k\lambda(1-\eta_r)+(2k-1)\beta}{2k+(2k-1)\beta}\right]^2$. If $\lambda < \frac{(2k-1)\beta}{2k\eta_r+(2k-1)\beta}$, then $\lambda < \frac{2k\lambda(1-\eta_r)+(2k-1)\beta}{2k+(2k-1)\beta}$, therefore, if $\lambda < \frac{(1-\lambda+\lambda\eta_r)(2k-1)\beta}{\eta_r[2k\eta_r+(2k-1)\beta]}$, that is $\lambda < \frac{(2k-1)\beta}{2k\eta_r^2+(2k-1)\beta}$, then $\lambda < \frac{(1-\lambda+\lambda\eta_r)[2k\lambda(1-\eta_r)+(2k-1)\beta]}{\eta_r[2k+(2k-1)\beta]}$. To sum up, if $\frac{(2k-1)\beta}{2k\eta_r+(2k-1)\beta} < \lambda < \frac{(2k-1)\beta}{2k\eta_r^2+(2k-1)\beta}$, then $\Pi^*_{s-s} > \Pi^*_{s-b}$ and $\Pi^*_{r-s} > \Pi^*_{r-b}$.

References

1. Sodhi, M.S., Tang, C.S.: Research opportunities in supply chain transparency. Prod. Oper. Manag. **28**(12), 2946–2959 (2019)
2. Babich, V., Hilary, G.: OM Forum-Distributed ledgers and operations: what operations management researchers should know about blockchain technology. Manuf. Serv. Oper. Manag. **22**(2), 223–240 (2020)
3. Fan, Z.P., Wu, X.Y., Cao, B.B.: Considering the traceability awareness of consumers: should the supply chain adopt the blockchain technology. Ann. Oper. Res. **309**, 837–860 (2020)
4. Galimberti, A., Labra, M., Sandionigi, A., Bruno, A., Mezzasalma, V., Mattia, F.D.: DNA barcoding for minor crops and food traceability. Adv. Agric. **2014**, 831875 (2014)
5. George, R.V., Harsh, H.O., Ray, P., Babu, A.K.: Food quality traceability prototype for restaurants using blockchain and food quality data index. J. Clean. Prod. **240**, 118021 (2019)
6. Wu, X.Y., Fan, Z.P., Cao, B.B.: An analysis of strategies for adopting blockchain technology in the fresh product supply chain. Int. J. Prod. Res. **61**(11), 3717–3734 (2023)
7. Dufwenberg, M., Kirchsteiger, G.: A theory of sequential reciprocity. Games Econ. Behav. **47**(2), 268–298 (2004)
8. Falk, A., Fischbacher, U.: A theory of reciprocity. Games Econ. Behav. **54**(2), 293–315 (2006)
9. Loch, C.H., Wu, Y.: Social preferences and supply chain performance: an experimental study. Manag. Sci. **54**(11), 1835–1849 (2008)
10. Wang, Y., Yu, Z., Jin, M., Mao, J.: Decisions and coordination of retailer-led low-carbon supply chain under altruistic preference. Eur. J. Oper. Res. **293**(3), 910–925 (2021)
11. Moon, I., Jeong, Y.J., Saha, S.: Investment and coordination decisions in a supply chain of fresh agricultural products. Oper. Res. Int. J. **20**, 2307–2331 (2020)
12. Bakos, Y., Halaburda, H.: Overcoming the coordination problem in new marketplaces via cryptographic tokens. Inf. Syst. Res. **33**(4), 1368–1385 (2022)

Approximation Algorithms for the Capacitated Min-Max and Minimum Graph Cover Problems

Jiafeng Xiong, Zhaohui Liu, and Wei Yu(✉)

School of Mathematics, East China University of Science and Technology, Shanghai 200237, China
y20170003@mail.ecust.edu.cn, {zhliu,yuwei}@ecust.edu.cn

Abstract. In this paper we obtain improved approximation algorithms for the Capacitated Min-Max Graph Cover Problems and the first constant-factor approximation algorithms for the Capacitated Minimum Graph Cover Problems. These problems are capacitated extension of the well-known min-max and minimum graph cover problems. We introduce several new ideas to bring down the approximation ratios for the Capacitated Min-Max Graph Cover Problems. For the Capacitated Minimum Graph Cover Problems, the constant-factor approximation algorithms are achieved by an approximation-preserving reduction to the corresponding uncapacitated problems.

Keywords: Approximation Algorithm · Tree Cover · Path Cover · Cycle Cover · Vehicle Routing

1 Introduction

Given a positive integer k and an undirected complete graph $G = (V, E)$ with vertex set V and edge set E, where each edge $e \in E$ is associated with a nonnegative weight $w(e)$ and the edge weights satisfy the triangle inequality, the Min-Max Tree Cover Problem (MMTCP) is to find a tree cover (of V), i.e. a set $\mathcal{T} = \{T_1, \ldots, T_p\}$ of trees in G collectively covering all the vertices in V, such that the number p of trees in $\mathcal{T}$ is at most k and the maximum weight of the trees in $\mathcal{T}$ is minimized. Here the weight of a tree means the total weight of its edges. In a related Minimum Tree Cover Problem (MTCP), the input consists of $B > 0$ and an undirected complete graph $G = (V, E)$ with nonnegative edge weights satisfying the triangle inequality, the objective is to find a tree cover $\mathcal{T} = \{T_1, \ldots, T_p\}$ (of V) such that the weight of each tree in $\mathcal{T}$ is at most B and the number p of trees in $\mathcal{T}$ is minimized. If we substitute paths/cycles for trees in the MMTCP/MTCP, we can define similarly the Min-Max Path/Cycle Cover Problem (MMPCP/MMCCP) and the Minimum Path/Cycle Cover Problem (MPCP/MCCP). These fundamental min-max and minimum graph cover problems and their variations have found numerous applications, including nurse

D. Du et al. (Eds.): COCOA 2024, LNCS 15434, pp. 18–30, 2025.
https://doi.org/10.1007/978-981-96-4445-2_2

station location [7], disaster relief efforts routing [5], data gathering/recharging in wireless sensor networks or IoT networks [18,20], data mule scheduling [3], multi-vehicle scheduling problem [2], distance-constrained vehicle routing [16], political districting [13], to name just a few.

The capacity constraint, which imposes an upper bound on the number of customers can be served by a single vehicle, is the most natural constraint in the classical vehicle routing literature [17]. This prompts us to consider the capacitated generalization of the MMTCP, called the Capacitated Min-Max Tree Cover Problem (CapMMTCP), where each vertex of G has a unit demand that must be served by exactly one tree of the tree cover $\mathcal{T} = \{T_1, \ldots, T_p\}$ with $p \leq k$ and the number of vertices served by each tree T_i $(i = 1, \ldots, p)$ cannot exceed a uniform capacity λ. When a vertex v is served by some tree T_i, we also say that it is covered by this tree and denote by $cov(T_i)$ the set of vertices covered by T_i. Analogously, one can define the Capacitated Min-Max Path/Cycle Cover Problem (CapMMPCP/CapMMCCP) and the Capacitated Minimum Tree/Path/Cycle Cover Problem (CapMTCP/CapMPCP/CapMCCP). Given the wide range of applications of the Min-Max and Minimum Tree/Path/Cycle Cover Problem, incorporating the capacity constraint makes these graph cover problems more realistic models, which deserves further investigation.

Since the (uncapacitated) graph cover problems are already NP-hard [9] and even APX-hard [16,19,24], we focus on the approximability results on their capacitated extension. Starting with the work of Even et al. [7] and Arkin et al. [1], approximation algorithms for the uncapacitated graph cover problems have been extensively studied. After a series of improvements, the state-of-the-art approximation ratios for the MMTCP, MMPCP and MMCCP stay at 3 [14], 4 [1,21] and 5 [8,23], respectively. Parallel to these results, the best-know approximation ratios for the MTCP, MPCP and MCCP are 2.5 [14], 3 [1] and 4 [20], respectively.

In contrast, it is somewhat surprising that few research has been conducted on the capacitated graph cover problems in the literature. Actually, it is until 2020 when the first constant-factor approximation algorithm was proposed for the CapMMTCP by Das et al. [6]. Previously, Guttmann-Beck and Hassin [11] only gave a $(2k-1)$-approximation algorithm for the partition variant of the special case of CapMMTCP where the total number of vertices in the graph equals $k\lambda$. Each feasible solution of this variant is a tree cover $\mathcal{T} = \{T_1, \ldots, T_p\}$ with $p = k$ such that $|V(T_i)| = |cov(T_i)| = \lambda$ for each i and $V(T_i) \cap V(T_j) = \emptyset$ for any $i \neq j$.

The algorithm of Das et al. consists of three phases $\mathcal{A}$, $\mathcal{B}$ and $\mathcal{C}$. For simplicity, we assume that the optimal value *opt* is known and G is connected containing no edges of weight greater than *opt*. Phase $\mathcal{A}$ constructs a set $\{T_1, \ldots, T_p\}$ of trees, which are called $\lceil \lambda/4 \rceil$-Steiner trees, such that $|cov(T_i)| = \lceil \lambda/4 \rceil$ for each i and $cov(T_i) \cap cov(T_j) = \emptyset$ for any $i \neq j$. These $\lceil \lambda/4 \rceil$-Steiner trees are generated one at a time by combining edge-doubling technique with a 2-approximation algorithm in [10] for a related q-MST problem, whose aim is to find a tree of minimum weight in an edge-weighted undirected graph that contains exactly q

vertices. Phase $\mathcal{A}$ ends when no $\lceil\lambda/4\rceil$-Steiner trees can be found. Then Phase $\mathcal{B}$ iteratively applies the algorithm of Khani and Salavatipour [14] and bipartite graph matching algorithm to cover the bad vertices, which are vertices not covered by the $\lceil\lambda/4\rceil$-Steiner trees in the previous phase. After adding a set V_{dummy} of at most $k\lambda - |V|$ dummy vertices, this phase ends up with a series of λ-good trees, i.e. trees T with $cov(T) \geq \lceil\lambda/4\rceil$, covering all the vertices in $V \cup V_{dummy}$. Finally, Phase $\mathcal{C}$ converts carefully these λ-good trees into a feasible tree cover of size at most k respecting the capacity constraint, using a result on the Hamiltonicity of the cube of a graph [12]. Das et al. [6] proved that this algorithm is a 260-approximation algorithm for the CapMMTCP. Using the standard edge-doubling strategy, this implies 520-approximation algorithms for both the CapMMPCP and CapMMCCP.

The main contribution of this paper is to devise better approximation algorithms for the CapMMTCP/CapMMPCP/CapMMCCP and the first constant factor approximation algorithms for the CapMTCP/CapMPCP/CapMCCP. For the CapMMTCP, we design an approximation algorithm following the three-phase framework of the algorithm in [6]. However, we introduce several new ideas in Phase $\mathcal{B}$ and Phase $\mathcal{C}$, which lead to an improved approximation ratio of $40 \ln 4 + 59 \approx 114.46$ (a detailed comparison between our algorithm and the algorithm in [6] is described at the end of Sect. 3). Moreover, we extend this approach to the CapMMPCP and CapMMCCP to obtain approximation algorithms with ratios 93 and 119, respectively, which are much better than the above ratios achieved by combining the algorithm for the CapMMTCP with the edge-doubling strategy. For the CapMTCP, CapMPCP and CapMCCP, we give an approximation-preserving reduction to the corresponding uncapacitated problems to derive the first constant-factor approximation algorithms with ratios 3.5, 4, 5, respectively.

The remainder of the paper is organized as follows. In Sect. 2, we define some notations and state formally the problems investigated in this paper. Section 3 presents the approximation algorithm for the CapMMTCP while Sect. 4 deals with the CapMMPCP and CapMMCCP. In Sect. 5, we describe an approximation-preserving reduction from the CapMTCP/CapMPCP/CapMCCP to the corresponding uncapacitated version.

2 Preliminaries

Given a capacity $\lambda \in \mathbb{Z}^+$ and an undirected complete graph $G = (V, E)$ with vertex set V and edge set E, each vertex has a unit demand to be served and each edge $e \in E$ is associated with a weight $w(e) \in \mathbb{Z}^+$ satisfying the triangle inequality. If $e = \{u, v\}$ with $u, v \in V$, we call u, v the endpoints of e and denote also by $w(u, v)$ the weight of e. Define $n = |V|$ to be the number of vertices in G. For an edge subset $E' \subseteq E$, we use $w(E') = \sum_{e \in E'} w(e)$ to indicate the total edge weight of the edges in E'. If H is a subgraph of G (e.g. tree, path, cycle), we denote by $V(H)$ and $E(H)$ the vertex set and edge set of H, respectively. The total weight of the edges in $E(H)$ is called the weight of H and is denoted by $w(H)$, i.e. $w(H) = w(E(H))$. If a vertex $v \in V(H)$ is served by H, we also say that v is covered by H. Let $cov(H)$ the set of vertices covered by H. The vertices in $V(H) \setminus cov(H)$ are passed (but not served) by H.

A tree cover (of V) is a set $\mathcal{T} = \{T_1, \ldots, T_p\}$ of trees in G collectively covering all the vertices in V, i.e., $V = \cup_{i=1}^{p} V(T_i)$. A capacitated tree cover (of V) is a tree cover $\mathcal{T} = \{T_1, \ldots, T_p\}$ along with a partition $V_1, \ldots, V_p$ of V such that $cov(T_i) = V_i \subseteq V(T_i)$ and $|cov(T_i)| \leq \lambda$ for each i. A Steiner tree cover of $R \subseteq V$ is a set $\mathcal{T} = \{T_1, \ldots, T_p\}$ of trees along with a partition $R_1, \ldots, R_p$ of R such that $cov(T_i) = R_i \subseteq V(T_i)$ for each i. If in addition $|cov(T_i)| = |R_i| \leq \lambda$ holds for each i, $\mathcal{T}$ is called a capacitated Steiner tree cover of R. The cost of $\mathcal{T}$, represented by $c(\mathcal{T})$, is defined as the maximum weight of the trees in $\mathcal{T}$, i.e. $\max_{1 \leq i \leq p} w(T_i)$. Analogously, we can define (capacitated) path/cycle covers and their cost by replacing trees with paths/cycles in the above definitions.

For $B > 0$, let $G[B]$ be the subgraph of G derived by deleting all edges $e \in E$ with $w(e) > B$. For $V_1 \subseteq V$ and $V_2 \subseteq V$, we define $d(V_1, V_2) = \min_{u \in V_1, v \in V_2} w(u, v)$ as the distance between V_1 and V_2. The distance $d(H_1, H_2)$ between two subgraphs H_1, H_2 of G is simply $d(V(H_1), V(H_2))$. In particular, for two trees T_1, T_2, the distance between T_1 and T_2 is defined as $d(T_1, T_2) = \min_{u \in V(T_1), v \in V(T_2)} w(u, v)$.

Throughout this paper, OPT denotes the optimal solution, i.e. a capacitated tree (path, cycle) cover, for the problem in discussion and *opt* is the corresponding optimal value. For the CapMMTCP (CapMMPCP, CapMMCCP), $opt = c(OPT)$. For the CapMTCP (CapMPCP, CapMCCP), $opt = |OPT|$.

The problems considered in this paper are described formally as follows:

Capacitated Min-Max Tree Cover Problem (CapMMTCP)
Input: *A positive integer k, a capacity $\lambda \in \mathbb{Z}^+$, an undirected complete graph $G = (V, E)$ with each vertex having a unit demand, and a weight function $w : E \to \mathbb{Z}^+$ satisfying the triangle inequality.*
Output: *A capacitated tree cover $\mathcal{T} = \{T_1, \ldots, T_p\}$ with $p \leq k$.*
Goal: *Minimize the cost of $\mathcal{T}$, i.e. $c(\mathcal{T}) = \max_{1 \leq i \leq p} w(T_i)$.*

Capacitated Minimum Tree Cover Problem (CapMTCP)
Input: *A positive integer B, a capacity $\lambda \in \mathbb{Z}^+$, an undirected complete graph $G = (V, E)$ with each vertex having a unit demand, and a weight function $w : E \to \mathbb{Z}^+$ satisfying the triangle inequality.*
Output: *A capacitated tree cover $\mathcal{T} = \{T_1, \ldots, T_p\}$ with $w(T_i) \leq B$ for each $i = 1, \ldots, p$.*
Goal: *Minimize the size of $\mathcal{T}$, i.e. $p = |\mathcal{T}|$.*

If we substitute paths/cycles for trees in the CapMMTCP and CapMTCP, we obtain the Capacitated Min-Max Path/Cycle Cover Problem (CapMM-PCP/CapMMCCP) and Capacitated Minimum Path/Cycle Cover Problem (CapMPCP/CapMCCP).

As mentioned in the introduction, the algorithm of Das et al. [6] uses an approximation algorithm for the following q-MST problem:

***q*-Minimum Spanning Tree Problem (*q*-MST)**
Input: *A positive integer* q, *an undirected graph* $G = (V, E)$ *with a root* $r \in V$ *and a weight function* $w : E \to \mathbb{Z}^+$.
Output: *A tree* T *in* G *rooted at* r *containing exactly* q *vertices.*
Goal: *Minimize the weight of* T, *i.e.* $w(T)$.

The q-MST can be generalized to the q-Minimum Steiner Tree Problem.

***q*-Minimum Steiner Tree Problem**
Input: *A positive integer* q, *an undirected graph* $G = (V, E)$ *with a root* $r \in V$, *a set* $R \subseteq V$ *of terminals with* $r \in R$ *and a weight function* $w : E \to \mathbb{Z}^+$.
Output: *A tree* T *in* G *rooted at* r *containing exactly* q *terminals in* R.
Goal: *Minimize the weight of* T, *i.e.* $w(T)$.

Garg [10] obtained a 2-approximation algorithm for the q-MST, which yields a 4-approximation algorithm for the q-Minimum Steiner Tree Problem using edge-doubling strategy[1].

Our algorithm also invokes an algorithm solving another generalization of the Min-Max Tree Cover Problem.

Min-Max Steiner Tree Cover Problem (MMSTCP)
Input: *A positive integer* $\hat{k}$, *an undirected complete graph* $G = (V, E)$ *with a set* $R \subseteq V$ *of terminals, and a weight function* $w : E \to \mathbb{Z}^+$.
Output: *A Steiner tree cover* $\mathcal{T} = \{T_1, \ldots, T_p\}$ *of* R *with* $p \leq \hat{k}$.
Goal: *Minimize the cost of* $\mathcal{T}$, *i.e.* $c(\mathcal{T}) = \max_{1 \leq i \leq p} w(T_i)$.

Yu [22] obtained a $4\alpha_{st}$-approximation algorithm for the MMSTCP, where α_{st} is the best-known approximation ratio for the Steiner Tree Problem. Currenly, $\alpha_{st} = \ln 4 \approx 1.387$ due to Byrka et al. [4].

To develop an α-approximation algorithm for the CapMMTCP (CapMMPCP, CapMMCCP), we give the following definition on α-subroutines.

Definition 1. *An* α-*subroutine for the CapMMTCP (CapMMPCP, CapMMCCP) is a polynomial time algorithm such that for any instance consisting of* $k, \lambda \in \mathbb{Z}^+$ *and* $G = (V, E)$ *with weight function* $w : E \to \mathbb{Z}^+$, *and any value* $0 \leq B \leq 2w(E)$, *the algorithm returns either failure or a capacitated tree (path, cycle) cover* $\mathcal{T}$ *such that* $|\mathcal{T}| \leq k$ *and* $c(\mathcal{T}) \leq \alpha B$ *and it always returns the latter output as long as* $opt \leq B$.

---

[1] Given an instance I of the q-Minimum Steiner Tree Problem defined on $G = (V, E)$ with terminal set R and root $r \in R$, we derive an instance I' of the q-MST defined on an undirected complete graph on R, whose edge weights equal the lengths of the shortest paths between the corresponding endpoints, and find a tree T' rooted in r such that $|V(T')| = q$ and $w(T') \leq 2opt(I')$ by the algorithm of Garg [10]. Replacing the edges of T' by the corresponding shortest paths in G derives a feasible solution T to I with $w(T) \leq w(T')$. Since doubling the edges of the optimal solution to I and then shortcut the non-terminal vertices derives a feasible solution to I', we have that $opt(I') \leq 2opt(I)$. Therefore, $w(T) \leq w(T') \leq 2opt(I') \leq 4opt(I)$.

It is well-known (see e.g. [14]) that an α-subroutine can be converted into an α-approximation algorithm by applying the standard binary search on the interval $[0, 2w(E)]$, which clearly contains the optimal value.

Lemma 1. *Any α-subroutine for the CapMMTCP (CapMMPCP, CapMM-CCP) that runs in $T(n)$ time yields an α-approximation algorithm for the same problem that runs in $O(T(n) \log w(E))$ time.*

3 Capacitated Min-Max Tree Cover

In this section, we present an improved approximation algorithm for the Cap-MMTCP with ratio $40 \ln 4 + 59 \approx 114.46$. This is achieved by designing an α-subroutine for the CapMMTCP with $\alpha = 40 \ln 4 + 59$. Similar to [6], the subroutine consists of three phases $\mathcal{A}$, $\mathcal{B}$ and $\mathcal{C}$. However, we refined their analysis and introduce some new ideas in phases $\mathcal{B}$ and $\mathcal{C}$ to bring down the approximation ratio.

We introduce some more definitions before the formal description of our algorithm.

Definition 2. *For any $\lambda \in \mathbb{Z}^+$, a tree T in $G = (V, E)$ is called λ-good if $|cov(T)| \geq \frac{\lambda}{3}$.*

Definition 3. *For a graph $G = (V, E)$ and a set $R \subseteq V$ of terminals, a tree T in G is called a q-Steiner tree if $|V(T) \cap R| = q$, i.e. T contains exactly q terminals in R.*

Given an instance of the CapMMTCP consisting of $k, \lambda \in \mathbb{Z}^+$ and $G = (V, E)$ with weight function $w : E \to \mathbb{Z}^+$, and any value $0 \leq B \leq 2w(E)$, Phase $\mathcal{A}$ first removes the edges of weight greater than B to obtain the graph $G[B]$. Then it compute as many as possible $\lceil \lambda/3 \rceil$-Steiner trees in each connected components of $G[B]$. Phase $\mathcal{A}$ terminates when no such trees can be produced and returns the set $\mathcal{T}_g$ of all generated $\lceil \lambda/3 \rceil$-Steiner trees, which are λ-good trees, and the set V_b of vertices not covered by the λ-good trees in $\mathcal{T}_g$. We call the vertices in V_b bad vertices. The formal description of Phase $\mathcal{A}$ is given as below.

Algorithm $\mathcal{A}$

Step 1. Delete all the edges e with $w(e) > B$ in G to generate the graph $G[B]$. Let $F_1, \ldots, F_p$ be the connected components of $G[B]$. If $p > k$, stop and return failure.

Step 2. For each $i = 1, 2, \ldots, p$,

Step 2.1. Initialize $V_c^i = \emptyset$, $\mathcal{T}_g^i = \emptyset$ and $U = V(F_i)$.

Step 2.2. Choose any $v \in U$. Run Garg's algorithm for the q-Minimum Steiner Tree Problem to find a $\lceil \lambda/3 \rceil$-Steiner tree T in F_i with terminal set $R = V(F_i) \setminus V_c^i$. Let $cov(T)$ be the set of terminals covered by T.

Step 2.3. If $w(T) \leq 4B$, add T to $\mathcal{T}_g^i$ and update $U = U \setminus cov(T)$, $V_c^i = V_c^i \cup cov(T)$. Otherwise, set $U = U \setminus \{v\}$.

Step 2.4. If $U \neq \emptyset$, go to Step 2.2.; otherwise, i.e. $U = \emptyset$, set $V_b^i = V(F_i) \setminus V_c^i$.

Step 3. Let $\mathcal{T}_g = \cup_{i=1}^p \mathcal{T}_g^i$, $V_c = \cup_{i=1}^p V_c^i$, $V_b = \cup_{i=1}^p V_b^i$. Return $\mathcal{T}_g$ and V_b.

For a tree T^* in OPT, it is called light if $|cov(T^*)| \leq \frac{2}{3}\lambda$; otherwise, it is called heavy. The number of light (heavy) trees in OPT is denoted by k_l (k_h). It is easy to make the following observation.

Observation 1. *If $opt \leq B$, then $w(e) \leq opt \leq B$ for each edge $e \in E(OPT) = \bigcup_{T^* \in OPT} E(T^*)$.*

This observation implies that $V(T^*)$ is contained entirely in exactly one of $V(F_1), \ldots, V(F_p)$ for each tree T^* in OPT whenever $opt \leq B$. We denote by $\mathcal{T}_i^*$ the set of trees in OPT that are contained in F_i and define $k_i^* = |\mathcal{T}_i^*|$. Since OPT is a feasible solution of cost $opt \leq B$, $\mathcal{T}_i^*$ is a Steiner tree cover of $V(F_i)$ of cost at most B. Moreover, $\mathcal{T}_1^*, \ldots, \mathcal{T}_p^*$ form a partition of OPT.

Observation 2. *If $opt \leq B$, $\mathcal{T}_i^*$ is a Steiner tree cover of $V(F_i)$ of cost at most B. Furthermore, any $\mathcal{T}' \subseteq \mathcal{T}_i^*$ is a Steiner tree cover of $\bigcup_{T^* \in \mathcal{T}'} cov(T^*)$ of cost at most B.*

By the construction of Algorithm $\mathcal{A}$, we can show the following conclusion.

Lemma 2. *When Algorithm $\mathcal{A}$ terminates, it holds that (i) $V(F_i) = cov(\mathcal{T}_g^i) \cup V_b^i$ with $cov(\mathcal{T}_g^i) = \bigcup_{T \in \mathcal{T}_g^i} cov(T)$ for each $i = 1, \ldots, p$; (ii) for each $T \in \mathcal{T}_g$, $|cov(T)| = \lceil \lambda/3 \rceil$ and $w(T) \leq 4B$; (iii) if $opt \leq B$, then for any heavy tree $T^* \in \mathcal{T}_i^*$ ($i = 1, \ldots, p$) at least half of the vertices in $cov(T^*)$ are covered collectively by the trees in $\mathcal{T}_g^i$.*

Proof. (i)(ii) are straightforward due to the steps of Algorithm $\mathcal{A}$.

It remains to show that (iii) holds. Suppose by contradiction that there is a heavy tree $T^* \in \mathcal{T}_i^*$ such that no more than half of the vertices in $cov(T^*)$ are covered collectively by the trees in $\mathcal{T}_g$. That is, there exists a heavy tree T^* contained in some F_i with $w(T^*) \leq opt \leq B$ covering at least $\frac{1}{2}|cov(T^*)| \geq \frac{1}{2} \cdot \frac{2}{3}\lambda = \frac{\lambda}{3}$ vertices in V_b^i, where the inequality follows from the definition of heavy trees. Therefore, Step 2.2 will indeed find a $\lceil \lambda/3 \rceil$-Steiner tree T in F_i with $w(T) \leq 4B$, which contradicts to the termination of Algorithm $\mathcal{A}$. □

As we have seen, after the first phase bad vertices are not covered by any λ-good trees in $\mathcal{T}_g$. So in the second phase, our main task is to cover the bad vertices without incurring too much cost. We do this component-by-component. The plan is to either construct new trees of large size but low cost or merge some bad vertices into the existing λ-good trees. This process is guided by the maximum matching of a bipartite graph. One part of this bipartite graph corresponds to trees covering the bad vertices and these trees are iteratively refined during the running of the algorithm. In contrast, the other part, corresponding to the λ-good trees in $\mathcal{T}_g^i$, keeps the same in each iteration. The algorithm of Phase $\mathcal{B}$ with respect to the component F_i is given as follows.

Algorithm $\mathcal{B}_i$

Step 1. Initialize $\mathcal{T}_b^i$ as a set of $|V_b^i|$ trivial trees each of which consists of a distinct vertex in V_b^i and covers this vertex.

Step 2. Construct a bipartite graph $H_i = (B_i \cup X_i, E_i)$, where each vertex $u \in B_i$ corresponds to a tree $T_{b,u} \in \mathcal{T}_b^i$ and each vertex in $x \in X_i$ corresponds to a tree $T_{g,x} \in \mathcal{T}_g^i$. $\{u, x\} \in E_i$ if and only if $d(T_{b,u}, T_{g,x}) \leq B$. Find a maximum matching M_i on H_i.

Step 2.1. If M_i matches all the vertices in B_i, i.e., $|M_i| = |B_i|$, go to Step 3.

Step 2.2. If M_i does not match all the vertices in B_i, i.e., $|M_i| < |B_i|$, there must exist a Hall set $S_i \subseteq B_i$ of maximum deficiency[2], i.e. $S_i = \arg\max_{S \subseteq B_i}(|S| - |N(S)|)$ (> 0), where $N(S)$ indicates the set of all neighbors of the vertices in S in the graph H_i.

Step 2.3. Run the $4\alpha_{st}$-approximation algorithm in [22] for the instance of the MMSTCP defined on F_i with terminal set $R = \bigcup_{u \in S_i} cov(T_{b,u})$ and $\hat{k}_i = |S_i| - 1$. This produces a Steiner tree cover $\mathcal{T}_R^i$ of R with $|\mathcal{T}_R^i| \leq \hat{k}_i$. If $c(\mathcal{T}_R^i) > 4\alpha_{st}B$, then go to Step 3; otherwise, update $\mathcal{T}_b^i$ as $(\mathcal{T}_b^i \setminus \bigcup_{u \in S_i}\{T_{b,u}\}) \cup \mathcal{T}_R^i$ and go to Step 2.

Step 3. Update $\mathcal{T}_g^i$ in the following way. For each edge $\{u, x\} \in M_i$, replace $T_{g,x}$ by a new λ-good tree that is generated by connecting $T_{b,u}$ with $T_{g,x}$ via an edge of $E(F_i)$ corresponding to $d(T_{b,u}, T_{g,x})$ $(\leq B)$. For each vertex $u \in B_i$ not covered by M_i, add to $T_{b,u}$ $\lceil \lambda/3 \rceil$ dummy vertices co-located with an arbitrary vertex of this tree and put the obtained λ-good tree into $\mathcal{T}_g^i$. Let V_d^i be the set of all dummy vertices added.

Phase $\mathcal{B}$ runs Algorithm $\mathcal{B}_i$ for each i. After that, if $\sum_{i=1}^p |V(F_i) \cup V_d^i|/\lambda > k \geq \sum_{i=1}^p k_i^*$, it returns failure and stop (this is correct by Lemma 4(iii)). Otherwise, the main algorithm proceeds to Phase $\mathcal{C}$.

Let us define the projection of OPT on S_i in Step 2.2 as follows:

$$OPT|_{S_i} = \{T^* \in OPT \mid \exists u \in S_i, cov(T^*) \cap cov(T_{b,u}) \neq \emptyset\}.$$

Lemma 3. *If opt $\leq B$ and Algorithm $\mathcal{B}_i$ goes from Step 2.3 to Step 3 and terminates, let S_i be the corresponding maximum deficiency Hall set, then $|S_i| - |N(S_i)| \leq k_l^i$, where k_l^i is the number of light trees in $OPT|_{S_i}$.*

Proof. Denote by k_h^i the number of heavy trees in $OPT|_{S_i}$. Then we have $|OPT|_{S_i}| = k_l^i + k_h^i$. Due to Observation 2, $OPT|_{S_i}$ is a Steiner tree cover of $R = \bigcup_{u \in S_i} cov(T_{b,u}) \subseteq \bigcup_{T^* \in OPT|_{S_i}} cov(T^*)$ of cost at most B. If $|S_i| > k_l^i + k_h^i$, which implies $\hat{k}_i = |S_i| - 1 \geq k_l^i + k_h^i$, then Step 2.3 will produce a Steiner tree

[2] Such Hall set can be computed in polynomial time, e.g. using the standard augmenting path algorithm for bipartite matching [15].

cover $\mathcal{T}_R^i$ of R with $|\mathcal{T}_R^i| \leq \hat{k}_i$ whose cost is at most $4\alpha_{st}B$. This contradicts the termination of Step 2.3 of Algorithm $\mathcal{B}_i$. So we conclude that $|S_i| \leq k_l^i + k_h^i$.

On the other hand, we next show that $|N(S_i)| \geq k_h^i$. By Lemma 2(iii), for each heavy tree $T^* \in OPT|_{S_i}$ at least $\frac{1}{2}|cov(T^*)| \geq \frac{1}{2} \cdot \frac{2}{3}\lambda = \frac{\lambda}{3}$ vertices in $cov(T^*)$ are covered collectively by the λ-good trees in $\mathcal{T}_g^i$. All such λ-good trees correspond to vertices in $N(S_i)$, since their distances to some tree $T_{b,u}$ with $u \in S_i$ is at most $opt \leq B^3$. Because each λ-good tree corresponding to a vertex in $N(S_i)$ covers precisely $\lceil \lambda/3 \rceil$ vertices, we obtain that $k_h^i \cdot \lceil \frac{\lambda}{3} \rceil \leq |N(S_i)| \cdot \lceil \lambda/3 \rceil$, which leads to $|N(S_i)| \geq k_h^i$.

This completes the proof of the lemma. □

By the construction of Algorithm $\mathcal{B}_i$ and analogous proofs in [6], we can show the following result.

Lemma 4. *When Algorithm $\mathcal{B}_i$ terminates, it holds that (i) $\mathcal{T}_g^i$ is a set of λ-good trees which constitute a Steiner tree cover of $V(F_i) \cup V_d^i$; (ii) $w(T) \leq (4\alpha_{st}+5)B$ for each $T \in \mathcal{T}_g^i$; (iii) if $opt \leq B$, then $|V(F_i) \cup V_d^i| \leq k_i^* \lambda$.*

Due to the above lemma, we can add some more dummy vertices such that $|V(F_i) \cup V_d^i| = t_i\lambda \leq k_i^*\lambda$ is a multiple of λ for some positive integer t_i. After this modification, Lemma 4 still holds.

In Phase $\mathcal{C}$, we still proceeds component-by-component to convert the obtained λ-good trees in $\mathcal{T}_g^i$ into a capacitated Steiner tree cover $\hat{\mathcal{T}}_i$ of $V(F_i)$ with $|\hat{\mathcal{T}}_i| \leq k_i^*$ when $opt \leq B$. This is achieved by first connecting the trees in $\mathcal{T}_g^i$ in a sequence generated by a result of Karaganis [12] and then partition the vertices according to this sequence into groups of size λ. Eventually, discarding the dummy vertices gives the desired capacitated Steiner tree cover $\hat{\mathcal{T}}_i$ of $V(F_i)$. The detailed steps of Phase $\mathcal{C}$ with respect to F_i is described as follows.

Algorithm $\mathcal{C}_i$

Step 1. Suppose that $\mathcal{T}_g^i = \{T_1, \ldots, T_{m_i}\}$ (by definition $|V(\mathcal{T}_g^i)| = \sum_{j=1}^{m_i} |V(T_j)| = |V(F_i) \cup V_d^i| = t_i\lambda$). Construct an unweighted graph $G_i = (V_g^i, E_g^i)$, where each vertex in $v \in V_g^i$ corresponds to a tree $T_v \in \mathcal{T}_g^i$ and $\{u, v\} \in E_g^i$ if and only if $d(T_u, T_v) \leq B$. Note that G_i is connected since F_i is a connected component of $G[B]$.

Step 2. Construct a new graph G_i^3, called the cube of G_i, with the same set of vertices as G_i and edge set E_g^{i3} such that $\{u, v\} \in E_g^{i3}$ if and only if the length of the shortest path between u and v is at most 3. Find a Hamiltonian path $P_i = v_1, \ldots, v_{m_i}$ in G_i^3. By renumber the vertices, we may assume that v_j corresponds to T_j for $j = 1, \ldots, m_i$.

Step 3. Keep choosing vertices according to the sequence $V(T_1), \ldots, V(T_{m_i})$ to partition $V(F_i) \cup V_d^i = \cup_{j=1}^{m_i} V(T_j)$ into t_i groups $R_1, \ldots, R_{t_i}$, each of which contains exactly λ vertices. For each R_j ($j = 1, \ldots, t_i$), let $V(T_{f_j})$ ($V(T_{l_j})$) be the vertex set among $V(T_1), \ldots, V(T_{m_i})$ from which the first (last) vertex

[3] By definition, there exists some edge $e \in E(T^*)$ connecting a vertex of such λ-good tree with some tree $T_{b,u}$ such that $u \in S_i$. Observation 1 implies that $w(e) \leq opt \leq B$.

of R_j is chosen. Generate a tree $\hat{T}_i^j$ by adding to $T_{f_j}, T_{f_j+1}, \ldots, T_{l_j}$ the paths of F_i corresponding to the edges $\{v_{f_j}, v_{f_j+1}\}, \{v_{f_j+1}, v_{f_j+2}\}, \ldots, \{v_{l_j-1}, v_{l_j}\}$ in G_i^3. Set $cov(\hat{T}_i^j) = R_j$ and put $\hat{T}_i^j$ into $\hat{\mathcal{T}}_i$.

Step 4. Remove all the dummy vertices from each tree $\hat{T}_i^j \in \hat{\mathcal{T}}_i$ and update $cov(\hat{T}_i^j) = cov(\hat{T}_i^j) \setminus V_d^i$. Return $\hat{\mathcal{T}}_i$.

By the construction of Algorithm $\mathcal{C}_i$, we may show the following result.

Lemma 5. *When Algorithm $\mathcal{C}_i$ terminates it holds that (i) $\hat{\mathcal{T}}_i$ is a capacitated Steiner tree cover of $V(F_i)$; (ii) If $opt \leq B$, then $|\hat{\mathcal{T}}_i| \leq k_i^*$. (iii) $c(\hat{\mathcal{T}}_i) \leq (40\alpha_{st} + 59)B$;*

Proof. (i) We have mentioned that for the connected graph G_i in Step 2 G_i^3 admits a Hamiltonian path that can be found in polynomial time [12]. After that Steps 3 and 4 of Algorithm $\mathcal{C}_i$ indeed produce a Steiner tree cover $\hat{\mathcal{T}}_i$ of $V(F_i)$ with $|\hat{\mathcal{T}}_i| \leq t_i$.

(ii) If $opt \leq B$, Lemma 4(iii) implies that $|\hat{\mathcal{T}}_i| \leq t_i \leq k_i^*$.

To prove (iii), it is sufficient to give the desired upper bound on $w(\hat{T}_i^j)$ for each $\hat{T}_i^j \in \hat{\mathcal{T}}_i$. By the construction of Step 3, $\hat{T}_i^j$ is obtained by connecting the trees $T_{f_j}, T_{f_j+1}, \ldots, T_{l_j}$ in $\mathcal{T}_g^i$ with the paths of F_i corresponding to the edges $\{v_{f_j}, v_{f_j+1}\}, \{v_{f_j+1}, v_{f_j+2}\}, \ldots, \{v_{l_j-1}, v_{l_j}\}$ in G_i^3. Since each tree in $\mathcal{T}_g^i$ is a λ-good tree, which covers at least $\lambda/3$ vertices, $\hat{T}_i^j$ can contain at most 4 trees in $\mathcal{T}_g^i$. That is, $l_j - f_j \leq 3$. As a result, at most 3 paths corresponding to $\{v_{f_j}, v_{f_j+1}\}, \{v_{f_j+1}, v_{f_j+2}\}, \ldots, \{v_{l_j-1}, v_{l_j}\}$ are added, each of which has a weight of no more than

$$3B + 2 \cdot \max_{T \in \mathcal{T}_g^i}(\text{diameter of } T) \leq 3B + 2w(T) \leq (3 + 2(4\alpha_{st} + 5))B,$$

where the last inequality follows from Lemma 4(ii).

So we can establish an upper bound on $w(\hat{T}_i^j)$ as follows:

$$w(\hat{T}_i^j) \leq 4 \cdot (4\alpha_{st} + 5)B + 3 \cdot (3 + 2(4\alpha_{st} + 5))B = (40\alpha_{st} + 59)B. \quad \square$$

Finally, Phase $\mathcal{C}$ ends up with a capacitated tree cover $\hat{\mathcal{T}} = \cup_{i=1}^p \hat{\mathcal{T}}_i$. If $|\hat{\mathcal{T}}| = \sum_{i=1}^p |\hat{\mathcal{T}}_i| > k$, it returns failure and stop (this is correct by Lemma 5(ii)). Otherwise, return $\hat{\mathcal{T}}$.

Lemma 5 implies that when $opt \leq B$, we have that $|\hat{\mathcal{T}}| = \sum_{i=1}^p |\hat{\mathcal{T}}_i| \leq \sum_{i=1}^p k_i^* \leq k$ and $c(\hat{\mathcal{T}}) \leq (40\alpha_{st} + 59)B$.

Combining Phases $\mathcal{A}$, $\mathcal{B}$ and $\mathcal{C}$, which clearly run in polynomial time, we obtain a $(40\alpha_{st} + 59)$-subroutine for the CapMMTCP. This implies the following result by Lemma 1.

Theorem 1. *There is a $(40\alpha_{st} + 59)$-approximation algorithm for the CapMMTCP. In particular, by taking $\alpha_{st} = \ln 4$, there is an approximation algorithm for the CapMMTCP with ratio $40 \ln 4 + 59 \approx 114.45$.*

As mentioned before, our algorithm for the CapMMTCP applies the same three-phase framework as the algorithm by Das et al. [6]. However, we refined their analysis and introduce several new ideas in Phase $\mathcal{B}$ and Phase $\mathcal{C}$. First, we change the threshold value from $\frac{\lambda}{2}$ to $\frac{2}{3}\lambda$ when defining light/heavy trees to suit our analysis. Second, in Phase $\mathcal{B}$ Das et al. adopt the 3-approximation algorithm for the MMTCP in [14] combined with edge-doubling technique to derive the Steiner tree cover $\mathcal{T}_R^i$ of R (Step 2.3 of Algorithm $\mathcal{B}_i$). In contrast, we apply directly the $4\alpha_{st}$-approximation algorithm for the MMSTCP to do the same thing, which incurs lower cost. Lastly, in Phase $\mathcal{C}$ when constructing the trees $\hat{T}_i^j$ in the final capacitated tree cover (Step 3 of Algorithm $\mathcal{C}_i$), Das et al. first merge the trees in $\mathcal{T}_g^i$ into trees covering at least λ vertices and then convert these trees into a feasible capacitated tree cover. Again, we use a more direct approach which connects at most four trees of $\mathcal{T}_g^i$ in a row along the sequence corresponding to the Hamiltonian path in G_i^3 to generate $\hat{T}_i^j$. It turns out our approach saves the cost indeed. Overall, our new methods manage to bring down the approximation ratio moderately.

4 Capacitated Min-Max Path/Cycle Cover

In this section we extend the approach for the CapMMTCP to solve the CapMMPCP/CapMMCCP and obtain the following results.

Theorem 2. *There is a 93-approximation algorithm for the CapMMPCP.*

Theorem 3. *There is a 119-approximation algorithm for the CapMMCCP.*

5 Capacitated Minimum Tree/Path/Cycle Cover

In this section, we deal with the Capacitated Minimum Tree/Path/Cycle Cover Problem. We give an approximation-preserving reduction from the CapMTCP/CapMPCP/CapMCCP to its uncapacitated version with a loss of one on the approximation ratio. Plugging in the best available ratios for the corresponding uncapacitated problems, as mentioned in the introduction, we are able to derive approximation algorithms for the three problems with ratios 3.5, 4 and 5, respectively.

Theorem 4. *Any α-approximation algorithm for the MTCP/MPCP/MCCP yields an $(\alpha + 1)$-approximation algorithm for the CapMTCP/CapMPCP/CapMCCP.*

Acknowledgements. This research is supported by the National Natural Science Foundation of China under grant number 11871213.

References

1. Arkin, E.M., Hassin, R., Levin, A.: Approximations for minimum and min-max vehicle routing problems. J. Algor. **59**, 1–18 (2006)
2. Bhattacharya, B., Hu, Y.: Approximation algorithms for the multi-vehicle scheduling problem. In: Cheong, O., Chwa, K.-Y., Park, K. (eds.) ISAAC 2010. LNCS, vol. 6507, pp. 192–205. Springer, Heidelberg (2010). https://doi.org/10.1007/978-3-642-17514-5_17
3. Chen, Z., Zhang, Z., Ran, Y., Shi, Y., Du, D.Z.: Data mule scheduling on a path with handling time and time span constraints. Optim. Lett. **14**, 1701–1710 (2020)
4. Byrka, J., Grandoni, F., Rothvob, T., Sanita, L.: Steiner tree approximation via iterative randomized rounding. J. ACM **60**(1), 6 (2013)
5. Campbell, A.M., Vandenbussche, D., Hermann, W.: Routing for relief efforts. Transp. Sci. **42**, 127–145 (2008)
6. Das, S., Jain, L., Kumar, N.: A constant factor approximation for capacitated min-max tree cover. In: Byrka, J., Meka, R. (eds.) The Proceedings of the 23rd International Conference on Approximation Algorithms for Combinatorial Optimization Problems (APPROX 2020), Article No. 55, pp. 55:1–55:13 (2020)
7. Even, G., Garg, N., Koemann, J., Ravi, R., Sinha, A.: Min-max tree covers of graphs. Oper. Res. Lett. **32**, 309–315 (2004)
8. Gao, X., Fan, J., Wu, F., Chen, G.: Approximation algorithms for sweep coverage problem with multiple mobile sensors. IEEE/ACM Trans. Network. **26**(2), 990–1003 (2018)
9. Garey, M., Johnson, D.: Computers and Intractability. Freeman, San Francisco (1979)
10. Garg, N.: Saving an epsilon: a 2-approximation for the k-mst problem in graphs. In: Proceedings of the Thirty-seventh Annual ACM Symposium on Theory of Computing (STOC 2005), pp. 396–402 (2005)
11. Guttmann-Beck, N., Hassin, R.: Approximation algorithms for min-max tree partition. J. Algor. **24**(2), 266–286 (1997)
12. Karaganis, J.J.: On the cube of a graph. Can. Math. Bull. **11**(2), 295–296 (1968)
13. Karakawa, S., Morsy, E., Nagamochi, H.: Minmax tree cover in the euclidean space. J. Graph Algor. Appl. **15**, 345–371 (2011)
14. Khani, M.R., Salavatipour, M.R.: Approximation algorithms for min-max tree cover and bounded tree cover problems. Algorithmica **69**, 443–460 (2014)
15. Lovasz, L., Plummer, M.D.: Mathcing Theory. AMS Chelsea Publishing, Providence (2009)
16. Nagarajan, V., Ravi, R.: Approximation algorithms for distance constrained vehicle routing problems. Networks **59**(2), 209–214 (2012)
17. Toth, P., Vigo, D.: Vehicle Routing: Problems, Methods, and Applications. SIAM, Philadelphia (2014)
18. Xu, W., Liang, W., Lin, X.: Approximation algorithms for min-max cycle cover problems. IEEE Trans. Comput. **64**(3), 600–613 (2015)
19. Xu, Z., Wen, Q.: Approximation hardness of min-max tree covers. Oper. Res. Lett. **38**, 408–416 (2010)
20. Xu, W., et al.: Minimizing the deployment cost of UAVs for delay-sensitive data collection in IoT networks. IEEE/ACM Trans. Network. **30**, 812–825 (2022)
21. Xu, Z., Xu, L., Li, C.-L.: Approximation results for min-max path cover problems in vehicle routing. Nav. Res. Logist. **57**, 728–748 (2010)

22. Yu, W.: Improved approximation algorithms for some min-max postmen cover problems with applications to the min-max subtree cover. Math. Methods Oper. Res. **97**, 135–157 (2023)
23. Yu, W., Liu, Z.: Improved approximation algorithms for some min-max and minimum cycle cover problems. Theor. Comput. Sci. **654**, 45–58 (2016)
24. Yu, W., Liu, Z.: Better approximability results for min-max tree/cycle/path cover problems. J. Comb. Optim. **37**(2), 563–578 (2019)

A Simple Algorithm for Scheduling Unit Jobs with Unknown Number of Machines

Lishi Yu(✉) and Zhiyi Tan

School of Mathematical Science, Zhejiang University, Hangzhou 310027, People's Republic of China
{12035036,tanzy}@zju.edu.cn

Abstract. This paper investigates a two-stage online scheduling problem for parallel machines with the goal of minimizing the makespan. The first stage requires packing a set of n unit jobs into M bags. In the second stage, these bags are allocated to $m \leq M$ machines, with the constraint that jobs within the same bag are processed on the same machine. By introducing an elaborate way of packing, we have developed an online algorithm based on the rule of the LPT and demonstrated that it is the optimal algorithm with a competitive ratio of $\frac{4}{3}$.

Keywords: Scheduling · Online algorithm · Competitive analysis

1 Introduction

In [6], Stein and Zhang introduced an intriguing two-stage scheduling problem with the goal of minimizing the makespan. In the first stage, we are provided with a set of n jobs that need to be scheduled on a collection of parallel and identical machines. Although the precise number of machines m is not known initially, an upper bound M is given. Our task is to pack all the jobs into M bags. In the second stage, when the number of machines, m, becomes known, we must schedule the M bags on these m machines, subject to the constraint that the jobs within the same bag must be processed on the same machine.

Since the number of machines is unknown at the time of packing the jobs, and the packing decision is irrevocable, the problem belongs to the category of online problems. Surprisingly, although online scheduling on parallel machines has been extensively studied [2,4,7], this particular problem has not been addressed before. In a sense, this problem has the least amount of missing information compared to other online scheduling problems. Both the first and second stages are inherently offline problems. Specifically, the second stage is equivalent to scheduling M jobs on m parallel machines, a problem that admits fully polynomial time approximation schemes [5]. However, since the critical information m is unknown, it is still not possible to design an algorithm that guarantees certain approximation performance.

Stein and Zhang [6] demonstrated that no online algorithm can achieve a competitive ratio better than $\frac{4}{3}$, and this lower bound remains valid even when

D. Du et al. (Eds.): COCOA 2024, LNCS 15434, pp. 31–43, 2025.
https://doi.org/10.1007/978-981-96-4445-2_3

all the jobs have unit processing time. They also developed an online algorithm with a competitive ratio of $\frac{5}{3}+\epsilon$. Even in the infinitesimal job case, where any job can be divided into arbitrarily small parts and different parts of the same job can be processed simultaneously, no online algorithm can achieve a competitive ratio better than $\max\left\{\frac{1}{\frac{a}{M-a}+\frac{M-2a}{M}} | a \leq \frac{M}{2} \text{ and } a \text{ is an integer}\right\}$. Later, Eberle et al. [1] provided two optimal algorithms with competitive ratios matching the lower bounds given in [6] for the unit job and infinitesimal job cases, respectively. They also generalized the problem to scenarios where the machines have different speeds. The time required for a machine with speed s to process a job with processing time p is $\frac{p}{s}$. They developed three algorithms with competitive ratios of $2-\frac{1}{M}$, $\frac{9}{5}$ and $\frac{M^M}{M^M-(M-1)^{M-1}}$ for the arbitrary job, unit job and infinitesimal job cases, respectively. Only the algorithm for the infinitesimal job case was proven to be optimal.

In this paper, we introduce a new optimal online algorithm for the unit job and identical machine case. Compared to the algorithm presented in [1], our algorithm is more succinct and easier to implement. For the first stage, our algorithm provides an explicit method for packing jobs into bags and reduces the number of special cases that require exceptional handling. For the second stage, the algorithm offers a universal and simple approach to assigning the bags to the machines. Our results indicate that even if only the LPT is used in the second stage, an optimal online algorithm can be obtained.

The overall structure of the algorithm is quite simple. In the first stage, all the jobs are packed into M bags, and in the second stage, these M bags are assigned to m machines according to the LPT rule. The LPT (Longest Processing Time first) rule was first proposed by Graham in his seminal paper [3]. It involves sorting all jobs in descending order of their processing times and then assigning the first unprocessed job in the sequence to the machine that can complete it the earliest. To ensure that this simple rule yields good performance, the packing process must be handled with great care.

W.l.o.g., we assume that all the jobs have processing time 1. Obviously, $C^*(n,m) = \lceil \frac{n}{m} \rceil$, where $C^*(n,m)$ is the optimum of scheduling n jobs to m machines. Let $\sigma(n,M,m)$ be the schedule produced by our algorithm, where M is the upper bound of the number of machines. Denote by $\mathcal{M}_i$ the ith machine and C_i the set of jobs scheduled on $\mathcal{M}_i$ in the schedule $\sigma(n,M,m)$, $i=1,\ldots,m$. Let $C(n,M,m) = \max_{i=1,\ldots,m} |C_i|$ be the makespan of $\sigma(n,M,m)$. The competitive ratio of the algorithm is $\max_{m \leq M} \frac{C(n,M,m)}{C^*(n,m)}$.

The structure of the paper is as follows. In Sect. 2, we presents several properties of the schedule generated by the LPT rule, and discuss scenarios with either M or n is small. In Sect. 3, we provide a detailed description of the algorithm, focusing particularly on the number of jobs included in each bag. Finally, Sect. 4 is dedicated to proving the competitive ratio of the algorithm. Due to space limitations, the proofs for some lemmas will be omitted.

2 Warm Up

Suppose that $\mathcal{M}_u$ determines the makespan of $\sigma(n, M, m)$, and there are w bags scheduled on $\mathcal{M}_u$, among which the last bag is B_v. Then $C(n, M, m) = |C_u|$, and B_v starts to be processed at time $|C_u| - |B_v|$.

Lemma 1. *(i)* $C^*(n, m) \geq |C_u| - |B_v| + 1$.
(ii) If $(4 - w)|B_v| \leq 4$, *then* $\frac{C(n,M,m)}{C^*(n,m)} \leq \frac{4}{3}$.

Proof. (i) By the rule of the LPT, the number of jobs that are assigned to any machine is at least $|C_v| - |B_v|$. Hence,

$$C^*(n, m) = \left\lceil \frac{n}{m} \right\rceil \geq \left\lceil \frac{m(|C_u| - |B_v|) + |B_v|}{m} \right\rceil \geq |C_u| - |B_v| + 1.$$

(ii) By the rule of the LPT, all the bags scheduled on $\mathcal{M}_u$ contain at least $|B_v|$ jobs. Hence, $|C_u| \geq w|B_v|$. By (i),

$$\frac{C(n, M, m)}{C^*(n, m)} \leq \frac{|C_u|}{|C_u| - |B_v| + 1} \leq \frac{w|B_v|}{(w - 1)|B_v| + 1} \leq \frac{4}{3}.$$

□

Obviously, the problem is nontrivial only when $n > M \geq m > 1$. If $\lceil \frac{n}{M} \rceil = \left\lceil \frac{n}{M-1} \right\rceil$, then just adopt the schedule for n jobs and $M - 1$ bags. If occasionally $m = M$, simply leave a machine empty. Thus $C(n, M, m) = C(n, M - 1, m)$ for any $m \leq M - 1$, and $C(n, M, M) = C(n, M - 1, M - 1)$. Clearly, for any $m \leq M - 1$,

$$\frac{C(n, M, m)}{C^*(n, m)} = \frac{C(n, M - 1, m)}{C^*(n, m)} \leq \max_{m \leq M-1} \frac{C(n, M - 1, m)}{C^*(n, m)}.$$

Even if $m = M$, then $C^*(n, m) = \lceil \frac{n}{M} \rceil = \left\lceil \frac{n}{M-1} \right\rceil = C^*(n, M - 1)$. Thus

$$\frac{C(n, M, m)}{C^*(n, m)} = \frac{C(n, M - 1, M - 1)}{C^*(n, M - 1)} \leq \max_{m \leq M-1} \frac{C(n, M - 1, m)}{C^*(n, m)}.$$

Therefore, $\max_{m \leq M} \frac{C(n,M,m)}{C^*(n,m)} \leq \max_{m \leq M-1} \frac{C(n,M-1,m)}{C^*(n,m)}$. Thus restricting to the case where $\left\lceil \frac{n}{M-1} \right\rceil > \lceil \frac{n}{M} \rceil$ is sufficient.

For $2 \leq M \leq 3$, we pack the jobs to M bags as even as possible. It results an optimal schedule when $M = m$. For the remaining case of $M = 3$ and $m = 2$, the LPT rule schedules the bag with the largest number of jobs separately, then

$$\frac{C(n, 3, 2)}{C^*(n, 2)} \leq \frac{n - \lceil \frac{n}{3} \rceil}{\lceil \frac{n}{2} \rceil} \leq \frac{\frac{2n}{3}}{\frac{n}{2}} = \frac{4}{3}.$$

We turn to the case of $qM < n \le (q+1)M$, where $q \in \{1,2,3\}$. Let $\gamma = \lfloor \frac{4}{3}(q+1) \rfloor$. Since

$$\gamma - q = \left\lfloor \frac{4}{3}(q+1) \right\rfloor - q = \begin{cases} 1, q = 1, \\ 2, q = 2, 3, \end{cases}$$

we have

$$\left\lceil \frac{\gamma M - n}{\gamma - q} \right\rceil + \left\lfloor \frac{\gamma M - n}{\gamma - q} \right\rfloor = \frac{2(\gamma M - n)}{\gamma - q} \tag{1}$$

for any $q \in \{1,2,3\}$, and

$$2\left\lfloor \frac{\gamma M - n}{\gamma - q} \right\rfloor + 1 = 2\left\lfloor \frac{(q+2)M - n}{2} \right\rfloor + 1 \ge 2\left\lfloor \frac{M}{2} \right\rfloor + 1 \ge M \tag{2}$$

for any $q \in \{2,3\}$.

Pack the n jobs into M bags such that there are $M - \left\lceil \frac{\gamma M - n}{\gamma - q} \right\rceil$ bags, each containing γ jobs, $\left\lfloor \frac{\gamma M - n}{\gamma - q} \right\rfloor$ bags, each containing q jobs, and $\left\lceil \frac{\gamma M - n}{\gamma - q} \right\rceil - \left\lfloor \frac{\gamma M - n}{\gamma - q} \right\rfloor \le 1$ bag containing $\frac{\gamma + q}{2}$ jobs. The bag which containing γ, q and $\frac{\gamma+q}{2}$ jobs is called *big*, *small* and *middle*, respectively. We will show below that $C(n, M, m) \le \frac{4}{3}C^*(n, m)$. By Lemma 1(ii), we only need to consider the case of $1 \le w \le 3$.

If $w = 1$, then

$$C(n, M, m) \le \gamma \le \frac{4}{3}(q+1) = \frac{4}{3}\left\lceil \frac{n}{M} \right\rceil \le \frac{4}{3}\left\lceil \frac{n}{m} \right\rceil = \frac{4}{3}C^*(n, m).$$

If $w \ge 2$, then $m \le M - 1$ and $C^*(n, m) = \lceil \frac{n}{m} \rceil \ge \left\lceil \frac{n}{M-1} \right\rceil > \lceil \frac{n}{M} \rceil = q + 1$.

First assume that B_v is a small bag. Since $|B_v| = q \le 3$, we only need to consider the case of $w = 2$ and $q = 3$ by Lemma 1(ii). If the bag scheduled on $\mathcal{M}_u$ other than B_v is also a small bag, then $\frac{C(n,M,m)}{C^*(n,m)} \le \frac{2q}{q+2} \le \frac{4}{3}$. Otherwise, by the rule of the LPT and the fact that there is at most one middle bag, each of the remaining machines processes at least a big bag or two small bags, and $|C_u| \ge 2q + 1 > \gamma$. Hence, $C^*(n, m) \ge \gamma + 1 = q + 3$ and $\frac{C(n,M,m)}{C^*(n,m)} \le \frac{q+\gamma}{q+3} \le \frac{2q+2}{q+3} = \frac{4}{3}$.

Next assume that B_v is not a small bag. Since B_v contains no more than γ jobs, we only need to consider the case of $q \ge 2$ by Lemma 1(ii). Since two big or middle bags are scheduled on $\mathcal{M}_u$, $|C_u| - |B_v| \ge \gamma$ and the number of big or middle bags is great than m. By (2),

$$\begin{aligned} \left\lfloor \frac{\gamma M - n}{\gamma - q} \right\rfloor &\ge M - \left\lfloor \frac{\gamma M - n}{\gamma - q} \right\rfloor - 1 \\ &= \left(M - \left\lceil \frac{\gamma M - n}{\gamma - q} \right\rceil \right) + \left(\left\lceil \frac{\gamma M - n}{\gamma - q} \right\rceil - \left\lfloor \frac{\gamma M - n}{\gamma - q} \right\rfloor \right) - 1 \ge m - 1. \end{aligned}$$

It follows that the number of small bags is no less than $m - 1$. By the rule of the LPT, at the time B_v is scheduled on $\mathcal{M}_u$, the number of jobs scheduled on any other machine is at least $|C_u| - |B_v|$ and no small bags have been assigned yet. Hence,

$$C^*(n,m) = \left\lceil \frac{n}{m} \right\rceil \geq \left\lceil \frac{m(|C_u| - |B_v|) + |B_v| + (m-1)q}{m} \right\rceil \geq |C_u| - |B_v| + q,$$

and

$$\frac{C(n,M,m)}{C^*(n,m)} \leq \frac{|C_u|}{|C_u| - |B_v| + q} = \frac{|C_u| - |B_v| + |B_v|}{|C_u| - |B_v| + q} \leq \frac{2\gamma}{\gamma + q} = \frac{2q+4}{2q+2} \leq \frac{4}{3}.$$

However, when $n > 4M$, the above approach may not yield the expected performance. For example, $q = 4$ and $\gamma = 6$ when $n = 84$ and $M = 17$. We will obtain 8 large bags, each containing 6 jobs, and 9 small bags, each containing 4 jobs. It is not hard to verify that $C(n, M, 12) = 10$ and $C^*(n, 12) = 7$. Thus $\frac{C(n,M,12)}{C^*(n,12)} = \frac{10}{7} > \frac{4}{3}$. Note that the schedule is already the best one for the M bags, indicating the need for a more sophisticated packing of the jobs.

3 Pack and Schedule

By the discussions in the Sect. 2, we assume that $M \geq 4$, $n > 4M$ and $\left\lceil \frac{n}{M-1} \right\rceil > \left\lceil \frac{n}{M} \right\rceil$ in the following. For any given n and M, let $\alpha, \beta, \gamma, \delta, \varepsilon$ be integers such that

$$\alpha = \left\lfloor \frac{M}{4} \right\rfloor, \beta = \left\lfloor \frac{n}{M-\alpha} \right\rfloor, \gamma = \left\lfloor \frac{4}{3} \left\lceil \frac{n}{M} \right\rceil \right\rfloor, \delta = M - 4\alpha, \varepsilon = n - (M-\alpha)\beta. \quad (3)$$

Obiviously, $\delta \in \{0, 1, 2, 3\}$ and $\varepsilon \in \{0, 1, \ldots, M - \alpha - 1\}$. Hence

$$n = (M - \alpha)\beta + \varepsilon \geq (M - \alpha)\beta. \quad (4)$$

and

$$4\alpha \leq M \leq 4\alpha + 3. \quad (5)$$

Moreover, under the previous assumption, $\alpha \geq 1$, $\beta \geq \left\lfloor \frac{n}{M-1} \right\rfloor \geq \left\lceil \frac{n}{M-1} \right\rceil - 1 > \left\lceil \frac{n}{M} \right\rceil - 1 = 4$ and $\gamma = \lfloor \frac{4}{3} \lceil \frac{n}{M} \rceil \rfloor \geq \lfloor \frac{4}{3} \cdot 5 \rfloor = 6$. In the following lemma, we give further properties of these numbers.

Lemma 2. *(i)* $\beta \leq \gamma$.
(ii) If $\beta = \gamma$ *and* $n > 2M$, *then* $\varepsilon < 2\alpha$.
(iii) If $\beta = \gamma$ *and* $\varepsilon > \alpha$, *then* $\gamma = \frac{4}{3} \lceil \frac{n}{M} \rceil - \frac{2}{3}$ *and* $\gamma \equiv 2 \pmod 4$.
(iv) If $\left\lceil \frac{n}{M-1} \right\rceil > \lceil \frac{n}{M} \rceil$ *and* $\alpha \geq 1$, *then* $\gamma \leq \frac{4}{3}\beta$.
(v) If $\beta \geq 5$, *then* $\lfloor \frac{2}{3}(\beta + 2) \rfloor + \lfloor \frac{1}{2}(\beta + 2) \rfloor \leq \frac{4}{3}(\beta + 1)$.

Next, we will describe the detailed steps involved in the two stages, with the first stage being further divided into two Procedures Batch and Divide. In the following Procedure Batch, we batch n jobs into $M - \alpha$ batches $A_1, \ldots, A_{M-\alpha}$. Among them, the first $M - 2\alpha$ batches are called *whole batches*, each of them becomes a bag in the Procedure Divide. The last α batches are called *dividable batches*, each of them will be divided into two bags in the Procedure Divide.

Procedure Batch

1. If $\varepsilon \leq \alpha$, then

$$|A_i| = \begin{cases} \beta, & i = 1, \ldots, M - 2\alpha, \\ \beta + 1, & i = M - 2\alpha + 1, \ldots, M - 2\alpha + \varepsilon, \\ \beta, & i = M - 2\alpha + \varepsilon + 1, \ldots, M - \alpha. \end{cases}$$

2. If $\varepsilon \geq \alpha + 1$ and $\beta < \gamma$, then

$$|A_i| = \begin{cases} \beta + 1, & i = 1, \ldots, \varepsilon - \alpha, \\ \beta, & i = \varepsilon - \alpha + 1, \ldots, M - 2\alpha, \\ \beta + 1, & i = M - 2\alpha + 1, \ldots, M - \alpha. \end{cases}$$

3. If $\varepsilon \geq \alpha + 1$ and $\beta = \gamma$, then

$$|A_i| = \begin{cases} \beta, & i = 1, \ldots, M - 2\alpha, \\ \beta + 2, & i = M - 2\alpha + 1, \ldots, M + \varepsilon - 3\alpha, \\ \beta + 1, & i = M + \varepsilon - 3\alpha + 1, \ldots, M - \alpha. \end{cases}$$

By $\varepsilon < M - \alpha$ and Lemma 2(ii), the procedure is well defined. Since

$$(M - \alpha - \varepsilon)\beta + \varepsilon(\beta + 1) = (M - \alpha)\beta + \varepsilon = \mathrm{n}$$

and

$$(M - 2\alpha)\beta + (2\alpha - \varepsilon)(\beta + 1) + (\varepsilon - \alpha)(\beta + 2) = (M - \alpha)\beta + \varepsilon = n,$$

each job belongs to exactly one batch. Moreover, we have the following lemma on the number of jobs in the batches.

Lemma 3. *(i)* $|A_{M-2\alpha+1}| \geq |A_{M-2\alpha+2}| \geq \cdots \geq |A_{M-\alpha}| \geq |A_1| \geq |A_2| \geq \cdots \geq |A_{M-2\alpha}| \geq \beta$.

(ii) The difference of the number of the jobs contained in any two whole (resp. dividable) batches is no more than 1.

(iii) Each whole batch contains no more γ *jobs.*

(iv) Each dividable batch contains no more than $\beta + 2$ *jobs. Furthermore, if* $\varepsilon \leq \alpha$ *or* $\beta < \gamma$, *each dividable batch contains no more than* $\beta + 1$ *jobs, and if* $\varepsilon > \alpha$, *each dividable batch contains no less than* $\beta + 1$ *jobs.*

In the Procedure Divide, we pack all the jobs into M bags based on the batches obtained in the Procedure Batch. Each whole batch automatically becomes a bag, while each dividable batch is divided into two bags.

Procedure Divide

1. Convert the whole batch A_i, $i = 1, \ldots, M - 2\alpha$ as it is into a bag B_i.

2. Partition the dividable batch $A_{M-2\alpha+i}$ into two bags $B_{M-2\alpha+i}$ and $B_{M-\alpha+i}$, such that

$$|B_{M-\alpha+i}| = \min\left\{\left\lfloor\frac{4}{3}\left\lceil\frac{n}{M-\alpha-\lceil\frac{i}{2}\rceil}\right\rceil\right\rfloor - |A_{M-2\alpha-i}|, \left\lfloor\frac{1}{2}|A_{M-2\alpha+i}|\right\rfloor\right\} \tag{6}$$

and

$$\begin{aligned}|B_{M-2\alpha+i}| &= |A_{M-2\alpha+i}| - |B_{M-\alpha+i}| \\ &= \max\left\{|A_{M-2\alpha+i}| + |A_{M-2\alpha-i}| - \left\lfloor\frac{4}{3}\left\lceil\frac{n}{M-\alpha-\lceil\frac{i}{2}\rceil}\right\rceil\right\rfloor, \left\lceil\frac{1}{2}|A_{M-2\alpha+i}|\right\rceil\right\}\end{aligned} \tag{7}$$

Denote by $\mathcal{B}_W = \{B_1, \ldots, B_{M-2\alpha}\}, \mathcal{B}_L = \{B_{M-2\alpha+1}, \ldots, B_{M-\alpha}\}$ and $\mathcal{B}_S = \{B_{M-\alpha+1}, \ldots, B_M\}$. The bags contained in $\mathcal{B}_W, \mathcal{B}_L$ and $\mathcal{B}_S$ are called *whole* bags, *large divided* bags and *small divided* bags, respectively. The following two lemmas provide properties on the number of jobs contained in each bag, as well as the relationship between the number of jobs contained in different bags.

Lemma 4. *(i) For any* $1 \le i \le \alpha$, $|B_{M-2\alpha+i}| \le \lfloor\frac{2}{3}|A_{M-2\alpha+i}|\rfloor$.

(ii) For any $1 \le i \le \alpha$, $|B_{M-2\alpha+i}| + |B_{M-4\alpha+i}| \le \left\lfloor\frac{4}{3}\left\lceil\frac{n}{M-2\alpha+\lceil\frac{i}{2}\rceil-1}\right\rceil\right\rfloor$.

(iii) If $\beta = \gamma \ge 6$ *and* $\varepsilon > \alpha$, *then for any* $M-2\alpha+1 \le i < 2m+1-i \le M-\alpha$ *such that* $|B_i| > \lceil\frac{1}{2}|A_i|\rceil$ *and* $|B_{2m+1-i}| > \lceil\frac{1}{2}|A_{2m+1-i}|\rceil$, $|B_i| + |B_{2m+1-i}| \le \lfloor\frac{4}{3}\lceil\frac{n}{m}\rceil\rfloor$.

Lemma 5. *(i)* $\beta \le |B_i| = |A_i| \le \gamma$, $i = 1, \ldots, M-2\alpha$.

(ii) Each small divided bag contains no less than $\beta - \left\lfloor\frac{2\beta}{3}\right\rfloor$ *and no more than* $\left\lfloor\frac{\beta+2}{2}\right\rfloor$ *jobs. Furthermore, if* $\varepsilon \le \alpha$ *or* $\beta < \gamma$, *each small divided bag contains no more than* $\left\lfloor\frac{\beta+1}{2}\right\rfloor$ *jobs, and if* $\varepsilon > \alpha$, *each small divided bag contains no less than* $\beta + 1 - \left\lfloor\frac{2(\beta+1)}{3}\right\rfloor$ *jobs.*

(iii) Each large divided bag contains no more than $\lfloor\frac{2}{3}(\beta+2)\rfloor$ *jobs. Furthermore, if* $\varepsilon \le \alpha$ *or* $\beta < \gamma$, *each large divided bag contains no more than* $\lfloor\frac{2}{3}(\beta+1)\rfloor$ *jobs.*

(iv) The number of jobs contained in any small divided bag is no more than those in any large divided bag. The number of jobs contained in any divided bag is no more than those in any whole bag.

Sort all the M bags in the nonincreasing order based on the number of jobs each contains. Suppose that $B_1', \ldots, B_M'$ is a permutation of $B_1, \ldots, B_M$ such that $|B_1'| \ge \cdots \ge |B_M'|$. By Lemmas 5(i)(iv) and 3(i), we have

$$\mathcal{B}_W = \{B_1, \ldots, B_{M-2\alpha}\}, \mathcal{B}_L = \{B'_{M-2\alpha+1}, \ldots, B'_{M-\alpha}\}, \mathcal{B}_S = \{B'_{M-\alpha+1}, \ldots, B'_M\}.$$

The following lemma provides the relationship of the number of jobs contained between different numbers of bags, which helps to characterize the schedule produced by the LPT rule.

Lemma 6. *(i) The total number of jobs contained in any two small divided bags is no less than those in any large divided bag.*

(ii) The total number of jobs contained in a whole bag and a small divided bag is no less than those in any whole bag.

(iii) The total number of jobs contained in two whole bags is no less than the total number of jobs contained in a whole bag and a divided bag.

(iv) The total number of jobs contained in the ith largest large divided bag and the ith smallest small divided bag is no less than those in any whole bag. That is, $|B'_{M-2\alpha+i}| + |B'_{M-i+1}| \geq |B_1|$ *for any* $1 \leq i \leq \alpha$.

In the second stage, we assign the bags $B'_1, \ldots, B'_M$ one by one to the machine that processes the smallest number of jobs among the m machines. By means of the relationship between the number of jobs contained in different bags, we can show the quantity and type of bags scheduled on different machines for different values of m.

When $m \geq M - 2\alpha$, $M \leq 2m$ by (5). The whole bags B_i, $i = 1, \ldots, M - 2\alpha$ must be the first bags scheduled on $\mathcal{M}_i$. If $m \geq M - \alpha$, all α large divided bags and the first $m - (M - \alpha)$ small divided bags $B'_{M-\alpha+1}, \ldots, B'_m$ are also the first bags scheduled on $\mathcal{M}_i$, $i = M - \alpha + 1, \ldots, m$. Regarding the schedule of the remaining $M - m$ small divided bags, B'_{2m+1-i}, $i = m, \ldots, 2m - M + 1$ is scheduled on $\mathcal{M}_i$ together with B'_i by Lemma 6(i). Hence,

$$C_i = \begin{cases} B_i, & i = 1, \ldots, M - 2\alpha, \\ B'_i, & i = M - 2\alpha + 1, \ldots, 2m - M, \\ B'_i \cup B'_{2m+1-i}, & i = 2m - M + 1, \ldots, m, \end{cases}$$

If $M - 2\alpha \leq m < M - \alpha$, the first $m - (M - 2\alpha)$ large divided bags $B'_{M-2\alpha+1}, \ldots, B'_m$ are also the first bag scheduled on $\mathcal{M}_i$, $i = M - 2\alpha + 1, \ldots, m$. Next, consider the schedule of the remaining $M - m - \alpha$ large divided bags and all α small divided bags. For any $M - 2\alpha + 1 \leq i \leq m$, write $k = i - (M - 2\alpha)$, then

$$|B'_i| + |B'_{2m+1-i}| \geq |B'_i| + |B'_{2M-2\alpha-i+1}| = |B'_{M-2\alpha+k}| + |B'_{M-k+1}| \geq |B_1|$$

by Lemma 6(iv). Together with Lemma 6(ii), it follows that B'_{2m+1-i}, $i = m, \ldots, 2m - M + 1$ is scheduled on $\mathcal{M}_i$ together with B_i or B'_i. Hence,

$$C_i = \begin{cases} B_i, & i = 1, \ldots, 2m - M, \\ B_i \cup B'_{2m+1-i}, & i = 2m - M + 1, \ldots, M - 2\alpha, \\ B'_i \cup B'_{2m+1-i}, & i = M - 2\alpha + 1, \ldots, m. \end{cases}$$

From the above description, it can be concluded that each machine processes at most two bags when $m \geq M - 2\alpha$.

When $m \leq M - 2\alpha - 1$, although the number of bags scheduled on each machine may be greater than two, the first two bags on each machine still adhere to certain rules with the help of Lemma 6(ii). The first m whole bags B_i must be the first bags scheduled on $\mathcal{M}_i$, $i = 1, \ldots, m$. The second bag scheduled on $\mathcal{M}_i$ together with B_i is the whole bag B_{2m+1-i} when $\max\{1, 2m+1-(M-2\alpha)\} \leq i \leq m$, or the large divided bag B'_{2m+1-i} when $\max\{1, 2m+1-(M-\alpha)\} \leq i < 2m+1-(M-2\alpha)$, or the small divided bag B'_{2m+1-i} when $1 \leq i < 2m+1-(M-\alpha)$. Hence, if B'_k is the third bag scheduled on a machine, then $k \geq 2m+1$. Consequently, there are at least three whole bags scheduled on a machine iff $2m+1 \leq M-2\alpha$. If there are at least three bags scheduled on a machine, among which the third one is a large divided bag, then $2m+1 \leq M-\alpha$. These facts are very helpful in analyzing the performance of the schedule $\sigma(n, M, m)$.

4 Competitive Analysis

In this section, we will prove that schedule the M bags obtained by Procedure Batch and Divide to m machines by the LPT rule meets the expected performance. Recall that $\mathcal{M}_u$ is the machine determines the makespan, and B_v is the wth and the last bag scheduled on $\mathcal{M}_u$. By Lemma 1(ii), we are safe to assume that $w \leq 3$. We divide the proof into several lemmas according to the value of m.

Lemma 7. *If $m \leq M - 2\alpha - 1$, $C(n, M, m) \leq \frac{4}{3} C^*(n, m)$.*

Proof. Note that any bag contains at most γ jobs by Lemma 5(i)(iv). If $w = 1$, then by (3),

$$C(n, M, m) \leq \gamma \leq \frac{4}{3}\left\lceil \frac{n}{M} \right\rceil \leq \frac{4}{3}\left\lceil \frac{n}{m} \right\rceil = \frac{4}{3} C^*(n, m).$$

If $w = 2$, both bags scheduled on $\mathcal{M}_u$ are whole bags by Lemmas 5(iv) and 6(iii). Then $C_u = B_u \cup B_{2m+1-u}$ and $M - 2\alpha \geq 2m+1-u$. By Lemmas 3(i)(ii) and 5(i),

$$\begin{aligned} n &= \sum_{l=1}^{u} |A_l| + \sum_{l=u+1}^{2m-u} |A_l| + |A_{2m+1-u}| + \sum_{l=2m-u+2}^{M-2\alpha} |A_l| + \sum_{l=M-2\alpha+1}^{M-\alpha} |A_l| \\ &\geq u|A_u| + |A_{2m+1-u}| + (M - 2\alpha - u - 1)(|A_u| - 1) + \alpha|A_u| \\ &= |B_{2m+1-u}| + (M - \alpha - 1)|B_u| - (M - 2\alpha - u - 1) \\ &\geq |B_{2m+1-u}| + (M - \alpha - 1)|B_u| - 2(M - 2\alpha - m - 1). \end{aligned}$$

Hence, by $|B_{2m+1-u}| \leq |B_u|$, $m \leq M - 2\alpha - 1$ and (5),

$$\begin{aligned} \frac{C(n, M, m)}{C^*(n, m)} &= \frac{|B_u| + |B_{2m+1-u}|}{\lceil \frac{n}{m} \rceil} \\ &\leq \frac{m(|B_{2m+1-u}| + |B_u|)}{|B_{2m+1-u}| + (M - \alpha - 1)|B_u| - 2(M - 2\alpha - m - 1)} \\ &\leq \frac{2m|B_u|}{(M - \alpha)|B_u| - 2(M - 2\alpha - m - 1)} \leq \frac{2(M - 2\alpha - 1)|B_u|}{(M - \alpha)|B_u|} \leq \frac{4}{3}. \end{aligned}$$

Finally, suppose that $w = 3$. If all the three bags are whole bags, then $M - 2\alpha \geq 2m + 1$. Combining it with Lemma 2(iv) and (5), (4), we have

$$C(n, M, m) \leq 3\gamma \leq 4\beta \leq \frac{4}{3}\frac{M-\alpha}{\frac{M-2\alpha-1}{2}}\beta \leq \frac{4}{3}\frac{M-\alpha}{m}\beta \leq \frac{4}{3}\left\lceil\frac{n}{m}\right\rceil = \frac{4}{3}C^*(n, m).$$

If only the first two bags are whole bags, then $|C_u| - |B_v| \geq 2\beta$ by Lemma 5(i). By Lemmas 1(i) and 5(iii)(iv),

$$\frac{C(n, M, m)}{C^*(n, m)} = \frac{|C_u|}{|C_u| - |B_v| + 1} = \frac{|C_u| - |B_v| + |B_v|}{|C_u| - |B_v| + 1} \leq \frac{2\beta + \frac{2}{3}\beta + \frac{4}{3}}{2\beta + 1} = \frac{4}{3}.$$

We are left with the case that B_u is the only whole bag scheduled on M_u. If B_v is a small divided bag, then $|C_u| \geq |B_u| + 2|B_v|$. By Lemmas 1(i) and 5(i)(ii),

$$\frac{C(n, M, m)}{C^*(n, m)} \leq \frac{|C_u|}{|C_u| - |B_v| + 1} \leq \frac{|B_u| + 2|B_v|}{|B_u| + |B_v| + 1} \leq \frac{\beta + 2|B_v|}{\beta + |B_v| + 1}$$
$$\leq \frac{\beta + \beta + 2}{\beta + \frac{\beta+2}{2} + 1} \leq \frac{4}{3}.$$

If B_v is a large divided bag, then $M - \alpha \geq 2m + 1$ and

$$C^*(n, m) = \left\lceil\frac{n}{m}\right\rceil \geq \left\lceil\frac{(M-\alpha)\beta}{\frac{M-\alpha-1}{2}}\right\rceil > 2\beta.$$

by (4). If $\beta = \gamma$ and $\varepsilon \geq \alpha + 1$, then by Lemma 5(iii),

$$C(n, M, m) \leq \beta + 2\left\lfloor\frac{2}{3}\beta + \frac{4}{3}\right\rfloor \leq \frac{7}{3}\beta + \frac{8}{3} \leq \frac{4}{3}(2\beta + 1) \leq \frac{4}{3}C^*(n, m).$$

Otherwise, by Lemma 5(iii),

$$C(n, M, m) \leq \beta + 1 + 2\left\lfloor\frac{2}{3}\beta + \frac{2}{3}\right\rfloor \leq \frac{7}{3}\beta + \frac{7}{3} \leq \frac{4}{3}(2\beta + 1) \leq \frac{4}{3}C^*(n, m).$$

□

Recall that each machine processes at most two bags when $m \geq M - 2\alpha$. This enables us to utilize the following well-known property of the LPT rule. For the classical parallel and identical machine scheduling problem with nonidentical job processing times, if each machine processes at most two jobs in the schedule produced by the LPT rule, then the schedule has the minimal makespan among all feasible schedules [3]. The property can be applied to our problem as long as bag B_i is considered to be a job with a processing time of $|B_i|$.

In the following, we will provide an auxiliary schedule σ^A of the M bags with a makespan of at most $\frac{4}{3}C^*(n, m)$. Then by using the above property, $C(n, M, m) \leq \frac{4}{3}C^*(n, m)$ as well. Although the proof for σ^A is simpler, the implementation of the LPT rule is more straightforward and convenient, and it maintains consistency with other cases. Denote by C_i^A, $i = 1, \ldots, m$, the set of jobs scheduled on $\mathcal{M}_i$ in the schedule σ^A.

Lemma 8. *If* $M-2\alpha \le m < M-\alpha$, $C(n,M,m) \le \frac{4}{3}C^*(n,m)$.

Proof. The schedule σ^A when $M-2\alpha \le m < M-\frac{3}{2}\alpha$ can be described as follows,

$$C_i^A = \begin{cases} B_i, & i = 1,\ldots,2m-M,\\ B_i \cup B_{i+2\alpha}, & i = 2m-M+1,\ldots,M-3\alpha,\\ B_i \cup B_{2M-3\alpha-i+1}, & i = M-3\alpha+1,\ldots,M-2\alpha,\\ B_i \cup B_{2m+1-i}, & i = M-2\alpha+1,\ldots,m. \end{cases}$$

The schedule σ^A when $M-\frac{3}{2}\alpha \le m < M-\alpha$ can be described as follows,

$$C_i^A = \begin{cases} B_i, & i = 1,\ldots,2m-M,\\ B_i \cup B_{2M-3\alpha-i+1}, & i = 2m-M+1,\ldots,M-2\alpha,\\ B_i \cup B_{2M-2\alpha-i+1}, & i = M-2\alpha+1,\ldots,2m-M+\alpha,\\ B_i \cup B_{2m+1-i}, & i = 2m-M+\alpha+1,\ldots,m. \end{cases}$$

Obviously, each bag is scheduled on one of the m machines. We will prove that $|C_i^A| \le \frac{4}{3}C^*(n,m)$ for any i.

Since $m < M-\alpha$, $C^*(n,m) = \lceil \frac{n}{m} \rceil \ge \left\lceil \frac{n}{M-\alpha-1} \right\rceil > \left\lfloor \frac{n}{M-\alpha} \right\rfloor = \beta$. Hence,

$$C^*(n,m) \ge \beta + 1. \tag{8}$$

For any $1 \le i \le 2m-M$, $|C_i^A| = |B_i| \le \gamma \le \frac{4}{3}\beta \le \frac{4}{3}C^*(n,m)$ by Lemmas 2(iv), 5(i) and (8).

For any $\max\{2m-M+1, M-3\alpha+1\} \le i \le M-2\alpha$, write $k = M-2\alpha-i+1$, then $k \le \min\{2M-2m-2\alpha, \alpha\}$ and thus $M-\alpha-\lceil\frac{k}{2}\rceil \ge m$. By Lemmas 5(i), 3(i) and (6),

$$\begin{aligned} |C_i^A| &= |B_i| + |B_{2M-3\alpha-i+1}| = |B_{M-2\alpha-k+1}| + |B_{M-\alpha+k}| \\ &= |A_{M-2\alpha-k+1}| + |B_{M-\alpha+k}| \le |A_{M-2\alpha-k}| + |B_{M-\alpha+k}| \\ &\le \left\lfloor \frac{4}{3}\left\lceil \frac{n}{M-\alpha-\lceil\frac{k}{2}\rceil} \right\rceil \right\rfloor \le \left\lfloor \frac{4}{3}\left\lceil \frac{n}{m} \right\rceil \right\rfloor \le \frac{4}{3}C^*(n,m). \end{aligned}$$

For any $2m-M+1 \le i \le M-3\alpha$, write $k = i-M+4\alpha$, then $2m-2M+4\alpha+1 \le k \le \alpha$ and thus $M-2\alpha+\lceil\frac{k}{2}\rceil - 1 \ge m$. By Lemma 4(ii),

$$\begin{aligned} |C_i^A| &= |B_i| + |B_{i+2\alpha}| = |B_{M-4\alpha+k}| + |B_{M-2\alpha+k}| \\ &\le \left\lfloor \frac{4}{3}\left\lceil \frac{n}{M-2\alpha+\lceil\frac{k}{2}\rceil-1} \right\rceil \right\rfloor \le \left\lfloor \frac{4}{3}\left\lceil \frac{n}{m} \right\rceil \right\rfloor \le \frac{4}{3}C^*(n,m). \end{aligned}$$

For any $M-2\alpha+1 \le i \le m$, $\mathcal{M}_i$ processes two divided bags in σ^A, one of them is the large divided bag B_i. If the other bag is a small divided bag, then by Lemmas 5(ii)(iii), 2(v) and (8),

$$|C_i^A| \le \left\lfloor \frac{2}{3}(\beta+2) \right\rfloor + \left\lfloor \frac{1}{2}(\beta+2) \right\rfloor \le \frac{4}{3}(\beta+1) \le \frac{4}{3}C^*(n,m).$$

If both bags are large divided bags, then $C_i^A = B_i \cup B_{2m+1-i}$. If both bags contain no more than $\lfloor \frac{2}{3}(\beta+1) \rfloor$ jobs, then $|C_i^A| \le 2\lfloor \frac{2}{3}(\beta+1) \rfloor \le \frac{4}{3}(\beta+1) \le \frac{4}{3}C^*(n,m)$ by (8). Otherwise, at least one bag contains $\lfloor \frac{2}{3}(\beta+1) \rfloor + 1 \ge \lfloor \frac{2}{3}(\beta+2) \rfloor$ jobs. Then $\beta = \gamma \ge 6$ and $\beta \equiv 2 \pmod 4$ by Lemmas 5(iii) and 2(iii). If at least one bag contains no more than $\lceil \frac{1}{2}(\beta+2) \rceil = \lfloor \frac{1}{2}(\beta+2) \rfloor$ jobs, then by Lemmas 5(iii), 2(v) and (8),

$$|C_i^A| = |B_i| + |B_{2m+1-i}| \le \left\lfloor \frac{2}{3}(\beta+2) \right\rfloor + \left\lfloor \frac{1}{2}(\beta+2) \right\rfloor \le \frac{4}{3}(\beta+1) \le \frac{4}{3}C^*(n,m).$$

Otherwise, $|B_i| > \lceil \frac{1}{2}|A_i| \rceil$ and $|B_{2m+1-i}| > \lceil \frac{1}{2}|A_{2m+1-i}| \rceil$ by Lemma 5(ii). Then by Lemma 4(iii),

$$|C_i^A| = |B_i| + |B_{2m+1-i}| \le \frac{4}{3}\left\lceil \frac{n}{m} \right\rceil = \frac{4}{3}C^*(n,m).$$

□

Lemma 9. *If* $M - \alpha \le m \le M$, $C(n,M,m) \le \frac{4}{3}C^*(n,m)$.

Proof. The schedule σ^A when $M - \alpha \le m \le M$ can be described as follows. That is,

$$C_i^A = \begin{cases} B_i, & i = 1, \dots, M-2\alpha, \\ B_i, & i = M-2\alpha+1, \dots, m-\alpha, \\ B_i \cup B_{i+\alpha}, & i = m-\alpha+1, \dots, M-\alpha, \\ B_i, & i = M-\alpha+1, \dots, m. \end{cases}$$

Obviously, each bag is scheduled on one of the m machines. We will prove that $|C_i^A| \le \frac{4}{3}C^*(n,m)$ for any i.

If $m = M$, each bag is processed on a machine. By (3),

$$|C_i^A| \le \gamma \le \frac{4}{3}\left\lceil \frac{n}{M} \right\rceil = \frac{4}{3}\left\lceil \frac{n}{m} \right\rceil = \frac{4}{3}C^*(n,m).$$

If $M - \alpha \le m < M$, each machine processes either a whole bag, or two divided bags comes from the same dividable batch, or a divided bag. Then if $\beta < \gamma$ or $\varepsilon \le \alpha$, by Lemma 3(iii)(iv) and (3),

$$|C_i^A| \le \gamma + 1 \le \frac{4}{3}\left\lceil \frac{n}{M} \right\rceil + 1 \le \frac{4}{3}\left(\left\lceil \frac{n}{M} \right\rceil + 1\right) \le \frac{4}{3}\left\lceil \frac{n}{M-1} \right\rceil \le \frac{4}{3}\left\lceil \frac{n}{m} \right\rceil = \frac{4}{3}C^*(n,m).$$

Otherwise, by Lemmas 3(iii)(iv) and 2(iii),

$$|C_i^A| \le \gamma + 2 = \frac{4}{3}\left(\left\lceil \frac{n}{M} \right\rceil + 1\right) \le \frac{4}{3}\left\lceil \frac{n}{M-1} \right\rceil \le \frac{4}{3}\left\lceil \frac{n}{m} \right\rceil = \frac{4}{3}C^*(n,m).$$

□

Through Lemmas 7, 8 and 9, we have finished the proof of our main result.

Acknowledgments. This study was funded by National Natural Science Foundation of China (12071427).

References

1. Eberle, F., Hoeksma, R., Megow, N., Nölke, L., Schewior, K., Simon, B.: Speed-robust scheduling: sand, bricks, and rocks. Math. Program. **197**(2), 1009–1048 (2023)
2. Epstein, L.: A survey on makespan minimization in semi-online environments. J. Sched. **21**(3), 269–284 (2018). https://doi.org/10.1007/s10951-018-0567-z
3. Graham, R.L.: Bounds on multiprocessing timing anomalies. SIAM J. Appl. Math. **17**, 416–429 (1969)
4. Pruhs, K., Sgall, J., Torng, E.: Online scheduling. In: Handbook of Scheduling: Algorithms, Models, and Performance Analysis. CRC, Boca Raton (2004)
5. Sahni, S.: Algorithms for scheduling independent tasks. J. ACM **23**, 116–127 (1976)
6. Stein, C., Zhong, M.: Scheduling when you do not know the number of machines. ACM Trans. Algor. **16**, Article No. 9 (2019)
7. Tan, Z., Zhang, A.: Online and semi-online scheduling. In: Pardalos, P.M., Du, D.-Z., Graham, R.L. (eds.) Handbook of Combinatorial Optimization, pp. 2191–2252. Springer, New York (2013). https://doi.org/10.1007/978-1-4419-7997-1_2

Faster Algorithms for Grid and Layered Drawings of Plane 3-Trees

Ahmed Hossain[1], Md. Hasanul Islam[1], Debajyoti Mondal[2](✉), and Md. Saidur Rahman[1]

[1] Graph Drawing and Information Visualization Laboratory, Department of Computer Science and Engineering (CSE), Bangladesh University of Engineering and Technology (BUET), Dhaka, Bangladesh
{1805016,1805045}@ugrad.cse.buet.ac.bd, saidurrahman@cse.buet.ac.bd
[2] Department of Computer Science, University of Saskatchewan, Saskatoon, Canada
d.mondal@usask.ca

Abstract. A grid drawing of a plane graph G is a planar drawing of G such that each vertex is mapped to an integer grid point and each edge is mapped to a straight line segment with the endpoints determined by its end vertices. The area of a grid drawing is the area of the smallest axis-aligned rectangle that encloses the drawing. A k-layer drawing of G is a planar drawing on a set L of k horizontal lines such that each vertex is mapped to a point on L and each edge is mapped to a straight line segment. A minimum area (minimum-layer) drawing of G is a straight-line grid (layered) drawing of G where the area (number of layers) is minimized. In this paper we examine the grid and layered drawing of a well-studied class of planar graphs called plane 3-tree. Given a n-vertex plane 3-tree, one can compute a straight line drawing with minimum number of layers or area in $O(n^5)$ and $O(n^9 \log n)$ time, respectively [Mondal *et al.*, JGAA, 15(2), 2011]. Here we improve the running times to $O(n^3)$ and $O(n^7 \log n)$, respectively. We then give a fast approximation algorithm for computing minimum-layer drawings for a nontrivial subclass of plane 3-trees.

1 Introduction

A *straight-line grid drawing* of a plane graph G is a planar drawing where every vertex is drawn as an integer grid point and every edge is drawn as a straight line segment between its corresponding end vertices. Throughout the paper we assume that the input is a plane graph (i.e., a fixed combinatorial planar embedding) and the output drawing must respect the given embedding. The *area* of a drawing is the area of its axis-aligned bounding box (Fig. 1(a)–(b)). Drawing plane graphs on a small area is a well-studied topic in graph drawing [21]. The objective of minimizing the drawing area is often motivated by the need for designing compact VLSI circuit layouts and for information visualization such as creating entity–relationship diagrams of software architecture or displaying phylogenetic trees, where the constraint on area is imposed naturally by the size of the display screen.

D. Du et al. (Eds.): COCOA 2024, LNCS 15434, pp. 44–55, 2025.
https://doi.org/10.1007/978-981-96-4445-2_4

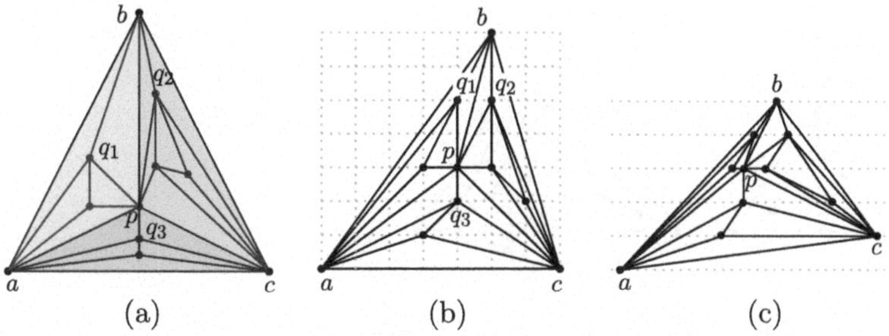

Fig. 1. (a) A plane 3-tree G. Its drawing (b) on a 7×7 grid and (c) on 6 layers.

Researchers have also attempted minimizing and approximating one dimension (i.e., height or, equivalently width), where the other dimension is unbounded [1,2,15,20]. This can be seen as a drawing of a graph on k horizontal lines or layers. Formally, a *k-layer drawing* of G is a planar straight-line drawing on a set L of k horizontal lines (i.e., *layers*) such that each vertex is mapped to a point (not necessarily a grid point) on L and each edge is drawn with a straight line segment (Fig. 1(c)).

In 1990, de Fraysseix et al. [5] and Schnyder [19] independently proved that every n-vertex planar graph admits a straight-line drawing with at most $(2n - 4) \times (n-2)$ and $(n-2) \times (n-2)$ area, respectively. Chrobak and Nakano [4] focused on minimizing the width (equivalently, height) of the drawing and gave an algorithm to find grid drawings in $\lfloor 2(n-1)/3 \rfloor \times (4\lfloor 2(n-1)/3 \rfloor - 1)$ area, which implies that every planar graph can be drawn on $2n/3$ layers. A lower bound of $\lfloor 2(n-1)/3 \rfloor \times \lfloor 2(n-1)/3 \rfloor$ area is determined by a class of graphs called nested triangles graphs [6,21]. The same graph class provides a lower bound of $2\lfloor (n-1)/3 \rfloor$ layers for layered drawings. For 4-connected planar graphs, Felsner [10] gave an upper bound of $\sqrt{2n}$ layers.

The problem of finding a minimum-area grid drawing of a plane graph is NP-hard [3,14]. Biedl [3] gave a polynomial-time algorithm for finding minimum-area grid drawings for planar graphs when both treewidth and face degrees are bounded, and showed that the problem becomes NP-hard if the bounded treewidth or bounded face degree restriction is dropped. Finding fast approximation algorithms appears to be challenging. Currently, a linear-time algorithm for computing minimum-layer drawings is known for trees [15], and a linear-time 2-approximation algorithm is known for maximal outerplanar graphs [1].

This paper examines a well-studied planar graph class called plane 3-trees. A *plane 3-tree* is a triangulated plane graph that can be constructed by starting with a triangle, and then repeatedly adding a vertex to some inner face of the current graph and triangulating that face. Figure 1(a) illustrates a plane 3-tree where the initial triangle is abc, the first vertex inserted is p, and the vertices inserted afterward into the resulting inner faces are q_1, q_2, q_3 and so on. Mondal *et al.* [16] showed that minimum-area grid drawings and minimum-layer drawings can be computed in polynomial time for the class of plane 3-trees. A rich body

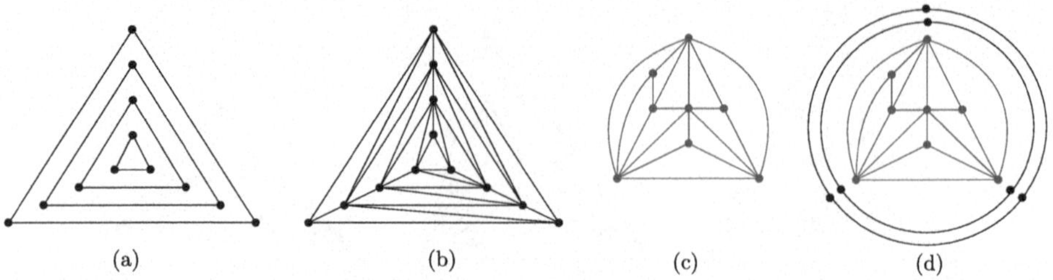

Fig. 2. (a)–(b) A nested sequence of triangles and its edge augmentation to a plane 3-tree. (c)–(d) Illustration for obtaining a general lower bound from a fixed size plane 3-tree.

of research attempts to compute upper bounds on various aesthetic parameters when drawing plane 3-trees [8,9,12,13,18].

In this paper we exclusively focus on computing grid and layered drawings for plane 3-trees. Plane 3-trees are interesting structures in the context of grid and layered drawings because the best-known lower bounds on the drawing area and the number of layers are determined by this graph class [6,11,16]. For example, a nested sequence of $n/3$ triangles is known to give a lower bound of $(2n/3) \times (2n/3)$ area or $(2n/3)$ layers (Fig. 2(a)). Such a sequence can be augmented to obtain a plane 3-tree (Fig. 2(b)). Furthermore, if we can establish a lower bound for a plane 3-tree with a fixed number of vertices, then one can make it nested inside a sequence of triangles to generalize the lower bound for graphs of arbitrary size. For example, the plane 3-tree of Fig. 2(c) is known to require a drawing area of 3×8 or 4×6 grid, and we can nest the graph into a sequence of triangles to obtain a general area bound (see e.g., [16] for the details). Therefore, computing the exact area or number of layers for drawing plane 3-trees is an interesting avenue for research. To this end, the high running time of the currently known algorithm is prohibitive.

Recently, Mondal and Rahman [17] have shown how every planar triangulation (or 3-connected graph) can be encoded into an edge-labelled plane 3-tree, which further justifies the importance of investigating this class of graphs.

Our Contribution: In this paper We give $O(n^3)$- and $O(n^7 \log n)$-time algorithms to find minimum-layer and minimum-area drawings for plane 3-trees, which improves the $O(n^5)$- and $O(n^9 \log n)$-time algorithms of Mondal *et al.* [16], respectively. Similar to Mondal *et al.* [16] we use dynamic program, but carefully leverage the idea of geometric invariants to achieve this significant speed up.

We then consider a non-trivial subclass of plane 3-trees, i.e., plane 3-trees with lineal distance γ. We show that for such plane 3-trees, one can find an $O(\gamma)$-approximation for minimum-layer drawing in linear time. This latter result is motivated by the structure of random plane 3-trees, which are likely to have small lineal distance.

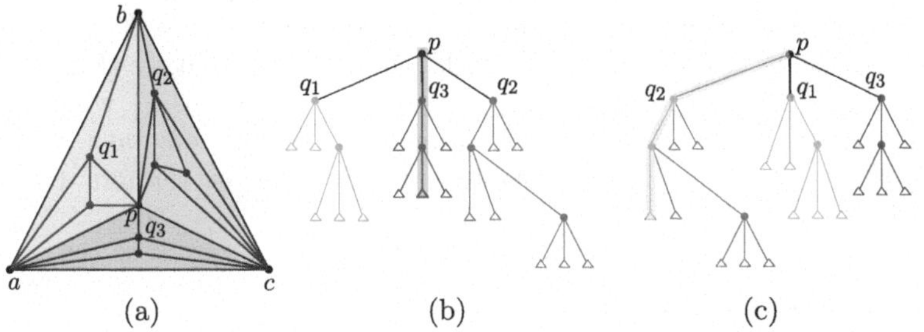

Fig. 3. (a) A plane 3-tree G with the representative vertex p. (b) A representative tree $\mathcal{T}$ of G with lineal distance three (determined by the path highlighted in blue). (c) Illustration for the lineal distance for a different ordering of the children around p.

2 Preliminaries

A *planar graph* is a graph that can be drawn in $\mathbb{R}^2$ without any edge crossing. A *plane graph* is a planar graph with a fixed combinatorial embedding. A plane graph delimits the plane into some connected regions called *faces*. The unbounded region is the *outer face* and the other regions are the *inner faces*. The vertices incident to the outer face are called *outer vertices* and the other vertices are called *inner vertices*.

Let G_{abc} be a n-vertex plane 3-tree with outer vertices a, b, c appearing in clockwise order on the outer face. Then G_{abc} is known to admit the following properties [16]. G_{abc} contains a unique inner vertex p that is incident to all three outer vertices a, b, c (Fig. 3(a)). The vertex p is called the *representative vertex* of G_{abc}. The subgraphs G_{abp}, G_{bcp} and G_{cap} enclosed by the triangles $\Delta_{abp}, \Delta_{bcp}$ and Δ_{cap} are also plane 3-trees. The *representative tree* of G_{abc} is an ordered rooted tree $\mathcal{T}$ (Fig. 3(b)) satisfying the following two conditions:

(a) If $n = 3$, then $\mathcal{T}$ consists of a single vertex.
(b) If $n > 3$, then the root p of $\mathcal{T}$ is the representative vertex of G_{abc} and the subtrees rooted at the three clockwise ordered children q_1, q_2 and q_3 of p in $\mathcal{T}$ are the representative trees of G_{abp}, G_{bcp} and G_{cap}, respectively.

Consider an *ordered upward planar drawing* of $\mathcal{T}$ where children are drawn below their parent maintaining the clockwise order around the parent. We refer to an edge of (u, v) of $\mathcal{T}$ as *left*, *middle*, or *right type* if u is a parent of v and v is a left, middle, or right child of u, respectively. By a *left* (*middle*, or *right*) *path* of $\mathcal{T}$ we denote a path where all edges are left (middle, or right) edges.

The *lineal distance* of G_{abc} is the length of the longest path P in $\mathcal{T}$ where all edges in P are of the same type. Figure 3(b) gives an example where the lineal distance is three and the distance is determined by a middle path. Note that the lineal distances for two different choices of left, right, and middle subtrees around the root may differ by at most one unit. The reason is that the clockwise ordering of the subtrees rooted at an internal node of $\mathcal{T}$ is determined starting at its parent node. However, for the root, we do not have any reference starting point to order its children. For example, if we redraw the tree $\mathcal{T}$ such that q_2, q_1 and q_3 are the left, middle, and right children of p, respectively (Fig. 3(c)), then the highlighted middle path of Fig. 3(b) loses the edge (p, q_3) which is now a right edge. Here the lineal distance is determined by the left path, whereas the ordering structure of the subtrees rooted at q_1, q_2 and q_3 remains the same as that of Fig. 3(b).

3 Layer Minimization

In this section we give an $O(n^3)$-time algorithm to find a layered drawing of a plane 3-tree G_{abc} on the minimum number of layers.

To check whether G_{abc} admits a layered drawing with a, b, c placed at layers a_y, b_y, c_y, respectively, Mondal *et al.* encoded the problem as $F[p, a_y, b_y, c_y]$. This was referred to as a feasibility checking problem. The problem $F[p, a_y, b_y, c_y]$ is set to true if there is a placement of p with $\min\{a_y, b_y, c_y\} < p_y < \max\{a_y, b_y, c_y\}$ such that the subgraphs G_{abp}, G_{bcp} and G_{cap} can be drawn with their outer vertices at layers $\{a_y, b_y, p_y\}, \{b_y, c_y, p_y\}$ and $\{c_y, a_y, p_y\}$, respectively, i.e., if $F[q_1, a_y, b_y, p_y], F[q_2, b_y, c_y, p_y], F[q_3, c_y, a_y, p_y]$ are all true. The reason that this provides a correct solution is that if the subgraphs can be drawn with the given placement of the outer vertices on layers, then the drawings can be stretched horizontally and patched together (keeping the layer numbers of the vertices the same) to obtain a layered drawing for G_{abc}. Specifically, it is straightforward to use the following lemma on the subgraphs to construct the drawing of G_{abc}.

Lemma 1 (Mondal et al. [16]). *Let Γ_{uvw} be a layered drawing of a plane 3-tree G_{uvw} and let Δ_{uvw} be the drawing of the outer face in Γ_{uvw}. Then for any other drawing Δ'_{uvw} of Δ_{uvw} that does not change the layer number of u, v, w, one can find a layered drawing of G_{abc} with Δ'_{uvw} as the outer face.*

Since p, a_y, b_y, c_y are bounded by $O(n)$, an $O(n^4)$ size table suffices to store subproblem solutions. Since each entry can be filled by checking for the $O(n)$ potential placements of the representative vertex, the total running time is $O(n^5)$.

We now describe our approach. Let Γ_{abc} be a minimum-layer drawing of G_{abc}. We guess a and c to be at the bottommost and topmost layers, respectively (Fig. 4(a)). We refer to the remaining outer vertex (i.e., b) as the middle outer vertex. We run the algorithm for all $O(1)$ choices for picking the top and bottom outer vertices among a, b, and c, but that does not change the asymptotic complexity of the algorithm.

Assume that the bottommost layer is 1, i.e., that $a_y = 1$. Since a and c are at the bottommost and topmost layers, respectively, the drawing of the outer face can be described with two parameters: first, the layer difference between b and a, i.e., $(b_y - a_y)$, and second, the position of c. Therefore, to minimize the number of layers, we can iterate b_y from 1 to n, and at each iteration ask for the lowest layer for c (with $b_y \leq c_y$) such that G_{abc} admits a layered drawing. We can then take the minimum over all iterations. We first prove an useful property of the layered drawing of a plane 3-tree.

Lemma 2. *Assume that a plane 3-tree G_{abc} admits a k-layer drawing with $c_y = k$. Then G_{abc} admits a layered drawing for every placement of c at a higher layer without changing the placement of a and b.*

Proof. The case when G_{abc} contains exactly three vertices is straightforward. Otherwise, let p, q_1, q_2 and q_3 be the representative vertices of G_{abc}, G_{abp}, G_{bcp} and G_{cap}, respectively. By induction, G_{bcp} (similarly, G_{cap}) admits a layered drawing for every placement of c at a higher layer without changing the placement of (b and p) and (p and a). Therefore, we can draw G_{bcp} and G_{cap} separately and then merge the drawings with the drawing of G_{abp} using Lemma 1 to obtain the required drawing for G_{abc} (Fig. 4(a)–(d)). □

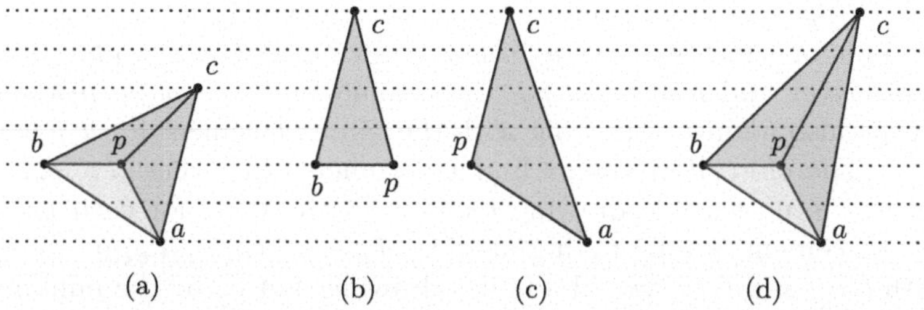

Fig. 4. (a) A drawing of G_{abc} with $c_y = 5$. (b)–(c) Drawings of G_{bcp} and G_{cap} with $c_y = 7$. (d) A new drawing of G_{abc}. (e) Case 1 and (f) Case 2.

We encode the problem of finding the lowest layer for c (given a placement for b), as $L_p[b_y, (a, b)]$. Here p is the representative vertex of G_{abc} and the tuple (a, b) indicates that a is the bottommost outer vertex at layer 1 and b is the middle outer vertex on layer b_y. To find the $L_p[b_y, (a, b)]$, we use the encoding for the drawings of the subgraphs G_{abp}, G_{bcp} and G_{cap} considering two cases depending on whether in the minimum-layer drawing, the layer number of p is higher than the layer number of b or not. We take the minimum over these two cases.

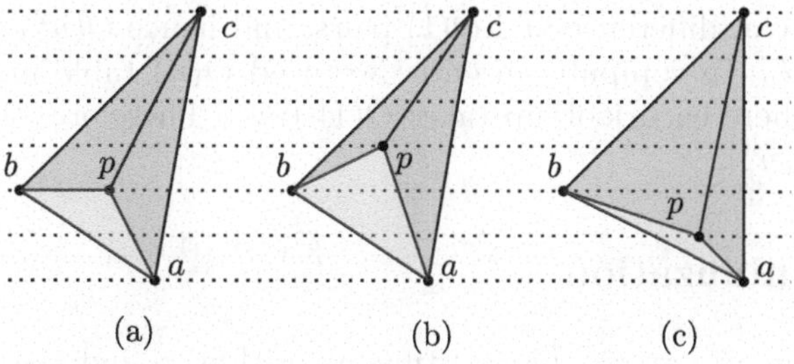

Fig. 5. (a)–(b) Case 1 and (c) Case 2.

Case 1 ($p_y \geq b_y$)**:** Here b and p can be considered to be the middle and top outer vertices of G_{abp} in the drawing of Δ_{abp}, respectively (Fig. 5(a)–(b)). Therefore, $L_{q_1}[b_y, (a, b)]$ would correspond to the lowest possible placement of p that allows for a layered drawing of G_{abp}. By Lemma 2, G_{abp} admits a drawing as long as $p_y \geq L_{q_1}[b_y, (a, b)]$. To find $L_p[b_y, (a, b)]$, we iterate p_y from $L_{q_1}[b_y, (a, b)]$ to n and at each step, find the lowest placement of c by taking $\max(L_{q_2}[p_y - b_y, (b, p)], L_{q_3}[p_y, (a, p)])$, i.e., where both G_{bcp} and G_{cap} would admit a drawing. Note that the term '$-b_y$' is needed as the notation $L_{q_2}[p_y - b_y, (b, p)]$ assumes b to be at layer 1. Let c_y be the minimum over all iterations. We set $L_p[b_y, (a, b)]$ to $\min\{c_y, L_p[b_y, (a, b)]\}$.

Case 2 ($p_y < b_y$)**:** Here p and b can be considered to be the middle and top outer vertices of G_{abp} in the drawing of Δ_{abp}, respectively (Fig. 5(c)). We iterate p_y from 2 to $(b_y - 1)$ and examine $L_{q_1}[p_y, (a, p)]$, which is the lowest possible placement of b for which we have a feasible drawing for G_{abp} by placing a at the bottommost layer and p at layer p_y. By Lemma 2, if $L_{q_1}[p_y, (a, p)]$ is at most b_y, then G_{abp} also admits a drawing with the same placement for a, p where b is placed at b_y. Therefore, if we cannot find a placement of p_y where $L_{q_1}[p_y, (a, p)] \leq b_y$, then there is no feasible drawing for G_{abp}. Otherwise, for each p_y, we find the lowest placement of c by taking $\max(L_{q_2}[b_y - p_y, (p, b)], L_{q_3}[p_y, (a, p)])$, i.e., where both G_{bcp} and G_{cap} would admit a drawing. Let c_y be the minimum over all iterations. We set $L_p[b_y, (a, b)]$ to $\min\{c_y, L_p[b_y, (a, b)]\}$.

The following theorem results from the observation that an $O(n^2)$-size table suffices to store the subproblem solutions, where each entry is filled by checking $O(n)$ candidate placements for the representative vertex.

Theorem 1. *Given a plane 3-tree G with n vertices, one can compute a minimum-layer grid drawing of G in $O(n^3)$ time.*

Proof. We employ a dynamic program and compute the solution to the subproblems using a bottom-up approach. A leaf r of $\mathcal{T}$ represents a triangular face in G_{abc} and two layers are necessary and sufficient to draw a triangle. For each leaf r with outer vertices u, v, w, we set $L_r[i, (\alpha, \beta)]$ to 2, where $0 \leq i \leq n$ and $\alpha, \beta \in \{u, v, w\}$. We then compute the entries for the internal vertices of $\mathcal{T}$ using Case 1 and Case 2 (taking the minimum value obtained from these two cases). There are $O(n)$ entries per vertex because for each vertex we have an index representing a layer difference and $O(1)$ pairs, and hence $O(n^2)$ entries over all the vertices of $\mathcal{T}$. To compute an entry we need $O(n)$ table look-up for Case 1 and Case 2 where each look-up takes $O(1)$ time. Therefore, the overall time complexity is $O(n^3)$. □

4 Area Minimization

In this section we give an $O(n^7 \log n)$ time algorithm to find a straight-line grid drawing of a plane 3-tree on a grid of minimum area.

To check whether G_{abc} admits a grid drawing with a fixed position for a, b, c, Mondal *et al.* encoded the problem as $A[p, a_x^y, b_x^y, c_x^y]$, where r_x^y denotes the x- and y-coordinates of a point r. This was referred to as a feasibility checking problem. The problem $A[p, a_x^y, b_x^y, c_x^y]$ is set to true if there is a placement of p at a grid point interior to Δ_{abc} such that $A[q_1, a_x^y, b_x^y, p_x^y]$, $A[q_2, b_x^y, c_x^y, p_x^y]$ and $A[q_3, c_x^y, a_x^y, p_x^y]$ are true. Although the first parameter of A can take $O(n)$ values and there are 6 coordinates, each bounded by $O(n^2)$, Mondal *et al.* showed that a careful analysis yields a running time of $O(n^9 \log n)$.

We now describe our approach to finding a minimum area drawing. We will use the following lemma.

Lemma 3. *Let Δ_{abc} be a triangle on an integer grid. One can apply at most two flip operations, and then a translation operation to bring a corner of Δ_{abc} at $(0,0)$ such that Δ_{abc} lies on the first quadrant.*

Proof. Without loss of generality assume that a is the leftmost corner (breaking ties arbitrarily). If the $a_y \leq \min\{b_y, c_y\}$, then translating Δ_{abc} to place a at $(0,0)$ would bring Δ_{abc} at the first quadrant. If $a_y \geq \max\{b_y, c_y\}$, then one can flip Δ_{abc} vertically and then translate Δ_{abc} to place a at $(0,0)$. Otherwise, either $\{b_y \leq a_y \leq c_y\}$ or $\{c_y \leq a_y \leq b_y\}$ which we process as follows.

Without loss of generality assume that b is the rightmost corner (breaking ties arbitrarily). If $\{b_y \leq a_y \leq c_y\}$, then one can flip Δ_{abc} horizontally and then translate Δ_{abc} to place b at $(0,0)$. If $\{c_y \leq a_y \leq b_y\}$, then one can flip Δ_{abc} first horizontally and then vertically, and finally, translate Δ_{abc} to place b at $(0,0)$. □

Assume that $f(\Delta_{rst})$ returns a transformation of triangle Δ_{rst} (i.e., three new coordinates for r, s, t using Lemma 3) where one corner is at $(0,0)$ and the other two corners at some points v_x^y and w_x^y. To check whether G_{abc} admits a grid drawing with a fixed position for a, b, c, we ask whether it admits a drawing inside $f(\Delta_{abc})$. We represent this feasibility checking problem as $\Gamma[p, f(\Delta_{abc})]$ and store the solution at a table $D[p, m, u_x^y, v_x^y]$, where m is the mapping of the triangle to $(0,0), u_x^y, v_x^y$. Since p correspond to the vertices of $\mathcal{T}$, $p = O(n)$. For a fixed vertex of $\mathcal{T}$ and for a fixed position of the triangle corresponding to this vertex, $m = O(1)$. We now bound the number of distinct triangles. Without loss of generality assume that the width W of the minimum-area drawing is at most its height H. Since $W \times H \leq n^2$, $W \leq n$. Therefore, we can iterate u_x and v_x from 1 to n. For every u_x and v_x, we can iterate the y-coordinates from 1 to $n^2/\max\{u_x, v_x\}$. Thus the number of distinct candidate triangles is at most

$$\sum_{u_x=1}^{n}\sum_{v_x=1}^{n}\left(\frac{n^2}{\max\{u_x, v_x\}}\right)^2 = O\left(\sum_{u_x=1}^{n}\sum_{v_x=1}^{u_x}\left(\frac{n^2}{u_x}\right)^2\right)$$
$$= O\left(\sum_{u_x=1}^{n}\frac{n^4}{u_x}\right)$$
$$= O(n^4 \log n).$$

We now show that an $O(n^7)$-size table suffices to store the subproblem solutions, but only $O(n^5 \log n)$ entries are relevant, where an entry can be filled by checking $O(n^2)$ placements for the representative vertex.

Theorem 2. *Given a plane 3-tree G with n vertices, one can compute a minimum-area grid drawing of G in $O(n^7 \log n)$ time.*

Proof. For each of the $O(n^4 \log n)$ candidate positions for Δ_{abc}, one can ask whether there is a placement for the representative vertex p inside Δ_{abc} such that the subgraphs G_{abp}, G_{bcp} and G_{cap} can be drawn inside Δ_{abp}, Δ_{bcp} and Δ_{cap}, respectively, i.e., if $\Gamma[q_1, f(\Delta_{abp})], \Gamma[q_2, f(\Delta_{bcp})]$ and $\Gamma[q_3, f(\Delta_{cap})]$ are feasible. This is done using a table lookup in $O(1)$ time per placement of p and $O(n^2)$ time overall.

We use a bottom-up approach using the representative tree to fill the dynamic programming table. The base case (with three vertices) is true as long as the three points of the triangle are non-colinear. We allocate a table $D[p, m, u_x^y, v_x^y]$, where $p = O(n)$, $u_x = O(n)$, $u_x = O(n)$, $u_y = O(n^2)$, $u_y = O(n^2)$, and for a fixed value of (p, u_x^y, v_x^y), $m = O(1)$. Therefore, $O(n^7)$ space and time is enough for the initialization.

Each internal vertex in the tree represents a subgraph that is also a plane 3-tree, and hence the outer face is a triangle. We have $O(n^4 \log n)$ candidate positions for this triangle and hence we need to fill in $O(n^5 \log n)$ entries in total over all the vertices of $\mathcal{T}$. Computing each entry of the table requires $O(n^2)$ time, and hence we obtain a running time of $O(n^7 \log n)$. □

5 Plane 3-Trees with Lineal Distance γ

We now show that for a plane 3-tree G with lineal distance γ, one can find an $O(\gamma)$-approximation for its layered drawing in linear time. The motivation behind computing an approximation with respect to the lineal distance comes from the case of random plane 3-trees. It is straightforward to show that with high probability the lineal distance of a random plane 3-tree, which is constructed by repeatedly inserting a vertex into a randomly chosen inner face, is bounded by $\log n$. Therefore, our result provides insights into the layered drawing of such randomly generated plane 3-trees.

Let $\Delta = (\Delta_1, \Delta_2, \ldots, \Delta_k)$ be an ordered sequence of triangles in G. We call Δ *a set of semi-nested triangles* if for every $1 \leq i < j \leq k$, the interior of Δ_i lies interior to Δ_j and Δ_i shares at most one vertex with Δ_j. Let $\mathcal{T}$ be a representative tree of G and consider an upward planar drawing of $\mathcal{T}$ as defined in Sect. 2. Let v be a vertex in $\mathcal{T}$. A *spider rooted at an internal vertex* v is a subtree of in $\mathcal{T}$ with the vertex v and the maximal left, middle, and right paths starting at v. Consider a tree $\mathcal{T}'$ with each vertex a' mapped to a distinct spider $S(a')$ in $\mathcal{T}$, where the decomposition of $\mathcal{T}$ into spiders is computed by recursively removing the spiders starting from the root. Here a vertex u' is a parent of a vertex v' (in $\mathcal{T}'$) if $S(u')$ contains the parent of the topmost vertex of $S(v')$. We refer to $\mathcal{T}'$ as a *folded representative tree* (Fig. 6(a)–(c)). The following lemma

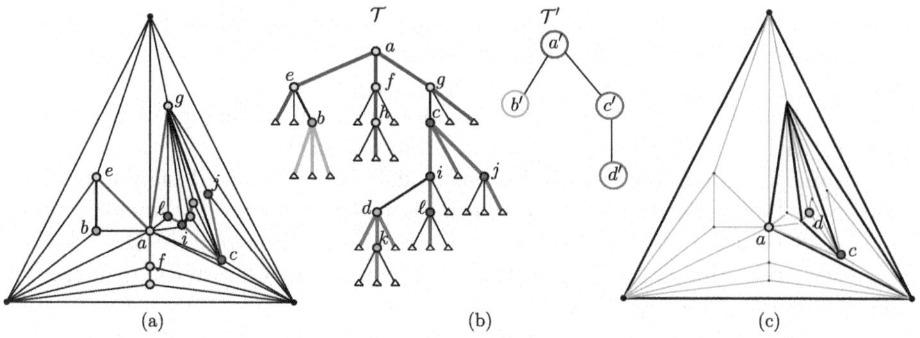

Fig. 6. (a) A plane 3-tree G. (b) The representative tree $\mathcal{T}$, where the spiders are highlighted in blue, orange, red, and green. The lineal distance of $\mathcal{T}$ is 4 (determined by the left path starting at g, i.e., the longest path $(g, c, i, d, \ldots)$ of the same type edges). $\mathcal{T}'$ is the corresponding folded representative tree. (c) Semi-nested triangles corresponding to the path (a', c', d') in $\mathcal{T}'$.

provides a lower bound on the number of layers whose proof is omitted due to space constraints.

Lemma 4. *Let $\mathcal{T}'$ be a folded representative tree of a plane 3-tree G and let $\mathcal{P}'$ be a path in $\mathcal{T}'$ from the root to a leaf. Then G contains at least $\Omega(|\mathcal{P}'|)$ semi-nested triangles and any layered drawing of G requires $\Omega(|\mathcal{P}'|)$ layers.*

The *depth* ρ of a plane 3-tree G is the number of vertices on the longest path from the root to a leaf in its representative tree. Hossain et al. [12] showed every plane 3-tree of depth ρ can be drawn on $(\rho + 2)$ layers in linear time. If the lineal distance is $O(\gamma)$, then the depth of G is at most $O(\gamma d')$, where d' is the depth of its folded representative tree $\mathcal{T}'$. The reason is that each spider can only contribute to $O(\gamma)$ amount towards the depth of the representative tree (equivalently, to the depth of G). Hence G can be drawn in $O(\gamma d')$ layers. By Lemma 4, G requires $\Omega(d')$ layers. Therefore, we obtain the following theorem.

Theorem 3. *Given a plane 3-tree G with lineal distance γ, we can find an $O(\gamma)$-approximation for minimum-layer drawing in linear time, i.e., we can compute a layered drawing in linear time where the number of layers is within an $O(\gamma)$ factor of the optimum.*

6 Directions for Future Research

Given a n-vertex plane 3-tree, we show how one can compute a straight line drawing with minimum number of layers or area in $O(n^5)$ and $O(n^9 \log n)$ time, respectively. We then give a fast approximation algorithm for computing minimum-layer drawings for a nontrivial subclass of plane 3-trees, e.g., plane 3-trees with bounded lineal distance. Here are some natural questions that arise from our work.

- Can we further improve the running time of our algorithms for computing minimum-layer or minimum-area drawings?
- Can we prove a non-trivial lower bound on running time?
- Can we find a constant-factor approximation for minimizing area for plane 3-trees with bounded lineal distance in polynomial time?
- A parameter called 'pathwidth' has previously been used to design algorithms for layered drawings [7,20]. Note that a plane 3-tree with $O(1)$ lineal distance may still contain $\Omega(n)$ semi-nested triangles and thus require $\Omega(n)$ layers. Can we find a linear-time algorithm to compute an $O(k)$-layer drawing for plane 3-trees of pathwidth k?

Acknowledgements. The work of Debajyoti Mondal is supported in part by NSERC. The work of Md. Saidur Rahman is done under RISE Internal Research Grant 2021-01-06 and partly supported by "Basic Research Grant" of BUET.

References

1. Biedl, T., Demontigny, P.: A 2-approximation for the height of maximal outerplanar graph drawings. In: WADS 2017. LNCS, vol. 10389, pp. 145–156. Springer, Cham (2017). https://doi.org/10.1007/978-3-319-62127-2_13
2. Biedl, T., Sheth, M.: Drawing Halin-graphs with small height. J. Graph Algor. Appl. **26**(1), 473–502 (2022)
3. Biedl, T.: On area-optimal planar graph drawings. In: Esparza, J., Fraigniaud, P., Husfeldt, T., Koutsoupias, E. (eds.) ICALP 2014. LNCS, vol. 8572, pp. 198–210. Springer, Heidelberg (2014). https://doi.org/10.1007/978-3-662-43948-7_17
4. Chrobak, M., Nakano, S.I.: Minimum-width grid drawings of plane graphs. Comput. Geom. Theory Appl. **11**(1), 29–54 (1998)
5. de Fraysseix, H., Pach, J., Pollack, R.: How to draw a planar graph on a grid. Combinatorica **10**(1), 41–51 (1990)
6. Dolev, D., Leighton, T., Trickey, H.: Planar embedding of planar graphs. Adv. Comput. Res. **2**, 147–161 (1984)
7. Dujmović, V., et al.: On the parameterized complexity of layered graph drawing. Algorithmica **52**(2), 267–292 (2008)
8. Durocher, S., Mondal, D.: On balanced -contact representations. In: Wismath, S., Wolff, A. (eds.) GD 2013. LNCS, vol. 8242, pp. 143–154. Springer, Cham (2013). https://doi.org/10.1007/978-3-319-03841-4_13
9. Durocher, S., Mondal, D.: Drawing planar graphs with reduced height. J. Graph Algor. Appl. **21**(4), 433–453 (2017)
10. Felsner, S.: 4-connected triangulations on few lines. J. Comput. Geom. 11(1), 476–492 (2020). https://doi.org/10.20382/JOCG.V11I1A18
11. Frati, F., Patrignani, M.: A note on minimum-area straight-line drawings of planar graphs. In: Hong, S.-H., Nishizeki, T., Quan, W. (eds.) GD 2007. LNCS, vol. 4875, pp. 339–344. Springer, Heidelberg (2008). https://doi.org/10.1007/978-3-540-77537-9_33
12. Hossain, M.I., Mondal, D., Rahman, M.S., Salma, S.A.: Universal line-sets for drawing planar 3-trees. J. Graph Algor. Appl. **17**(2), 59–79 (2013)
13. Jelínek, V., Jelínková, E., Kratochvíl, J., Lidický, B., Tesar, M., Vyskocil, T.: The planar slope number of planar partial 3-trees of bounded degree. Graphs Comb. **29**(4), 981–1005 (2013)

14. Krug, M., Wagner, D.: Minimizing the area for planar straight-line grid drawings. In: Hong, S.-H., Nishizeki, T., Quan, W. (eds.) GD 2007. LNCS, vol. 4875, pp. 207–212. Springer, Heidelberg (2008). https://doi.org/10.1007/978-3-540-77537-9_21
15. Mondal, D., Alam, M.J., Rahman, M.S.: Minimum-layer drawings of trees. In: Katoh, N., Kumar, A. (eds.) WALCOM 2011. LNCS, vol. 6552, pp. 221–232. Springer, Heidelberg (2011). https://doi.org/10.1007/978-3-642-19094-0_23
16. Mondal, D., Nishat, R.I., Rahman, M.S., Alam, M.J.: Minimum-area drawings of plane 3-trees. J. Graph Algor. Appl. **15**(2), 177–204 (2011)
17. Mondal, D., Rahman, M.S.: On the 3-tree core of plane graphs. In: Proceedings of the 18th Annual Conference on Theory and Applications of Models of Computation (TAMC). LNCS, vol. 14637. Springer, Heidelberg (2024). https://doi.org/10.1007/978-981-97-2340-9_13
18. Nishat, R.I., Mondal, D., Rahman, M.S.: Visibility drawings of plane 3-trees with minimum area. Math. Comput. Sci. **5**(1), 119–132 (2011)
19. Schnyder, W.: Embedding planar graphs on the grid. In: Proceedings of the 1st Annual ACM-SIAM Symposium on Discrete Algorithms (SODA), San Francisco, California, USA, pp. 138–148. ACM (1990)
20. Suderman, M.: Pathwidth and layered drawing of trees. Int. J. Comput. Geom. Appl. **14**(3), 203–225 (2004)
21. Tamassia, R. (ed.): Handbook of Graph Drawing and Visualization. CRC Press, Boca Raton (2013)

Reinforcement Learning for Airline Continuous Dynamic Pricing

Zhicheng Yao and Wenguo Yang(✉)

School of Mathematical Sciences, University of Chinese Academy of Sciences, Beijing 100049, China
yaozhicheng21@mails.ucas.ac.cn, yangwg@ucas.ac.cn

Abstract. The airline industry's New Distribution Capability has spurred interest in continuous dynamic pricing, where fares aren't restricted to fixed price points. Forecasting demand across a continuous price range is difficult, and no optimal solution algorithm exists for general demand functions. This study employs the Soft Actor-Critic (SAC) algorithm for continuous action spaces in reinforcement learning to tackle these challenges in continuous dynamic pricing. In the context of dynamic pricing, we found that due to its inherent model limitations, the Soft Actor-Critic algorithm requires a substantial number of training iterations to address the issue of local convergence to the boundary when learning optimal actions near the boundary. Therefore, utilizing the structure of the existing approximate solutions, we propose the Reward Shaping Soft Actor-Critic (RSAC) algorithm, which is based on a penalty for low prices. The experimental results indicate that RSAC significantly improved the convergence speed and performance of the SAC algorithm.

Keywords: airline dynamic pricing · revenue management · reinforcement learning · Soft Actor-Critic · Continuous pricing

1 Introduction

Dynamic pricing is generally considered a part of revenue management, which originated and was first applied in the airline business in the 1980s. After the 1970s, with the abolition of aviation regulations and the opening up of the airline market, airlines were finally able to determine their own prices and flight schedules. Following this, Revenue Management Systems (RMSs) have undergone rapid development, with seat control and price selection theories based on demand forecasting being widely applied in airline pricing models, resulting in significant revenue enhancements. In order to maximize revenue and profit, airlines began to study how to select prices based on limited seat availability and sales time to maximize total revenue.

Traditional RMS methods have three distinct drawbacks:

- Parametric demand models assumptions for customer behavior: Customer behavior is divided into two stages, arrival and purchase. Customer arrivals

D. Du et al. (Eds.): COCOA 2024, LNCS 15434, pp. 56–68, 2025.
https://doi.org/10.1007/978-981-96-4445-2_5

are assumed to follow certain specified distributions (e.g., Poisson distribution), and whether a customer purchases a ticket after arrival depends on their Willingness to Pay (WTP).
- Curse of dimensionality: When applied to the joint pricing problem of multiple fare classes, RMS methods must contend with an exponential increase in the number of states as the number of classes increases.
- Dependent on historical data, but the information obtained solely from historical data is often insufficient to describe the future.

Reinforcement learning, a model-free machine learning approach, effectively addresses dynamic pricing challenges by learning policies to maximize returns in environmental interactions, approximating value functions to combat dimensionality issues. Traditional research in dynamic pricing, or Assortment Optimization, focused on discrete price points, but with IATA's NDC, airlines can now offer prices in a continuous range. To adapt, we propose using reinforcement learning for continuous action spaces and introduce an enhanced algorithm that speeds up convergence for continuous dynamic pricing.

The main contributions are summarized as follows:

- Continuous pricing algorithms
  We have introduced Soft Actor-Critic algorithms adapted for continuous action spaces, overcoming the computational challenges faced with continuous action intervals and thereby achieving continuous dynamic pricing.
- Reward Shaping method
  We propose a reward shaping method specifically tailored for airline dynamic pricing based on the structural properties of dynamic pricing. It expands the difference between the action-value functions of the optimal strategy and the boundary actions by penalizing the rewards for low-price actions, thereby helping the agent to quickly learn the optimal strategy.

The structure of this paper is as follows: Sect. 2 reviews past research. Section 3 first describes the dynamic pricing problem, as well as the dynamic programming method used for optimizing discrete pricing. Section 4 introduces the Soft Actor-Critic algorithm and proposes the Reward Shaping Soft Actor-Critic algorithm. Section 5 is dedicated to presenting the detailed results of the numerical experiments. Finally, Sect. 6 concludes the study and offers future research directions.

2 Literature Review

We review the existing research in two parts: the first part is dedicated to traditional pricing methods, and the second part focuses on reinforcement learning approaches.

2.1 Traditional Methods

The study of dynamic pricing can be traced back to Gallego and Van Ryzin [1], who addressed the dynamic pricing problem for single flights and single products by introducing what would later be termed the fluid model approach. They provided asymptotically optimal fixed-price heuristics for the scenario where revenue is a concave function of demand and the demand function is completely known. Subsequently, Gallego and Van Ryzin [2] extended their research to the network level.

Belobaba [3] introduced the Expected Marginal Seat Revenue (EMSRb) heuristic, which employs a seat protection mechanism. Lautenbacher and Stidham [4] employed dynamic programming to devise a dynamic pricing method.

Following the establishment of the mathematical programming and dynamic programming frameworks, the problem of customer demand forecasting for dynamic pricing has been extensively studied.

Talluri and van Ryzin [5] employed methods such as Ordinary Least Squares (OLS) or Maximum Likelihood Estimation (MLE) to predict the parameters of unknown demand models using historical databases. Kumar et al. [6] introduced a combined MLE and EM approach. Otero and Akhavan-Tabatabaei [7] used phase-type distributions, which offer a better fit, to model the distribution of customer arrival times and purchase probabilities. Akçay [8] el al. introduced a comprehensive consumer choice model based on linear stochastic consumer utility, yielding monotonicity results for vertically differentiated products. Selçuk et al. [9] examined optimal strategies for four distinct price-demand relationships.

2.2 Reinforcement Learning

The model-free nature of reinforcement learning algorithms and their ability to interact directly with the environment to learn optimal policies make RL highly suitable for industries like aviation, where demand is complex and variable.

Gosavii et al. [10] introduced the λ-SMART algorithm to solve the single-flight dynamic pricing problem, and the results showed that their method significantly outperformed the widely used EMSR-b nesting algorithm in the industry. Gosavi [11] present a Reinforcement Learning (RL) algorithm based on policy iteration for solving average reward Markov and semi-Markov decision problems. Lawhead and Gosavi [12] addressed the issue of excessive values of the actor in the Actor-critic algorithm, which can lead to overflow, by improving the iterative steps.

Bondoux et al. [13] applied Deep Q-Learning to dynamic pricing and used RMSprop for supervised learning to initialize and accelerate the training process of the DQN. In their experiments, they investigated the changes in revenue under competitive environments.

Jo et al. [14] ensured the Markov property of the pricing process in the presence of patient customers by incorporating historical pricing into the state space. Zhu et al. [15] addressed the Multi-Flight Dynamic Pricing (MFDP) problem by designing an innovative utility function for the Multinomial Logit (MNL) model.

They implemented several reinforcement learning algorithms (DQN, A2C, PPO, TRPO) on the pricing problem.

The reinforcement learning algorithms for dynamic pricing mentioned above have shown superior performance to traditional methods in experiments. However, these methods are all discrete pricing strategies, and there is a lack of research on reinforcement learning algorithms for continuous dynamic pricing.

3 Model Formulation

Before describing the model in detail, we assume that the airline operates in a monopolistic market, free from competition. It is important to emphasize that this study considers continuous pricing, where the set of prices is a continuous interval rather than a set of pre-defined discrete price points. The airline's focus is on maximizing revenue for a single flight and a single class of tickets, without taking into account issues such as overbooking or cancellations. Let's now consider a single-resource optimization problem with finite capacity C and a finite time horizon T. At time t (where $t = 0, 1, \ldots, T{-}1$), we need to select a positive number a_t as the ticket price based on the current state s_t, with the goal of maximizing total revenue. Here, s_t represents the current remaining number of tickets, t is the time that has elapsed, and T denotes the departure time.

3.1 MDP

If we assume that customer demand is solely dependent on the current price, we can represent the dynamic pricing problem as a Markov Decision Process (MDP). In the context of a Markov Decision Process (MDP), a finite tuple typically refers to a well-defined sequence of elements that together specify a component of the MDP. For the dynamic pricing problem, a key tuple is the one that defines the MDP itself, which consists of the following five components:

- A finite set of states $\mathcal{S}$: Each state s_t is typically represented the remaining number of tickets s_t at the time t.
- A infinite set of actions $\mathcal{A}$: This includes all the possible prices that the airline can set for the tickets.
- A transition probability function $\mathcal{P}$: This function describes the probabilities of transitioning from one state to another when a particular action is taken. It involves modeling how likely customers are to purchase tickets at different prices.
- A reward function $\mathcal{R}$: This function specifies the immediate reward that the airline receives for taking an action in a given state s_t. It equals the number of tickets sold at time t multiplied by the ticket price p_t, that is $r_t{=}a_t{\cdot}(s_t{-}s_{t+1})$.
- A discount factor γ: This factor is used to discount future rewards, In the context of dynamic pricing problems, the discount factor γ is often set to 1.

Together, these components form the MDP tuple: MDP $= <\mathcal{S}\ ,\mathcal{A},\ \mathcal{P},\ \mathcal{R},\ \gamma>$.

3.2 DP for RMS

Traditional revenue management methods have to assume that customer behavior satisfies certain specific models. We consider a widely used assumption in this regard. At time t, the number of arriving customers follows a Poisson distribution with mean λ_t (We use $\mathcal{P}(\lambda)$ to represent a Poisson distribution with a mean of λ):

$$\begin{aligned} u_t(i) &= \Pr(i \text{ arrivings at time } t) \\ &= \Pr\{X = i | X \sim \mathcal{P}(\lambda_t)\} \\ &= \frac{\lambda_t^i e^{-\lambda_t}}{i!}. \end{aligned} \tag{1}$$

And the arriving customers purchase tickets with a probability that is an exponential function of the negative of the price f, i.e.:

$$\begin{aligned} p_t &= \Pr(\text{booking at time t}) \\ &= min(e^{(-\beta_t(\frac{f}{f_0}-1))},\ 1), \end{aligned} \tag{2}$$

β_t is known as the price sensitivity coefficient, which represents the degree to which customers' willingness to pay is influenced by price changes. And f_0 is a special price point, if the published price is below this point, then arriving customers will definitely purchase a ticket.

Once we have determined the transition probabilities, we can obtain the optimal state-value function $V_t^*(s_t)$ and the optimal policy $\pi^*(a_t|s_t)$ by solving the Bellman equation, which is:

$$V_t^*(s_t) = \max_{a_t \in \mathcal{A}}\{\sum_{i=0}^{s_t} \Pr(i \text{ bookings} \mid a_t)(i \cdot a_t + V_{t+1}^*(s_t - i))\}. \tag{3}$$

When all the tickets are sold out or the departure time arrives, the value of the state is 0. Therefore, the boundary condition is:

$$V_T^*(s_t) = 0, \tag{4}$$

and

$$V_t^*(0) = 0. \tag{5}$$

If the action set $\mathcal{A}$ is finite, we can solve the Bellman equation using dynamic programming. However, when the action set is a continuous interval, dynamic programming methods cannot easily find the optimal solution. Therefore, to achieve continuous pricing, it is necessary to overcome this difficulty.

4 Continuous Pricing Algorithm

4.1 Soft Actor-Critic

The Soft Actor-Critic algorithm falls under the category of maximum entropy reinforcement learning. It incorporates the entropy of the policy π as a regularization term into the learning objective, which ensures that while maximizing cumulative returns, the learned policy maintains a certain level of randomness:

$$\pi^* = \arg\max_{\pi} \mathbb{E}_{\pi}\left[\sum_t r(s_t, a_t) + \alpha H(\pi(\cdot \mid s_t))\right]. \tag{6}$$

Here, the entropy $H(X)$ is a measure of the randomness of a random variable X, which is defined as:

$$H(X) = \mathbb{E}_{x \sim p}[-\log p(x)]. \tag{7}$$

Due to the inclusion of a regularization term in the objective function, the state-value function is accordingly modified to:

$$V(s_t) = \mathbb{E}_{a_t \sim \pi}[Q(s_t, a_t) - \alpha \log \pi(a_t|s_t)]. \tag{8}$$

To be applicable to continuous action and state spaces, the SAC algorithm uses two neural networks to approximate the action-value function $Q_{\theta_1}(s_t, a_t)$,$Q_{\theta_2}(s_t, a_t)$ and one neural network to output the mean and variance of a Gaussian distribution as the action distribution $\pi_\phi(a_t|s_t)$. To make the policy function differentiable, the reparameterization trick is employed, which involves sampling from the standard Gaussian distribution $\mathcal{N}$, multiplying the sample by the network's output variance, and adding the mean. The algorithm employs a method of stochastic gradient descent to update the parameters θ_1,θ_2,ϕ,α. The loss function of the action-value network is constructed to minimize the soft Bellman residual, which is:

$$J_Q(\theta) = \mathbb{E}_{(s_t,a_t)\sim\mathcal{D}}[\frac{1}{2}(Q_\theta(s_t, a_t) - (r(s_t, a_t) + \gamma \mathbb{E}_{s_{t+1}\sim p}[V_{\bar{\theta}}(s_{t+1})]))^2]. \tag{9}$$

The loss function for the policy network parameters is derived from the Kullback-Leibler divergence, and the SAC algorithm proposes an automatic adjustment method for the regularization term coefficient α. We present these two loss functions directly below, with the specific details available in the referenced Haarnoja [16].

$$J_\pi(\phi) = \mathbb{E}_{s_t\sim\mathcal{D},\varepsilon_t\sim\mathcal{N}}[\alpha \log(\pi_\phi(f_\phi(\varepsilon_t; s_t)|s_t) - Q_\theta(s_t, f_\phi(\varepsilon_t; s_t)))], \tag{10}$$

$$L(\alpha) = \mathbb{E}_{(s_t,a_t)\sim\mathcal{D}}[-\alpha \log \pi(a_t|s_t) - \alpha\mathcal{H}_0]. \tag{11}$$

ε_t is a noise vector sampled from a standard Gaussian distribution, $f_\phi(\varepsilon_t; s_t)$ represents the specific action output based on the current state and the sampled vector, and $\mathcal{H}_0$ is the minimum threshold for the entropy of the policy.

4.2 Reward Shaping

Assuming that customer's arrivals are related to the price and they will purchase a ticket upon arrival. The customer's demand-rate is a monotonically decreasing function of the price, that is, $\lambda = f(p)$. In the research by Gallego and Van

Ryzin [1], they derived an approximation of the optimal revenue by the fluid model under the assumption that the demand-rate function is known:

$$T \times (\max\{p^0, p^c\}) f(\max\{p^0, p^c\}) = T \times (\min\{\lambda^0, \lambda^c\}) f^{-1}(\min\{\lambda^0, \lambda^c\}). \quad (12)$$

p^0 is the price that maximizes revenue ($p^o = \text{argmax}_p pf(p)$), and p^c is the clearance price that ensures the initial inventory is sold out throughout the entire sales period, $C = T\lambda^c$, $p^c = f^{-1}(\lambda^c)$.

During the sales process, the true demand-rate function $f(\cdot)$ is unknown, making it difficult to determine the exact value of λ^0. However, we can easily determine λ^c, as $C = T\lambda^c$. The assumption that the demand rate monotonically decreases with respect to price allows the correctness of the set price to be reflected by the sales rate. Since $\min\{\lambda^0, \lambda^c\} \leq \lambda^c$, the optimal demand-rate should not exceed the clearance rate. If the number of tickets sold exceeds the clearance rate λ_t^c, it will lead to a decrease in the clearance rate λ_{t+1}^c at the next time $t+1$, i.e. $\lambda_t^c - \lambda_{t+1}^c > 0$. Therefore, we can choose the penalty term to be $\lambda_t^c - \lambda_{t+1}^c$, such that the shaping of the reward takes the form of:

$$r_{t,\mu} = r_t - \mu(\lambda_t^c - \gamma\lambda_{t+1}^c). \quad (13)$$

μ is an adjustable penalty term coefficient,and γ is discount factor. At time T, since sales are halted, we can consider the clearance rate to be equal to 0. Therefore:

$$\lambda_t^c = \begin{cases} \frac{s_t}{T-t}, t = 0, 1, ..., T-1; \\ 0, t = T. \end{cases} \quad (14)$$

Using a method similar to that of Devlin et al. [17], we can prove that adding a penalty term for low prices does not alter the optimal policy. We assume that the optimal value function for the new Markov Decision Process with the given parameter μ are represented by $V_{t,\mu}^*(s_t)$, respectively. Let us consider the return U experiencing sequence $\bar{s}$ in a discounted framework without shaping:

$$U(\bar{s}) = \sum_{j=0}^{T-1} \gamma^j r_j. \quad (15)$$

And with shaping:

$$\begin{aligned} U_\mu(\bar{s}) &= \sum_{j=0}^{T-1} \gamma^j (r_j - \mu(\lambda_j^c - \gamma\lambda_{j+1}^c)) \\ &= \sum_{j=0}^{T-1} \gamma^j r_j - \mu(\sum_{j=0}^{T-1} \gamma^j \lambda_j^c - \sum_{j=0}^{T-1} \gamma^{j+1} \lambda_{j+1}^c) \\ &= \sum_{j=0}^{T-1} \gamma^j r_j - \mu(\sum_{j=0}^{T-1} \gamma^j \lambda_j^c - \sum_{j=1}^{T} \gamma^j \lambda_j^c) \\ &= U(\bar{s}) - \mu\lambda_0^c + \mu\gamma^T\lambda_T^c \\ &= U(\bar{s}) - \mu\lambda_0^c, \end{aligned} \quad (16)$$

where r_j is the reward received at time j following the sequence $\bar{s}$.

$$\begin{aligned} V_\mu^*(s,a) &= \sum_{\bar{s}} Pr(\bar{s}|s,a)U_\mu(\bar{s}) \\ &= \sum_{\bar{s}} Pr(\bar{s}|s,a)U(\bar{s}) - \mu \sum_{\bar{s}} Pr(\bar{s}|s,a)\lambda_0^c \\ &= \sum_{\bar{s}} Pr(\bar{s}|s,a)U(\bar{s}) - \mu\lambda_0^c \\ &= V^*(s,a) - \mu\lambda_0^c. \end{aligned} \tag{17}$$

Since the difference term λ_0^c in the reward shaping is independent of the action, we can conclude that the optimal policy remains consistent.

However, such reward shaping terms may not be easy to control in experiments. Due to the time-varying factor '$1/(T-t-1)$' in the penalty term, the denominator for the first and last day is different by a factor of T. Such issues make it difficult for us to adjust the value of μ to an appropriate level in experiments.

In the airline scenario, the discount factor γ is typically set to 1, which allows us to directly ignore γ, and we note that:

$$\begin{aligned} \lambda_t^c - \lambda_{t+1}^c > 0 &\Leftrightarrow \frac{s_t}{T-t} - \frac{s_{t+1}}{T-t-1} > 0 \\ &\Leftrightarrow \frac{s_t}{T-t} - (s_t - s_{t+1}) < 0. \end{aligned} \tag{18}$$

$s_t/(T-t)$ represents the clearance rate,$s_t - s_{t+1}$represents the number of airline tickets sold at time t, (18) indicates that a penalty is required when the number of tickets sold exceeds the clearance rate. So we can incorporate a penalty term into the reward function, which is:

$$r_{t,\mu} = r_t - \mu \text{min}^2\{0, \frac{s_t}{T-t} - (s_t - s_{t+1})\}. \tag{19}$$

When the price is set too low, which is equivalent to having a high demand-rate, the penalty term $s_t/t - (s_t - s_{t+1}) < 0$, which punishes the agent for being on the wrong strategy, encourages the agent to learn the optimal strategy faster. A larger μ encourages the agent to learn a pricing strategy that is higher. The use of a squared penalty term amplifies the impact of extremely unreasonable low-price strategies and reduces the punishment for smaller deviations, thereby producing a more pronounced reward signal to facilitate learning.

5 Numerical Experiments

In this section, the reinforcement learning algorithms used, Soft Actor-Critic and Reward Shaping for Soft Actor-Critic, have their hyperparameters set in accordance with Haarnoja et al. [16] except for the unique hyperparameter μ specific to RSAC, $\mu = (\frac{100}{100+\text{episode}})^2$.

The revenue obtained during each training session is smoothed by taking the average of the last one hundred training sessions to make the curve more smooth. The experiments are repeated on five different random seeds to test the robustness of the results. After learning is complete, we will conduct 500 simulation experiments in the same environment to approximate the agent's expected revenue.

5.1 Comparison with Discrete Pricing

In this part of the experiment, we compare the results of SAC, RSAC, and dynamic programming algorithm in Sect. 3.2 with common price granularity in a small-scale setting. The experimental setup is as follows:

- $T = 20$, $C = 20$.
- Discrete price pool $\mathcal{F} = \{1, 2, 3, 4, 5\}$, Continuous price set $\mathcal{F} = [1, 5]$.
- Customer behavior is divided into two stages: arrival and purchase. The arrival rate $\lambda = 2$, and the price sensitivity coefficient $\beta = 1.2$.

From Table 1, it can be observed that although reinforcement learning algorithms with model-free properties are unaware of customer demand, the rewards of SAC (except for seed 4) and RSAC surpass the theoretical optimal upper bound of the DP algorithm where customer demand is completely known as training progresses. The failure of the SAC algorithm on seed 4 is because the agent was initialized near the boundary action of $a_t = 1$, and a large number of optimal actions are relatively close to the boundary. However, due to the use of the $tanh(\cdot)$ activation function in the network, gradient vanishing occurred, preventing the network from updating quickly near the boundary. This can be observed from Fig. 1, where the reward for seed 4 remains at 20 for an extended period, indicating that the pricing of 1 was adopted for most states.

Table 1. Revenue evaluating after training.

Seed	Revenue ratio to DP	
	SAC	RSAC
Seed 0	106.185%	106.996%
Seed 1	106.854%	106.271%
Seed 2	105.495%	104.166%
Seed 3	105.875%	105.822%
Seed 4	72.132%	107.128%

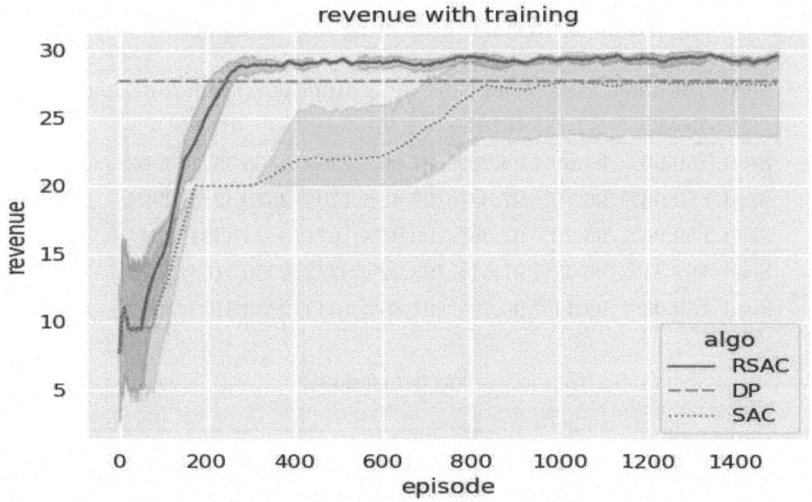

Fig. 1. The training process compared with assortment optimization for small-scale problems.

5.2 Comparison with Continus Pricing

To accurately measure the learning capability of the SAC algorithm on continuous pricing, we consider customer behavior to follow a Poisson distribution with a demand rate of $\lambda(p) = ae^{-\alpha p}$,where $a > 0, \alpha > 0$. The problem under this scenario has been addressed in the research by Gallego and Van Ryzin [1], where analytical solutions for the optimal value and optimal action are provided as: $V^*(n,t) = \log \sum_{i=1}^{n}((\lambda^* t)^i/i!)$, and $p^*(n,t) = V^*(n,t) - V^*(n-1,t) + 1$.

The experiment is divided into two demand-rate, corresponding to the off-season and peak season of sales, respectively:

- $T = 100$, $C = 100$.
- RL's Continuous price set $\mathcal{F} = [1, 9]$.
- low demand-rate: $a = 2e$, $\alpha = 1$, high demand-rate $a = 4e^{0.4}$, $\alpha = 0.4$.

Figure 2 illustrates the training process of the SAC and RSAC algorithms under the two scenarios, while Table 2 presents the ratio of the average reward to the optimal reward expectation after 500 simulation runs, as well as the training time, for the algorithms after training. OV represents the expected value of optimal revenue. Due to the large number of optimal actions being close to the boundary under the low demand-rate scenario, the SAC algorithm failed to train successfully in four out of the five seeds. In the high demand-rate scenario, although the SAC algorithm eventually converges, it requires a significant number of iterations to escape the boundary actions. The RSAC algorithm converges quickly under both scenarios across all five random seeds, and in comparison to the SAC results that do converge, the optimality of RSAC is not inferior to SAC. Therefore, we can conclude that in the context of continuous dynamic pricing, the RSAC algorithm significantly outperforms the SAC algorithm.

Table 2. Revenue evaluating after training.

Seed	Revenue for low/high		Runtime (s) for low /high	
	SAC	RSAC	SAC	RSAC
Seed 0	59.96%/97.60%	98.20%/97.29%	1760/2838	3371/3062
Seed 1	59.96%/96.79%	97.33%/97.43%	1705/2579	3238/2994
Seed 2	59.96%/96.20%	97.38%/96.95%	1671/3024	3281/2944
Seed 3	97.31%/96.47%	97.51%/98.22%	3222/2650	3164/3012
Seed 4	59.96%/98.11%	96.72%/97.85%	1643/2333	3125/3012

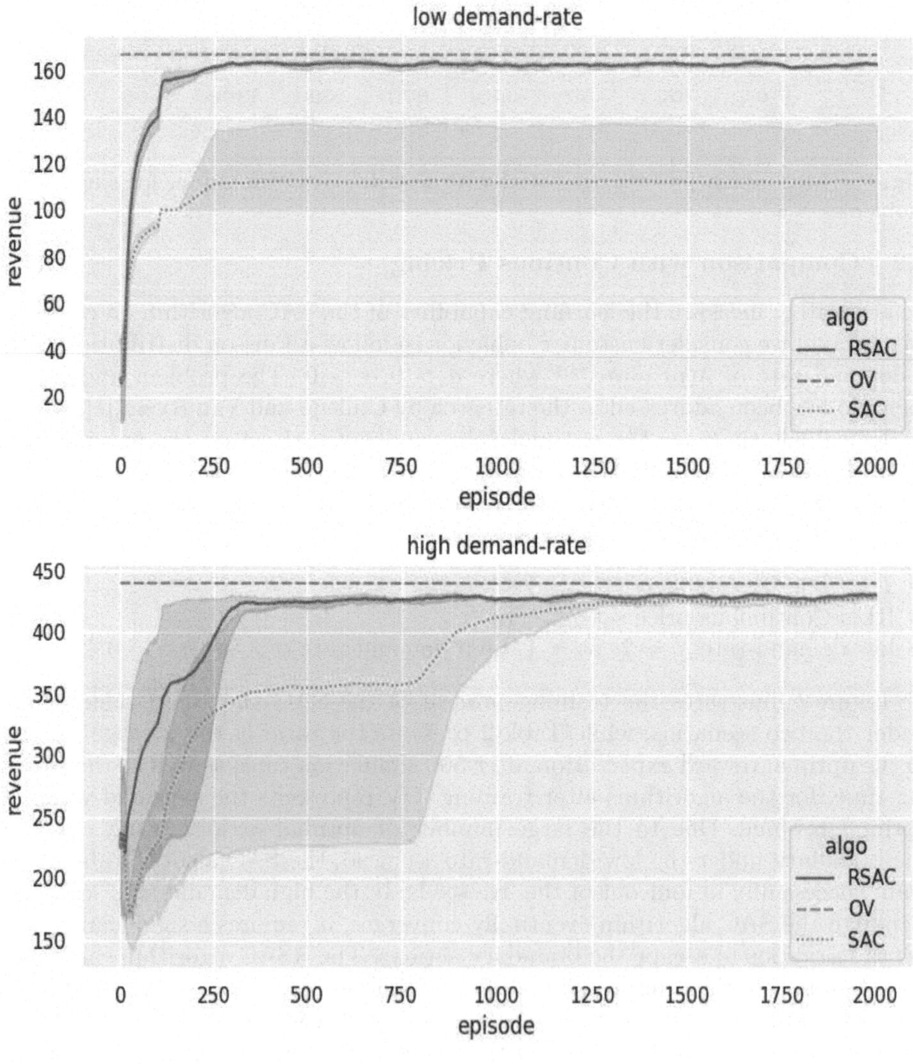

Fig. 2. Training process for low/high demand-rate.

6 Conclusion

With the introduction of IATA's new policy, continuous dynamic pricing has become feasible. However, there is currently limited research on continuous dynamic pricing in the aviation industry, especially regarding the application of reinforcement learning methods to continuous dynamic pricing. This study applies the Soft Actor-Critic algorithm, suitable for continuous action spaces, to address the continuous dynamic pricing problem and proposes a method to accelerate learning. This method smoothly amplifies the value difference between boundary actions and optimal actions, thereby helping the agent to learn more effectively.

We compared the SAC and RSAC algorithms with the dynamic programming method with known demand on a small-scale problem. The results indicate that the RSAC algorithm can exceed the optimal expected revenue of discrete dynamic pricing by at least 4.166%, while the SAC algorithm, under the premise of successful learning, achieves 5.495%. This fully demonstrates that compared to discrete dynamic pricing, continuous dynamic pricing using reinforcement learning methods offers a significant improvement in benefits. To clarify the learning capability of RL algorithms in continuous dynamic pricing, we considered a demand rate function in exponential form: when demand is known, this problem can be solved analytically. The SAC algorithm performs poorly in learning under the low demand-rate scenario, with four out of five random seeds stuck at the boundary actions for a long time, and its convergence speed for the high demand-rate scenario is also inferior to that of RSAC. The RSAC algorithm is able to converge quickly and stabilize under both scenarios, achieving at least 96.20%/96.95% of the optimal expected revenue for the low/high demand-rate cases.

In this study, the main focus was on implementing continuous dynamic pricing for single-flight single-class tickets using reinforcement learning algorithms, and numerical experiments were conducted to analyze algorithm performance with an exponential demand function. For future research directions, the first priority is to develop continuous dynamic pricing algorithms suitable for multi-class ticketing within flight networks. The second is to study more general demand functions that can accurately describe customer behavior. Finally, is to consider how to describe and model the competitive relationships between airlines.

References

1. Gallego, G., van Ryzin, G.: Optimal dynamic pricing of inventories with stochastic demand over finite horizons. Manag. Sci. **40**(8), 999–1020 (1994)
2. Gallego, G., van Ryzin, G.: A multiproduct dynamic pricing problem and its applications to network yield management. Oper. Res. **45**(1), 24–41 (1997)
3. Belobaba, P.P.: Optimal vs. heuristic methods for nested seat allocation. In: Presentation at ORSA/TIMS Joint National Meeting (1992)
4. Lautenbacher, C.J., Stidham, S., Jr.: The underlying Markov decision process in the single-leg airline yield-management problem. Transport. Sci. **33**(2), 136–146 (1999)

5. Talluri, K., van Ryzin, G.: Revenue management under a general discrete choice model of consumer behavior. Manag. Sci. **50**(1), 15–33 (2004)
6. Kumar, R., Li, A., Wang, W.: Learning and optimizing through dynamic pricing. J. Revenue Pricing Manag. **17**, 63–77 (2018). https://doi.org/10.1057/s41272-017-0120-2
7. Otero, D.F., Akhavan-Tabatabaei, R.: A stochastic dynamic pricing model for the multiclass problems in the airline industry. Eur. J. Oper. Res. **242**, 188–200 (2015)
8. Akçay, Y., Natarajan, H.P., Xu, S.H.: Joint dynamic pricing of multiple perishable products under consumer choice. Manag. Sci. **56**(8), 1345–1361 (2010). https://doi.org/10.1287/mnsc.1100.1178
9. Selcuk, A.M., Avsar, Z.M.: Dynamic pricing in airline revenue management. J. Math. Anal. Appl. **478**, 1191–1217 (2019). https://doi.org/10.1016/j.jmaa.2019.06.012
10. Gosavii, A., Bandla, N., Das, T.K.: A reinforcement learning approach to a single leg airline revenue management problem with multiple fare classes and overbooking. IIE Trans. **34**, 729–742 (2002)
11. Gosavi, A.: A reinforcement learning algorithm based on policy iteration for average reward: empirical results with yield management and convergence analysis. Mach. Learn. **55**, 5–29 (2004)
12. Lawhead, R.J., Gosavi, A.: A bounded actor-critic reinforcement learning algorithm applied to airline revenue management. Eng. Appl. Artif. Intell. **82**, 252–262 (2019)
13. Bondoux, N., Nguyen, A.Q., Fiig, T., Acuna-Agost, R.: Reinforcement learning applied to airline revenue management. J. Revenue Pricing Manag. **19**, 332–348 (2020)
14. Jo, S., Lee, G.M., Moon, I.: Airline dynamic pricing with patient customers using deep exploration-based reinforcement learning. Eng. Appl. Artif. Intell. **133**, 108073 (2024)
15. Zhu, X., Jian, L., Chen, X., Zhao, Q.: Reinforcement learning for multi-flight dynamic pricing. Comput. Ind. Eng. **193**, 110302 (2024)
16. Haarnoja, T., Zhou, A., Abbeel, P., Levine, S.: Soft actor-critic: off-policy maximum entropy deep reinforcement learning with a stochastic actor. ArXiv arxiv:1801.01290 (2018)
17. Devlin, S., Kudenko, D.: Dynamic potential-based reward shaping. In: Adaptive Agents and Multi-Agent Systems (2012)

Approximation Algorithms for the Combination of Prize-Collecting Vertex Cover and Scheduling

Li Guan, Hongli Liu, and Xin Liu(✉)

School of Mathematics and Statistics, Yunnan University,
Kunming 650504, People's Republic of China
guanli@ynu.edu.cn, xiaolan2409839160@163.com

Abstract. Given an undirected graph $G = (V, E)$ with weighted vertices and edges, a set J of $n = |V|$ independent jobs and m unrelated machines, each vertex $v \in V$ corresponds to a job $J_j \in J$. The combination of prize-collecting vertex cover and scheduling (CPVCS) is to select a subset $C \subseteq V$ of vertices and schedule the jobs corresponding to the vertices in C on m unrelated machines, edges that are not covered incur a penalty cost, such that the sum of makespan and the total weight of the uncovered edges by C is minimized, where the makespan is defined as the maximum load of the machines. When the number of machines is an input number, we design a 4-approximation algorithm based on LP-rounding method. When the number of machines is a constant, we design a 3-approximation algorithm by guessing the job with maximum processing time on each machine.

Keywords: Prize-collecting vertex cover · unrelated parallel scheduling · approximation algorithm

1 Introduction

Both vertex cover [11] and parallel machine scheduling [17] are classical problems in combinatorial optimization and theoretical computer science. Wang and Cui [19] introduced the combination of vertex cover and parallel machine scheduling, denoted by $P|VC|C_{max}$, which is to select a subset of vertices of a given graph, and schedule the corresponding jobs on the parallel machines, such that the makespan is minimized. If the given graph is complete, $P|VC|C_{max}$ is exactly the parallel machine scheduling problem [2]. If the number of machines is 1, $P|VC|C_{max}$ is exactly the vertex cover [11]. Based on the local ratio method and the LPT rule, Wang and Cui [19] developed a 3-approximation algorithm. Epstein et al. [7] gave an improved 2-approximation algorithm and proposed 3-approximation algorithm for the combination of vertex cover and unrelated machine scheduling, denoted by $R|VC|C_{max}$, which is a generalized version of $P|VC|C_{max}$.

D. Du et al. (Eds.): COCOA 2024, LNCS 15434, pp. 69–76, 2025.
https://doi.org/10.1007/978-981-96-4445-2_6

Prize-collecting vertex cover [10] is an extension of the vertex cover problem, where it is permitted to not cover some edges, but there is a nonnegative penalty for the uncovered edges. The framework for solving integer programs over monotone inequalities in three variables in [10] implies that prize-collecting vertex cover can be approximated within 2. Recently, prize-collecting receives more and more attentions in combinatorial optimization community, such as prize-collecting Steiner forest [1], prize-collecting TSP [4], and prize-collecting power cover [6].

In the scheduling problems, the prize-collecting version implies that a job can be rejected, which is first proposed in [3]. Kones and Levin [12] design a unified framework for designing efficient polynomial time approximation scheme (EPTAS) for load balancing on parallel machines, which implies that scheduling with rejection [3] admits an EPTAS. Recently, Li and Ou [16] considered a more general case of scheduling with rejection where the total weight of rejected jobs is bounded by a given upper bound, and gave a polynomial time approximation scheme. More related results can be found in [5,14,15,18].

In [20], by combing prize-collecting vertex cover and parallel machine scheduling, the authors proposed the problem $P|PCVC|C_{max}$, and designed a 3-approxi-mation algorithms, where the jobs corresponding to the vertices in the cover and the uncovered edges must be scheduled. To the best of our knowledge, breaking the 3-approximation is a very challenging problem both for $P|PCVC|C_{max}$ [20] and $R|VC|C_{max}$ [7].

In this paper, we consider the combination of prize-collecting vertex cover [10] and unrelated machine scheduling [13], which generalizes $P|PCVC|C_{max}$ [20] and [7]. We design a 4-approximation algorithm for the general case, and a 3-approximation algorithm when the number of machines is fixed.

2 Preliminaries

For any positive integer k, let $[k] = \{1, 2, \ldots, k\}$. We are given a weighted undirected graph $G = (V, E, \pi)$ with $|V| = n$, a set $J = \{J_1, J_2, \ldots, J_n\}$ of n independent jobs, and a set $M = \{M_1, M_2, \ldots, M_m\}$ of m unrelated parallel machines. Each vertex $v_j \in V$ is corresponding to the job $J_j \in J$ for $j \in [n]$. Let $P = (p_{ij})$ is a $m \times n$ matrix corresponding to the processing times. If J_j (or vertex v_j) is processed on machine M_i, the processing time is p_{ij}.

The combination of prize-collecting vertex cover and scheduling (CPVCS) problem is to find a subset $C \subseteq V$ of vertices, and schedule the corresponding jobs in $J_C = \{J_j \in J | v_j \in C\}$ on the m machines, such that the maximum load (makespan) plus the total penalty of uncovered edges $E_C = \{e = (u, v) \in E | u \notin V, v \notin V\}$ is minimized, where the penalty of the uncovered edge $e \in E_C$ is $\pi(e)$. Formally, let $\sigma : J_C \mapsto M$ be the schedule which means that the job $J_j \in J_C$ is scheduled on machine $\sigma(J_j)$, and $\sigma^{-1}(M_i)$ is the set of jobs scheduled on machine M_i. The load of machine M_i is defined as $L_i = \sum_{j:J_j \in \sigma^{-1}(M_i)} p_{ij}$, and the makespan of schedule σ is defined as the maximum load of the machines, *i.e.*, $L_{max}(C, \sigma) = \max_i L_i$. The objective of CPVCS is to minimize $L_{max}(C, \sigma) + \pi(E_C)$, where $\pi(E_C) = \sum_{e:e \in E_C} \pi(e)$.

For each vertex $v_j \in V$, we introduce $m+1$ binary variables $x_{ij} \in \{0,1\}$, $i \in \{0,1,2,\ldots,m\}$, where $x_{ij} = 1$ if and only if v_j is selected in C and the corresponding job J_j is processed on machine M_i for each $i \in [m]$, and $x_{0j} = 1$ if and only if v_j is not selected in C. For each edge $e \in E$, we introduce a binary variable $z_e \in \{0,1\}$ where $z_e = 1$ if and only if the edge e is uncovered by C. The CPVCS problem can be formulated as the following integer linear program (ILP).

$$\begin{aligned}
\min \quad & L_{max} + \sum_{e:e\in E} \pi(e) z_e \\
\text{s.t.} \quad & \sum_{i=0}^{m} x_{ij} = 1, \forall j \in [n], \\
& L_i = \sum_{j=1}^{n} p_{ij} x_{ij} \le L_{max}, \forall i \in [m], \\
& \sum_{i=1}^{m} x_{ij_1} + \sum_{i=1}^{m} x_{ij_2} + z_e \ge 1, \forall e = (v_{j_1}, v_{j_2}) \in E, \\
& x_{ij} \in \{0,1\}, z_e \in \{0,1\}, \forall i \in \{0\} \cup [m], \forall j \in [n], \forall e \in E.
\end{aligned}$$

When $\pi(e) = +\infty$ implying that all the edges should be covered, CPVCS is exactly the vertex cover that meets scheduling on unrelated machines [7], which admits a 3-approximation algorithm. Under the identical parallel machine, let p_j as the processing time of J_j. When $p_{ij} = p_j$ for any i, j, CPVCS is very close to the problem $P|PCVC|C_{max}$ [20], which admits a 3-approximation algorithm, too. The main difference between CPVCS and $P|PCVC|C_{max}$ is whether the uncovered edges should be processed on the machines.

Let $X = (x_{ij})$ be the $(m+1) \times n$ decision matrix, and (X, z) be a feasible solution for ILP corresponding to (C, σ), whose makespan is $L_{max}(X) = \max_i \sum_{j=1}^{n} p_{ij} x_{ij}$ and total penalty is $\pi(z) = \sum_{e:e\in E} \pi(e) z_e$. As in [7,13], we guess a parameter T which denotes an upper bound on the maximum processing time of the jobs in J_C scheduled on the machines. Consider the following relaxation $LP(T)$ of ILP.

$$\begin{aligned}
\min \quad & L_{max} + \sum_{e:e\in E} \pi(e) z_e \\
\text{s.t.} \quad & \sum_{i=0}^{m} x_{ij} = 1, \forall j \in [n], \\
& L_i = \sum_{j=1}^{n} p_{ij} x_{ij} \le L_{max}, \forall i \in [m], \\
& \sum_{i=1}^{m} x_{ij_1} + \sum_{i=1}^{m} x_{ij_2} + z_e \ge 1, \forall e = (v_{j_1}, v_{j_2}) \in E, \\
& x_{ij} = 0, \text{ if } p_{ij} > T, \\
& x_{ij} \in [0,1], z_e \in [0,1], \forall i \in \{0,1,2,\ldots,m\}, \forall j = \{1,\cdots,n\}, \forall e \in E.
\end{aligned}$$

As in [7], It is easy to prove that

Lemma 1. *For any feasible solution* (C, σ) *to CPVCS with makespan* $L_{max}(C, \sigma)$ *and total penalty* $\pi(E_C)$*, there is a feasible solution* (X, z) *for* $LP(T)$ *with objective value at most* $L_{max}(C, \sigma) + \pi(E_C)$*, where* $T = L_{max}(C, \sigma)$*.*

Moreover, using the classical method in [13], we have

Lemma 2 [13]**.** *For any feasible solution* (X, z) *to* $LP(T)$ *with objective value* $L_{max}(X) + \pi(z)$ *where* $z_e \in \{0, 1\}$ *for any* $e \in E$*, there is integer feasible solution* $(\bar{X}, z)$ *for* $LP(T)$ *with objective value at most* $L_{max}(X) + \tau_i + \pi(z)$*, where* $\tau_i = \max\{p_{ij} | x_{ij} > 0\}$ *and* $\bar{x}_{ij} \in \{0, 1\}$ *for any* i, j*.*

Let (C^*, σ^*) be an optimal solution for CPVCS and (X^*, z^*) be the corresponding optimal solution for ILP. For simplicity, by using bisection method, we can assume that $T^* = L_{max}(C^*, \sigma^*)$ is known.

3 The Case m Is an Input Number

In this section, we assume that m is an input number, which implies that the running time of the algorithm must be polynomial in m. Let (X, z) be the fractional optimal solution for $LP(T^*)$. Thus, we have

$$L_{max}(X) + \pi(z) \le L_{max}(X^*) + \pi(z^*) = L_{max}(C^*, \sigma^*) + \pi(E_{C^*}) = OPT, \quad (1)$$

where OPT is the objective value of the optimal solution, and $L_{max}(X^*) = \max_i \sum_{j=1}^{n} p_{ij} x^*_{ij}$. It is easy to verify that

$$z_e = 1 - \min\{1, \sum_{i=1}^{m} x_{ij_1} + \sum_{i=1}^{m} x_{ij_2}\}, \forall e = (v_{j_1}, v_{j_2}) \in E, \quad (2)$$

following from the optimality of (X, z).

Based on the fractional optimal solution (X, z), we construct an integer solution as follows. For each vertex $v_j \in V$, if $\sum_{i=1}^{m} x_{ij} \in [\frac{1}{3}, 1]$, set

$$x'_{ij} = \frac{x_{ij}}{\sum_{i=1}^{m} x_{ij}} \in [0, 1], \text{ and } x'_{0j} = 0.$$

Otherwise, set $x'_{ij} = 0$ for $i = 1, 2, \ldots, m$, and $x'_{0j} = 1$. For each $e = (v_{j_1}, v_{j_2}) \in E$, if $\sum_{i=1}^{m} x_{ij_1} \in (0, \frac{1}{3})$ and $\sum_{i=1}^{m} x_{ij_2} \in (0, \frac{1}{3})$, set $z'_e = 1$. Otherwise, set $z'_e = 0$.

Lemma 3. (X', z') *is a feasible solution for* $LP(T^*)$ *with objective value at most* $3(L_{max}(X) + \pi(z))$*.*

Proof. For each vertex $v_j \in V$, if $\sum_{i=1}^{m} x_{ij} \in [\frac{1}{3}, 1]$, we have

$$\sum_{i=0}^{m} x'_{ij} = \sum_{i=1}^{m} \frac{x_{ij}}{\sum_{i=1}^{m} x_{ij}} = 1, \text{ and } x'_{ij} \le 3x_{ij}, \text{ for } i = 0, 1, \ldots, m.$$

If $\sum_{i=1}^{m} x_{ij} \in [0, \frac{1}{3})$, we have $x_{0j} = 1 - \sum_{i=1}^{m} x_{ij} > \frac{1}{3}$ and

$$\sum_{i=0}^{m} x'_{ij} = 1, \text{ and } x'_{ij} \leq 3x_{ij}, \text{ for } i = 0, 1, \ldots, m.$$

For each $e = (v_{j_1}, v_{j_2}) \in E$, if there is a vertex v_j, $j \in \{j_1, j_2\}$, such that $\sum_{i=1}^{m} x_{ij} \in [\frac{1}{3}, 1]$, we have $z'_e = 0 \leq 3z_e$. Otherwise, we have $z_e = 1 - \sum_{i=1}^{m} x_{ij_1} - \sum_{i=1}^{m} x_{ij_2} \in (\frac{1}{3}, 1]$ and $z'_e = 1 \leq 3z_e$. It implies that (X', z') is a feasible solution for $LP(T^*)$. Moreover, we have

$$L_{max}(X') + \pi(z') = \max_i \sum_{j=1}^{n} p_{ij} x'_{ij} + \sum_{e:e \in E} \pi(e) z'_e \leq 3(L_{max}(X) + \pi(z)).$$

Thus, the lemma holds. ∎

By Lemma 2, we can find an integer feasible solution $(\bar{X}, z')$ for $LP(T^*)$ with objective value

$$\begin{aligned} L_{max}(\bar{X}) + \pi(z') &\leq L_{max}(X') + \tau_i + \pi(z') \\ &\leq L_{max}(X') + T^* + \pi(z') \\ &\leq 3(L_{max}(X) + \pi(z)) + T^* \\ &\leq 4OPT, \end{aligned}$$

where the third inequality follows from Lemma 3, and the last inequality follows from (1).

Constructing a feasible solution (C, σ) based on $(\bar{X}, z)$, we have

Theorem 1. *The CPVCS problem admits a 4-approximation algorithm.*

4 The Case m is Fixed

In this section, we assume that m is a constant. In the optimal solution (X^*, z^*), let $l_i^* = \arg\max_{j:x_{ij}^*=1} p_{ij}$ be the job with largest processing time scheduled on machine M_i, where $l_i^* = 0$ if $x_{ij}^* = 0$ for every $j \in [n]$. Since each machine has at most $n+1$ choices, by trying at most $(n+1)^m$ possibilities, we can assume that l_i^* are known.

Construct the following integer linear program $ILP(l_1^*, l_2^*, \ldots, l_m^*)$:

$$\begin{aligned} \min \quad & L_{max} + \sum_{e:e \in E} \pi(e) z_e \\ \text{s.t.} \quad & \sum_{i=0}^{m} x_{ij} = 1, \forall j \in [n], \\ & p_{il_i^*} + \sum_{j \in [n]: j \neq l_i^*} p_{ij} x_{ij} \leq L_{max}, \forall i \in [m], \end{aligned}$$

$$\sum_{i=1}^{m} x_{ij_1} + \sum_{i=1}^{m} x_{ij_2} + z_e \geq 1, \forall e = (v_{j_1}, v_{j_2}) \in E,$$
$$x_{il_i^*} = 1, \forall i \in \{i | l_i^* \neq 0, i \in [m]\},$$
$$x_{ij} = 0, \text{ if } p_{ij} > p_{il_i^*},$$
$$x_{ij} \in \{0,1\}, z_e \in \{0,1\}, \forall i \in \{0,1,2,\ldots,m\}, \forall j \in [n], \forall e \in E,$$

where $p_{il_i^*} = 0$, if $l_i^* = 0$.

As in the last section, we can find a feasible solution (X', z') with $z'_e \in \{0,1\}$ based on the fractional optimal solution (X, z) for the relaxation of $ILP(l_1^*, l_2^*, \ldots, l_m^*)$ such that

$$\pi(z') \leq 3\pi(z), \text{ and } L'_i = p_{il_i^*} + \sum_{j\in[n]:j\neq l_i^*} p_{ij}x'_{ij} \leq p_{il_i^*} + 3 \sum_{j\in[n]:j\neq l_i^*} p_{ij}x_{ij},$$

for each $i \in [m]$. Note that in the solution (X', z'), we have $x'_{il_i^*} = x_{il_i^*} = x^*_{il_i^*} = 1$ for each $i \in [m]$, implying that $\tau_i = \max\{p_{ij} | x'_{ij} > 0\} = p_{il_i^*} \leq T^*$.

By Lemma 2, we can find an integer feasible solution $(\bar{X}, z')$ for $LP(T^*)$ with objective value

$$\begin{aligned}
L_{max}(\bar{X}) + \pi(z') &= \max_i\{p_{il_i^*} + \sum_{j\in[n]:j\neq l_i^*} p_{ij}\bar{x}_{ij}\} + \pi(z') \\
&\leq \max_i\{p_{il_i^*} + \sum_{j\in[n]:j\neq l_i^*} p_{ij}x'_{ij} + p_{il_i^*}\} + 3\pi(z) \\
&\leq \max_i\{3 \sum_{j\in[n]} p_{ij}x'_{ij}\} + 3\pi(z) \\
&= 3(L_{max}(X) + \pi(z)) \\
&\leq 3OPT,
\end{aligned}$$

where the last inequality follows from (1).

Constructing a feasible solution (C, σ) based on $(\bar{X}, z)$, we have

Theorem 2. *The CPVCS problem admits a 3-approximation algorithm when m is a constant.*

5 Discussion

We have designed a 4-approximation algorithm for the CPVCS problem. Since the best approximation ratio for a special case of CPVCS [7] is 3, it is challenging to design 3-approximation algorithm for CPVCS. It is also interesting to consider other combined problems of graph optimization [8,20] and unrelated machine scheduling [13] or fair allocation [9].

References

1. Ahmadi, A., Gholami, I., Hajiaghayi, M.T.: 2-approximation for prize-collecting Steiner forest. In: Proceedings of the 2024 Annual ACM-SIAM Symposium on Discrete Algorithms (SODA), pp. 669–693 (2024). https://doi.org/10.1137/1.9781611977912.25
2. Alon, N., Azar, Y., Woeginger, G.J.: Approximation schemes for scheduling on parallel machines. J. Sched. **1**(1), 55–66 (1998). https://doi.org/10.1002/(SICI)1099-1425(199806)1:1<55::AID-JOS2>3.0.CO;2-J
3. Bartal, Y., Leonardi, S., Marchetti-Spaccamela, A., Sgall, J., Stougie, L.: Multiprocessor scheduling with rejection. SIAM J. Disc. Math. **13**(1), 64–78 (2000). https://doi.org/10.1137/S0895480196300522
4. Blauth, J., Klein, N., Nagele, M.: A better-than-1.6-approximation for prize-collecting TSP. In: International Conference on Integer Programming and Combinatorial Optimization, pp. 28–42 (2024). https://doi.org/10.1007/978-3-031-59835-7_3
5. Dai, B., Li, W.: Vector scheduling with rejection on two machines. Int. J. Comput. Math. **97**(12), 2507–2515 (2020). https://doi.org/10.48550/arXiv.2308.06254
6. Dai, H., Li, W., Liu, X.: An approximation algorithm for the H-Prize-Collecting power cover problem. In: International Workshop on Frontiers in Algorithmics, pp. 89–98 (2022). https://doi.org/10.1007/978-3-031-20796-9_7
7. Epstein, L., Levin, A., Woeginger, G.J.: Vertex cover meets scheduling. Algorithmica **74**(3), 1148–1173 (2015). https://doi.org/10.1007/s00453-015-9992-y
8. Guan, L., Li, J., Li, W., Lichen, J.: Improved approximation algorithms for the combination problem of parallel machine scheduling and path. J. Comb. Optim. **38**(3), 689–697 (2019). https://doi.org/10.1007/s10878-019-00406-0
9. Guo, H., Li, W., Deng, B.: A survey on fair allocation of chores. Mathematics **11**(16), 3616 (2023). https://doi.org/10.3390/math11163616
10. Hochbaum, D.S.: Solving integer programs over monotone inequalities in three variables: a framework for half integrality and good approximations. Eur. J. Oper. Res. **140**(2), 291–321 (2002). https://doi.org/10.1016/S0377-2217(02)00071-1
11. Karakostas, G.: A better approximation ratio for the vertex cover problem. ACM Trans. Algor. **5**(4), 1–8 (2009). https://doi.org/10.1145/1597036.1597045
12. Kones, I., Levin, A.: A unified framework for designing EPTAS for load balancing on parallel machines. Algorithmica **81**(7), 3025–3046 (2019). https://doi.org/10.1007/s00453-019-00566-9
13. Lenstra, J.K., Shmoys, D.B., Tardos, E.: Approximation algorithms for scheduling unrelated parallel machines. Math. Program. **46**, 259–271 (1990). https://doi.org/10.1007/BF01585745
14. Li, W., Li, J., Zhang, X., Chen, Z.: Penalty cost constrained identical parallel machine scheduling problem. Theor. Comput. Sci. **607**, 181–192 (2015). https://doi.org/10.1016/j.tcs.2015.10.007
15. Li, W., Cui, Q.: Vector scheduling with rejection on a single machine. 4OR **16**(1), 95–104 (2017). https://doi.org/10.1007/s10288-017-0356-0
16. Li, W., Ou, J.: Machine scheduling with restricted rejection an application to task offloading in cloud-edge collaborative computing. Eur. J. Oper. Res. **314**(3), 912–919 (2024). https://doi.org/10.1016/j.ejor.2023.11.002
17. Mokotoff, E.: Parallel machine scheduling problems: a survey. Asia-Pac. J. Oper. Res. **18**(2), 193 (2001)

18. Shabtay, D., Gaspar, N., Kaspi, M.: A survey on offline scheduling with rejection. J. Sched. **16**, 3–28 (2013). https://doi.org/10.1007/s10951-012-0303-z
19. Wang, Z., Cui, Z.: Combination of parallel machine scheduling and vertex cover. Theor. Comput. Sci. **460**, 10–15 (2012). https://doi.org/10.1016/j.tcs.2012.06.003
20. Wang, Z., Hong, W., He, D.: Combination of parallel machine scheduling and the covering problem. Pac. J. Optim. **10**(3), 577–591 (2013)

Randomized Mechanisms for Improved Approximation Ratios in Heterogeneous Two-Facility Location

Meng Xu, Qian Liu, Min Li, and Yang Zhou(✉)

School of Mathematics and Statistics, Shandong Normal University, Jinan 250358, People's Republic of China
{liminEmily,zhouyang}@sdnu.edu.cn

Abstract. This study explores heterogeneous two-facility location mixed mechanisms, aiming to develop an approach for positioning two facilities that ensures agent strategyproofness while minimizing social costs. We introduce a mixed mechanism that achieves an approximation ratio of $\frac{25}{8}$, demonstrating a significant improvement over the latest deterministic mechanism, which has an approximation ratio of $\frac{17}{4}$.

Keywords: mechanism design · facility location game · strategyproofness · approximation ratio

1 Introduction

Facility location games represent a significant area of study within game theory, operations research, and mechanism design. In these games, players or agents strategically select locations for facilities with the goal of minimizing costs, maximizing utilities, or achieving specific social objectives. The decisions made by individual agents can greatly impact the overall efficiency and effectiveness of the system. Researchers have explored various factors within these games, including player preferences, distances between facilities, strategic behavior, and mechanisms to ensure fairness and stability [4,5,11,13,14,18]. In this paper, we focus on the strategic mechanism design problem in two-facility location games, considering that players will report their location information truthfully but may misrepresent their preferences regarding facility locations.

Research on mechanism design for facility location games has gained significant attention over the past decade. Procaccia and Tennenholtz [19] were the first to study the problem of facility location games. In the context of proving one or more facilities, they derived the approximation ratio and lower bounds for two objective functions, thereby opening the door for further exploration of this issue. Subsequently, Serafino and Ventre (2016) [21] developed an anti-lying deterministic mechanism that enables each agent to simultaneously report

This work is supported by National Science Foundation of China (No. 12371099) and Natural Science Foundation of Shandong Province of China (No. ZR2024MA015).

D. Du et al. (Eds.): COCOA 2024, LNCS 15434, pp. 77–87, 2025.
https://doi.org/10.1007/978-981-96-4445-2_7

their location and preferences based on their individual facility preferences. They achieved an $(n-1)$-approximate mechanism for social cost and a 3-approximate mechanism for maximum cost. Ordinary dual facilities are continuously evolving into facilities with specific preferences. Shortly thereafter, Fong et al. (2018) [12] examined a model where each agent has a fractional preference for two facilities that sum to 1, designing various mechanisms based on agents' false reporting. Anastasiadis and Deligkas (2020) [1] made further advancements, suggesting that agents' preferences could include indifference, approval, and disapproval for k facilities. There are many studies on the location of heterogeneous facilities [7,17,20].

In addition to changing preferences, many researchers have proposed various adaptations and extensions of the facility location games. For example, Duan et al. (2019) [6] proposed that all agents are within the interval $[0, 1]$, and the cost is the sum of the distance between the agents and the two facilities.They investigated heterogeneous facilities, proving a lower bound of $\frac{7-d}{6}$ for the approximation ratio achieved by deterministic strategyproof mechanisms with respect to social cost, where d represents the smallest distance between the facilities. Kyropoulou et al. [16] made two changes on the initial basis: the preference of agents is only for and against, while the cost of the second is not only related to the preference, but also to the location of the two facilities. Zhao et al. (2022) [23] studied the costs incurred by each agent in k heterogeneous facilities and related these costs to the farthest distances of the facilities serving them. Most people mainly study that each agent will misreport about his location or preference. Moreover, we can consider other situations for agents. For example, each agent may have multiple locations, which will cause agents to hide their locations [3,14,22].

In the context of the approximation ratio of mechanisms under the premise of strategyproofness, in addition to the more common social cost and maximum cost, researchers also focus on various definitions of objective functions as measurement criteria [8,9]. Anshelevich et al. [2] divided agents into disjoint districts and examined the sum, maximum, and minimum costs within each district. Subsequently, Filos-Ratsikas et al. [10] studied another objective focused on maximizing the sum within each district, providing tight upper and lower bounds for this measure.

In this paper, for the heterogeneous two-facility location game described above, we propose a randomized mechanism with the property of strategyproofness. This mechanism achieves an approximation ratio of $\frac{25}{8}$, and we provide an illustrative example demonstrating the tightness of this ratio under our mechanism.

The structure of the remaining content in this paper is as follows. Section 2 mainly expounds on the research background of facility location games, as well as the contents and achievements of related research both domestically and internationally. Section 3 introduces the mathematical model and related definitions of this problem in detail, designs a strategyproof mechanism for social cost, and proves its approximation ratio. Section 4 summarizes the results and challenges encountered in this paper, and provides some insights into future work.

2 Preliminaries

We consider a scenario on a line with m points, which are often set as integers ranging from 1 to m for comparison purposes. Among these points, there are $n \leq m$ agents distributed across various locations. Additionally, we aim to select two distinct positions to place two facilities. Together, these agents and facilities form the basic elements of this line graph.

Each agent is assigned to a distinct node x_i along the line. The collective positions of all agents, denoted as $\mathbf{x}$, encompass the nodes they occupy along with potential empty node positions, constituting common knowledge among all stakeholders.

Furthermore, each agent i has private approval preferences for the two facilities, ensuring that at least one of the facilities is approved. Agent i approves facility j (where $j = 1$ or 2) when $t_{ij} = 1$; otherwise, she remains indifferent. Based on this condition, we have that for any $i \in [n]$, it holds that $t_{i1} + t_{i2} \geq 1$. The preference profile $\mathbf{t} = (\mathbf{t}_i)_{i \in N}$ encapsulates the preferences of all agents. Notably, we denote $N_j = \{i \mid t_{ij} = 1\}$. With $\mathbf{x}$ and $\mathbf{t}$ encompassing all pertinent information, we denote this collective information as $I = (\mathbf{x}, \mathbf{t})$.

Ensuring that the locations of the two facilities are different, we define a feasible solution $\mathbf{z} = (z_1, z_2)$, where z_j represents the location of facility $j \in \{1, 2\}$.

According to the solution, the cost for each agent i depends on the sum of her distances to the facilities she wishes to connect with, defined as:

$$\text{cost}_i(\mathbf{z}|I) = \sum_{j \in \{1,2\}} t_{ij} \cdot d(i, j),$$

in which the distance is defined as $d(i, j) := |x_i - z_j|$.

For the randomized mechanism, the cost of each agent is equal to the distance to the facility which she approves multiplied by the probability of the facility. Consequently, a mechanism receives an instance I as input and yields a feasible solution as output.

To prevent agents from misreporting about their preferences, we need to design a strategyproof mechanism.

Definition 1 (strategyproof). *A mechanism M is strategyproof if she misreports about her preferences $\mathbf{t}_i^{'}$, her cost would not be lower than when she does not misreport about it, i.e.,*

$$\text{cost}_i(M(I)|I) \leq \text{cost}_i(M(\mathbf{x}, (\mathbf{t}_i^{'}, \mathbf{t}_{-i}))|I).$$

Here, $(\mathbf{t}_i^{'}, \mathbf{t}_{-i})$ represents the reporting scenario in which agent i has misreported her information as $\mathbf{t}_i^{'}$ with other agents remaining the same as in $\mathbf{t}_{-i}$.

The social cost objective is considered in this paper as a function of any feasible solution.

Definition 2 (social cost). *The social cost associated with a feasible solution* $\mathbf{z}$ *for a given instance* I *is determined by the overall cost incurred by all agents, i.e.,*

$$\mathrm{SC}(\mathbf{z}|I) := \sum_{i \in N} \mathrm{cost}_i(\mathbf{z}|I).$$

For a given instance I, we define the optimal social cost obtained through all feasible solutions as $\mathrm{SC}^*(I) = \min_{\mathbf{z}} \mathrm{SC}(\mathbf{z}|I)$. Our focus is on designing randomized mechanisms that are strategyproof, aiming for an as low approximation ratio as possible. The definition of the approximation ratio for a mechanism is as follows.

Definition 3 (approximation ratio). *The approximation ratio of mechanism* M *with respect to a measurement* f *denotes the maximum ratio (across all instances) between function values of the output produced by* M *and the minimum result attainable from feasible solutions, expressed as*

$$\rho_f(M) = \sup_I \frac{f(M(I)|I)}{f^*(I)}.$$

3 The Mixed Mechanism

We initiate our discussion by outlining the social cost objective as the foundation of our approach. Our primary focus centers on general instances involving n agents, where we have devised a randomized strategyproof mechanism boasting an impressive approximation ratio of 25/8. Because there are m nodes and n agents in total, the following two situations will occur. When $m = n$, each agent can only occupy one position, that is to say, there will be one agent at each node. When $m > n$, there are only n agents in total, which will lead to some nodes that are not occupied by agents. These nodes are called empty nodes. This mechanism capitalizes on the available information regarding the position profile and adapts its operations based on whether the given instance includes empty nodes. When empty nodes are present, the mechanism strategically positions one of the facilities at the median location of all agents endorsing facility j with probability $\frac{1}{2}$, while placing facility $3-j$ at the empty node nearest to the median location of agents approving facility.

Mechanism 1. Mixed mechanism for heterogeneous two-facility location

Arguments: an instance I
Return: location of facilities $\mathbf{z} = (z_1, z_2)$

```
if m = n then
    z1 ← ⌊n/2⌋
    z2 ← ⌊n/2⌋ + 1
else
    for j ∈ {1, 2} do
        yj ← position of the leftmost median of Nj
    end for
    p ← a random number uniformly chosen in [0, 1]
    if p > 0.5 then
        z1 ← y1
        z2 ← nearest empty node to y2
    else
        z1 ← nearest empty node to y1
        z2 ← y2
    end if
end if
```

The following theorem establishes the strategyproofness of Mechanism 1.

Theorem 1. *Mechanism 1 is strategyproof.*

Proof. We know where the agent is, so the agent can only change her cost by lying about preferences.

The preference of any agent has nothing to do with the location of the facility in the situation of no empty nodes, so lying doesn't change the cost. The following proof only considers the case with empty nodes.

Given an arbitrary instance I. Suppose that facility 1 is positioned at y_1 with a probability $\frac{1}{2}$ and at some empty node a with $a > y_1$ with a probability $\frac{1}{2}$, facility 2 is positioned at y_2 with a probability $\frac{1}{2}$ and at b with $b > y_2$ with a probability $\frac{1}{2}$. Suppose that agent i is the one who lies. We classified three scenarios based on preference:

- $i \in N_1 \setminus N_2$.
  - If agent i misreports that she approves both facilities, then $y_1^{'} = y_1$ and $a^{'} = a$, and the cost of agent i does not decrease.

  Next, we consider that agents use the median y_1 to split agents into two parts if they falsely report the preference for facility 2.
  - If $x_i \leq y_1$, then the leftmost median of $N_1^{'}$ either doesn't change, or it gets bigger to the right. Thus, the closest empty node to $y_1^{'}$ is either still a, keeping the cost of i the same and the cost of i remains exactly the same, or a different empty node $a^{'}$ where $a^{'} > a$ and the cost of i strictly increases.

- If $x_i > y_1$, then $y_1^{'} \leq y_1$. Thus, either a remains the closest empty node to $y_1^{'}$, keeping the cost unchanged, or a different empty node $a^{'}$ where $a^{'} < y_1 < a$ becomes the nearest empty node to $y_1^{'}$.
When $y_1 < x_i < a$,

$$\text{cost}_i^{'} = \frac{1}{2}(x_i - y_1^{'}) + \frac{1}{2}(x_i - a^{'}),$$

because the nearest empty node to y_1 is a, so $a - y_1 < y_1 - a^{'}$, $y_1 > \frac{1}{2}(a + a^{'})$, the cost of agent i increases, so she does not misreport.
when $x_i > a$,

$$\text{cost}_i^{'} = \frac{1}{2}(x_i - y_1^{'}) + \frac{1}{2}(x_i - a^{'}) \geq \text{cost}_i.$$

- $i \in N_2 \setminus N_1$. Similarly, we can show that these agents do not lie about their preferences.
- $i \in N_1 \cap N_2$. By definition, $\text{cost}_i = |x_i - y_1| + |x_i - y_2|$. When the agent falsely claims that she approves only one of the facilities j, the change in the cost is only related to the distance of the facilities $3 - j$, so one of the above two situations is satisfied, that is, the agent of this type will not lie. □

In the proof below, we use the symbol $\mathbf{1}\{X\}$, where X represents an event, which is 1 if the event occurs, otherwise $\mathbf{1}\{X\} = 0$.

Lemma 1. *[15] Given any instance I that there are n agents approving facility $j \in \{1, 2\}$, the following holds:*

$$\begin{aligned} \text{SC}^*(I) &\geq \frac{1}{4}\left(n_1^2 + n_2^2 - \mathbf{1}\{n_1 \ odd\} - \mathbf{1}\{n_2 \ odd\}\right) \\ &\geq \frac{1}{4}(n_1^2 + n_2^2 - 2). \end{aligned} \tag{1}$$

Proof. We independently discuss each facility $j \in \{1, 2\}$. The discussion is based on whether n_j is odd or even. When n_j is odd, the optimal location for facility j is located at one of these n_j nodes and get the distance i between the facility for each agent, for $i \in \left\{1, \ldots, \frac{n_j - 1}{2}\right\}$ from j. The sum over the cost of all agents yields $\frac{n_j^2 - 1}{4}$. Similarly, when n_j is even, the optimal allocation can have at most an agent at distance $\frac{n_j}{2}$, besides, The distance from the remaining agents to the facility is i, for $i \in \{1, \ldots, \frac{n_j}{2} - 1\}$ from j; in this case, the total cost for facility j is calculated $\frac{n_j^2}{4}$. □

We will use this inequality in the subsequent demonstration of approximation ratios.

Lemma 2. *[16] For any scenario involving $n \geq 5$ agents and with no vacant nodes, the SC-approximation ratio does not exceed 3.*

In the case of empty nodes, and $n \geq 6$, the approximation ratio of this mechanism is 25/8. Our next proof of the approximation ratio requires the help of a previous lemma.

Lemma 3. *Assume that* $x, y \in \mathbb{Z}_+$ *and* $f(x,y) = \frac{4xy+2x+2y+4}{x^2+y^2-2}$. *When* $x + y \geq 6$, *it satisfies* $f(x,y) \leq 13/4$.

Proof. We focus only on values of positive integers x, y that satisfy $f(x,y) > 3$ i.e., by simplifying x, y such that $4xy+2x+2y+4 > 3(x^2+y^2-2)$.

By rearranging, we obtain

$$2x+2y+10 > 2(x-y)^2 + x^2 + y^2 \geq x^2 + y^2.$$

According to $x^2 + y^2 \geq \frac{(x+y)^2}{2}$, so we can proof that

$$2x+2y+10 \geq \frac{(x+y)^2}{2},$$

therefore, $(x+y)^2 - 4(x+y) - 20 < 0$, $(x+y-2)^2 - 24 < 0$.

Clearly, for $x + y - 2 \geq 5$, the left-hand-side term is still non-negative, so the inequality never holds. Hence, when $x + y \geq 7$, it is a contradiction. For the remaining values of x and y, i.e., $x + y = 6$, $x \in \{0, 1, 2, 3, 4, 5, 6\}$. By the symmetry of the function, we can only consider $x \in \{0, 1, 2, 3\}$. When $x = 0$, then $y = 6$, $f(0,6) = \frac{16}{34} < \frac{13}{4}$. When $x = 1$, then $y = 5$, $f(1,5) = \frac{36}{24} < \frac{13}{4}$. When $x = 2$, then $y = 4$, $f(2,4) = \frac{48}{18} < \frac{13}{4}$. When $x = 3$, then $y = 3$, $f(3,3) = \frac{13}{4}$. □

We have completed all necessary preparations and are fully equipped to demonstrate the bound for cases involving empty nodes.

Theorem 2. *The SC-approximation ratio of Mechanism 1 is* 25/8 *for any scenario with* $n \geq 6$ *and there exists at least one empty node.*

Proof. Given any instance I, we first consider the optimal social cost of I. To minimize social costs, a solution locates facility $j \in \{1, 2\}$ to y_j which is the median agent of N_j. However, if $y_1 = y_2$, this might not be a feasible solution, and also get larger than the optimal social cost. We have that

$$\mathrm{SC}^*(\mathrm{I}) \geq \sum_{i \in N_1} d(x_i, y_1) + \sum_{i \in N_2} d(x_i, y_2). \tag{2}$$

According to the facilities $\mathbf{z}$ obtained by the mechanism, we then calculate the social cost. Suppose two facilities be placed at (y_1, b) with probability 0.5, and at (y_1, b) with probability 0.5. Without loss of generality, we can assume that $a > y_1$, $b > y_2$. By definition we can obtain the following formula,

$$\mathrm{SC}(\mathbf{z}|I) = \frac{1}{2}\Big(\sum_{i \in N_1} d(x_i, y_1) + \sum_{i \in N_2} d(x_i, b)\Big) + \frac{1}{2}\Big(\sum_{i \in N_1} d(x_i, a) + \sum_{i \in N_2} d(x_i, y_2)\Big).$$

According to [15], we similarly partition the set N_l, $l \in \{1, 2\}$ into three sets L_l, M_l and R_l as follows:

- $L_l = \{i \in N_l : x_i \leq y_l\}$;
- $M_l = \{i \in N_l : x_i \in (y_l, a \text{ or } b)\}$;
- $R_l = \{i \in N_l : x_i > a \text{ or } b\}$.

By the definition of median, we have that $|L_l| \geq |M_l| + |R_l|$, where $l \in \{1, 2\}$; in particular, if $n_l = |N_l|$ is even, then $|L_l| = |M_l| + |R_l|$. Otherwise if n_l is odd, then $|L_l| > |M_l| + |R_l|$.

If we only consider when placing two facilities in one case, can be obtained the following inequalities:

$$\sum_{i \in N_1} d(x_i, y_1) + \sum_{i \in N_2} d(x_i, b) \leq \sum_{i \in N_1} d(x_i, y_1) + \sum_{i \in N_2} d(x_i, y_2) + \frac{1}{4}(n_2^2 + 4n_1n_2 + 2n_2 + 1).$$

For agents who approve facility 1 and all agents who approve facility 2 cannot be included in each other. So when $d(y_1, a) \leq 1 + \frac{1}{2}(n_1 - \mathbf{1}\{n_1\ odd\})$, then $d(y_2, b) \leq n_1 + 1 + \frac{1}{2}(n_2 - \mathbf{1}\{n_2\ odd\})$. By adding two inequalities together, we can get the following inequalities:

$$d(y_2, b)\mathbf{1}\{n_2\ odd\} + 2\sum_{i \in M_1} d(x_i, b) \leq \frac{1}{4}(n_1^2 + 2n_2 + 1).$$

So we can calculate the sum cost,

$$\begin{aligned} \mathrm{SC}(\mathbf{z}|I) &= \frac{1}{2}(\sum_{i \in N_1} d(x_i, y_1) + \sum_{i \in N_2} d(x_i, b)) + \frac{1}{2}(\sum_{i \in N_1} d(x_i, a) + \sum_{i \in N_2} d(x_i, y_2)) \\ &\leq \sum_{i \in N_1} d(x_i, y_1) + \sum_{i \in N_2} d(x_i, y_2) + \frac{1}{4}(n_1^2 + n_2^2 + 4n_1n_2 + 2n_1 + 2n_2 + 2). \end{aligned} \tag{3}$$

According to Lemma 1, $\mathrm{SC}^*(I) \geq \frac{1}{4}(n_1^2 + n_2^2 - 2)$,

$$\frac{\mathrm{SC}(\mathbf{z}|I)}{\mathrm{SC}^*(I)} \leq 1 + \frac{1}{2}(1 + \frac{4n_1n_2 + 2n_1 + 2n_2 + 4}{n_1^2 + n_2^2 - 2}).$$

We assign n_1 to x and n_2 to y, respectively, then $f(x, y) \leq \frac{13}{4}$, so we can get $\rho_{SC}(M) = \frac{25}{8}$. □

Lemma 4. *There exists an instance with $n \geq 6$ and only one empty node such that the approximation ratio of this mechanism is tight.*

Proof. The following instance I with 6 agents with only one empty node, with $N_1 = \{1, 2, 3\}$, $N_2 = \{4, 5, 6\}$, and the empty node is 7. When we put facilities 1 and 2 in positions 2 and 5 respectively, we get the optimal solution of 4 with $\mathrm{SC}^*(I) = 4$. This mechanism puts facility 1 at node 2 and facility 2 at node 7 with a one-in-half probability, and also puts facility 1 in 7 and facility 2 in 5 with a one-in-half probability. The social cost of this mechanism is $\mathrm{SC}(\mathbf{z}|I) = \frac{25}{2}$. The approximation ratio is $\frac{25}{8}$.

If there is no empty node, for the social cost of a single facility, putting the facility in the median of all agents is the optimal solution. So we consider another mechanism for no empty nodes. In the case of $y_1 \neq y_2$, we expect facility i, $i \in \{1, 2\}$ to be in the position of y_i; When $y_1 = y_2$, facility 1 should be placed in the position of y_1 and facility 2 should be placed in the position of $y_2 + 1$.

Lemma 5. *When there is no empty node,this mechanism is not strategyproof.*

Proof. The following instance I with 8 agents with no empty nodes. Agents that approve facility 1 are 1,2,4,6,8. Agents that approve facility 2 are 1,3,4,5,7. We can get that $y_1 = y_2 = 4$, so facility 1 is placed at node 4, facility 2 is placed at node 5. For agent at node 3, his cost is $5 - 3 = 2$. But if he misreports that he is only interested in facility, it causes the median of N_1 and N_2 to change, so $y_1^{'} = 3$, $y_2^{'} = 4$. This satisfies the case where the two nodes are different, so $(z_1^{'}, z_2^{'}) = (3, 4)$. Now, the cost of agent 3 is 1, which is lower than it would have been if he hadn't lied.

Therefore, in the case of no empty points, even though this mechanism may get a better approximation ratio, it will cause agents to lie about their preferences.

4 Conclusions

In this paper, we primarily focus on a scenario where the locations of all agents are known, but their preferences are unknown. All agents and facilities are on a straight line, and there are m nodes in total, the distance between each node is 1. Both facilities and agents are situated along this line. Among$m$ nodes, the mechanism is designed to minimize social cost, considering both scenarios with and without empty nodes. The approximation ratios obtained are 3 and $\frac{25}{8}$, respectively. An example is provided to demonstrate that the result is tight under this mechanism.

Future research could delve into scenarios involving two facilities where the locations and preferences of agents are not only unknown but also prone to misreporting. This includes studying settings with varied preferences, like fractional or clear preferences, or more complex location maps, such as trees or rule-based maps. Additionally, we might place facilities in multi-dimensional spaces and design strategyproof mechanisms for preferences or locations. If it is determined that the strategyproof mechanism depends on the number of facilities, specifically showing that for $k = 2$, the proved approximation ratio is achieved, then similar results can be extended to cases with different numbers of facilities. In addition, the facilities proposed by Zhao et al. [23] can only be placed in some given candidate locations. We can consider designing a strategyproof mechanism if both facilities have their own candidate locations.

Besides social cost, another important aspect to consider is the maximum cost. In 2016, Serafino and Ventre [21] demonstrated that the approximation ratio of the TwoExtremes mechanism is upper-bounded by 3, while any strategyproof mechanism is lower-bounded by $\frac{3}{2}$. Building on this foundation, Kanellopoulos et al. [15] developed a deterministic maximum mechanism and improved the approximation ratio to 2. They put the facilities in different positions according to three situations. At the beginning of this question, for the single facility randomized strategyproof mechanism, this facility is placed in the positions of the leftmost agent, the center and the rightmost agent with different probabilities and get an approximation ratio of $\frac{3}{2}$. We can try to put this mechanism in three situations to prove whether we can get a better approximation ratio than 2.

References

1. Anastasiadis, E., Deligkas, A.: Heterogeneous facility location games. arXiv preprint arXiv:2005.03095 (2020)
2. Anshelevich, E., Filos-Ratsikas, A., Voudouris, A.A.: The distortion of distributed metric social choice. Artif. Intell. **308**, 103713 (2022)
3. Babaioff, M., Feldman, M., Tennenholtz, M.: Mechanism design with strategic mediators. ACM Trans. Econ. Comput. (TEAC) **4**(2), 1–48 (2016)
4. Cai, Q., Filos-Ratsikas, A., Tang, P.: Facility location with minimax envy. In: AAAI Press/International Joint Conferences on Artificial Intelligence (2016)
5. Cheng, Y., Yu, W., Zhang, G.: Strategy-proof approximation mechanisms for an obnoxious facility game on networks. Theor. Comput. Sci. **497**, 154–163 (2013)
6. Duan, L., Li, B., Li, M., Xu, X.: Heterogeneous two-facility location games with minimum distance requirement. In: AAMAS, pp. 1461–1469 (2019)
7. Feigenbaum, I., Sethuraman, J.: Strategyproof mechanisms for one-dimensional hybrid and obnoxious facility location models. In: Workshops at the Twenty-Ninth AAAI Conference on Artificial Intelligence (2015)
8. Feigenbaum, I., Sethuraman, J., Ye, C.: Approximately optimal mechanisms for strategyproof facility location: minimizing LP norm of costs. Math. Oper. Res. **42**(2), 434–447 (2017)
9. Feldman, M., Wilf, Y.: Strategyproof facility location and the least squares objective. In: Proceedings of the Fourteenth ACM Conference on Electronic Commerce, pp. 873–890 (2013)
10. Filos-Ratsikas, A., Kanellopoulos, P., Voudouris, A.A., Zhang, R.: Settling the distortion of distributed facility location. arXiv preprint arXiv:2301.01604 (2023)
11. Filos-Ratsikas, A., Li, M., Zhang, J., Zhang, Q.: Facility location with double-peaked preferences. Auton. Agent. Multi-Agent Syst. **31**(6), 1209–1235 (2017). https://doi.org/10.1007/s10458-017-9361-0
12. Fong, C.K.K., Li, M., Lu, P., Todo, T., Yokoo, M.: Facility location games with fractional preferences. In: Proceedings of the AAAI Conference on Artificial Intelligence, vol. 32 (2018)
13. Fotakis, D., Tzamos, C.: Winner-imposing strategyproof mechanisms for multiple facility location games. Theor. Comput. Sci. **472**, 90–103 (2013)
14. Hossain, S., Micha, E., Shah, N.: The surprising power of hiding information in facility location. In: Proceedings of the AAAI Conference on Artificial Intelligence, vol. 34, pp. 2168–2175 (2020)
15. Kanellopoulos, P., Voudouris, A.A., Zhang, R.: On discrete truthful heterogeneous two-facility location. SIAM J. Discret. Math. **37**(2), 779–799 (2023)
16. Kyropoulou, M., Ventre, C., Zhang, X.: Mechanism design for constrained heterogeneous facility location. In: Fotakis, D., Markakis, E. (eds.) SAGT 2019. LNCS, vol. 11801, pp. 63–76. Springer, Cham (2019). https://doi.org/10.1007/978-3-030-30473-7_5

17. Li, M., Lu, P., Yao, Y., Zhang, J.: Strategyproof mechanism for two heterogeneous facilities with constant approximation ratio. arXiv preprint arXiv:1907.08918 (2019)
18. Menon, V., Larson, K.: Mechanism design for locating a facility under partial information. In: Fotakis, D., Markakis, E. (eds.) SAGT 2019. LNCS, vol. 11801, pp. 49–62. Springer, Cham (2019). https://doi.org/10.1007/978-3-030-30473-7_4
19. Procaccia, A.D., Tennenholtz, M.: Approximate mechanism design without money. ACM Trans. Econ. Comput. (TEAC) **1**(4), 1–26 (2013)
20. Serafino, P., Ventre, C.: Truthful mechanisms without money for non-utilitarian heterogeneous facility location. In: Proceedings of the AAAI Conference on Artificial Intelligence, vol. 29 (2015)
21. Serafino, P., Ventre, C.: Heterogeneous facility location without money. Theor. Comput. Sci. **636**, 27–46 (2016)
22. Yan, X., Chen, Y.: Strategyproof facility location mechanisms with richer action spaces. arXiv preprint arXiv:2002.07889 (2020)
23. Zhao, Q., Liu, W., Fang, Q., Nong, Q.: Constrained heterogeneous two-facility location games with max-variant cost. In: International Workshop on Frontiers in Algorithmics, pp. 25–43. Springer (2022)

Comprehensive Multi-view Subspace Clustering with Global-and-Local Representation Learning

Bin Xiao, Jipeng Guo, Juntao Hu, Yifan Dong, and Youqing Wang(✉)

College of Information Science and Technology, Beijing University of Chemical Technology, Beijing 100029, China
{2023200757,guojipeng,2022030551,2022030024}@buct.edu.cn,
wang_youqing@mail.buct.edu.cn

Abstract. Multi-view subspace clustering (MSC) has attracted significant attention due to its superiority in unsupervised multi-source data analysis. For MSC methods, the key is to explore the local view-specific and global view-shared information embedded in multi-view data. However, most existing MSC methods only explore view-specific information or shared information across multi-view data, ignoring the interaction between global and local information. This paper proposes a Comprehensive Multi-view Subspace Clustering (CMSC) with global-and-local representation collaborative learning. Specially, view-specific subspace representations are learned in the original multi-view feature space for exploring complementary local description information. And, view-shared subspace representation learned in shared latent space could explore global consistency among multi-view data. To achieve the interaction of global-and-local information and fully explore the complex correlation among multi-views, local view-specific and global view-shared representations are uniformly merged into a low-rank tensor regularized by non-convex and more compact tensor logarithmic nuclear norm. Thus, the proposed CMSC can sufficiently explore the complementarity-consistency among multi-view data and capture more comprehensive semantic correlation. Experimental results on several benchmark datasets verify the performance superiority of CMSC over some classic MSC methods.

Keywords: Multi-view Subspace Clustering · Local View-specific Representation · Global View-shared Representation · Global-and-Local Information Interaction · Non-convex Low-rank Tensor

1 Introduction

The data collected in reality exhibits new characteristics: 1) the complex structure, 2) the lack of label information, and 3) rich forms of data. Especially with

D. Du et al. (Eds.): COCOA 2024, LNCS 15434, pp. 88–99, 2025.
https://doi.org/10.1007/978-981-96-4445-2_8

the development of data sensors and collection technology, multi-view data is becoming increasingly common [1]. Multi-view data is usually collected from different fields or generated by various feature extractors, resulting in complementary and comprehensive descriptions [2]. The forms of multi-view data usually include different modalities, different features, and data from different sources. Recently, multi-view subspace clustering methods with representation learning have attracted widespread attention, exploring data intrinsic structure in an unsupervised way [3–5]. There are two important principles for multi-view representation learning [6]: (1) the consensus principle, which maximizes the consistency among various views; (2) the complementarity principle, which maximizes the exploring of view-specific information.

To better explore the complementarity in multi-view data, Cao et al. [7] proposed a Diversity-induced MSC (DiMSC) using Hilbert-Schmidt Independence Criterion (HSIC) to model the mutually independence correlation among multiple subspace representations [8]. Some MSC methods also embed multi-view data into the shared low-dimensional space to learn the common subspace representation among multi-view data [9–11]. Although these methods have achieved significant performance improvements, they have the following limitations: a) such matrix-oriented methods ignore the high-order correlations between multi-view representations. b) these multi-view subspace clustering models lack effective perception of global structural information during the multi-view subspace representation learning.

To this end, some tensor-based MSC methods with low-rank tensor structure were proposed to explore complex cross-view correlations. Low-rank Tensor constrained MSC (LTMSC) constructed the tensor by simply merging multi-view representations and utilized the tensor unfolding-based nuclear norm to characterize low-rank structure [12]. Further, the tensor Singular Value Decomposition (t-SVD) based MSC (t-SVD-MSC) [13] was proposed, using 3-order low-rank tensor merged and rotated by multiple subspace representations to model high-order correlations among multi-view data. The t-SVD-MSC significantly improved clustering performance on multi-view clustering tasks and demonstrated the strong robustness and stability. Subsequently, on basic of t-SVD-MSC, several improved tensor-based MSC versions were further proposed to improve performance.

However, most existing tensor-based MSC methods still face some crucial limitations: 1) They only learn view-specific representations or view-shared representation, and thus cannot model effective interactions between global and local information. Hence, the existing methods are unable to deeply explore structural information. 2) The tensor nuclear norm is widely used as the low-rank tensor approximation. However, convex tensor nuclear norm regularizes each singular value equally when modeling the low-rank structure of target tensor. When solving the tensor nuclear norm minimization problem, different singular values are shrunk by same threshold, which results in important information loss and noise redundancy [14]. To address these issues, this study proposes a novel MSC method, Comprehensive Multi-view Subspace Clustering (CMSC), which simultaneously learns local and global representations and achieves the interaction between them. The main contributions of this paper are summarized as follows:

- In a unified representation leaning framework, CMSC learns view-specific representation in original feature and view-shared representation in latent feature fitted by non-linear HSIC constraint. Specifically, the HSIC is used to characterize the non-linear correlation between latent feature and multi-view feature, resulting in powerful and comprehensive latent feature shared by multi-view data.
- To achieve the interaction and mutual guidance of global and local representations, CMSC integrates view-specific representations and view-shared representation into a unified tensor space and models the discriminative high-order correlations using non-convex tensor Logarithmic norm constraint.
- This study formulates the proposed CMSC as an optimization problem and proposes a simple and effective algorithm to solve it. And, extensive experiments on various real datasets verify the advantage and effectiveness of proposed CMSC method.

2 Related Work

Low-rank tensor based MSC methods achieved the satisfactory clustering performance due to its strong ability to effectively model high-order cross-view correlations and explore structural information, and the representative method was t-SVD-MSC [13]. Given the N observed samples collected from V views, i.e., $\{\mathbf{X}^{(1)}, \mathbf{X}^{(2)}, \cdots, \mathbf{X}^{(V)}\}$ where $\mathbf{X}^{(v)} = \{\mathbf{X}_1^{(v)}, \mathbf{X}_2^{(v)}, \ldots, \mathbf{X}_N^{(v)}\} \in \mathbb{R}^{d^{(v)} \times N}$, the t-SVD-MSC model could be formulated as follows

$$\begin{aligned} \min_{\mathbf{Z}^{(v)}, \mathbf{E}^{(v)}} \quad & \|\mathcal{Z}\|_{\circledast} + \lambda \|\mathbf{E}\|_{2,1} \\ \text{s.t.} \quad & \mathbf{X}^{(v)} = \mathbf{X}^{(v)}\mathbf{Z}^{(v)} + \mathbf{E}^{(v)}, \quad v = 1, 2, \cdots, V \\ & \mathcal{Z} = \phi(\mathbf{Z}^{(1)}, \mathbf{Z}^{(2)}, \ldots, \mathbf{Z}^{(V)}) \\ & \mathbf{E} = \left[\mathbf{E}^{(1)}; \mathbf{E}^{(2)}; \cdots; \mathbf{E}^{(V)}\right] \end{aligned} \tag{1}$$

where $\|\mathcal{Z}\|_{\circledast}$ is the tensor nuclear norm pursuing the low-rank structure, and $\phi(\cdot)$ is a tensor construction operator. The t-SVD-MSC learned the view-specific subspace representation $\mathbf{Z}^{(v)}$ and merged them into a 3-order low-rank tensor space $\mathcal{Z}$ to explore complex correlations. Although t-SVD-MSC has achieved excellent performance, there are still several limitations:

- The t-SVD-MSC method learns view-specific representations from a local perspective, which can capture complementary information but ignores the global consistency of multi-view data. Furthermore, global consistency information and local information cannot effectively interact, making it impossible to fully explore clustering structure and model complex correlations.

- The t-SVD-MSC utilizes simple and convex tensor nuclear norm to characterize low-rank tensor structure, resulting in biased threshold shrinkage and sub-optimal low-rank tensor representation. Therefore, cross-view correlation redundancy cannot be effectively removed.

3 The Proposed Approach

To fully explore clustering structure from multi-view data, this paper develops a novel MSC method dubbed Comprehensive Multi-view Subspace Clustering (CMSC), which mainly consists of Local View-specific Representation Learning (LVRL), Global View-shared Representation Learning (GVRL), and Global-and-Local Collaborative Interaction (GLCI) modules. Next, this section introduces these modules in detail and provides formulation for proposed CMSC model. And, an effective optimization algorithm is further designed for formulated objective problem of CMSC.

3.1 Formulation of CMSC

Local View-Specific Representation Learning. Each view provides specific descriptive information for target samples, and thus multi-view data can provide complementary information. To this end, CMSC learns multiple view-specific subspace representations in original feature space. Thus, given view-specific feature $\mathbf{X}^{(v)}$, the LVRL module is formulated as follows

$$\mathbf{X}^{(v)} = \mathbf{X}^{(v)}\mathbf{Z}^{(v)} + \mathbf{E}^{(v)} \tag{2}$$

where $\mathbf{Z}^{(v)} \in \mathbb{R}^{N\times N}$ is view-specific subspace representation, which reflects the clustering structure from local perspective. Due to complementarity of original multi-view data, the representations $\{\mathbf{Z}^{(v)}\}_{v=1}^{V}$ could provide rich and complementary structural information. In addition, $\mathbf{E}^{(v)}$ is view-specific reconstruction error fitting existing noise in original feature. There are some suitable structural constraints on $\mathbf{Z}^{(v)}$ and $\mathbf{E}^{(v)}$, such as low-rank nuclear norm constraint on $\mathbf{Z}^{(v)}$ and $l_{2,1}$ sparsity constraint on $\mathbf{E}^{(v)}$. The used constraints would be shown in overall problem formulation.

Global View-Consistency Representation Learning. Although multi-view data provides different descriptive information from various aspects, the sample they described is consistent. Thus, modeling and exploring the consistency among multiple subspace representations is necessary and beneficial for accurately understanding data and capturing clustering structure. The consistent information is partially shared by multi-view data and embedded in latent space shared by original multi-view feature. Extracting discriminative and powerful latent representation from original multi-view data is crucial.

The Hilbert-Schmidt Independence Criterion (HSIC) [15] is a nonlinear method that could model the complex correlation between two variables. Assume that the common feature representation of multi-view data is $\mathbf{H}$, CMSC uses HSIC constraint to model its nonlinear relationship with multi-view data $\{\mathbf{X}^{(v)}\}_{v=1}^{V}$, which is formulated as follows

$$\max_{\mathbf{H}} \sum_{v=1}^{V} \mathrm{HSIC}(\mathbf{X}^{(v)}, \mathbf{H}) \tag{3}$$

where $\mathrm{HSIC}(\mathbf{A}, \mathbf{B}) = \mathrm{tr}(\mathbf{K}_A \mathbf{L} \mathbf{K}_B \mathbf{L})$ is an approximate HSIC constraint, $\mathbf{K}_A$ and $\mathbf{K}_B$ are Gram matrices in the feature space with centered means, and $\mathbf{L}_{ij} = \delta_{ij} - 1/N$ centers the Gram matrix to have zero mean in the feature space. The smaller value of $\mathrm{HSIC}(\mathbf{X}^{(v)}, \mathbf{H})$, the smaller independence of the variables. Thus, maximizing the objective function in (3) can ensure that the latent representation fully contains consistent information embedded in multi-view data. Thus, the GVRL module could learn view-shared subspace representation from latent feature $\mathbf{H}$, i.e.,

$$\mathbf{H} = \mathbf{H}\mathbf{Z} + \mathbf{E}, \tag{4}$$

where $\mathbf{Z} \in \mathbb{R}^{N \times N}$ is consistent subspace representation, which explores the clustering structure from a comprehensive perspective. Thus, the representation shared by multi-view data consists of global information. In addition, $\mathbf{E}$ is also used to fit noise.

Global-and-Local Representation Collaborative Interaction. Through the LVRL and GVRL modules, CMSC successfully learns both the view-specific representations corresponding to each view and the global representation capturing consistency across multi-views. These representations $\{\mathbf{Z}^{(1)}, \cdots, \mathbf{Z}^{(V)}, \mathbf{Z}\}$ contain structural information at different levels. Thus, to further facilitate the information interaction and mutual guidance among global and local representations, CMSC integrates them into a unified tensor space and pursues low-rank structure, which models the complex correlations and reduces the redundant information. The GLCI module is mathematically formulated as follows

$$\min_{\mathcal{Z}} \|\mathcal{Z}\|_{\circledast} \quad s.t.\ \mathcal{Z} = \phi(\mathbf{Z}^{(1)}, \mathbf{Z}^{(2)}, \cdots, \mathbf{Z}^{(V)}, \mathbf{Z}) \tag{5}$$

Nonetheless, the tensor nuclear norm treats all singular values equally and shrinks them with same threshold for the optimization of tensor nuclear norm minimization problem, which is a biased and relaxed approximation of low-rank tensor constraint.

Inspired by [16,17], CMSC utilizes the tensor Logarithmic nuclear norm as effective low-rank structure constraint, penalizing all singular values by the non-convex penalty function $f(x) = \log(1 + x)$. Thus, given any 3-order tensor $\mathcal{X} \in \mathbb{R}^{n_1 \times n_2 \times n_3}$, the tensor Logarithmic nuclear norm is defined as follows

$$\|\mathcal{X}\|_L = \frac{1}{n_3} \sum_{k=1}^{n_3} \|\mathcal{X}_f^{(k)}\|_L = \frac{1}{n_3} \sum_{k=1}^{n_3} \sum_{i=1}^{h} \log\left(1 + \mathcal{S}_f^{(k)}(i,i)\right) \tag{6}$$

where $\mathcal{S}_f$ is the diagonal singular value tensor of $\mathcal{X}$ in the Fourier domain; $h = \min(n_1, n_2)$. Due to the non-convex property of penalty function $f(x) = \log(1 + x)$, the tensor Logarithmic nuclear norm offers several advantages:

- the tensor Logarithmic nuclear norm is a more tighter approximation of the original tensor rank function.
- the tensor Logarithmic nuclear norm is an unbiased approximation of the original tensor rank function. This is mainly attributed that the derivative of function $f(x)$ monotonically decreases, which could achieve heavier penalty for smaller singular values representing noise and mild penalty for large singular values representing discriminative information.

Thus, the GLCI module is further revised as follows

$$\min_{\mathcal{Z}} \|\mathcal{Z}\|_L \quad s.t. \ \mathcal{Z} = \phi(\mathbf{Z}^{(1)}, \mathbf{Z}^{(2)}, \cdots, \mathbf{Z}^{(V)}, \mathbf{Z}) \tag{7}$$

With this optimization framework, the more redundancies could be fully remove, resulting in discriminative correlations.

Objective Function. By combining GVLR, LVRL, and GLCI modules, the proposed CMSC could obtain comprehensive subspace representations. The final formulation of CMSC is as follows:

$$\begin{aligned} \min_{\Theta} & -\sum_{v=1}^{V} \mathrm{HSIC}(\mathbf{X}^{(v)}, \mathbf{H}) + \lambda_1 \|\mathcal{Z}\|_L + \lambda_2 \left(\|\mathbf{E}\|_{2,1} + \sum_{v=1}^{V} \|\mathbf{E}^{(v)}\|_{2,1} \right) \\ \text{s.t. } & \mathbf{X}^{(v)} = \mathbf{X}^{(v)}\mathbf{Z}^{(v)} + \mathbf{E}^{(v)}, \\ & \mathbf{H} = \mathbf{H}\mathbf{Z} + \mathbf{E}, \\ & \mathcal{Z} = \phi(\mathbf{Z}^{(1)}, \mathbf{Z}^{(2)}, \cdots, \mathbf{Z}^{(V)}, \mathbf{Z}) \end{aligned} \tag{8}$$

where $\Theta = \{\mathbf{H}, \mathbf{Z}^{(1)}, \mathbf{Z}^{(2)}, \cdots, \mathbf{Z}^{(V)}, \mathbf{Z}, \mathbf{E}_1^{(v)}, \cdots, \mathbf{E}_v^{(v)}, \mathbf{E}\}$ are unknown variables; λ_1 and λ_2 are non-negative balanced parameters. Clearly, the CMSC learns both local view-specific representations and global view-shared representation, which could simultaneously explores complementary information and consistency structure. The designed information interaction module ensures that the global representation could provide beneficial guidance for local subspace representation learning and removing the multi-view redundancy; conversely, complementary view-specific representations could provide discriminative information, which is integrated into the global representation. After the model is optimized, a comprehensive and discriminative representation can be constructed, i.e., $\mathbf{Z}_c = \frac{1}{V+1}\left(\sum_{v=1}^{V} \mathbf{Z}^{(v)} + \mathbf{Z}\right)$. The clustering labels could be contained by conducting the spectral clustering on the similarity matrix $\mathbf{S} = (|\mathbf{Z}_c| + |\mathbf{Z}_c^T|)/2$.

3.2 Optimization

To optimize the above objective function with multiple variables, the Alternating Direction Method of Multipliers (ADMM) [18] is employed as optimization framework and provides an easy-to-implement optimization algorithm. To make the variables separable, a splitting variable $\mathcal{J}$ is introduced and then its unconstrained augmented Lagrangian function is provided as follows

$$\begin{aligned}\mathcal{L} = &-\sum_{v=1}^{V}\mathrm{HSIC}(\mathbf{X}^{(v)},\mathbf{H}) + \lambda_1\|\mathcal{J}\|_L + \lambda_2(\sum_{v=1}^{V}\|\mathbf{E}^{(v)}\|_{2,1} + \|\mathbf{E}\|_{2,1})\\ &+\sum_{v=1}^{V}\Omega(\mathbf{Y}_1^{(v)},\mathbf{X}^{(v)} - \mathbf{X}^{(v)}\mathbf{Z}^{(v)} - \mathbf{E}^{(v)}) + \Omega\left(\mathbf{Y}_2,\mathbf{H} - \mathbf{HZ} - \mathbf{E}\right)\\ &+\Omega\left(\mathcal{W},\mathcal{Z} - \mathcal{J}\right)\end{aligned} \tag{9}$$

where $\{\mathbf{Y}_1^{(v)} \in \mathbb{R}^{d_v\times n}\}_{v=1}^{V}$, $\mathbf{Y}_2 \in \mathbb{R}^{d\times n}$, and $\mathcal{W} \in \mathbb{R}^{n\times(V+1)\times n}$ are Lagrange multipliers, and $\Omega(\mathbf{A},\mathbf{B}) = \langle\mathbf{A},\mathbf{B}\rangle + \frac{\mu}{2}\|\mathbf{B}\|_F^2$ for any matrix or tensor, where μ is the penalty parameter. Therefore, in each iteration, the variables are alternately optimized as follows:

Update $\mathbf{Z}^{(v)}$: Fixing the other variables, the sub-problem involved $\mathbf{Z}^{(v)}$ is as follows

$$\begin{aligned}&\min_{\mathbf{Z}^{(v)}}\Omega(\mathbf{Y}_1^{(v)},\mathbf{X}^{(v)} - \mathbf{X}^{(v)}\mathbf{Z}^{(v)} - \mathbf{E}^{(v)}) + \Omega(\mathbf{W}^{(v)},\mathbf{Z}^{(v)} - \mathbf{J}^{(v)})\\ \Leftrightarrow&\min_{\mathbf{Z}^{(v)}}\frac{\mu}{2}\left\|\mathbf{X}^{(v)} - \mathbf{X}^{(v)}\mathbf{Z}^{(v)} - \mathbf{E}^{(v)} + \frac{1}{\mu}\mathbf{Y}_1^{(v)}\right\|_F^2 + \frac{\mu}{2}\left\|\mathbf{Z}^{(v)} - \mathbf{J}^{(v)} + \frac{1}{\mu}\mathbf{W}^{(v)}\right\|_F^2\end{aligned} \tag{10}$$

Taking the derivative with respect to $\mathbf{Z}^{(v)}$ and setting it to zero, we have

$$\mathbf{Z}^{(v)} = \left(\mu\left(\mathbf{X}^{(v)}\right)^T\mathbf{X}^{(v)} + \mu\mathbf{I}\right)^{-1}\widehat{\mathbf{Z}}^{(v)} \tag{11}$$

where $\widehat{\mathbf{Z}}^{(v)} = \mu(\mathbf{X}^{(v)})^T(\mathbf{X}^{(v)} + \mathbf{Y}_1^{(v)} - \mathbf{E}^{(v)}) - \mathbf{W}^{(v)} + \mu\mathbf{J}^{(v)}$, $\mathbf{W}^{(v)} = \phi^{-1}(\mathcal{W})$, and $\mathbf{J}^{(v)} = \phi^{-1}(\mathcal{J})$.

Update $\mathbf{E}^{(v)}$: Fixing the other variables, the sub-problem involved $\mathbf{E}^{(v)}$ is as follows

$$\begin{aligned}&\min_{\mathbf{E}^{(v)}}\lambda_2\|\mathbf{E}^{(v)}\|_{2,1} + \Omega(\mathbf{Y}_1^{(v)},\mathbf{X}^{(v)} - \mathbf{X}^{(v)}\mathbf{Z}^{(v)} - \mathbf{E}^{(v)})\\ \Leftrightarrow&\min_{\mathbf{E}^{(v)}}\frac{\lambda_2}{\mu}\|\mathbf{E}^{(v)}\|_{2,1} + \frac{1}{2}\left\|\mathbf{X}^{(v)} - \mathbf{X}^{(v)}\mathbf{Z}^{(v)} - \mathbf{E}^{(v)} + \frac{1}{\mu}\mathbf{Y}_1^{(v)}\right\|_F^2\end{aligned} \tag{12}$$

This optimization problem with closed-form solution can be easily optimized by sparse thresholding algorithm [19].

Update $\mathbf{E}$: Fixing the other variables, the sub-problem involved $\mathbf{E}$ is as follows

$$\begin{aligned}&\min_{\mathbf{E}}\lambda_2\|\mathbf{E}\|_{2,1} + \Omega(\mathbf{Y}_2,\mathbf{H} - \mathbf{HZ} - \mathbf{E})\\ \Leftrightarrow&\min_{\mathbf{E}}\frac{\lambda_2}{\mu}\|\mathbf{E}\|_{2,1} + \frac{1}{2}\left\|\mathbf{H} - \mathbf{HZ} - \mathbf{E} + \frac{1}{\mu}\mathbf{Y}_2\right\|_F^2\end{aligned} \tag{13}$$

Similarly, it could be solved by sparse thresholding algorithm.

Update Z: Fixing the other variables, the sub-problem involved **Z** is as follows

$$\begin{aligned} &\min_{\mathbf{Z}} \Omega(\mathbf{Y}_2, \mathbf{H}-\mathbf{H}\mathbf{Z}-\mathbf{E}) + \Omega(\mathbf{W}, \mathbf{Z}-\mathbf{J}) \\ \Leftrightarrow &\min_{\mathbf{Z}} \frac{\mu}{2}\left\|\mathbf{H}-\mathbf{H}\mathbf{Z}-\mathbf{E}+\frac{1}{\mu}\mathbf{Y}_2\right\|_F^2 + \frac{\mu}{2}\left\|\mathbf{Z}-\mathbf{J}+\frac{1}{\mu}\mathbf{W}\right\|_F^2 \end{aligned} \tag{14}$$

where $\mathbf{W} = \phi^{-1}(\mathcal{W})(V+1)$, and $\mathbf{J} = \phi^{-1}(\mathcal{J})(V+1)$. Taking the derivative with respect to **Z** and setting it to zero, we have

$$\mathbf{Z} = \left(\mu\mathbf{H}^T\mathbf{H} + \mu\mathbf{I}\right)^{-1}\left(\mu\mathbf{H}^T\mathbf{H} + \mathbf{H}^T\mathbf{Y}_2 - \mu\mathbf{H}^T\mathbf{E} - \mathbf{W} + \mu\mathbf{J}\right) \tag{15}$$

Update $\mathcal{J}$: Fixing the other variables, the sub-problem involved $\mathcal{J}$ is as follows

$$\min_{\mathcal{J}} \lambda_1\|\mathcal{J}\|_L + \Omega(\mathcal{W}, \mathcal{Z}-\mathcal{J}) \Leftrightarrow \min_{\mathcal{J}} \frac{\lambda_1}{\mu}\|\mathcal{J}\|_L + \frac{1}{2}\left\|\mathcal{Z}-\mathcal{J}+\frac{1}{\mu}\mathcal{W}\right\|_F^2 \tag{16}$$

This is a tensor Logarithmic nuclear norm minimization problem, which could be solved directly by the existing low-rank thresholding algorithm [16].

Update H: Fixing the other variables, the sub-problem involved **H** is as follows

$$\begin{aligned} &\min_{\mathbf{H}} -\sum_{v=1}^{V} \mathrm{HSIC}(\mathbf{X}^{(v)}, \mathbf{H}) + \Omega(\mathbf{Y}_2, \mathbf{H}-\mathbf{H}\mathbf{Z}-\mathbf{E}) \\ \Leftrightarrow &\min_{\mathbf{H}} -\mathrm{tr}(\mathbf{K}_X\mathbf{L}\mathbf{K}_H\mathbf{L}) + \frac{\mu}{2}\left\|\mathbf{H}-\mathbf{H}\mathbf{Z}-\mathbf{E}+\frac{1}{\mu}\mathbf{Y}_2\right\|_F^2 \end{aligned} \tag{17}$$

where $\mathbf{K}_X = \sum_{v=1}^{V}(\mathbf{X}^{(v)})^T\mathbf{X}^{(v)}$, $\mathbf{K}_H = \mathbf{H}^T\mathbf{H}$.

Taking the derivative with respect to **H** and setting it to zero, we obtain

$$\mathbf{H} = \left(\mu(\mathbf{I}-\mathbf{Z})(\mathbf{I}-\mathbf{Z})^T - 2\mathbf{L}\mathbf{K}_X\mathbf{L}\right)^{-1}(\mu\mathbf{E}-\mathbf{Y}_2)(\mathbf{I}-\mathbf{Z})^T \tag{18}$$

Update Lagrange Multipliers $\mathbf{Y}_1^{(v)}$, $\mathbf{Y}_2$, $\mathcal{W}$, and Penalty Parameter μ:

$$\begin{aligned} &\mathbf{Y}_1^{(v)} : \mathbf{Y}_1^{(v)} = \mathbf{Y}_1^{(v)} + \mu\left(\mathbf{X}^{(v)} - \mathbf{X}^{(v)}\mathbf{Z}^{(v)} - \mathbf{E}^{(v)}\right) \\ &\mathbf{Y}_2 : \mathbf{Y}_2 = \mathbf{Y}_2 + \mu(\mathbf{H}-\mathbf{H}\mathbf{Z}-\mathbf{E}) \\ &\mathcal{W} : \mathcal{W} = \mathcal{W} + \mu(\mathcal{Z}-\mathcal{J}) \\ &\mu : \mu = \min(\rho\mu, \mu_{\max}) \end{aligned} \tag{19}$$

where $\rho > 1$ is a constant to accelerate the convergence.

4 Experiment

To demonstrate the superiority of our CMSC, this section conducts comprehensive experiments on several benchmark datasets.

4.1 Experimental Setting

Datasets: Five different types of real-world datasets are utilized in this experiment, including **UCI-Digits**, **Caltech101-7**, **ORL**, **BBC**, and **MSRCv1**. Statistics and more details of these datasets can be found in [6] .

Competitors: This experiment compares proposed CMSC against several representative MSC methods, including **RMSC** [20], **DiMSC** [7], **LTMSC** [12], **t-SVD-MSC** [13], and **ETLMSC** [21]. Please refer to the Sect. 1 or original papers for the detailed descriptions of these methods.

Clustering Metrics: To scientifically evaluate the clustering performance of all MSC methods, several commonly used clustering metrics are employed, including Accuracy (ACC), Normalized Mutual Information (NMI), F-score, and Precision (PRE).

Table 1. Clustering results on the all datasets.

Datasets	Methods	RMSC	DiMSC	LTMSC	t-SVD-MSC	ETLMSC	CMSC
UCI-Digits	ACC	0.915	0.703	0.803	0.955	0.997	**0.999**
	NMI	0.822	0.772	0.775	0.932	0.993	**0.999**
	F-score	0.811	0.695	0.775	0.932	0.995	**0.999**
	PRE	0.797	0.673	0.739	0.933	0.995	**0.999**
Caltech101-7	ACC	0.453	0.426	0.853	0.907	0.715	**0.987**
	NMI	0.351	0.311	0.564	0.841	0.633	**0.967**
	F-score	0.332	0.335	0.582	0.835	0.611	**0.974**
	PRE	0.259	0.255	0.644	0.823	0.578	**0.973**
ORL	ACC	0.723	0.838	0.795	0.897	0.765	**0.987**
	NMI	0.872	0.940	0.930	0.979	0.903	**0.997**
	F-score	0.654	0.807	0.768	0.902	0.896	**0.987**
	PRE	0.607	0.764	0.766	0.846	0.851	**0.979**
BBC	ACC	0.775	0.892	0.591	0.995	0.997	**0.998**
	NMI	0.616	0.728	0.442	0.989	0.817	**0.993**
	F-score	0.656	0.810	0.546	0.789	0.833	**0.996**
	PRE	0.703	0.811	0.525	0.800	0.886	**0.997**
MSRCv1	ACC	0.596	0.593	0.842	0.995	0.695	**1.000**
	NMI	0.421	0.476	0.755	0.989	0.622	**1.000**
	F-score	0.556	0.584	0.736	0.990	0.585	**1.000**
	PRE	0.711	0.883	0.725	0.990	0.571	**1.000**

4.2 Experimental Results

The experimental results of all MSC methods on five real-world datasets are presented in Table 1, where the best results are highlighted with the bold. Some interesting observations can be clearly summarized as follows.

- Overall, the proposed method CMSC consistently outperforms all compared MSC baselines over all clustering metrics on all datasets. Interestingly, CMSC achieves the best clustering performance in unsupervised scenarios. And, the performance improvement of proposed CMSC on some benchmark datasets is remarkable. For instance, on ORL dataset, CMSC outperforms the sub-optimal t-SVD-MSC with improvements of 9.0%, 1.8%, 8.5%, and 13.3% in terms of ACC, NMI, F-score, and Precision, respectively. These experimental phenomenons intuitively demonstrate the superiority of proposed CMSC method on unsupervised clustering task.
- Among all competitors, t-SVD-MSC is the most critical baseline. Compared with t-SVD-MSC, the corresponding improvement of proposed CMSC is 4.44% on ACC, 4.52% on NMI, 10.16% on F-score, and 11.20% on Precision averagely. This is mainly attributed that CMSC method simultaneously explores local view-specific and global information and achieves the collaborative interaction between them. Thus, CMSC could capture comprehensive clustering structure and deeply understand multi-view data.
- Under various clustering task scenarios (including UCI-Digits, Caltech101-7, MSRCv1 object datasets; ORL face dataset; BBC text dataset), the performance of proposed CMSC is always satisfactory, verifying its stability and robustness.

5 Conclusion

This paper proposed a novel multi-view subspace clustering method, CMSC, which learns comprehensive subspace representations from local to global. To capture complementary information embedded in multi-view data, CMSC learned view-specific subspace representations in original feature space. Further, CMSC learned view-shared subspace representation in latent space to explore consistency information across multi-views. Due to the fact that interaction between global and local information is beneficial for understanding multi-view data, CMSC explored high-order and complex correlations among view-specific and view-shared representations in a unified tensor space with low-rank structure. The experimental results highlighted the superior clustering performance of proposed CMSC method.

Acknowledgments. This research was supported by the Beijing Natural Science Foundation under Grant 4244085; in part by National Science Fund for Distinguished Young Scholars of China under Grant 62225303; in part by National Natural Science Foundation of China under Grant 62403043; in part by the Postdoctoral Fellowship Program of CPSF under Grant GZC20230203; in part by the China Postdoctoral Science Foundation under Grant 2023M740201; in part by Fundamental Research Funds for the Central Universities under Grant ZY2414.

Disclosure of Interests. The authors have no competing interests to declare that are relevant to the content of this article.

References

1. Xu, C., Tao, D., Xu, C.: A survey on multi-view learning. arXiv preprint arXiv:1304.5634 (2013)
2. Zhao, J., Xie, X., Xu, X., Sun, S.: Multi-view learning overview: recent progress and new challenges. Information Fusion **38**, 43–54 (2017)
3. Han, L., Wang, C., Xu, D., Zhang, D.: Algorithms for the prize-collecting k-Steiner tree problem. Tsinghua Sci. Technol. **27**(5), 785–792 (2022)
4. Chen, M.-S., et al.: Representation learning in multi-view clustering: a literature review. Data Sci. Eng. **7**(3), 225–241 (2022)
5. Wang, H., et al.: Learning task-driving affinity matrix for accurate multi-view clustering through tensor subspace learning. Inf. Sci. **563**, 290–308 (2021)
6. Guo, J., Sun, Y., Gao, J., Hu, Y., Yin, B.: Rank consistency induced multiview subspace clustering via low-rank matrix factorization. IEEE Trans. Neural Netw. Learn. Syst. **33**(7), 3157–3170 (2021)
7. Cao, X., Zhang, C., Fu, H., Liu, S., Zhang, H.: Diversity-induced multi-view subspace clustering. In: CVPR, pp. 586–594 (2015)
8. Du, G., Zhou, L., Lü, K., Wu, H., Xu, Z.: Multiview subspace clustering with multilevel representations and adversarial regularization. IEEE Trans. Neural Netw. Learn. Syst. **34**(12), 10279–10293 (2022)
9. Zhang, C., Hu, Q., Fu, H., Zhu, P., Cao, X.: Latent multi-view subspace clustering. In: CVPR, pp. 4333–4341 (2017)
10. Zhang, C., et al.: Generalized latent multi-view subspace clustering. IEEE Trans. Pattern Anal. Mach. Intell. **42**(1), 86–99 (2020)
11. Li, R., Zhang, C., Hu, Q., Zhu, P., Wang, Z.: Flexible multi-view representation learning for subspace clustering. In: IJCAI 2019, pp. 2916–2922 (2019)
12. Zhang, C., Fu, H., Liu, S., Liu, G., Cao, X.: Low-rank tensor constrained multiview subspace clustering. In: ICCV, pp. 1582–1590 (2015)
13. Xie, Y., Tao, D., Zhang, W., Liu, Y., Zhang, L., Qu, Y.: On unifying multi-view self-representations for clustering by tensor multi-rank minimization. Int. J. Comput. Vis. **126**, 1157–1179 (2018)
14. Chen, Y., Wang, S., Peng, C., Hua, Z., Zhou, Y.: Generalized nonconvex low-rank tensor approximation for multi-view subspace clustering. IEEE Trans. Image Process. **30**, 4022–4035 (2021)
15. Greenfeld, D., Shalit, U.: Robust learning with the Hilbert-Schmidt independence criterion. In: ICML, pp. 3759–3768 (2020)
16. Guo, J., Sun, Y., Gao, J., Hu, Y., Yin, B.: Logarithmic schatten-p norm minimization for tensorial multi-View subspace clustering. IEEE Trans. Pattern Anal. Mach. Intell. **45**(3), 3396–3410 (2022)
17. Guo, J., Sun, Y., Gao, J., Hu, Y., Yin, B.: Multi-attribute subspace clustering via auto-weighted tensor nuclear norm minimization. IEEE Trans. Image Process. **31**, 7191–7205 (2022)
18. Lin, Z., Chen, M., Ma, Y.: The augmented Lagrange multiplier method for exact recovery of corrupted low-rank matrices. arXiv preprint arXiv:1009.5055 (2010)
19. Elhamifar, E., Vidal, R.: Sparse subspace clustering: algorithm, theory, and applications. IEEE Trans. Pattern Anal. Mach. Intell. **35**(11), 2765–2781 (2013)

20. Xia, R., Pan, Y., Du, L., Yin, J.: Robust multi-view spectral clustering via low-rank and sparse decomposition. In: AAAI, pp. 2149–2155 (2014)
21. Wu, J., Lin, Z., Zha, H.: Essential tensor learning for multi-view spectral clustering. IEEE Trans. Image Process. **28**(12), 5910–5922 (2019)

On the Twin-width of Outerplanar Graphs

Muhammad Anwarul Azim(✉), Sk Ruhul Azgor, Sadia Sharmin, and Md. Saidur Rahman

Graph Drawing and Information Visualization Laboratory, Department of Computer Science and Engineering, Bangladesh University of Engineering and Technology (BUET), Dhaka 1205, Bangladesh
0422052044@grad.cse.buet.ac.bd, 1805091@ugrad.cse.buet.ac.bd, {sadiasharmin,saidurrahman}@cse.buet.ac.bd

Abstract. Twin-width is a recently introduced graph parameter that measures the similarity of a graph to a cograph. It is analogous to treewidth, which measures how similar a graph to a tree is. Many NP-hard problems are fixed-parameter tractable on graphs with bounded treewidth or twin-width. Twin-width is proven as a more general width parameter as it is bounded for all bounded treewidth graphs and beyond. For example, the treewidth and twin-width for planar graphs are $O(\sqrt{n})$ and at most 8, respectively. Nonetheless, the twin-width bound for planar graphs is yet to be proven tight, and very little is known about the subclasses of planar graphs. In this paper, we focus on obtaining the tight bound of the twin-width of outerplanar graphs, a well-known subclass of planar graphs. We show that every outerplanar graph has twin-width at most 3, and there is an outerplanar graph with twin-width at least 3.

Keywords: Twin-width · Fixed-parameter tractable · Outerplanar graphs

1 Introduction

A cograph is a graph without a path of four vertices as an induced subgraph. Many NP-hard problems, such as the maximum clique problem, the minimum coloring problem, hamiltonicity testing, etc., can be efficiently solved in cographs [9]. Recently, researchers have been interested in a new graph parameter introduced by Bonnet et al. called twin-width [7], which measures how far a graph is from being a cograph using the notion of "twins." Two vertices u and v in a graph are twins if they have the same set of neighbors (excluding themselves if adjacent). A cograph always contains twins, and "contracting" them results in a smaller cograph. A graph that is not a cograph does not have this property, but still, we can find vertex pairs that are "almost twins", that is, the intersection of their neighborhood is the maximum. Twin-width is defined via iterated contraction of vertices that are almost twins. Twin-width resembles treewidth, a widely-studied graph parameter that measures how "tree-like" a graph is [15].

D. Du et al. (Eds.): COCOA 2024, LNCS 15434, pp. 100–112, 2025.
https://doi.org/10.1007/978-981-96-4445-2_9

Also, similar to treewidth, twin-width has been proven successful in designing parameterized algorithms for several NP-hard problems, which include the independent set problem, dominating set problem, induced subgraph isomorphism, etc. [5].

Although computing twin-width for arbitrary graphs is NP-hard [3], many important graph classes have bounded twin-width, including all graphs with bounded known width parameters (e.g., treewidth, clique-width, rank-width) [6] and beyond, such as planar graphs. See [4] for classes of bounded twin-width graphs. However, such bounds for many graph classes were proven using general arguments, making them non-explicit with astronomical value. For example, a non-explicit twin-width bound for planar graphs was proven in [7], which is at least 2^{1000} as mentioned in [8]. Therefore, a series of works has been devoted to improving the twin-width of planar graphs from 2^{1000} to 583 [8], 183 [12], 37 [2], and finally, 8 [11]. Moreover, Král and Lamison [13] have constructed a graph with twin-width exactly 7, which is very close to the best upper bound. Nonetheless, the tight bound remains open, and further improvement seems challenging as both the current lower and upper bound demand complex technical proofs.

Fortunately, there are well-studied subclasses of planar graphs, such as outerplanar graphs, and studying their twin-width bound may provide valuable insights for finding the tight twin-width bound for planar graphs. Studying a restricted class has previously proven to be helpful for solving problems in general class; see [1,10]. Yet, except for grid graphs and planar bipartite graphs, which have twin-width at most 4 [7,16], and at most 6 [11], respectively; we could not find any published result on the twin-width upper bound for any nontrivial subclass of planar graphs. Note that we consider trees and cycles as trivial as it is easy to see that their twin width is not more than two, and examples can be constructed easily where the bound is tight.

In this paper, we show that the twin-width of any outerplanar graph is at most three. We first prove the bound for maximal outerplanar graphs, where we construct a contraction sequence of width at most 3 based on a bottom-up search on a rooted weak dual of the maximal outerplanar graph. Later, we show this holds for any outerplanar graph. Finally, we construct a maximal outerplanar graph with 12 vertices with twin-width at least 3, proving the bound as tight.

The rest of the paper is organized as follows. In Sect. 2, we give the necessary terminologies used throughout the paper. Section 3 deals with the twin-width of outerplanar graphs, first the upper bound and then the lower bound. Finally, Sect. 4 is a conclusion.

2 Preliminaries

In this section, we define the basic terminologies that are used throughout the paper. For basic graph theoretic definitions, we refer to [14].

Let $G = (V, E)$ be a simple graph with vertex set V and edge set E. We denote the set of vertices by $V(G)$ and the set of edges by $E(G)$. We usually denote by G_n a graph of n vertices. An edge joining vertices u and v is denoted by uv. If $e = uv \in E(G)$, we call u a *neighbor* of v, and v is a neighbor of u in G. Also, we say a and b are *adjacent*, and the edge e is *incident* to u and v. We denote by $N(v)$ the set of neighbors, also referred to as the *neighborhood*, of a vertex v, and the cardinality of $N(v)$ is called the *degree* of v. We denote by $deg(v)$ the degree of a vertex v. Two vertices u, v in G are *twins* if $N(u) - \{v\} = N(v) - \{u\}$.

A connected graph without a cycle is a *tree* whose vertices we refer to as *nodes*. In a tree, a node is an *internal node* if it has a degree of at least two; otherwise, it is called a *leaf*. A *path graph* is a tree, denoted as P_n, where n is the number of nodes, such that exactly two nodes are leaves and the rest of the nodes have degrees exactly two. For any tree T, any single node can be selected as the *root* of T, say r, and a *parent/child* relationship can be constructed such that every node u, except for r, has exactly one node as a *parent* and $deg(u) - 1$ nodes as *children*. Thus, the root has no parent, and nodes with no children are called the *leaves* of T.

A *subgraph* of a graph $G = (V, E)$ is a graph $G^{'} = (V^{'}, E^{'})$ such that $V^{'} \subseteq V$ and $E^{'} \subseteq E$. For a graph G, if $S \subseteq V(G)$, then the graph obtained by deleting $V(G) - S$ is known as the subgraph of G *induced* by S, denoted as $G[S]$. Also, instead of deleting vertices, which simultaneously deletes incident edges, subgraphs can be obtained by deleting edges only, defined as *spanning* subgraphs. Graphs where every pair of vertices are mutually adjacent are known as the *complete graphs*, denoted as K_n, where n represents the number of vertices. It is easy to see that any pair of vertices in a complete graph K_n are twins. For a graph G, if $S \subseteq V(G)$ such that $G[S]$ is a complete graph, then S is known as a k-clique of G, where k denotes the cardinality of S. An *isomorphism* is a bijection between the vertex sets of two graphs that preserves the adjacency relation. Any graph without any induced subgraph isomorphic to P_4 is defined as a *cograph*, and every cograph contains at least a pair of twins. Also, cographs are closed under deleting one vertex from a twin; thus, repeating this process reduces any cograph to a single vertex.

A graph is *planar* if it can be embedded in the plane such that no two edges intersect except for their common vertex. A *plane graph* is a planar graph with a fixed planar embedding. A plane graph divides the plane into connected regions called *faces*. The unbounded region is called the *outer face*; all other faces are *inner faces*. Any edge in the common boundary of two inner faces is an *inner edge*; otherwise, it is an *outer edge*. A graph is *outerplanar* if it has a plane embedding where all vertices are on the outer face. An outerplanar embedding of an outerplanar graph is called an *outerplane* graph. An outerplanar graph is *maximal* if it has the maximum possible edges for the given number of vertices to be outerplanar. It is known that every maximal outerplanar graph $G_n, n > 2$ is biconnected, has $n-2$ inner faces, and each inner face contains exactly 3 vertices in its boundary, inducing a 3-clique of G_n [14]. Also, it is easy to observe that, for any vertex v in a maximal outerplanar graph G, the subgraph induced by the neighbors of v, $G[N(v)]$, is a path graph.

A graph G is a *trigraph* if its edges are partitioned into two sets, the set of *red edges* $E_R(G)$, and the set of *black edges* $E_B(G)$, such that $E_R(G) \cup E_B(G) =$

$E(G)$, and $E_R(G) \cap E_B(G) = \varnothing$. Note that either of $E_R(G)$ and $E_B(G)$ can be empty. For any $v \in V(G)$, if uv is a red edge, then v is a *red neighbor* of u, and u is a red neighbor of v. The number of red neighbors of a vertex v is the *red degree* of v. Similarly, we define a *black neighbor* and the *black degree* of a vertex. A *d-trigraph* is a trigraph that contains vertices with red degrees at most d.

Let u and v be two vertices in a trigraph G_n of $n \geq 2$ vertices. We obtain another trigraph G_{n-1} from G_n by (i) deleting u and v of G_n, and adding a new vertex w, and (ii) adding edges between w and all $z \in (N(u)) \cup N(v)) - \{u, v\}$. We regard the edge wz in G_{n-1} as a black edge if both uz and vz exist in G_n as black edges; otherwise, we regard wz as a red edge. We call the operation above a *contraction* and denote it by $< u, v > \rightarrow w$. We call $\{u, v\}$ the *contraction pair of* G_n and w the *contracted vertex* of G_{n-1}. We also call $\{u, v\}$ and w the contraction pair and the contracted vertex of the contraction $< u, v > \rightarrow w$, respectively. The *width* of the contraction is d if G_{n-1} is a d-trigraph. We call a contraction of width d a *d-contraction.*

Figure 1 illustrates a 4-contraction, $< u, v > \rightarrow w$, where Fig. 1(a) depicts a 2-trigraph G_7 and Fig. 1(b) depicts the 4-trigraph G_6 obtained by the contraction. The black and red edges in the graphs in Figs. 1(a), (b) are represented with solid and dash-dotted lines, respectively.

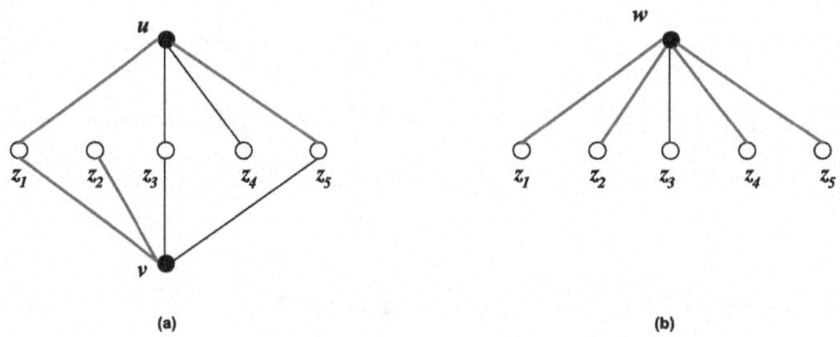

Fig. 1. Illustration of (a) 2-trigraph G_7 with contraction pairs $\{u, v\}$ and (b) 4-trigraph G_6, obtained from G_7 by the contraction $< u, v > \rightarrow w$ (Color figure online)

For an input trigraph G_n with no red edge, if a sequence of trigraphs $G_{n-1}, ..., G_i, n > i > 0$ is obtained from a sequence of contractions S, such that no contraction in S has width more than d, then we say S is a *d-sequence* that reduces G_n to G_i. Also, we say S is a d-sequence of G_n if it reduces G_n to G_1. Observe that the width of a contraction depends on the choice of its contraction pairs, and there are total $\binom{n}{2}$ different choices of contraction pairs in G_n for performing a single contraction. Consequently, there are total $\binom{n}{2}\binom{n-1}{2}...\binom{2}{2} = \frac{n!\,(n-1)!}{2^{(n-1)}}$ ways of generating contraction sequence of G_n. The *twin-width* of G_n is the minimum integer d such that a d-sequence of G_n exists. It is easy to see that any trigraph G_n trivially admits a $n-1$-sequence. Also, if G_n admits a d-

sequence and $G_{n'}$ is an induced subgraph of G_n, then $G_{n'}$ admits a d'-sequence such that $d' \leq d$.

Figure 2 illustrates two different contraction sequences for the path graph with four vertices, P_4, where Fig. 2(a) shows a 2-sequence and Fig. 2(b) shows a 1-sequence. It is easy to observe that P_4 does not admit a 0-sequence; hence, the twin-width of P_4 is one.

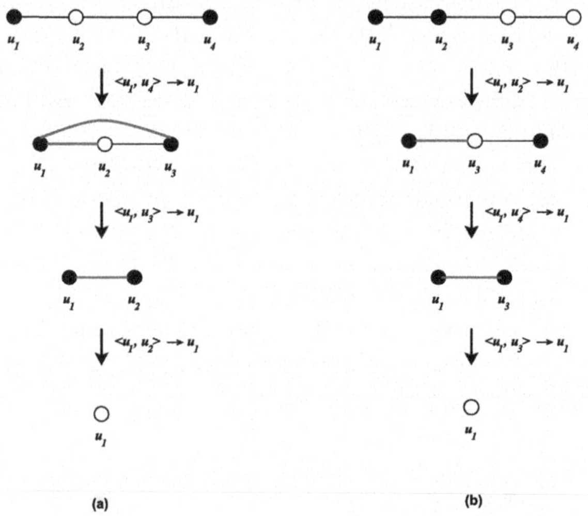

Fig. 2. Illustration of (a) 2-sequence, and (b) 1-sequence, of P_4

3 Twin-width of Outerplanar Graphs

In this section, we first show that the twin-width of a maximal outerplanar graph is at most three by constructing a contraction sequence of width at most three. We then show that this bound holds for any outerplanar graph. Before proceeding further, we prove some properties of a maximal outerplanar graph.

Let G be a maximal outerplanar graph. We consider any plane embedding of G and construct another graph G^* as follows: we place a vertex v' in each inner face of G, and for each inner edge e of G, we draw an edge e' which crosses e and joins two vertices of G^* corresponding to the two faces of G on which e lies. We call G^* a *weak dual* of G. It is known that the weak dual G^* is a tree [14]. We consider a vertex of degree one as the root and make G^* a rooted tree T. See Fig. 3(a) for an example. Throughout this section, the nodes of T are labeled using $'$ (prime) notation to distinguish them from the vertices of G. We denote the face of G corresponding to the node u' of T by $F(u')$, and we denote the set of vertices that belong to the boundary of $F(u')$ by $V(F(u'))$.

Lemma 1. *Let T be the rooted weak dual of a maximal outerplanar graph G_n where $n \geq 5$. Then there exists $u' \in V(T)$ such that u' has at most two children which are all leaves of T. Also, for any leaf v' of T, the face $F(v')$ in G_n has a vertex of degree 2.*

Proof. As G_n has at least 5 vertices, the number of faces in G_n is at least 3, and T has at least 3 vertices. Since every inner face of G_n contains exactly three edges on its boundary, the degree of any node of T is at most 3. As we considered, T is rooted at r' where $deg(r') = 1$; hence, every internal node of T, except r', has one parent and, at most, two children. Therefore, $u' \in V(T) - \{r'\}$ exists such that u' is the parent at most two leaves of T. For the second claim, consider v' to be a leaf of T, thus $deg(v') = 1$. Therefore, $F(v')$ is adjacent to one inner face of G_n and thus has two outer edges on its boundary. Since these two outer edges are incident to a vertex in $F(v')$, say v, it follows that $deg(v) = 2$. □

Lemma 2. *Let T be the rooted weak dual of a maximal outerplanar graph G_n, $n \geq 5$, and u' be a node with at most two children which are leaves of T. Then, the face $F(u')$ contains a vertex u of degree 3 or 4, and for each of the child v' of u', there exists a vertex v of degree 2 in face $F(v')$ such that v is adjacent to u. If u' has two children, then u has degree 4; otherwise, u has degree 3.*

Proof. We first consider the case where u' has two children, v' and w'. Since both v' and w' are leaves of T, by Lemma 1, each of $F(v')$ and $F(w')$ in G_n contains a vertex of degree 2. Let v and w be the vertices of degree 2 on $F(v')$ and $F(w')$, respectively. Observe that, $N(v) \neq N(w)$. Since v' is a child of u', $F(v')$ and $F(u')$ has exactly one common inner edge e of G_n. Furthermore, $F(v')$ does not have a common inner edge with any other inner face of G_n since v' is a leaf of T. Therefore, the end vertices of e are the two neighbors of v in G_n. Similarly, $F(w')$ and $F(u')$ have exactly one common inner edge of G_n whose end vertices are the two neighbors of w in G_n. Then $N(v) \cup N(w) \subset V(F(u'))$. Since $|V(F(u'))| = 3$, $N(v) \cap N(w)$ contains a single vertex u, which clearly belongs to $F(u')$. Let the other two vertices of $F(u')$ are p and q. Therefore, $\{p, q, v, w\} \subseteq N(u)$, and $deg(u)$ is at least 4.

We now show that $deg(u)$ cannot be larger than 4. Assume for a contradiction, $deg(u) > 4$, then $N(u) = R \cup S$ such that, $R = \{p,\ q,\ v,\ w\}$, $S \subset V(G)$, and S is nonempty. Let $x \in S$, then, x is adjacent to at least one vertex from R since the subgraph induced by $N(u)$ is a path graph. Now, x can not be adjacent to p or q because, in that case, G_n will not remain outerplanar. Again, x can not be adjacent to v or w because that will make v' or w' a non-leaf in T; a contradiction. Therefore, S is empty, and the degree of u is exactly 4.

We now consider the case where u' has exactly one child v'. Following the similar argument in the proof of the previous case (by omitting the leaf node w' from T and the vertex w from G_n), we can show that there exists $u \in F(u')$ with degree exactly three which is adjacent to the vertex of degree 2 in $F(v')$. □

We now define a graph as *red-bounded* if it has a maximal outerplane embedding such that all of its outer edges and, at most one inner edge is red, see

Fig. 3(a). Therefore, if a red-bounded graph has no red inner edge, each vertex has a red degree exactly 2. Otherwise, every vertex, except for the two endpoints of the red inner edge, has a red degree exactly 2, and two endpoints of the red inner edge have a red degree exactly 3. Therefore, every red-bounded graph is a 3-trigraph.

Lemma 3. *Let $G_n, n \geq 5$ be a red-bounded graph with no red inner edge; then there exists a 3-sequence S that reduces G_n to G_4 such that each contraction of S obtains a red-bounded graph.*

Proof. Let T be the rooted weak dual of G_n. By Lemma 1, a node u' in T exists with at most two children such that the children of u' are leaves of T. If u' has one child v', then by Lemma 2, there exists u in $F(u')$ and v in $F(v')$ such that, $deg(u) = 3$, $deg(v) = 2$, and $uv \in E(G_n)$. Now, if we perform $< u, v >\rightarrow u$ to obtain G_{n-1}, it is easy to see that G_{n-1} is red-bounded with no red inner edge. We define this type of contraction as Type-0; see Fig. 3(b). Therefore, a Type-0 contraction obtains G_{x-1} from G_x, $x > 4$, such that both graphs are red bounded with no red inner edge.

Now, if u' has two children v' and w', then by Lemma 2, there exists u in $F(u')$, v in $F(v')$, and w in $F(w')$ such that, $deg(u) = 4$, $deg(w) = deg(v) = 2$, and u is adjacent to both v and w in G_n. Therefore, w.l.o.g., if we perform $< u, v >\rightarrow u$ to obtain G_{n-1}, it is easy to see that G_{n-1} is red bounded with exactly one red inner edge. We define this type of contraction as Type-1; see Fig. 3(c). Therefore, a Type-1 contraction is performed on a red-bounded graph with no red inner edge, and a red-bounded graph with exactly one red inner edge is obtained. Observe that, if we perform $< u, w >\rightarrow u$ on G_{n-1} and obtain G_{n-2} then, G_{n-2} is red bounded with no red inner edge. We define this type of contraction as Type-2, see Fig. 3(d). Therefore, a Type-2 contraction is performed on a red-bounded graph with exactly one red inner edge, which resulted from the previous Type-1 contraction, and a red-bounded graph with no red inner edge is obtained. Note that, every Type-1 contraction is always followed by a Type-2 contraction.

Now, assume S is a 3-sequence consisting of Type-0, Type-1 and Type-2 contractions, and every trigraph G_i, $4 < i < n$ obtained from S are red bounded. We have to show that we can obtain G_{i-1} from G_i via contractions of the defined types such that G_{i-1} remains red-bounded. Now, if G_i has no red inner edge, then we can perform either Type-0 or Type-1 contraction to obtain G_{i-1}. We thus consider G_i has one red inner edge. Then the contraction that obtained G_i is of Type-1, and we can perform a Type-2 contraction to obtain G_{i-1}. Thus, in all cases, we can obtain the next red-bounded graph, proving the existence of S. □

We now have the following corollary from Lemma 3.

Corollary 1. *Let $G_n, n > 4$ be a maximal outerplanar graph; then a 3-sequence exists that reduces G_n to G_4.*

Proof. The claim of Lemma 3 is stronger than the claim of Corollary 1. □

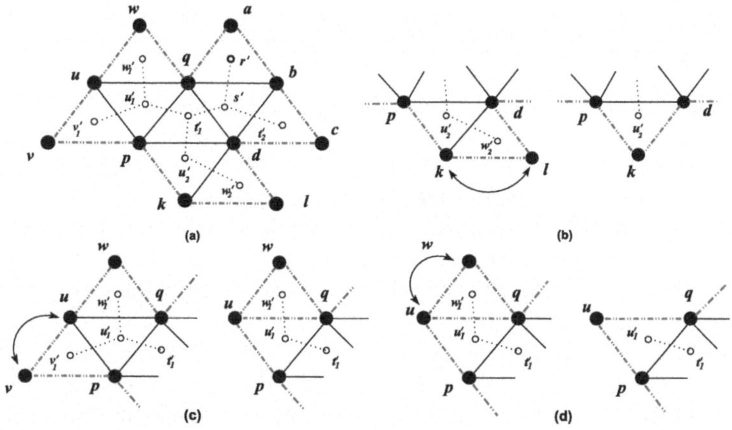

Fig. 3. Red-bounded graph and contractions (a) Red-bounded graph G_{11} with its weak dual rooted at r' (b) Type-0 contraction $< k, l > \rightarrow k$ (c) Type-1 contraction $< v, u > \rightarrow u$ (d) Type-2 contraction $< w, u > \rightarrow u$

Theorem 1. *The twin width of maximal outerplanar graphs is at most 3.*

Proof. Let G_n be a maximal outerplanar graph. If $n = 4$, then no contraction sequence for G_n has a width of more than three, even if all edges are red. This is because G_4 has the maximum degree of three, and any trigraph obtained by performing contraction on G_4 has the maximum degree of less than three. Therefore, we assume $n > 4$ for G_n. Let S_1 be a 3-sequence that reduces G_n to G_4. We know S_1 exists from Corollary 1. Let S_2 be a contraction sequence that reduces G_4 to a single vertex. Clearly, S_2 has a width of no more than 3. Therefore, $S = S_1 + S_2$ is a 3-sequence for G_n. □

In our next theorem, we generalize the result of Theorem 1 for any outerplanar graph.

Theorem 2. *The twin width of outerplanar graphs is at most 3.*

Proof. Let G be an outerplanar graph. We consider G as a connected graph since the twin-width of a graph is the maximum of the twin-width of its connected components. Let H be a maximal outerplanar graph obtained from G by introducing the minimum number of edges in G. Since, H is a maximal outerplanar graph, there exists a 3-sequence S of H by Theorem 1. Now, following the proof of Lemma 3, for any contraction $C \in S$, if u and v are the contraction pair of C, then w.l.o.g., $deg(v) = 2$, $deg(u) \in \{3, 4\}$, and u and v have exactly one common neighbor in H. Observe that, G is a spanning subgraph of H, therefore, $deg(v) \leq 2$, $deg(u) \leq 4$, and u and v have at most one common neighbor in G. Now, if we perform the contraction sequence S on G, all the trigraphs, obtained by the contractions of S, have a red degree of at most three. Therefore, the theorem holds. □

We now show that the upper bound of 3 for the twin-width of outerplanar graphs is indeed tight by the following theorem.

Theorem 3. *There exists a maximal outerplanar graph with a twin width of at least 3.*

Proof. We consider the maximal outerplanar graph G_{12} as shown in Fig. 4. We regard the vertices $D = \{u_1, u_2, ..., u_6\}$ as *deep* vertices and the vertices $S = \{v_1, v_2, ..., v_6\}$ as *shallow* vertices. Let, $\mathcal{S} = C_{12}, C_{11}, ..., C_2$, be the d-sequence with the minimum d, that obtains the sequence of trigraphs $G_{11}, G_{10}, ..., G_1$, starting from G_{12}. To keep the proof simple, w.l.o.g. we assume for any $C_i \in \mathcal{S}$ which reduces G_i to G_{i-1}, if $\{u, v\}$ is the contraction pair of C_i, then the contracted vertex has the label u in G_{i-1} only if $deg(u) \geq deg(v)$ in G_i. Since any shallow vertex in G_i can have a degree at most 3 (possibly by 2-contraction of two shallow vertices) and any deep vertex has a degree at least 4, for any G_i obtaind from $\mathcal{S}$, if $v \in V(G_i)$ is a deep vertex, then $v \in D$.

Observe that, for any two distinct vertices $u, v \in V(G_{12})$, $|N(u) \Delta N(v)| \geq 2$. Here Δ denotes the symmetric difference of sets such that, $N(u) \Delta N(v) = (N(u) - N(v)) \cup (N(v) - N(u))$. Therefore, $\mathcal{S}$ is at least a 2-sequence, and we have to show that there exists $C_i \in \mathcal{S}$ of width at least 3.

Let, $C_i =< d_x, d_y >\rightarrow d_x$ be the contraction in $\mathcal{S}$, with the maximum i, such that, $d_x, d_y \in D$. In other words, no C_j, $j > i$ exists in $\mathcal{S}$ where both the vertices of the contraction pair of C_j are deep vertices. Also, let G' is the subgraph of G_{12} induced by D (see Fig. 4). Therefore, C_i reduces G_i to G_{i-1}, and for $u, v \in V(G')$ if $uv \in E(G')$ then $uv \in E(G_i)$; Otherwise, there must be a contraction $C_j \in \mathcal{S}, j > i$ removing uv, but that requires both the vertices of the contraction pair of C_j to be deep vertices, a contradiction. We assume G_i is not a 3-trigraph, as otherwise, the theorem holds trivially.

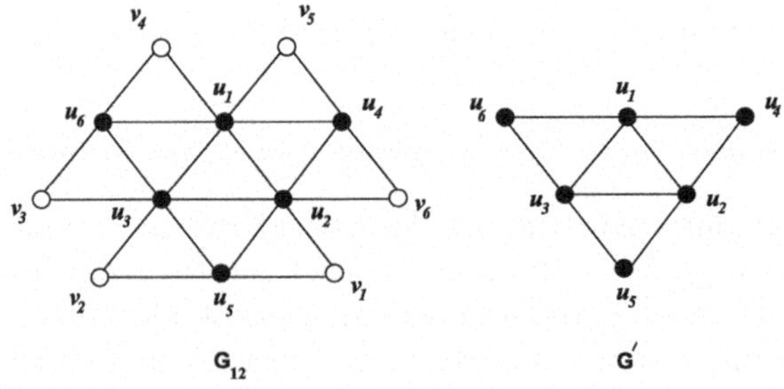

Fig. 4. Outerplanar graph G_{12} and its subgraph G' induced by the deep vertices only

Observe that, both d_x and d_y of C_i also belong to $V(G')$, and $V(G')$ only contains 2 types of vertices; vertices of degree 2 and vertices of degree degree 4. Therefore, C_i can be of at most four types based on $N(d_x)$ and $N(d_y)$ in G'; *Type A:* $deg(d_x) = deg(d_y) = 4$, *Type B:* $deg(d_x) = deg(d_y) = 2$, *Type C:* $deg(d_x) = 4$,

$deg(d_y) = 2$, and $d_x d_y \in E(G')$, and *Type D:* $deg(d_x) = 4$, $deg(d_y) = 2$, and $d_x d_y \notin E(G')$. We categorize C_i with respect to G' so that the number of types remains low due to the symmetry of G'. Thus, it suffices to prove that none of the above-mentioned types are 2-contraction.

Before proceeding further, we present four crucial observations. (I) For $u, v \in V(G_{12})$, if $uv \notin E(G_{12})$ but $uv \in E(G_i)$, where G_i is obtained by some contraction of $\mathcal{S}$, then uv is a red edge in G_i. This is because uv is introduced later via some contraction, and they are red by definition. (II) Given $C_i =< d_x, d_y >\rightarrow d_x$, if $p \in N(d_x) \cap N(d_y)$ in G_i, and any of pd_x or pd_y is a red edge in G_i, then, p is a red neighbor of d_x in G_{i-1}. (III) For $u, v, w \in V(G_{12})$, where $w \in N(u) \Delta N(v)$ in G_{12}, if $w \in N(u) \cup N(v)$ in G_i, $i < 12$, and $< u, v >\rightarrow u$ obtains G_{i-1} from G_i, then w is a red neighbor of u in G_{i-1}. The reason is, w is not adjacent to exactly one of u and v in G_{12}, w.l.o.g. let $wu \notin E(G_{12})$. Therefore, if $w \in N(u) \cup N(v)$ in G_i, then either $wu \notin E(G_i)$ or, by observation I, wu is a red edge in G_i. Thus, regardless of the case, $< u, v >\rightarrow u$ makes w a red neighbor of u in G_{i-1}. (IV) Given $C_i =< d_x, d_y >\rightarrow d_x$, if $q \in N(d_x) \Delta N(d_y)$ in G' and $q \in N(d_x) \cup N(d_y)$ in G_i, then q is a red neighbor of d_x in G_{i-1}. As G' is an induced subgraph of G_{12}, observation IV is a special case of observation III.

Now, we complete the proof by case analysis.

Type A: Let w.l.o.g., $C_i =< u_1, u_3 >\rightarrow u_1$ be a 2-contraction of Type A in G_i. As $u_4, u_5 \in N(u_1) \Delta N(u_3)$ in G', they are the two red neighbors of u_1 in G_{i-1} by observation IV. Then, any third red neighbor of u_1 in G_{i-1} makes C_i a 3-contraction. Now, w.l.o.g. consider the black edge $u_1 v_4$ of G_{12} and assume $v_4 \in N(u_1) \cup N(u_3)$ in G_i. As $v_4 \in N(u_1) \Delta N(u_3)$ in G_{12}, hence, by observation III, v_4 is the third red neighbor of u_1 in G_{i-1}. Therefore, C_i can not be a 2-contraction if $v_4 \in N(u_1) \cup N(u_3)$ in G_i. Then, there exists a 2-contraction $C_j \in \mathcal{S}$, where $j > i$, such that, C_j obtains G_{j-1} where $v_4 \notin N(u_1)$ in G_{j-1}. It is easy to see that no 2-contraction, except for $< u_6, v_4 >\rightarrow u_6$, can satisfy such condition. However, $C_j =< u_6, v_4 >\rightarrow u_6$ makes the edge $u_3 u_6$ red. As $E(G') \subseteq E(G_i)$ and $u_3 u_6 \in E(G')$, the red edge $u_3 u_6$ remains in G_i. Nonetheless, $u_6 \in N(u_3) \cap N(u_1)$ in G_i, and by observation II, u_6 is the third red neighbor of u_1 in G_{i-1}. Therefore, C_i is at least a 3-contraction if it is of Type A.

Type B: Let w.l.o.g., $C_i =< u_4, u_6 >\rightarrow u_6$ be a 2-contraction of Type B in G_i. As $u_2, u_3 \in N(u_4) \Delta N(u_6)$ in G', they are the two red neighbors of u_6 in G_{i-1} by observation IV. Now, w.l.o.g. consider the black edge $u_6 v_3$ of G_{12} and assume $v_3 \in N(u_4) \cup N(u_6)$ in G_i. Then, by observation III, v_3 is the third red neighbor of u_6 in G_{i-1}. Therefore, assume $v_3 \notin N(u_4) \cup N(u_6)$ in G_i. Then, there exists a 2-contraction $C_j \in \mathcal{S}$, where $j > i$, such that, C_j obtains G_{j-1} where $v_3 \notin N(u_6)$ in G_{j-1}. It is easy to see that no 2-contraction, except for $< u_6, v_3 >\rightarrow u_6$, can satisfy such condition. However, $C_j =< u_6, v_3 >\rightarrow u_6$ makes the edge $u_1 u_6$ red which remains in G_i. Nonetheless, $u_1 \in N(u_4) \cap N(u_6)$ in G_i, and by observation II, u_1 is the third red neighbor of u_6 in G_{i-1}. Therefore, C_i is at least a 3-contraction if it is of Type B.

Type C: Let w.l.o.g., $C_i =< u_1, u_6 >\rightarrow u_1$ be a 2-contraction of Type C in G_i. As $u_2, u_4 \in N(u_1) \Delta N(u_6)$ in G', they are the two red neighbors of u_1 in G_{i-1}

by observation IV. Now, w.l.o.g. consider the black edge u_6v_3 of G_{12} and assume $v_3 \in N(u_1) \cup N(u_6)$ in G_i. Then, by observation III, v_3 is the third red neighbor of u_1 in G_{i-1}. Therefore, assume $v_3 \notin N(u_1) \cup N(u_6)$ in G_i. Then, there exists a 2-contraction $C_j \in \mathcal{S}$, where $j > i$, such that, C_j obtains G_{j-1} where $v_3 \notin N(u_6)$ in G_{j-1}. It is easy to see that no 2-contraction, except for $< u_6, v_3 >\rightarrow u_6$, can satisfy such condition. We thus assume $C_j =< u_6, v_3 >\rightarrow u_6$ obtains G_{j-1}. As $v_4 \in N(u_6) \cap N(u_1)$ in G_{12}, if $v_4 \in V(G_j)$, then u_6v_4 becomes a red edge in G_{j-1} due to C_j. As u_6v_4 remains red in G_i, where $v_4 \in N(u_1) \cap N(u_6)$ in G_i, by observation II, v_4 becomes the third red neighbor of u_1 in G_{i-1}. Therefore, assume $v_4 \notin V(G_j)$, then there exists $C_k \in \mathcal{S}$, $k > j$, that obtains G_{k-1} such that, $v_4 \notin N(u_6) \cup N(u_1)$ in $V(G_{k-1})$. It is easy to see that, any valid choice of contraction pair for C_k makes u_3 a red neighbor of either u_6 or u_1, and C_i eventually makes u_3 the third red neighbor of u_1. This concludes C_i is at least a 3-contraction if it is of Type C.

Type D: Let w.l.o.g., $C_i =< u_1, u_5 >\rightarrow u_1$ be a 2-contraction of Type D in G_i. Then, u_4 and u_6 are the two red neighbors of u_1 in G_{i-1}. Observe that, $v_1 \in N(u_2) \cap N(u_5)$ and $v_2 \in N(u_3) \cap N(u_5)$ in G_{12}. As $v_1, v_2 \notin N(u_1)$ in G_{12}, both vertices must be removed from $N(u_5)$ before performing C_i; otherwise they will be the third and fourth red neighbor of u_1 in G_{i-1}. Let w.l.o.g., $C_j \in \mathcal{S}, j > 1$ obtains G_{j-1} such that, $v_1 \notin N(u_5)$ in G_{j-1}. It is easy to see that only $C_j =< u_5, v_1 >\rightarrow u_1$ can satisfy such a condition. However, then u_3u_5 becomes red. As, $u_3 \in N(u_1) \cap N(u_5)$ in G_i, by observation II, C_i makes u_3 the third red neighbor of u_1 in G_{i-1}. Therefore, C_i is at least a 3-contraction if it is of Type D.

This concludes the proof. □

4 Conclusion

In this paper, we proved the twin-width of outerplanar graphs is at most three, and this bound is tight. The next step in our quest to find the tight twin-width bound of planar graphs and their subclasses is to solve this problem for the plane 3-tree, a well-structured subclass of maximal planar graphs.

Acknowledgments. This work has been done under the project "Development of Efficient Graph Algorithms for Big Data Analytics and Visualization" funded by Research and Innovation Centre for Science and Engineering (RISE), Internal Research Grant 2021-01-006 of Bangladesh University of Engineering and Technology (BUET). The work is also partially supported by the basic research grants of BUET.

References

1. Alam, J.M., Bekos, M.A., Gronemann, M., Kaufmann, M., Pupyrev, S.: Queue layouts of planar 3-trees. Algorithmica **82**, 2564–2585 (2020)

2. Bekos, M.A., Lozzo, G.D., Hliněný, P., Kaufmann, M.: Graph product structure for h-framed graphs. In: 33rd International Symposium on Algorithms and Computation (ISAAC 2022), pp. 23:1–23:15. Leibniz International Proceedings in Informatics (2022). https://doi.org/10.4230/LIPIcs.ISAAC.2022.23
3. Bergé, P., Bonnet, É., Déprés, H.: Deciding twin-width at most 4 is np-complete. In: 49th International Colloquium on Automata, Languages and Programming (ICALP 2022) (2022). https://doi.org/10.4230/LIPIcs.ICALP.2022.18
4. Bonnet, É., Geniet, C., Kim, E.J., Thomassé, S., Watrigant, R.: Twin-width ii: small classes. In: Proceedings of the 2021 ACM-SIAM Symposium on Discrete Algorithms (SODA), pp. 1977–1996. SIAM (2021). https://doi.org/10.1137/1.9781611976465.118
5. Bonnet, É., Geniet, C., Kim, E.J., Thomassé, S., Watrigant, R.: Twin-width iii: max independent set, min dominating set, and coloring. In: 48th International Colloquium on Automata, Languages, and Programming (ICALP 2021) (2021). https://doi.org/10.4230/LIPIcs.ICALP.2021.35
6. Bonnet, É., Kim, E.J., Reinald, A., Thomassé, S.: Twin-width vi: the lens of contraction sequences. In: Proceedings of the 2022 Annual ACM-SIAM Symposium on Discrete Algorithms (SODA), pp. 1036–1056. SIAM (2022). https://doi.org/10.1137/1.9781611977073.45
7. Bonnet, É., Kim, E.J., Thomassé, S., Watrigant, R.: Twin-width i: tractable FO model checking. ACM J. ACM (JACM) **69**(1), 1–46 (2021). https://doi.org/10.1145/3486655
8. Bonnet, É., Kwon, O.J., Wood, D.R., et al.: Reduced bandwidth: a qualitative strengthening of twin-width in minor-closed classes (and beyond). arXiv preprint arXiv:2202.11858 (2022). https://doi.org/10.48550/arXiv.2202.11858
9. Corneil, D.G., Lerchs, H., Burlingham, L.S.: Complement reducible graphs. Discret. Appl. Math. **3**(3), 163–174 (1981). https://doi.org/10.1016/0166-218x(81)90013-5
10. Dujmović, V., Joret, G., Micek, P., Morin, P., Ueckerdt, T., Wood, D.R.: Planar graphs have bounded queue-number. J. ACM (JACM) **67**(4), 1–38 (2020)
11. Hliněný, P., Jedelský, J.: Twin-width of planar graphs is at most 8, and at most 6 when bipartite planar. In: 50th International Colloquium on Automata, Languages, and Programming (ICALP 2023) (2023). https://doi.org/10.4230/LIPIcs.ICALP.2023.75
12. Jacob, H., Pilipczuk, M.: Bounding twin-width for bounded-treewidth graphs, planar graphs, and bipartite graphs. In: International Workshop on Graph-Theoretic Concepts in Computer Science, pp. 287–299. Springer (2022). https://doi.org/10.1007/978-3-031-15914-5
13. Král', D., Lamaison, A.: Planar graph with twin-width seven. Eur. J. Comb. 103749 (2023). https://doi.org/10.1016/j.ejc.2023.103749
14. Rahman, M.S.: Basic Graph Theory. UTCS, Springer, Cham (2017). https://doi.org/10.1007/978-3-319-49475-3

15. Robertson, N., Seymour, P.D.: Graph minors. ii. Algorithmic aspects of tree-width. J. Algorithms **7**(3), 309–322 (1986). https://doi.org/10.1016/0196-6774(86)90023-4
16. Schidler, A., Szeider, S.: A sat approach to twin-width. In: 2022 Proceedings of the Symposium on Algorithm Engineering and Experiments (ALENEX), pp. 67–77. SIAM (2022). https://doi.org/10.1137/1.9781611977042.6

Maximize an Approximate k-submodular Function under a Knapsack Constraint

Fanqing Meng[1], Qingqin Nong[1(✉)], Suning Gong[1,2], and Xiaoying Qu[1]

[1] Ocean University of China, Qingdao Shandong 266100, People's Republic of China
qqnong@ouc.edu.cn

[2] Qingdao University, Qingdao 266071, Shandong, People's Republic of China

Abstract. In this paper, we consider the problem of maximizing a monotone approximate k-submodular function under a knapsack constraint. For monotone approximate k-submodular functions, we characterize the approximately k-submodular degree by k-submodularity ratio and present a $\frac{\tilde{\gamma}}{1+\tilde{\gamma}} \cdot \left(1 - e^{-1}\right)$ greedy approximation algorithm under a knapsack constraint.

Keywords: Approximate k-submodular function · Knapsack · Greedy algorithm

1 Introduction

Submodular function $f: 2^E \to \mathbb{R}_+$, which satisfies $f(S) + f(T) \geq f(S \cap T) + f(S \cup T)$ for any subset S and T, is a key concept in operations research and combinatorial optimization [1–5]. The submodularity also implies the property of diminishing marginal utility which can be denoted by $f(S \cup \{j\}) - f(S) = f_S(j) \geq f_T(j)$ for any subset $S \subseteq T \subseteq E$ and any element $j \in E \backslash T$. It is found that submodular function maximization has many practical applications, such as machine learning, computer vision, data mining and other fields [7,8].

Because the submodular maximization problem is NP-hard [9], the existing literatures focus on the performance of approximation algorithms under different constraints. However, in many practical problems, the set function may not have strictly submodular property, such as sparse Gaussian process [10], which prompts scholars to propose the concept of approximate submodular function. In 2011, Das and Kempe [11] proposed submodularity ratio to characterize the approximately submodular degree of a set function $\beta = \min\limits_{T,S \subseteq E} \frac{\sum_{j \in T \backslash S} [f(S \cup \{j\}) - f(S)]}{f(S \cup T) - f(S)}$. In 2018, Elenberg [12] proposed a weakly submodularity ratio based on a given subset S and an integer M which is the upper bound of set size. In the same year, Gong *et al.* [13] proposed a general submodularity ratio from the perspective of diminishing marginal benefit and proved greedy

This research was supported by the National Natural Science Foundation of China [Grant Numbers 12171444, 12301415].

D. Du et al. (Eds.): COCOA 2024, LNCS 15434, pp. 113–126, 2025.
https://doi.org/10.1007/978-981-96-4445-2_10

algorithm can return $1-e^{-\gamma}$, $1-e^{-\gamma}$ and $\frac{\gamma}{(k+\gamma)}$ approximate solutions under cardinality constraint, knapsack constraint and k-matroid constraint respectively, where $\gamma = \min_{T,S,j} \{\frac{f(S\cup\{j\})-f(S)}{f(T\cup\{j\})-f(T)} : S \subseteq T \subseteq E, j \in E\backslash T\}$.

On the other hand, functions in the practical problems are no longer limited to single-set functions. A universal extension of submodular function was introduced by Lovász [6] in 1983. He introduced submodularity into 2-dimensional set function. And the corresponding optimization problem is applied to solve the bicooperative games. Many practical problems imply that set functions derived from multiple disjoint sets are more realistic than single-set functions, such as the sensor placement problem [14].

In 2012, Huber and Kolmogorov [15] first proposed the conception of k-submodular function. For an integer $k \geq 1$ and a non-empty finite set as the ground set E, k-submodular function $f : (k+1)^E \rightarrow \mathbb{R}_+$ is defined on the set of a k-tuple which composed of k-disjoint subsets, and satisfy $f(\boldsymbol{s}) + f(\boldsymbol{t}) \geq f(\boldsymbol{s} \sqcap \boldsymbol{t}) + f(\boldsymbol{s} \sqcup \boldsymbol{t})$ for any k-tuples $\boldsymbol{s} = (S_1, \cdots, S_k)$ and $\boldsymbol{t} = (T_1, \cdots, T_k)$.

An approximation algorithm for the k-submodular function maximization problem with non-constraints was first proposed by Ward and Živný [16]. At present, existing literatures mainly focus on the construction of approximate algorithms under different constraints such as cardinality constraint, knapsack constraint and matroid constraint. Ohsaka and Yoshida [14] proposed a construction method from a current solution to the optimal solution and presented a greedy $\frac{1}{2}$-approximation algorithm for monotone k-submodular maximization under a cardinality constraint with $O(k|E|B)$ queries. Sakaue [17] solved monotone k-submodular maximization under a matroid constraint with $O(M|E|(IO + kEO))$ time complexity and achieved $\frac{1}{2}$-approximation, where M is the upper bound of the size of optimal solution, IO and EO represent the calculated matroid independence and the complexity required for the k-submodular function. For k-submodular maximization with knapsack constraint, Tang *et al.* [18] presented a greedy $\frac{(1-e^{-1})}{2}$-approximation algorithm with $O(|E|^4 k^3)$ time complexity. Xiao *et al.* [19] improved this result to $\frac{(1-e^{-2})}{2}$ and proved that the approximation of $\frac{(1-e^{-3})}{3}$ can be achieved when the function is non-monotone. By applying the theory of multilinear extension, Huang *et al.* [20] obtained a similar conclusion to Xiao *et al.* [19].

The problem of k-submodular function maximization has been extensively studied in the last decade, but the k-submodulity may not be strictly guaranteed in practical applications. About this situation, Zheng *et al.* [21] constructed $\epsilon - AS$ and $\epsilon - ADR$ approximations for k-submodular function and they also proved that the greedy algorithm can achieve $\frac{(1-\epsilon)}{2(1+\epsilon)}$ and $\frac{(1-\epsilon)}{3(1+\epsilon)}$ approximation ratio with entire cardinality constraint and individual cardinality constraint. From the perspective of diminishing marginal utility, Wang *et al.* [22] constructed orthant submodular ratio α to characterize the approximately k-submodular degree of a monotone function and presented a $\frac{\alpha}{(1+\alpha)}$-approximation algorithm with time complexity of $O(M|E|(IO + kEO))$ under single matroid constraint.

Jiang *et al.* [23] integrated the above two concepts [21,22] and proposed two classes of approximations: $\epsilon - A - \alpha - WS$ and $\epsilon - A - \alpha - WDR$, which has been proved that the greedy algorithm is $\frac{\alpha(1-\epsilon)^2}{(1+\epsilon)[(\alpha+1)(1-\epsilon)+2B\epsilon]}$ and $\frac{(1-\epsilon)\alpha}{1+\epsilon+\alpha-\epsilon\alpha}$ approximation when the function is monotone under single matroid constraint.

In this paper, we will explore the problem of maximizing a monotone approximate k-submodular function f under a knapsack constraint (denoted by MAkSK). The standard greedy algorithm cannot obtain a bounded approximation ratio. We apply an technique introduced by Sahni [25], named partial enumeration to improve the performance of our algorithm. We provide a $\frac{\widetilde{\gamma}}{1+\widetilde{\gamma}} \cdot (1-e^{-1})$-approximation algorithm for the MAkSK, where the $\widetilde{\gamma}$ is the k-submodularity ratio of the objective function.

2 Preliminaries

Let E be a ground set and k be a positive integer. The set of k-tuples on E is defined as

$$(k+1)^E = \{\boldsymbol{x} = (X_1, \cdots, X_k) \,|X_i \subseteq E, \forall i \in [k], X_i \cap X_j = \emptyset, \forall i,j \in [k], i \neq j\}.$$

and define two operations on $(k+1)^E$ as follows:

$$\begin{aligned}
&\boldsymbol{x} \sqcap \boldsymbol{y} = (X_1 \cap Y_1, \cdots, X_k \cap Y_k),\\
&\boldsymbol{x} \sqcup \boldsymbol{y} = ((X_1 \cup Y_1) \setminus \cup_{i\neq 1} (X_i \cup Y_i), \cdots, (X_j \cup Y_j) \setminus \cup_{i \neq j} (X_i \cup Y_i),\\
&\qquad\qquad \cdots, (X_k \cup Y_k) \setminus \cup_{i \neq k} (X_i \cup Y_i)).
\end{aligned}$$

k-submodularity is a generalization of submodularity in the domain of definition.

Definition 1 (k-submodular set function). *Given a positive integer k and a set E, set function $f: (k+1)^E \to \mathbb{R}$ is k-submodular if it satisfies the following inequality for any k-tuples $\boldsymbol{x}$ and $\boldsymbol{y}$*

$$f(\boldsymbol{x}) + f(\boldsymbol{y}) \geq f(\boldsymbol{x} \sqcap \boldsymbol{y}) + f(\boldsymbol{x} \sqcup \boldsymbol{y}).$$

We introduce a partial order $\preccurlyeq$ on $(k+1)^E$: $\forall \boldsymbol{x}, \boldsymbol{y} \in (k+1)^E$, $\boldsymbol{x} \preccurlyeq \boldsymbol{y}$ if and only if $X_i \subseteq Y_i$ for each $i \in [k]$. And we call a k-submodular set function f is monotone if for any $\boldsymbol{x} \preccurlyeq \boldsymbol{y}$, we have $f(\boldsymbol{x}) \leq f(\boldsymbol{y})$.

Let $P(\boldsymbol{x}) = \cup_{i \in [k]} X_i$ represents the support set of the k-tuple $\boldsymbol{x}$. Define a function $\boldsymbol{x}(\cdot): E \to \{0, 1, \cdots, k\}$ to represent the subset where e is located in $\boldsymbol{x}$

$$\boldsymbol{x}(e) = \begin{cases} i & if\ e \in P(\boldsymbol{x})\ and\ e \in X_i \\ 0 & otherwise \end{cases}.$$

A special class of k-tuples is the support set only contains one element e,

$$\boldsymbol{I}_{[e,i]} = (\emptyset, \cdots, \underbrace{\{e\}}_{i}, \cdots \emptyset) \in (k+1)^E.$$

Let $\boldsymbol{x}\backslash \boldsymbol{I}_{[e,\boldsymbol{x}(e)]}$ represent the situation where e is removed from $P(\boldsymbol{x})$ without changing the subset index of other elements. For an element $e \notin P(\boldsymbol{x})$, the marginal benefit of adding e to the i-th subset X_i of $\boldsymbol{x}$ is

$$\Delta_{e,i} f(\boldsymbol{x}) = f\left(\boldsymbol{x} \sqcup \boldsymbol{I}_{[e,i]}\right) - f(\boldsymbol{x}).$$

And if $e \in P(\boldsymbol{x})$, the marginal benefit of moving e from $\boldsymbol{x}(e)$-th subset $X_{\boldsymbol{x}(e)}$ to the i-th subset X_i of $\boldsymbol{x}$ is

$$\Delta_{e,i} f\left(\boldsymbol{x}\backslash \boldsymbol{I}_{[e,\boldsymbol{x}(e)]}\right) - \Delta_{e,\boldsymbol{x}(e)} f\left(\boldsymbol{x}\backslash \boldsymbol{I}_{[e,\boldsymbol{x}(e)]}\right).$$

Similar to submodularity, k-submodular functions also have the property of diminishing marginal benefit. That is $\Delta_{e,i} f(\boldsymbol{x}) \geq \Delta_{e,i} f(\boldsymbol{y})$ for any $e \in E\backslash P(\boldsymbol{y})$ and $x \preccurlyeq y$.

When function f cannot strictly guarantee the k-submodularity, we follow the research of Wang *et al.* [22] and approximate the k-submodularity of the function by k-submodularity ratio.

Definition 2 ($\widetilde{\gamma}$-approximate k-submodular function). *A nonnegative monotone k-dimension set function $f : (k+1)^E \to \mathbb{R}_+$ is called $\widetilde{\gamma}$-approximate k-submodular if*

$$\widetilde{\gamma} = \min_{\boldsymbol{x},\boldsymbol{y} \in (k+1)^E, e \in E} \left\{\frac{\Delta_{e,i} f(\boldsymbol{x})}{\Delta_{e,i} f(\boldsymbol{y})} : \boldsymbol{x} \preccurlyeq \boldsymbol{y}, e \in E\backslash P(\boldsymbol{y}), i \in [k]\right\}.$$

Clearly, $\widetilde{\gamma} \in [0,1]$, and for $\boldsymbol{x},\boldsymbol{y} \in (k+1)^E$ with $\boldsymbol{x} \preccurlyeq \boldsymbol{y}$, $\forall e \in E\backslash P(\boldsymbol{y})$ and $\forall i \in [k]$, the following holds $\Delta_{e,i} f(\boldsymbol{x}) \geq \widetilde{\gamma} \cdot \Delta_{e,i} f(\boldsymbol{y})$. We can draw an important result about $\widetilde{\gamma}$-approximate k-submodular function.

Proposition 1. *A $\widetilde{\gamma}$-approximate k-submodular function $f : (k+1)^E \to \mathbb{R}$ satisfies the following inequality for any pair of $\boldsymbol{x} \preccurlyeq \boldsymbol{y}$,*

$$f(\boldsymbol{y}) - f(\boldsymbol{x}) \leq \frac{1}{\widetilde{\gamma}} \cdot \sum_{e \in P(\boldsymbol{y})\backslash P(\boldsymbol{x})} \Delta_{e,\boldsymbol{y}(e)} f(\boldsymbol{x}).$$

Proof. Let $P(\boldsymbol{y})\backslash P(\boldsymbol{x}) = \{e_1, e_2, \cdots, e_m\}$, we can define a sequence $\{\boldsymbol{z}^{(j)}\}_{j=0}^{m}$ such that (1) $\boldsymbol{z}^{(0)} = \boldsymbol{x}$ and $\boldsymbol{z}^{(m)} = \boldsymbol{y}$; (2) $\boldsymbol{z}^{(j)} = \boldsymbol{z}^{(j-1)} \sqcup \boldsymbol{I}_{[e_j,\boldsymbol{y}(e_j)]}$ for $j \in [m]$. Then we have

$$\begin{aligned} f(\boldsymbol{y}) - f(\boldsymbol{x}) &= \sum_{j=1}^{m} \Delta_{e_j,\boldsymbol{y}(e_j)} f(\boldsymbol{z}^{(j-1)}) \leq \frac{1}{\widetilde{\gamma}} \cdot \sum_{j=1}^{m} \Delta_{e_j,\boldsymbol{y}(e_j)} f(\boldsymbol{z}^{(0)}) \\ &= \frac{1}{\widetilde{\gamma}} \cdot \sum_{j=1}^{m} \Delta_{e_j,\boldsymbol{y}(e_j)} f(\boldsymbol{x}) = \frac{1}{\widetilde{\gamma}} \cdot \sum_{e \in P(\boldsymbol{y})\backslash P(\boldsymbol{x})} \Delta_{e,\boldsymbol{y}(e)} f(\boldsymbol{x}), \end{aligned}$$

where the inequality holds from $\boldsymbol{z}^{(0)} \preccurlyeq \boldsymbol{z}^{(1)} \preccurlyeq \cdots \preccurlyeq \boldsymbol{z}^{(m)}$ and $f(\cdot)$ is $\widetilde{\gamma}$-approximate k-submodular.

The following lemma also plays an important role.

Lemma 1 **([24]).** *Let A and B be arbitrary positive integers, $\rho_1, \rho_2, \cdots, \rho_A$ be arbitrary nonnegative real values and $\rho_1 > 0$, then the following inequality holds*

$$\frac{\sum_{i=1}^{A} \rho_i}{\min_{j=1,\cdots,A}\left(\sum_{i=1}^{j-1} \rho_i + B\rho_j\right)} \geq 1 - \left(1 - \frac{1}{B}\right)^A \geq 1 - e^{-\frac{A}{B}}.$$

3 Greedy Algorithms for MAkSK

This section focuses on maximizing a monotone $\widetilde{\gamma}$-approximate k-submodular function under a knapsack constraint (referred to as MAkSK). Given a ground set E, a monotone $\widetilde{\gamma}$-approximate k-submodular function f, a capacity budget $B \in \mathbb{N}_+$, and a cost function $c : E \to \mathbb{N}_+$, we can formulate our problem as

$$\max\{f(\boldsymbol{x}) : \boldsymbol{x} \in (k+1)^E \textit{ and } c(P(\boldsymbol{x})) \leq B\},$$

where $c(P(\boldsymbol{x})) = \sum_{e \in P(\boldsymbol{x})} c(e)$.

Algorithm 1. Greedy algorithm for MAkSK

Require: A ground set E, a monotone approximate k-submodular function f, a capacity budget B and a cost function c
Ensure: solution $\boldsymbol{s}_G$ with $c(P(\boldsymbol{s}_G)) \leq B$

$\boldsymbol{s}_G \leftarrow \emptyset$
while $E \neq \emptyset$ **do**
  $[\tilde{e}, \tilde{i}] \leftarrow \arg\max_{e \in E, i \in [k]} \frac{\Delta_{e,i} f(\boldsymbol{s}_G)}{c(e)}$
  if $c(P(\boldsymbol{s}_G)) + c(\tilde{e}) \leq B$ **then**
    $\boldsymbol{s}_G = \boldsymbol{s}_G \sqcup \boldsymbol{I}_{[\tilde{e}, \tilde{i}]}$
  end if
  $E \leftarrow E \backslash \{\tilde{e}\}$
end while
return $\boldsymbol{s}_G$

Unfortunately, Algorithm 1 has unbounded approximation ratio, problems arise when there are high value items with high cost. Consider the case where $E = \{e_1, e_2\}$ with $c(e_1) = B$ and $c(e_2) = 1$, and for any $i \in [k]$ there has

$$f(\boldsymbol{I}_{[e_1,i]}) = B \text{ and } f(\boldsymbol{I}_{[e_2,i]}) = 2.$$

The best solution is clearly to pick e_1 into any one of X_i for a value of B. Since $\frac{\Delta_{e_2,i} f(\emptyset)}{c(e_2)} = 2 > 1 = \frac{\Delta_{e_1,i} f(\emptyset)}{c(e_1)}$ for any $i \in [k]$, Algorithm 1 picks e_2 into one of X_i and obtains a value of 2. As B gets close to infinity, the approximation ratio becomes arbitrarily large.

Algorithm 2. Greedy algorithm for MAkSK with partial enumeration

Require: A ground set E, a monotone approximate k-submodular function f, a capacity budget B and a cost function c
Ensure: solution $\boldsymbol{s}$ with $c(P(\boldsymbol{s})) \leq B$

$\boldsymbol{s}_1 \leftarrow \arg\max_{P(\boldsymbol{s})\subseteq E, c(\boldsymbol{s})\leq B, |P(\boldsymbol{s})|<4} f(\boldsymbol{s})$
$\boldsymbol{s}_2 \leftarrow \emptyset$
for all $\boldsymbol{s}^{(0)} \in \{\boldsymbol{s} | P(\boldsymbol{s}) \subseteq E, |P(\boldsymbol{s})| = 4, c(P(\boldsymbol{s})) \leq B\}$ **do**
  $E' \leftarrow E \backslash P(\boldsymbol{s}^{(0)})$
  $\boldsymbol{s}_G \leftarrow$ Algorithm 1 for E' with initialization $\boldsymbol{s}_G \leftarrow \boldsymbol{s}^{(0)}$
  if $f(\boldsymbol{s}_G > f(\boldsymbol{s}_2))$ **then**
    $\boldsymbol{s}_2 \leftarrow \boldsymbol{s}_G$
  end if
end for
return $\arg\max \{f(\boldsymbol{s}_1), f(\boldsymbol{s}_2)\}$

An approach to improve approximation ratio is partial enumeration. In the following part of section, we present an algorithm with better approximation ratio by adding partial enumeration before greedy which is shown by Algorithm 2. And to ensure inequality (11), the bound of $|P(\boldsymbol{s})|$ requires to be set no less than 4.

Let $\boldsymbol{o}$ be the optimal solution and $|P(\boldsymbol{o})| = r$. Nothing that if $r \leq 3$, the algorithm can search the optimal solution in the traversal process, so we assume $r \geq 4$. In order to analyze the approximation ratio of the greedy algorithm, it is necessary to set an order for $P(\boldsymbol{o}) = \{e_1^*, \cdots, e_r^*\}$, and define a new k-tuple sequence set $\{\boldsymbol{q}^{(j)}\}_{j=0}^{r}$. Let $\boldsymbol{q}^{(0)} = \boldsymbol{\emptyset}, (e_1^*, i_1^*) = \underset{e\in P(\boldsymbol{o}), i\in[k]}{\arg\max}\ f(\boldsymbol{I}_{[e,i]})$ and iteratively, for $j = 1, \cdots, r$

$$\boldsymbol{q}^{(j)} = \boldsymbol{q}^{(j-1)} \sqcup \boldsymbol{I}_{[e_j^*, i_j^*]} \text{ and } (e_{j+1}^*, i_{j+1}^*) = \underset{e\in P(\boldsymbol{o})\backslash P(\boldsymbol{q}^{(j)}), i\in[k])}{\arg\max} f(\boldsymbol{q}^{(j)} \sqcup \boldsymbol{I}_{[e,i]}).$$

It is obvious that $\boldsymbol{q}^{(j-1)} \preccurlyeq \boldsymbol{q}^{(j)}$.

To analyze our algorithm, we construct another sequence of k-tuples $\{\bar{\boldsymbol{o}}^{(j)}\}_{j=0}^{r}$. Let $\bar{\boldsymbol{o}}^{(j)} = \left(\bar{\boldsymbol{o}}^{(j-1)} \sqcup \boldsymbol{q}^{(j)}\right) \sqcup \boldsymbol{q}^{(j)},\ j \in [r]$. One can see that for any $j \in [r]$, $P(\bar{\boldsymbol{o}}^{(j)}) = P(\boldsymbol{o})$, $\bar{\boldsymbol{o}}^{(0)} = \boldsymbol{o}$ and

$$\bar{\boldsymbol{o}}^{(j)}(e_u^*) = \begin{cases} \boldsymbol{q}^{(j)}(e_u^*) & if\ u \leq j \\ \boldsymbol{o}(e_u^*) & if\ u > j. \end{cases}$$

From the definition of $\boldsymbol{q}^{(j)}$ and $\bar{\boldsymbol{o}}^{(j)}$, $\boldsymbol{q}^{(j)} \preccurlyeq \bar{\boldsymbol{o}}^{(j)}$ for any $j \in [r]$. The following result holds for any monotone $\widetilde{\gamma}$-approximate k-submodular functions.

Lemma 2. *Let f be a $\widetilde{\gamma}$-approximate k-submodular function. If f is monotone,*

$$f\left(\bar{\boldsymbol{o}}^{(j-1)}\right) - f\left(\bar{\boldsymbol{o}}^{(j)}\right) \leq \frac{1}{\widetilde{\gamma}}\left(f\left(\boldsymbol{q}^{(j)}\right) - f\left(\boldsymbol{q}^{(j-1)}\right)\right). \tag{1}$$

Proof. Recall that $P(\boldsymbol{q}^{(j)}) = P(\boldsymbol{q}^{(j-1)}) \cup \{e_{j^*}\}$ and $\boldsymbol{q}^{(j)} = \boldsymbol{q}^{(j-1)} \sqcup \boldsymbol{I}_{[e_j^*, i_j^*]}$. We have $f(\boldsymbol{q}^{(j)}) - f(\boldsymbol{q}^{(j-1)}) = \Delta_{e_j^*, i_j^*} f(\boldsymbol{q}^{(j-1)})$. On the other hand

$$\bar{\boldsymbol{o}}^{(j-1)} = \left(\bar{\boldsymbol{o}}^{(j)} \backslash \boldsymbol{I}_{[e_j^*, i_j^*]}\right) \sqcup \boldsymbol{I}_{[e_j^*, \boldsymbol{o}(e_j^*)]}.$$

Thus

$$\begin{aligned} f\left(\bar{\boldsymbol{o}}^{(j-1)}\right) &= f(\bar{\boldsymbol{o}}^{(j)} \backslash \boldsymbol{I}_{[e_j^*, i_j^*]}) + \Delta_{e_j^*, \boldsymbol{o}(e_j^*)} f(\bar{\boldsymbol{o}}^{(j)} \backslash \boldsymbol{I}_{[e_j^*, i_j^*]}) \\ &\leq f(\bar{\boldsymbol{o}}^{(j)}) + \Delta_{e_j^*, \boldsymbol{o}(e_j^*)} f(\bar{\boldsymbol{o}}^{(j)} \backslash \boldsymbol{I}_{[e_j^*, i_j^*]}) \leq f(\bar{\boldsymbol{o}}^{(j)}) + \Delta_{e_j^*, i_j^*} f(\bar{\boldsymbol{o}}^{(j)} \backslash \boldsymbol{I}_{[e_j^*, i_j^*]}) \\ &\leq f(\bar{\boldsymbol{o}}^{(j)}) + \frac{1}{\tilde{\gamma}} \Delta_{e_j^*, i_j^*} f(\boldsymbol{q}^{(j-1)}), \end{aligned}$$

where the first inequality holds by monotonity, the second holds by the definition of $\boldsymbol{q}^{(j)}$ and the last holds by $\boldsymbol{q}^{(j-1)} \preccurlyeq \bar{\boldsymbol{o}}^{(j)} \backslash \boldsymbol{I}_{[e_j^*, i_j^*]}$.

Summing the left and right sides of the inequality (1) for $j \in [4]$, we can obtain

$$f(\boldsymbol{o}) - \frac{1}{\tilde{\gamma}} f\left(\boldsymbol{q}^{(4)}\right) \leq f\left(\bar{\boldsymbol{o}}^{(4)}\right). \tag{2}$$

Assuming that $\boldsymbol{s}$ is a feasible solution given by the greedy algorithm beginning with some $\boldsymbol{s}^{(0)}$. Let $e^{(j)}$ is the element which is selected by the greedy algorithm in j-th iteration and $\boldsymbol{s}(e^{(j)}) = i^{(j)}$. For $j \in \left[|P(\boldsymbol{s}) \backslash P(\boldsymbol{s}^{(0)})|\right]$, the iterations of Algorithm 2 can be defined as follows

$$\boldsymbol{s}^{(j)} = \boldsymbol{s}^{(j-1)} \sqcup \boldsymbol{I}_{[e^{(j)}, i^{(j)}]}.$$

Without loss of generality, we assume that the instance only consists of elements in $P(\boldsymbol{s}) \cup P(\boldsymbol{o})$. It is because even removing an element which is not in $P(\boldsymbol{s}) \cup P(\boldsymbol{o})$ does not affect the optimal solution and any steps of the algorithm. For an element $e^{(p+1)}$ belonging to $P(\boldsymbol{o}) \backslash P(\boldsymbol{s})$, without loss of generality, we assumes that it is the first element in the above situation which is considered by Algorithm 2 in the $p+1$ iteration, but it is not put into the partial solution $\boldsymbol{s}^{(p)}$ for the sake of $c(\boldsymbol{s}^p) + c(e^{(p+1)}) > B$.

Let $\boldsymbol{o}^{(0)} = \bar{\boldsymbol{o}}^{(4)} \backslash \boldsymbol{I}_{[e^{(p+1)}, \bar{\boldsymbol{o}}^{(4)}(e^{(p+1)})]}$. For $j \in \left[|P(\boldsymbol{s}) \backslash P(\boldsymbol{s}^{(0)})|\right]$, we define

$$\boldsymbol{o}^{(j)} = \left(\boldsymbol{o}^{(j-1)} \sqcup \boldsymbol{I}_{[e^{(j)}, i^{(j)}]}\right) \sqcup \boldsymbol{I}_{[e^{(j)}, i^{(j)}]}.$$

Because $\boldsymbol{s}^{(j)}$ is a partial solution obtained by the algorithm after running j iterations, it is not difficult to find that $P\left(\boldsymbol{o}^{(j)}\right) = P\left(\boldsymbol{o}^{(0)}\right) \cup P\left(\boldsymbol{s}^{(j)}\right)$. If $e \in P(\boldsymbol{s}^{(j)})$, $\boldsymbol{o}^{(j)}(e) = \boldsymbol{s}^{(j)}(e)$ and if $e \in P(\boldsymbol{o}^{(0)}) \backslash P(\boldsymbol{s}^{(j)})$, $\boldsymbol{o}^{(j)}(e) = \boldsymbol{o}^{(0)}(e)$.

And it can be observed that the above k-tuples have the following partial order relationships $\boldsymbol{s}^{(j)} \preccurlyeq \boldsymbol{o}^{(j)}$. In addition, for $t = 0, \cdots, p$, there is an equality relationship on the support set

$$P(\boldsymbol{o}^{(0)}) \backslash P(\boldsymbol{s}^{(t)}) = P(\boldsymbol{o}^{(t)}) \backslash P(\boldsymbol{s}^{(t)}). \tag{3}$$

Considering the above analysis, the following lemma can be obtained.

Lemma 3. *For any $j \in [k]$, the following inequality holds*

$$\Delta_{e^{(p+1)},j} f\left(\boldsymbol{o}^{(0)}\right) \leq \frac{1}{4\widetilde{\gamma}} f\left(\boldsymbol{s}^{(0)}\right).$$

Proof. Due to $\boldsymbol{q}^{(4)} = \boldsymbol{s}^{(0)} \preccurlyeq \boldsymbol{o}^{(0)}$, and function f is $\widetilde{\gamma}$-approximate k-submodular, we have

$$\Delta_{e^{(p+1)},j} f\left(\boldsymbol{o}^{(0)}\right) \leq \frac{1}{\widetilde{\gamma}} \Delta_{e^{(p+1)},j} f\left(\boldsymbol{q}^{(4)}\right).$$

Note that $\boldsymbol{q}^{(0)} \leq \boldsymbol{q}^{(1)} \leq \cdots \boldsymbol{q}^{(4)}$. Combining with $f(\emptyset) \geq 0$, we have

$$\begin{aligned}
\Delta_{e^{(p+1)},j} f(\boldsymbol{o}^{(0)}) &\leq \frac{1}{\widetilde{\gamma}} \cdot \min\{\Delta_{e^{(p+1)},j} f(\boldsymbol{q}^{(4)}), \cdots, \Delta_{e^{(p+1)},j} f(\boldsymbol{q}^{(0)})\} \\
&\leq \frac{1}{\widetilde{\gamma}} \cdot \min\{f\left(\boldsymbol{q}^{(4)}\right) - f\left(\boldsymbol{q}^{(4-1)}\right), \cdots, f\left(\boldsymbol{q}^{(1)}\right) - f\left(\boldsymbol{q}^{(0)}\right)\} \\
&\leq \frac{1}{\widetilde{\gamma}} \cdot \frac{f\left(\boldsymbol{q}^{(4)}\right) - f\left(\boldsymbol{q}^{(4-1)}\right) + \cdots + f\left(\boldsymbol{q}^{(1)}\right) - f\left(\boldsymbol{q}^{(0)}\right)}{4} \\
&\leq \frac{1}{4 \cdot \widetilde{\gamma}} f\left(\boldsymbol{q}^{(4)}\right),
\end{aligned}$$

where the second inequality holds from $\boldsymbol{q}^{(4)} = \boldsymbol{s}^{0}$, $e^{(p+1)} \in P(\boldsymbol{o}) \backslash P(\boldsymbol{s}^{(0)})$ and the definition of $\boldsymbol{q}^{(j)}$ $(j \in [r])$.

By analyzing the first p iterations of the greedy procedure, the following lemma can be obtained.

Lemma 4. *For $t = 0, \cdots, p$, if the $\widetilde{\gamma}$-approximate k-submodular function f is monotone, there is*

$$f\left(\boldsymbol{o}^{(0)}\right) \leq \left(1 + \frac{1}{\widetilde{\gamma}}\right) f\left(\boldsymbol{s}^{(t)}\right) - \frac{1}{\widetilde{\gamma}} f\left(\boldsymbol{s}^{(0)}\right) + \frac{1}{\widetilde{\gamma}} \sum_{e \in P(\boldsymbol{o}^{(0)}) \backslash P(\boldsymbol{s}^{(t)})} \Delta_{e, \boldsymbol{o}^{(0)}(e)} f(\boldsymbol{s}^{(t)}).$$

Proof. Obviously, the above inequality holds at $t = 0$ and therefore we only consider the cases with $t = 1, \cdots, p$. To obtain the inequality, we first demonstrate that the inequality below holds for any $j \in [p]$ based on whether $e^{(j)}$ is in $P\left(\boldsymbol{o}^{(j-1)}\right)$ or not,

$$f\left(\boldsymbol{o}^{(j-1)}\right) - f\left(\boldsymbol{o}^{(j)}\right) \leq \frac{1}{\widetilde{\gamma}} \cdot \left[f\left(\boldsymbol{s}^{(j)}\right) - f\left(\boldsymbol{s}^{(j-1)}\right)\right]. \tag{4}$$

Recall that $P(\boldsymbol{o}^{(j)}) = P(\boldsymbol{o}^{(j-1)}) \cup \{e^{(j)}\}$ and $\boldsymbol{s}^{(j)} = \boldsymbol{s}^{(j-1) \sqcup \boldsymbol{I}_{[e^{(j)}, i^{(j)}]}}$, we have

$$\begin{aligned}
\boldsymbol{o}^{(j-1)} &= \left(\boldsymbol{o}^{(j)} \backslash \boldsymbol{I}_{[e^{(j)}, i^{(j)}]}\right) \sqcup \boldsymbol{I}_{[e^{(j)}, \boldsymbol{o}^{(j-1)}(e^{(j)})]}, \\
f(\boldsymbol{o}^{(j-1)}) &= f(\boldsymbol{o}^{(j)} \backslash \boldsymbol{I}_{[e^{(j)}, i^{(j)}]}) + \Delta_{e^{(j)}, \boldsymbol{o}^{(j-1)}(e^{(j)})} f(\boldsymbol{o}^{(j)} \backslash \boldsymbol{I}_{[e^{(j)}, i^{(j)}]}).
\end{aligned}$$

Case 1, $e^{(j)} \in P\left(\boldsymbol{o}^{(j-1)}\right)$, then $\boldsymbol{o}^{(j-1)}(e^{(j)}) \neq 0$.

If $\boldsymbol{o}^{(j-1)}(e^{(j)}) = i^{(j)}$, the subset index of $e^{(j)}$ in the optimal solution is the same as that in $\boldsymbol{s}^{(j)}$, then there is $\boldsymbol{o}^{(j-1)} = \left(\boldsymbol{o}^{(j)} \backslash \boldsymbol{I}_{[e^{(j)}, \boldsymbol{o}^{(j-1)}(e^{(j)})]}\right) \sqcup \boldsymbol{I}_{[e^{(j)}, \boldsymbol{o}^{(j-1)}(e^{(j)})]} = \boldsymbol{o}^{(j)}$, implying

$$f\left(\boldsymbol{o}^{(j-1)}\right) - f\left(\boldsymbol{o}^{(j)}\right) = 0 \leq \frac{1}{\tilde{\tilde{\gamma}}} \cdot \left[f\left(\boldsymbol{s}^{(j)}\right) - f\left(\boldsymbol{s}^{(j-1)}\right)\right].$$

If $\boldsymbol{o}^{(j-1)}(e^{(j)}) \neq i^{(j)}$, similar to the analysis in Lemma 2, we have

$$\begin{aligned}
f(\boldsymbol{o}^{(j-1)}) &= f(\boldsymbol{o}^{(j)} \backslash \boldsymbol{I}_{[e^{(j)}, i^{(j)}]}) + \Delta_{e^{(j)}, \boldsymbol{o}^{(j-1)}(e^{(j)})} f(\boldsymbol{o}^{(j)} \backslash \boldsymbol{I}_{[e^{(j)}, i^{(j)}]}) \\
&\leq f(\boldsymbol{o}^{(j)}) + \Delta_{e^{(j)}, \boldsymbol{o}^{(j-1)}(e^{(j)})} f(\boldsymbol{o}^{(j)} \backslash \boldsymbol{I}_{[e^{(j)}, i^{(j)}]}) \\
&\leq f(\boldsymbol{o}^{(j)}) + \Delta_{e^{(j)}, i^{(j)}} f(\boldsymbol{o}^{(j)} \backslash \boldsymbol{I}_{[e^{(j)}, i^{(j)}]}) \\
&\leq f(\boldsymbol{o}^{(j)}) + \frac{1}{\tilde{\tilde{\gamma}}} \Delta_{e^{(j)}, i^{(j)}} f(\boldsymbol{s}^{(j-1)}) \\
&= f(\boldsymbol{o}^{(j)}) + \frac{1}{\tilde{\tilde{\gamma}}} \left[f\left(\boldsymbol{s}^{(j)}\right) - f\left(\boldsymbol{s}^{(j-1)}\right)\right].
\end{aligned}$$

where the first inequality holds by monotonity, the second holds by the definition of $\boldsymbol{s}^{(j)}$ and the last holds by $\boldsymbol{s}^{(j-1)} \preccurlyeq \boldsymbol{o}^{(j)} \backslash \boldsymbol{I}_{[e_j, i_j]}$.

Case 2, $e^{(j)} \notin P\left(\boldsymbol{o}^{(j-1)}\right)$, namely $\boldsymbol{o}^{(j-1)}(e^{(j)}) = 0$.

It yields that $\boldsymbol{I}_{[e^{(j)}, \boldsymbol{o}^{(j-1)}(e^{(j)})]} = \emptyset$ and implies

$$f(\boldsymbol{o}^{(j-1)}) = f(\boldsymbol{o}^{(j)} \backslash \boldsymbol{I}_{[e^{(j)}, i^{(j)}]}) = f(\boldsymbol{o}^{(j)}) - \Delta_{e^{(j)}, i^{(j)}} f(\boldsymbol{o}^{(j-1)}),$$

we have

$$f(\boldsymbol{o}^{(j-1)}) - f(\boldsymbol{o}^{(j)}) = -\Delta_{e^{(j)}, i^{(j)}} f(\boldsymbol{o}^{(j-1)}) \leq 0 \leq \frac{1}{\tilde{\tilde{\gamma}}} \cdot \left[f(\boldsymbol{s}^{(j)}) - f(\boldsymbol{s}^{(j-1)})\right].$$

By adding the two sides of the inequality (4) to the first t iterations, we have

$$f\left(\boldsymbol{o}^{(0)}\right) - f\left(\boldsymbol{o}^{(t)}\right) \leq \frac{1}{\tilde{\tilde{\gamma}}} \cdot \left[f\left(\boldsymbol{s}^{(t)}\right) - f\left(\boldsymbol{s}^{(0)}\right)\right]. \tag{5}$$

Using Proposition 1 and the fact that $\boldsymbol{s}^t \preccurlyeq \boldsymbol{o}^t$, we obtain

$$f\left(\boldsymbol{o}^{(t)}\right) \leq f\left(\boldsymbol{s}^{(t)}\right) + \frac{1}{\tilde{\tilde{\gamma}}} \cdot \sum_{e \in P(\boldsymbol{o}^{(t)}) \backslash P(\boldsymbol{s}^{(t)})} \Delta_{e, \boldsymbol{o}^{(t)}(e)} f(\boldsymbol{s}^{(t)}). \tag{6}$$

Combining equations (3), (5) and (6), we have

$$\begin{aligned}
f\left(\boldsymbol{o}^{(0)}\right) &\leq f\left(\boldsymbol{o}^{(t)}\right) + \frac{1}{\tilde{\tilde{\gamma}}} \cdot \left[f\left(\boldsymbol{s}^{(t)}\right) - f\left(\boldsymbol{s}^{(0)}\right)\right] \\
&\leq f\left(\boldsymbol{s}^{(t)}\right) + \frac{1}{\tilde{\tilde{\gamma}}} \cdot \sum_{e \in P\left(\boldsymbol{o}^{(t)}\right) \backslash P\left(\boldsymbol{s}^{(t)}\right)} \Delta_{e, \boldsymbol{o}^{(t)}(e)} f(\boldsymbol{s}^{(t)}) + \frac{1}{\tilde{\tilde{\gamma}}} \cdot \left[f\left(\boldsymbol{s}^{(t)}\right) - f\left(\boldsymbol{s}^{(0)}\right)\right]
\end{aligned}$$

$$= f\left(\boldsymbol{s}^{(t)}\right) + \frac{1}{\widetilde{\gamma}} \cdot \sum_{e \in P\left(\boldsymbol{o}^{(0)}\right) \backslash P\left(\boldsymbol{s}^{(t)}\right)} \Delta_{e,\boldsymbol{o}^{(0)}(e)} f(\boldsymbol{s}^{(t)}) + \frac{1}{\widetilde{\gamma}} \cdot \left[f\left(\boldsymbol{s}^{(t)}\right) - f\left(\boldsymbol{s}^{(0)}\right)\right]$$

$$= \left(1 + \frac{1}{\widetilde{\gamma}}\right) f\left(\boldsymbol{s}^{(t)}\right) - \frac{1}{\widetilde{\gamma}} f\left(\boldsymbol{s}^{(0)}\right) + \frac{1}{\widetilde{\gamma}} \sum_{e \in P\left(\boldsymbol{o}^{(0)}\right) \backslash P\left(\boldsymbol{s}^{(t)}\right)} \Delta_{e,\boldsymbol{o}^{(0)}(e)} f(\boldsymbol{s}^{(t)}).$$

Theorem 1. *If the $\widetilde{\gamma}$-approximate k-submodular function f is monotone, the query complexity of Algorithm 2 is $O\left(k^5 |E|^6\right)$ and it returns a $\frac{\widetilde{\gamma}}{1+\widetilde{\gamma}} \cdot (1 - e^{-1})$-approximate solution.*

Proof. Firstly, we make the query complexity analysis of the algorithm. Before the greedy loop, we need to randomly enumerate 4 elements in E and allocate subset index to these elements, which up to $O(k|E|)^4$ cases. The number of iterations of the greedy process is up to $|E| - 4$ times, and in each iteration, the number of function value queries is $O(k|E|)$. Thus the query complexity of Algorithm 2 is $O\left(k^5 |E|^6\right)$.

Secondly, we make the approximate ratio analysis for the algorithm. For $j \in [|P(\boldsymbol{s})|]$, let $\theta_j = \frac{\Delta_{e^{(j)},i^{(j)}} f(\boldsymbol{s}^{(j-1)})}{c(e)}$, the marginal benefit of $e^{(j)}$, we have

$$\begin{aligned}
& f(\boldsymbol{o}^{(0)}) - f(\boldsymbol{s}^{(0)}) \\
& \leq \left(1 + \frac{1}{\widetilde{\gamma}}\right) (f(\boldsymbol{s}^{(t)}) - f(\boldsymbol{s}^{(0)})) + \frac{1}{\widetilde{\gamma}} \cdot \sum_{e \in P\left(\boldsymbol{o}^{(0)}\right) \backslash P\left(\boldsymbol{s}^{(t)}\right)} \Delta_{e,\boldsymbol{o}^{(0)}(e)} f(\boldsymbol{s}^{(t)}) \\
& = \left(1 + \frac{1}{\widetilde{\gamma}}\right) (f(\boldsymbol{s}^{(t)}) - f(\boldsymbol{s}^{(0)})) + \frac{1}{\widetilde{\gamma}} \cdot \sum_{e \in P\left(\boldsymbol{o}^{(0)}\right) \backslash P\left(\boldsymbol{s}^{(t)}\right)} c(e) \cdot \frac{\Delta_{e,\boldsymbol{o}^{(0)}(e)} f(\boldsymbol{s}^{(t)})}{c(e)} \\
& \leq \left(1 + \frac{1}{\widetilde{\gamma}}\right) (f(\boldsymbol{s}^{(t)}) - f(\boldsymbol{s}^{(0)})) + \frac{1}{\widetilde{\gamma}} \cdot \sum_{e \in P\left(\boldsymbol{o}^{(0)}\right) \backslash P\left(\boldsymbol{s}^{(t)}\right)} c(e) \theta_{t+1} \\
& = \left(1 + \frac{1}{\widetilde{\gamma}}\right) (f(\boldsymbol{s}^{(t)}) - f(\boldsymbol{s}^{(0)})) + \frac{1}{\widetilde{\gamma}} \cdot c\left(P\left(\boldsymbol{o}^{(0)}\right) \backslash P\left(\boldsymbol{s}^{(t)}\right)\right) \theta_{t+1} \\
& \leq \left(1 + \frac{1}{\widetilde{\gamma}}\right) (f(\boldsymbol{s}^{(t)}) - f(\boldsymbol{s}^{(0)})) + \frac{1}{\widetilde{\gamma}} \cdot c\left(P\left(\boldsymbol{o}^{(0)}\right) \backslash P\left(\boldsymbol{s}^{(0)}\right)\right) \theta_{t+1} \\
& = \left(1 + \frac{1}{\widetilde{\gamma}}\right) \left[f(\boldsymbol{s}^{(t)}) - f(\boldsymbol{s}^{(0)}) + \frac{c\left(P\left(\boldsymbol{o}^{(0)}\right) \backslash P\left(\boldsymbol{s}^{(0)}\right)\right)}{1 + \widetilde{\gamma}} \cdot \theta_{t+1}\right].
\end{aligned}$$

The inequality implies

$$\begin{aligned}
& f(\boldsymbol{o}^{(0)}) - f(\boldsymbol{s}^{(0)}) \\
& \leq \left(1 + \frac{1}{\widetilde{\gamma}}\right) \min_{t=0,\cdots,p-1} \left[f(\boldsymbol{s}^{(t)}) - f(\boldsymbol{s}^{(0)}) + \frac{c\left(P\left(\boldsymbol{o}^{(0)}\right) \backslash P\left(\boldsymbol{s}^{(0)}\right)\right)}{1 + \widetilde{\gamma}} \cdot \theta_{t+1}\right], \quad (7)
\end{aligned}$$

where the first inequality holds from Lemma 4, the second inequality holds from the definition of θ_j and the greedy of Algorithm 2, the last inequality holds from $c\left(P\left(\boldsymbol{o}^{(0)}\right) \backslash P\left(\boldsymbol{s}^{(t)}\right)\right) \leq c\left(P\left(\boldsymbol{o}^{(0)}\right) \backslash P\left(\boldsymbol{s}^{(0)}\right)\right)$ for any $t=0, \cdots, p$.

Let $B_t=c\left(P\left(\boldsymbol{s}^{(t)}\right) \backslash P\left(\boldsymbol{s}^{(0)}\right)\right)=\sum_{j=1}^t c(e^{(j)})$ for any $t=0, \cdots, p$. Recall that each $c(e)$ for $e \in E$ is a non-negative integer. Then $B_t \in \mathbb{Z}_+$. For $l \in [B_p]$, we define $\rho_l=\theta_j$ if $l=B_{j-1}+1, \cdots, B_j$, That is $\rho_1, \rho_2, \cdots, \rho_{B_p}$ can be partitioned into p categories and the t-th category contains $c(e^{(t)})$ integers and each of them is equal to θ_t. From this definition, we obtain $f(\boldsymbol{s}^{(t)})-f(\boldsymbol{s}^{(0)})=\sum_{j=1}^t\left[f\left(\boldsymbol{s}^{(j)}\right)-f\left(\boldsymbol{s}^{(j-1)}\right)\right]=\sum_{j=1}^t c(e^j)\theta_j=\sum_{l=1}^{B_p} \rho_l$. Furthermore, it holds that

$$c\left(P\left(\boldsymbol{o}^{(0)}\right) \backslash P\left(\boldsymbol{s}^{(0)}\right)\right) \leq c\left(P\left(\boldsymbol{s}^{(p)}\right) \backslash P\left(\boldsymbol{s}^{(0)}\right)\right)=B_p.$$

The above inequality can be proved by contradiction: if $c\left(P\left(\boldsymbol{o}^{(0)}\right) \backslash P\left(\boldsymbol{s}^{(0)}\right)\right) > c\left(P\left(\boldsymbol{s}^{(p)}\right) \backslash P\left(\boldsymbol{s}^{(0)}\right)\right)$, then $c\left(P\left(\boldsymbol{s}^{(p)}\right) \backslash P\left(\boldsymbol{s}^{(0)}\right)\right)+c\left(e^{(p+1)}\right) < c\left(P\left(\boldsymbol{o}^{(0)}\right) \backslash P\left(\boldsymbol{s}^{(0)}\right)\right)+c\left(e^{(p+1)}\right) \leq B-c\left(P\left(\boldsymbol{s}^{(0)}\right)\right)$. It means that $e^{(p+1)}$ can be put into $\boldsymbol{s}^{(p)}$, a contradiction to $e^{(p+1)} \in P(\boldsymbol{o}) \backslash P(\boldsymbol{s})$. Thus we have

$$\begin{aligned}
&\min_{t=0,\cdots,p-1}\left\{f(\boldsymbol{s}^{(t)})-f(\boldsymbol{s}^{(0)})+\frac{c(P(\boldsymbol{o}^{(0)})\backslash P(\boldsymbol{s}^{(0)}))}{1+\widetilde{\gamma}}\theta_{t+1}\right\}\\
&\leq \min_{t=0,\cdots,p-1}\left\{f(\boldsymbol{s}^{(t)})-f(\boldsymbol{s}^{(0)})+\frac{B_p}{1+\widetilde{\gamma}}\theta_{t+1}\right\}=\min_{t=0,\cdots,p-1}\left\{\sum_{l=1}^{B_t}\rho_l+\frac{B_p}{1+\widetilde{\gamma}}\rho_{B_t+1}\right\}\\
&=\min_{v=1,\cdots,B_p}\left\{\sum_{l=1}^{v-1}\rho_l+\frac{B_p}{1+\widetilde{\gamma}}\rho_v\right\},
\end{aligned} \tag{8}$$

where the last equality holds by the fact that $\min_{v=1,\cdots,B_p}\left\{\sum_{l=1}^{v-1}\rho_l+\frac{B_p}{1+\widetilde{\gamma}}\rho_v\right\}$ achieves its minimum at some B_t+1 $(t=0,\cdots,p-1)$. To see this, we assume that $\arg\min_{v=1,\cdots,B_p}\left\{\sum_{l=1}^{v-1}\rho_l+\frac{B_p}{1+\widetilde{\gamma}}\rho_v\right\}=v^*$ and $B_{t-1}+2 \leq v^* \leq B_t$. Note that $\rho_l=\theta_t$ for each $B_{t-1}+1 \leq l \leq B_t$, we have $\sum_{l=1}^{v^*}\rho_l+\frac{B_p}{1+\widetilde{\gamma}}\rho_{v^*} > \sum_{l=1}^{B_{t-1}}\rho_l+\frac{B_p}{1+\widetilde{\gamma}}\rho_{v^*}=\sum_{l=1}^{B_{t-1}}\rho_l+\frac{B_p}{1+\widetilde{\gamma}}\rho_{B_{t-1}+1}$, a contradiction. Combining ineqalities (7) and (8), we have a further conclusion of Lemma 4:

$$f(\boldsymbol{o}^{(0)})-f(\boldsymbol{s}^{(0)}) \leq \left(1+\frac{1}{\widetilde{\gamma}}\right)\min_{v=1,\cdots,B_p}\left\{\sum_{l=1}^{v-1}\rho_l+\lceil\frac{B_p}{1+\widetilde{\gamma}}\rceil\rho_v\right\}. \tag{9}$$

Combining Lemma 1 and relation (9), we have

$$\begin{aligned}
\frac{f(\boldsymbol{s}^{(p)})-f(\boldsymbol{s}^{(0)}))}{f(\boldsymbol{o}^{(0)})-f(\boldsymbol{s}^{(0)})} &\geq \frac{\sum_{l=1}^{B_p}\rho_l}{\left(1+\frac{1}{\widetilde{\gamma}}\right)\cdot\min_{v=1,\cdots,B_p}\left\{\sum_{l=1}^{v-1}\rho_l+\lceil\frac{B_p}{1+\widetilde{\gamma}}\rceil\rho_s\right\}}\\
&\geq \frac{\widetilde{\gamma}}{1+\widetilde{\gamma}}\cdot\left(1-e^{-\left(\frac{B_p}{\lceil\frac{B_p}{1+\widetilde{\gamma}}\rceil}\right)}\right)=\frac{\widetilde{\gamma}}{1+\widetilde{\gamma}}\cdot\left(1-e^{-\left(\frac{B_p}{\lfloor\frac{B_p+\widetilde{\gamma}}{1+\widetilde{\gamma}}\rfloor}\right)}\right)
\end{aligned}$$

$$\geq \frac{\widetilde{\gamma}}{1+\widetilde{\gamma}} \cdot \left(1 - e^{-\left(\frac{B_p(1+\widetilde{\gamma})}{B_p+\widetilde{\gamma}}\right)}\right) = \frac{\widetilde{\gamma}}{1+\widetilde{\gamma}} \cdot \left(1 - e^{-\left(1+\widetilde{\gamma}-\frac{\widetilde{\gamma}(1+\widetilde{\gamma})}{B_p+\widetilde{\gamma}}\right)}\right)$$
$$\geq \frac{\widetilde{\gamma}}{1+\widetilde{\gamma}} \cdot \left(1 - e^{-\left(1+\widetilde{\gamma}-\frac{\widetilde{\gamma}(1+\widetilde{\gamma})}{1+\widetilde{\gamma}}\right)}\right) = \frac{\widetilde{\gamma}}{1+\widetilde{\gamma}} \cdot \left(1-e^{-1}\right),$$

where the first equation is because $\lceil \frac{n}{m} \rceil = \lfloor \frac{n+m-1}{m} \rfloor$ for any $n, m \in \mathbb{R}_+$and the last inequality holds from $B_p \geq \min_{e \in E} c(e) \geq 1$. From inequality (2) and Lemma 3, we have

$$\frac{4 \cdot \widetilde{\gamma}+1}{4 \cdot \widetilde{\gamma}} f\left(\boldsymbol{s}^{(0)}\right) = f\left(\boldsymbol{s}^{(0)}\right) + \frac{1}{4 \cdot \widetilde{\gamma}} f\left(\boldsymbol{s}^{(0)}\right) \geq f\left(\boldsymbol{s}^{(0)}\right) + \Delta_{e^{(p+1)}, \bar{o}^{(4)}(e^{(p+1)})} f(\boldsymbol{o}^{(0)})$$
$$= f\left(\bar{\boldsymbol{o}}^{(4)}\right) - \left[f\left(\boldsymbol{o}^{(0)}\right) - f\left(\boldsymbol{s}^{(0)}\right)\right]$$
$$\geq f\left(\boldsymbol{o}\right) - \frac{1}{\widetilde{\gamma}} f\left(\boldsymbol{s}^{(0)}\right) - \left[f\left(\boldsymbol{o}^{(0)}\right) - f\left(\boldsymbol{s}^{(0)}\right)\right].$$

This further reveals the relationship among $f(\boldsymbol{o})$, $f\left(\boldsymbol{o}^{(0)}\right)$ and $f\left(\boldsymbol{s}^{(0)}\right)$

$$\frac{4 \cdot \widetilde{\gamma}+5}{4 \cdot \widetilde{\gamma}} f\left(\boldsymbol{s}^{(0)}\right) \geq f\left(\boldsymbol{o}\right) - \left(f\left(\boldsymbol{o}^{(0)}\right) - f\left(\boldsymbol{s}^{(0)}\right)\right). \tag{10}$$

Finally, considering and (10), we obtain a lower bound on

$$f(\boldsymbol{s}) \geq f\left(\boldsymbol{s}^{(p)}\right) = f\left(\boldsymbol{s}^{(0)}\right) + f(\boldsymbol{s}^{(p)}) - f(\boldsymbol{s}^{(0)})$$
$$\geq f\left(\boldsymbol{s}^{(0)}\right) + \frac{\widetilde{\gamma}}{1+\widetilde{\gamma}} \cdot \left(1-e^{-1}\right) \cdot \left(f(\boldsymbol{o}^{(0)}) - f(\boldsymbol{s}^{(0)})\right)$$
$$\geq \frac{4 \cdot \widetilde{\gamma}}{4 \cdot \widetilde{\gamma}+5} \cdot \left(f\left(\boldsymbol{o}\right) - \left(f\left(\boldsymbol{o}^{(0)}\right) - f\left(\boldsymbol{s}^{(0)}\right)\right)\right)$$
$$+ \frac{\widetilde{\gamma}}{1+\widetilde{\gamma}} \cdot \left(1-e^{-1}\right) \cdot \left(f(\boldsymbol{o}^{(0)}) - f(\boldsymbol{s}^{(0)})\right)$$
$$\geq \min\left\{\frac{4 \cdot \widetilde{\gamma}}{4 \cdot \widetilde{\gamma}+5}, \frac{\widetilde{\gamma}}{1+\widetilde{\gamma}} \cdot \left(1-e^{-1}\right)\right\} \cdot f\left(\boldsymbol{o}\right).$$

By calculation, the following inequality holds for arbitrary $\widetilde{\gamma} \in [0, 1]$

$$\frac{4 \cdot \widetilde{\gamma}}{4 \cdot \widetilde{\gamma}+5} \geq \frac{\widetilde{\gamma}}{1+\widetilde{\gamma}} \cdot \left(1-e^{-1}\right). \tag{11}$$

Therefore, it yields that $f(\boldsymbol{s}) \geq \frac{\widetilde{\gamma}}{1+\widetilde{\gamma}} \cdot (1-e^{-1}) f(\boldsymbol{o})$.

4 Conclusion

In this work, we investigate the problem of maximizing a monotone approximate k-submodular function subject to a knapsack constraint. By greedy technique, our algorithms provide a $\frac{\widetilde{\gamma}}{1+\widetilde{\gamma}} \cdot (1-e^{-1})$-approximate solution, where $0 < \widetilde{\gamma} \leq 1$ is the k-submodularity ratio. In the future, one may discuss approximate k-submodular maximization under other general constraints, which needs to create new analysis techniques.

References

1. Narayanan, H.: Submodular Functions and Electrical Networks. Elsevier, New Brunswick (1997)
2. Topkis, D.M.: Supermodularity and Complementarity. Princeton University Press, Princeton (1998)
3. Fujishige, S.: Submodular Functions and Optimization. Elsevier, Amsterdam (2005)
4. Bernhard Korte, J.V.: Combinatorial Optimization. Springer, Heidelberg (2018)
5. Iwata, S.: Submodular function minimization. Math. Program. **112**, 45–64 (2008)
6. Lovász, L.: Submodular functions and convexity. In: Mathematical Programming The State of the Art: Bonn 1982, pp. 235–257. Springer, Berlin, Heidelberg (1983)
7. Balcan, M.-F., Harvey, N.J.: Learning submodular functions. Proceedings of the Forty-third Annual ACM Symposium on Theory of Computing, pp. 793–802 (2011)
8. Hochbaum, D.S.: An efficient algorithm for image segmentation, Markov random fields and related problems. J. ACM **48**(4), 686–701 (2001)
9. Feige, U.: A threshold of ln n for approximating set cover. J. ACM **45**(4), 634–652 (1998)
10. Seeger, M., Herbrich, R.: Fast sparse gaussian process methods: the informative vector machine. Adv. Neural Inf. Process. Syst. **15**, 625–632 (2003)
11. Das, A., Kempe, D.: Submodular meets spectral: greedy algorithms for subset selection, sparse approximation and dictionary selection. arXiv preprint, arXiv:1102.3975 (2011). https://api.semanticscholar.org/CorpusID:16480285
12. Elenberg, E.R., Khanna, R., Dimakis, A.G., Negahban, S.: Restricted strong convexity implies weak submodularity. Ann. Stat. **46**(6B), 3539–3568 (2018)
13. Gong, S., Nong, Q., Sun, T., Fang, Q., Shao, X.: Maximize a monotone function with a generic submodularity ratio. Theor. Comput. Sci. **853**, 16–24 (2020)
14. Ohsaka, N., Yoshida, Y.: Monotone k-submodular function maximization with size constraints. In: Proceedings of the 28th International Conference on Neural Information Processing Systems, vol. 1, pp. 694–702 (2015)
15. Huber, A., Kolmogorov, V.: Towards minimizing k-submodular functions. In: Mahjoub, A.R., Markakis, V., Milis, I., Paschos, V.T. (eds.) ISCO 2012. LNCS, vol. 7422, pp. 451–462. Springer, Heidelberg (2012). https://doi.org/10.1007/978-3-642-32147-4_40
16. Ward, J., Živný, S.: Maximizing k-submodular functions and beyond. ACM Trans. Algorithms **12**, 1–26 (2016)
17. Sakaue, S.: On maximizing a monotone k-submodular function subject to a matroid constraint. Discret. Optim. **23**, 105–113 (2017)
18. Tang, Z., Wang, C., Chan, H.: On maximizing a monotone k-submodular function under a knapsack constraint. Oper. Res. Lett. **50**(1), 28–31 (2022)
19. Xiao, H., Liu, Q., Zhou, Y., Li, M.: Approximation algorithms for k-submodular maximization subject to a knapsack constraint. CoRR, ArXiv abs/2306.14520 (2023). https://arxiv.org/abs/2306.14520
20. Huang, L., Wang, B., Zhou, H.: On optimal approximations for k-submodular maximization via multilinear extension. arXiv preprint arXiv:2107.07103 (2021). https://arxiv.org/abs/2107.07103
21. Zheng, L., Chan, H., Loukides, G., Li, M.: Maximizing approximately k- submodular functions. In: Proceedings of the 2021 SIAM International Conference on Data Mining, pp. 414–422 (2021)

22. Wang, Y., Zhang, D., Zhang, Y., Zhang, Z.: Weakly k-submodular maximization under matroid constraint. In: Theory and Applications of Models of Computation, pp. 393–40 (2022)
23. Jiang, Y., Wang, Y., Yang, R., Ye, W.: Maximizing approximately non-k- submodular monotone set function with matroid constraint. In: International Conference on Theory and Applications of Models of Computation, pp. 11–20 (2022)
24. Wolsey, L.A.: Maximising real-valued submodular functions: primal and dual heuristics for location problems. Math. Oper. Res. **7**(3), 410–425 (1982)
25. Sahni, S.S.: Approximate algorithms for the 0–1 knapsack problem. J. ACM **22**, 115–124 (1975)

A Short Proof and Experimental Study of the Approximation Algorithm for Label s-t Cut

Peng Zhang(✉)

School of Software, Shandong University, Jinan 250101, China
algzhang@sdu.edu.cn

Abstract. Labeled graph is a graph with labels defined on edges. Given a labeled graph $G = (V, E)$ with label set $L = \{\ell_1, \ell_2, \ldots, \ell_q\}$, a source $s \in V$ and a sink $t \in V$, the Label s-t Cut problem asks to find a minimum size label subset $L' \subseteq L$ such that the removal of all edges from G with labels in L' disconnects s and t. This problem is NP-hard. The current best approximation ratio of Label s-t Cut is $O(\frac{n^{2/3}}{\mathrm{OPT}^{1/3}})$, where OPT is the size of an optimal solution to the problem. This ratio is achieved by a simple purely combinatorial two-phase approximation algorithm. In this paper, we give a new but short (maybe the shortest) proof for the two-phase approximation algorithm. We also perform experiments for the algorithm. Experimental results show that the two-phase approximation algorithm has actually very good practical performance in random graphs.

Keywords: Label s-t Cut · Approximation Algorithm · Algorithm Analysis · Performance Evaluation · Combinatorial Optimization

1 Introduction

The Label s-t Cut problem is a well known combinatorial optimization problem arising in the application areas such as system security [5,9,10] and computer networks [1,2]. The problem is defined as in Definition 1.

Definition 1 (Instance). *We are given a graph $G = (V, E)$ with labels defined on E. Let $L = \{\ell_1, \ell_2, \ldots, \ell_q\}$ be the set of all the labels on edges. We are also given a source vertex s and a sink vertex t.*

(Goal) *The* Label s-t Cut *problem asks to find a minimum size label subset $L' \subseteq L$ such that the removal of edges from G with labels in L' can disconnect s and t.*

The Label s-t Cut problem is NP-hard [5]. The difficulty of the problem comes from the fact that many edges may share the same label. When each edge in the input graph has a distinct label, the Label s-t Cut problem degenerates to the classical Label s-t Cut problem, which can be solved in polynomial time. This

D. Du et al. (Eds.): COCOA 2024, LNCS 15434, pp. 127–138, 2025.
https://doi.org/10.1007/978-981-96-4445-2_11

means that Label s-t Cut is a non-trivial generalization of Label s-t Cut (assuming that P is not equal to NP).

A graph with labels defined on edges is called a *labeled graph.* Label s-t Cut is a cut problem defined on labeled graph. In this meaning, the optimal solution value of Label s-t Cut is a generalization of s-t connectivity, which is equal to the optimal value of Min s-t Cut by Menger's Theorem [7] and is an important concept in graph theory.

We would like to briefly introduce labeled graph via examples coming from computer networks [1]. Most computer networks can be conceptually divided into two layers, i.e., the logical layer and the physical layer. For example, in a VPN (Virtual Private Network) network computers are connected by the logical VPN connections. A VPN connection is actually realized by a path of links in the underlying ądphysical network. (We use link to denote a direct physical connection between two sites in the physical network.) We can use a graph to model the VPN network, in which each vertex denotes a computer and each edge denotes a VPN connection. To reflect the correspondences of edges in the graph and the path of links in the underlying network, we can assign each link with a different label and then use these labels to mark the edges in the graph. In this way, we get a graph with labels defined on edges.

Note that an edge would obtain several labels since it corresponds to a physical path of links. It is reasonable to replace such an edge by a path in which each edge has only one label (since links in a physical path are connected in series). Therefore, we finally obtain a graph with each edge having one associated label, namely, a labeled graph.

A key observation is that in a labeled graph many edges may have the same label. This reflects the fact that many VPN connections in the logical layer may share the same link in the physical layer. Once a link in the physical layer fails, all the VPN connections share this link fail. This is why in the Label s-t Cut problem, once a label is removed, all edges with this label are removed from the graph. So, it is important to calculate the "labeled s-t connectivity" in a labeled graph, which is precisely the task of the Label s-t Cut problem.

1.1 Related Work

To the best of our knowledge, currently most approximation algorithms for Label s-t Cut are of two-phase type, including the algorithms in [3,8,11,13], and [14]. The first phase of these algorithms is to make the input graph sparse by using some label selecting strategy. Once a label is selected, all the edges in the graph with this label are removed. When the graph becomes sparse to some extent, the second phase computes a minimum s-t cut to separate s and t directly. Since the graph has been sparse in the second phase, it is expected that the minimum s-t cut contains small number of edges, and hence even small number of labels on these edges.

The algorithm in [13] gives an $O(m^{1/2})$-approximation for weighted Label s-t Cut, where m is the number of edges in the input graph, and by "weighted" we mean that each label has a nonnegative weight. This is the first approximation algorithm for Label s-t Cut whose ratio is given in terms of m. Its first phase is a greedy approach that selects labels by repeatedly solving the instances of the Budgeted Max Coverage problem [6], while its second phase is to compute a minimum s-t cut. The ratio $O(m^{1/2})$ was improved to $O(\frac{m^{1/2}}{\mathrm{OPT}^{1/2}})$ in [11, 14].

The algorithm in [8] gives an $O(n^{2/3})$-approximation for (un-weighted) Label s-t Cut, where n is the number of vertrices in the graph. This is the first approximation algorithm for Label s-t Cut (whose ratio is given in terms of n). Its first phase removes s-t paths not very long, and its second pahse divides vertices into layers by depth-first search. The ratio $O(n^{2/3})$ was improved to $O(\frac{n^{2/3}}{\mathrm{OPT}^{1/3}})$ in [11, 14].

The algorithm in [11] gives an $O(\frac{m^{1/2}}{\mathrm{OPT}^{1/2}})$-approximation and an $O(\frac{n^{2/3}}{\mathrm{OPT}^{1/3}})$-approximation for (un-weighted) Label s-t Cut. The first phase of the algorithms in [11] consists of solving a linear program for Label s-t Cut and then selecting labels by rounding the optimal solution of the linear program. The second phase is still to compute a minimum s-t cut. The approximation ratios $O(\frac{m^{1/2}}{\mathrm{OPT}^{1/2}})$ and $O(\frac{n^{2/3}}{\mathrm{OPT}^{1/3}})$ are currently the best for the Label s-t Cut problem. They are incomparable due to different parameters.

Thereafter, [14] (journal version of [11]) gives two purely combinatorial approximation algorithms for Label s-t Cut with the same ratios as in [11]. These two algorithms have the same two-phase structure. The first phase of the algorithms selects labels by repeatedly finding the shortest s-t path in the current graph, while the second phase computes a minimum s-t cut. The two algorithms only differ in a specific parameter used in the first phase.

Independently, [3] gives an $O(n^{2/3})$-approximation algorithm for (un-weighted) Label s-t Cut. The first phase of the algorithm is to solve a linear program for the problem that is different to the one in [11] using the Ellipsoid method. The second phase is to compute a minimum s-t cut.

On the aspect of computational hardness, it is known that Label s-t Cut cannot be approximated within $2^{\log^{1-o(1)}|I|}$ unless P is equal to NP, where $|I|$ is the input length of the problem [13]. In addition, a natural linear program for Label s-t Cut is known having integrality gap $\Omega(m^{1/3-\epsilon})$ for any small arbitrary constant $\epsilon > 0$ [15], implying that obtaining approximation ratio better than $\Omega(m^{1/3-\epsilon})$ via pure linear programming techniques on this LP is impossible.

1.2 Our Results

In this paper, we give a new but short (maybe the shortest) proof for the two-phase $O(\frac{n^{2/3}}{\mathrm{OPT}^{1/3}})$-approximation algorithm (see Algorithm 1) for Label s-t Cut. The ratio $O(\frac{n^{2/3}}{\mathrm{OPT}^{1/3}})$ is somewhat more attractive than the ratio $O(\frac{m^{1/2}}{\mathrm{OPT}^{1/2}})$ (up to the factor OPT) since in dense graphs the edge number m is easily larger than $n^{4/3}$. Our new proof is inspired by a result in [4]. This result (see Theorem 1) is a short length version of Menger's theorem. With the help of Theorem 1, the new proof is only of no more than half page.

We also perform experiments to study the practical performance of Algorithm 1. As there is no benchmark data set for Label s-t Cut, we test Algorithm 1 on random data sets generated by the Erdős-Rényi model (also known as the G(n, p) model) and the configuration model. In the experiments, each data set contains many groups of random instances. For each group, the average approximation ratio is only a bit above one. Even the worst ratio of each group never exceed two. The actual approximation ratios on random instances form sharp contrast with their analytical counterpart $O(\frac{n^{2/3}}{\text{OPT}^{1/3}})$, whose value will be much higher than one for moderate and large n. This shows that Algorithm 1 has actually very good practical performance in random graphs.

2 A Short Proof for the Two-Phase Approximation Algorithm

It is known that the Label s-t Cut problem can be approximated within $O(\frac{n^{2/3}}{\text{OPT}^{1/3}})$ [14]. The algorithm in [14] is shown as Algorithm 1. This algorithm applies to both directed and undirected graphs. We give a new, but short proof for this result. To the best of our knowledge, the new proof should be the shortest one known so far for the $O(\frac{n^{2/3}}{\text{OPT}^{1/3}})$-approximation result.

Algorithm 1 Two-phase Approximation Algorithm for Label s-t Cut

1: Guess OPT.
2: $L_1 \leftarrow \emptyset$.
3: **while** the s-t distance $\leq \frac{n^{2/3}}{\text{OPT}^{1/3}}$ **do**
4: Find a shortest s-t path P.
5: Remove all the edges from G whose labels are in $L(P)$.
6: $L_1 \leftarrow L_1 \cup L(P)$.
7: **end while**
8: Denote by R the remaining graph. Find a minimum size s-t cut $E' \subseteq E(R)$ of R.
9: $L_2 \leftarrow L(E')$.
10: **return** $L_1 \cup L_2$.

In step 1 of Algorithm 1, by "guess OPT" we actually mean repeating the algorithm for each possible value of OPT in $\{1, 2, \ldots, q\}$, and returning the solution of minimum size ever found. The guess technique is common in design of approximation algorithms. The range of OPT is polynomially bounded, so the whole algorithm still runs in polynomial time. In the algorithm $L(P)$ denotes the set of labels appearing in path P, and similarly $L(E')$ denotes the set of labels appearing in the edge set E'.

Theorem 1 ([4]). *For a simple n-vertex (undirected or directed) graph G, if there are k edge-disjoint paths between two vertices s and t, then the average length (number of edges) of these paths is $O(n/\sqrt{k})$.*

Theorem 2. *Algorithm 1 is an $O(\frac{n^{2/3}}{\text{OPT}^{1/3}})$-approximation algorithm for the* Label s-t Cut *problem on (directed or undirected) simple graphs.*

Proof. First note that the algorithm obviously gives a feasible solution in polynomial time. Every path found in stage one has length $\leq \frac{n^{2/3}}{\mathrm{OPT}^{1/3}}$. All these paths are label-disjoint (i.e., no two paths of them share the same label), and hence edge-disjoint. It follows that the optimal solution must select at least one label from each path, implying that the number of these paths is $\leq$ OPT. Therefore, we get that

$$|L_1| \leq \frac{n^{2/3}}{\mathrm{OPT}^{1/3}}\mathrm{OPT}. \tag{1}$$

By Menger's theorem (see, e.g., [7]), the size (number of edges) $|E'|$ of the minimum s-t cut is equal to the number edge disjoint s-t paths in graph R. By Theorem 1, the average length of these paths is $O(n/\sqrt{|E'|})$. Let $d_R(s,t)$ be the length of the shortest s-t path in graph R. When Algorithm 1 enters stage two, it holds that $d_R(s,t) \geq \frac{n^{2/3}}{\mathrm{OPT}^{1/3}}$. We thus have $\frac{n^{2/3}}{\mathrm{OPT}^{1/3}} \leq d_R(s,t) \leq O(n/\sqrt{|E'|})$, implying that

$$|L_2| \leq |E'| = O(n^{2/3}\mathrm{OPT}^{2/3}) = O\left(\frac{n^{2/3}}{\mathrm{OPT}^{1/3}}\right)\mathrm{OPT}. \tag{2}$$

The theorem then follows by (1) and (2). □

The Menger's theorem we use in the proof of Theorem 2 is its edge version, since the minimum s-t cut is a set of edges. It is known that Menger's theorem has also the vertex version, which says that in a (directed or undirected) graph the size of its minimum s-t separator is equal to the number of vertex (openly) disjoint s-t paths. An s-t separator is a set of vertices whose removal from the graph disconnects s and t. The difficulty of the proof of Theorem 2 is how to relate the size (number of edges) of the minimum s-t cut to the vertex number n, where we insist on n since the approximation ratio is given in terms of it. Theorem 1 helps to overcome this difficulty. Although there is also the vertex version of Menger's theorem, currently we do not know how to use this theorem to analyze Algorithm 1.

3 Experiments

We test Algorithm 1 for Label s-t Cut on random instances. The input graphs in the instances are generated by the G(n, p) model and the configuration model. Algorithm 1 is implemented in C++ and tested using a DELL T7920 workstation with Intel Xeon 24-core CPU and 96GB RAM. As an implementation of Algorithm 1, it is reasonable to remove redundant labels from its solution. This is done in polynomial time by checking for each label whether it can be removed from the solution. The program code can be accessed by query email to the author.

3.1 Experiments on the G(n, p) Model

Overview of the Experiments. The G(n, p) model (also known as the Erdős-Rényi model) is a standard generating model of random graphs. Given vertex number n and probability p, the G(n, p) model puts an edge between vertex u and vertex v in an n-vertex graph G with probability p, for all possible vertex pairs $u \neq v$.

We perform 31 experiments for the G(n, p) model with different label numbers and edge-generating probabilities. The 31 experiments are

$$\text{ER-}q10 - p0.1, \text{ER-}q10 - p0.2, \ldots, \text{ER-}q10 - p0.9,$$
$$\text{ER-}q12 - p0.1, \text{ER-}q12 - p0.2, \ldots, \text{ER-}q12 - p0.9,$$
$$\text{ER-}q14 - p0.1, \text{ER-}q14 - p0.2, \ldots, \text{ER-}q14 - p0.9,$$
$$\text{ER-}q16 - p0.1, \text{ER-}q16 - p0.2, \text{ER-}q16 - p0.3, \text{ER-}q16 - p0.4.$$

The nine experiments ER-q10-p0.1, ER-q10-p0.2, ..., ER-q10-p0.9 are about the tests of Algorithm 1 on the G(n, p) model with $q = 10$ labels and edge generating probability p ranges over $\{0.1, 0.2, \ldots, 0.9\}$. Similarly, experiments ER-q12-$p\cdot$, ER-q14-$p\cdot$, ER-q16-$p\cdot$ are about $q = 12$ labels, $q = 14$ labels, and $q = 16$ labels, respectively.

Each experiment ER-$q\cdot$-$p\cdot$ consists of six groups of random instances. For example, Table 1 presents the test results of experiment ER-q12-p0.2. The vertex number n of these groups ranges over $\{10, 20, 30, 40, 50, 60\}$. Each group in turn contains ten random instances (G, s, t, L) of Label s-t Cut with the specified vertex number and label number. Graph G is generated using the G(n, p) model. For simplicity, we always designate the first vertex v_1 in G as the source s and the last vertex v_n as the sink t. (The vertices obtain their indices when graph G is generated.) Each edge in G is given a label picked uniformly at random from the set L containing q labels. Therefore, each experiment actually contains 60 random instances of Label s-t Cut.

Due to space limitation, we have to select five representative experiments to explain their results. The five experiments are

$$\text{ER-}q12\text{-}p0.2, \text{ER-}q12\text{-}p0.5, \text{ER-}q12\text{-}p0.8, \text{ER-}q16\text{-}p0.2, \text{ER-}q16\text{-}p0.4, \quad (3)$$

whose test results are given in Tables 1, 2, 3, 4, and 5, respectively. The test results of all the 31 experiments will be given in the full version of the paper.

Table Structure Showing the Test Results. All the tables representing the test results in this paper have the same structure. We introduce the structure by using Table 1 as an example. Let us focus on group i ($1 \le i \le 6$) in Table 1. Group i consists of ten random instances. We compute the optimal value (i.e., the number of labels in an optimal solution) of each instance by enumerating, obtaining ten optimal values. The column "avg. OPT" in Table 1 is the average of these ten optimal values. We run Algorithm 1 on each of the ten instances, obtaining ten approximate solutions. The column "avg. SOL" is the average of these ten approximation values. The column "avg. ratio" is then the ratio of the

Table 1. Results for experiment ER-q12-p0.2

group	n	avg. OPT	avg. SOL	avg. ratio	worst ratio
1	10	0.6	0.6	1	1
2	20	1.8	1.9	1.05556	1.5
3	30	4	4.6	1.15	1.66667
4	40	4.9	4.9	1	1
5	50	6.1	6.7	1.09836	1.5
6	60	6.5	6.9	1.06154	1.4

Table 2. Results for experiment ER-q12-p0.5

group	n	avg. OPT	avg. SOL	avg. ratio	worst ratio
1	10	3.2	3.3	1.03125	1.33333
2	20	6.4	6.7	1.04688	1.25
3	30	7.5	7.6	1.01333	1.11111
4	40	9.2	10.1	1.09783	1.25
5	50	10.3	10.4	1.00971	1.1
6	60	10.4	10.4	1	1

Table 3. Results for experiment ER-q12-p0.8

group	n	avg. OPT	avg. SOL	avg. ratio	worst ratio
1	10	5.1	5.1	1	1
2	20	8.2	8.4	1.02439	1.14286
3	30	10	10.2	1.02	1.11111
4	40	11	11.1	1.00909	1.1
5	50	11.2	11.2	1	1
6	60	11.8	11.8	1	1

Table 4. Results for experiment ER-q16-p0.2

group	n	avg. OPT	avg. SOL	avg. ratio	worst ratio
1	10	0.6	0.6	1	1
2	20	1.9	1.9	1	1
3	30	4.6	5.4	1.17391	1.66667
4	40	5.5	6	1.09091	1.4
5	50	6.5	7.2	1.10769	1.5
6	60	6.8	7.1	1.04412	1.33333

Table 5. Results for experiment ER-q16-p0.4

group	n	avg. OPT	avg. SOL	avg. ratio	worst ratio
1	10	2.5	2.5	1	1
2	20	5	5.1	1.02	1.16667
3	30	7.7	8.5	1.1039	1.5
4	40	9	9.6	1.06667	1.33333
5	50	10.4	11.8	1.13462	1.36364
6	60	11.2	11.7	1.04464	1.25

average optimal value and the average solution value. To the aim of evaluating Algorithm 1 more comprehensively, we also compute the worst ratio, which is the largest ratio of the optimal value and the approximation value over the ten instances. So we have the column "worst ratio" in Table 1.

To compute the optimal value of an instance by enumerating is time consuming, even for mild value of label number q and somewhat large vertex number n. For example, we spent 47.22 min to finish the calculation of optimal values for group 5 in Table 5 (about experiment ER-q16-p0.4), and 71.72 min for group 6 in the table. When the label number and edge generating probability get larger, the time spent on the experiments can not be endured. This is why we stop the test for the G(n, p) model at experiment ER-q16-p0.4.

Test Results of the Experiments on the $\mathrm{G}(n, p)$ **Model.** (i) From Tables 1, 2, 3, 4, and 5 of the experiments in (3), we see that the approximation ratios of Algorithm 1 on these random instances are actually very good. The average ratios are only about 1.1. Even the worst ratios do not exceed 1.67. The experiments show that the approximation ratios of Algorithm 1 on random instances are much better than the worst-case analysis result $O(\frac{n^{2/3}}{\mathrm{OPT}^{1/3}})$. For example, for group 6 of Table 5 the analytical ratio will be $6.85c_0$, where c_0 is the hidden constant in the big-O of the approximation ratio and its value would be conservatively at least five. The average ratio of group 6 is only 1.04464.

Since approximation ratios are always at least one, if the average ratio is one, then the worst ratio is also one, and vice versa. We can see that there are nine groups in the tables for the experiment in (3) that the average ratio is equal to one for each of them. This means that Algorithm 1 finds optimal solutions for all the ten random instance in each of the nine groups.

(ii) We also notice that in each experiment, the average optimal value increases as the vertex number n increases, and so does the average approximate solution value. This is because according to the $\mathrm{G}(n, p)$ model, the expected edge number in each random instance is $\frac{1}{2}n(n-1)p$. As n grows, the expected edge number increases quadratically. So in expectation, more labels will be used in the optimal solutions, as well as the approximate solutions.

(iii) On the other hand, we observe that as the edge generating probability p grows, the average optimal values in experiments with high value of p are overall larger than the ones in experiments with low value of p. This observation also applies to the average approximate value. For example, the average optimal values of Table 5 are pairwise larger than the average optimal values of Table 4. This is in general because according to the $\mathrm{G}(n, p)$ model, the larger the edge generating probability, the denser the resulting graph.

Remarks. By the same reasons as in above (ii) and (iii), the resulting $\mathrm{G}(n, p)$ graph will be more dense for large n and p. So, for random instances with these dense graphs, no matter the optimal solutions and the approximate solutions would use almost all the labels to disconnect s and t, as shown by, e.g., groups 4, 5, and 6 of Table 3 for the experiment ER-q12-p0.8.

Note that for group 1 in Table 1 (experiment ER-q12-p0.2) the average optimal value is equal to 0.6. (The average approximate solution value is also equal to 0.6.) This is because in this experiment the edge generating probability is small (only 0.2), and especially for group 1, the vertex number n is also small (only 10). Consequently, in group 1 there are several random instances in which s and t are disconnected. For these instances, both the optimal value and the approximate value are zero. So, we get the average optimal value being less than one. The same thing happens for group 1 in Table 4 (experiment ER-p16-q0.2).

3.2 Experiments on the Configuration Model

Overview of the Experiments. In the G(n, p) model, the vertex degree has the binomial distribution with parameters $n-1$ (number of trials) and p (success probability). The vertices tend to have degrees close to the expected degree $(n-1)p$. Consequently, the large graphs generated by the G(n, p) model with high probability p have no enough diversity in vertex degrees. One cannot specify the degrees of vertices in the G(n, p) model.

In contrast, using the configuration model one can generate a random graph with the specified vertex degree sequence. It works as follows. Firstly, we generate a random degree sequence $(d_1, d_2, \ldots, d_n)$. Each degree d_i is a random number sampling from $\{1, 2, \ldots, n-1\}$. The sum of the degrees should be even. If it is not even, we just abandon the sequence and generate the next one. Then, we generate edges on n vertices satisfying the degree sequence. One can image that vertex v_i has d_i line ends attaching to it. We pick two ones uniformly at random from the ends of all the vertices, and put an edge between the two vertices corresponding to these two ends. If the two ends randomly picked happen to come from the same vertex, this round is canceled (so that no self loop occurs) and the next round starts. In this way, a random graph with the specified vertex degree sequence $(d_1, d_2, \ldots, d_n)$ will be obtained.

We perform four experiments using the configuration model, namely,

CM-q10, CM-q12, CM-q14, and CM-q16.

The test results for these four experiments are given in Tables 6, 7, 8, and 9, respectively. The four tables (experiments) have the same structure as the tables in Sect. 3.1.

Test Results of the Experiments on the Configuration Model. (i) From Tables 6, 7, 8 and 9, we can see that the approximation ratios of Algorithm 1 on the random instances are actually very good. The average ratios wave only about 1.1. Even the worst ratios do not exceed 2. We can see that there are eight groups whose average ratio is one. This means that Algorithm 1 finds the optimal solutions for all the ten instances in each group.

(ii) For each table of the four experiments, we can see that the average optimal value, as well as the average approximate value, (roughly) increases as the vertex number n increases. This is because according to the graph generating method of the configuration model, the expected degree of a vertex is $\frac{(1/2)n(n-1)}{n-1} = \frac{1}{2}n$, and thus the expected number of edges in the graph is equal to $\frac{1}{4}n^2$. So, as n increases, the expected number of edges increases quadratically, resulting in a possibly large solution value (number of labels) to the random instance.

Table 6. Results for experiment CM-$q10$

group	n	avg. OPT	avg. SOL	avg. ratio	worst ratio
1	10	2.2	2.2	1	1
2	20	5	5.2	1.04	1.16667
3	30	4.9	5.3	1.08163	1.33333
4	40	6.6	6.7	1.01515	1.14286
5	50	6.1	6.9	1.13115	1.75
6	60	5.9	5.9	1	1

Table 7. Results for experiment CM-$q12$

group	n	avg. OPT	avg. SOL	avg. ratio	worst ratio
1	10	2.3	2.3	1	1
2	20	5.1	5.6	1.09804	2
3	30	5.3	6	1.13208	2
4	40	6.6	6.8	1.0303	1.14286
5	50	7.2	7.5	1.04167	1.22222
6	60	6.8	6.8	1	1

Table 8. Results for experiment CM-$q14$

group	n	avg. OPT	avg. SOL	avg. ratio	worst ratio
1	10	2.3	2.3	1	1
2	20	5.2	5.3	1.01923	1.14286
3	30	5.9	6	1.01695	1.1
4	40	7	7.7	1.1	1.33333
5	50	7.6	7.7	1.01316	1.14286
6	60	7.7	7.8	1.01299	1.08333

Table 9. Results for experiment CM-$q16$

group	n	avg. OPT	avg. SOL	avg. ratio	worst ratio
1	10	2.3	2.3	1	1
2	20	5.3	5.4	1.01887	1.125
3	30	6.1	6.1	1	1
4	40	7.4	7.4	1	1
5	50	7.9	8.3	1.05063	1.25
6	60	7.9	8.5	1.07595	1.55556

4 Discussions and Conclusions

From the experimental results about the Label s-t Cut problem, we learn that the actual approximation ratios of Algorithm 1 on the tested instances are much better than the analytical counterpart $O(\frac{n^{2/3}}{\text{OPT}^{1/3}})$. This shows that Algorithm 1 has very good practical performance on random instances generated by the $G(n, p)$ model and the configuration model.

However, one should not be excessively surprising to the good practical performance of Algorithm 1 on random instances. In fact, we have an example to show the approximation ratio of Algorithm 1 is at least $\Omega(n^{1/3})$, as shown in Fig. 1. Since the probability to generate such an instance is very low, it is hard to meet such instance by the $G(n, p)$ model or the configuration model.

The graph in Fig. 1 consists of $h = a^{1/3}$ openly vertex-disjoint s-t paths for some $a > 0$ (i.e., these paths only have common end vertices s and t), each of which has length (i.e., number of edges) $f = a^{2/3} + 1$. So, the total number of vertices in the graph is $n = h(f - 1) + 2 = a + 2$. The edges out-going from s have labels $\ell_1, \ell_2, \ldots, \ell_h$, respectively. All the remaining edges have the same label ℓ_{h+1}. In the instance the number of labels is $q = h + 1$. Denote by $\mathcal{I}$ the constructed instance. It is easy to see that the optimal solution to $\mathcal{I}$ is $\{\ell_{h+1}\}$ and thus OPT $= 1$.

When running Algorithm 1 on instance $\mathcal{I}$, the first stage will remove nothing since each s-t path has length $f = a^{2/3} + 1 > (a + 2)^{2/3} = \frac{n^{2/3}}{\text{OPT}^{1/3}}$. The second stage will compute an minimum s-t cut. We can reasonably assume that this minimum s-t cut is the set of all the edges out-going from s. Therefore, the solution to instance $\mathcal{I}$ found by the algorithm is $\{\ell_1, \ell_2, \ldots, \ell_h\}$. Consequently, the approximation ratio of Algorithm 1 on instance $\mathcal{I}$ is $h/1 = a^{1/3} = \Omega(n^{1/3})$.

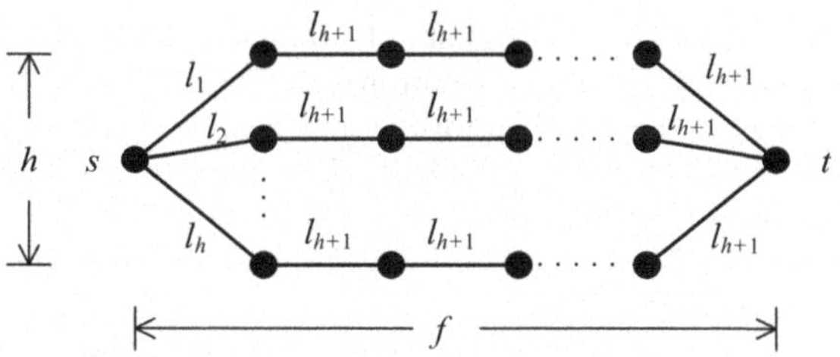

Fig. 1. A bad example for Algorithm 1

So far, we arrive at a situation that Algorithm 1 has approximation ratio between $\Omega(n^{1/3})$ and $O(\frac{n^{2/3}}{\mathrm{OPT}^{1/3}})$ on the one hand, and has very good practical performance on the other hand. Actually, this phenomenon has already been pointed out by Vazirani in his textbook [12, Preface]. We would like to cite some statements from the book. The preface of [12] says the following. "*The approximation ratio guarantee only reflects the performance of the algorithm on the most pathological instances.*" "*Experiments have confirmed that these and other sophisticated algorithms do have error bounds of the desired magnitude, 2% to 5%, on typical instances, even though their worst case error bounds are much higher.*" Our algorithms and experiments give strong support to the statements.

The approximation ratio only reveals the extreme behaviours of an approximation algorithm. This suggests that in practice, many approximation algorithms qualify as routines solving the corresponding problems. These algorithms have not only rigorously proved theoretical approximation guarantees, but also good practical performance. We conclude the paper by what Vazirani pointed out in [12, Preface]: "*Perhaps it is more appropriate to view the approximation guarantee as a measure that forces us to explore deeper into the combinatorial structure of the problem and discover more powerful tools for exploiting this structure.*"

Acknowledgements. This work is supported by the National Natural Science Foundation of China under Grant 62272280, Grant 61972228, and Grant 61672323, and by the Natural Science Foundation of Shandong Province, China under Grant ZR2021ZD15.

References

1. Coudert, D., Datta, P., Pérennes, S., Rivano, H., Voge, M.-E.: Shared risk resource group: complexity and approximability issues. Parallel Process. Lett. **17**, 169–184 (2007)
2. Coudert, D., Pérennes, S., Rivano, H., Voge, M.E.: Combinatorial optimization in networks with shared risk link groups. Discret. Math. Theor. Comput. Sci. **18**(3), 8:1–8:25 (2016)
3. Dutta, T., Heath, L.S., Kumar, V.A., Marathe, M.V.: Labeled cuts in graphs. Theor. Comput. Sci. **648**, 34–39 (2016)

4. Galil, Z., Yu, X.: Short length versions of Menger's theorem. In: Proceedings of the 27th Symposium on Theory of Computing (STOC), pp. 499–508, 1995
5. Jha, S., Sheyner, O., Wing, J.: Two formal analyses of attack graphs. In: Proceedings of the 15th IEEE Computer Security Foundations Workshop (CSFW), pp. 49–63. IEEE Computer Society, 2002
6. Khuller, S., Moss, A., Naor, J.: The budgeted maximum coverage problem. Inf. Process. Lett. **70**(1), 39–45 (1999)
7. Korte, B., Vygen, J.: Combinatorial Optimization: Theory and Algorithms. Springer, 5th edition, 2012
8. Rohloff, K., Khuller, S., Kortsarz, G.: Approximating the minimal sensor selection for supervisory control. Discret. Event Dyn. Syst. **16**(1), 143–170 (2006)
9. Sheyner, O., Haines, J., Jha, S., Lippmann, R., Wing, J.M.: Automated generation and analysis of attack graphs. In: Proceedings of the IEEE Symposium on Security and Privacy, pp. 273–284, 2002
10. Sheyner, O., Wing, J.M.: Tools for generating and analyzing attack graphs. In: Proceedings of Formal Methods for Components and Objects, Second International Symposium (FMCO), pp. 344–371 (2004)
11. Tang, L., Zhang, P.: Approximating minimum label s-t cut via linear programming. In: Fernández-Baca, D. (ed.), Proceedings of the 10th Latin American Theoretical Informatics Symposium (LATIN), pp. 655–666, 2012
12. Vazirani, V.V.: Approximation Algorithms. Springer (2001)
13. Zhang, P., Cai, J.-Y., Tang, L., Zhao, W.: Approximation and hardness results for label cut and related problems. J. Comb. Optim. **21**(2), 192–208 (2011)
14. Zhang, P., Bin, F., Tang, L.: Simpler and better approximation algorithms for the unweighted label s-t cut problem. Algorithmica **80**(1), 398–409 (2018)
15. Zhang, P., Tang, L.: Minimum label s-t cut has large integrality gaps. Inf. Comput. **275**, 104543:1–104543:15 (2020)

Risk-Embedded Scheduling Optimization for a Virtual Power Plant Under Carbon Emission Trading Constraints

Jiaming Hu[1], Boon-Han Lim[2], Maowen Lu[1], Xiaoyun Tian[1(✉)], Dongzhao Wang[1], and Wenqing Xu[1]

[1] Institute of Operations Research and Information Engineering, Beijing University of Technology, Beijing 100124, People's Republic of China
xiaoyun_txy@163.com

[2] Department of Electrical and Electronic Engineering, Lee Kong Chian Faculty of Engineering and Science, Universiti Tunku Abdul Rahman, 43000 Kajang, Selangor, Malaysia

Abstract. Virtual power plants (VPPs), as a novel integrated energy aggregation model, can effectively realize peak shaving and valley filling in the power system. This paper mainly investigates the model of a VPP consisting of distributed energy resources (DERs) as energy producers and converters, and an electric vehicle parking lot (EVPL) serving as both a battery storage system and an energy consumer. To mitigate the risks associated with uncertainties, a downside risk constraint method is concerned, which has not yet been applied to this model with multiple uncertainties. We concentrate on the risk-embedded scheduling of the VPP with EVPL, taking into account key uncertain factors, such as, the arrival and departure times of the EVs. Numerical results reveal that the primary factor affecting VPP profits is the instability of renewable energy. In our model, compared to risk-neutral strategies, risk-averse strategies lead to a reduction in average profits by 5.66%, highlighting the trade-off between minimizing risk and maximizing profits.

Keywords: Virtual power plant · Scheduling · Carbon trading mechanism · Downside risk constraint

1 Introduction

In recent years, exceedingly high summer temperatures have significantly impacted human capital productivity and daily life. The excessive use of fossil fuels and the resultant global warming not only degrade quality of life but also pose substantial threats to economic development. To address these challenges, sustainable energy solutions have become paramount. With the increasing societal emphasis on low-carbon and environmental protection, there is an intensified focus on expediting the construction of new power systems and developing microgrids that primarily depend on renewable energy for interactions with the

D. Du et al. (Eds.): COCOA 2024, LNCS 15434, pp. 139–150, 2025.
https://doi.org/10.1007/978-981-96-4445-2_12

upstream grid [4]. The virtual power plant (VPP), as a novel integrated aggregation model, is progressively receiving public recognition.

The VPP, with its ability to aggregate electricity from various sources over time, embodies the concept of "accumulating grains of sand to form a tower", first introduced by Shimon Awerbuch [2,12]. Technically, a VPP is an advanced information and energy interaction system capable of aggregating and synergistically optimizing distributed energy, energy storage systems, and controllable loads [10,13,16,18]. Fang et al. [6] introduce the concept of edge intelligence (EI) architecture to address the economic dispatch problem (EDP) within VPPs in the internet of energy context. In contrast to traditional edge computing, the proposed EI structure inherits the benefit of reduced cloud computing burden, while concurrently bolstering the computing capacity of edge computing. Liu et al. [9] establish a VPP scheduling model of energy sharing and interaction among multiple regions, and propose a coordinated optimal scheduling strategy of distributed energy multiple regions. Zhang et al. [21] incorporate the carbon trading mechanism and the green certificate trading mechanism into the optimal dispatch model of VPP including wind power generation, photovoltaic power generation, gas turbines and energy storage device.

The downside risk method is commonly used in financial management, portfolio optimization, and decision analysis to limit potential losses that exceed an investor's or decision-maker's risk tolerance [3]. In the energy sector, the uncertainty of renewable energy and the immediacy of demand means such potential risk. Saberi-Beglar et al. [14] propose a downside risk constraint to mitigate risks arising from uncertainties in the risk-embedded scheduling of a residential energy hub model. Recently, approaches have been proposed to improve resilience and reduce risk in distribution grids using power-to-hydrogen technology to optimize day-ahead scheduling [20]. These approaches employ downside risk constraints to ensure the robustness of resilience-oriented outcomes, significantly impacted by uncertainties.

In this paper, we focus on the risk-embedded scheduling of a VPP with an electric vehicle parking lot (EVPL), considering various uncertain factors. We consider the involvement of VPP in carbon emission trading, contributing to promoting system-wide emission reduction. The main contributions of this paper compared to existing work are threefold. 1.) We propose a downside risk constraint to characterize the uncertainty risk faced in economic dispatch. 2.) We analyze the change in the operator's expected profit under risk-neutral strategies and risk-averse conditions, as well as analyze the reasons for the change in profit based on the performance of different power sources in each time period. 3.) Numerical results reveal that, compared to risk-neutral strategies, risk-averse strategies are expected to reduce average profits by 5.66% in our model.

The organization of this paper is as follows: Sect. 2 presents the model and the mathematical formulation of the VPP under consideration. The objective function of the model, encompassing profit and operating and carbon emission costs, is delineated in Sect. 3. Section 4 describes our algorithm for handling risk-embedded scheduling. Section 5 illustrates a numerical study of the proposed model, discussing the input data and numerical results. Finally, Sect. 6 presents the conclusion.

2 Model of the VPP

In this section, we predominantly investigate the VPP concept, which comprises distributed energy resources (DERs) and an EVPL. By sequentially presenting the constraints at each stage from energy generation to consumption, we construct the overall framework required for day-ahead scheduling.

2.1 Model Characterization

To explore the change in returns associated with uncertainty-related risks, we use a downside risk constraint approach. Our study centers on the risk-embedded scheduling of a VPP with EVPL, accounting for various uncertainties. This model integrates several main features by introducing a carbon emission mechanism that effectively limits the reliance on fuel engines. It also incorporates a robust framework for modeling uncertainties, capturing the variability of renewable energy sources and the dynamic charging and discharging behavior of EV owners. Furthermore, the downside risk constraints are designed to safeguard target profits in Subsect. 2.5, offering a balanced approach to risk management and profitability.

Formally, let N_s denote a scenario set, where the VPP includes N_g gas turbine units, N_q renewable energy units (wind turbine and PV generator), and one EVPL that can accommodate N_v EVs. We consider a scheduling period of one day (T), divided into 24 time periods with each interval t lasting one hour. The parameters are assumed to remain constant in each interval.

2.2 The Carbon Emission Trading Mechanism

In order to achieve low-carbon output, we consider the carbon emissions generated by gas turbines and calculate the carbon emission cost as follows:

$$C_{t,s}^{\mathrm{C}} = K_{\mathrm{C}}(M_{\mathrm{C},t,s} - M_{\mathrm{D},t,s}), \quad \forall s,t \tag{1}$$

$$M_{\mathrm{C},t,s} = \sum_{g=1}^{N_G} Q_g P_{g,t,s}^{\mathrm{gas}}, \quad \forall s,t \tag{2}$$

$$M_{\mathrm{D},t,s} = \sum_{m=1}^{N_m} \varepsilon P_{m,t,s}, \quad \forall s,t \tag{3}$$

Equation (1) shows the carbon emission cost of VPP at time t in scenario s, where K_{C} is the transaction price of the per unit carbon emission. The carbon emission credit $M_{\mathrm{D},t,s}$ and the actual carbon emission $M_{\mathrm{C},t,s}$ is determined by Eqs. (2) and (3). The term $P_{g,t,s}^{\mathrm{gas}}$ is the generation output of gas turbine g at time

t, and the term $P_{m,t,s}$ is the generation output of generator m at time t. The notation Q_g is used to characterize carbon emission intensity of the generation output, and ε is the carbon emission distribution coefficient, which is determined by the "National Grid Baseline Emission Factor [11]".

2.3 Uncertainty Modeling

Models for uncertain parameters, such as the arrival and departure times of electric vehicles (EVs) and the variability of renewable energy production, are frequently explored. A widely adopted approach to model EV behavior involves using a truncated Gaussian distribution. The probability distribution function ($\mathrm{PDF}_a/\mathrm{PDF}_d$) of the arrival/departure times parameters $t^a_{v,s}/t^d_{v,s}$ for each EV's owner v in the scenario s in Eqs. (4–6).

$$\mathrm{PDF}_a(x) = \frac{1}{\sqrt{2\pi\sigma^a}} \cdot \exp\left[\frac{(x-\mu^a)^2}{2(\sigma^a)^2}\right],$$

$$\mathrm{PDF}_d(x) = \frac{1}{\sqrt{2\pi\sigma^d}} \cdot \exp\left[\frac{(x-\mu^d)^2}{2(\sigma^d)^2}\right], \tag{4}$$

$$t^a_{v,s} = F(x; \mu^a, \sigma^a, (t^a_{-v}, t_v^{-a})), \quad \forall v, s \tag{5}$$

$$t^d_{v,s} = F(x; \mu^d, \sigma^d, (\max\{t^a_v, t^d_{-v}\}, t_v^{-d})), \quad \forall v, s \tag{6}$$

The terms μ and σ are the corresponding mean and standard variance value, respectively, and X is the uncertain parameter. Let (t^a_{-v}, t_v^{-a}) in Eq. (5) denote the minimum and maximum arrival time. Similarly, the departure time of each EV is restricted in a range of $(\max\{t^a_v, t^d_{-v}\}, t_v^{-d})$ in Eq. (6). Renewable energy mainly includes wind turbines and photovoltaic generators in the VPP. The output power of wind turbines based on wind speed is illustrated in Eq. (7).

$$P_t^{\mathrm{wind}} = \begin{cases} 0 & \text{if } v_t < v_{\mathrm{in}} \text{ or } v_t > v_{\mathrm{out}}, \\ P_{\mathrm{rate}}^{\mathrm{wind}} \frac{v_t^3 - v_{\mathrm{in}}^3}{v_{\mathrm{rate}}^3 - v_{\mathrm{in}}^3} & \text{if } v_{\mathrm{in}} \le v_t \le v_{\mathrm{rate}}, \\ P_{\mathrm{rate}}^{\mathrm{wind}} & \text{if } v_{\mathrm{rate}} < v_t \le v_{\mathrm{out}}, \end{cases} \tag{7}$$

where $P_{\mathrm{rate}}^{\mathrm{wind}}$ represents the power corresponding to rated wind speed v_{rate}; v_{in} and v_{out} denote the cut-in wind speed and cut-out wind speed, respectively, when the wind speed exceeds the cut-out wind speed, the unit must be halted to prevent damage. Base on the analysis of historical data, wind speed follows the Weibull distribution [8] given by the following PDF:

$$f(v) = \frac{k}{c}\left(\frac{v}{c}\right)^{k-1} \exp\left[-\left(\frac{v}{c}\right)^k\right].$$

The output power of PV based on solar irradiance is illustrated in Eq. (8).

$$P_{t,s}^{\mathrm{pv}} = \eta^{\mathrm{pv}} \cdot A^{\mathrm{pv}} \cdot r_{t,s}^{\mathrm{pv}} \cdot (1 - 0.005 \cdot (T_{t,s}^{\mathrm{out}} - 25)), \quad \forall s, t \tag{8}$$

The term η^{pv} is the power generation efficiency of PV, A^{pv} is the surface area of PV, and $r^{\mathrm{pv}}_{t,s}$ is the amount of solar radiation. Similarly, the solar irradiance follows the Beta distribution [8] given by the following PDF:

$$f(r) = \frac{\Gamma(a+b)}{\Gamma(a)\Gamma(b)}(\frac{r}{r_{\max}})^{a-1}(1-\frac{r}{r_{\max}})^{b-1},$$

where $a = (\mu_r \cdot b)/(1-\mu_r)$, $b = \left(\mu_r(1+\mu_r)/\sigma_r^2 - 1\right)(1-\mu_r)$.

2.4 Other Constraints

The output and ramping constraints of gas turbines are illustrated in Eqs. (9–10).

$$u_{g,t,s}P^{\mathrm{gas}}_{g,\min} \leq P^{\mathrm{gas}}_{g,t,s} \leq u_{g,t,s}P^{\mathrm{gas}}_{g,\max}, \quad \forall g,s,t \tag{9}$$

$$-R^{\mathrm{D}}_{g}\Delta t \leq P^{\mathrm{gas}}_{g,t,s} - P^{\mathrm{gas}}_{g,(t-1),s} \leq R^{\mathrm{U}}_{g}\Delta t, \quad \forall g,s,t \tag{10}$$

where $P^{\mathrm{gas}}_{g,\min}$ and $P^{\mathrm{gas}}_{g,\max}$ are the minimum and maximum power outputs of gas turbine g; R^{D}_{g} and R^{U}_{g} are the ramping-down and ramping-up limits of gas turbine g. The constraints of EVPL including the power and state of charging at time t are presented in Eqs. (11–17).

$$P^{\mathrm{ch}}_{v,t,s} = u^{\mathrm{ch}}_{v,t,s} \cdot P^{\mathrm{ch}}_{v,s}, \quad \forall v,s,t \tag{11}$$

$$P^{\mathrm{dch}}_{v,t,s} = u^{\mathrm{dch}}_{v,t,s} \cdot P^{\mathrm{dch}}_{v,s}, \quad \forall v,s,t \tag{12}$$

$$u^{\mathrm{ch}}_{v,t,s} + u^{\mathrm{dch}}_{v,t,s} \leq u^{\mathrm{in}}_{v,t,s}, \quad \forall v,s,t \tag{13}$$

$$S_{v,s,\min} \leq S_{v,t,s} \leq S_{v,s,\max}, \quad \forall v,s,t \tag{14}$$

$$S_{v,t^{a}_{v,s},s} = S_{v,a,s}, \quad \forall v,s \tag{15}$$

$$S_{v,d,s} \leq S_{v,t^{d}_{v,s},s} \leq S_{v,s,\max}, \quad \forall v,s \tag{16}$$

$$S_{v,t,s} = S_{v,t-1,s} + P^{\mathrm{ch}}_{v,t,s}\eta^{\mathrm{ch}}_{v,s}\Delta t - \frac{P^{\mathrm{dch}}_{v,t,s}}{\eta^{\mathrm{dch}}_{v,s}}\Delta t, \forall v,s,t \tag{17}$$

The terms $u^{\mathrm{ch}}_{v,t,s}$ and $u^{\mathrm{dch}}_{v,t,s}$ are binary decision variables separately used to denote the charging and discharging status of each EV; $u^{\mathrm{in}}_{v,t,s} \in \{0,1\}$ is whether EV stay in EVPL; $S_{v,t,s}$ denotes the state of charging of each EV; $P^{\mathrm{ch}}_{v,s,t}$ and $P^{\mathrm{dch}}_{v,s,t}$ are separately used to represent the rated charging and discharging power; $\eta^{\mathrm{ch}}_{v,s}$ and $\eta^{\mathrm{dch}}_{v,s}$ are respectively the charging and discharging efficiencies.

The constraints of renewable energies including wind power and PV power are depicted in Eqs. (18–20).

$$0 \leq P^{\mathrm{type}}_{q,t,s} \leq P^{\mathrm{type}}_{q,t,\max}, \forall q,s,t \tag{18}$$

$$\sum_{t=1}^{T} P^{\mathrm{wind}}_{q,t,s} \leq \sum_{t=1}^{T}(1-\alpha)P^{\mathrm{wind}}_{q,t,\max}, \quad \forall q,s \tag{19}$$

$$\sum_{t=1}^{T} P^{\mathrm{PV}}_{q,t,s} \leq \sum_{t=1}^{T}(1-\beta)P^{\mathrm{PV}}_{q,t,\max}, \quad \forall q,s \tag{20}$$

where α and β are the maximum curtailment ratio of wind and PV; $P_{q,t,\max}^{\text{type}}$ is the expected maximum output power of the type set (type $\in$ {wind, PV}). The term $P_{t,s}^{\text{sell}}$ refers to the power sold at time t in scenario s. The power balance constraint is presented in Eq. (21).

$$\sum_{g=1}^{N_G} P_{g,t,s}^{\text{gas}} + \sum_{q=1}^{N_q} \sum_{\text{type}} P_{q,t,s}^{\text{type}} + \sum_{v=1}^{N_v} P_{v,t,s}^{\text{dch}} = P_{t,s}^{\text{sell}} + \sum_{v=1}^{N_v} P_{v,t,s}^{\text{ch}}, \forall t, s \tag{21}$$

2.5 Downside Risk Constraint

Considering the numerous uncertainties inherent in real-world energy systems, VPP operators aim to minimize economic losses while ensuring responsibility toward all participants. In this paper, we seek to maximize the VPP operating profit, denoted by $M(s)$, through a risk-embedded scheduling problem, as outlined in Sect. 3. The operator may set a target profit M_0. If the profit of scenario s is more than the target profit, the difference between these profits is the downside risk (RISK(s)). The amounts of expected downside risk (EDR(M_0)) refers to the profit in excess of the target profit, therefore the small amounts will be better for the operator. The term EDR_0 is expected downside without risk constraints. The downward risk constraint is formalized in Eqs. (22–25).

$$\text{RISK}(s) = \begin{cases} M_0 - M(s), & \text{if } M(s) > M_0, \\ 0, & \text{if } M(s) \leq M_0, \end{cases} \tag{22}$$

$$\text{EDR}(M_0) = \exp\left[\text{RISK}(s)\right] = \sum_{s=1}^{N_s} \pi(s) \times \text{RISK}(s), \tag{23}$$

$$\text{EDR}(M_0) \leq \theta \text{EDR}_0, \tag{24}$$

$$\text{EDR}_0 = \sum_{s=1}^{N_s} \pi(s)(M_0 - M(s)), \tag{25}$$

where $\pi(s)$ represents the probability of each scenario in N_s; the parameter θ represents the operator's risk tolerance. When $\theta = 1$, it indicates a risk-neutral strategy, while a value closer to 0 reflects a risk-averse strategy.

3 Objective Function

Our goal is to maximize the total profit $\sum_{s=1}^{N_s} M(s)$ generated by VPP's energy interaction with the upstream grid and EVPL and participation in carbon trading within T, which is presented in Eq. (26).

$$\max \sum_{s=1}^{N_s} \sum_{t=1}^{T} \left[\lambda_t^{\text{sell}} P_{t,s}^{\text{sell}} - (C_{t,s}^{\text{gas}} + C_{t,s}^{\text{C}} + C_{t,s}^{\text{evpl}})\right]. \tag{26}$$

The first term is the profit of power transactions with the upstream grid where λ_t^{sell} is the spot price. The second term is the cost of the gas turbines, and the third term is the carbon emission cost. And the last term is the operating cost of EVPL. The costs of gas turbines (including switching costs) are shown in Eqs. (27–29).

$$C_{t,s}^{\mathrm{gas}} = \sum_{g=1}^{N_G} \left[\left(a_g (P_{g,t,s}^{\mathrm{gas}})^2 + b_g P_{g,t,s}^{\mathrm{gas}} + c_g \right) + C_{g,t,s}^{\mathrm{SU}} + C_{g,t,s}^{\mathrm{SD}} \right], \forall s, t \tag{27}$$

$$C_{g,t,s}^{\mathrm{SU}} = \max \left\{ \lambda_g^{\mathrm{SU}} \left(u_{g,t,s} - u_{g,(t-1),s} \right), 0 \right\}, \forall g, s, t \tag{28}$$

$$C_{g,t,s}^{\mathrm{SD}} = \max \left\{ \lambda_g^{\mathrm{SD}} \left(u_{g,(t-1),s} - u_{g,t,s} \right), 0 \right\}, \forall g, s, t \tag{29}$$

where $u_{g,t,s}$ is a binary variable showing the status (open-close) of each gas turbine g at time t; a_g, b_g and c_g are the coefficients of the quadratic cost function; $P_{g,t,s}^{\mathrm{gas}}$ is the power output of gas turbine g at time t; λ_g^{SU} and λ_g^{SD} are the fixed start-up and shut down cost. The total costs of EVPL (including capacity, penalty and degradation costs) are represented in Eqs. (30–33).

$$C_{t,s}^{\mathrm{evpl}} = \sum_{v=1}^{N_v} (\lambda_t^{\mathrm{dch}} P_{v,t,s}^{\mathrm{dch}} - \lambda_t^{\mathrm{ch}} P_{v,t,s}^{\mathrm{ch}}) + C_{t,s}^{\mathrm{evpl,\ cap}} + C_{t,s}^{\mathrm{evpl,\ pen}} + C_{t,s}^{\mathrm{evpl,\ deg}}, \quad \forall s, t \tag{30}$$

$$C_{t,s}^{\mathrm{evpl,\ cap}} = \sum_{v=1}^{N_v} (S_{v,s,\max} - S_{v,s,\min}) C^{\mathrm{ev}} \lambda_t^{\mathrm{cap}}, \quad \forall s; t = t_{v,s}^d \tag{31}$$

$$C_{t,s}^{\mathrm{evpl,\ pen}} = \sum_{v=1}^{N_v} \left| S_{v,t_v^d,s} - S_{v,t_v^a,s} \right| C^{\mathrm{ev}} \lambda_t^{\mathrm{pen}}, \quad \forall s; t = t_{v,s}^d \tag{32}$$

$$C_{t,s}^{\mathrm{evpl,\ deg}} = \sum_{v=1}^{N_v} (P_{v,t,s}^{\mathrm{ch}} + P_{v,t,s}^{\mathrm{dch}}) \cdot \frac{C_v^{\mathrm{rep,ev}}}{\mathrm{Cap}_v^{\mathrm{ev}}}, \quad \forall s, t \tag{33}$$

where λ_t^{ch} and λ_t^{dch} represent the prices of charging/discharging power of each EV; λ_t^{cap} represents the price of using EV's battery capacity and λ_t^{pen} represents the penalty pirce of charging insufficiently at time t; $C_v^{\mathrm{rep,ev}}$ is the the cost of replacing of each EV's battery; $\mathrm{Cap}_v^{\mathrm{ev}}$ denotes the charging/discharging capacity of EV's battery during its lifetime.

4 Methods

Algorithm 1 explains how we can find feasible scheduling solutions. Firstly, we generate initial 1000 scenarios through a Monte-Carlo simulation [15], which is employed for all uncertain parameters. Then, the k-means++ algorithm [1] is used to cluster a large number of initial scenarios and reduce the scenario number to 10. The basic idea is to divide the dataset into k clusters, so that the data points within each cluster are as similar as possible, while the data points between different clusters are as dissimilar as possible [5,7,17,22]. The problem is solved without considering a downside risk constraint to attain a target profit $M(s)$ and expected downside risk EDR_0. After achieving acceptable risk, the model is solved by considering a downside risk constraint to calculate the optimal solution for operating profit.

Algorithm 1. Risk post-processing algorithm

Require: All model parameters.
1: Initializate time t, scenario s;
2: Generate uncertainty modeling scenarios based on Monte Carlo Simulation;
3: Reduce scenarios number from 1000 to 10 through k-means++ method;
4: Solve scenario-based problem without downside risk constraint Eqs. (1–21, 26–33);
5: Obtain the target profit and expected downside risk;
6: Solve the optimization problem with downside risk constraint (24), if the expected downside risk is acceptable, obtain the scheduling solution for VPP, otherwise return step 4 and decrease previous risk value by 10%.

---

5 Numerical Study

In this section, to assess the impact of the downside risk constraint on the decision-making of risk-averse investors, we conduct simulation experiments using Algorithm 1. Our problem is modeled with mixed-integer quadratic programming and solved using the DOCPLEX/CPLEX solver in Python 3.8.

5.1 Input Data

The VPP input data include fuel generator parameters, carbon emission index, and some parameters that need to be used in the EVPL. Based on the probability distribution function and Monte Carlo simulation process, we randomly generate the forecasted wind and photovoltaic power, and the arrival and departure times of the EVs for ten scenarios.

5.2 Result Analysis

To provide a more intuitive visualization of the numerical effects, we simulate the day-ahead scheduling and revenue of the VPP system. The DRC approach is used to study the system's unpredictable behavior under risk-averse ($\theta = 0$) and risk-neutral ($\theta = 1$) models. Two conditions are provided in the following to analyze the impacts of different conditions on the proposed model.

Condition 1: A VPP with and without EV's forced charging in summer.

Condition 2: Implementing DRC to the proposed VPP to assess the risk level and maximize it for the conservative operator.

For each of the above two conditions, we provide the discussions of the energy transactions and the risk decision as follows.

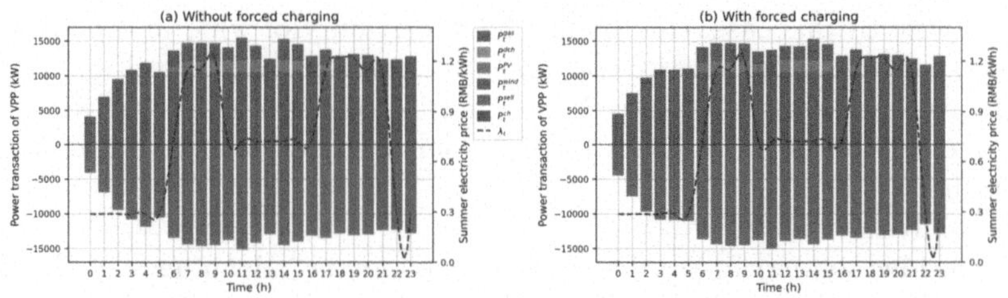

Fig. 1. VPP transaction with/without forced charging in one day (the dotted line indicates the summer electricity price).

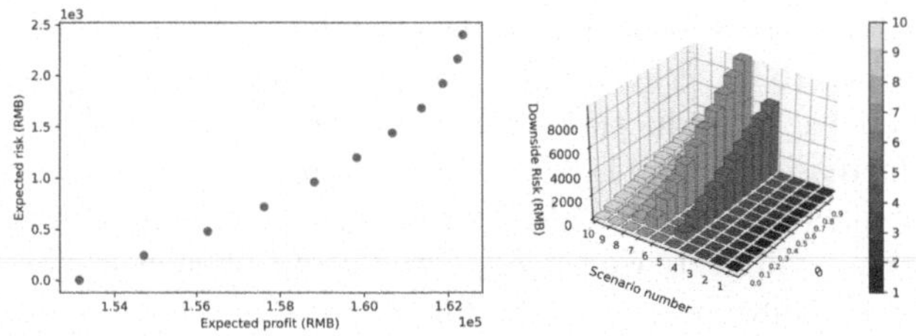

Fig. 2. Comparison chart of expected profit and expected risk (left); Distribution of downside risks under different scenarios (right).

1. Energy transactions of the VPP
   Figure 1 displays the average profits and segmented electricity price for ten scenarios without risk constraints. Since demand-side responses other than EVs are not considered in this model, all our electrical energy can be injected into the upstream grid for consumption. In this model, the gas turbines, as the primary power generators, operate at their total capacity from 3:00-23:00. VPP injects the most power into the upstream grid at 11:00, where renewable energy achieves maximum electric power input into the upstream grid, leading to profit gain. In the case without forced charging, the role of EVPL has weakened. After comparing electric vehicle storage and discharge with direct electricity sales, we find that selling electricity directly will bring more benefits to the operator. Moreover, during this period, electricity prices are relatively low, which attracts many vehicle owners to charge their vehicles. The VPP average profit is 162, 336.48 (CNY) calculated by Eq. (26).
   EVs can maximize total profit by preferentially discharging during parking periods. While this assumption is consistent with the profit-maximizing objective function, it may not reflect real-life situations. We therefore make the assumption that arriving vehicles are forced to charge. Considering the initial charging level of the EV and its battery capacity, we added Eq. (16) to ensure that the EV is charged upon arrival until it reaches the minimum drive-off charge. This ensures the normal travel of EV owners. By exporting the values of the decision variables, we find that the charge level of EVs upon departure exceeds 17, except for a few EVs that didn't stay long enough to charge for

four hours. In this case, combining the uncertainty of wind and photovoltaic power, we obtain the energy conversion diagram as shown in Fig. 1(b). The VPP average profit with forced charging is 160, 983.19 (CNY), which is an average reduction of 1, 353.29 (CNY) compared to the case without forced charging.

2. Risk analysis
We consider numerical results to demonstrate the advantages of the downside risk constraints method. According to the results obtained in Fig. 2, the expected profit of the VPP without downside risk constraints is 162, 336.48 (CNY), and the expected risk is 2, 401.91 (CNY). All scenarios will reach a certain amount of expected cost when assigning the risk level θ to zero. As the expected risk decreases according to the risk level, our expected profit gradually decreases to 153, 148.90 (CNY). In the worst scenario 7, operation profit does not decrease by reduction of θ, while the profit of the most favorable scenario 1 decreases to 153, 148.90 (CNY) in $\theta = 0$.
In summary, it is evident that decreasing θ or increasing conservativeness in decision-making will decrease the expected profit of each scenario. Due to our risk experiment being based on non forced charging, the impact of uncertain EV owner behavior on the expected profits is relatively small. Through the experiments, we found that when wind power generation is not stable enough, our profits will decrease. This reflects that risk-averse investors conservatively estimate profits, and they will not rely on unstable portfolios for profits.

6 Conclusion

In this paper, we present a risk-embedded scheduling model for a VPP with EVPL, taking carbon emission trading into account. Our focus is on maximizing profits, rather than minimizing losses, to offer a more beneficial perspective for decision-makers. We provide a set of scheduling schemes without downside risk constraints and analyze the scheduling situation at each time step. Additionally, we examine how operators' expected profits shift under different risk levels, comparing risk-neutral and risk-averse strategies. The numerical results indicate that risk-averse strategies reduce average profits by 5.66% compared to risk-neutral strategies, emphasizing the trade-off between risk and reward.

Additionally, we identify potential limitations and challenges associated with this model. When the electricity usage in EVPL is less than 1% of the production power, it is related to the size of EVPL and the battery capacity of the EVs we are considering. If the problem is considered as a game between multiple VPPs, the appropriate selling price and the limited selling quantity will be considered to be added to the model. Moreover, the accuracy of renewable energy predictions significantly impacts the model's profitability, a topic that has been extensively studied and remains a critical area for further improvement.

Acknowledgments. The research is supported by National Key R&D Program of China (No. 2022-YFE0196100), and Fundamental Research Grant Scheme (FRGS) of

Malaysia (FRGS/1/2022/TK0/UTAR/02/8). The above two grants are also parked under China-Malaysia Intergovernmental Science, Technology and Innovation Cooperative Programme 2023 (offer letter MOSTI.700-2/2/8(39), dated 9 February 2023).

References

1. Arthur, D., Vassilvitskii, S. *k*-means++: the advantages of careful seeding. In: Proceedings of the Eighteenth Annual ACM-SIAM Symposium on Discrete Algorithms, pp. 1027–1035 (2007)
2. Awerbuch, S., Preston, A. (eds.) The virtual utility: Accounting, technology and competitive aspects of the emerging industry (Vol. 26). Springer Science and Business Media (2012)
3. Cao, Y., Wang, Q., Cheng, W., Nojavan, S., Jermsittiparsert, K.: Risk-constrained optimal operation of fuel cell/photovoltaic/battery/grid hybrid energy system using downside risk constraints method. Int. J. Hydrogen Energy **45**(27), 14108–14118 (2020)
4. Chen, C., Chen, C.C., Hsiao, C.T.: Determinants of installing solar power generation equipment in Taiwan: a viewpoint of integrating market environment and government policy. Int. J. Green Energy **19**(6), 664–675 (2022)
5. Chen, J., Han, J., Meng, X., Li, Y., Li, H.: Graph convolutional network combined with semantic feature guidance for deep clustering. Tsinghua Sci. Technol. **27**(5), 855–868 (2022)
6. Fang, D., Guan, X., Lin, L., Peng, Y., Sun, D., Hassan, M.M.: Edge intelligence based economic dispatch for virtual power plant in 5G internet of energy. Comput. Commun. **151**, 42–50 (2020)
7. Ji, S., Xu, D., Du, D., Gai, L., Zhao, Z.: Approximation algorithm for the balanced 2-correlation clustering problem. Tsinghua Sci. Technol. **27**(5), 777–784 (2022)
8. Karaki, S.H., Chedid, R.B., Ramadan, R.: Probabilistic performance assessment of autonomous solar-wind energy conversion systems. IEEE Trans. Energy Convers. **14**(3), 766–772 (1999)
9. Liu, X.: Research on optimal dispatch method of virtual power plant considering various energy complementary and energy low carbonization. Int. J. Electr. Power Energy Syst. **136**, 107670 (2022)
10. Liu, Z., et al.: Optimal dispatch of a virtual power plant considering demand response and carbon trading. Energies **11**(6), 1488 (2018)
11. Lou, S., Hu, B., Wu, Y., Lu, S.: Optimal dispatch of power system integrated with large scale photovoltaic generation under carbon trading environment. Autom. Electric Power Syst. **38**(17), 91–97 (2014)
12. Morales, J. M., Conejo, A. J., Madsen, H., Pinson, P., Zugno, M.: Integrating renewables in electricity markets: operational problems, vol. 205. Springer Science and Business Media (2013)
13. Naval, N., Yusta, J.M.: Virtual power plant models and electricity markets-a review. Renew. Sustain. Energy Rev. **149**, 111393 (2021)
14. Saberi-Beglar, K., Zare, K., Seyedi, H., Marzband, M., Nojavan, S.: Risk-embedded scheduling of a CCHP integrated with electric vehicle parking lot in a residential energy hub considering flexible thermal and electrical loads. Appl. Energy **329**, 120265 (2023)
15. Wealer, B., Bauer, S., Hirschhausen, C.V., Kemfert, C., Göke, L.: Investing into third generation nuclear power plants-review of recent trends and analysis of

future investments using monte Carlo simulation. Renew. Sustain. Energy Rev. **43**, 110836 (2021)
16. Wu, Y., et al.: Energy management for PV prosumers inside microgrids based on Stackelberg-Nash game considering demand response. Sustainable Energy Technol. Assess. **68**, 103856 (2024)
17. Xu, Y., Möhring, R.H., Xu, D., Zhang, Y., Zou, Y.: A constant FPT approximation algorithm for hard-capacitated k-means. Optim. Eng. **21**, 709–722 (2020)
18. Yang, X., Zhang, Y.: A comprehensive review on electric vehicles integrated in virtual power plants. Sustain. Energy Technol. Assess. **48**, 101678 (2021)
19. Ying, Z., Yu, T., Huang, Y.: Research on clearing the operation market of Shanghai virtual power plant. J. Electric Power **35**(2), 129–134 (2020)
20. Yuan, Z., Li, J.: Photovoltaic-penetrated power distribution networks' resiliency-oriented day-ahead scheduling equipped with power-to-hydrogen systems: a risk-driven decision framework. Energy **299**, 131115 (2024)
21. Zhang, L., Liu, D., Cai, G., Lyu, L., Koh, L.H., Wang, T.: An optimal dispatch model for virtual power plant that incorporates carbon trading and green certificate trading. Int. J. Electr. Power Energy Syst. **144**, 108558 (2023)
22. Zhou, R., Zhang, Y., Feng, S., Luktarhan, N.: A novel hierarchical clustering algorithm based on density peaks for complex datasets. Complexity **1**, 2032461 (2018)

On the Constrained Steiner Strong Connectivity Augmentation Problem

Junran Lichen[1(✉)], Shuwen Ge[1], Runtao Xie[2], and Ping Yang[2]

[1] School of Mathematics and Physics, Beijing University of Chemical Technology, North Third Ring East Road, Chaoyang District, Beijing 100029, People's Republic of China
J.R.Lichen@buct.edu.cn, 15237571096@163.com

[2] School of Mathematics and Statistics, Yunnan University, East Outer Ring South Road, University Town, Kunming 650504, People's Republic of China
xieruntao7@163.com, 1573395725@qq.com

Abstract. In this paper, we consider the constrained Steiner strong connectivity augmentation (CStSCA) problem. Concretely, given a weighted digraph $D = (V, A; w)$, where $w : A \to R^+$ is a weight function, K ($\subseteq V$) is a vertex-set consisting of k fixed vertices (called as terminals) and $T_k = (V_k, A_k)$ is a strongly connected Steiner subgraph (of D), *i.e.*, T_k contains at least one directed path from each terminal s to other each terminal t, where s, $t \in K$, we are asked to find an arc-set $A' \subseteq A \backslash A_k$ to satisfy the constraints that, given each arc $e \in A_k$, the subgraph $D[A' \cup A_k \backslash \{e\}]$ is still a strongly connected Steiner subgraph (of D), the objective is to minimize the summation of weights of all arcs in A', where the summation is taken among all arc-sets to satisfy the constraints as mentioned-above. In particular, given a vertex-set $K = \{x, y\}$, we refer the CStSCA problem as to the constrained directed circuit augmentation (CDCA) problem, and given a vertex-set $K = V$, we refer the CStSCA problem as to the constrained strong connectivity augmentation (CSCA) problem. We obtain the following three main results. (1) We design an 2-approximation algorithm to solve the CDCA problem in time $O(mn)$; (2) Using the preceding algorithm in (1) as a subroutine for many times, we present an $2k$-approximation algorithm to solve the CStSCA problem in time $O(kmn)$; (3) We provide an 2-approximation algorithm to solve the CSCA problem in time $O(n(m + n \log n) \log n)$.

Keywords: Combinatorial optimization · Constrained Steiner strong connectivity augmentation · Constrained strong connectivity augmentation · Approximation algorithms · Complexity

---

This paper is supported by the National Natural Science Foundation of China [Nos.12361066, 12101593]. Junran Lichen and Shuwen Ge are also supported by Fundamental Research Funds for the Central Universities [No. buctrc202219], Runtao Xie and Ping Yang are also supported by Project of Yunling Scholars Training of Yunnan Province [No.K264202011820].

D. Du et al. (Eds.): COCOA 2024, LNCS 15434, pp. 151–162, 2025.
https://doi.org/10.1007/978-981-96-4445-2_13

1 Introduction and Problem Descriptions

The Steiner tree problem on weighted graphs [18] is one fundamental and classical problem in combinatorial optimization, and it has many wide important applications in scientific research and reality life. Concretely, given a weighted graph $G = (V, E; w)$, where $w : E \to R^+$ is a weight function, and a set K of k fixed vertices of G, it is asked to find a tree T (of G) to interconnect all vertices in K, the objective is to minimize the summation of weights of all edges of that tree T. The Steiner tree problem is *NP*-complete even when the graph is induced by some points in the Euclidean plane R^2 [6], it has been extensively studied, and there are many approximation algorithms for solving the Steiner tree problem and related problems [1,3,11,18].

We are interested in the directed Steiner tree problem [10,14]. Concretely, given a weighted digraph $D = (V, A; w)$, where $w : A \to R^+$ is a weight function, a fixed vertex $r \in V$ (called as a root), a set K of k fixed vertices in $V \backslash \{r\}$ (called as k terminals), it is asked to find a directed tree T to connect from this root r to each terminal in K, the objective is to minimize the summation of weights of all arcs of that directed tree T, where T is called as a directed Steiner tree rooted at r, and T may be permitted to contain other vertices not in K. Especially, given a set $K = V \backslash \{r\}$, the directed Steiner tree problem becomes the minimum arborescence problem [14], and this directed Steiner tree T rooted at r is called as an arborescence rooted at r. The directed Steiner tree problem and its variations have lots of applications in both network design and network routing. There are some famous but more difficult approximation algorithms for solving the directed Steiner tree problem, for example, Charikar et al. [4] in 1999 designed an $O(k^{\epsilon})$-approximation algorithm in polynomial time, where $\epsilon > 0$, and an $O(\log^3 k)$-approximation algorithm in quasi-polynomial-time, *i.e.*, in time $O(n^{3 \log k})$. Grandoni et al. [9] in 2022 proposed a tight $O(\log^2 k / \log \log k)$-approximation algorithm in quasi-polynomial time, *i.e.*, in time $O(n^{poly \log k})$.

Since Eswaran and Tarjan [6] in 1976 began to propose a theoretical framework for studying augmentation problems on weighted complete graphs, Frederickson and JáJá [7] in 1981 proposed graph augmentation problems on general weighted graphs or digraphs. Concretely, given a weighted graph or digraph $G = (V, E; w)$, a spanning subgraph $H = (V, E')$ of G, and a cost function w on E, it is asked to find a set of edges, *i.e.*, $E'' \subseteq E \backslash E'$, such that the (reduced) subgraph $H \cup E'' = (V, E' \cup E'')$ satisfies a required property $\mathcal{P}$, the objective is to minimize the total cost of edges in E'', *i.e.*, $\min w(E'') = \sum_{e \in E''} w(e)$. Frederickson and JáJá [7] reconsidered three properties proposed by Eswaran and Tarjan [6], *i.e.*, strong connectivity for digraphs, and biconnectivity and bridge-connectivity for graphs, and they designed three 2-approximation algorithms for solving these three augmentation problems, respectively. Khuller and Thurimella [13] in 1993 presented a simple 2-approximation algorithm for resolving the biconnectivity augmentation problem in time $O(m + n \log n)$.

The tree augmentation problem [17] is modelled in the following setting. Given a weighted graph $G = (V, E; w)$ with cost function $w(\cdot)$ on E and a spanning tree $T = (V, E')$ of G, it is asked to find a minimum cost edge-set E''

$\subseteq E \backslash E'$ such that the (reduced) subgraph $T \cup E'' = (V, E' \cup E'')$ becomes 2-edge-connected. Given any constant $\varepsilon > 0$, Nagamochi [17] in 2003 presented a $(1.875 + \varepsilon)$-approximation algorithm based on the maximum matching algorithm and a tree decomposition for solving the tree augmentation problem in time $O(n^{1/2}m + n^2)$. Cohen and Nutov [5] in 2013 designed a $(1 + \ln 2)$-approximation algorithm for solving the tree augmentation problem on trees with radius bounded by a constant in time $n^{3^{h(T)}} \cdot poly(n)$, where n is the number of vertices and $h(T)$ is the radius of an initial tree as an input.

In the recent years, it is interesting for many researchers to consider the survivable network construction problems, which can survive certain edge/vertex failures. Georgiadis et al. [8] in 2020 investigated some basic connectivity problems on digraphs, and Hommelsheim et al. [12] in 2021 considered the problem of how to secure matchings against edge failures, those problems are not only theoretically interesting, but also arise in a variety of application areas, including network analysis and computational biology. Li et al. [15] in 2024 considered the constrained directed Steiner tree augmentation (CDStTA) problem, they designed some approximation or exact algorithms for solving the CDStTA problem and related special versions.

Motivated by these situations and many augmentation problems as mentioned-above, we consider the constrained Steiner strong connectivity augmentation (CStSCA) problem. Concretely, given a weighted digraph $D = (V, A; w)$, where $w : A \rightarrow R^+$ is a weight function, K ($\subseteq V$) is a vertex-set of k fixed vertices (called as terminals) and $T_k = (V_k, A_k)$ is a strongly connected Steiner subgraph (of D), *i.e.*, T_k contains at least one directed path $P_{s,t}$ from each terminal s to other each terminal t, where s, $t \in K$, we are asked to find an arc-set $A' \subseteq A \backslash A_k$ to satisfy the constraints that, given each arc $e \in A_k$, the (induced) subgraph $D[A' \cup A_k \backslash \{e\}]$ is still a strongly connected Steiner subgraph (of D), the objective is to minimize the summation of weights of all arcs in A', where the summation is taken among all arc-sets to satisfy the constraints as mentioned-above. In particular, given a vertex-set $K = \{x, y\}$, we refer the CStSCA problem as to the constrained directed circuit augmentation (CDCA) problem, and given a vertex-set $K = V$, we refer the CStSCA problem as to the constrained strong connectivity augmentation (CSCA) problem.

The CStSCA problem is different from some graph augmentation problems [6,7,13,17], and it is different from the connectivity augmentation problem [5, 17]. However, the CStSCA problem has a little relation corresponding to the connectivity problem [8] and to the CDStTA problem [15]. As far as we know, the CStSCA problem has not been considered in the literature, and we hope that there are more important applications about this new augmentation problem and other related problems in theory and/or in practice. We intend to design approximation algorithms to solve the CStSCA problem and the CSCA problem.

The remainder of this paper is organized into the following sections. In Sect. 2, we provide some notations and lemmas to ensure the correctness of our algorithms. In Sect. 3, we present an 2-approximation algorithm to solve the CDCA problem in time $O(mn)$, and using this preceding algorithm as a subroutine for

k times, we design an $2k$-approximation algorithm to solve the CStSCA problem in time $O(kmn)$; In Sect. 4, we present an 2-approximation algorithm to solve the CSCA problem in time $O(n(m+n\log n)\log n)$. In Sect. 5, we summarize our conclusions.

2 Notations and Key Lemmas

In this section, we provide some notations and lemmas such that we present proofs of algorithms to solve the CStSCA problem and the CSCA problem. Other ones not defined in the sequel will be found in the literature [13,14,19].

Given a weighted digraph $D = (V, A; w)$, we denote an arc $e^r = (y, x)$ to present the reverse of an arc $e = (x, y)$ in A, where $(e^r)^r$ is also the same arc e, *i.e.*, $e = (e^r)^r$, and for convenience, we denote a weighted digraph $H = (V, A^r; h)$ to present the reverse of $D = (V, A; w)$, where $h : A^r \rightarrow R^+$ is a weight function, satisfying $h(e) = w(e^r)$ for each arc $e \in A^r$.

Given a weighted digraph $D = (V, A; w)$ with a vertex-set K $(\subseteq V)$ consisting of k fixed terminals and a weight function $w : A \rightarrow R^+$, we still use a notation $D = (V, A; w; T_k)$ to present an instance of the CStSCA problem, where $T_k = (V_k, A_k)$ is a strongly connected Steiner subgraph (of D) containing all terminals in K, *i.e.*, T_k contains one directed path $P_{s,t}$ from each terminal s to other each terminal t. Similarly, we use $D = (V, A; w; C_{x,y})$ to denote an instance of the CDCA problem, where $C_{x,y} = (V_{x,y}, A_{x,y})$ is a strongly connected subgraph (of D) containing the two fixed vertices x and y, and $D = (V, A; w; T)$ to denote an instance of the CSCA problem, where $T = (V, A_T)$ is a strongly connected spanning subgraph of D, respectively. In addition, given an arc-set $A' \subseteq A$, we denote $w(A') = \sum_{e \in A'} w(e)$.

We have known that Edmonds [20] in 1973 obtained the two main results, one of which provides a necessary and sufficient condition to determine whether exists k arc-disjoint arborescences rooted at r, and another of which provided an exact algorithm [20,21] in time $O(k^2m^2)$ to produce k arc-disjoint arborescences rooted at r with minimum cost, or there are no such k arc-disjoint arborescences rooted at r. Using these two results, we immediately present the following

Lemma 1 [20]. Given a digraph $D = (V, A)$, a vertex $r \in V$ and a positive integer k, D contains k arc-disjoint arborescences rooted at r if and only if there are at least k arc-disjoint directed paths from r to each other vertex x of D.

We intend to use the following key result to produce a feasible solution for the CStSCA problem, whose full proof can be found in the reference [16].

Lemma 2. Given an instance $D = (V, A; w; T_k)$ of the CStSCA problem and an arc-set $A' \subseteq A \backslash A_k$, then this arc-set A' is a feasible solution for the instance D if and only if, for a pair of any two distinct terminals $s, t \in K$, there are at least two arc-disjoint directed paths from s to t in $D[A' \cup A_k]$. In addition, given a vertex set $K = \{x, y\}$, if A^* $(\subseteq A \backslash A_{x,y})$ is an optimal solution for the instance $D = (V, A; w; C_{x,y})$ of the CDCA problem, then there are exactly two arc-disjoint directed paths from x to y and another exactly two arc-disjoint directed paths from y to x in the (induced) subgraph $D[A^* \cup A_{x,y}]$.

We have known that there are many exact algorithms for optimally solving the minimum cost s-t flow (MCF) problem [14], and especially the minimum cost s-t integer flow (MCIF) problem. We provide an exact algorithm $\mathcal{A}_{MCIF}$, due to Busacker and Gowen [2], for optimally solving the MCIF problem.

Lemma 3 [2,14]. Given a network $N = (V, A; b, w; s, t)$, where $b : A \rightarrow Z^+$ is a capacity function, $w : A \rightarrow R^+$ is a unit cost function, and k is a positive integer, the buildup algorithm $\mathcal{A}_{MCIF}$, which is also called as the successive shortest path algorithm, can solve the minimum cost s-t integer flow problem, and this algorithm runs in time $O(knm)$.

3 A 2k-Approximation Algorithm to Solve the CStSCA Problem

In this section, we first consider the CDCA problem, then we can present an approximation algorithm to solve the CStSCA problem.

We present the following algorithm $\mathcal{A}_{CDCA}$ to solve the CDCA problem.

Algorithm: $\mathcal{A}_{CDCA}$
INPUT: An instance $D = (V, A; w; C_{s,t})$ of the CDCA problem, where $C_{s,t} = (V_{s,t}, A_{s,t})$;
OUTPUT: A subset A_0 and a value $w(A_0)$.
Begin
Step 1 Based on an instance $D = (V, A; w; C_{s,t})$, construct a network $N_1' = (V, A; b_1', w_1'; s, t)$ as follows: a capacity function $b_1' : A \rightarrow \{1\}$ by $b_1'(e) = 1$ for each arc $e \in A$, and a unit cost function $w_1' : A \rightarrow R^+$ by $w_1'(e) = w(e)$ if $e \notin A_{s,t}$ and $w_1'(e) = 0$ if $e \in A_{s,t}$.
Step 2 Use the buildup algorithm $\mathcal{A}_{MCIF}$ [2] to produce a minimum cost s-t integer flow f_1' of value 2 (in N_1'); If there exists no such flow f_1', then output "no feasible solution for the instance D of the CDCA problem", and STOP.
Step 3 Based on the network $N_1' = (V, A; b_1', w_1'; s, t)$ and an s-t integer flow f_1' (in N_1'), construct a network $N_1'' = (V, A; b_1'', w_1''; t, s)$ as follows: a capacity function $b_1'' : A \rightarrow \{1\}$ with $b_1''(e) = 1$ for each arc $e \in A$, and a unit cost function $w_1'' : A \rightarrow R^+$ by $w_1''(e) = w(e)$ if $e \notin A_{s,t}$ and $f_1'(e) = 0$, $w_1''(e) = 0$ if either $e \notin A_{s,t}$ and $f_1'(e) = 1$ or $e \in A_{s,t}$.
Step 4 Use the buildup algorithm $\mathcal{A}_{MCIF}$ [2] to find a minimum cost t-s integer flow f_1'' of value 2 (in N_1''); If there exists no such flow f_1'', then output "no feasible solution for the instance D of the CDCA problem", and STOP.
Step 5 Denote $A_1 = \{e \in A \mid f_1'(e) = 1\} \cup \{e \in A \mid f_1''(e) = 1\}$;
Step 6 Based on an instance $D = (V, A; w; C_{s,t})$, construct a network $N_2' = (V, A; b_2', w_2'; t, s)$ as follows: a capacity function $b_2' : A \rightarrow \{1\}$ by $b_2'(e) = 1$ for each arc $e \in A$, and a unit cost function $w_2' : A \rightarrow R^+$ by $w_2'(e) = w(e)$ if $e \notin A_{s,t}$ and $w_2'(e) = 0$ if $e \in A_{s,t}$.

Step 7 Use the buildup algorithm $\mathcal{A}_{MCIF}$ [2] to find a minimum cost t-s integer flow f_2' of value 2 (in N_2'); If there exists no such flow f_2', then output "no feasible solution for the instance D of the CDCA problem", and STOP.

Step 8 Based on the network $N_2' = (V, A; b_2', w_2'; t, s)$ and a t-s integer flow f_2' (in N_2'), construct a network $N_2'' = (V, A; b_2'', w_2''; s, t)$ as follows: a capacity function $b_2'' : A \to \{1\}$ by $b_2''(e) = 1$ for each arc $e \in A$, and a unit cost function $w_2'' : A \to R^+$ by $w_1''(e) = w(e)$ if $e \notin A_{s,t}$ and $f_2'(e) = 0$, $w_1''(e) = 0$ if either $e \notin A_{s,t}$ and $f_2'(e) = 1$ or $e \in A_{s,t}$.

Step 9 Use the buildup algorithm $\mathcal{A}_{MCIF}$ [2] to produce a minimum cost s-t integer flow f_2'' of value 2 (in N_2''); If there exists no such flow f_2'', then output "no feasible solution for the instance D of the CDCA problem", and STOP.

Step 10 Denote $A_2 = \{e \in A \mid f_2'(e) = 1\} \cup \{e \in A \mid f_2''(e) = 1\}$;

Step 11 Output "a best arc-set A_0 as either A_1 or A_2 to satisfy $w(A_0) = \min\{w(A_1), w(A_2)\}$".

End

Using the algorithm $\mathcal{A}_{CDCA}$, we determine the following result.

Theorem 1. The algorithm $\mathcal{A}_{CDCA}$ is an 2-approximation algorithm to solve the CDCA problem, and this algorithm runs in time $O(mn)$, where n is the number of vertices, m is the number of arcs of a weighted digraph $D = (V, A; w; C_{s,t})$.

Proof. Given an instance $D = (V, A; w; C_{s,t})$ of the CDCA problem, we may suppose that there is an optimal solution for the instance D, *i.e.*, there is an optimal arc-set A^* with optimal value $w(A^*) = \min_{A' \subseteq A} \sum_{e \in A'} w(e)$. And for the same instance D, the algorithm $\mathcal{A}_{CDCA}$ can produce an arc-set A_0 with its output value $w(A_0)$. We may assume, with loss of generality, that $w(A_0) = w(A_1)$ holds. We shall prove $w(A_0) \leq 2w(A^*)$ in the sequel. For convenience, we may denote $B_1' = \{e \in A \mid f_1'(e) = 1\}$ and $B_1'' = \{e \in A \mid f_1''(e) = 1\}$, then we have $A_1 = B_1' \cup B_1''$ and $B_1' \cap B_1'' = \phi$.

We first show that A_1 $(= B_1' \cup B_1'')$ is a feasible solution for the instance D of the CDCA problem, *i.e.*, for each arc $e \in C_{s,t}$, there are a directed path from s to t and another directed path from t to s in $D[C_{s,t} \cup A_1 \backslash \{e\}]$. Since N_1' is constructed from D and the buildup algorithm $\mathcal{A}_{MCIF}$ [2] produces a minimum cost s-t integer flow f_1' of value 2 (in N_1'), we obtain that $D[C_{s,t} \cup B_1]$ contains two arc-disjoint directed paths P_1, P_2 from s to t in $D[C_{s,t} \cup B_1]$. For each arc $e \in C_{s,t}$, to prove that there is still a directed path from s to t in $D[C_{s,t} \cup B_1 \backslash \{e\}]$, we only need to consider the following three cases. (1) For $e \notin A(P_1) \cup A(P_2)$, there are two directed paths P_1, P_2 from s to t in $D[C_{s,t} \cup B_1 \backslash \{e\}]$; (2) For $e \in A(P_1)$, there is still a directed path P_2 from s to t in $D[C_{s,t} \cup B_1 \backslash \{e\}]$; (3) For $e \in A(P_2)$, there is still a directed path P_1 from s to t in $D[C_{s,t} \cup B_1 \backslash \{e\}]$. With the preceding similar arguments, for each arc $e \in C_{s,t}$, we can prove that there is still a directed path from t to s in $D[C_{s,t} \cup A_1 \backslash \{e\}]$. These arguments show that an arc-set A_1 produced by the algorithm $\mathcal{A}_{CDCA}$ is also a feasible solution for an instance D of the CDCA problem.

We now prove $w(A_0) = w(A_1) \leq 2w(A^*)$. Since since A^* is an optimal solution for the instance D of the CDCA problem, using Lemma 2, we obtain two arc-disjoint directed paths from s to t in the (induced) subgraph $D[A^* \cup A_{x,y}]$, and executing the buildup algorithm $\mathcal{A}_{MCIF}$ [2] at Step 2 of the algorithm $\mathcal{A}_{CDCA}$, we have $w(B_1') \leq w(A^*)$. Again using Lemma 2 and the preceding similar arguments, we obtain $w(B_1'') \leq w(A^*)$. This shows that $w(A_0) = w(A_1) = w(B_1') + w(B_1'') \leq 2w(A^*)$ holds. Thus, we conclude that the algorithm $\mathcal{A}_{CDCA}$ is an 2-approximation algorithm to solve the CDCA problem.

The complexity of the algorithm $\mathcal{A}_{CDCA}$ can be determined as follows. (1) Step 1 needs at most time $O(m)$ to construct a network N_1'; (2) Step 2 needs at most time $O(mn)$ to compute an s-t integer flow f_1' of value 2 (in N_1'); (3) Similarly, Step 3 needs at most time $O(m)$ to construct a network N_1'', Step 4 needs at most time $O(mn)$ to compute a t-s integer flow f_1'' of value 2 (in N_1''); (4) Similarly, it needs at most time $O(mn)$ to execute Step 6 to Step 9; (5) It needs at most time $O(m)$ to execute other steps. Thus, the running time of the algorithm $\mathcal{A}_{CDCA}$ is in total time $O(mn)$.

This completes the proof of this theorem. ∎

Using the algorithm $\mathcal{A}_{CDCA}$ as a subroutine for many times, we design the following algorithm $\mathcal{A}_{CStSCA}$ to solve the CStSCA problem.

Algorithm: $\mathcal{A}_{CStSCA}$
INPUT: An instance $D = (V, A; w; T_k)$ of the CStSCA problem, where $T_k = (V_k, A_k)$;
OUTPUT: An arc-subset A_0 and a value $w(A_0)$.
Begin
Step 1 For each terminal $s \in K$ do:
– 1.1 For other each terminal $t \in K \backslash \{s\}$ do:
– – Construct an instance $D_{s,t} = (V, A; w; T_k)$ of the CDCA problem as mentioned-above, and use the algorithm $\mathcal{A}_{CDCA}$ to find a minimum cost arc-set $A_{s,t} \subseteq A \backslash A_k$.
– – If (there exists no such subset $A_{s,t}$) then
– – Output "no feasible solution for the instance D of the CStSCA problem", and STOP.
– 1.2 Denote $A_s := \cup_{t \in K \backslash \{s\}} A_{s,t}$;
Step 2 Choose A_0 as one element in $\{A_s \mid s \in K\}$ to satisfy $w(A_0) = \min\{w(A_s) \mid s \in K\}$;
Step 3 Output "the arc-subset A_0 and the value $w(A_0)$".
End

Using the algorithm $\mathcal{A}_{CStSCA}$, we can obtain the following

Theorem 2. The algorithm $\mathcal{A}_{CStSCA}$ is an $2k$-approximation algorithm to solve the CStSCA problem, and this algorithm runs in time $O(kmn)$, where k is the number of terminals, n is the number of vertices and m is the number of arcs of $D = (V, A; w; T_k)$ with $T_k = (V_k, A_k)$.

Proof. Given an instance $D = (V, A; w; T_k)$ of the CStSCA problem, we may suppose that there is an optimal solution for the instance D, *i.e.*, there is an optimal arc-set A^* with optimal value $w(A^*) = \min_{A' \subseteq A} \sum_{e \in A'} w(e)$ such that, for each arc $e \in A_k$, the reduced subgraph $D[A^* \cup A_k \backslash \{e\}]$ is a strongly connected Steiner subgraph (of D), *i.e.*, $D[A^* \cup A_k \backslash \{e\}]$ contains at least one directed path from each terminal s to other each terminal t, and that $w(A^*)$ is minimized among all values in $\{w(A') \mid D[A' \cup A_k \backslash \{e\}]$ is still a strongly connected Steiner subgraph (of D), where $A' \subseteq A$, $\forall e \in A_k\}$. And for the same instance D, the algorithm $\mathcal{A}_{CStSCA}$ can produce a subset A_0 with output value $w(A_0)$.

Given each terminal t ($\in K$) of the same instance D, by using the algorithm $\mathcal{A}_{CDCA}$ (seeing Step 1 of the algorithm $\mathcal{A}_{CDCA}$) and Theorem 1, the algorithm $\mathcal{A}_{CDCA}$ produces an arc-subset A_s such that, for each arc $e \in A_k$, there is still a directed path from s to each terminal t and a directed path from each terminal t to s in $D[A_k \cup A_s \backslash \{e\}]$ and that $w(A_{s,t})$ is minimized for each terminal t in K. Step 1.2 of the algorithm $\mathcal{A}_{CDCA}$ produces $A_s := \cup_{t \in K \backslash \{s\}} A_{s,t}$ such that, for each arc $e \in A_k$, $D[A_k \cup A_s \backslash \{e\}]$ contains a directed path from each terminal s to other each terminal t ($\in K$), *i.e.*, A_s is a feasible solution for the instance $D = (V, A; w; T_k)$ of the the CDCA problem. By executing Step 2, A_0 is a feasible solution for the instance $D = (V, A; w; T_k)$ of the the CDCA problem.

At Step 1 of the algorithm $\mathcal{A}_{CDCA}$ for per iteration, Theorem 1 implies that $w(A_{s,t})$ is minimized for a pair of any two terminals $s, t \in K$ at per iteration. Since A^* is an optimal solution for an instance of the CStSCA problem, we obtain that, for each fixed two terminals s, t in K, A^* is a feasible solution for the instance $D_{s,t} = (V, A; w; T_k)$ of the CDCA problem in the algorithm $\mathcal{A}_{CDCA}$, *i.e.*, $D[A_s \cup A^* \backslash \{e\}]$ contains a directed path from s to t and a directed path from t to s in K, implying that we can obtain $w(A_s) \leq 2w(A^*)$ for each terminal s in K.

Since K contains exactly k terminals of D, Step 1 of the algorithm $\mathcal{A}_{CDCA}$ is exactly executed k iterations. Thus, using Step 2, we have $w(A_0) = \min\{w(A_s) \mid s \in K\} \leq 2kw(A^*)$, *i.e.*, $w(A_0) \leq 2kw(A^*)$. This shows that the algorithm $\mathcal{A}_{CStSCA}$ is a $2k$-approximation algorithm to solve the CStSCA problem.

The complexity of the algorithm $\mathcal{A}_{CStSCA}$ can be determined as follows. (1) Given a terminal $s \in K$, for each terminal $t \in K$, the algorithm $\mathcal{A}_{CDCA}$ needs time $O(mn)$ to find an arc-set $A_{s,t}$ such that $w(A_{s,t})$ is minimized by Theorem 1, implying that Step 1.1 needs at most time $O(kmn)$ to execute k iterations; (2) Step 1.2 needs at most time $O(m)$ to produce a union A_s of arc-sets as mentioned-above; (3) Step 2 needs time $O(n)$ to choose A_0 as one element in $\{A_s \mid s \in K\}$ to satisfy $w(A_0) = \min\{w(A_s) \mid s \in K\}$; (4) Step 3 needs at most time $O(m)$ to compute the weight $w(A_0)$. Thus, the running time of the algorithm $\mathcal{A}_{CDCA}$ is in total $O(kmn)$.

This completes the proof of this theorem. ∎

4 An 2-Approximation Algorithm to Solve the CSCA Problem

In this section, we consider the constrained strong connectivity augmentation (CSCA) problem, where $K = V$, and we intend to design an approximation algorithm to solve the CSCA problem.

Motivated by the strategy of an algorithm $\mathcal{A}_{CAA}$ [15] to solve the constrained arborescence augmentation (CAA) problem, we hope to use the following result, due to Gabow [20], to find two arc-disjoint arborescences rooted at r.

Lemma 4 [20]. Given a weighted digraph $D = (V, A; w)$, where $w : A \to R^+$ is a weight function, r is a fixed vertex in D, k is a positive integer, the algorithm $\mathcal{A}_{Gabow}$ can produce the minimum cost k arc-disjoint arborescences rooted at r, and this algorithm $\mathcal{A}_{Gabow}$ runs in time $O(kn(m + n\log n)\log n)$.

Our approximation algorithm to solve the CSCA problem is presented in the following setting.

Algorithm: $\mathcal{A}_{CSCA}$
INPUT: An instance $D = (V, A; w; T)$ of the CSCA problem, where $T = (V, A_T; w)$;
OUTPUT: An arc-set A_0 and a weight $w(A_0)$.
Begin
Step 1 For each vertex $s \in V$ do:

- 1.1 Based on an instance $D = (V, A; w; T)$ of the CSCA problem, construct a weighted digraph $G_s = (V, A; g_s)$ as follows: a weight function $g_s : A \to R^+$ by $g_s(e) = w(e)$ if $e \notin A_T$ and $g_s(e) = 0$ if $e \in A_T$.,
- 1.2 Use the algorithm $\mathcal{A}_{Gabow}$ [20] on $G_s = (V, A; g_s)$ to produce a minimum cost directed subgraph $G'_s = (V, A_{G'_s})$ to satisfy that G'_s is exactly composed of two arc-disjoint arborescences rooted at s; If there exists no such directed subgraph G'_s, then output "no feasible solution for the instance D of the CSCA problem", and STOP.
- 1.3 Based on the weighted digraph $G_s = (V, A; g_s)$ and a directed subgraph $G'_s = (V, A_{G'_s})$, denote an arc-set $A^r = \{(x, y) \mid (y, x) \in A\}$, *i.e.*, A^r is composed of the reverses of all arcs in A, and construct a weighted digraph $G''_s = (V, A^r; g'_s)$ as follows: a weight function $g'_s : A^r \to R^+$ by $g'_s(e) = w(e^r)$ if $e^r \notin A_T$ and $e^r \notin A_{G'_s}$, $g'_s(e) = 0$ if $e^r \notin A_T$ and $e^r \in A_{G'_s}$ or $e^r \in A_T$.
- 1.4 Use the algorithm $\mathcal{A}_{Gabow}$ [20] on $G'_s = (V, A^r; g'_s)$ to produce a minimum cost directed subgraph $G''_s = (V, A^r_{G''_s})$ to satisfy that G''_s is exactly composed of two arc-disjoint arborescences rooted at s (in G'_s); If there exists no such directed subgraph G''_s, then output "no feasible solution for the instance D of the CSCA problem", and STOP.
- 1.5 Denote $A'_s := A_{G'_s} \cup A_{G''_s} \backslash A_T$, where $A_{G''_s}$ is composed of all reverses of arcs in $A^r_{G''_s}$, *i.e.*, $A_{G''_s} = (A^r_{G''_s})^r$.

Step 2 For each vertex $s \in V$ do:

- 2.1 Based on an instance $D = (V, A; w; T)$ of the CSCA problem, denote an arc-set $A^r = \{(x, y) \mid (y, x) \in A\}$, and construct a weighted digraph

$H_s = (V, A^r; h_s)$ as follows: a weight function $h_s : A^r \to R^+$ by $h_s(e) = w(e^r)$ if $e^r \notin A_T$, and $h_s(e) = 0$ if $e^r \in A_T$, where e^r $(\in A)$ is the reverse of an arc e in A^r.

– 2.2 Use the algorithm $\mathcal{A}_{Gabow}$ [20] on $H_s = (V, A^r; h_s)$ to produce a minimum cost directed subgraph $H'_s = (V, A^r_{H'_s})$ to satisfy that H'_s is exactly composed of two arc-disjoint arborescences rooted at s; If there exists no such directed subgraph H'_s, then output "no feasible solution for the instance D of the CSCA problem", and STOP.

– 2.3 Based on the weighted digraph $H_s = (V, A^r; h_s)$ and a directed subgraph $H'_s = (V, A^r_{H'_s})$, construct a weighted digraph $H''_s = (V, A; h'_s)$ as follows: a weight function $h'_s : A \to R^+$ by $h'_s(e) = w(e)$ if $e \notin A_T$ and $e^r \notin A_{H'_s}$, and $h'_s(e) = 0$ if $e \notin A_T$ and $e^r \in A_{H'_s}$ or $e \in A_T$.

– 2.4 Use the algorithm $\mathcal{A}_{Gabow}$ [20] on $H''_s = (V, A; h'_s)$ to produce a minimum cost directed subgraph $H''_s = (V, A_{H''_s})$ to satisfy that H''_s is exactly composed of two arc-disjoint arborescences rooted at s; If there exists no such directed subgraph H''_s, then output "no feasible solution for the instance D of the CSCA problem", and STOP.

– 2.5 Denote $A''_s := A_{H'_s} \cup A_{H''_s} \backslash A_T$, where $A_{H'_s}$ is composed of all reverses of arcs in $A^r_{H'_s}$, *i.e.*, $A_{H'_s} = (A^r_{H'_s})^r$.

– 2.6 Denote A_s as either A'_s or A''_s to satisfy $w(A_s) = \min\{w(A'_s), w(A''_s)\}$;

Step 3 Denote A_0 as one element in $\{A_s \mid s \in V\}$ to satisfy $w(A_0) = \min\{w(A_s) \mid s \in V\}$.

Step 4 Output "the arc-subset A_0, and compute value $w(A_0) = \sum_{e \in A_0} w(e)$".

End

Using the algorithm $\mathcal{A}_{CSCA}$, we can obtain the following

Theorem 3. The algorithm $\mathcal{A}_{CSCA}$ is an 2-approximation algorithm to solve the CSCA problem, and this algorithm runs in time $O(n(m + n \log n) \log n)$, where n is the number of vertices, m is the number of arcs.

Proof. Given an instance $D = (V, A; w; T)$ of the CSCA problem, we may suppose that there is an optimal solution for the instance D, *i.e.*, there is an optimal subset A^* with optimal value $w(A^*) = \min\{w(A') = \sum_{e \in A'} w(e) \mid D[A' \cup A_T \backslash \{e\}]$ still contains a strongly connected spanning subgraph, where $A' \subseteq A$, $\forall e \in A_T\}$. And for the same instance D, the algorithm $\mathcal{A}_{CSCA}$ can produce an arc-set A_0 with output value $w(A_0)$. For convenience, we may assume $w(A_0) = w(A_s)$ for some vertex $s \in V$. In addition, we may assume that $w(A_s)$ $(= w(A_0)) = w(A'_s)$ holds, *i.e.*, A_s is produced at Step 1.5 for some vertex $s \in V$.

At Step 1.2 of the algorithm $\mathcal{A}_{CSCA}$, using Lemma 4 [20], we obtain the fact that the subgraph $G'_s = (V, A_{G'_s})$ exactly contains two arc-disjoint arborescences rooted at s. Since we have $A_s = A_{G'_s} \backslash A_T$ (Step 1.2 of the algorithm $\mathcal{A}_{CSCA}$), we may consider the following two cases: (1) If $A_T \subseteq A_{G'_s}$ holds, we have $A_s \cap A_T = \phi$, implying that $A_{G'_s} = A_T \cup A_s$ and $A_s \cap A_T = \phi$, then we obtain $G'_s = (V, A_{G'_s}) = D[A_T \cup A_s]$; (2) If $A_T \not\subset A_{G'_s}$ holds, *i.e.*, $A_T \backslash A_{G'_s} \neq \phi$, we denote $A'_{G'_s} = A_T \backslash A_{G'_s}$, implying that $A_{G'_s} \cup A'_{G'_s} = A_T \cup A_s$, then we obtain that

$G'_s = (V, A_{G'_s})$ is a subgraph of $D[A_{G'_s} \cup A'_{G'_s}] = D[A_T \cup A_s]$. This shows that, for each arc $e \in A_T$, $D[A_T \cup A_{G'_s} \backslash \{e\}]$ still contains one directed path from s to every other vertex in V. Similarly, at Step 1.4 of the algorithm $\mathcal{A}_{CSCA}$, using Lemma 4 [20], we obtain the fact that, for each arc $e \in A_T$, $D[A_T \cup A_{G''_s} \backslash \{e\}]$ still contains one directed path from every other vertex in V to s. Thus, we conclude that A_s is a feasible solution for the instance D of the CSCA problem.

After executing Step 1.2 of the algorithm $\mathcal{A}_{CSCA}$, using Lemma 4 [20], we obtain that the cost of the subgraph $G'_s = (V, A_{G'_s})$ is minimized, *i.e.*, $w'(A_{G'_s}) \leq w'(A''^*) \leq w(A^*)$, where A''^* is an optimal solution for the instance H, corresponding to weight function $w(\cdot)$, to find the minimum cost two arc-disjoint arborescences rooted at r. Similarly, after executing Step 1.4 of the algorithm $\mathcal{A}_{CSCA}$, using Lemma 4 [20], we obtain that the cost of the subgraph $G''_s = (V, A^r_{G''_s})$ is minimized, *i.e.*, $w'(A^r_{G''_s}) \leq w(A^*)$. This shows that $w(A_0) = w(A_s) \leq w'(A_{G'_s}) + w'(A^r_{G''_s}) \leq 2w(A^*)$ holds. Thus, we conclude that $\mathcal{A}_{CSCA}$ is an 2-approximation algorithm to solve the CSCA problem.

The complexity of the algorithm $\mathcal{A}_{CSCA}$ can be determined as follows. (1) Step 1.1 needs at most time $O(m)$ to contract a new weighted digraph $G_s = (V, A; g_s)$; (2) Step 1.2 needs at most time $O(n(m + n\log n)\log n)$ to find a minimum cost subgraph $G'_s = (V, A_{G'_s})$ by Lemma 4; (3) Step 1.3 needs at most time $O(m)$ to construct a new weighted digraph $G''_s = (V, A^r; g'_s)$; (4) Step 1.4 needs at most time $O(n(m + n\log n)\log n)$ to find a minimum cost subgraph $G''_s = (V, A^r_{G''_s})$ by Lemma 4; This shows that Step 1 needs at most $O(n(m+n\log n)\log n)$ to be executed. Similarly, Step 2 needs at most $O(n(m+n\log n)\log n)$ to be executed, and Steps 3–4 need at most time $O(m)$. Thus, the running time of the algorithm $\mathcal{A}_{CSCA}$ is in total $O(n(m + n\log n)\log n)$.

This completes the proof of this theorem. ∎

5 Conclusions

In this paper, we consider the CStSCA problem and its two special versions, *i.e.*, the CDCA problem and the CSCA problem. We obtain the following main results.

(1) We design an 2-approximation algorithm $\mathcal{A}_{CDCA}$ to solve the CDCA problem, and this algorithm $\mathcal{A}_{CDCA}$ runs in time $O(mn)$;
(2) Using the algorithm $\mathcal{A}_{CDCA}$ as subroutine for many times, we present an $2k$-approximation algorithm to solve the CStSCA problem, and that algorithm $\mathcal{A}_{CStSCA}$ runs in time $O(kmn)$;
(3) We provide an 2-approximation algorithm $\mathcal{A}_{CSCA}$ to solve the CSCA problem, and that that algorithm $\mathcal{A}_{CSCA}$ runs in time $O(n(m + n\log n)\log n)$.

Acknowledgment. These authors are grateful to the three anonymous reviewers for their insightful comments and for their suggested changes that improve the current presentation greatly.

Conflict of interest. The authors declare that they have no conflict of interest.

References

1. Bern, M.W., Plassmann, P.E.: The Steiner problem with edge lengths 1 and 2. Inf. Process. Lett. **32**(4), 171–176 (1989)
2. Busacker, R.G., Gowen, P.J.: A procedure for determining a family of minimum cost network flow patterns. ORO Technical Paper 15, Operational Research Office, Johns Hopkins University, Baltimore (1961)
3. Byrka, J., Grandoni, F., Rothvoss, T., Sanità, L.: Steiner tree approximation via iterative randomized rounding. J. ACM **60**, 1–33 (2013)
4. Charikar, M., et al.: Approximation algorithms for directed Steiner problems. J. Algorithms **33**(1), 73–91 (1999)
5. Cohen, N., Nutov, Z.: A $(1 + \ln 2)$-approximation algorithm for minimum-cost 2-edge-connectivity augmentation of trees with constant radius. Theoret. Comput. Sci. **489**, 67–74 (2013)
6. Eswaran, K.P., Tarjan, R.E.: Augmentation problems. SIAM J. Comput. **5**, 653–665 (1976)
7. Frederickson, G., JáJá, J.: Approximation algorithms for several graph augmentation problems. SIAM J. Comput. **10**(2), 270–283 (1981)
8. Georgiadis, L., Italino, G.F., Parotsidia, N.: Strong connectivity in directed graphs under failures, with applications. SIAM J. Comput. **49**(5), 865–926 (2020)
9. Grandoni, F., Laekhanukit, B., Li, S.: $O(\log^2 k/\log\log k)$-approximation algorithm for directed steiner tree: a tight quasi-polynomial time algorithm. SIAM J. Comput. **52**(2), 298–322 (2022)
10. Hartmanis, J., Computers and intractability: a guide to the theory of NP-completeness (Michael R. Garey and David S. Johnson). SIAM Rev. **24**, 90–91 (1982)
11. Han, L., Wang, C.J., Xu, D.C., Zhang, D.M.: Algorithms for the prize-collecting k-Steiner tree problem. Tsinghua Sci. Technol. **27**(5), 785–792 (2022)
12. Hommelsheim, F., Muehlenthaler, M., Schaudt, D.: How to secure matchings against edge failures. SIAM J. Discret. Math. **35**(3), 2265–2292 (2021)
13. Khuller, S., Thurimella, R.: Approximation algorithms for graph augmentation. J. Algorithms **14**(2), 214–225 (1993)
14. Korte, B., Vygen, J.: Combinatorial Optimization: Theory and Algorithms, 6th edn. Springer-Verlag, Berlin (2018)
15. Li, J., Xie, R., Lichen, J.: Approximation algorithms to solve the constrained directed Steiner tree augmentation problem, submitted (2024)
16. Lichen, J., Ge, S., Xie, R.: Approximation algorithms to solve the constrained Steiner strong connectivity augmentation problem (Full version), prepared (2024)
17. Nagamochi, H.: An approximation for finding a smallest 2-edge connected subgraph containing a specified spanning tree. Discret. Appl. Math. **126**, 83–113 (2003)
18. Robins, G., Zelikovsky, A.: Tighter bounds for graph Steiner tree approximation. SIAM J. Discret. Math. **19**, 122–134 (2005)
19. Rosenthal, A., Goldner, A.: Smallest augmentations to biconnect a graph. SIAM J. Comput. **6**(1), 55–66 (1977)
20. Schrijver, A.: Combinatorial Optimization: Polyhedra and Efficiency, Volume A-C, Springer-Verlag Berlin Heidelberg (2004)
21. Tarjan, R.E.: A good algorithm for edge-disjoint branching. Inf. Process. Lett. **3**, 51–53 (1974)

Competitive Algorithms for Online Traveling Salesman Problem on a Semi-line

Qian Liu[1], An Zhang[1(✉)], Kai Wang[1], Yong Chen[1], and Guangting Chen[2]

[1] Department of Mathematics, Hangzhou Dianzi University, Hangzhou 310018, China
{qianliu,anzhang,kaiwang,chenyong}@hdu.edu.cn
[2] Zhejiang University of Water Resources and Electric Power, Hangzhou 310018, Zhejiang, China
gtchen@hdu.edu.cn

Abstract. We study the online traveling salesman problem on a semi-line, where a server (salesman) starts from the origin point of the semi-line and moves either at unit speed or 'waits' (with zero speed) at somewhere, so as to serve the requests that are released online along the semi-line. The objective is to minimize the total time for serving all requests. We distinguish the homing version and the nomadic version by whether the server is required to return to the origin point or not. When no information of the requests is disclosed, the best online algorithms for two versions are $\frac{3}{2}$- and 2-competitive respectively. When the locations of all requests are known in advance, the online algorithm for the homing version could be 1-competitive. We present a $\frac{10}{7}$-competitive algorithm for the nomadic version, improving the previous $\frac{13}{9}$-competitive algorithm. We also improve the lower bound of this version from $\frac{4}{3}$ to 1.356. If machine-learned predictions are adopted for the locations, then we introduce learning-augmented algorithms that have competitive ratios of $\min\{1+\frac{\eta}{2},2\}$ and $\min\{\frac{3}{2}+\frac{\eta}{2},2\}$ for the homing version and the nomadic version respectively, where $\eta \geq 0$ denotes the prediction error, i.e., the maximum deviation between the predicted location and the real location normalized by the length of the shortest path serving all requests.

Keywords: Online traveling salesman problem · Competitive ratio · Learning-augmented algorithms · Prediction error

1 Introduction

The online traveling salesman problem (OLTSP) was introduced by Ausiello et al. [6], in which the requests are presented in an online manner and located at points of a metric space. The objective is to minimize the time for serving all requests by a single server (the salesman). In [6], Ausiello et al. introduced two versions of this problem. In the homing one it is required that the server (salesman) has to return to the departure point after serving all the requests,

D. Du et al. (Eds.): COCOA 2024, LNCS 15434, pp. 163–175, 2025.
https://doi.org/10.1007/978-981-96-4445-2_14

while in the nomadic one it is not required. For the homing version of OLTSP (H-OLTSP), they presented an optimal 2-competitive algorithm (in non-polytime) on a wide class of metric spaces. In particular, they provided both a lower bound of 1.64 and a 1.75-competitive algorithm for the special metric space of a line (H-OLTSP-L). For the nomadic version of OLTSP (N-OLTSP), they presented a 2.5-competitive algorithm (in non-polytime) on the same class of general metric spaces. They also provided a lower bound of 2 and a $\frac{7}{3}$-competitive algorithm for the line metirc (N-OLTSP-L). For H-OLTSP-L, the optimal 1.64-competitive algorithm matching the lower bound [6] is given by Bjelde et al. [8]. An optimal randomized 1.5-competitive algorithm is given by Chen et al. [3]. In [8], Bjelde et al. also improved the lower bound of N-OLTSP-L from 2 to 2.04, and provided an optimal algorithm for that version. Jawgal et al. [5] proved that the lower bound of OLTSP on a cycle (OLTSP-C) is at least 2, for both homing and nomadic version. Blom et al. [2] considered the homing version of OLTSP where the server starts at the origin point of a semi-line (H-OLTSP-SL), they showed that a very natural strategy can give an optimal $\frac{3}{2}$-competitive algorithm. Note in all the above studies, no information (such as release times, locations) about the requests is disclosed to the server in advance.

The case when the server initially learns the exact locations of all requests is denoted as OLTSPL. Bampis et al. [1] gave non-polytime algorithms with competitive ratio $\frac{3}{2}$ for both H-OLTSPL and N-OLTSPL on general metric spaces, and provided a lower bound of $\frac{3}{2}$ for N-OLTSPL on a cycle and a lower bound of $\frac{4}{3}$ on a semi-line. Results in Gouleakis et al. [4] can imply the same lower bound of $\frac{13}{9}$ for N-OLTSPL on a line or a star, and imply the same lower bound of $\frac{3}{2}$ for H-OLTSPL on a line, a cycle or a star. In [7], Bampis et al. further improved the lower bounds of N-OLTSPL on a line or a star to $\frac{1+\sqrt{61}}{6} \approx 1.468$. Gouleakis et al. [4] presented an optimal $\frac{3}{2}$-competitive algorithm for H-OLTSPL-L (see also [7]) and a $\frac{5}{3}$-competitive algorithm for N-OLTSPL-L. Bampis et al. [7] provided optimal $\frac{3}{2}$-competitive algorithms for H-OLTSPL-C, N-OLTSPL-C and N-OLTSPL-L. In [1], Bampis et al. gave a $(\frac{7}{4} + \epsilon)$-competitive algorithm for H-OLTSPL on a star, an optimal 1-competitive algorithm for H-OLTSPL-SL, and a $\frac{13}{9}$-competitive algorithm for N-OLTSPL-SL.

A most recent line of research focuses on the design and analysis of learning-augmented algorithms for OLTSP, denoted as OLTSPP. In OLTSPP, the locations of all requests are given by machine-learned predictions with a prediction error $\eta \geq 0$ measured by the maximum deviation between the predicted location and the real location normalized by the length of the shortest path serving all requests. Note the prediction error cannot be known in advance and hence cannot be used by an algorithm. The goal of learning-augmented algorithms is to utilize predictions of the future input to improve the competitive ratio in the case that the predictions are accurate enough (consistency), while maintaining worst-case guarantees even if the prediction error is arbitrarily large (robustness). Such algorithms are also expected to be smooth, i.e., their performance degrades gracefully as the prediction error increases. For OLTSPP, the best known results are given in [7], where Gouleakis et al. presented a

$\min\{3/2 + 5\eta, 11/4\}$-competitive algorithm for the homing version on a cycle, a $\min\{3/2 + 5\eta, 5/2\}$-competitive algorithm for the homing version on a line or a star, a $\min\{3/2 + 5\eta, 17/6\}$-competitive algorithm for the nomadic version on a cycle, and a $\min\{3/2 + 5\eta, 8/3\}$-competitive algorithm for the nomadic version on a line or a star. Among them, only the algorithms on a line or a cycle are in polytime.

In this paper, we study online TSP on a semi-line, which might be the simplest metric space lack of careful treatment in the literature. For N-OLTSP-SL and N-OLTSPL-SL, we present an optimal 2-competitive algorithm and an improved $\frac{10}{7}$-competitive algorithm respectively. And for the latter problem, we improve the lower bound of $\frac{4}{3}$ given by Bampis et al. [1] to $\frac{1+\sqrt{97}}{8} \approx 1.356$ as well. For H-OLTSPP and N-OLTSPP, we provide a $\min\{1 + \frac{\eta}{2}, 2\}$-competitive algorithm and a $\min\{\frac{3}{2} + \frac{\eta}{2}, 2\}$-competitive algorithm respectively.

2 Preliminary

The input of OLTSP on a semi-line (OLTSP-SL in short) consists of a server initially located at the origin point 0, and a sequence $Q = \{q_1, q_2, \ldots, q_n\}$ of n requests that are located at the positive side of a real line and released online over time. Let $rel(q_i))$ denote the release time of the request q_i, and with a little confusion, let q_i denote its real location. Note the server cannot know the real location of a request until its release time. The server can either move at unit speed or 'waits' (with zero speed) at somewhere. The problem of OLTSP-SL aims to find a route of the server so as to visit each request at/after its release time. In the homing variant (H-OLTSP-SL) we require the server to return to the origin point after serving all the requests, while in the nomadic variant(N-OLTSP-SL) returning to the origin point is not required. The objective is to minimize the total time for serving all requests. If the location point q_i for all i is disclosed to the server initially, then this is the problem of OLTSPL-SL. If one employs a machine-learned sequence $P = \{p_1, p_2, \ldots, p_n\}$ to predict the locations of all requests, then this is the problem of OLTSPP-SL. Note in OLTSPP-SL, each p_i attempts to approximate the real location q_i. The quality of predictions may vary and is unknown to the algorithm. We define a measure $\eta \geq 0$ to evaluate the quality or the prediction error. Essentially, η is the maximum deviation between the predicted location and the real location, normalized by the length of the shortest path serving all requests. That is,

$$\eta = \max_i \frac{|p_i - q_i|}{L},$$

where $L = \max(Q)$. A learning-augmented algorithm is supposed to utilize the predictions to improve the competitive ratio with the prediction error. We say that such an online algorithm is:

α-consistent, if it is α-competitive when there is no error, i.e., $\eta = 0$.
β-robust, if it is β-competitive regardless of prediction error, i.e., $\eta = \infty$.

γ-smooth for a continuous function $\gamma(\eta)$, if it is $\gamma(\eta)$-competitive.

In the reminder of the paper, let t_f denote the time when the last request is released. For an online algorithm ALG and an optimal offline route/solution OPT, let $|ALG|$ and $|OPT|$ represent the total time spent by ALG and OPT respectively. At any time t, we use $loc(t)$ (resp., $loc^*(t)$) to denote the position of the server at time t in algorithm ALG (resp., in OPT), e.g., $loc(0) = loc^*(0) = 0$. The notation 'move(u) $\bigoplus$ move(v)' is used to indicate a move to u followed by a move to v in an algorithm. Let $W \subseteq Q$ contain all released but unserved requests and in the prediction model, let $P' \subseteq P$ contain all the predicted locations that are not yet released.

3 An Optimal Algorithm for OLTSP-SL

For the homing version of OLTSP-SL, Blom et al. [2] have proved a lower bound of $\frac{3}{2}$ and given a very natural strategy (algorithm) that matches the lower bound. In fact, the server only needs to visit the current rightmost request as quickly as possible; all the missing requests are served during its moving back to the origin point. We now show a lower bound for the nomadic version and provide an algorithm matching it.

Theorem 1. *No online algorithm for N-OLTSP-SL can have a competitive ratio smaller than* 2.

Proof. The proof is derived from the following arguments. If the online server waits at the origin point from time 0 until T, the adversary releases a request at point $x = T$. Thus, the online server needs a total time of at least $2T$ while the adversary can complete at time T. If the server starts moving at the very beginning, the adversary releases a request at the origin point when the online server reaches location x. Thus, the online server needs at least a time of $2x$ while the adversary can complete at time x. This proves the theorem.

The algorithm NEAREST FIRST computes the set W of released but unserved requests, and keeps moving towards $\min(W)$ followed by $\max(W)$ unless a new request is released (so that W is updated).

Algorithm 1. NEAREST FIRST update function.

Input: nothing is known in advance.
Output: a series of (unit speed) moves to carry out until the next request is released.

```
W ← the released but unserved requests in Q.
if W is not empty then
    return move(min(W))⊕move(max(W));
else
    stay at the current location.
end if
```

Theorem 2. *The algorithm NEAREST FIRST has a competitive ratio of* 2 *for N-OLTSP-SL and is the best possible.*

Proof. At time t_f when the last request is released, let $q_x = \min(W)$ and $q_y = \max(W)$. If $loc(t_f) \leq q_x$, then $|ALG| = t_f + q_y - loc(t_f) \leq t_f + q_y \leq 2|OPT|$. If $loc(t_f) \geq q_y$, then $|ALG| = t_f + loc(t_f) - q_x \leq t_f + loc(t_f) \leq 2|OPT|$, where $|OPT| \geq loc(t_f)$ holds because by the rule of NEAREST FIRST, there must be requests at or farther than $loc(t_f)$. Now suppose that $q_x < loc(t_f) < q_y$. If OPT serves q_y followed by serving q_x, then $|OPT| \geq 2q_y - q_x$. Accordingly, $|ALG| = t_f + loc(t_f) - q_x + q_y - q_x \leq t_f + 2(q_y - q_x) \leq 2|OPT|$. Otherwise, OPT must serve q_x followed by serving q_y. Then we have $|OPT| \geq rel(q_x) + q_y - q_x$. Suppose that the server is located at loc at the time $rel(q_x)$ when q_x is released. By the rule of NEAREST FIRST, the server has to move towards q_x during the time interval $[rel(q_x), t_f]$, thus $q_x < loc \leq q_y$. Consequently, $|ALG| = rel(q_x) + loc - q_x + q_y - q_x \leq 2|OPT|$. This proves the theorem.

4 Improved Algorithm and Lower Bound for N-OLTSPL-SL

Now we consider the case when the locations of all requests are disclosed to the server initially. For the homing version, the optimal strategy is to serve the requests from the farthest to the nearest as quickly as possible (wait if the request is not released), and finally move back to the origin point. This algorithm is 1-competitive [1]. For the nomadic version of OLTSPL-SL, Bampis et al. [1] presented a 13/9-competitive algorithm and a lower bound of $\frac{4}{3}$. In this section, we improve both the lower and upper bound for this problem.

We say an algorithm sequentially serves from a to b if it serves from a to b without missing any request between a and b, and say an algorithm moves directly to b if it immediately terminates the current movement and performs moving to b at unit speed. And we may frequently use trival lower bounds on OPT for the competitive analysis of the algorithm, e.g., $|OPT|$ is no smaller than the completion time of each request, $|OPT| \geq t_f$, $|OPT| \geq rel(q_i)$ for any request q_i, $|OPT| \geq t + |loc^*(t) - q_i|$ for some request q_i unserved by time t, and $|OPT| \geq t + |q_i - q_j|$ for any two requests q_i and q_j such that q_i is served prior to q_j, and is completed at/after time t.

Theorem 3. *No online algorithm for N-OLTSPL-SL can have a competitive ratio smaller than* $\frac{1+\sqrt{97}}{8} \approx 1.356$.

Assume without loss of the generality that $0 \leq q_1 \leq q_2 \leq \cdots \leq q_n = \max\{Q\} = 1$. We now present an online algorithm SMART WAIT for N-OLTSP-SL. In fact, the server sequentially serves from 0 to 1 unless there is a request $q_x < 1/4$ with $rel(q_x) > q_x + 1/2$. Then it skips q_x at time $q_x + 1/2$ and moves towards 1/2. If requests located in the segment $[q_x, l_x]$ are all released before the server arrives at 1/2, then it moves directly to q_x and sequentially serves from q_x to 1. Otherwise, the server waits at 1/2 until one segment of requests in $[q_x, l_x]$

or [3/4, 1] are all released. Then it moves directly to an endpoint of that segment and sequentially serves the remaining requests along the semi-line. Here, we let

$$l_x = \min\{1/6 + q_x, 1/3\} = \begin{cases} 1/6 + q_x, & \text{if } q_x \in [0, 1/6), \\ 1/3, & \text{if } q_x \in [1/6, 1/4). \end{cases} \tag{1}$$

It's clear that $l_x - q_x \leq 1/6$ and $1/6 \leq l_x \leq 1/3$ are valid for all $q_x \in [0, 1/4)$.

Algorithm 2. SMART WAIT.

Input: set Q of requests; location of each request $q_i \in Q$.
Output: a series of (unit speed) moves to carry out until all requests are served.
  Sequentially serve from 0 to 1 unless there exists a request $q_x < 1/4$ with $rel(q_x) > q_x + 1/2$ (let q_x be the first such request).
  Do leave q_x at time $q_x + 1/2$ and move towards 1/2.
  if all requests in $[q_x, l_x]$ are released before the server reaches 1/2 **then**
    move directly to q_x after all requests in $[q_x, l_x]$ are released, and then **sequentially serve** from q_x to 1.
  else
    stay at 1/2 **until** one segment of requests in $[q_x, l_x]$ or [3/4, 1] are all released.
    if requests in $[q_x, l_x]$ are all released **then**
      move directly to q_x, and **sequentially serve** from q_x to 1.
    else
      move directly to 1, and **sequentially serve** from 1 to q_x.
    end if
  end if

First, we note the following observation was given in [7].

Lemma 1. *If OPT turns around at least once, then the first time it turns occurs at the endpoint* 1.

Lemma 2. *If q_x does not exist, then $|ALG| \leq \frac{10}{7}|OPT|$.*

Proof. When q_x does not exist, the algorithm must sequentially serve all requests from 0 to 1. If OPT never turns around either, then $|OPT| = |ALG|$. Otherwise, OPT has to fetch some missing request when it first arrives at the endpoint 1 by Lemma 1. Let t (≥ 1) be the time when OPT first arrives at 1, and let s_t^* be the farthest position such that all requests in $[0, s_t^*)$ have been served by OPT before/at time t. If no requests in $[0, loc(t)]$ are missed by OPT, then $loc^*(t) \leq loc(t)$ must hold since ALG takes the minimum possible time for serving reqeusts therein. However, this means that OPT does not need to turn around as we have $loc^*(t) = 1$. Thus it cannot be the case. If OPT misses some request(s) in $[0, loc(t)]$, then $s_t^* \leq loc(t)$. By Lemma 1,

$$|OPT| \geq t + 1 - s_t^* \geq t + 1 - loc(t). \tag{2}$$

If ALG never waits for a request after time t, then $|ALG| = t+1-loc(t) \leq |OPT|$. Otherwise, let q_y be the last request for which ALG waits. Thus

$$|ALG| = rel(q_y) + 1 - loc(rel(q_y)) \leq |OPT| + 1 - loc(t), \tag{3}$$

where the last inequality is because $|OPT| \geq rel(q_y)$ and $rel(q_y) > t$. Since q_x does not exist, $rel(q_j) \leq q_j + 1/2$ holds for any $q_j < 1/4$. Consequently, the server must have arrived at $1/4$ no later than $1/4 + 1/2 = 3/4 < t$. This implies that $loc(t) \geq 1/4$, which, together with (2) and (3), follows that

$$\frac{|ALG|}{|OPT|} \leq 1 + \frac{1 - loc(t)}{t + 1 - loc(t)} \leq 1 + \frac{1 - loc(t)}{2 - loc(t)} \leq \frac{10}{7}.$$

This proves the lemma.

Lemma 3. *If all requests in $[q_x, l_x]$ are released before the server reaches $1/2$, then $|ALG| \leq \frac{10}{7}|OPT|$.*

Proof. Suppose that q_z is the last request in $[q_x, l_x]$ and it is released at time $t = rel(q_z)$ before the server reaches $1/2$. Since the server leaves q_x at time $q_x + 1/2$ and would reach $1/2$ at time $(q_x + 1/2) + (1/2 - q_x) = 1$ if such q_z does not exist, we must have $1/2 + q_x < t < 1$ and $loc(t) = t - 1/2$.

Case 1. The algorithm never waits for a request after time t. Then

$$|ALG| = t + (loc(t) - q_x) + (1 - q_x) = 2t + 1/2 - 2q_x < 5/2 - 2q_x. \tag{4}$$

Consider the optimal solution. If $loc^*(t) \geq loc(t)$, then OPT cannot leave q_x later than $t - (loc^*(t) - q_x) \leq q_x + 1/2 < rel(q_x)$. Thus q_x is missed by OPT at that time. This means that OPT has to turn around. By Lemma 1, $|OPT| \geq 2 - q_x$. Together with (4),

$$\frac{|ALG|}{|OPT|} < \frac{5/2 - 2q_x}{2 - q_x} \leq \frac{5}{4} < \frac{10}{7}.$$

Otherwise, $loc^*(t) < loc(t)$. If $q_z < loc^*(t)$, then q_z is missed at time t. Accordingly, $|OPT| \geq t+1-loc^*(t)+1-q_z > t+1-loc(t)+1-l_x = 5/2-l_x > 2-q_x$ by $l_x - q_x \leq 1/6 < 1/2$. From (4), we still have $|ALG| \leq \frac{5}{4}|OPT|$. If $q_z \geq loc^*(t)$, then $|OPT| \geq t + 1 - loc^*(t) \geq t + 1 - q_z \geq t + 1 - l_x$. Combining it with the second to last term of (4) and $t < 1$, we obtain

$$\frac{|ALG|}{|OPT|} \leq 1 + \frac{t - 1/2 + l_x - 2q_x}{t + 1 - l_x} \leq 1 + \frac{t - 1/3 - q_x}{t + 1 - l_x} \leq 1 + \frac{t - 1/3}{t + 2/3} < \frac{7}{5} < \frac{10}{7}.$$

Case 2. The algorithm has to wait for some request(s) after time t. Let q_y be the last request for which ALG waits. Clearly $q_y > l_x$. Then

$$|ALG| = rel(q_y) + 1 - q_y \leq |OPT| + 1 - q_y < |OPT| + 1 - l_x. \tag{5}$$

Consider the optimal route. If OPT serves q_y prior to q_n $(= 1)$, then $|OPT| \geq rel(q_y) + 1 - q_y \geq |ALG|$. Otherwise, OPT must serve q_n prior to q_y. If OPT also serves q_n prior to q_x, then $|OPT| \geq 2 - q_x$, which, together with (5) and (1), follows

$$\frac{|ALG|}{|OPT|} < 1 + \frac{1 - l_x}{2 - q_x} = 1 + \frac{1 - \min\{1/6 + q_x, 1/3\}}{2 - q_x} \leq \max\{\frac{17}{12}, \frac{29}{21}\} < \frac{10}{7}.$$

Now the only remaining case is that OPT serves first q_x, then q_n, and finally q_y. This follows that

$$|OPT| \geq rel(q_x) + (1 - q_x) + (1 - q_y) > 5/2 - q_y,$$

where the last inequality is due to $rel(q_x) > q_x + 1/2$. By (5) and $q_y > l_x \geq 1/6$,

$$\frac{|ALG|}{|OPT|} \leq 1 + \frac{1 - q_y}{5/2 - q_y} < \frac{19}{14} < \frac{10}{7}.$$

This completes the proof.

Lemma 4. *If the last request in* $[q_x, l_x]$ *is released after the server reaches* $1/2$, *then* $|ALG| \leq \frac{10}{7}|OPT|$.

From Lemma 2, Lemma 3 and Lemma 4, we directly obtain the competitive ratio of of SMART WAIT.

Theorem 4. *The algorithm SMART WAIT has a competitive ratio of* $10/7$ *for N-OLTSPL-SL.*

We remark that the competitive analysis is tight. Consider the following instance. There are 5 requests located at points $0, 1/4 - \epsilon, 1/3, 3/4$ and 1, and are released at time $0, 3/4 + \epsilon, 5/4$ and 1, respectively. Here we let $\epsilon > 0$ be sufficiently small. SMART WAIT will find the request $q_x = 1/4 - \epsilon$ and compute $l_x = 1/3$, and leave q_x at time $3/4 - \epsilon$. Then it arrives $1/2$ and stays there until time $5/4$, when the segment $[3/4, 1]$ of requests are all released. Finally it moves directly to 1, and then serves sequentially from 1 to q_x. The total time spent by SMART WAIT equals exactly $|ALG| = 5/2 + \epsilon$. However, the optimal solution moves directly to 1 at the very beginning, and serves sequentially from 1 to q_x, implying that $|OPT| = 7/4 + \epsilon$. Thus $\frac{|ALG|}{|OPT|} \to \frac{10}{7}$ as $\epsilon \to 0$.

5 Learning-Augmented Algorithms for OLTSPP-SL

In this section, we consider the OLTSPP-SL, we will provide the FARTHEST and the NEAREST algorithm for the homing version and the nomadic version respectively. We introduce both algorithm via an update function that gets triggered whenever a request is released. This update function returns the plan of moves to be executed until the next release of a request. When all the moves are executed, the server waits for the next release. This occurs solely while waiting on a prediction. Note that the server moves at either zero or unit speed, thus

$$loc(t) \leq t. \tag{6}$$

5.1 H-OLTSPP-SL

The algorithm FARTHEST is designed for the homing version of OLTSPP-SL, where the server has to return to the origin point after serving all requests. In fact, it always moves towards or stays at the fartherst prediction $\max(P')$ unless a request with real location farther than $\max(P')$ is released, or the request determining $\max(P')$ is released and hence $\max(P')$ is updated. Then it serves all the requests farther than the current $\max(P')$ (if any), and returns to $\max(P')$ when finished. At the moment when all requests are released, it serves the unserved requests along an optimal route.

Algorithm 3. FARTHEST update function.

Input: number of requests n, current position of the server loc, set P of predictions.
Output: A series of (unit speed) moves to carry out until the next request is released.

```
W ← the released but unserved requests in Q;
P' ← the predictions of those unreleased requests in P;
if P' is not empty then
    return move(max(W ∪ P'))⊕move(max(P'));
else
    return move(max(W))⊕move(0).
end if
```

According to the FARTHEST algorithm, we define two special movements, 'waiting' and 'pickup', as follows. 'Waiting' occurs at $\max(P')$ and ends when a request farther than $\max(P')$ is released, or $\max(P')$ is updated (due to the release of the request determining the current $\max(P')$). Whenever a request q_x is released such that $\max(P') \leq loc(rel(q_x)) < q_x$, where $\max(P')$ is the farthest prediction right before $rel(q_x)$, the algorithm has to move to q_x. We call this movement 'pickup'.

Lemma 5. *The algorithm FARTHEST is 2-robust.*

Proof. Recall that t_f is the last release time of all requests. Since after t_f, the algorithm will move to $\max(W)$ (assume it is determined by q_w) and then to the origin point. Thus, we observe that $|ALG| = t_f + |q_w - loc(t_f)| + q_w$. Let q_f be the farthest request such that $rel(q_f) = t_f$, then it is clear that

$$|OPT| \geq \max\{t_f + q_f, rel(q_w) + q_w, 2q_w, 2q_f\}. \tag{7}$$

Case 1. $loc(t_f) < q_w$. If $q_f = q_w$, then by (7), we have

$$|ALG| = t_f + 2q_w - loc(t_f) \leq t_f + 2q_f \leq 3/2|OPT|.$$

Otherwise, $q_f < q_w$. By the definition of q_f, $rel(q_w) < t_f$. As $loc(t_f) < q_w$ and the algorithm has not served q_w by time t_f, the server cannot be located at or after q_w during the time period $[rel(q_w), t_f]$. Then $loc(rel(q_w)) < q_w$. According to the rule of FARTHEST, it must move towards q_w after time $rel(q_w)$, and arrive at $loc(t_f)$ at time t_f. That is, $loc(rel(q_w)) < loc(t_f)$. Therefore,

$$|ALG| = rel(q_w) + (q_w - loc(rel(q_w))) + q_w \leq rel(q_w) + 2q_w \leq 3/2|OPT|,$$

where the last inequality is due to (7).

Case 2. $loc(t_f) > q_w$. Then by (6) and (7), $|ALG| = t_f + loc(t_f) \leq 2t_f \leq 2|OPT|$. This proves the lemma.

Lemma 6. *The algorithm FARTHEST is* $(1 + \frac{\eta}{2})$*-smooth.*

Proof. Recall that $L = \max(Q)$ and $\eta L = \max_i\{|q_i - p_i|\}$. Then $|OPT| \geq 2L$. We distinguish three cases according to the special movements performed by the algorithm.

Case 1. The algorithm FARTHEST has never performed any special movements. Suppose that the server arrives at the farthest point $loc(t)$ at time t. Since neither 'waiting' nor 'pickup' is performed, we must have $loc(t) = t$ and $|ALG| = 2loc(t)$. If this point happens to be the current maximum prediction $\max(P')$, i.e., $loc(t) = \max(P')$, then the request determining $\max(P')$, denoted as q_m, must be released at t so that 'waiting' is unavailable, and it cannot be farther than $\max(P')$ so that 'pickup' is unavailable either. That is, we have $q_m \leq p_m = \max(P')$ and $rel(q_m) = t = loc(t)$. Accordingly,

$$|ALG| \leq rel(q_m) + p_m = (rel(q_m) + q_m) + (p_m - q_m) \leq |OPT| + \eta L \leq (1 + \frac{\eta}{2})|OPT|.$$

If the farthest point $loc(t) = q_x$ for some request q_x, then $|ALG| = 2q_x \leq |OPT|$. Otherwise, the server must be on the way moving towards $\max(P')$ and at time t, the request determining $\max(P')$, denoted as q_m, is released and its real location is nearer than $loc(t)$. That is, $t = rel(q_m)$ and $q_m < loc(t) < p_m = \max(P')$. Thus we also have $|ALG| \leq rel(q_m) + p_m \leq (1 + \frac{\eta}{2})|OPT|$.

Case 2. The last special movement the algorithm FARTHEST performed is 'waiting'. Then it must be 'waiting' at the current $\max(P')$, and must be 'waiting' for the request determining $\max(P')$ whose real location cannot be farther than $\max(P')$ so that 'pickup' is unavailable. Let q_x be this request after which no special movement occurs any more. Then similarly, $|ALG| = rel(q_x) + p_x \leq (1 + \frac{\eta}{2})|OPT|$.

Case 3. The last special movement the algorithm FARTHEST performed is 'pickup'. Then suppose that the last 'pickup' begins at time t and is to serve the request q_x with $rel(q_x) = t$. By the definition of the 'pickup' movement, we know that $p_x \leq \max(P') \leq loc(t) < q_x$. Since after this 'pickup' no special movements are available, we have

$$|ALG| = rel(q_x) + q_x - loc(t) + q_x \leq |OPT| + q_x - p_x \leq (1 + \frac{\eta}{2})|OPT|.$$

This proves the lemma.

By Lemma 5 and Lemma 6, we obtain the competitive ratio of FARTHEST.

Theorem 5. *The algorithm FARTHEST is $f(\eta)$-competitive, where*

$$f(\eta) = \begin{cases} 1 + \eta/2, & \eta \leq 2, \\ 2,, & \eta > 2. \end{cases}$$

5.2 N-OLTSPP-SL

The algorithm NEAREST is designed for N-OLTSPP-SL, for which returning to the origin point is no longer required. According to NEAREST, the server always moves towards and stays at the nearest prediction in P' unless a new released request closer to the origin point than that prediction arrives, or the request that determines $\min(P')$ is released (so that $\min(P')$ is updated). If the former occurs, then it will turn back to serve the new request. If P' becomes empty, it will follow the best route of the remaining unserved requests. Accordingly, we also say the algorithm is performing 'waiting' if it stays at $\min(P')$, and is performing 'turnback' if it moves towards the negative direction to fetch a new released request nearer than $\min(P')$. 'Waiting' and 'turnback' are two special movements of the NEAREST algorithm. Note before time t_f, unless a special movement is being performed, the server is always moving towards the positive direction. At any time t when P' is not empty,

$$loc(t) \leq \min(P'). \tag{8}$$

Theorem 6. *The algorithm NEAREST is $f(\eta)$-competitive, where*

$$f(\eta) = \begin{cases} \frac{3}{2} + \frac{\eta}{2}, & \eta \leq 1, \\ 2, & \eta > 1. \end{cases}$$

Algorithm 4. NEAREST update function.

Input: number of requests n, current position of the server loc, set P of predictions.
Output: a series of (unit speed) moves to carry out until the next request is released.
  $W \leftarrow$ the released but unserved requests in Q;
  $P' \leftarrow$ the predictions of those unreleased requests in P;
  if P' is not empty **then**
    return move(min($W \cup P'$))$\oplus$move(min(P'));
  else
    if $loc \leq \frac{\max(W)+\min(W)}{2}$ **then**
      return move(min(W))$\oplus$move(max(W));
    else
      return move(max(W))$\oplus$move(min(W)).
    end if
  end if

6 Conclusion

We focused on the online traveling salesman problem on the metric space of a semi-line (OLTSP-SL), one fundamental issue worthy of in-depth exploration. We first presented an optimal 2-competitive algorithm for the nomadic version of OLTSP-SL (N-OLTSP-SL) in which the server is not required to return to the departure point, compared to the optimal 3/2-competitive algorithm for homing version of OLTSP (H-OLTSLP-SL) given by Bloem et al. [2]. Then we revisited the OLTSP-SL with known locations. In [1], Bampis et al. provided a 1-competitive algorithm for the homing version, a 13/9-competitive algorithm and a lower bound of 4/3 for the nomadic version. We improved the latter version by giving a 10/7-competitive algorithm and a lower bound of $\frac{1+\sqrt{97}}{8} \approx 1.356$. We also studied OLTSP with machine-learned predictions. For the homing version and the nomadic version, we provided online algorithms with competitive ratios $\min\{1 + \frac{\eta}{2}, 2\}$ and $\min\{\frac{3}{2} + \frac{\eta}{2}, 2\}$ respectively, where $\eta \geq 0$ denotes the prediction error. We note that even with the inclusion of our contribution, the online traveling salesman problem on this specific metric deserves a further look. Take for example, there is a narrow gap between the upper and lower bound of the nomadic problem with known locations; the algorithm for N-OLTSP-SL with predictions only achieves a 3/2-consisteness, far away from the algorithm with known locations. Best possible online algorithms are of common interest but challenging.

Acknowledgements. This research is supported by the NSFC Grants 12371316, 12471301, and the Zhejiang Provincial NSF Grant LZ25A010001.

References

1. Bampis, E., Escoffier, B., Hahn N., Xefteris, M.: Online TSP with known locations. In: 18th Algorithms and Data Structures Symposium, pp. 65–78 (2023)

2. Blom, M., Krumke, S.-O., de Paepe, W., Stougie, L.: The online TSP against fair adversaries. INFORMS J. Comput. **13**(2), 138–148 (2001)
3. Chen, P.-C., Demaine, Erik D., Liao, C.-S., Wei, H. -T.: Waiting is not easy but worth it: the online TSP on the line revisited (2019). arXiv:1907.00317
4. Gouleakis, T., Lakis, K., Shahkarami, G.: Learning-augmented algorithms for online TSP on the line. In: Proceedings of the AAAI Conference on Artificial Intelligence, vol. 37, no. 10, pp. 11989–11996 (2023)
5. Jawgal, V.A., Muralidhara, V.N., Srinivasan, P.S.: Online travelling salesman problem on a circle. In: Gopal, T.V., Watada, J. (eds.) TAMC 2019. LNCS, vol. 11436, pp. 325–336. Springer, Cham (2019). https://doi.org/10.1007/978-3-030-14812-6_20
6. Ausiello, G., Feuerstein, E., Leonardi, S., Stougie, L., Talamo, M.: Algorithms for the on-line travelling salesman[1]. Algorithmica **29**, 560–581 (2001)
7. Bampis, E., et al.: Learning-augmented online TSP on rings, trees, flowers and (Almost) everywhere else. In: 31st Annual European Symposium on Algorithms, LIPIcs 274, pp. 1–17 (2023)
8. Bjelde, A., et al.: Tight bounds for online TSP on the line. ACM Trans. Algorithms **17**(1), 1–58 (2021)

On the Inapproximability of Two-machine Open Shop Scheduling with Exact Delays

Shunzhang Lu[1], An Zhang[1(✉)], Mengyuan Hu[1], Yong Chen[1], and Guangting Chen[2]

[1] Department of Mathematics, Hangzhou Dianzi University, Hangzhou 310018, China
{shunzhanglu,anzhang,mengyuanhu,chenyong}@hdu.edu.cn

[2] Zhejiang University of Water Resources and Electric Power, Hangzhou 310018, Zhejiang, China
gtchen@hdu.edu.cn

Abstract. In two-machine open shop scheduling with exact delays, each job consists of two operations with an arbitrary process order, provided that the second operation must start exactly a certain time after the first operation is completed. It is known that the problem of minimizing makespan cannot be approximated in polynomial time to within a factor of $2 - \epsilon$ for any $\epsilon > 0$, unless $P = NP$. In this paper, we show that similar inapproximability results hold even if all delays are equal. Specifically, it is shown that the problem with equal exact delays cannot be approximated within $\frac{4}{3} - \epsilon$ unless $P = NP$, and even if each job has equal-length operations, it is NP-hard to approximate the problem to within $\frac{5}{4} - \epsilon$. When there exist single-operation jobs, we show that the problem cannot be approximated within $\frac{7}{6} - \epsilon$ unless $P = NP$. On the positive side, we first observe that the $O(n \log n)$-time exact algorithm for the flow shop problem behaves as a 2-approximation algorithm for the open shop problem, and then we present an improved $\frac{5}{3}$-approximation algorithm for the case with equal-length operations for each job.

Keywords: Open shop · exact delay · polynomial time reduction · inapproximability · approximation algorithm

1 Introduction

Scheduling problems that involve process delays have been studied for decades, attracting considerable attention of the researchers [14,18]. The concept of delay duration is motivated by various applications, such as the transportation time from one production facility to another, or the transportation, communication, or cooling-down time between two consecutive tasks/operations of the same job (e.g., [25,26,29]). Hence the delays are often referred to as time-lags or transportation times. The scheduling problem that accounts for delays is also known as the coupled task/job scheduling problem in the literature. However, we notice

D. Du et al. (Eds.): COCOA 2024, LNCS 15434, pp. 176–189, 2025.
https://doi.org/10.1007/978-981-96-4445-2_15

some slight differences among these terms. Time lags and transportation times are essentially equivalent to the minimum delays, while coupled jobs are mostly associated with exact delays. For the sake of accuracy, we follow the definition given by Fondrevelle et al. [5,6] that the exact (resp. minimum, maximum) delay describes the situation where a succeeding operation should be started after the completion of its preceding operation, plus an exact (resp. minimum, maximum) amount of time. Let this time be l_j for a specific job j, and place "+" and "−" above it to further indicate the minimum and maximum delay respectively. In the shop-machine environments, it is generally assumed that each job must visit all machines; however, it is sometimes relaxed by allowing jobs to visit only a subset of machines, for which, as suggested by Leung et al. [17], a "g" is added above the term of machine environments, e.g., F^g is to denote flow shop scheduling containing single-operation jobs.

In this paper, we focus on two-machine open shop scheduling with exact delays. Formally, we are given a set of jobs to be non-preemptively processed on two machines, denoted by M_1 and M_2. Each machine can process one job at a time, and each job can be processed by one machine at a time. A job j consists of two operations, one with processing time or length p_j on M_1 and the other with processing time or length q_j on M_2; the process order of these two operations could be arbitrary, while the second operation must start exactly l_j units of time later than the completion of the first operation. We say a job starts if one of its operations starts processing, and is completed if both of its operations are finished. The problem is to seek for a feasible schedule so that the maximum completion time of all jobs or so-called the length/makespan of the schedule is minimized. According to the notations introduced by Graham et al. [10], this problem is denoted as $O2|(p_j, l_j, q_j)|C_{max}$.

Ageev [1] noted that the problem $O2|(p_j, l_j, q_j)|C_{max}$ must be strongly NP-hard since the strongly NP-hard no-wait scheduling problem $O_2|(p_j, q_j)|C_{max}$ [22] can be viewed as a special case with all zero delays, i.e., $l_j = 0$ for all j. They further proved that for any $\epsilon > 0$, the existence of a polynomial-time $(\frac{3}{2} - \epsilon)$-approximation algorithm for the special case where each job has equal length for both operations must imply $P = NP$. And even when the delays take only two distinct values of 0 and L, the problem $O2|(p_j, l_j, q_j)|C_{max}$ cannot be approximated within a factor of $\frac{5}{4} - \epsilon$ unless $P = NP$ [1]. This inapproximability proof also implies that the problem $O2^g|(p_j, l_j, p_j)|C_{max}$ with $l_j \in \{0, L\}$ cannot be approximated within $\frac{5}{4} - \epsilon$ either, unless $P = NP$. Recently, when studying the inapproximability of the parallel-machine scheduling problem with exact delays, Khatami et al. [13] observed that it is NP-hard to approximate the general problem $O2|(p_j, l_j, q_j)|C_{max}$ to within $2 - \epsilon$.

As a special case of the problem with exact delays, the no-wait problem $O2|(p_j, q_j)|C_{max}$ is strongly NP-hard [22], while it can be solved in $O(n \log n)$ time if $p_j - q_j$ becomes a constant for all j [20]. In particular, the problem $O2|(p_j, p_j)|C_{max}$ is polynomially solvable. When preemption of operations is allowed, Strusevich [24] proved that the no-wait problem $O2|(p_j, q_j)|C_{max}$ remains strongly NP-hard (see also [12]). Sidney and Sriskandarajah [23] used the exact solution for the corresponding two-machine no-wait flow shop problem $F2|(p_j, q_j)|C_{max}$ (proposed by Gilmore and Gomory [7] and implemented in $O(n \log n)$ time) as a heuristic solution for the open shop problem $O2|(p_j, q_j)|C_{max}$. They showed that this heuristic algorithm has an approximation ratio of $\frac{3}{2}$ and this ratio is tight.

In contrast, Leung et al. [17] proved that both $F2^g|(p_j, l_j, q_j)|C_{max}$ (directly from Sahni and Cho [22]) and $F2|(p_j, l_j, q_j)|C_{max}$ are strongly NP-hard even if $l_j \in \{L_1, L_2\}$, while $F2|(p_j, L, q_j)|C_{max}$ can be solved in $O(n \log n)$ time. They proposed an $O(n \log n)$-time 3-approximation algorithms for $F2^g|(p_j, l_j, q_j)|C_{max}$, and proved that the ratio can be reduced to 2 if either $p_j \leq q_j$ or $p_j \geq q_j$. Ageev [2] developed an $O(n \log n)$-time 2-approximation algorithm for the problem $F2|(p_j, l_j, q_j)|C_{max}$ with $l_j \in \{L_1, L_2\}$, and showed that the existence of a $(\frac{5}{4} - \epsilon)$-approximation algorithm for the problem implies $P = NP$, even if $L_1 = 0$. For the problem $F2|(p_j, l_j, p_j)|C_{max}$, Ageev and Kononov [4] proved that it cannot be approximated within $\frac{3}{2} - \epsilon$ unless $P = NP$. They also provided an $O(n \log n)$-time SPT-based algorithm with approximation ratio 3 for the general problem $F2|(p_j, l_j, q_j)|C_{max}$, and reduce the ratio to 2 when $p_j \leq q_j$ or $p_j \geq q_j$. When restricted further to the unit-length case that $p_j = q_j = 1$, the result in [28] (see also [27]) implies that the problem remains strongly NP-hard, and an $O(n^2 \log n)$-time $\frac{3}{2}$-approximation algorithm was later proposed by Ageev and Baburin [3].

The no-wait flow shop problem $F2|(p_j, q_j)|C_{max}$ seems essentially different. It can be formulated as an asymmetric traveling salesman problem (ATSP) where the distance matrix has a very special combinatorial structure, so that the famous subtour patching technique of Gilmore and Gomory [7] can yield an $O(n \log n)$ time exact algorithm. However, the result of Sahni and Cho [22] establishes that the problem $F2^g|(p_j, q_j)|C_{max}$ with missing operations at the first/second machines remains NP-hard in the strong sense. Glass et al. [8] presented an $O(n \log n)$-time heuristic with a worst-case ratio of $\frac{4}{3}$ for the problem $F2^g|(p_j, q_j)|C_{max}$ where some jobs require processing on the first machine only.

Vaessens and Dell'Amico [26] proved that even the open shop problem with equal-length operations for each job, i.e., the problem $O2|(p_j, l_j^+, p_j)|C_{max}$, is strongly NP-hard. Yu [28] confirmed the strong NP-hardness for a more special case where only two distinct minimal delays are involved. If all delays are equal, Rayward-Smith and Rebaine [21] proved that $O2|(p_j, L^+, q_j)|C_{max}$ is ordinary NP-hard, while Yu [28] confirmed that the NP-hardness holds even for the special case where $p_j = q_j$ for all j. Yu [28] showed that $O2|(1, l_j^+, 1)|C_{max}$ is strongly NP-hard (see also [27]). Using an extension of the Gonzalez-Sahni approach [11], Strusevich [25] presented an algorithm for the problem $O2|(p_j, l_j^+, q_j)|C_{max}$ that requires $O(n \log n)$-time and provides an approximation ratio of $\frac{3}{2}$. Munier-Kordon and Rebaine [19] proposed an $O(n \log n)$-time algorithm for two solvable cases of the problem $O2|(1, l_j^+, 1)|C_{max}$: all distinct delays and only two distinct delays. They also presented a heuristic with asymptotic approximation ratio $\frac{5}{4}$ for this unit-length open shop problem.

The strong NP-hardness of the flow shop problem $F2|(p_j, l_j^+, q_j)|C_{max}$ is early established in [15,16]. Vaessens and Dell'Amico [26] showed that the special case $F2|(p_j, l_j^+, p_j)|C_{max}$ is strongly NP-hard, and Yu [28] further confirmed that it remains strongly NP-hard even if $l_j \in \{0, L\}$. Yu [28] proved that the unit-length problem $F2|(1, l_j^+, 1)|C_{max}$ is strongly NP-hard (see also [27]).

We explore the two-machine open shop scheduling problem with exact delays by showing that even for the case where the delays are equal for all jobs, similar inapproximability results are valid. Specifically, we first show that $O2|(p_j, L, q_j)|C_{max}$ cannot be approximated within $\frac{4}{3} - \epsilon$ unless $P = NP$, compared to the known inappproximability ratios of $2 - \epsilon$ and $\frac{5}{4} - \epsilon$ respectively for $O2|(p_j, l_j, q_j)|C_{max}$ [13] and its special case where $l_j \in \{0, L\}$ [1]. Then we prove that it is NP-hard to approximate the case $O2|(p_j, L, p_j)|C_{max}$ to within $\frac{5}{4} - \epsilon$, compared to the known inappproximability ratio of $\frac{3}{2} - \epsilon$ for $O2|(p_j, l_j, p_j)|C_{max}$ [1], and that it is NP-hard to approximate the problem $O2^g|(p_j, L, q_j)|C_{max}$ that contains single-operation jobs to within $\frac{7}{6} - \epsilon$, compared to the known inapproximability ratio of $\frac{5}{4} - \epsilon$ for $O2^g|(p_j, l_j, p_j)|C_{max}$ where $l_j \in \{0, L\}$ [1]. Our hardness results cannot be directly derived from the existing proofs in the literature, but instead require new and perhaps more complicated reductions. On the positive side, we observe that the $O(n \log n)$-time exact algorithm for the flow shop problem $F2|(p_j, L, q_j)|C_{max}$ [7,14,17] is actually a 2-approximation algorithm for the open shop problem, and then present an improved $\frac{5}{3}$-approximation algorithm for the case with equal-length operations for each job.

2 The Inapproximability Results

All the inapproximability results are obtained by polynomial time reductions from the NP-complete PARTITION problem [9] defined as follows.

PARTITION (Without loss of generality, let $n > 2$, and $e_j < S$ for all $j \in X$.)
Input: $n + 1$ positive integers $e_1, e_2, \cdots, e_n$ and S such that $\sum_{j=1}^{n} e_j = 2S$.
Query: Is there a subset $Y \subset X = \{1, 2, \cdots, n\}$ such that $\sum_{j \in Y} e_j = S$?

In the two-machine open shop, each job is processed either in the order $M_1 \to M_2$ or in the order $M_2 \to M_1$, depending on which machine handles the first operation of the job. In the reminder of the paper, we may say a job starts with an operation or is started by an operation if this is the first operation of the job. We say a job starts (resp. is completed) at time t if the first (resp. second) operation of this job starts (resp. finishes) processing at t. The start time and the completion time of a job j are denoted by s_j and c_j respectively. Clearly, a feasible schedule must satisfy $c_j = s_j + p_j + L + q_j$ for all j. The time interval $[s_j, c_j]$ is also called the process interval of a job j. A gap on a machine is referred to as a time interval during which no job is processed. We may also say a job starts (resp. is completed) within a gap on a machine if the first (resp. second) operation of the job starts (resp. is completed) on the machine and at a time during the gap. We call two jobs *cross-processed* if their process orders are opposite and their process intervals overlap. Note whenever cross-processing a pair of jobs, a gap appears on each machine and the total length of the gaps on the two machines equals exactly $2L$.

Given any instance $\mathcal{I}$ of PARTITION, we construct in polynomial time an instance $\mathcal{I}'$ of the scheduling problem $O2|(p_j, L, q_j)|C_{max}$ with a set of $n+6$ jobs:

- n small jobs with $p_j = q_j = e_j$, $j \in X = \{1, 2, \cdots, n\}$.
- four large jobs with $(p_{n+1}, q_{n+1}) = (R+1, R)$, $(p_{n+2}, q_{n+2}) = (R, R+1)$, $(p_{n+3}, q_{n+3}) = (R-S, S)$ and $(p_{n+4}, q_{n+4}) = (S, R-S)$.
- two medium jobs with $(p_{n+5}, q_{n+5}) = (S, S+1)$ and $(p_{n+6}, q_{n+6}) = (S+1, S)$.

Let the exact delay of each job be equal to $L = R + S$ and set $R > 5S + 1$ (or $\gg S$).

Hereafter, a set of small jobs with index in $Z \subseteq X$ will be referred to as small jobs in Z. The job $n+i$ will be referred to as the i-th large job for $i = 1, 2, 3, 4$.

Lemma 1. *If the* PARTITION *instance $\mathcal{I}$ answers YES, then the scheduling instance $\mathcal{I}'$ admits a feasible schedule of length no greater than $3R + 5S + 1$.*

Proof. Given a YES solution Y to the PARTITION instance, i.e., $\sum_{j \in Y} e_j = S$, we can construct a feasible schedule with $C_{max} \leq 3R + 5S + 1$ (See Fig. 1). In fact, the four large jobs are scheduled in pairs, with the first pair of jobs $n+1$ and $n+2$ starting simultaneously on different machines, followed by the second pair of jobs $n+3$ and $n+4$ starting simultaneously as well. In addition, the second pair starts exactly S units of time after the first pair starts. Both pairs of jobs are cross-processed.

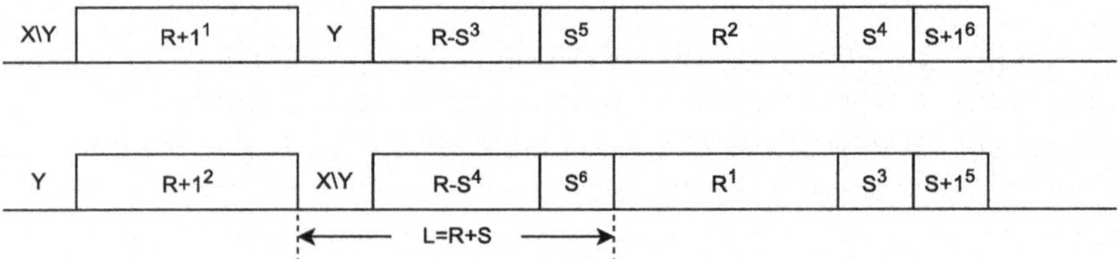

Fig. 1. A feasible schedule with makespan no greater than $3R + 5S + 1$.

Note that $L = R + S$, the arrangement of the four large jobs must produce two gaps of length S on each machine. Then the medium jobs $n+5$ and $n+6$ are cross-processed, starting within the second gap of M_1 and M_2 respectively. Subsequently, the small jobs in Y are started by the their operations on M_2 in non-increasing order of the processing times, while the small jobs in $X \setminus Y$ are started by the their operations on M_1 in non-increasing order of the processing times too. One sees that the arrangement of medium and small jobs is feasible, and all the four gaps are filled by them. This yields a feasible schedule of length no more than $3R + 5S + 1$ and hence proves the lemma.

Lemma 2. *If the* PARTITION *instance $\mathcal{I}$ answers NO, then any feasible schedule of the instance $\mathcal{I}'$ must achieve a length of at least $4R$.*

Proof. We prove by contradiction. Suppose that the PARTITION instance $\mathcal{I}$ answers NO and the scheduling instance $\mathcal{I}'$ admits a feasible solution with length $C_{max} < 4R$. Then we have the following claims.

Claim 1. The first two large jobs must be cross-processed.

In fact, if the large jobs $n+1$ and $n+2$ are processed in the same order, then it is clear that $C_{max} \geq \max\{c_{n+1}, c_{n+2}\} \geq q_{n+1}+q_{n+2}+L+p_{n+2} = 4R+S+1 > 4R$. And if their process intervals are disjoint, then $C_{max} \geq \max\{c_{n+1}, c_{n+2}\} \geq 2(p_{n+1} + L + q_{n+1}) = 6R + 2S + 2 > 4R$. This implies that the first two large jobs must be cross-processed.

By Claim 1, suppose without loss of generality that job $n+2$ is processed in the order $M_2 \rightarrow M_1$ and starts exactly α (≥ 0) units of time later than job $n+1$, creating (at most) two gaps of length $\delta_1 = L + \alpha$ and $\delta_2 = L - \alpha$ on machine M_1 and machine M_2 respectively (See Fig. 2 for an illustration). Henceforth, we refer to a gap solely by its length or replace the term "gap" with just its length for simplification.

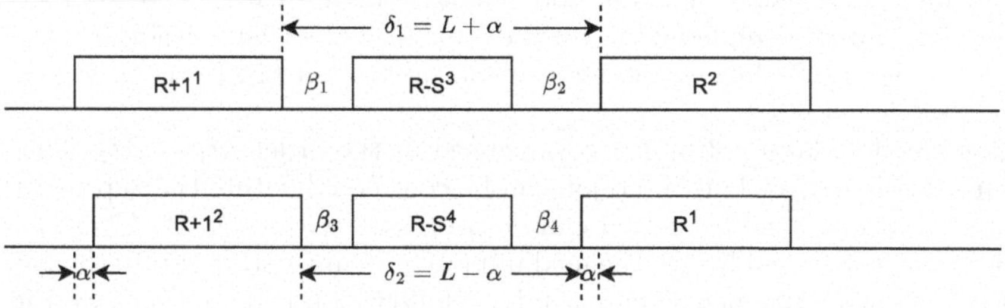

Fig. 2. Illustration for Claim 1 and Claim 2.

Claim 2. The larger operations p_{n+3} and q_{n+4} of the two remaining large jobs must be processed in the gaps δ_1 and δ_2 respectively.

Because the process interval of either job $n+1$ or $n+2$ already achieves a length of $p_{n+1}+L+q_{n+1} = 3R+S+1$, processing the larger operation $p_{n+3} = R-S$ (resp. $q_{n+4} = R-S$) of job $n+3$ (resp. $n+4$) outside the gap δ_1 (resp. δ_2) must lead to a schedule of length at least $3R+S+1+(R-S) = 4R+1 > 4R$. Thus Claim 2 is valid.

Due to the process of job $n+3$ and $n+4$ as demonstrated in Claim 2, (at most) four smaller gaps, denoted as β_i, $i = 1, 2, 3, 4$, are produced (See Fig. 2). Note that β_1 is on machine M_1 followed by β_2, while β_3 is on machine M_2 followed by β_4, and the total length of them equals exactly

$$\sum_{i=1}^{4} \beta_i = \delta_1 + \delta_2 - p_{n+3} - q_{n+4} = 2L - 2(R-S) = 4S. \tag{1}$$

Claim 3. All the gaps β_i, $i = 1, 2, 3, 4$, must be filled with the medium and small jobs. In particular, exactly one operation of each small job and the smaller operation of each medium job are processed in these gaps.

If there is a medium or small job j completely excluded by the gaps β_i, $i = 1, 2, 3, 4$, then the length of the schedule would be at least $p_{n+1} + L + q_{n+1} + p_j + L + q_j > 4R + 2S + 1 > 4R$. Thus each medium and small job must have at least one operation processed in these gaps. And if there are two operations of the same job or at least one larger operation of the medium jobs, then the total length of operations processed in the gaps would achieve $C_{max} \geq \sum_{j=1}^{n} e_j + 2S + 1 = 4S + 1$, exceeding the total length of the gaps calculated in (1), which yields a contradiction! This implies that only the smaller operation of each medium job and exactly one operation of each small job are processed in these gaps. Since the total length of these operations equals exactly $4S$, they definitely fill up the gaps. This confirms Claim 3.

Claim 4. Only the second operation of a job can be processed in the gap β_3, and only the first operation of a job can be processed in the gap β_4. If there is a job starting within the gap β_3, then its second operation must start on M_1, where the earliest available time is at the completion of job $n+2$. However, this would lead to a minimum of $q_{n+4} + p_{n+2} = (R - S) + R > L$ units of time for delay, which is infeasible! Thus no job can start within the gap β_3. In a similar way, no job can be completed within the gap β_4. Thus Claim 4 holds.

For further discussion, we distinguish two cases according to the process order of the large job $n+4$.

Case 1. The large job $n+4$ is processed in the order $M_1 \rightarrow M_2$. Then by similar arguments as Claim 4, no job can be completed within the gap β_3 either, thus indicating that no operation can be processed within β_3. By Claim 3, this implies that $\beta_3 = 0$ and hence, the medium job $n+6$ must start within the gap β_4, creating (at most) two new gaps γ_1 and γ_2 (follows after γ_1) on M_2 (See Fig. 3). Since a job starting within γ_2 (if any) must have its second operation starting on M_1, where the earliest available time is at the completion of job $n+6$, the delay of this job would be as large as $\alpha + p_{n+2} + p_{n+6} = \alpha + R + S + 1 > L$, resulting in a contradiction! This implies that no job can start within γ_2. Together with Claim 4, it further indicates that no operation can be processed in the gap γ_2, which means that $\gamma_2 = 0$ by Claim 3.

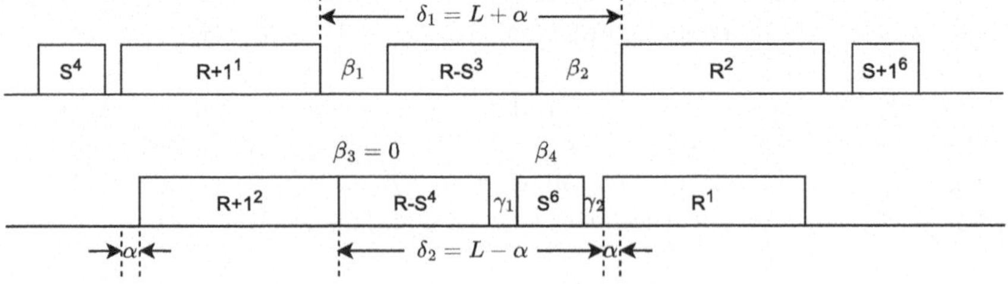

Fig. 3. Illustration for Case 1.

If the gap $\gamma_1 > 0$ (thus must be filled with small jobs), then by $\gamma_2 = 0$, a job that finishes its first operation in γ_1 has to wait at least $c_{n+2} - s_{n+6} =$

$R+S+\alpha = L+\alpha$ units of time for its second operation. Such specific case occurs only when $\alpha = 0$, and only if exactly one job starts within that gap. However, this cannot be the case since the gap γ_1 is strictly larger than any small job, i.e., $\gamma_1 = \delta_2 - q_{n+4} - q_{n+6} = L - (R - S) - S = S > e_j$ for any $j \in X$. Thus $\gamma_1 = 0$.

By $\beta_3 = \gamma_1 = \gamma_2 = 0$, we can calculate the exact value of α directly,

$$\alpha = (c_{n+1} - s_{n+1}) - (q_{n+2} + q_{n+4} + q_{n+6} + q_{n+1}) = (2R+1+L) - (2R+1+R) = S, \tag{2}$$

and from which, the delay of the large job $n+4$ must be as large as $(c_{n+4} - q_{n+4}) - s_{n+1} = R + 1 + \alpha = R + S + 1 > L$, implying a contradiction!

Case 2. The large job $n+4$ is processed in the order $M_2 \rightarrow M_1$. If $\beta_4 > 0$, which must be filled with medium or small jobs by Claim 3, then the minimum delay between the first and second operation of a job that starts within the gap β_4 (recall Claim 4 that no job is completed within β_4) achieves $\alpha + p_{n+2} + (c_{n+4} - c_{n+2}) \geq \alpha + R + S = \alpha + L$. This occurs only when $\alpha = 0, c_{n+4} = c_{n+2} + p_{n+4}$, and exactly one job starts within β_4. In addition, since $\beta_4 = L - p_{n+2} - \alpha = L - R = S$, it must be the job $n+6$ filling up the gap β_4. As consequence, we obtain $\beta_3 = \delta_2 - q_{n+4} - \beta_4 = L - R = S$, which consists only of small jobs by Claim 3. Let Y denote the set of these jobs. Then $\sum_{j \in Y} e_j = \beta_3 = S$, giving a YES answer to the Partition instance. However, this is a contradiction too.

Otherwise, $\beta_4 = 0$. Then together by Claim 3 and Claim 4, the medium job $n+6$ must be completed within β_3, creating (at most) two new gaps γ_3 and γ_4 (follows after γ_3) on M_2 (See Fig. 4). By similar arguments as Case 1, no small job can be completed within γ_3 or γ_4, which follows by Claim 3 that $\gamma_3 = \gamma_4 = 0$. Consequently, we compute $\alpha = S$ the same as that in (2). Then the minimum delay of job $n+6$ would be as large as $p_{n+1} + \alpha = R + S + 1 > L$, which yields a contradiction as well. This finally completes the proof of Lemma 2.

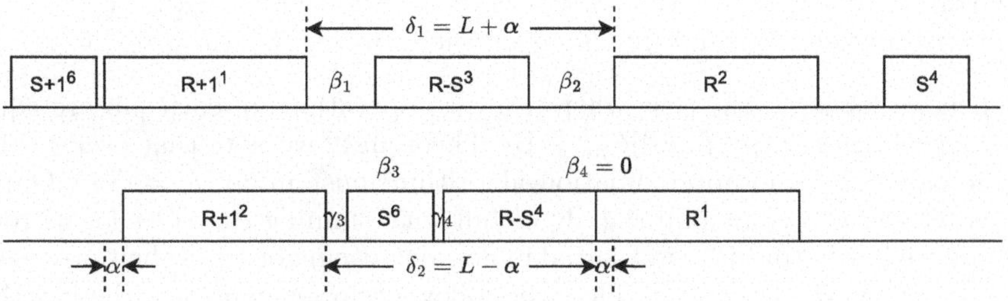

Fig. 4. Illustration for Case 2.

In conclusion, Lemma 1, 2 directly follow Theorem 1. And two additional inapproximability results can be derived by reductions from Partition as well.

Theorem 1. *The problem* $O2|(p_j, L, q_j)|C_{max}$ *does not admit a* $(\frac{4}{3} - \epsilon)$-*approximation algorithm unless* $P = NP$.

Theorem 2. *The problem* $O2|(p_j, L, p_j)|C_{max}$ *does not admit a* $(\frac{5}{4} - \epsilon)$-*approximation algorithm unless* $P = NP$.

Theorem 3. *The problem* $O2^g|(p_j, L, q_j)|C_{max}$ *does not admit a* $(\frac{7}{6} - \epsilon)$-*approximation algorithm unless* $P = NP$.

3 Approximation Algorithms

As we have shown, the open shop problem $O2|(p_j, L, q_j)|C_{max}$ is NP-hard to approximate within a factor of $\frac{4}{3}$. On the contrary, Leung et al. [17] noticed that the flow shop problem $F2|(p_j, L, q_j)|C_{max}$ can be solved in $O(n \log n)$ time. Khatami et al. [14] further observed that the optimal algorithm for the two-machine no-wait flow shop problem proposed by Gilmore and Gomory [7] turns out to be an optimal algorithm for $F2|(p_j, L, q_j)|C_{max}$ too. Since any solution to the flow shop problem also serves as a feasible solution to the open shop problem, applying the optimal algorithm for $F2|(p_j, L, q_j)|C_{max}$ to the open shop problem $O2|(p_j, L, q_j)|C_{max}$ can result in a feasible schedule of length no more than twice the optimal length. In fact, let C^*_{max} be the length of the optimal schedule for the open shop problem, in which the partial schedule consisting only of jobs with the process order $M_1 \to M_2$ (resp. $M_2 \to M_1$) has length $C^*_{max}(M_1 \to M_2) \le C^*_{max}$ (resp. $C^*_{max}(M_2 \to M_1) \le C^*_{max}$), and let C_{max} be the length of the optimal schedule for the flow shop problem, then it is clear that

$$C_{max} \le C^*_{max}(M_1 \to M_2) + C^*_{max}(M_2 \to M_1) \le 2C^*_{max}.$$

We summarize this observation as the following theorem.

Theorem 4. *The optimal* $O(n \log n)$-*time algorithm for* $F2|(p_j, L, q_j)|C_{max}$ *is a 2-approximation algorithm for the open shop problem* $O2|(p_j, L, q_j)|C_{max}$.

Then we focus on the case where the two operations of each job are equal, i.e., the problem $O2|(p_j, L, p_j)|C_{max}$. By Theorem 2, we note that even for this specific case a $(\frac{5}{4} - \epsilon)$-approximation algorithm must imply $P = NP$. Our $\frac{5}{3}$-approximation is a combinatorial algorithm with running time $O(n \log n)$ too.

Suppose firstly that jobs are sorted in non-decreasing order of their processing times, i.e., $p_1 \le p_2 \le \cdots \le p_n$. Then the following lower bound on the length of any feasible schedule holds obviously.

$$C^*_{max} \ge \max\{\sum_{j=1}^{n} p_j, 2p_n + L\}. \quad (3)$$

Next, we refer to a SPT (Shortest Processing Time first) rule as a strategy/algorithm for scheduling jobs along the above sequence, and we define two basic SPT rules, SPT_1 and SPT_2 as follows.

SPT$_1$. Sequentially start all the jobs on machine M_1 as early as possible, introducing idle times on machine M_2 whenever necessary (See Fig. 5).
SPT$_2$. Cross-process job 1 and job 2 so that two gaps of length $\delta_1 = L + p_2 - p_1$ and $\delta_2 = L - p_2 + p_1$ are created on M_1 and M_2 respectively; sequentially start the remaining jobs within the gaps δ_1 and δ_2 as early as possible with idle times introduced for the second operation of each job if necessary (See Fig. 6).

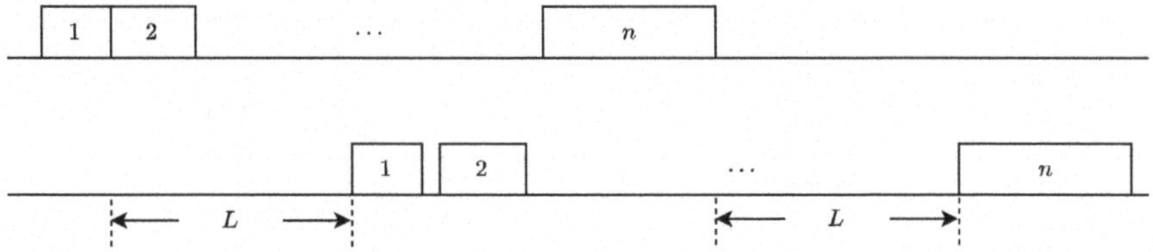

Fig. 5. The schedule produced by SPT$_1$.

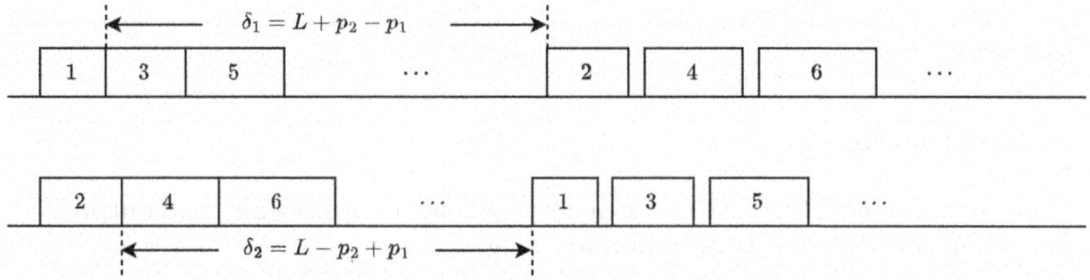

Fig. 6. The schedule produced by SPT$_2$.

Note that SPT$_1$ is always feasible, while SPT$_2$ is applicable only if $p_2 \leq p_1 + L$ (so that the two gaps are well-defined) and all the remaining jobs, $3, \cdots, n$, are successfully processed within the gaps. We give some further observations regarding the two SPT rules.

Lemma 3. *In the schedule produced by* SPT$_1$*, there is no idle time on machine* M_1 *and the length achieves* $C^1_{max} = \sum_{j=1}^{n} p_j + L + p_n$.

Lemma 4. *Suppose that* $p_2 \leq p_1 + L$*, then in the schedule produced by* SPT$_2$*,*

(a) no idle time occurs among the operations processed in the gaps δ_1 *and* δ_2*;*
(b) no fewer jobs are processed in the gap δ_1 *than in the gap* δ_2*, but up to a maximum of* 2 *additional jobs in* δ_1*;*
(c) the length achieves $C^2_{max} \leq \frac{1}{2} \sum_{j=1}^{n} p_j + 2p_n + L$ *if all the remaining jobs,* $3, \cdots, n$*, can be processed within the gaps* δ_1 *and* δ_2 *by* SPT$_2$*;*
(d) $\sum_{j=3}^{n} p_j > 2L + (i-2)p_n$ *if the last* i $(i = 1, 2, 3)$ *jobs fail to be processed within the gaps* δ_1 *and* δ_2 *by* SPT$_2$.

Our algorithm CombApprox applies both SPT_1 and SPT_2, and then a modified SPT_1 (m-SPT_1 in short) if neither SPT_1 nor SPT_2 can give the $\frac{5}{3}$-approximation. See Fig. 8 for the description.
m-SPT_1. Apply SPT_1 on the first $n-1$ jobs, and start job n on M_2 as early as possible before the second operation of job 1 (See Fig. 7 for an illustration).

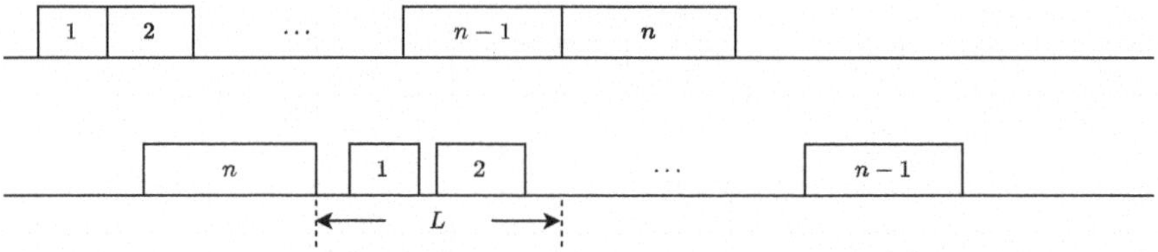

Fig. 7. A schedule produced by m-SPT_1.

Lemma 5. *If* $p_n < p_1 + L$ *and* $\sum_{j=2}^{n-1} p_j \leq 2L$*, then m-*$\text{SPT}_1$ *can output a feasible schedule with length* $C_{max}^3 = \max\{2p_n + L, \sum_{j=1}^{n} p_j + L\}$.

> **CombApprox.**
>
> 1. Run each of SPT_1, SPT_2, or m-SPT_1 whenever applicable, and output the schedule with the minimum length.

Fig. 8. A high-level description of CombApprox.

Theorem 5. *The algorithm* CombApprox *is an* $O(n \log n)$*-time* $\frac{5}{3}$*-approximation algorithm for the problem* $O2|(p_j, L, p_j)|C_{max}$.

Proof. If SPT_2 is applicable, then by Lemma 4 (c) and the lower bound (3), the length of SPT_2 achieves $C_{max}^2 \leq \frac{1}{2}\sum_{j=1}^{n} p_j + 2p_n + L \leq \frac{3}{2}C_{max}^*$. If at least three jobs fail to be processed by SPT_2, then by Lemma 4 (d), we have $\sum_{j=3}^{n} p_j > 2L + p_n$. Together with (3), $L + p_n = \frac{1}{3}(2L + p_n) + \frac{1}{3}(2p_n + L) < \frac{1}{3}\sum_{j=3}^{n} p_j + \frac{1}{3}C_{max}^* \leq \frac{2}{3}C_{max}^*$. By Lemma 3, the length of SPT_1 achieves $C_{max}^1 = \sum_{j=1}^{n} p_j + L + p_n \leq \frac{5}{3}C_{max}^*$. If $\sum_{j=2}^{n-1} p_j > 2L$, then we still have $C_{max}^* \geq \sum_{j=2}^{n} p_j > 2L + p_n$ and hence $C_{max}^1 \leq \frac{5}{3}C_{max}^*$ can be derived in the same way. If $p_n \geq L + p_1$, then $L + p_n \leq \frac{2}{3}(L + 2p_n) \leq \frac{2}{3}C_{max}^*$, which directly follows that $C_{max}^1 \leq \frac{5}{3}C_{max}^*$.

Now suppose that $\sum_{j=2}^{n-1} p_j \leq 2L$, $p_n < L + p_1$, and SPT_2 misses either one or two jobs. Then Lemma 5 gives that m-SPT_1 is applicable and its length equals exactly $C_{max}^3 = \max\{2p_n + L, \sum_{j=1}^{n} p_j + L\}$. If only one job fails to be processed by SPT_2, then by Lemma 4 (d), $\sum_{j=3}^{n} p_j > 2L - p_n$. Combining it with (3), we derive $L = \frac{1}{5}(2p_n + L) + \frac{2}{5}(2L - p_n) < \frac{1}{5}C^*_{max} + \frac{2}{5}\sum_{j=3}^{n} p_j \le \frac{3}{5}C^*_{max}$. Consequently, $C^3_{max} \le \frac{8}{5}C^*_{max}$. If exactly two jobs fail to be processed by SPT_2, then by Lemma 4 (d), $\sum_{j=3}^{n} p_j > 2L$. Then by (3), $L < \frac{1}{2}\sum_{j=3}^{n} p_j \le \frac{1}{2}C^*_{max}$. Thus we have $C^3_{max} \le \frac{3}{2}C^*_{max}$. This proves the theorem.

4 Conclusion

We investigated two-machine open shop scheduling with exact delays and established various inapproximatibility results for the case with all equal delays. We also provided positive results for this specific case. Our hardness results significantly strengthened previous results on the open shop problem with multiple exact delays [1,13,22]. In a recent survey on open shop problems [18], Ahmadian et al. concluded that coupled task scheduling in the open-shop setting "is still in the early stage of development", and they believed that "further research in this area is worthy". In another survey focusing on coupled task scheduling [14], the authors raised the point that "the flow shop coupled task scheduling problem has received fairly good attention, research in other shop environments, e.g., the flexible shop settings, needs further development", and they observed "investigation into the job shop and open shop settings is evolving; however, it is still in the nascent stage". In summary, our inapproximatibility and approximation results can be considered a preliminary progress on these problems. It definitely suggests lots of further research on open shop problems with exact (resp. minimum, maximum) delays, not only complexity-oriented, but the design of approximation algorithms for these problems is also challenging.

Acknowledgements. This research is supported by the NSFC Grants 12371316, 12471301, and the Zhejiang Provincial NSF Grant LZ25A010001.

References

1. Ageev, A.: Inapproximability lower bounds for open shop problems with exact delays. In: International Conference on Optimization Problems and Their Applications, pp. 45–55. Springer (2018)
2. Ageev, A.A.: Approximating the 2-machine flow shop problem with exact delays taking two values. J. Glob. Optim. **76**(3), 491–497 (2020)
3. Ageev, A.A., Baburin, A.E.: Approximation algorithms for UET scheduling problems with exact delays. Oper. Res. Lett. **35**(4), 533–540 (2007)
4. Ageev, A.A., Kononov, A.V.: Approximation algorithms for scheduling problems with exact delays. In: International Workshop on Approximation and Online Algorithms (WAOA), pp. 1–14. Springer (2007)
5. Fondrevelle, J., Oulamara, A., Portmann, M.-C.: Permutation flow shop scheduling problems with maximal and minimal time lags. Comput. Oper. Res. **33**, 1540–1566 (2006)

6. Fondrevelle, J., Oulamara, A., Portmann, M.-C., Allahverdi, A.: Permutation flowshops with exact time lags to minimize maximum lateness. Int. J. Prod. Res. **47**, 6759–6775 (2009)
7. Gilmore, P.C., Gomory, R.E.: Sequencing a one state-variable machine: a solvable case of the traveling salesman problem. Oper. Res. **12**(5), 655–679 (1964)
8. Glass, C.A., Gupta, J., Potts, C.N.: Two-machine no-wait flow shop scheduling with missing operations. Math. Oper. Res. **24**(4), 911–924 (1999)
9. Garey, M.R., Johnson, D.S.: Computers and Intractability: A Guide To The Theory of NP-Completeness. Freeman, San Francisco (1979)
10. Graham, R.-L., Lawler, E.-L., Lenstra, J.-K., Kan, A.: Optimization and approximation in deterministic sequencing and scheduling: a survey. Ann. Discrete Math. **5**, 287–326 (1979)
11. Gonzalez, T., Sahni, S.: Open shop scheduling to minimize finish time. J. Assoc. Comput. Mach. **23**, 665–679 (1976)
12. Hall, N.G., Sriskandarajah, C.: A survey of machine scheduling problems with blocking and no-wait in process. Oper. Res. **44**(3), 510–525 (1996)
13. Khatami, M., Oron, D., Salehipour, A.: Scheduling coupled tasks on parallel identical machines. Optim. Lett. **18**(4), 991–1003 (2024)
14. Khatami, M., Salehipour, A., Cheng, T.: Coupled task scheduling with exact delays: literature review and models. Eur. J. Oper. Res. **282**(1), 19–39 (2020)
15. Lenstra, J.-K.: Private communication (1991)
16. Lawler, E.L., Lenstra, J.K., Kan, A.H.R., Shmoys, D.B.:. Sequencing and scheduling: algorithms and complexity. In: Graves, S.C., et al. (eds.) Handbooks in Operations Research and Management Science, vol.4, pp. 445–522. Logistics of Production and Inventory, North-Holland, Amsterdam (1993)
17. Leung, J.Y.-T., Li, H., Zhao, H.: Scheduling two-machine flow shops with exact delays. Int. J. Found. Comput. Sci. **18**(2), 341–359 (2007)
18. Ahmadian, M.M., Khatami, M., Salehipour, A., Cheng, T.: Four decades of research on the open-shop scheduling problem to minimize the makespan. Eur. J. Oper. Res. **295**(2), 399–426 (2021)
19. Munier-Kordon, A., Rebaine, D.: The two-machine open-shop problem with unit-time operations and time delays to minimize the makespan. Eur. J. Oper. Res. **203**(1), 42–49 (2010)
20. Panwalkar, S.S., Koulamas, C.: The two-machine no-wait general and proportionate open shop makespan problem. Eur. J. Oper. Res. **238**(2), 471–475 (2014)
21. Rayward-Smith, V.J., Rebaine, D.: Open shop scheduling with delays. RAIRO Theoret. Inform. Appl. **26**, 439–447 (1992)
22. Sahni, S., Cho, Y.: Complexity of scheduling shops with no wait in process. Math. Oper. Res. **4**(4), 448–457 (1979)
23. Sidney, J.B., Sriskandarajah, C.: A heuristic for the two-machine no-wait openshop scheduling problem. Nav. Res. Logist. **46**(2), 129–145 (1999)
24. Strusevich, V.A.: Complexity aspects of shop scheduling problems. Erasmus Universiteit Rotterdam Ph.D. thesis
25. Strusevich, V.A.: A heuristic for the two-machine open-shop scheduling problem with transportation times. Discret. Appl. Math. **93**(2–3), 287–304 (1999)
26. Vaessens, R.J.M., Dell'Amico, M.: Flow and open shop scheduling on two machines with transportation times and machine-independent processing times is NP-hard, Preprint (1995)
27. Yu, W., Hoogeveen, H., Lenstra, J.-K.: Minimizing makespan in a two-machine flow shop with delays and unit-time operations is NP-Hard. J. Sched. **7**(5), 333–348 (2004)

28. Yu, W.: The two-machine flow shop problem with delays and the one-machine total tardiness problem. Ph.D. dissertation. Eindhoven University of Technology, The Netherlands (1996)
29. Zhang, X., van de Velde, S.: Polynomial-time approximation schemes for scheduling problems with time lags. J. Sched. **13**(5), 553–559 (2010)

Minimum Power Partial Cover with Fairness Constraint

Chensheng Ma and Zhao Zhang(✉)

Zhejiang Normal University, Jinhua 321004, Zhejiang, China
hxhzz@sina.com

Abstract. In wildlife conservation, monitoring species' activities is essential for understanding habitats, migration, and population dynamics. Sensors like detectors and cameras collect data, which is gathered by base stations within their signal ranges. Base stations can adjust their power, where the power consumption $p(s)$ relates to the sensing radius $r(s)$ as $p(s) = c \cdot r(s)^\alpha$. The minimum power partial cover problem (MinPPC) seeks to minimize power while covering at least k sensors. Addressing fairness concerns, we propose the MinPPC with fairness constraints (MinPPCF), ensuring each sensor category meet a coverage threshold. We present a polynomial-time algorithm with an approximation ratio of $O(\alpha + T)$, where T is the number of categories.

Keywords: Minimum power cover · Partial cover · Fairness constraint · Approximation algorithm

1 Introduction

Wireless sensor networks (WSNs) have emerged as a crucial technology in various applications, ranging from environmental monitoring to industrial automation [1–3]. For example, in a wildlife conservation, it is necessary to monitor the activities and behaviors of various species to understand their habitats, migration patterns, population dynamics, etc. To collect data from the environment, a variety of sensors, such as motion detectors, temperature sensors and cameras, are deployed at points of interest (PoIs). Then, data collected by these sensors are relayed to base stations for further processing. Suppose a base station s equipped with power $p(s)$ can collect data from those sensors located at distance at most $r(s)$ from s, where

$$p(s) = c \cdot r(s)^\alpha. \quad (1)$$

In other words, we say that s can cover all those PoIs at distance at most $r(s)$ from s. In a *minimum power cover* (MinPC) problem, it is desired to use the minimum power to cover all PoIs [4]. It has been show in [5] that compromising a little coverage may bring significant benefits in costs. In a *minimum power partial cover* (MinPPC) problem, it is desired to use the minimum power to cover at least k PoIs.

D. Du et al. (Eds.): COCOA 2024, LNCS 15434, pp. 190–199, 2025.
https://doi.org/10.1007/978-981-96-4445-2_16

Fairness has been an important concern in recent studies [6–10]. Consider the example of wildlife conservation again, it is desired that certain degree of coverage can be guaranteed across different categories of sensors. Incorporating such a fairness consideration into the minimum power cover problem, we propose the *minimum power cover with fairness constraint* (MinPCF) problem as follows. Suppose there are T different colors. Each color class C_i (consisting of those PoIs with color i) is associated with an integer r_i with $0 \leq r_i \leq |C_i|$. For each color class C_i, it is required that at least r_i PoIs are covered. A power assignment meeting all these requirements is said to be a *fair power assignment.* The goal of the MinPCF problem is to find a fair power assignment with the minimum total power. We also propose the *minimum power partial cover with fairness constraint* (MinPPCF) problem, which is a MinPFC problem with an additional constraint that, in total, at least k PoIs are covered.

In this paper, we study approximation algorithm for MinPFC and MinPPFC.

1.1 Related Work

In this section, we review key literature related to our study on optimizing power coverage in WSNs. Before our work, Freund and Rawitz [11] studied the minimum power coverage problem and obtained a 9 approximation factor that was only obtained for $\alpha = 2$ and $T = 1$. When $\alpha > 1$, Alt et al. [4] proved that the minimum power cover problem is NP-hard.

Li et al. [12,13] used a primal-dual approach and a local ratio method to obtain the same 3^{α}-approximation factor when α is general and $T = 1$. Note that they studied the partial version and their algorithm runs in time $O(km^2n^2)$. Later, Dai et al. [14] used a ρ-relaxed independence set instead of maximal independence set in their algorithm, and obtained an $O(\alpha)$-approximation algorithm which runs in time $O(km^2n^2)$. Ran et al. [15] presented a polynomial time approximation scheme (PTAS) for the minimum power partial multicover problem based on a plane subdivision technique. This was done under the assumption that the maximum number of times that a point must be covered, denoted as $cr_{\max}$, is bounded by a constant. There are also some variants. Liu et al. [16] considered the k-prize-collecting minimum power cover problem and obtained the 3^{α}-approximation factor by using the primal-dual scheme. And then Dai et al. [17] studied the H-prize-collecting power cover problem and obtained $(4 \cdot 3^{\alpha-1} + \varepsilon)$-approximation factor. Liu et al. [18] presented a $3^{\alpha} + 1$ approximation algorithm for the minimum power cover problem with submodular penalties. Although there are many works on minimum power cover, there is no existing study considering fairness constraint.

Bera et al. [19] studied the vertex cover problem with fairness constraints and presented an $O(\log T)$-approximation algorithm. Zhou et al. [20] first proposed the prize-collecting vertex cover problem with fairness constraints. With probability at least $1 - 1/T^{\delta}$, their algorithm achieves approximation ratio at most $\left(\frac{9(\delta+1)}{2} \ln T + 3\right)$, where δ is a constant.

1.2 Our Contributions

This paper introduces the minimum power cover problem with fairness constraints (MinPCF) and the minimum power partial cover problem with fairness constraints (MinPPCF), addressing the trade-off between fair coverage and power consumption in wireless sensor networks (WSNs). The motivation stems from real-world applications where different groups require balanced coverage.

Our algorithm employs a primal-dual strategy. Note that multi-color scenario poses new challenges on the effectiveness of this strategy. Due to the interlacing of different color classes through disks, the contribution of a disk might be counted repeatedly, which incurs uncontrollable increase in the approximation ratio. So, primal-dual methods in the previous works [12,16,17] for the partial cover or prize-collecting cover problems cannot be readily employed here.

The paper presents a new primal-dual method that mitigates these issues, achieving an approximation ratio of at most $\max\{e^{2\pi}\alpha, T+1\}$ for MinPCF, where α is the power stretch factor and T is the number of color classes. For MinPPCF, the approximation ratio becomes $\max\{2e^{2\pi}\alpha, e^{2\pi}\alpha+T+1\}$, and the algorithm runs in time $O((\sum_{i\in[T]} r_i + k)mn\log mn)$, where m is the number of points of interest (PoIs), n is the number of base stations, and k is the partial cover parameter. Note that MinPPC is a special case of MinPCF, in which there is only one color class. Our algorithm, applied on MinPPC, runs in time $O(mnk\log mn)$, which significantly decreases the time complexity $O(km^2n^2)$ of the previous algorithm presented in [14].

1.3 Organization

The rest of this paper is organized as follows: Sect. 2.1 describes the MinPPCF problem and the MinPCF problem, and proves their NP-hardness. In Sect. 2.2, algorithms for MinPCF and MinPPCF are proposed. Section 2.3 presents theoretical analysis. And the paper is concluded in Sect. 3.

2 The Algorithm and Its Analysis

2.1 Preliminaries

The MinPPCF problem is formally defined as follows.

Definition 1. [Minimum Power Partial Cover with Fairness Constraint problem (MinPPCF)] *Suppose X is a set of m PoIs and S is a set of n base stations on the plane. Denote $[T]=\{1,\ldots,T\}$. PoIs are divided into T disjoint color classes $C_1,\ldots,C_T$. Each color class C_i is associated with an integer r_i with $0\le r_i\le |C_i|$. A PoI $x\in X$ is covered by a base station $s\in S$ with power $p(s)$ if x belongs to the disk supported by $p(s)$, that is $x\in Disk(s,r(s))$ centered at s with radius $r(s)$, where $p(s)=c\cdot r(s)^\alpha$. A PoI is covered by a power assignment $\mathcal{F}: S\rightarrow \mathbb{R}^+$ if it is covered by some disk supported by $\mathcal{F}$. Suppose k is an integer with $k\ge\sum_{i\in[T]} r_i$. The goal of MinPPCF is to find a power assignment*

$\mathcal{F}$ to cover at least k PoIs such that the fairness constraints are satisfied, where the fairness constraint for color class C_i requires at least r_i PoIs from C_i being covered, subject to these conditions, the total power $\sum_{s\in\mathcal{F}} p(s)$ is as small as possible. Here, we assume that there is no limit on the power at a base station.

When the MinPPCF problem is not required to cover at least k PoIs, we call this problem the Minimm Power Cover with Fairness Constraint (MinPCF).

When $T = 1$ and $k = r_1 = m$, the MinPPCF problem becomes the *minimum power cover* problem, which was shown to be NP-hard in [4]. So, both the MinPPCF problem and the MinPCF problem are also NP-hard.

Note that although a station s can adjust its power $p(s)$ (and thus its transmission radius $r(s)$) continuously, in an optimal solution, $r(s)$ can only take a value from a discrete set $\{\|sx\|: x \in X\} \cup \{0\}$, where $\|sx\|$ is the Euclidean distance between station s and PoI x. To be more concrete, denote

$$\mathcal{D}_s = \{Disk(s, \|sx\|): x \in X\}, \text{let } \mathcal{D} = \bigcup_{s\in\mathcal{S}} \mathcal{D}_s, \tag{2}$$

and define the weight of disk D as

$$w_D = c \cdot r_D^{\alpha}, \tag{3}$$

where r_D is the radius of disk $D \in \mathcal{D}$, then the MinPPCF problem can be viewed as picking a minimum weight set of disks from $\mathcal{D}$ to cover at least k PoIs such that the fairness constraints are satisfied and

$$\text{for each base station } s \in \mathcal{S}, \text{ at most one disk from } \mathcal{D}_s \text{ can be picked.} \tag{4}$$

Constraint (4) indicates that every station can only set its power at one fixed level. Note that any *minimal* solution satisfies condition (4).

2.2 The Algorithm

The main algorithm is presented in Algorithm 1. It combines a solution $\mathcal{F}^1$ to the MinPPC problem and a solution $\mathcal{F}^2$ to the MinPCF problem. There have been many studies on MinPPC [12–14]. Our algorithm is a modification of the one in [12,14] that achieves approximation factor $e^{2\pi}\alpha$. Different from [12,14], our algorithm, denoted as MinPPC^A, adds a binary search to reduce the time complexity without compromising accuracy. We also designed an algorithm called MinPCF^A to compute an approximate solution to MinPCF. For $i = 1, 2$, denote by $p_{\mathcal{F}^i}(s)$ the power assigned to station s under $\mathcal{F}^i$. In the final solution $\mathcal{F}$, the power assigned to station s is $p_{\mathcal{F}}(s) = \max\{p_{\mathcal{F}^1}(s), p_{\mathcal{F}^2}(s)\}$, this operation is denoted as $\mathcal{F} = \mathcal{F}^1 \vee \mathcal{F}^2$. Clearly, such an assignment satisfies both the total covering requirement k and the fairness constraints for all groups.

Algorithm 1. MinPPCF$^A(X, \mathcal{S}, \{r_i\}_{i\in[T]}, k)$

Input: A set of points X, a set of base stations $\mathcal{S}$, fairness requirements $\{r_i\}_{i\in[T]}$ and total covering requirement k.
Output: A power assignment $\mathcal{F}$.
1: $\mathcal{F}^1 \leftarrow$ MinPPC$^A(X, \mathcal{S}, k)$;
2: $\mathcal{F}^2 \leftarrow$ MinPCF$^A(X, \mathcal{S}, \{r_i\}_{i\in[T]})$;
3: $\mathcal{F} \leftarrow \mathcal{F}^1 \vee \mathcal{F}^2$;
4: **return** $\mathcal{F}$

Next, we present the algorithm MinPCF^A for the MinPCF problem. It is based on the following (relaxed) LP formulation.

$$\begin{aligned} \min \quad & \sum_{D\in\mathcal{D}} w_D z_D && (5)\\ \text{s.t.} \quad & \sum_{D: x\in D} z_D \geq y_x, \forall x \in X, \\ & \sum_{x\in C_i} y_x \geq r_i, \quad \forall i \in [T], \\ & z_D \geq 0, \qquad \forall D \in \mathcal{D}, \\ & y_x \geq 0, \qquad \forall x \in X. \end{aligned}$$

To control the approximation ratio of the algorithm, MinPCF^A will *guess* a largest disk in an optimal solution. Suppose $D_{\max}$ is such a disk. Denote by $\mathcal{D}_{\leq D_{\max}} = \{D \in \mathcal{D} \colon r_D \leq r_{D_{\max}}\} \setminus \{D_{\max}\}$ the subset of disks in $\mathcal{D}$ that are no larger than $D_{\max}$, excluding $D_{\max}$ itself. Then, we may restrict our attention on those disks in $\mathcal{D}_{\leq D_{\max}}$. The concrete meaning is stated in the following observation. For a disk D, denote by $X(D)$ the set of PoIs covered by D, and denote by $C_i(D) = X(D) \cap C_i$ the set of points in color class C_i that are covered by D.

Observation 1. *Given a MinPCF instance* $(X, \mathcal{D}, \{r_i\}_{i\in[T]})$, *suppose* $D_{\max}$ *is a largest disk in an optimal solution. Let* $X^{res}_{D_{\max}} = X \setminus X(D_{\max})$ *be the residual set of PoIs that are not covered by* $D_{\max}$ *and* $r^{res}_{i,D_{\max}} = \max\{0, r_i - |C_i(D_{\max})|\}$ *be the residual fairness requirement for color class* C_i *after* $D_{\max}$ *is picked. Then,* $opt\left(X, \mathcal{D}, \{r_i\}_{i\in[T]}\right) = opt\left(X^{res}_{D_{\max}}, \mathcal{D}_{\leq D_{\max}}, \{r^{res}_{i,D_{\max}}\}_{i\in[T]}\right) + w_{D_{\max}}$, *where the left opt and the right opt refer to the optimal power of the MinPCF instances* $(X, \mathcal{D}, \{r_i\}_{i\in[T]})$ *and* $\left(X^{res}_{D_{\max}}, \mathcal{D}_{\leq D_{\max}}, \{r^{res}_{i,D_{\max}}\}_{i\in[T]}\right)$, *respectively.*

The algorithm MinPCF^A, described in Algorithm 2, uses a binary search to guess the largest disk $D_{\max}$. In the binary search, *low*, *high*, and *mid* represent the left, right, and middle indices, respectively, and Ψ records those indices for which a feasible solution can be found. For each guessed disk D, the algorithm calls a sub-procedure Cov to compute an approximate solution for the residual instance $\left(X^{res}_D, \mathcal{D}_{\leq D}, \{r^{res}_{i,D}\}_{i\in[T]}\right)$. A boolean variable Γ is used to indicate whether Cov has found a feasible solution for the currently guessed disk D. If not, the algorithm guesses a larger disk. Otherwise, the guessed radius is reduced. The final output $\mathcal{F}^2$ is the best solution from all guesses that yield a feasible result. For simplicity, we do not distinguish between a power assignment and the set of disks determining this power assignment in the following discussion.

Algorithm 2. MinPCF$^A(X, \mathcal{S}, \{r_i\}_{i\in[T]})$

Input: A set of PoIs X, a set of base stations $\mathcal{S}$, fairness requirements $\{r_i\}_{i\in[T]}$.
Output: A power assignment $\mathcal{F}^2$ satisfying fairness constraints.

construct disk set $\mathcal{D}$ as in (2); determine weights of disks as in (3);
sort disks as $D_1, \dots, D_{|\mathcal{D}|}$ in increasing order of power;
$low \leftarrow 1, high \leftarrow |\mathcal{D}|, \Psi \leftarrow \emptyset$;
while $low < high$ **do**
    $mid = \lfloor low + high \rfloor /2$;
    $(\mathcal{M}_{D_{mid}}, \Gamma) \leftarrow \text{Cov}\left(X^{res}_{D_{mid}}, \mathcal{D}_{\le D_{mid}}, \{r^{res}_{i,D_{mid}}\}_{i\in[T]}, \rho\right)$;
    if $\Gamma =$ True **then**
        $\mathcal{F}_{D_{mid}} \leftarrow \mathcal{M}_{D_{mid}} \cup \{D_{mid}\}$;
        $\Psi \leftarrow \Psi \cup \{mid\}$;
        $high \leftarrow mid$;
    else
        $low \leftarrow mid$;
    end if
end while
$\tilde{D} \leftarrow \arg\min_{i\in\Psi} \{w(\mathcal{F}_{D_i})\}$;
$\mathcal{F}^2 \leftarrow$ the power assignment corresponding to $\mathcal{F}_{\tilde{D}}$;
return $\mathcal{F}^2$

Sub-procedure Cov, described in Algorithm 3, is used to solve the residual instance $\left(X^{res}_D, \mathcal{D}_{\le D}, \{r^{res}_{i,D}\}_{i\in[T]}\right)$. It first finds a feasible solution $\bar{\mathcal{D}}$ by a primal-dual approach (Step 2 to Step 10), based on the following dual program of (5):

$$\begin{aligned} \max \quad & \sum_{i\in[T]} r_i\gamma_i \\ \text{s.t.} \quad & \sum_{x\in D} \beta_x \le w_D, \forall D \in \mathcal{D}, \\ & \sum_{i:x\in C_i} \gamma_i \le \beta_x, \forall x \in X, \\ & \beta_x \ge 0, \qquad \forall x \in X, \\ & \gamma_i \ge 0, \qquad \forall i \in [T]. \end{aligned} \tag{6}$$

Note that in the second constraint, there is only one term in the left summation, because the color classes are disjoint.

Initially, all dual variables are set to be zeros and considered to be *active*. As long as a color class's fairness constraint is unsatisfied, all active dual variables are increased until a disk D meets its dual constraint $\sum_{x\in D} \beta_x = w_D$. This disk is added into $\bar{\mathcal{D}}$, and corresponding β_x variables are deactivated. When a color class C_i meets its coverage requirement, we deactivate γ_i and reset all $\beta_x = 0$ for $x \in C_i \setminus C_i(\bar{\mathcal{D}})$. If the algorithm exits the while loop with some color class

Algorithm 3. Cov($X', \mathcal{D}', \{r'_i\}_{i\in[T]}, \rho$)

Input: A set of PoIs X', a set of disks $\mathcal{D}'$, fairness requirements $\{r'_i\}_{i\in[T]}$ and a parameter $\rho > 0$.
Output: A power assignment $\mathcal{M}$.

1: $\bar{\mathcal{D}} \leftarrow \emptyset, \beta_x \leftarrow 0$ for each $x \in X$, $\gamma_i \leftarrow 0$ for each $i \in [T]$;
2: **while** there is an index $i \in [T]$ with $|C_i(\bar{\mathcal{D}})| < r'_i$ and $\mathcal{D} \setminus \bar{\mathcal{D}} \neq \emptyset$ **do**
3: simultaneously increase active variables until a disk D becomes tight;
4: $\bar{\mathcal{D}} \leftarrow \bar{\mathcal{D}} \cup \{D\}$; $X' \leftarrow X' \setminus D$;
5: deactivate those variables β_x with $x \in D$;
6: **if** $\exists i \in [T]$, s.t. $|C_i(\bar{\mathcal{D}})| \geq r'_i$ **then**
7: $\beta_x \leftarrow 0\ \forall x \in C_i \setminus C_i(\bar{\mathcal{D}})$;
8: deactivate γ_i and $\beta_x, \forall x \in C_i$;
9: **end if**
10: **end while**
11: **if** $\exists$ an index $i \in [T]$ with $|C_i(\bar{\mathcal{D}})| < r'_i$ and $\mathcal{D} \setminus \bar{\mathcal{D}} = \emptyset$ **then**
return ($\emptyset$, False);
12: **end if**
13: $D_i^{cri} \leftarrow$ critical disk for color class C_i ($\forall i \in [T]$);
14: $\mathcal{I} \leftarrow \bar{\mathcal{D}} \setminus \{\bigcup_{i\in[T]} D_i^{cri}\}$; $l \leftarrow |\mathcal{I}|$;
15: sort $\mathcal{I}$ as $D_1, D_2, \ldots, D_l$ such that $r_{D_1} \geq r_{D_2} \geq \cdots \geq r_{D_l}$;
16: **for** $l' = 1$ to l with $D_{l'} \in \mathcal{I}$ **do**
17: **for** every $l'' > l'$ with $X(D_{l''}) \cap X(D_{l'}) \neq \emptyset$ and $\|s_{D_{l''}} s_{D_{l'}}\| \leq \rho r_{D_{l''}}$ **do**
18: $\mathcal{I} \leftarrow \mathcal{I} - \{D_{l''}\}$;
19: **end for**
20: **end for**
21: $\mathcal{M} \leftarrow \{Disk(s_D, (1+\rho)r_D) : D \in \mathcal{I}\} \cup \{D_i^{cri} | i \in [T]\}$;
22: **return** ($\mathcal{M}$, True).

still has its fairness requirement unsatisfied, then the guessed disk is too small, and thus Γ is set to False (see line 11). Otherwise, the algorithm proceeds.

For each color class C_i, suppose D_i^{cri} is the disk added into $\bar{\mathcal{D}}$ that pushes the number of covered PoIs in C_i across threshold r'_i. Call D_i^{cri} the *critical disk* for color class C_i. Note that after guessing the largest disk $D_{\max}$ in an optimal solution, a critical disk has its weight upper bounded by *opt*.

The algorithm then sparsifies the non-critical disks in $\bar{\mathcal{D}}$. If two non-critical disks are too close, the smaller one is removed. As a result, no station has more than one disk remaining (line 17). After this step, if the solution becomes infeasible, the radii of the non-critical disks are expanded by a factor of $(1+\rho)$, and critical disks are added back into the final solution $\mathcal{M}$.

2.3 Theoretical Analysis

We first analyze the performance of MinPCF^A.

Lemma 2. *For an integer $t \in \{2, 3, \ldots\}$, let $\rho_t = 2\sin\frac{\pi}{t}$, and $\mathcal{I}_t$ be the set of disks computed at the end of line 20 of Algorithm 3 using ρ_t as the parameter.*

Then $p(\mathcal{I}_t) \le (t-1)\cdot p(\mathcal{D}_D^*)$, *where* $\mathcal{D}_D^*$ *is an optimal solution to the MinPCF reinstance* $\left(X_D^{res}, \mathcal{D}_{\le D}, \{r_{i,D}^{res}\}_{i\in[T]}\right)$, *when the algorithm guesses* D.

Proof sketch. Since only tight disks are included in $\bar{\mathcal{D}}$ and $\mathcal{I} \subseteq \bar{\mathcal{D}}$, for each disk $D \in \mathcal{I}$, we have $\sum_{x:x\in D}\beta_x = w_D$. By Lemma 1 in [14], any point x is covered by at most $t-1$ disks. Denote by $C_i^{(e)} = \{x \in C_i\colon \beta_x > 0\}$ the set of PoIs effectively covered by some disks selected into $\bar{\mathcal{D}}$. Thus, $p(\mathcal{I}) = \sum_{D\in\mathcal{I}} w_D = \sum_{D\in\mathcal{I}}\sum_{x:x\in D}\beta_x \le (t-1)\sum_{i\in[T]}\sum_{x\in C_i^{(e)}(\mathcal{I})}\beta_x$. Since $\beta_x = \gamma_i$ for any $x \in C_i^{(e)}\left(\bar{\mathcal{D}}\setminus\bigcup_{i\in[T]}\{D_i^{cri}\}\right)$ and $\left|C_i^{(e)}\left(\bar{\mathcal{D}}\setminus\bigcup_{i\in[T]}\{D_i^{cri}\}\right)\right| < r_i^{res}$, we have: $\sum_{x\in C_i^{(e)}(\mathcal{I})}\beta_x \le r_i^{res}\gamma_i$. Thus $p(\mathcal{I}) \le (t-1)\sum_{i\in[T]} r_i^{res}\gamma_i$. Then the result follows from primal-dual theory.

For a guessed disk D, denote by $\mathcal{M}_D$ the set of disks $\mathcal{M}$ produced by Algorithm 3. It is straightforward from Lemma 2 that the power of $\mathcal{M}_D$ could be estimated as in the next lemma.

Lemma 3. *For an integer* $t \in \{2,3,\ldots\}$, *let* $\rho_t = 2\sin\frac{\pi}{t}$, *we have* $p(\mathcal{M}_D) \le (t-1)(1+\rho_t)^\alpha p(\mathcal{D}_D^*) + \sum_{i\in[T]} p(D_i^{cri})$.

The feasibility of the algorithm is given in the following lemma.

Lemma 4. *If* $\Gamma_D =$ *True, then* $\mathcal{M}_D$ *is a feasible solution, where* Γ_D *denotes the indicator* Γ *when disk* D *is guessed.*

Now we are ready to prove the approximation ratio of Algorithm 2.

Theorem 1. *MinPCF*^{A} *is a* $\max\{e^{2\pi}\alpha, T+1\}$*-approximation algorithm for the MinPCF problem, which runs in time* $O\left(\left(\sum_{i\in[T]} r_i\right)mn\log mn\right)$, *where* α *is the stretch factor of power and* T *is the number of color classes.*

Proof. Suppose $OPT^{(1)}$ is an optimal solution to the MinPCF problem, $opt^{(1)}$ is its power, and $D_{\max}$ is the largest disk in $OPT^{(1)}$. By Lemma 3 and the fact that for any $i \in [T]$, $p(D_i^{cri,\max}) \le p(D_{\max})$, where $D_i^{cri,\max}$ is the critical disk for color class C_i when the guessed disk is $D_{\max}$, we have

$$\begin{aligned} p(\mathcal{F}_{D_{\max}}) &= p(\mathcal{M}_{D_{\max}}) + p(D_{\max}) \\ &\le (t-1)(1+\rho_t)^\alpha \cdot p(\mathcal{D}_{D_{\max}}^*) + \sum_{i\in[T]} p(D_i^{cri,\max}) + p(D_{\max}) \\ &\le (t-1)(1+\rho_t)^\alpha \cdot p(\mathcal{D}_{D_{\max}}^*) + (T+1)\cdot p(D_{\max}) \\ &\le \max\{(t-1)(1+\rho_t)^\alpha, T+1\}\cdot opt^{(1)}. \end{aligned} \tag{7}$$

Note that (7) holds for any $t = 2,3,\ldots$. By setting $t^* = \arg\min_{t\in\{2,3,\ldots\}}(t-1)(1+\rho_t)^\alpha$, we have $(t^*-1)(1+2\cdot\sin\frac{\pi}{t^*})^\alpha \approx e^{2\pi}\alpha$. The claimed approximation ratio then follows from (7).

The time complexity is the result of the following observations: the binary search is executed $O(\log mn)$ rounds; each round of the while loop newly covers at least one PoI and takes time $O(mn)$; the while loop is executed at most $\sum_{i\in[T]} r_i$ times.

The performance of Algorithm 1 is given as follows.

Theorem 2. *MinPPCF^A is a* $\max\{2e^{2\pi}\alpha, e^{2\pi}\alpha + T + 1\}$*-approximation algorithm for the MinPPCF problem, which runs in time* $O((\sum_{i\in[T]} r_i + k)mn \log mn)$.

Proof. Let $opt^{(1)}$ and $opt^{(2)}$ be the optimal powers for the MinPPC and the MinPCF instances, respectively. Let $apx^{(1)}$ and $apx^{(2)}$ be the approximate powers computed by MinPPC^A and MinPCF^A, respectively. For the MinPPCF instance, denote by opt and apx the optimal and the approximate powers computed by MinPPCF^A. Since $opt \geq \max\{opt^{(1)}, opt^{(2)}\}$, and $p(\mathcal{F}) \leq p(\mathcal{F}^{(1)}) + p(\mathcal{F}^{(2)})$, we can estimate:

$$\begin{aligned} apx &\leq apx^{(1)} + apx^{(2)} \\ &\leq \max\{e^{2\pi}\alpha, T + 1\} \cdot opt^{(1)} + e^{2\pi}\alpha \cdot opt^{(2)} \\ &\leq \max\{e^{2\pi}\alpha, T + 1\} \cdot opt + e^{2\pi}\alpha \cdot opt \\ &= \max\{2e^{2\pi}\alpha, e^{2\pi}\alpha + T + 1\} \cdot opt. \end{aligned}$$

The time complexity can be obtained by a similar analysis as in Theorem 1.

3 Conclusion

This paper initiated the study on the minimum power cover problem with fairness constraints (MinPCF), and designed a $\max\{e^{2\pi}\alpha, T + 1\}$-approximation algorithm. We also proposed the minimum power partial cover problem with fairness constraints, and designed a $\max\{2e^{2\pi}\alpha, e^{2\pi}\alpha + T + 1\}$-approximation algorithm. It is the first time to consider different categories of PoIs and fairness constraints in the minimum power cover problem. In addition, we have successfully reduced the original quadratic time complexity to a more efficient linear-logarithmic complexity. This improvement significantly improves the performance of the algorithm, making it more practical for large-scale problems.

References

1. Pinar, Y., Zuhair, A., Hamad, A., Resit, A., Shiva, K., Omar, A.: Wireless Sensor Networks (WSNs). In: 2016 IEEE Long Island Systems, Applications and Technology Conference, pp. 1–8. IEEE, New York, NY, USA (2016)
2. Jaiswal, S.K., Dwivedi, A.K.: A security and application of wireless sensor network: a comprehensive study. In: 2023 International Conference on IoT, Communication and Automation Technology, pp. 1–5. IEEE, Gorakhpur, India (2023)
3. Rawat, P., Singh, K.D., Chaouchi, H., Bonnin, J.M.: Wireless sensor networks: a survey on recent developments and potential synergies. J. Supercomput. **68**, 1–48 (2014)
4. Alt, H., et al.: Minimum-cost coverage of point sets by disks. In: Proceedings of 22nd ACM Symposium on Computational Geometry, pp. 449–458. ACM, New York, NY, USA (2006)

5. Liu, P., Huang, X.: Approximation algorithm for partial set multicover versus full set multicover. Discret. Math. Algorithms Appl. **10**(2), 1850026 (2018)
6. Zafar, M.B., Valera, I., Gomez-Rodriguez, M., Gummadi, K.P.: Fairness constraints: a flexible approach for fair classification. J. Mach. Learn. Res. **20**(75), 1–42 (2019)
7. Tang, S., Yuan, J., Mensah-Boateng, T.: Achieving long-term fairness in submodular maximization through randomization. In: Brieden, A., Pickl, S., Siegle, M. (eds) Graphs and Combinatorial Optimization: From Theory to Applications, AIRO Springer Series, vol. 13, pp. 161–173. Springer, Cham (2024). https://doi.org/10.1007/978-3-031-46826-1_13
8. Yuan, J., Tang, S.: Group fairness in non-monotone submodular maximization. J. Comb. Optim. **45**(88) (2023)
9. Asudeh, A., Jagadish, H.V., Stoyanovich, J., Das, G.: Designing fair ranking schemes. In: Proceedings of the 2019 International Conference on Management of Data, pp. 1259–1276. ACM, Amsterdam, Netherlands (2019)
10. Beutel, A., et al.: Fairness in recommendation ranking through pairwise comparisons. In: Proceedings of the 25th ACM SIGKDD International Conference on Knowledge Discovery & Data Mining, pp. 2212–2220. ACM, AK, USA (2019)
11. Freund, A., Rawitz, D.: Combinatorial interpretations of dual fitting and primal fitting. In: Solis-Oba, R., Jansen, K. (eds.) Approximation and Online Algorithms. LNCS, vol. 2909, pp. 137–150. Springer, Heidelberg (2004). https://doi.org/10.1007/978-3-540-24592-6_11
12. Li, M., Ran, Y., Zhang, Z.: A primal-dual algorithm for the minimum power partial cover problem. J. Comb. Optim. **44**, 1913–1923 (2022)
13. Li, M., Ran, Y., Zhang, Z.: Approximation algorithms for the minimum power partial cover problem. In: Du, D.Z., Li, L., Sun, X., Zhang, J. (eds.) Algorithmic Aspects in Information and Management. LNCS, vol. 11640, pp. 1–13. Springer, Cham (2019). https://doi.org/10.1007/978-3-030-27195-4_17
14. Dai, H., Deng, B., Li, W., Liu, X.: A note on the minimum power partial cover problem on the plane. J. Comb. Optim. **44**, 970–978 (2022)
15. Ran, Y., Huang, X., Zhang, Z., Du, D.-Z.: Approximation algorithm for minimum power partial multi-coverage in wireless sensor networks. J. Glob. Optim. **80**(3), 661–677 (2021). https://doi.org/10.1007/s10898-021-01033-y
16. Liu, X., Li, W., Xie, R.: A primal-dual approximation algorithm for the k-prize-collecting minimum power cover problem. Optim. Lett. **16**, 2373–2385 (2022)
17. Dai, H., Li, W., Liu, X.: An approximation algorithm for the H-prize-collecting power cover problem. In: Li, M., Sun, X. (eds.) Frontiers of Algorithmic Wisdom. LNCS, vol. 13461, pp. 1–13. Springer, Cham (2022). https://doi.org/10.1007/978-3-031-20796-9_7
18. Liu, X., Li, W., Dai, H.: Approximation algorithms for the minimum power cover problem with submodular/linear penalties. Theor. Comput. Sci. **923**, 256–270 (2022)
19. Bera, S., Gupta, S., Kumar, A., Roy, S.: Approximation algorithms for the partition vertex cover problem. Theor. Comput. Sci. **555**, 2–8 (2014)
20. Zhou, M., Zhang, Z., Du, D.: Approximation algorithm for prize-collecting vertex cover with fairness constraints. J. Comb. Optim. (to appear)

Differentially Private Counting Queries on Approximate Shortest Paths

Jesse Campbell[1] and Chunjiang Zhu[2](✉)

[1] Duke Kunshan University, Kunshan, Jiangsu, China
jesse.campbell@duke.edu
[2] University of North Carolina at Greensboro, Greensboro, NC, USA
chunjiang.zhu@uncg.edu

Abstract. Given a weighted graph $G = (V, E, \mathbf{w})$ with some private edge attribute function $f : E \to \mathbb{R}^+$, the differentially private counting query problem aims to release the sum of the private edge attributes along a desired path (or set of paths) in G with differential privacy. In this work, we introduce the differentially private counting query problem on approximate shortest paths. A nearly-optimal algorithm with $\widetilde{\Theta}(n^{1/4})$ error is known for releasing the counting queries along all true shortest paths in G [6,10]. We extend this result by showing if G admits a *t-collective tree spanner* with η_t spanning trees, then for each pair of vertices in V, we can release the counting query over a t-approximate shortest path with $\widetilde{O}(\eta_t/\epsilon)$ additive error in the ε-DP case, and $\widetilde{O}(\sqrt{\eta_t}/\epsilon)$ in the (ε, δ)-DP case. As an example of this, we use the collective tree $O(k \log\log n)$-spanner of [1] to give an algorithm which, for each pair of vertices, releases counting queries over a $O(k \log\log n)$-approximate shortest path in G with $\widetilde{O}(\sqrt{kn^{1/k}}/\epsilon)$ error. Our result is based on the polylogarithmic error in the differential private release of attributes along paths in trees, which is much lower than the $\widetilde{\Omega}(n^{1/4})$ error in the general case.

1 Introduction

Differential privacy (DP) is a mathematical framework which quantifies the leakage of sensitive user information by an algorithm. The definition of differential privacy requires that the probability of a certain outcome being attained from a dataset without the contribution of any single individual is very close to the probability the same outcome is attained from a dataset with the individual's contribution restored. Differential privacy has seen applications in industry and government agencies, with algorithms satisfying differential privacy being adopted by organizations such as Google [3,5], Microsoft [11], and the US Census Bureau [22,23].

The *counting query* problem takes on input a weighted, connected graph $G = (V, E, \mathbf{w})$ and some edge attribute function $f : E \to \mathbb{R}^+$. The goal is to release the sum of the attributes along a set of paths in G. Naturally, the *differentially private* counting query on a set of paths aims to release the counting

D. Du et al. (Eds.): COCOA 2024, LNCS 15434, pp. 200–212, 2025.
https://doi.org/10.1007/978-981-96-4445-2_17

queries over the set of paths with differential privacy, where the edge attributes are kept private (and are not dependent on the public edge weights). Previous work on this problem has shown it to be $\widetilde{\Theta}(n^{1/4})$-accurate for the set of (unique) shortest paths on general graphs (where $n = |V|$ and $\widetilde{\Theta}(\cdot)$ hides a polylogarithmic factor), for sufficiently small $\varepsilon, \delta > 0$ [6,10].

Calculating private estimates of counting queries over shortest path systems is motivated by many real-world applications. For instance, consider a graph where edges represent transactions between financial entities, and the edge attributes model the private value of the transactions. Fraud researchers could plausibly query this graph along certain paths to check for abnormal activity. When considering the *shortest path* between two entities, however, the additive error lower bound of $\Omega(n^{1/4})$ may be too high. As another example, consider a graph which models a network of healthcare facilities for the application of patient transfers, with the edge weights corresponding to public travel times between facilities. Suppose the private edge attributes correspond to the number of patients *en route* between facilities. The lower bound of $\widetilde{\Omega}(n^{1/4})$ additive error in releasing the private edge attributes may be too high. It would be useful to have a way of determining a more accurate estimate of patients in-transfer between facilities to get their expected capacities towards a successful patient transfer, at the cost of taking a slightly longer path between facilities. Therefore, it is well-motivated to improve the additive error for applications where the importance of being on a shortest path is small compared to the importance of having an accurate estimate of the private edge attributes. We believe this relaxed problem shares many common practical applications as the original problem [10], e.g., in financial transaction networks, transportation networks, and supply chain networks.

In this setting, we provide a new privacy framework which gives the user a way of releasing counting queries over *approximate* shortest paths while greatly improving the additive error bounds. In specific, we provide an explicit way of releasing counting queries over $O(k \log\log n)$-approximate shortest paths with $\widetilde{O}(kn^{1/k}/\epsilon)$ additive error in the ε-DP case, and $\widetilde{O}(\sqrt{kn^{1/k}}/\epsilon)$ additive error in the (ε, δ)-DP case.

Our results will make use of the notion of *collective tree spanners*. A graph spanner is a sparse subgraph of a graph which preserves approximate pairwise distances. In specific, if H is an α-spanner of G, then the distance between any two vertices in H is at most α times their distance in G, where α is referred to as the *stretch*. In 2004, the notion of *collective tree spanners* was introduced as an extension of *tree t-spanners*, which consist of a single tree and are the sparsest type of graph spanner [2]. A collective tree spanner is a collection of spanning trees of G such that for every pair of vertices, there is a single tree which approximates their distance [14]. We note that the idea of collective tree spanners are highly related to the notion of *tree covers*, as introduced by [24]. A tree cover is a collection of tree subgraphs whose union spans the entire vertex set. Unlike a collective tree spanner, the individual trees in a tree cover are not required to span the entire vertex set.

1.1 Our Results

In this work, we introduce the notion of the *differentially private counting query on approximate shortest paths* (DPCQ on ASP) problem. In particular, we show that we are able to greatly improve upon the lower bound of $\widetilde{\Omega}(n^{1/4})$ for the additive error in general graphs by allowing a counting query to be on a $O(k \log \log n)$ *approximate* shortest path. Due to the key observation that we can release differentially private counting query estimates on tree graphs with $O(\text{polylog}(n))$ error via a recursive algorithm [10,29], we can release a private estimate of the counting queries over all trees in a collective tree spanner with additive error proportional to the number of trees. Consequently, we can release an estimate over an approximate shortest path for every pair of vertices in V with low error.

Theorem 1. *Let* $\mathbf{T}$ *be a* t*-collective tree spanner of* G *such that* $|\mathbf{T}| = \eta_t$. *For* $\gamma \in (0, 0.5]$ *and* $\varepsilon \in (0, 1]$, *there is an* ε*-DP algorithm for releasing the counting query between* $u, v \in V$ *on a* t*-approximate shortest path in* G *that is* $O(\eta_t \cdot \log^{2.5}(n) \cdot \log(1/\gamma)/\varepsilon)$*-accurate with probability* $1 - \gamma$.

Theorem 2. *Let* $\mathbf{T}$ *be a* t*-collective tree spanner of* G *such that* $|\mathbf{T}| = \eta_t$. *For* $\gamma \in (0, 0.5]$ *and* $\varepsilon, \delta \in (0, 1]$, *there is an* (ε, δ)*-DP algorithm for releasing the counting query between* $u, v \in V$ *on a* t*-approximate shortest path that is* $O(\sqrt{\eta_t} \cdot \log^2(n) \cdot \log(1/\delta) \cdot \sqrt{\log(1/\gamma)}/\varepsilon)$*-accurate with probability* $1 - \gamma$.

Based on Ramsey spanning trees which are a collective tree $O(k \log \log n)$-spanner with $kn^{1/k}$ trees [1], we immediately arrive at the following results.

Corollary 1. *For* $\gamma \in (0, 0.5]$ *and* $\varepsilon \in (0, 1]$, *there exists an* ε*-DP algorithm for releasing the counting query between* $u, v \in V$ *on a* $O(k \log \log n)$*-approximate shortest path in* G *that is* $O(kn^{1/k} \cdot \log^{2.5}(n) \cdot \log(1/\gamma)/\varepsilon)$*-accurate with probability* $1 - \gamma$.

Corollary 2. *For* $\gamma \in (0, 0.5]$ *and* $\varepsilon, \delta \in (0, 1]$, *there exists an* (ε, δ)*-DP algorithm for releasing the counting query between* $u, v \in V$ *on a* $O(k \log \log n)$*-approximate shortest path that is* $O(\sqrt{kn^{1/k}} \cdot \log^2(n) \cdot \log(1/\delta) \cdot \sqrt{\log(1/\gamma)}/\varepsilon)$*-accurate with probability* $1 - \gamma$.

1.2 Related Work

Differentially Private Range Query. In 2023, Deng et al. [10] were the first to study the application of differential privacy to the range query problem on shortest paths. We follow their DP framework where two graphs are neighboring if they share the graph topology and edge weights, and the vectors of edge attributes differ by a most 1 in the l_1-distance. In particular, [10] showed an algorithm for DP counting queries on shortest paths that is $O(n^{1/3} \cdot \log^{5/6}(n)/\varepsilon)$-accurate in the ε-DP case and $O(n^{1/4} \cdot \log^{2/3}(n) \cdot \log^{1/4}(1/\delta)/\varepsilon)$-accurate in the (ε, δ)-DP case.

The application of differential privacy to graph problems was first studied by Hay et al. [25] in 2009, by giving a mechanism to release a private estimate of the degree distribution in a graph. In their work, they introduce the notions of node- and edge-differential privacy, which defines neighboring graphs based on their underlying graph topology. Algorithms with node- and edge-differential privacy have been studied considerably in recent literature [26–28]. Adam Sealfon [29] was the first to study the application of differential privacy to the all-pairs shortest distances (APSD) problem [9,20,21], defining neighboring graphs based on the l_1 distance of their private edge weight vectors.

Furthermore, Chen et al. [9] proved a lower bound for the additive error of $\Omega(n^{1/6})$ in any algorithm for DP all-pairs distances release on general graphs, for a sufficiently small $\varepsilon, \delta > 0$, by exploiting a reduction from the *linear queries* problem to APSD, and a hereditary discrepancy lower bounded provided by Chazelle and Lvov [8]. It was noted by Deng et al. [10] that this lower-bound technique also applies to the DP range query on shortest paths problem. Soon after, Bodwin et al. [6] improved this lower bound to $\Omega(n^{1/4}/\sqrt{\log n})$ as a result of a better discrepancy lower bound, proving, together with the upper bound by [10], that the DP range query on shortest paths problem is $\widetilde{\Theta}(n^{1/4})$-accurate for general graphs.

Collective Tree Spanners. The notion of collective tree spanners was first introduced in 2004 by Dragan et al. [14], who gave constructions of *additive* collective tree spanners for chordal graphs, co-comparability graphs, and more. Since then, additive collective tree spanners have been developed in the literature, with many results on specific families of graphs such as graphs with bounded parameters [13], bounded tree-breadth graphs [12], and more [15,16]. *Multiplicative collective tree spanners* were first (informally) studied in 2001 by Gupta et al. [24] as an application to a graph routing problem, showing that all *planar* graphs have a collective multiplicative tree 1-spanner with $O(\sqrt{n})$ trees, and a 3-spanner with $O(\log(n))$ trees. Recently, the topic of multiplicative collective tree spanners has seen an increase in the literature due to the notion of *tree padding spanners*, which is an instance of the collective tree spanner problem such that each vertex has a single tree for which all the α-approximate paths are contained. Abraham et al. [1] gave an algorithm which constructs a tree padding $(k \log\log n)$-spanner which is a collection of $kn^{1/k}$ spanning trees. Similarly, Bartal et al. [4] gave constructions of tree covers for metric spaces with a small amount of trees for general, doubling, and planar metrics in the normal and Ramsey (analogous to the tree padding spanner case for graphs) cases.

2 Preliminaries

First, we present the definition of *collective tree spanners*. For $x, y \in V$, we let $d_G(x, y)$ to be the shortest path distance between x and y in G.

Definition 1. *A collection of spanning trees* $\mathbf{T}$ *of* G *is said to be an* α*-collective tree spanner of* G *if for every* $u, v \in V$*, there is a tree* $T \in \mathbf{T}$ *such that* $d_T(u,v) \leq \alpha \cdot d_G(u,v)$.

Next, we define the notion of what it means for a dataset to differ in the contribution of one individual. In our case, as in [10], isomorphic, undirected, weighted graphs are neighbors if their edge attribute vectors differ by at most 1 in the l_1 norm.

Definition 2. *Two isomorphic graphs* $G_1, G_2 = (V, E, \mathbf{w})$ *with edge attribute functions* $f_1, f_2 : E \rightarrow \mathbb{R}^+$ *are said to be neighboring if*

$$\sum_{e \in E} |f_1(e) - f_2(e)| \leq 1$$

Moreover, we define the notion of the sensitivity of a function as the most the function can differ over neighboring datasets.

Definition 3. *The* l_1 *sensitivity of* $\mathcal{F} : \mathcal{X} \rightarrow \mathbb{R}^D$ *is defined as*

$$\Delta_1(f) :- \max_{X,X'} \|\mathcal{F}(X) - \mathcal{F}(X')\|_1$$

where X, X' *are neighboring datasets.*

We are now ready to present the formal definition of *differential privacy.*

Definition 4. *An algorithm* $\mathcal{M} : \mathcal{X} \rightarrow \mathbb{R}^D$ *is said to be* (ε, δ)*-differentially private if, for all outcomes* $S \subseteq \mathbb{R}^D$ *and neighboring datasets* X, X',

$$\mathbb{P}[\mathcal{M}(X) \in S] \leq e^{\varepsilon} \cdot \mathbb{P}[\mathcal{M}(X') \in S] + \delta$$

Furthermore, we present the Laplace mechanism, which is one of the primary tools in the theory of differential privacy. In specific, the Laplace mechanism describes how much one needs to perturb a released statistic in order to guarantee it is ε-DP.

Lemma 1 (Laplace Mechanism, [19]). *Given any function* $f : \mathcal{X} \rightarrow \mathbb{R}^k$*, the Laplace mechanism on input* $X \in \mathcal{X}$ *independently samples* $Y_1, ..., Y_k$ *according to Lap*$(\Delta_1(f)/\varepsilon)$ *and outputs,*

$$\mathcal{M}_{f,\varepsilon}(X) = f(X) + (Y_1, ..., Y_k)$$

The Laplace mechanism is ε*-differentially private.*

We also present a useful result about the concentration of independent Laplace random variables.

Lemma 2 ([7]). *Let $X_1, ..., X_t$ be independent random variables distributed according to Lap(b), and let $X = X_1 + ... + X_t$. Then for all $\gamma \in (0,1)$, with probability at least $1-\gamma$ we have,*

$$|X| < O(b\sqrt{t}\log(1/\gamma))$$

Next, we present here two key results about the composition of differentially private systems.

Lemma 3 (Basic Composition, [17,19]). *Let $\varepsilon, \delta \in [0,1]$ and $k \in \mathbb{N}$. If we run k mechanisms where each mechanism is $(\varepsilon/k, \delta/k)$-differentially private, then the entire algorithm is (ε, δ)-differentially private.*

Lemma 4 (Advanced Composition, [18,19]). *Let $\varepsilon, \delta \in (0,1]$ and $k \in \mathbb{N}$. If we run k mechanisms where each mechanism is $\left(\frac{\varepsilon}{2\sqrt{2k\log(2/\delta)}}, \frac{\delta}{2k}\right)$-DP, then the entire algorithm is (ε, δ)-DP.*

3 DP Counting Queries on Approximate Shortest Paths

In this section, we show how we are able to release all-pairs counting queries much more accurately by allowing them to be on t-approximate shortest paths. We achieve a much lower error than the DP range query problem on true shortest paths by leveraging a result which gives an ε-DP recursive algorithm for releasing the counting queries over all-pairs shortest paths in trees with $O(\text{polylog}(n))$ additive error. Previous similar results were on DP releasing all-pairs shortest distances in trees by Sealfon [29], which were adapted to the (ε, δ)-DP and range query setting by Deng et al. [10]. In this paper, we explicitly provide the proof for releasing DP counting queries on all pairs of vertices in a tree under the ε-DP setting. We then apply the result to the studied problem of DP counting queries over approximate shortest paths carefully in order to save a $\log n$ term. Throughout this section, for any two vertices $x, y \in V$, we let $P_G(x,y)$ to be the shortest path between x and y and $\omega_G(x,y) = \sum_{e\in P_G(x,y)} f(e)$ to be the accumulation of edge attributes along the shortest path between x and y, or the *counting query*.

We will first prove the result for releasing counting queries from the root of a rooted tree to all other vertices, and then extend this result to counting queries for all pairs of vertices in a tree.

Lemma 5. *Let $\mathcal{T} = (V, E, \mathbf{w})$ be a tree rooted at z with $\varepsilon \in (0,1]$ and $\gamma \in (0, 0.5]$. Then there is an ε-DP algorithm for releasing counting queries from the root to all other vertices on $\mathcal{T}$ that is $O(\log^{1.5}(n) \cdot \log(n/\gamma)/\varepsilon)$-accurate with probability $1-\gamma$.*

Proof. The algorithm works by splitting the tree into subtrees, each with fewer than $n/2$ vertices, computing counting queries between the root and the roots of the subtrees, and then applying the algorithm recursively on each subtree.

It is a well-known result that every tree has either one or two vertices known as *centroids* such that their removal disconnects the tree into subtrees, each with fewer than $n/2$ vertices (see, for example, [31]). If $\mathcal{T}$ has a single centroid, we let z^* be that vertex. If $\mathcal{T}$ has two centroids, we let z^* be the centroid that is closer to the root node z. By choosing it in this way, we guarantee that z^* is the unique vertex such that the subtree rooted at z^* has more than $n/2$ vertices, but the subtree rooted at each of z^*'s children has at most $n/2$ vertices. Let $z_1, ..., z_t$ be the children of z^* and $\mathcal{T}_i = (V_i, E_i)$, $i \in \{1, ..., t\}$, their corresponding subtrees. Additionally, we define $\mathcal{T}_0$ to be the subtree rooted at z with vertex set $V_0 = V - \{V_1, ..., V_t\}$.

Next, we release the counting queries between z and z^*, as well as between z^* and each of it's children by adding Laplace noise from the distribution $\text{Lap}(\log(n)/\varepsilon)$. In particular, we note that since each tree is disjoint by construction, the function which releases these counting queries has sensitivity 1. Hence, by Lemma 1, releasing each subtree is $(\varepsilon/\log(n))$-DP. Then, we recursively call this algorithm for each subtree until each subtree consists only of a single vertex.

Since each subtree has at most $n/2$ vertices, the maximum recursion depth is bounded by $\log(n)$. Hence, by basic composition (Lemma 3), since the released queries in our algorithm are a composition of $\log(n)$, $(\varepsilon/\log(n))$-DP mechanisms, the total algorithm is ε-DP.

It remains to be shown that the error in each released estimate is bounded above by $O(\text{polylog}(n))$. We note that for every $u \in V$, the estimate of $\omega(z, u)$ is the composition of released queries from root nodes to centroids, and from centroids to their children. By bounding the number of released estimates in the composition, we can bound the total error incurred from each noisy random variable.

We will prove by induction on the number of vertices that the number of released estimates used to calculate an estimate of $\omega(z, u)$ is bounded above by $2\log(n)$. Firstly, the base case of $n = 1$ is vacuously true. Suppose $n > 1$. In the first iteration of the algorithm, the vertex sets $V_0, V_1, ..., V_t$ are a partition of V. Hence, $u \in V_i$ for some $i \in [t]$. By the induction hypothesis, we can release the distance from z_i to u by the composition of at most $2\log(n/2) = 2\log(n) - 2$ noisy distances. Then, we have that $\omega(z, z_i) = \omega(z, z^*) + \omega(z^*, z_i)$, giving a total of $2\log(n)$ noisy estimates.

It follows that the error in each released estimate is the concatenation of $2\log(n)$ random variables distributed according to $\text{Lap}(\log(n)/\varepsilon)$. By Lemma 2, with probability at most γ, we have that the error exceeds $O(\log^{1.5}(n) \cdot \log(1/\gamma)/\varepsilon)$. Hence, by a union bound, with probability at least $1 - \gamma$, the error for the estimate from z to all vertices $u \in V$ is bounded above by $O(\log^{1.5}(n) \cdot \log(n/\gamma)/\varepsilon)$, as desired. □

Next, we will prove another result by extending Lemma 5 to the all-pairs case. We intentionally keep the term $O(\log^{1.5}(n) \cdot \log(n/\gamma)/\varepsilon)$ instead of $O(\log^{2.5}(n) \cdot \log(1/\gamma)/\varepsilon)$, in order to facilitate a more careful analysis in Theorem 1 to save a $\log n$ term through Proposition 1.

Proposition 1 states a few basic results about asymptotics.

Proposition 1. *Let* $\alpha \geq 2$, $\gamma \in (0, 0.5]$, *and* $\lambda > 0$, *then,*

1. $\log(\alpha^{1+\lambda}/\gamma) = O(\log(\alpha/\gamma))$, *and,*
2. $\log(\alpha/\gamma) = O(\log(\alpha)\log(1/\gamma))$.

Lemma 6. *Let* $\mathcal{T} = (V, E, \mathbf{w})$ *be a tree with* $\varepsilon \in (0, 1]$ *and* $\gamma \in (0, 0.5]$. *Then there is an* ε*-DP algorithm for releasing counting queries on all pairs of vertices in* $\mathcal{T}$ *that is* $O(\log^{1.5}(n) \cdot \log(n/\gamma)/\varepsilon)$*-accurate with probability* $1 - \gamma$.

Proof. Choose a vertex $z \in V$ arbitrarily as a root vertex of $\mathcal{T}$, and call the rooted tree $\mathcal{T}_z$. Then, using Lemma 5, we can privately release the counting query from z to every other vertex in $\mathcal{T}_z$. We will show that these estimate suffice to compute the estimate for any pair $x, y \in V$ of vertices in V.

Let $v_{x,y} \in V$ be the lowest common ancestor of x and y in the rooted tree $\mathcal{T}_z$. Then, $\omega_{\mathcal{T}}(x, y) = \omega_{\mathcal{T}}(x, v_{x,y}) + \omega_{\mathcal{T}}(v_{x,y}, y) = \omega_{\mathcal{T}_z}(z, x) + \omega_{\mathcal{T}_z}(z, y) - 2\omega_{\mathcal{T}_z}(z, v_{x,y})$. By the result in Lemma 5, we condition on the event that for all $w \in V$, $\omega(z, w)$ is $O(\log^{1.5}(n) \cdot \log(n/\gamma)/\varepsilon)$-accurate with probability $1 - \gamma$. Since our query estimate is the sum of four estimates from z, the total error is at most four times this, which is still $O(\log^{1.5}(n) \cdot \log(n/\gamma)/\varepsilon)$. By a union bound, with probability $1 - \gamma$ each counting query among all n^2 pairs counting queries is at most $O(\log^{1.5}(n) \cdot \log(n^3/\gamma)/\varepsilon) = O(\log^{1.5}(n) \cdot \log(n/\gamma)/\varepsilon)$, by Proposition 1. □

Applying Lemma 6 directly to a collective tree spanner of a graph gives the desired result.

Proof (of **Theorem** 1*).* Since $\mathbf{T}$ is constructed only based on the public edge weights and topology of G, each tree $\mathcal{T} \in \mathbf{T}$ is also public. For each tree $\mathcal{T}$ we run the $\varepsilon/|\mathbf{T}|$-DP algorithm from Lemma 6 to release the counting query over $\mathcal{T}$. By basic composition (Lemma 3), releasing the counting queries in all the trees is ε-DP.

By the definition of a collective tree spanner, for every pair of vertices $u, v \in V$, there is a tree $\mathcal{T} \in \mathbf{T}$ such that $P_{\mathcal{T}}(u, v)$ is a t-approximate shortest path between u and v. We let $\widetilde{\omega}(u, v)$ to be the private counting query estimate over the path $P_{\mathcal{T}}(u, v)$. The result of Lemma 6 ensures that, with probability $1-\gamma/n^2$, the additive error in the released counting query is,

$$\begin{aligned}\max_{u,v \in V} |\omega_{\mathcal{T}}(u, v) - \widetilde{\omega}(u, v)| &\leq O(|\mathbf{T}| \cdot \log^{1.5}(n) \cdot \log(n^3/\gamma)/\varepsilon) \\ &= O(\eta_t \cdot \log^{1.5}(n) \cdot \log(n/\gamma)/\varepsilon) \qquad (1) \\ &= O(\eta_t \cdot \log^{2.5}(n) \cdot \log(1/\gamma)/\varepsilon)\end{aligned}$$

where the two equalities hold because of Proposition 1. By a union bound, the upper bound (1) holds for all n^2 pairs of vertices with probability at least $1 - \gamma$. □

For DPCQ on ASP, the (ε, δ) case is identical to the ε-DP case except for two differences. Firstly, we use *advanced composition* to release the private counting query estimates for each tree $\mathcal{T} \in \mathbf{T}$ instead of basic composition (Lemma 3).

The second difference is that, instead of applying Lemma 6 to compute the distance estimates in each tree $\mathcal{T}$, we apply the following lemma by Deng et al. which adds Gaussian noise to the edge weights instead of Laplacian noise to reduce the final error by a factor of $\log(n)$ in the (ε, δ)-DP case. We state the result here but refer the reader to [10] for a detailed proof.

Lemma 7. *Let $\mathcal{T} = (V, E, \mathbf{w})$ be a tree with $\varepsilon, \delta \in (0, 1]$ and $\gamma \in (0, 0.5]$. Then there is an (ε, δ)-DP algorithm for releasing counting queries on all pairs of vertices in $\mathcal{T}$ that is $O(\log(n) \cdot \sqrt{\log(1/\delta)} \cdot \sqrt{\log(n/\gamma)}/\varepsilon)$-accurate with probability $1 - \gamma$.*

The proof of Theorem 2 is similar to the proof of Theorem 1 and is therefore deferred to Appendix A.

To give an explicit instance of Theorem 1 and 2, we note that Abraham et al. [1] gave a construction based on Ramsey spanning trees for a collective tree $O(k \log\log n)$-spanner with $kn^{1/k}$ trees. By the result of Theorem 2, choosing $k \geq 3$ immediately gives improvement on the additive approximate over the $\widetilde{O}(n^{1/4})$ algorithm on true shortest paths from [10].

Runtime Analysis. Not only does releasing differentially private counting queries over approximate shortest paths allow us to greatly improve the additive error in the released estimates, but also potentially the runtime of the algorithm to release the estimates. In particular, the $\widetilde{\Theta}(n^{1/4})$ algorithm from [10] to release counting queries over exact shortest paths requires first sampling a *hitting set* of vertices of size $\widetilde{O}(\sqrt{n})$, and then computing the shortest path trees spanning from each vertex in the hitting set. Using Dijkstra's algorithm, this time complexity of this process is $\widetilde{O}(n^{2.5})$.

By allowing the counting queries to be along approximate shortest paths, we bypass the need to calculate exact shortest paths from a hitting set of vertices. In specific, since the Laplace noise can be sampled and added to the edges of a tree in $O(1)$ time and the centroid can be found in linear time, the recursive process described in the proof of Lemma 5 takes $O(n \log n)$ time. It follows that the total algorithm to release the counting queries on approximate paths can run in $O(S(n) + \eta_t \cdot n \log n)$ time, where $S(n)$ is the amount of time needed to construct the collective tree spanner on the graph G.

4 Conclusion and Future Work

In this work, we introduce the problem of differentially private counting queries on approximate shortest paths in graphs. In specific, we show that the additive error lower bound of $\widetilde{\Omega}(n^{1/4})$ for counting queries on true shortest paths can be greatly improved to $\widetilde{O}(\sqrt{kn^{1/k}}/\epsilon)$ at the cost of querying along $\widetilde{O}(k)$-approximate shortest paths. It is still an open question what the optimal trade-off between the quality of shortest paths and the additive error is. We discuss the future work from the perspective of collective tree spanners.

Collective Tree Spanners

The method we provide in this work is based on the existence of *collective tree spanners* for general graphs. Tree graphs have many convenient properties such as being well-suited to recursive algorithms and having unique path systems, emphasizing their convenience in algorithm design. Despite this, until recently there have been few results about multiplicative collective tree spanners for general graphs. Even the collective tree $O(k \log\log n)$-spanner mentioned in this work is an instance of a *tree padding spanner*, making it optimal for routing problems, but sacrificing an optimal tradeoff between stretch and size for a property that is not necessary for applications like ours.

In specific, we note that one clear gap in the current literature is the construction of a constant stretch collective tree spanner with optimal size and stretch tradeoff, namely a collective tree $(2k-1)$-spanner with $O(n^{1/k})$ trees. In what follows, we give an outline of the construction of such a spanner.

In particular, consider the celebrated distance oracle of Thorup and Zwick [30]. Their distance oracle samples nested subsets of vertices of decreasing size $V = A_0 \supseteq A_1 \supseteq ... \supseteq A_{k-1} \supseteq A_k = \emptyset$. For each vertex $w \in A_i$, they define a set of vertices called a *cluster* by $C(w) :- \{v \in V \mid d(w,v) < d(v, A_{i+1})\}$. Then, they compute the shortest path tree rooted at w spanning each cluster. The shortest path distances within each cluster are then used to output the distance approximations, inducing a *tree cover* of the graph. In expectation, the number of edges in the union of spanning trees of clusters centered at vertices in A_i is $O(n^{1+1/k})$, meaning that the clusters at each level could plausibly be *packed* into $O(n^{1/k})$ spanning trees of G, giving a final collective tree $(2k-1)$-spanner with $O(kn^{1/k})$ trees.

One issue with this approach is that the clusters at each level may have some intersections with one another, and hence we cannot guarantee that packing the clusters together will produce a spanning tree of G. Instead, we can equally partition each set A_i randomly into $\beta \in \mathbb{N}$ subsets A_i^r, $r \in [\beta]$ and then define the clusters $C^*(w) = \{v \in V \mid d(w,v) < d(v, A_i^r - \{w\})\}$. By this definition, clearly the clusters at each level are disjoint. As long as β is chosen to be large enough such that, with high probability $d(v, A_{i+1}) < d(v, A_i^r - \{w\})$ for every $v \in V$, $i \in [k-1]$, and $r \in [\beta]$, we have that $C(w) \subseteq C^*(w)$, implying that the $(2k-1)$ stretch result still holds for the modified definitions of clusters. Furthermore, since for each partition of A_i we construct a single spanning tree of G, the final collective tree spanner would consist of $k \cdot \beta$ trees. Future work can formalize how large β must be in-terms of n to give an explicit construction of a collective tree $(2k-1)$-spanner using this method.

Acknowledgements. This material is based upon work supported by NSF Grant CNS-2349369. This work was completed during the UNCG GraLNA 2024 REU, where Jesse Campbell was working under the mentorship of Chunjiang Zhu.

A Proof of Theorem 2

Theorem 2. *Let* $\mathbf{T}$ *be a t-collective tree spanner of* G *such that* $|\mathbf{T}| = \eta_t$. *For* $\gamma \in (0, 0.5]$ *and* $\varepsilon, \delta \in (0, 1]$, *there is an* (ε, δ)*-DP algorithm for releasing the counting query between* $u, v \in V$ *on a t-approximate shortest path that is* $O(\sqrt{\eta_t} \cdot \log^2(n) \cdot \log(1/\delta) \cdot \sqrt{\log(1/\gamma)}/\varepsilon)$*-accurate with probability* $1 - \gamma$.

Proof. For each tree $\mathcal{T} \in \mathbf{T}$ we run the $(\varepsilon/(2\sqrt{2|\mathbf{T}|\log(2/\delta)}), \delta/(2|\mathbf{T}|))$-DP algorithm from Lemma 7 to release the counting query over $\mathcal{T}$. By advanced composition (Lemma 4), releasing the counting queries in all the trees is (ε, δ)-DP.

By the definition of collective tree spanners, for every pair of vertices $u, v \in V$, there is a tree $\mathcal{T} \in \mathbf{T}$ such that $P_{\mathcal{T}}(u, v)$ is a t-approximate shortest path between u and v. We let $\widetilde{\omega}(u, v)$ to be the private counting query estimate over the path $P_{\mathcal{T}}(u, v)$. The result of Lemma 7 ensures that, with probability $1 - \gamma/n^2$, the additive error in the released counting query is,

$$\max_{u,v \in V} |\omega_{\mathcal{T}}(u, v) - \widetilde{\omega}(u, v)| \leq O(\sqrt{|\mathbf{T}|} \cdot \sqrt{\log(2/\delta)} \cdot \log(n) \cdot \sqrt{\log(1/\delta)} \cdot \sqrt{\log(n^3/\gamma)}/\varepsilon) = O(\sqrt{\eta_t} \cdot \log^2(n) \cdot \log(1/\delta) \cdot \sqrt{\log(1/\gamma)}/\varepsilon) \quad (2)$$

by Proposition 1. By a union bound, the upper bound (2) holds for all n^2 pairs of vertices with probability at least $1 - \gamma$. □

B Proof of Proposition 1

Proposition 1. *Let* $\alpha \geq 2$, $\gamma \in (0, 0.5]$, *and* $\lambda > 0$, *then,*

1. $\log(\alpha^{1+\lambda}/\gamma) = O(\log(\alpha/\gamma))$, *and,*
2. $\log(\alpha/\gamma) = O(\log(\alpha)\log(1/\gamma))$.

Proof (of 1).

$$\log(\alpha^{1+\lambda}/\gamma) = \log((\alpha/\gamma^{1/(1+\lambda)})^{1+\lambda}) = (1+\lambda) \cdot \log(\alpha/\gamma^{1/(1+\lambda)}) \leq (1+\lambda) \cdot \log(\alpha/\gamma) = O(\log(\alpha/\gamma))$$

Proof (of 2). Suppose the base of our logarithm is $b = \{2, e\}$. Then,

$$\log(\alpha/\gamma) = \log(\alpha) + \log(1/\gamma) = O(\max\{\log(\alpha), \log(1/\gamma)\}) = O(\log(\alpha)\log(1/\gamma))$$

as $\alpha, 1/\gamma \geq 2$. □

References

1. Abraham, I., Chechik, S., Elkin, M., Filtser, A., Neiman, O.: Ramsey spanning trees and their applications. ACM Trans. Algorithms **16**(2) (2020). https://doi.org/10.1145/3371039. ISSN 1549-6325
2. Ahmed, R., et al.: Graph spanners: a tutorial review. Comput. Sci. Rev. **37** (2020)
3. Aktay, A., et al.: Google Covid-19 community mobility reports: anonymization process description (version 1.1). arXiv preprint arXiv:2004.04145 (2020)
4. Bartal, Y., Fandina, O.N., Neiman, O.: Covering metric spaces by few trees. J. Comput. Syst. Sci. **130**, 26–42 (2022)
5. Bavadekar, S., et al.: Google Covid-19 search trends symptoms dataset: anonymization process description (version 1.0). arXiv preprint arXiv:2009.01265 (2020)
6. Bodwin, G., Deng, C., Gao, J., Hoppenworth, G., Upadhyay, J., Wang, C.: The discrepancy of shortest paths. arXiv preprint arXiv:2401.15781 (2024)
7. Chan, H., Shi, E., Song, D.: Private and continual release of statistics. ACM Trans. Inf. Syst. Secur. **14**(3) (2011). https://doi.org/10.1145/2043621.2043626. ISSN 1094-9224
8. Chazelle, B., Lvov, A.: A trace bound for the hereditary discrepancy. Discret. Comput. Geom. **26**(2), 221–231 (2001). https://doi.org/10.1007/s00454-001-0030-2
9. Chen, J.Y., et al.: Differentially private all-pairs shortest path distances: improved algorithms and lower bounds. In: Proceedings of the 2023 Annual ACM-SIAM Symposium on Discrete Algorithms (SODA), pp. 5040–5067 (2023). https://doi.org/10.1137/1.9781611977554.ch184
10. Deng, C., Gao, J., Upadhyay, J., Wang, C.: Differentially private range query on shortest paths. In: Algorithms and Data Structures - 18th International Symposium, WADS 2023, Proceedings, pp. 340–370. Springer, Cham (2023)
11. Ding, B., Kulkarni, J., Yekhanin, S.: Collecting telemetry data privately. In: Guyon, I., et al. (eds.) Advances in Neural Information Processing Systems, vol. 30. Curran Associates, Inc. (2017)
12. Dragan, F.F., Abu-Ata, M.: Collective additive tree spanners of bounded tree-breadth graphs with generalizations and consequences. Theor. Comput. Sci. **547**, 1–17 (2014)
13. Dragan, F.F., Yan, C.: Collective tree spanners in graphs with bounded parameters. Algorithmica **57**(1), 22–43 (2010)
14. Dragan, F.F., Yan, C., Lomonosov, I.: Collective tree spanners of graphs. In: Hagerup, T., Katajainen, J. (eds.) SWAT 2004. LNCS, vol. 3111, pp. 64–76. Springer, Heidelberg (2004). https://doi.org/10.1007/978-3-540-27810-8_7
15. Dragan, F.F., Yan, C., Corneil, D.G.: Collective tree spanners and routing in AT-free related graphs. In: Hromkovič, J., Nagl, M., Westfechtel, B. (eds.) WG 2004. LNCS, vol. 3353, pp. 68–80. Springer, Heidelberg (2004). https://doi.org/10.1007/978-3-540-30559-0_6
16. Dragan, F.F., Corneil, D.G., Köhler, E., Xiang, Y.: Collective additive tree spanners for circle graphs and polygonal graphs. Discret. Appl. Math. **160**(12), 1717–1729 (2012)
17. Dwork, C., Lei, J.: Differential privacy and robust statistics. In: Proceedings of the Forty-First Annual ACM Symposium on Theory of Computing, STOC 2009, pp. 371–380. Association for Computing Machinery, New York (2009). https://doi.org/10.1145/1536414.1536466. ISBN 9781605585062

18. Dwork, C., Roth, A.: The algorithmic foundations of differential privacy. Found. Trends Theor. Comput. Sci. **9**(3–4), 211–407 (2014). https://doi.org/10.1561/0400000042. ISSN 1551-305X
19. Dwork, C., McSherry, F., Nissim, K., Smith, A.: Calibrating noise to sensitivity in private data analysis. In: Halevi, S., Rabin, T. (eds.) Theory of Cryptography, pp. 265–284. Springer, Heidelberg (2006). ISBN 978-3-540-32732-5
20. Fan, C., Li, P., Li, X.: Breaking the linear error barrier in differentially private graph distance release. In: NeurIPS (2022)
21. Fan, C., Li, P., Li, X.: Private graph all-pairwise-shortest-path distance release with improved error rate. In: Proceedings of the 36th International Conference on Neural Information Processing Systems, NIPS 2022. Curran Associates Inc. (2024). ISBN 9781713871088
22. Foote, A.D., Machanavajjhala, A., McKinney, K.: Releasing earnings distributions using differential privacy: disclosure avoidance system for post-secondary employment outcomes (PSEO). J. Privacy Confidentiality **9**(2) (2019). https://doi.org/10.29012/jpc.722
23. Garfinkel, S.L., Abowd, J.M., Powazek, S.: Issues encountered deploying differential privacy. In: Proceedings of the 2018 Workshop on Privacy in the Electronic Society, WPES 2018, pp. 133–137. Association for Computing Machinery, New York (2018). https://doi.org/10.1145/3267323.3268949. ISBN 9781450359894
24. Gupta, A., Kumar, A., Rastogi, R.: Traveling with a Pez dispenser (or, routing issues in MPLS). In: Proceedings 42nd IEEE Symposium on Foundations of Computer Science, pp. 148–157 (2001)
25. Hay, M., Li, C., Miklau, G., Jensen, D.: Accurate estimation of the degree distribution of private networks. In: 2009 Ninth IEEE International Conference on Data Mining, pp. 169–178 (2009). https://doi.org/10.1109/ICDM.2009.11
26. Jian, X., Wang, Y., Chen, L.: Publishing graphs under node differential privacy. IEEE Trans. Knowl. Data Eng. **35**(4), 4164–4177 (2023). https://doi.org/10.1109/TKDE.2021.3128946
27. Liu, G., Ma, X., Li, W.: Publishing node strength distribution with node differential privacy. IEEE Access **8**, 217642–217650 (2020). https://doi.org/10.1109/ACCESS.2020.3040077
28. Liu, W., Liu, B., Qiang, X., Lei, H.: Graph node strength histogram publication method with node differential privacy. In: Journal of Physics: Conference Series, vol. 1757, no. 1, p. 012186 (2021). https://doi.org/10.1088/1742-6596/1757/1/012186
29. Sealfon, A.: Shortest paths and distances with differential privacy. In: PODS 2016, pp. 29–41. Association for Computing Machinery, New York (2016). https://doi.org/10.1145/2902251.2902291. ISBN 9781450341912
30. Thorup, M., Zwick, U.: Approximate distance oracles. J. ACM **52**(1), 1–24 (2005)
31. Wang, H.: Centroid, leaf-centroid, and internal-centroid. Graphs Comb. **31**(3), 783–793 (2013). https://doi.org/10.1007/s00373-013-1401-1

Semi-online Multiprocessor Scheduling with Known Largest Job Processing Time

Mingyang Gong[1], Guohui Lin[1(✉)], and Zhiyi Tan[2(✉)]

[1] Department of Computing Science, University of Alberta, Edmonton, Canada
{mgong4,guohui}@ualberta.ca
[2] School of Mathematical Science, Zhejiang University, Hangzhou, China
tanzy@zju.edu.cn

Abstract. We study a semi-online variant of the online multiprocessor scheduling problem, in which the largest job processing time is known. We show affirmatively that such a known value can be explored to design better online algorithms, for a large number of machines in particular. Specifically, we design the first 1.8686-competitive algorithm for the semi-online variant, compared against the long-standing competitive ratio of 1.9201 for the classic online problem. When there are only three machines, we present a 1.4655-competitive algorithm, improving the previous best known ratio of 1.5. Our algorithm design deviates from the well-known list scheduling that always assigns the current job to a least loaded machine, by incorporating a novel idea to assign the current job to a maximum loaded "safe" machine, where "safe" is defined using the largest job processing time.

Keywords: Scheduling · semi-online · multiprocessor · largest processing time · approximation algorithm

1 Introduction

Online scheduling over list models many real-life applications and has received extensive studies. It can be described as follows: There is a set $\mathcal{J}$ of jobs $J_1, J_2, \ldots$ arriving one by one in sequence and to be processed non-preemptively on a set $\mathcal{M}$ of parallel identical machines. The scheduler knows the processing time p_j of J_j upon its arrival, and makes irrevocable assignment decision on which machine to process J_j, with the objective to minimize the makespan, i.e., the largest job completion time. Such an online problem is denoted as $P \mid\mid C_{\max}$, or $Pm \mid\mid C_{\max}$ when the number of machines m is fixed.

For an instance I of the problem, let $C^A(I)$ denote the makespan of the schedule produced by a specific *polynomial time online* algorithm A and $C^*(I)$ the optimal *offline* makespan, respectively. Here offline means all the jobs arrive, and their processing times are known, at the beginning of the scheduling. The performance of A is measured in its *competitive ratio* defined as $r_A = \sup_I \frac{C^A(I)}{C^*(I)}$, and the algorithm is said r_A*-competitive*. If for an online problem one can prove

D. Du et al. (Eds.): COCOA 2024, LNCS 15434, pp. 213–225, 2025.
https://doi.org/10.1007/978-981-96-4445-2_18

that no algorithm has a competitive ratio less than ρ, then ρ is called a *lower bound* for the problem. An algorithm A is *optimal* (or best possible) for the problem if $r_A = \rho$.

1.1 Results on Online Multiprocessor Scheduling

For $P \mid\mid C_{\max}$, Graham [12] proposed the well-known *list scheduling* (LS for short) algorithm, which always schedules the current job on a least loaded machine for processing, and proved it to be $(2 - \frac{1}{m})$-competitive, where m is the number of machines. When $m = 2, 3$, the LS algorithm is shown optimal [9]. For $P4 \mid\mid C_{\max}$, Chen et al. [7] presented an improved $\frac{26}{15}$-competitive algorithm, which remains the currently best; a lower bound of $\sqrt{3}$ is given in [18], leaving a tiny gap of 0.0013 for future improvement.

Many efforts have also been made to improve the competitive ratios and lower bounds for large m's. When m is sufficiently large, Fleischer and Wahl [10] proposed the currently best 1.9201-competitive algorithm after a long sequence of improvements; the best known lower bound is 1.8535 [11].

1.2 Results on Semi-online Multiprocessor Scheduling

An important characteristic of the online problem $P \mid\mid C_{\max}$ is that when the job J_j arrives, the scheduler does not know whether there are more jobs coming and their processing times. One thus might wonder what partial information about the future jobs could help the scheduler make better decisions, leading to provably better algorithms. In the literature, there are several *semi-online* variants been studied, in each of which a certain piece of extra information about the jobs is provided to the scheduler at the beginning of the scheduling. We review four typical ones below.

The first variant is denoted as $P \mid decr \mid C_{\max}$, in which the jobs arrive in the non-increasing order of their processing times. The competitive ratio of LS is $\frac{4}{3} - \frac{1}{3m}$, the same as the *longest processing time first* algorithm for the offline multiprocessor scheduling problem [13]. Seiden et al. [19] showed that LS is actually optimal for $P2 \mid decr \mid C_{\max}$. Cheng et al. [8] presented an optimal $\frac{1+\sqrt{37}}{6}$-competitive algorithm for $P3 \mid decr \mid C_{\max}$, and a $\frac{5}{4}$-competitive algorithm, which is strictly better than LS, when there are 5 or more machines.

In the second problem $P \mid sum \mid C_{\max}$, the total job processing time is known. We remark that in this variant, one can actually tell whether or not there are more jobs coming after a certain job, though not their processing times. For $P2 \mid sum \mid C_{\max}$, Kellerer et al. [16] presented an optimal $\frac{4}{3}$-competitive algorithm. When the number of machines is arbitrarily large, the competitive ratio lower bound for the problem had been improved from 1.5652 [2] to 1.5850 [1], and Kellerer et al. [15] presented an optimal 1.5850-competitive algorithm. This way, $P \mid sum \mid C_{\max}$ is the first semi-online variant that admits an optimal algorithm when the number of machines is sufficiently large.

The problem $P \mid opt \mid C_{\max}$, where the optimal offline makespan is known to the scheduler, has also been studied. For $P2 \mid opt \mid C_{\max}$, Azar and Regev [3] presented an optimal $\frac{4}{3}$-competitive algorithm. For $P3 \mid opt \mid C_{\max}$, Böhm et al. [4] contributed the current best $\frac{11}{8}$-competitive algorithm and found a lower bound of $\frac{15}{11}$ through a computer program; for an arbitrary number of machines, the current best $\frac{3}{2}$-competitive algorithm is proposed by Böhm et al. [5].

In the last problem denoted as $P \mid \max \mid C_{\max}$, the largest job processing time $p_{\max}$ is known in advance. We remark that, similarly to the above second variant, one can tell there are more jobs coming after a certain job if the largest job processing time is not seen yet. He and Zhang [14] presented an optimal $\frac{4}{3}$-competitive algorithm for $P2 \mid \max \mid C_{\max}$; for $P3 \mid \max \mid C_{\max}$, a lower bound of $\sqrt{2}$ is given by Cai [6] and the current best $\frac{3}{2}$-competitive algorithm is due to Wu et al. [20]. For $m = 4, 5$, Lee and Lim [17] presented a $\frac{1+\sqrt{5}}{2}$- and $\frac{5}{3}$-competitive algorithms, respectively, and showed a lower bound of $\frac{3+\sqrt{33}}{6} \approx 1.4574$ for every $m \geq 4$. For sufficiently large numbers of machines, the 1.9201-competitive algorithm [10] for the classic online problem $P \mid\mid C_{\max}$ remains the current best, that is, prior to our work, the value $p_{\max}$ is unknown helpful or not for designing better algorithms.

1.3 Our Contributions

We study the last semi-online variant $P \mid \max \mid C_{\max}$ reviewed in the above, which has interesting applications such as video-on-demand services [17]. We address affirmatively that the largest job processing time $p_{\max}$ is helpful, by presenting a 1.8686-competitive algorithm for $m \geq 4$; extending the design idea, we also present an improved 1.4655-competitive algorithm for $P3 \mid \max \mid C_{\max}$. Recall that the LS algorithm always assigns the current job to a least loaded machine; our algorithms deviate from it to assign the current job to a maximum loaded "safe" machine, where "safe" is a novel concept defined using $p_{\max}$. When there is no safe machine, our algorithms reduce to LS to assign the current job to a least loaded machine.

In the next section, we present the concept of safe machine and the 1.8686-competitive algorithm; and we deal with $P3 \mid \max \mid C_{\max}$ in Sect. 3. We conclude the paper in Sect. 4 with some remarks.

2 Safe Machines and a 1.8686-Competitive Algorithm

We scale down all the numerical such that the largest job processing time $p_{\max} = 1$. For an instance I of $P \mid \max \mid C_{\max}$, $C^A(I)$ denotes the makespan of the schedule produced by an online algorithm A and $C^*(I)$ denotes the optimal offline makespan, respectively. We also drop I from the notations $C^A(I)$ and $C^*(I)$, when the instance I is clear from the context. Note the job arrival sequence $J_1, J_2, \ldots$; we use $P_j = \sum_{k=1}^{j} p_k$ to denote the total job processing time up to the job J_j. We have two lower bounds on C^* when J_j arrives:

$$C^* \geq \max\left\{p_{\max}, \frac{P_j}{m}\right\} = \max\left\{1, \frac{P_j}{m}\right\}. \tag{1}$$

The following parameter $\beta(m)$ is used to define *safe* machines, where $m \geq 4$ is the number of machines.

$$\beta(m) = \frac{\sqrt{13m^2 - 8m + 4} - m - 2}{3m}. \tag{2}$$

When the job J_j arrives and is assigned for processing, let L_i^j be the load of the machine M_i for $i = 1, 2, \ldots, m$. Since $C^* \geq 1$, if the current makespan is at most $1 + \beta(m)$, then the current competitive ratio $\frac{C^A}{C^*} \leq 1 + \beta(m)$. This motivates the definition of safe machines.

Definition 1. *The machine M_i is* safe *for J_j if $L_i^{j-1} + p_j \leq 1 + \beta(m)$.*

Using $\beta(m)$, we classify the machines into three subsets with respect to J_j as follows:

$$\begin{aligned} \text{(double safe)}\ \mathcal{D}_j &= \{M_i \mid L_i^{j-1} + p_j \leq \beta(m)\}, \\ \text{(safe)}\ \mathcal{S}_j &= \{M_i \mid \beta(m) < L_i^{j-1} + p_j \leq 1 + \beta(m)\}, \\ \text{(unsafe)}\ \mathcal{N}_j &= \mathcal{M} \setminus \{\mathcal{D}_j \cup \mathcal{S}_j\}. \end{aligned} \tag{3}$$

In Eq. (3), a machine of $\mathcal{D}_j$ is *double safe* in that it is safe not only for J_j but also for the next job J_{j+1}, if any, since $p_{j+1} \leq 1$. In our algorithm, double safe machines have the top priority to process J_j; but unlike the LS algorithm to assign J_j to a least loaded machine, our algorithm assigns J_j to a maximum loaded double safe machine. If there is no double safe machine, i.e., $\mathcal{D}_j = \emptyset$, then our algorithm assigns J_j to a maximum loaded safe machine; and further if $\mathcal{D}_j \cup \mathcal{S}_j = \emptyset$, then our algorithm assigns J_j to a least loaded machine. We remark that the novel idea deviating from the LS algorithm is, when $\mathcal{D}_j \neq \emptyset$ or $\mathcal{S}_j \neq \emptyset$, to assign the current job to a maximum loaded double safe or safe machine while leaving larger rooms to future jobs as we do not know their processing times yet. A pseudo-code of our algorithm, denoted GLTm, is depicted in Algorithm 1.

Algorithm 1. GLTm

Input: A sequence of jobs with $p_{\max} = 1$ and $m \geq 4$ parallel identical machines;
Output: A feasible schedule with makespan at most $(1 + \beta(m))C^*$.
Initialize $L_i^0 = 0$ for $i = 1, 2, \ldots, m$;
for J_j with the processing time p_j, $j = 1, 2, \ldots$ **do**
    Classify machines into $\mathcal{D}_j$, $\mathcal{S}_j$, $\mathcal{N}_j$ as in Eq. (3);
    if $\mathcal{D}_j \neq \emptyset$ **then**
        Schedule J_j on the largest load machine of $\mathcal{D}_j$;
    else if $\mathcal{S}_j \neq \emptyset$ **then**
        Schedule J_j on the largest load machine of $\mathcal{S}_j$;
    else
        Schedule J_j on the least-loaded machine (of $\mathcal{N}_j$).

We will show that the competitive ratio of our algorithm GLTm is at most $1 + \beta(m) = \frac{\sqrt{13m^2-8m+4}+2m-2}{3m}$, which approaches to $\frac{\sqrt{13}+2}{3} < 1.8686$ when m tends to infinity. To this purpose, we define

$$\alpha(m) = \frac{\sqrt{13m^2 - 8m + 4} - 3m - 2}{4m} \tag{4}$$

such that $\alpha(m)$ and $\beta(m)$ satisfy

$$3\beta(m) = 2 + 4\alpha(m), \text{ and } \frac{m-1}{m(1+\alpha(m))+1} = \beta(m). \tag{5}$$

Lemma 1. *If $m \geq 4$, then $\beta(m) - \alpha(m) > \frac{1+\alpha(m)}{2} + \frac{1}{2m}$.*

Proof. By Eq. (5), $\beta(m) = \frac{4\alpha(m)+2}{3}$ and by $m \geq 4$, we have

$$\beta(m) - \alpha(m) - \frac{1+\alpha(m)}{2} - \frac{1}{2m} = \frac{1-\alpha(m)}{6} - \frac{1}{2m} \geq \frac{1-4\alpha(m)}{24}.$$

One can check from Eq. (4) that $\alpha(m) < \frac{1}{4}$, and thus the lemma follows. □

We prove the competitive ratio by a *minimal counterexample* to assume that J_ℓ is the first job such that $\max_{j=1}^{\ell} p_j = 1$ and $\frac{C^A}{C^*} > 1 + \beta(m)$ when J_ℓ is scheduled. That is, for every $k < \ell$, the job sequence $(J_1, J_2, \ldots, J_k)$ either is invalid (i.e., no job of processing time $p_{\max} = 1$) or is valid but $\frac{C^A}{C^*} \leq 1 + \beta(m)$ when J_k is scheduled. We assume the job J_n with $n \leq \ell$ determines the makespan C^A of the schedule by GLTm for this minimal counterexample $(J_1, J_2, \ldots, J_\ell)$. Let b be the least load before J_n arrives, and thus $b + p_n \leq C^A$ by the algorithm description.

Lemma 2. *$L_i^{n-1} \geq b$ for every $i = 1, 2, \ldots, m$, and $C^A = b + p_n$.*

Proof. The first half of the lemma holds by the definition of b being the least load before J_n arrives.

We next argue that J_n is scheduled on a least loaded machine by the algorithm. If $\mathcal{D}_n \neq \emptyset$ ($\mathcal{S}_n \neq \emptyset$, respectively), then J_n is scheduled on a machine of $\mathcal{D}_n$ ($\mathcal{S}_n$, respectively). This means $C^A \leq 1 + \beta(m)$. By Eq. (1), we have $\frac{C^A}{C^*} \leq 1 + \beta(m)$, a contradiction. Therefore,

$$\mathcal{D}_n = \mathcal{S}_n = \emptyset, \tag{6}$$

and subsequently by the algorithm description J_n is scheduled on a least loaded machine. □

Lemma 3. *$\beta(m) < b < 1 + \alpha(m)$ and $p_n > \beta(m) - \alpha(m)$.*

Proof. If $b \leq \beta(m)$, then by Lemma 2, $C^A \leq \beta(m) + 1$, a contradiction.

Lemma 2 also tells that $P_n \geq \sum_{i=1}^{m} L_i^{n-1} + p_n \geq mb + p_n$. If follows from Eq. (1) that $C^* \geq \frac{P_n}{m} \geq b + \frac{p_n}{m}$. If $b \geq 1 + \alpha(m)$, then by Eq. (5) we have

$$\frac{C^A}{C^*} \leq \frac{b+p_n}{b+p_n/m} \leq \frac{1+\alpha(m)+1}{1+\alpha(m)+1/m} = 1 + \frac{m-1}{m(1+\alpha(m))+1} = 1 + \beta(m),$$

a contradiction too. This proves the first half of the lemma.

For the second half, if $p_n \leq \beta(m) - \alpha(m)$, then by $b < 1 + \alpha(m)$ we have $C^A \leq b + p_n \leq 1 + \beta(m)$, a contradiction again. □

Definition 2. *A job J_j is* large *if $p_j \geq \beta(m) - \alpha(m)$.*

Definition 3. *A job is a* filling job *if its starting time is before or at time point $\beta(m)$ and its finishing time is strictly after $\beta(m)$.*

From Lemmas 2 and 3 we know that J_n is large but not a filling job. Also, since $L_i^{n-1} \geq b > \beta(m)$ for every i, every machine processes exactly one filling job.

Recall from Lemma 2 and Eq. (6) that there is no double safe or safe machine for J_n, i.e., $\mathcal{M} = \mathcal{N}_n$. We define

$$\mathcal{T}_n = \{M_i \mid L_i^{n-1} < 1 + \alpha(m)\}. \tag{7}$$

Lemma 3 states that $b < 1 + \alpha(m)$, indicating $\mathcal{T}_n$ is not empty. In fact, there are at least three machines in $\mathcal{T}_n$.

Lemma 4. $|\mathcal{T}_n| \geq 3$.

Proof. Suppose to the contrary that $|\mathcal{T}_n| \leq 2$. By Eq. (7), if $M_i \notin \mathcal{T}_n$, then $L_i^{n-1} \geq 1 + \alpha(m)$. Together with Lemma 2, we have

$$\begin{aligned} P_n &\geq |\mathcal{T}_n| b + (m - |\mathcal{T}_n|)(1 + \alpha(m)) + p_n \\ &= |\mathcal{T}_n|(b - 1 - \alpha(m)) + m(1 + \alpha(m)) + p_n. \end{aligned} \tag{8}$$

Since $b < 1 + \alpha(m)$ by Lemma 3, the right side of Eq. (8) is decreasing in $|\mathcal{T}_n|$, and thus

$$P_n \geq 2(b - 1 - \alpha(m)) + m(1 + \alpha(m)) + p_n.$$

It follows from Eq. (1) that

$$\begin{aligned} \frac{C^A}{C^*} \leq \frac{b + p_n}{P_n/m} &= \frac{m(b + p_n)}{2(b - 1 - \alpha(m)) + m(1 + \alpha(m)) + p_n} \\ &= \frac{mb + mp_n}{2b + p_n + (m - 2)(1 + \alpha(m))} \\ &= \frac{m}{2} \times \left(1 - \frac{(m - 2)(1 + \alpha(m)) - p_n}{2b + p_n + (m - 2)(1 + \alpha(m))}\right) \end{aligned} \tag{9}$$

Using $m \geq 4$, $(m - 2)(1 + \alpha(m)) - p_n \geq 2(1 + \alpha(m)) - 1 > 0$, and Eq. (9) is increasing in b. Since $b < 1 + \alpha(m)$ by Lemma 3, we have

$$\frac{C^A}{C^*} \leq \frac{m(1 + \alpha(m) + p_n)}{m(1 + \alpha(m)) + p_n} = m\left(1 - \frac{(m - 1)(1 + \alpha(m))}{m(1 + \alpha(m)) + p_n}\right). \tag{10}$$

The right side of Eq. (10) is increasing in p_n, and thus by Eq. (5) $\frac{C^A}{C^*} \leq \frac{m(2+\alpha(m))}{m(1+\alpha(m))+1} = 1 + \beta(m)$, a contradiction. □

Remark 1. Recall that every machine processes exactly one filling job. Let J_{f_i} denote the filling job on the machine M_i, for every i. We assume w.l.o.g. that $\mathcal{T}_n = \{M_1, M_2, \ldots, M_{|\mathcal{T}_n|}\}$ and J_{f_i} arrives before $J_{f_{i+1}}$ in the job sequence, i.e., $f_i < f_{i+1}$, for $i = 1, 2, \ldots, |\mathcal{T}_n| - 1$.

Lemma 5. *$M_i \in \mathcal{S}_{f_i}$ for every filling job J_{f_i}.*

Proof. By Definition 3 of the filling job J_{f_i}, its starting time is $L_i^{f_i-1} \leq \beta(m)$ and its finishing time is $L_i^{f_i-1} + p_{f_i} > \beta(m)$. Since $p_{f_i} \leq p_{\max} = 1$, we have $\beta(m) < L_i^{f_i-1} + p_{f_i} \leq \beta(m) + 1$; that is, $M_i \in \mathcal{S}_{f_i}$ by Eq. (3). □

Lemma 6. *The filling job J_{f_i} is large for every $i = 2, 3, \ldots, |\mathcal{T}_n|$.*

Proof. Since J_{f_1} arrives before J_{f_i} for every $i = 2, 3, \ldots, |\mathcal{T}_n|$, by Definition 3 the loads of the machines M_1 and M_i satisfy $L_1^{f_i-1} \geq L_1^{f_1} > \beta(m) \geq L_i^{f_i-1}$. That is, when J_{f_i} arrives, the load of M_1 is strictly larger than that of M_k. By Lemma 5, M_i is safe for J_{f_i}, and thus M_1 is unsafe for J_{f_i} as otherwise the algorithm would assign J_{f_i} to M_1, i.e., $L_1^{f_i-1} + p_{f_i} > 1 + \beta(m)$.

On the other hand, filling jobs all arrive before J_n; and since $M_1 \in \mathcal{T}_n$ we have $L_1^{f_i-1} \leq L_1^{n-1} < 1+\alpha(m)$. Therefore, $p_{f_i} > \beta(m)-\alpha(m)$, i.e., J_{f_i} is large. □

Lemma 7. *Every filling job is a large job.*

Proof. We continue from Lemma 6 to prove that J_{f_i} is large for every $i = 1, |\mathcal{T}_n|+1, |\mathcal{T}_n| + 2, \ldots, m$. Recall from Lemma 4 that $|\mathcal{T}_n| \geq 3$, and thus $M_2, M_3 \in \mathcal{T}_n$ and J_{f_2}, J_{f_3} are proven large in Lemma 6.

Consider the two machines M_2 and M_3, for which $L_2^{n-1} < 1 + \alpha(m)$ and $L_3^{n-1} < 1 + \alpha(m)$. Using Eq. (5), $L_2^{f_2-1} \leq L_2^{n-1} - p_{f_2} < 1 + \alpha(m) - (\beta(m) - \alpha(m)) = \frac{\beta(m)}{2}$, and the same $L_3^{f_2-1} \leq L_3^{f_3-1} < \frac{\beta(m)}{2}$. Therefore, $L_2^{f_2-1} + L_3^{f_2-1} < \beta(m)$. That is, both machines are double safe for each of the jobs $J_1, J_2, \ldots, J_{f_2-1}$ assigned to M_2 or M_3, indicating by the algorithm description that either $L_2^{f_2-1} = 0$ or $L_3^{f_2-1} = 0$. We assume w.l.o.g. $L_2^{f_2-1} = 0$.

Consider a filling job J_{f_i} with $i \neq 2$. We distinguish two cases on the relative positions of J_{f_i} and J_{f_2} in the job sequence. In the first case, J_{f_i} arrives before J_{f_2}. Since $L_2^{f_i-1} \leq L_2^{f_2-1} = 0$, when J_{f_i} arrives, M_2 is an empty machine but does not receive J_{f_i}; by Lemma 5 stating that $M_i \in \mathcal{S}_{f_i}$, M_2 is not a double safe machine for J_{f_i} and thus $p_{f_i} > \beta(m)$. In the second case, J_{f_i} arrives after J_{f_2}. Since $L_2^{f_i-1} \geq L_2^{f_2} > \beta(m) \geq L_i^{f_i-1}$, when J_{f_i} arrives, M_2 has a greater load but does not receive J_{f_i}; by Lemma 5 stating that $M_i \in \mathcal{S}_{f_i}$, M_2 is an unsafe machine for J_{f_i} and thus $L_2^{f_i-1} + p_{f_i} > 1 + \beta(m)$, which by $M_2 \in \mathcal{T}_n$ implies $p_{f_i} > 1 + \beta(m) - L_2^{n-1} > \beta(m) - \alpha(m)$.

In summary, in either case we have $p_{f_i} > \beta(m) - \alpha(m)$, i.e., J_{f_i} is large. □

Theorem 1. *The algorithm* GLTm *is $(1 + \beta(m))$-competitive, where $m \geq 4$, $\beta(m)$ is defined in Eq. (2), and $1 + \beta(m) < 1.8686$.*

Proof. Assuming the existence of a minimal counterexample $(J_1, J_2, \ldots, J_\ell)$ in which the job J_n determines the makespan generated by GLTm. Lemmas 3 and 7 together say that there are $m+1$ large jobs which are the m filling jobs and J_n. This way, at least two large jobs are assigned to the same machine in an optimal offline schedule, of which the makespan $C^* \geq 2(\beta(m) - \alpha(m))$. On the other hand, from Lemma 3 we have $C^A = b + p_n < 2 + \alpha(m)$. It follows from Lemma 1 and Eq. (5) that

$$\frac{C^A}{C^*} < \frac{2+\alpha(m)}{2(\beta(m)-\alpha(m))} \leq \frac{m(2+\alpha(m))}{m(1+\alpha(m))+1} = 1 + \beta(m).$$

This means the minimal counterexample actually is not a counterexample, and so proves the theorem. □

3 An Improved Algorithm for $P3 \mid \max \mid C_{\max}$

Our earlier algorithm GLTm requires $m \geq 4$ for defining large jobs, and thus does not work for $P3 \mid \max \mid C_{\max}$. Nevertheless, one sees from Lemma 4 the existence of at least three machines of load less than $1+\alpha(m)$, suggesting possible room for improvement at the presence of only three machines.

Indeed, we present next an improved algorithm for $P3 \mid \max \mid C_{\max}$ by slightly modifying the parameters. Let $\alpha(3)$ be the unique positive root to the equation $12x^2 + 19x - 2 = 0$ and $\beta(3) = \frac{4\alpha(3)+1}{3}$ (cf. Eq. (5)). One sees that

$$\beta(3) = \tfrac{4\alpha(3)+1}{3} = \tfrac{2}{3\alpha(3)+4},$$
$$\alpha(3) = \tfrac{\sqrt{457}-19}{24} \approx 0.099, \ \beta(3) = \frac{\sqrt{457}-13}{18} \approx 0.4654 < 0.4655; \qquad (11)$$

Since there exists at least one job of processing time $p_{\max} = 1$, we can assume w.l.o.g. that $p_1 = 1$ and J_1 is scheduled on M_1 in our algorithm (or otherwise we simply reserve the processing interval of length 1 on M_1 for the first arriving job of processing time 1). We use L_i^{j-1} to denote the load of M_i before J_j arrives.

For now we identify only *safe* and *unsafe* machines defined in Eq. (3). The key idea in our algorithm, denoted as GLT3, is to "sacrifice" the machine M_1 when it is safe for J_j, by assigning J_j on M_1. Otherwise, we distinguish the following three cases: If exactly one of M_2, M_3 is safe for J_j, then assign J_j to the safe machine. If both M_2, M_3 are safe for J_j, then assign J_j to one of them by checking the two conditions described below in the pseudo-code (Line 10). In the last case where no machine is safe for J_j, assign J_j on a least loaded machine. A pseudo-code of GLT3 is presented in Algorithm 2.

Lemma 8. *The following two inequalities hold:*

(i) $(1 + \beta(3) - \alpha(3))(1 + \beta(3)) > 2$.
(ii) $(1 + \beta(3))^2 > 2 + \alpha(3)$.

Algorithm 2. GLT3

Input: A sequence of jobs with $p_1 = p_{\max} = 1$ and three parallel identical machines;
Output: A feasible schedule with makespan at most $(1+\beta(3))C^*$.
Schedule J_1 on M_1; set $L_1^1 = 1$, $L_2^1 = L_3^1 = 0$;
for J_j with processing time p_j, $j = 2, 3, \ldots$ **do**
  if M_1 is safe for J_j **then**
    Schedule J_j on M_1;
  else if exactly one of M_2, M_3 is safe for J_j **then**
    Schedule J_j on the safe machine;
  else if both M_2, M_3 are safe for J_j **then**
    if $\min\{L_2^{j-1}, p_j\} < \beta(3)$ and $L_2^{j-1} + p_j > 1$ **then**
      Schedule J_j on M_3;
    else
      Schedule J_j on M_2;
  else
    Schedule J_j on the least loaded machine;

Proof. These two inequalities can be proven using the definition that $\alpha(3)$ is the positive root to the equation $12x^2 + 19x - 2 = 0$ and Eq. (11). □

To prove the competitive ratio of $1 + \beta(3) = \frac{\sqrt{457}+5}{18} < 1.4655$, we again use a minimal counterexample. Slightly different from the proof of Theorem 1, since we have $p_1 = p_{\max} = 1$, we can assume J_n (certainly, $\neq J_1$) is the last job in the counterexample and it determines the makespan of the schedule generated by GLT3. Let b the least load of the three machines when the job J_n arrives.

The next Lemma 9 is similar to Lemma 2 and half of Lemma 3.

Lemma 9. *$L_i^{n-1} \geq b$ for every $i = 1, 2, 3$, $C^A = b + p_n$, and $\beta(3) < b$.*

Proof. By Eq. (1), for this counterexample we have $C^A > 1 + \beta(3)$.

Since b is the least load of the three machines when the job J_n arrives, $L_i^{n-1} \geq b$ for every $i = 1, 2, 3$. We show next J_n is scheduled on the least load machine. If at least one of the machines is safe for J_n, then J_n is scheduled on a safe machine M_i by the algorithm and by safe it means $C^A = L_i^{n-1} + p_n \leq 1 + \beta(3)$, a contradiction. Therefore, there is no safe machine for J_n, and thus the algorithm assigns J_n on the least-loaded machine.

Lastly, combining $C^A = b + p_n > 1 + \beta(3)$ and $p_n \leq 1$, we have $b > \beta(3)$. □

The next Lemma 10 is similar to Lemma 4 and the second half of Lemma 3.

Lemma 10. *$L_i^{n-1} < 1 + \alpha(3)$ for every $i = 1, 2, 3$, and $p_n > \beta(3) - \alpha(3)$.*

Proof. Assume $\max\{L_1^{n-1}, L_2^{n-1}, L_3^{n-1}\} \geq 1 + \alpha(3)$. By Lemma 9, $C^A = b + p_n$. We distinguish two cases about the value of b.

Case 1. $b \geq 1 + \alpha(3)$. By Lemma 9, $P_n \geq 3b + p_n$. By Eq. (1), we know $C^* \geq \frac{P_n}{3} \geq b + \frac{p_n}{3}$. Therefore by $p_n \leq 1$ and Eq. (11), we have

$$\begin{aligned}\frac{C^A}{C^*} &\leq \frac{3b+3p_n}{3b+p_n} = 1 + \frac{2p_n}{3b+p_n}\\ &\leq 1 + \frac{2p_n}{3\alpha(3)+3+p_n}\\ &\leq 1 + \frac{2}{3\alpha(3)+4} = 1+\beta(3).\end{aligned}$$

Case 2. $b < 1 + \alpha(3)$. So, by Lemma 9 and $\max\{L_1^{n-1}, L_2^{n-1}, L_3^{n-1}\} \geq 1 + \alpha(3)$, we have $P_n \geq 2b + 1 + \alpha(3) + p_n$. By Eq. (1) again, $C^* \geq \frac{P_n}{3} \geq \frac{1}{3}(2b + 1 + \alpha(3) + p_n)$. Therefore, by $p_n \leq 1$ and Eq. (11), we have

$$\begin{aligned}\frac{C^A}{C^*} &\leq \frac{3b+3p_n}{2b+1+\alpha(3)+p_n} = \frac{3}{2}\left(1 + \frac{p_n - 1 - \alpha(3)}{2b+1+\alpha(3)+p_n}\right)\\ &< \frac{3}{2}\left(1 + \frac{p_n - 1 - \alpha(3)}{3(1+\alpha(3))+p_n}\right) = 1 + \frac{2p_n}{3(1+\alpha(3))+p_n}\\ &\leq 1 + \frac{2}{3\alpha(3)+4} = 1+\beta(3).\end{aligned}$$

It follows that $\frac{C^A}{C^*} \leq 1+\beta(3)$ for these two cases, which contradicts the definition of J_n. Hence $\max\{L_1^{n-1}, L_2^{n-1}, L_3^{n-1}\} < 1 + \alpha(3)$.

Since $b \leq \max\{L_1^{n-1}, L_2^{n-1}, L_3^{n-1}\} < 1 + \alpha(3)$, we have $C^A < 1 + \alpha(3) + p_n$. Recall that $C^A > 1 + \beta(3)$ and in turn $p_n > \beta(3) - \alpha(3)$. Then the lemma is proved. □

Definition 4. *Let J_{i_t} denote the t-th job scheduled on M_i, for $i = 2, 3$. The job set $\mathcal{F}$ includes all the jobs scheduled on M_2, M_3 when J_n arrives.*

From the algorithm description, one sees that J_{2_1} is the first arriving job among the jobs of $\mathcal{F}$.

Lemma 11. *If $J_j \in \mathcal{F}$, then $p_j > \beta(3) - \alpha(3)$. Moreover, $2 \leq |\mathcal{F}| \leq 4$.*

Proof. Lemma 10 says $L_1^{n-1} < 1 + \alpha(3)$. If $J_j \in \mathcal{F}$ and $p_j \leq \beta(3) - \alpha(3)$, then J_j would be assigned to M_1 by the algorithm, a contradiction.

We next prove that $2 \leq |\mathcal{F}| \leq 4$. Lemma 9 says $L_i^{n-1} \geq b > \beta(3)$, for $i = 2, 3$. That is, at least one job has been scheduled on M_2, M_3 when J_n arrives, indicating $|\mathcal{F}| \geq 2$. On the other hand, if $|\mathcal{F}| \geq 5$, then one of M_2, M_3 has at least three jobs when J_n arrives. By Eq. (11), this means $\max\{L_2^{n-1}, L_3^{n-1}\} > 3(\beta(3) - \alpha(3)) = 1 + \alpha(3)$, contradicting Lemma 10. Therefore, $|\mathcal{F}| \leq 4$. □

Lemma 12. *If $C^A \leq 2$, then $|\mathcal{F}| = 2$ and the three jobs in $\mathcal{F} \cup \{J_n\}$ are scheduled on two machines in the optimal offline schedule.*

Proof. By Lemmas 10 and 11, the processing time of each job in $\mathcal{F} \cup \{J_n\}$ is greater than $\beta(3) - \alpha(3)$.

If $C^* \geq 1 + \beta(3) - \alpha(3)$, then by Lemma 8(i), we have $\frac{C^A}{C^*} \leq \frac{2}{1+\beta(3)-\alpha(3)} < 1 + \beta(3)$, a contradiction. Therefore, $C^* < 1 + \beta(3) - \alpha(3)$. Subsequently, since $p_1 = 1$, in the optimal offline schedule, J_1 cannot be scheduled on the same machine with any job in $\mathcal{F} \cup \{J_n\}$. That is, all the jobs of $\mathcal{F} \cup \{J_n\}$ are scheduled on two machines in the optimal schedule.

Assume to the contrary that $|\mathcal{F}| \geq 3$. Then, one of M_2, M_3 has been scheduled at least two jobs when J_n comes, implying by Lemma 9 that $L_2^{n-1} + L_3^{n-1} > b + 2(\beta(3) - \alpha(3))$. Therefore,

$$C^* \geq \frac{1}{2}\left(L_2^{n-1} + L_3^{n-1} + p_n\right) > \frac{1}{2}(b + 2(\beta(3) - \alpha(3)) + p_n) = \frac{C^A}{2} + \beta(3) - \alpha(3).$$

By Lemma 8(i) and $C^A \leq 2$, we have

$$\frac{C^A}{C^*} < \frac{2C^A}{C^A + 2(\beta(3) - \alpha(3))} \leq \frac{2}{1 + \beta(3) - \alpha(3)} < 1 + \beta(3), \text{ a contradiction.}$$

This proves $|\mathcal{F}| = 2$. □

Definition 5. *A job J_j is* large *if $p_j > \beta(3)$ (cf. Definition 2).*

Lemma 13. $|\mathcal{F}| = 4$.

Proof. By Lemma 11, $2 \leq |\mathcal{F}| \leq 4$; Assume to the contrary $|\mathcal{F}| \leq 3$. Then, when J_n arrives, one of M_2, M_3 has been scheduled with at most one job, implying by Lemma 9 that $C^A = b + p_n \leq 2$. By Lemma 12, this in turn implies $|\mathcal{F}| = 2$.

Since $L_i^{n-1} \geq b > \beta(3)$, for $i = 2, 3$ (Lemma 9), there is exactly one job, which is large, on each of M_2, M_3 when J_n arrives. That is, $\mathcal{F} = \{J_{2_1}, J_{3_1}\}$ by Definition 4 and both are large jobs.

Apparently none of J_{2_1}, J_{3_1} is scheduled with J_n on the machine in the optimal offline schedule, since otherwise $C^* \geq b + p_n = C^A$. By Lemma 12, J_{2_1} and J_{3_1} have to be scheduled on the same machine in the optimal offline schedule, and thus $C^* \geq p_{2_1} + p_{3_1}$. On the other hand, in the computed schedule, when J_{3_1} arrives, the load of M_2 is $p_{2_1} > \beta(3)$ since J_{2_1} arrives before J_{3_1}. since $p_{3_1} > \beta(3)$ and J_{3_1} is scheduled on M_3, M_3 is the only machine safe for J_{3_1}, i.e., $p_{2_1} + p_{3_1} > 1 + \beta(3)$. Therefore, $C^* > 1 + \beta(3)$. Recall that $C^A \leq 2$ and by Lemma 8(ii), we have $\frac{C^A}{C^*} < 1 + \beta(3)$, a contradiction.

This contradiction proves that $|\mathcal{F}| = 4$. □

Lemma 14. $L_i^{n-1} > 1$ *for $i = 2, 3$, and at most two jobs of $\mathcal{F}$ are large.*

Proof. By Lemmas 12 and 13, $C^A = b + p_n > 2$, and thus $b > 1$ (which is stronger than $b > \beta(3)$ stated in Lemma 9). By Lemma 9, $L_i^{n-1} \geq b > 1$, for every $i = 1, 2, 3$. It follows that the three largest of the four jobs of $\mathcal{F}$ (Lemma 13) have their total job processing time greater than $\frac{3}{4}(1 + 1) = \frac{3}{2}$.

Using $b < 1 + \alpha(3)$ by Lemma 10, we have $2 < C^A = b + p_n < 2 + \alpha(3)$ and $1 \geq p_n > 1 - \alpha(3) > \beta(3)$ (which is stronger than $p_n > \beta(3) - \alpha(3)$ stated in Lemma 10).

If $C^* \geq 1 + \beta(3)$, then by Lemma 8(ii), $\frac{C^A}{C^*} < \frac{2+\alpha(3)}{1+\beta(3)} < 1 + \beta(3)$, a contradiction. Therefore, $C^* < 1 + \beta(3)$, and thus no large job can be scheduled on the same machine with J_1 in the optimal offline schedule.

Note that the three largest of the four jobs of $\mathcal{F}$ cannot be scheduled on the same machine in the optimal schedule, since otherwise $C^* > \frac{3}{2} > 1 + \beta(3)$, a contradiction. It follows that if there are three or more large jobs in $\mathcal{F}$, then at least one of them is scheduled on the same machine with J_n, which is also large, in the optimal schedule. Therefore, $C^* > \beta(3) + p_n$. By Lemma 8(i) and $p_n > 1 - \alpha(3)$, we have

$$\frac{C^A}{C^*} < \frac{1 + \alpha(3) + p_n}{\beta(3) + p_n} < \frac{2}{1 + \beta(3) - \alpha(3)} < 1 + \beta(3), \text{ a contradiction.}$$

Therefore, there are at most two large jobs in $\mathcal{F}$. □

Theorem 2. *The algorithm* GLT3 *is* $1+\beta(3)$*-competitive, where* $\beta(3)$ *is defined in Eq. (11) and* $1 + \beta(3) < 1.4655$.

Proof. This is done by showing for this minimal counterexample $(J_1, J_2, \ldots, J_n)$, there are at least three large jobs in $\mathcal{F}$, contradicting Lemma 14. The details are available in the full version of this extended abstract.

4 Conclusion

In this paper, we studied the semi-online multiprocessing scheduling problem $P \mid \max \mid C_{\max}$, in which the largest job processing time is known to the scheduler in advance. We presented the first 1.8686-competitive algorithm for any number of machines, confirming that the largest job processing time enables designing better online algorithms. The novel idea in the design is to identify double safe or safe machines, using the largest job processing time, and to assign the current job to one of them with maximum load. If no such double safe or safe machine exists, then the algorithm assigns the current job to the least loaded machine, the same as the classic LS algorithm. Extending the design idea, we also presented an improved 1.4655-competitive algorithm for $P3 \mid \max \mid C_{\max}$. For future research, besides designing better algorithms, one might want to show improved lower bounds for $P \mid \max \mid C_{\max}$. It would be interesting to see an optimal online algorithm when there are three machines.

Acknowledgments. The research is supported by the NSERC Canada, the National Natural Science Foundation of China Grant 12071427, and the Ministry of Science and Technology of China Grants G2022040004L and G2023016016L.

References

1. Albers, S., Hellwig, M.: Semi-online scheduling revisited. Theor. Comput. Sci. **443**, 1–9 (2012)
2. Angelelli, E., Nagy, A.B., Speranza, M.G., Tuza, Z.: The on-line multiprocessor scheduling problem with known sum of the tasks. J. Sched. **7**, 421–428 (2004)
3. Azar, Y., Regev, O.: On-line bin-stretching. Theor. Comput. Sci. **268**, 17–41 (2001)
4. Böhm, M., Sgall, J., van Stee, R., Veselý, P.: Online bin stretching with three bins. J. Sched. **20**(6), 601–621 (2017). https://doi.org/10.1007/s10951-016-0504-y
5. Böhm, M., Sgall, J., van Stee, R., Vesely, P.: A two-phase algorithm for bin stretching with stretching factor 1.5. J. Comb. Optim. **34**, 810–828 (2017)
6. Cai, S.Y.: Semi online scheduling on three identical machines. J. Wenzhou Normal Coll. (Nat. Sci. Ed.) **23**, 1–3 (2002). (in Chinese)
7. Chen, B., Vliet, A., Woeginger, G.: New lower and upper bounds for on-line scheduling. Oper. Res. Lett. **16**, 221–230 (1994)
8. Cheng, T., Kellerer, H., Kotov, K.: Algorithms better than LPT for semi-online scheduling with decreasing processing times. Oper. Res. Lett. **40**, 349–352 (2012)
9. Faigle, U., Kern, W., Turan, G.: On the performance of on-line algorithms for partition problems. Acta Cybern. **9**, 107–119 (1989)
10. Fleischer, R., Wahl, M.: On-line scheduling revisited. J. Sched. **3**, 343–353 (2000)
11. Gormley, T., Reingold, N., Torng, E., Westbrook, J.: Generating adversaries for request-answer games. In: Proceeding of the 11th ACM-SIAM Symposium on Discrete Algorithms, pp. 564–565 (2000)
12. Graham, R.L.: Bounds for certain multiprocessing anomalies. Bell Labs Tech. J. **45**, 1563–1581 (1966)
13. Graham, R.L.: Bounds on multiprocessing timing anomalies. SIAM J. Appl. Math. **17**, 263–269 (1969)
14. He, Y., Zhang, G.: Semi online scheduling on two identical machines. Computing **62**, 179–187 (1999)
15. Kellerer, H., Kotov, V., Gabay, M.: An efficient algorithm for semi-online multiprocessor scheduling with given total processing time. J. Sched. **18**(6), 623–630 (2015). https://doi.org/10.1007/s10951-015-0430-4
16. Kellerer, H., Kotov, V., Speranza, M.G., Tuza, Z.: Semi on-line algorithms for the partition problem. Oper. Res. Lett. **21**, 235–242 (1997)
17. Lee, K., Lim, K.: Semi-online scheduling problems on a small number of machines. J. Sched. **16**, 461–477 (2013)
18. Rudin, J.F., III., Chandrasekaran, R.: Improved bound for the online scheduling problem. SIAM J. Comput. **32**, 717–735 (2003)
19. Seiden, S.S., Sgall, J., Woeginger, G.: Semi online scheduling with decreasing job sizes. Oper. Res. Lett. **27**, 215–221 (2000)
20. Wu, Y., Huang, Y., Yang, Q.F.: Semi-online multiprocessor scheduling with the longest given processing time. J. Zhejiang Univ. Sci. Ed. **35**, 23–26 (2008). (in Chinese)

Acyclically Edge Color Triangle-free Toroidal Graphs in $\Delta + 2$ Colors

Qiaojun Shu[1] and Guohui Lin[2(✉)]

[1] Department of Mathematics, Hangzhou Dianzi University, Hangzhou, China
qjshu@hdu.edu.cn

[2] Department of Computing Science, University of Alberta, Edmonton, Canada
guohui@ualberta.ca

Abstract. An acyclic edge coloring of a graph G colors the edges in G such that adjacent edges receive distinct colors and the subgraph induced by the edges of any two colors is acyclic. It is conjectured that every simple graph with maximum degree Δ can be acyclically edge $(\Delta + 2)$-colored. The conjecture has been proven true on triangle-free planar graphs, but is unknown true or not for general planar graphs. We make a step forward to acyclically edge color triangle-free toroidal graphs in $\Delta + 2$ colors. This improves the latest $(\Delta + 3)$-coloring scheme by Chen and Hou, and completely resolves affirmatively the conjecture on such a super class of triangle-free planar graphs. We achieve the result by first showing that every triangle-free toroidal graph has one of the five specified groups of local structures, and then inductively coloring the edges at the presence of each local structure.

Keywords: Acyclic edge coloring · triangle-free toroidal graphs · discharging method · induction

1 Introduction

Only simple and connected graphs are considered in this paper. Let G be such a graph with vertex set $V(G)$ and edge set $E(G)$. For an integer $k \geq 2$, an *edge k-coloring* is a mapping $c : E(G) \to \{1, 2, \ldots, k\}$ such that any two adjacent edges receive distinct colors. G is *edge k-colorable* if G has an edge k-coloring. The *chromatic index* $\chi'(G)$ of G is the smallest integer k such that G is edge k-colorable. An edge k-coloring c of G is called *acyclic* if there are no bichromatic cycles in G, i.e., the subgraph of G induced by the edges of any two colors is a forest. The *acyclic chromatic index* of G, denoted by $a'(G)$, is the smallest integer k such that G is acyclically edge k-colorable.

Let Δ denote the maximum degree of the graph G. One sees that $\Delta \leq \chi'(G) \leq a'(G)$. Note that $\chi'(G) \leq \Delta + 1$ by Vizing's theorem [20] while $a'(K_4) = \Delta + 2$. Fiamčik [10] and Alon, Sudakov, and Zaks [2] independently made the *acyclic edge coloring conjecture* (AECC) that for any graph G, $a'(G) \leq \Delta + 2$.

D. Du et al. (Eds.): COCOA 2024, LNCS 15434, pp. 226–239, 2025.
https://doi.org/10.1007/978-981-96-4445-2_19

The following milestones on AECC have been achieved: Alon, McDiarmid and Reed [1] proved that $a'(G) \leq 64\Delta$ by a probabilistic argument. This upper bound was improved to 16Δ [13], to $\lceil 9.62(\Delta - 1)\rceil$ [14], to $4\Delta - 4$ [9], and most recently in 2017 to $\lceil 3.74(\Delta - 1)\rceil + 1$ by Giotis et al. [11] using the Lovász local lemma. On the other hand, the AECC has been confirmed true for graphs with $\Delta \in \{3, 4\}$ [3,4,18,22], respectively, and true for graphs with $\Delta \geq 4.2 \times 10^{14}$ [8].

Planar graphs have genus 0, that is, they are embeddable in the plane without any crossings (i.e., edges meet only at the vertices of the graph). They can be a starting class of graphs to investigate the AECC. For any planar graph G, Basavaraju et al. [5] showed that $a'(G) \leq \Delta + 12$; and this upper bound was improved to $\Delta + 7$ by Wang et al. [24] and to $\Delta + 6$ by Wang and Zhang [21]. The conjecture has been confirmed true for planar graphs without i-cycles for each $i \in \{3, 4, 5, 6\}$ in [16,17,23,26], respectively. Recently, Shu et al. [15] showed that AECC is true on planar graphs without intersecting triangles.

A graph G is *toroidal* if G can be embedded on a torus. Toroidal graphs have genus at most 1, on which there is an equal amount of study to its subclass of planar graphs, including numerous literature on their various parameters such as the odd chromatic number, the list vertex arboricity, and the acyclically choosability [7,12,19,25]. In particular, Xu [27] examined the acyclic edge coloring of toroidal graphs and showed several sufficient conditions for a toroidal graph G to have $a'(G) = \Delta$.

A *triangle* is synonymous for a 3-cycle. A *triangle-free* planar/toroidal graph is a planar/toroidal graph which contains no triangles. Recall that the AECC holds true for triangle-free planar graphs [17]. For a triangle-free toroidal graph G, Chen and Hou [6] have showed that $a'(G) \leq \Delta + 3$, which is only 1 off the conjectured upper bound of $\Delta + 2$. In this paper, we completely settle the AECC on triangle-free toroidal graphs, by showing the following main theorem.

Theorem 1. *The AECC is true for triangle-free toroidal graphs.*

The rest of the paper is organized as follows. In Sect. 2, we characterize five local configurations or local structures, and by discharging methods we prove that any triangle-free toroidal graph must contain one of these local structures. Incorporating a known property of edge colorings and bichromatic cycles (Lemma 6, proven by Shu and her coauthors in [18]), in Sect. 3 we prove by an induction on the number of edges that any triangle-free toroidal graph admits an acyclic edge $(\Delta + 2)$-coloring. We conclude the paper in Sect. 4.

2 The Local Configurations

Note that in the context of edge coloring, we can assume without loss of generality that graphs are 2-connected. We continue to use most of the notations introduced in [15]. Given a graph G, let $d_G(v)$ denote the degree of the vertex v in G (and the subscript is omitted when the graph is clear from the context). A vertex of degree k (at least k, at most k, respectively) is called a *k-vertex* (*k^+-vertex*, *k^--vertex*, respectively). Let $n_k(v)$ ($n_{k^+}(v)$, $n_{k^-}(v)$, respectively) denote the number of k-vertices (k^+-vertices, k^--vertices, respectively) adjacent to v in G, abbreviated as k-neighbors (k^+-neighbors, k^--neighbors, respectively).

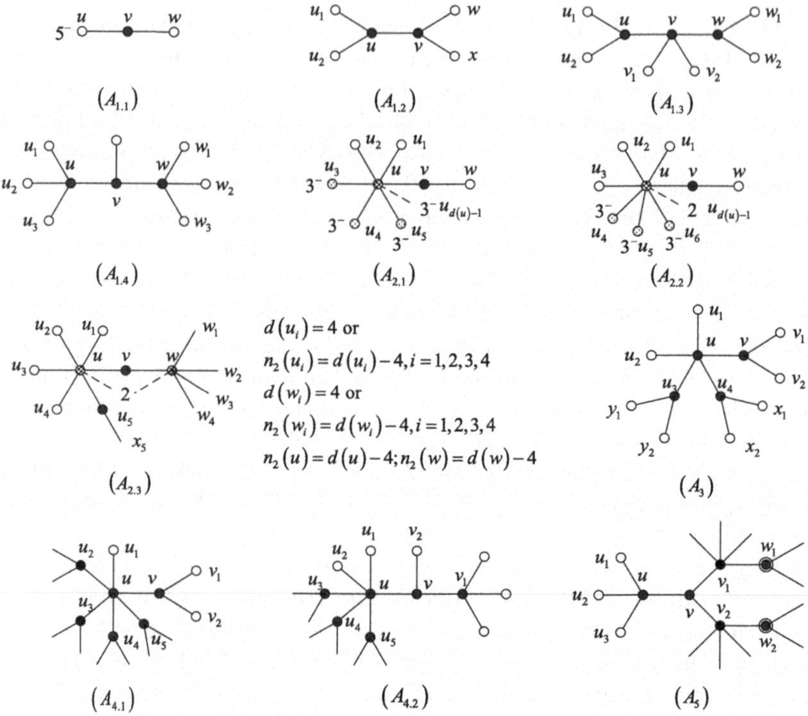

Fig. 1. The five groups of local configurations (A_1)–(A_5) specified in Theorem 2. In each local structure, filled vertices have all their neighbors determined while the non-filled ones could be adjacent to other vertices outside of the local structure. Each dashed edge represents zero, one, or more edges in the graph.

Theorem 2. *A 2-connected triangle-free toroidal graph G with $\Delta \geq 5$ contains one of the following local configurations (A_1)–(A_5) illustrated in Fig. 1:*

(A_1) *This configuration has a path uvw and includes four cases in each of which one of the following four holds:*
($A_{1.1}$) $d(v) = 2$ *and* $d(u) \leq 5$;
($A_{1.2}$) $d(u) = d(v) = 3$;
($A_{1.3}$) $d(v) = 4$ *and* $d(u) = d(w) = 3$;
($A_{1.4}$) $d(v) = 3$ *and* $d(u) = d(w) = 4$.
(A_2) *This configuration has a vertex u with $n_2(u) \geq 1$. Let $u_1, u_2, \ldots, u_{d(u)-1}$ and $v = u_{d(u)}$ be the sorted neighbors of u such that $d(v) = 2$ and $d(u_i)-n_2(u_i) \geq d(u_{i+1})-n_2(u_{i+1})$ for every $i = 1, 2, \ldots, d(u)-2$. The configuration includes three cases in each of which one of the three following holds:*

($A_{2.1}$) $n_2(u) + n_3(u) \geq d(u) - 2$;
($A_{2.2}$) $n_2(u) + n_3(u) = d(u) - 3$, *and* $n_3(u) \leq 3$;
($A_{2.3}$) $n_2(u) = d(u) - 4$, $n_2(w) = d(w) - 4$, $d(x) \in \{4, n_2(x) + 4\}$ *for each* $x \in \{u_1, u_2, u_3, u_4, w_1, w_2, w_3, w_4\}$, *where* w *is the other neighbor of* v *and* w_1, w_2, w_3, w_4 *are non-2-neighbors of* w.

(A_3) *This configuration hasa* 5*-vertex* u *adjacent to* u_1, u_2, *and three* 3*-vertices* u_3, u_4, v.

(A_4) *This configuration hasa* 6*-vertex* u *adjacent to* u_1, u_2 *and four* 3*-vertices* u_3, u_4, u_5, v. *It includes two cases in each of which one of the following two holds:*
($A_{4.1}$) $d(u_2) = 3$ *too;*
($A_{4.2}$) v *has a* 4*-neighbor* v_1.

(A_5) *This configuration hasa path* $u_1uvv_1w_1$ *with* $d(u) = 4$, $d(v) = 3$, $d(v_1) = 5$ *and* $d(w_1) = 3$, *and another path* $u_1uvv_2w_2$ *with* $d(v_2) = 5$ *and* $d(w_2) = 3$, *where* w_2 *could collide into* w_1.

The rest of the section is devoted to the proof of Theorem 2, by contradiction. To this purpose, we assume to the contrary that G contains none of the local configurations (A_1)–(A_5). We employ several known local structural properties for a non-crossing embedding of a graph from [15,16]. These properties are summarized in Lemmas 1–3.

2.1 Definitions and Notations

Since G is 2-connected, $d_G(v) \geq 2$ for any $v \in V(G)$. Let G' be the graph obtained from G by deleting all the 2-vertices of G, and let H be a connected component of G'. Clearly, every vertex $v \in V(H)$ has $d_G(v) \geq 3$, and H is also triangle-free toroidal. In what follows, we assume that H is embedded in torus such that its edges intersect only at their ending vertices. We use the following notations with respect to H:

- $N_H(v) = \{u \mid uv \in E(H)\}$ and $d_H(v) = |N_H(v)|$—the set of neighbors of v and the degree of v;
- similarly define what a k*-vertex*, a k^+*-vertex* and a k^-*-vertex* are;
- $n'_k(v)$ ($n'_{k^+}(v)$, $n'_{k^+}(v)$, respectively)—the number of k-neighbors (k^+-neighbors, k^--neighbors, respectively) of v;
- $F(H)$—the face set;
- $V(f)$—the set of vertices on (the boundary of) the face f;
- a vertex v and a face f are *incident* if $v \in V(f)$;
- $n_{k^+}(f)$—the number of k^+-vertices in $V(f)$;
- $d(f)$—the degree of the face f, which is the number of edges on the face f with cut edges counted twice, and similarly define what a k*-face*, a k^+*-face*, and a k^-*-face* are;
- $F(v) = \{f \in F(H) \mid v \in V(f)\}$;
- $m_k(v)$ ($m_{k^+}(v)$, $m_{k^-}(v)$, respectively)—the number of k-faces (k^+-faces, k^--faces, respectively) in $F(v)$;

- for a k-vertex v in H, let $u_1, u_2, \ldots, u_k$ be all the neighbors of v indexed clockwise, and denote the face containing u_1vu_2 (u_2vu_3, ..., $u_{k-1}vu_k$, u_kvu_1, respectively) as f_1 ($f_2, \ldots, f_{k-1}, f_k$, respectively). (Note that these faces might not be distinct from each other.)

2.2 Structural Properties

Assuming the in-existences of some of (A_1)–(A_5), the next two lemmas are proven for planar graphs without 5-cycles in [16] or for planar graphs without intersecting triangles in [15], but in fact they hold for all graphs.

Lemma 1. [15] *The minimum degree of the vertices in H is $\delta(H) \geq 3$. Furthermore, if the degree of a vertex u in G is $3 \leq d_G(u) \leq 5$, then $d_H(u) = d(u)$; if $d_G(u) \geq 6$, then $d_H(u) \geq 4$.*

Lemma 2. [16] (1) *For every vertex $u \in V(H)$, $n_3'(u) = n_3(u)$.*
(2) *If a vertex u has $d_H(u) = 4$ and $n_3'(u) \geq 1$, then $d_G(u) = 4$.*

Lemma 3. [15] (1) *Let u be a vertex with $d_H(u) = 5$. Then $n_3(u) \leq 2$; and if $n_3(u) = 2$ then $d_G(u) = 5$.*
(2) *Let u be a vertex with $d_H(u) = 6$. Then $n_3(u) \leq 4$; if $n_3(u) \geq 3$, then $d_G(u) = 6$; and if $n_3(u) = 4$, then $n_{5+}'(v) \geq 2$ for every 3-neighbor v of u in H.*

Lemma 4. *Let v be a vertex with $d_H(v) = 3$ and $n_4'(v) \geq 1$, and u be a 4-neighbor of v in H. Then $d_G(u) = d_H(u) = 4$ and $n_4'(v) = n_4(v) = 1$ (i.e., u is the only 4-neighbor of v).*

Proof. By Lemma 1, $d_G(v) = d_H(v) = 3$ and thus v is a 3-neighbor of u in H (i.e., $n_3'(u) \geq 1$); and then by Lemma 2, $d_G(u) = d_H(u) = 4$.

Let x be another neighbor of v. Since G contains no $(A_{1.4})$ or $(A_{1.2})$, $d_G(x) \geq 5$ and thus by Lemma 1 and the last paragraph we have $d_H(x) \geq 5$ too. That is, $n_4'(v) = n_4(v) = 1$. □

2.3 Discharging to Show Contradictions

To derive a contradiction, we make use of the discharging method, which is analogous to amortized analysis, on H. First, by Euler's formula for toroidal graphs $|V(H)| - |E(H)| + |F(H)| = 0$ and the relations $\sum_{u\in V(H)} d_H(u) = \sum_{f\in F(H)} d(f) = 2|E(H)|$, we have the following equality:

$$\sum_{u\in V(H)} (d_H(u) - 4) + \sum_{f\in F(H)} (d(f) - 4) = 0. \quad (1)$$

The High-Level Proof Scheme: We define an *initial* weight function $\omega(\cdot)$ on $V(H) \cup F(H)$, where

$$\begin{cases} \omega(u) = d_H(u) - 4, & \text{for each vertex } u \in V(H); \\ \omega(f) = d(f) - 4, & \text{for each face } f \in F(H). \end{cases} \tag{2}$$

It follows from Eq. (1) that the total weight of the elements of $V(H) \cup F(H)$ is $\omega(V(H) \cup F(H)) = 0$.

Below we will define two discharging rules (R1) and (R2) to move a portion of weight from one vertex to another (the function $\tau(u \to v)$) or from a face to a vertex (the function $\tau(f \to u)$). That is, $\tau(x \to y)$ denotes the amount of weight redistributed from x to y. At the end of discharging, we achieve a *new* weight function $\omega'(\cdot)$ and we will show two statements that 1) $\omega'(x) \geq 0$ for every element $x \in V(H) \cup F(H)$ and 2) there is a vertex u for which $\omega'(u) > 0$. In other words, the total weight after discharging becomes positive, which contradicts Eq. (1).

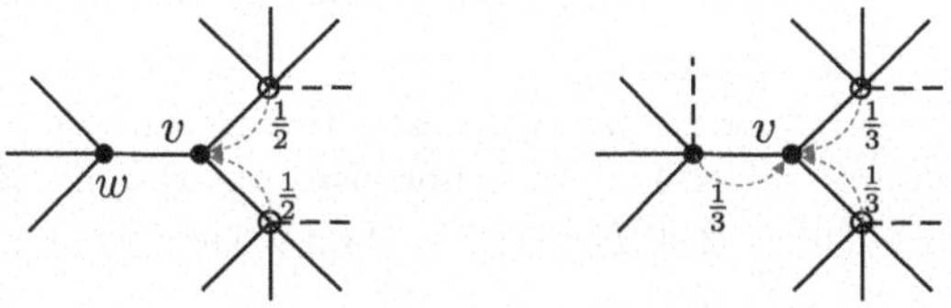

(a) A bad 3-vertex v. (b) A regular 3-vertex v.

Fig. 2. The discharging rule (R1) $\tau(u \to v)$ applying to a 5^+-vertex u and its 3-neighbor v: (a) v is bad and $\tau(u \to v) = \frac{1}{2}$; (b) v is not bad and $\tau(u \to v) = \frac{1}{3}$.

The Discharging Rules and Proofs of the Two Statements:
A 3-vertex v is called *bad* if v is incident to a 4-vertex w, see Fig. 2 for an illustration. For each vertex u, let $b_3'(u)$ denote the number of bad 3-vertices adjacent to u. The two discharging rules are as follows.

(R1) When a 5^+-vertex u is adjacent to a 3-vertex v, $\tau(u \to v) = \frac{1}{2}$ if v is bad, or otherwise $\tau(u \to v) = \frac{1}{3}$.
(R2) When a 5^+-face f is adjacent to a 5^+-vertex u, $\tau(f \to u) = \frac{1}{5}$.

We proceed to validate the two statements, using Eq. (2), (R1) and (R2):

1) $\omega'(x) \geq 0$ for every element $x \in V(H) \cup F(H)$. There are eight separate cases.
   1. For every face $f \in F(H)$, if $d(f) = 4$, then $\omega'(f) = \omega(f) = d(f) - 4 = 0$. Otherwise, $d(f) \geq 5$. Then $\omega'(f) = \omega(f) - \frac{1}{5} \times n_{5+}(f) \geq d(f) - 4 - \frac{1}{5} \times d(f) = \frac{4}{5} \times d(f) - 4 \geq \frac{4}{5} \times 5 - 4 = 0$.
   Recall from Lemma 1 that $d_H(u) \geq 3$.
   2. If $d_H(u) = 3$, then $\omega(u) = d_H(u) - 4 = -1$. By Lemma 4 and (R1), $\omega'(u) = -1 + \frac{1}{2} \times 2 = 0$ if u is bad, or $\omega'(u) = -1 + \frac{1}{3} \times 3 = 0$ otherwise.

3. If $d_H(u) = 4$, then $\omega'(u) = \omega(u) = d_H(u) - 4 = 0$ by (R1).
4. If $d_H(u) \geq 9$, then $\omega'(u) \geq d_H(u) - 4 - \frac{1}{2} \times b_3'(u) - \frac{1}{3} \times (n_3'(u) - b_3'(u)) \geq d_H(u) - 4 - \frac{1}{2} \times d_H(u) = \frac{1}{2} \times d_H(u) - 4 \geq \frac{1}{2}$.
5. If $d_H(u) = 8$, then $\omega(u) = d_H(u) - 4 = 4$. When $n_3'(u) \leq 7$, $\omega'(u) \geq 4 - \frac{1}{2} \times n_3'(u) \geq 4 - \frac{1}{2} \times 7 = \frac{1}{2}$; when ($n_3'(u) = 8$ and) $b_3'(u) \leq 7$, $\omega'(u) \geq 4 - \frac{1}{2} \times 7 - \frac{1}{3} \times 1 = \frac{1}{6}$; in the other case, $b_3'(u) = n_3'(u) = n_3(u) = 8$. Since G does not contain ($A_{1.4}$), $m_{5^+}(u) \geq 1$ and $\omega'(u) \geq 4 - \frac{1}{2} \times 8 + \frac{1}{5} \times m_{5^+}(u) \geq \frac{1}{5}$.
6. If $d_H(u) = 7$, then $\omega(u) = d_H(u) - 4 = 3$. When $n_3'(u) \leq 5$, $\omega'(u) \geq 3 - \frac{1}{2} \times n_3'(u) \geq 3 - \frac{1}{2} \times 5 = \frac{1}{2}$; when ($n_3'(u) \geq 6$ and) $b_3'(u) \leq 3$, $\omega'(u) \geq 3 - \frac{1}{2} \times 3 - \frac{1}{3} \times 4 = \frac{1}{6}$; in the other case, $n_3'(u) \geq 6$ and $b_3'(u) \geq 4$. Since G does not contain ($A_{1.4}$), $m_{5^+}(u) \geq 1$. If $n_3'(u) = 6$, then $\omega'(u) \geq 3 - \frac{1}{2} \times 6 + \frac{1}{5} \times m_{5^+}(u) \geq \frac{1}{5}$. If $n_3'(u) = 7$, then $\omega'(u) \geq 3 - \frac{1}{2} \times b_3'(u) - \frac{1}{3} \times (7 - b_3'(u)) + \frac{1}{5} \times m_{5^+}(u) = \frac{2}{3} - \frac{1}{6} \times b_3'(u) + \frac{1}{5} \times m_{5^+}(u)$. It follows that when $b_3'(u) \leq 5$, $\omega'(u) \geq \frac{2}{3} - \frac{1}{6} \times 5 + \frac{1}{5} \times 1 = \frac{1}{30}$; when $b_3'(u) \geq 6$, $m_{5^+}(u) \geq 3$ and $\omega'(u) \geq 3 - \frac{1}{2} \times 7 + \frac{1}{5} \times 3 = \frac{1}{10}$.
7. If $d_H(u) = 6$, then $\omega(u) = d_H(u) - 4 = 2$. When $n_3'(u) \leq 3$, $\omega'(u) \geq 2 - \frac{1}{2} \times n_3'(u) \geq 2 - \frac{1}{2} \times 3 = \frac{1}{2}$; otherwise, $n_3'(u) \geq 4$. By Lemma 3(2), $d_G(u) = 6$ and $b_3'(u) = 0$, and consequently $\omega'(u) \geq 2 - \frac{1}{2} \times 0 - \frac{1}{3} \times 4 = \frac{2}{3}$.
8. Lastly, if $d_H(u) = 5$, then $\omega(u) = d_H(u) - 4 = 1$. When $n_3'(u) \leq 1$, $\omega'(u) \geq 1 - \frac{1}{2} \times n_3'(u) \geq 1 - \frac{1}{2} \times 1 = \frac{1}{2}$; otherwise, $n_3'(u) \geq 2$. By Lemma 3(1), $d_G(u) = 5$ and $n_3'(u) = 2$. If $b_3'(u) \leq 1$, then $\omega'(u) \geq 1 - \frac{1}{2} \times 1 - \frac{1}{3} \times 1 = \frac{1}{6}$; if $b_3'(u) = 2$, then $\omega'(u) \geq 1 - \frac{1}{2} \times 2 = 0$.

2) There is a vertex u for which $\omega'(u) > 0$.
From the above analysis, when there is a vertex u with $d_H(u) \geq 6$ (Cases 4–7), its $\omega'(u) > 0$ and the statement is proved. We thus assume the remaining case where $\Delta(H) \leq 4$ (Cases 2–3), or $\Delta(H) = 5$ such that every 5-vertex u in H is adjacent to exactly two 3-vertices both of which are bad (i.e., $n_3'(u) = b_3'(u) = 2$, the last scenario of Case 8).
Since $\Delta \geq 5$, $\Delta(H) \leq 4$ implies the existence of (A_2); $\Delta(H) = 5$ implies either a bad 3-vertex has a positive new weight, or it is adjacent to two 5-vertices and thus by Lemma 4 (A_5) exists. Therefore, either the statement is proved or we derive a contradiction.

The above two statements together say that the total weight of the elements of $V(H) \cup F(H)$ is positive, but not 0 as in Eq. (1), a contradiction.

We remark that for planar graphs one only needs to prove the non-negativity of the total weight of the vertices and faces (for example, as in [15]), while here for toroidal graphs we have to locate a vertex or face of positive weight, because toroidal graphs have genus up to 1.

3 Acyclic Edge Coloring

We now show how to derive an acyclic edge $(\Delta+2)$-coloring by an induction on $|E(G)|$. The following lemma gives some base cases for the induction.

Lemma 5. [3,4,18,22] *If* $\varDelta \in \{3, 4\}$, *then* $a'(G) \le \varDelta + 2$.

Let c denote a partial acyclic edge k-coloring of the graph G using the color set $C = \{1, 2, \ldots, k\}$. For a vertex $v \in V(G)$, let $C(v)$ denote the set of colors assigned to the edges incident with v under c. If the edges of a path $P = ux \ldots v$ are alternatively colored by colors i and j, we call P an $(i, j)_{(u,v)}$-path. We remark that, if $uv \in E(G)$, then it is colored by some color $k \ne i, j$, or it is uncolored yet, under c. Furthermore, if uv is uncolored yet, then coloring it by i or j would give rise to an $(i, j)_{(u,v)}$-cycle $ux \ldots vu$. Such a bichromatic cycle is infeasible, and its edge coloring needs to be revised.

For simplicity, we use $\{e_1, e_2, \ldots, e_m\} \to a$ to state that all the edges e_1, e_2, ..., e_m are colored by a (or re-colored by a if any of them was colored under c previously), $e_1 \to S$ ($S \ne \emptyset$) to state that e_1 is colored by a color in S, $e_1 \to a$ to state that e_1 is colored by a, and $(e_1, e_2, \ldots, e_m) \to (a_1, a_2, \ldots, a_m)$ to state that e_j is colored by a_j, for each j, respectively. We also use $(e_1, e_2, \ldots, e_m)_c = (a_1, a_2, \ldots, a_m)$ to denote that $c(e_j) = a_j$, for each j, respectively. This way, when $(uv, xy)_c = (a, b)$, switching their colors is represented as $(uv, xy) \to (b, a)$.

Lemma 6. [18] *Suppose G has an acyclic edge coloring c, and $P = uv_1v_2$-...-v_kv_{k+1} is a maximal $(a, b)_{(u,v_{k+1})}$-path with $c(uv_1) = a$ and $b \notin C(u)$. Then there is no $(a, b)_{(u,w)}$-path for any vertex $w \notin V(P)$.*

Proof of Theorem 1. The induction is on the number $|E(G)|$ of edges in the graph G. For the base cases, when $|E(G)| \le \varDelta+2$, each edge is colored distinctly leading to a valid edge $(\varDelta + 2)$-coloring; when $\varDelta \le 4$, an acyclic edge $(\varDelta + 2)$-coloring exists by Lemma 5.

In the sequel we fix a 2-connected graph G for discussion, and assume that $\varDelta \ge 5$ and $|E(G)| \ge \varDelta + 3$.

Theorem 2 tells that G contains at least one of the five configurations (A_1)–(A_5). In each of these configurations, there is an edge uv such that $d(u)+d(v) \le 8$ or $d(v) \in \{2, 3\}$. We pick such an edge uv and let $H = G - uv$ (which is the shorthand for $H = (V(G), E(G) \setminus \{uv\})$). The graph H is also triangle-free toroidal and $\varDelta \ge \varDelta(H) \ge \varDelta - 1 \ge 4$. By the inductive assumption, H has an acyclic edge $(\varDelta + 2)$-coloring c using the color set $C = \{1, 2, \ldots, \varDelta + 2\}$. This c is the partial acyclic edge $(\varDelta + 2)$-coloring we use for G, in which the only uncolored edge is uv.

Since $d_H(u) + d_H(v) \le 6$, or $d_H(u) \le \varDelta - 1$ and $d_H(v) \in \{1, 2\}$, we conclude from $|C| = \varDelta + 2 > \max\{6, \varDelta + 1\}$ that $C \setminus (C(u) \cup C(v)) \ne \emptyset$.

If $C(u) \cap C(v) = \emptyset$, then $uv \to C \setminus (C(u) \cup C(v))$ results in an acyclic edge $(\varDelta + 2)$-coloring for G and we are done. (In the sequel, we say *we are done* to mean that an acyclic edge $(\varDelta + 2)$-coloring for G is achieved.) Similarly, if there exists some $j \in C \setminus (C(u) \cup C(v))$ such that H contains no $(i, j)_{(u,v)}$-path for every $i \in C(u) \cap C(v)$, then $uv \to j$ and we are done. In the remaining case, for each $i \in C(u) \cap C(v)$, let $B_i = \{j \mid \text{there is an } (i, j)_{(u_s,v_t)}\text{-path}\}$, where $c(uu_s) = c(vv_t) = i$. One clearly sees that $B_i \subseteq C(u_s) \cap C(v_t)$. In the sequel we continue the proof with the following assumption (AS#1) and let $S_i = \{c(uu_j) \mid d(u_j) = i\}$ for every $i \in \{2, 3, 4, 5\}$.

(AS#1) $C(u) \cap C(v) \neq \emptyset$; for every $j \in C \setminus (C(u) \cup C(v))$, H contains an $(i,j)_{(u,v)}$-path for some $i \in C(u) \cap C(v)$; and $C \setminus (C(u) \cup C(v)) \subseteq \bigcup_{i \in C(u) \cap C(v)} B_i$.

It is not hard to see that if G contains (A_1) or $(A_{2.1})$, then the proof for the cases $(A_{1.1})$–$(A_{1.4})$ and $(A_{2.1})$ of the planar graphs in [16] can be adopted; if G contains $(A_{2.2})$, (A_3), $(A_{4.1})$, or $(A_{4.2})$, then the proof for the cases $(A_{2.2})$, $(A_{5.1})$, $(A_{6.1})$, and $(A_{3.2})$ of the planar graphs in [15] can be adopted.

3.1 Configuration $(A_{2.3})$

Let us recall the configuration $(A_{2.3})$, and see Fig. 1 for an illustration.

This configuration has a vertex u with $n_2(u) = d(u) - 4 \geq 1$. Let $u_1, u_2, \ldots, u_{d(u)-1}$ and $v = u_{d(u)}$ be the sorted neighbors of u such that $d(v) = 2$ and $d(u_i) - n_2(u_i) \geq d(u_{i+1}) - n_2(u_{i+1})$ for every $i = 1, 2, \ldots, d(u) - 2$. $n_2(w) = d(w) - 4$, and $d(x) \in \{4, n_2(x) + 4\}$ for each $x \in \{u_1, u_2, u_3, u_4, w_1, w_2, w_3, w_4\}$, where w is the other neighbor of v and w_1, w_2, w_3, w_4 are non-2-neighbors of w.

For each $1 \leq i \leq d(u)$, if $d(u_i) = 2$, then let t_i denote the other neighbor of u_i different from u. By (AS#1), $c(vw) \in C(u)$ and H contains a $(c(vw), i)_{(u,w)}$-path for every $i \in C \setminus C(u)$ and $(C \setminus C(u)) \subseteq C(w)$. Note that $|C \setminus C(u)| \geq 3$. If there exists an $i = c(uu_i) \in S_2 \setminus (C(w) \setminus \{c(vw)\})$, then $vw \to i$ and $uv \to (C \setminus C(u)) \setminus C(u_i)$, and we are done. Otherwise, we can assume that:

(AS#2) $c(vw) \in C(u) \setminus S_2$ and $S_2 \cup (C \setminus C(u)) \subseteq C(w) \setminus \{c(vw)\}$.

Suppose $c(uu_i) = i$ for $1 \leq i \leq d(u) - 1$. By (AS#1) and (AS#2), we may further assume that $c(vw) = 1$ and $\{5, 6, \cdots, \Delta + 2\} \subseteq C(w)$. By (AS#1), $C \setminus C(u) \subseteq B_1$.

If there exists an $\alpha \in C \setminus C(u)$ such that $\alpha \notin B_i$ for some $i \in \{2,3,4\} \setminus C(w)$, then $(vw, uv) \to (i, \alpha)$ and we are done. Hence we consider below that $(C \setminus C(u)) \subseteq B^* \triangleq B_1 \cap B_i \subseteq C(u_1) \cap C(u_i)$, where $i \in \{2,3,4\} \setminus C(w)$. Note that $|\{2,3,4\} \setminus C(w)| \geq 2$, and without losing generality, assume that $3, 4 \notin C(w)$.

Lemma 7. *For each $j = c(uu_j) \in S_2$, if (i) $c(u_jt_j) \in C \setminus \{2\} = \{1,3,4\} \cup (C \setminus C(u)) \cup S_2$, or (ii) $c(u_jt_j) = 2$ and, $t_j \neq w$ or $t_j = w$ but G contains no $(2,k)_{(u,w)}$-path for some $k \in C \setminus C(u)$, then*

(1) re-coloring uu_j (and u_jt_j if needed) with some color in $C \setminus C(u)$ gives another acyclic edge $(\Delta + 2)$-coloring of H with $j \notin C(u)$, and
(2) $j \in B_i$ for each $i \in \{1\} \cup (\{2,3,4\} \setminus C(w))$.

Proof. When condition (i) holds, if $c(u_jt_j) \in \{1,3,4\} \cup (C \setminus C(u))$, then $uu_j \to (C \setminus C(u)) \setminus \{c(u_jt_j)\}$; if $c(u_jx_j) = c(uu_s) \in S_2$, then $uu_j \to (C \setminus C(u)) \setminus C(u_s))$. Note that $t_j \neq w$ if $c(u_jt_j) \in \{1,3,4\}$, and thus no bichromatic cycles are produced. That is, the new edge coloring is still acyclic. When condition (ii) holds, if $c(u_jt_j) = 2$ and $t_j \neq w$, then recolor u_jt_j with a color in $(C \setminus \{2, j\}) \setminus C(t_j)$ enables condition (i). If $t_j = w$ and G contains no $(2,k)_{(u,w)}$-path for some $k \in C \setminus C(u)$, then $uu_j \to k$ results in an acyclic edge coloring. This proves (1).

To prove (2), assume $j \notin B_i$ for some $i \in \{1\} \cup (\{2,3,4\} \setminus C(w))$. We first re-color uu_j (and u_jt_j if needed) as in (1), and then $(uv, vw) \to (j, i)$ results in an acyclic edge coloring of G and we are done. □

Lemma 7 basically states that the edge uu_j can be colored by any color in $\{j\} \cup (C \setminus C(u))$, and consequently when the premise holds, we consider j and the colors in $C \setminus C(u)$ to be *equivalent* to each other. Hence, we assume that:

(AS#3) $S_2 \subseteq B^*$, or there exists $j \in S_2$ such that $S_2 \setminus \{j\} \subseteq B^*$, and $t_j = w$, $c(u_jt_j) = 2$ and G contains a $(2,k)_{(u,w)}$-path for each $k \in C \setminus C(u)$.

We continue to explore the configuration under the assumptions (AS#1)–(AS#3). When $d(u_1) = 4$, it follows from (AS#2) and (AS#3) that $d(u) = \Delta = 6$ and $t_5 = w$, $c(u_5t_5) = 2$. Then $(uu_1, uu_5, uv) \to (5,1,6)$ and we are done. Hence, we consider in the following $d(u_1) > 4$ and, for the same reason, $d(u_3) > 4$ and $d(u_4) > 4$.

Let $y_1, y_2, y_3, y_4, \cdots, y_{d(u_1)-1}$ be the neighbors of u_1 such that $d(y_i) = 2$, for $i \in \{4, \cdots, d(u_1)-1\}$, and $S_2^1 = \{c(u_1y_j) \mid d(y_j) = 2\}$.

Firstly, when $|(C \setminus C(u)) \cap S_2^1| \geq 2$, without loss of generality, assume that $c(u_1y_4) = \Delta+2$ and $c(u_1y_5) = \Delta+1$. Swapping the colors of u_1y_4 and u_1y_5, $uv \to \Delta+2$, and we are done.

Secondly, when $|(C \setminus C(u)) \cap S_2^1| = 1$ and $|(S_2 \cap B)^* \cap S_2^1| \geq 1$, without loss of generality, assume that $c(u_1y_4) = \Delta+2$ and $c(u_1y_5) = 5$. Swapping the colors of u_1y_4 and u_1y_5, $uv \to \Delta+2$, $uu_5 \to \Delta+1$ if $c(u_5t_5) = 1$, and we are done.

Thirdly, when $|(S_2 \cap B^*) \cap S_2^1| \geq 2$, w.l.o.g., assume that $c(u_1y_4) = 6$ and $c(u_1y_5) = 5$. Swap the colors of u_1y_4 and u_1y_5 first, and if $c(u_6t_6) = 1$ then $(uu_6, uv) \to (\Delta+2, 6)$, and next if $c(u_5t_5) = 1$ then $uu_5 \to \Delta+1$, and we are done. Otherwise, $c(u_5t_5) \neq 1$ and $c(u_6t_6) \neq 1$, and we assume that $u_6 \neq w$ or $u_6 = w$ and $c(u_6t_6) \neq 2$. Using the same proof for Lemma 7(1), we can recolor uu_6 (and u_6t_6 if needed) and $uv \to 6$ to obtain an acyclic edge coloring of G.

Lastly, we may assume that:

(AS#4) $|((C \setminus C(u)) \cup (S_2 \cap B^*)) \cap S_2^1| \leq 1$.

Note that $|C \setminus C(u)| = \Delta+2-(d(u)-1) = \Delta - d(u) + 3$. It follows that $d(u) \geq \Delta - 1$. When $d(u) = \Delta - 1$, it follows from (AS#4) and (AS#3) that $d(u) = 6$, $c(u_1y_i) = i+5$, $u_5 = w$ and G contains a $(2,k)_{(u,w)}$-path for each $k \in C \setminus C(u)$. Then $(uu_1, uv) \to (9,6)$, $u_1y_4 \to \{2,3,4\} \setminus C(u_1)$, and we are done.

When $d(u) = \Delta$, we can furthermore assume that each vertex adjacent to a 2-vertex is a Δ-vertex. From now no, let $C(w) = C \setminus \{3,4\}$. And it follows from (AS#4) that $\Delta \leq 7$ and it suffices to consider the following two situations:

Case 1. For each $j \in S_2$, $S_2 \subseteq B^*$. It follows from (AS#4) that $\Delta = 6$ and $\{c(u_1y_1), c(u_1y_2), c(u_1y_3), c(u_1y_4)\} = \{5,6,7,8\}$. Note that $\{3,4\} \setminus C(u_1) \neq \emptyset$. Assume that $3 \notin C(u_1)$. By Lemma 7, we assume that $c(u_1y_4) = 8$ and $C(y_4) = \{8,1\}$. Then $(u_1y_4, uu_1, uv) \to (3,8,7)$. Recall that $8 \in B_3 \cap B_4$. If G contains neither a $(5,8)_{(u_1,u_5)}$-path nor a $(2,8)_{(u_1,u_2)}$-path, then we are done.

Otherwise, if G contains a $(5,8)_{(u_1,u_5)}$-path, then recolor $uu_5 \to 1$; if G contains a $(2,8)_{(u_1,u_2)}$-path, then $c(u_1y_5) = 2$, $C(y_5) = \{2,8\}$, and recolor $uu_5 \to 4$. In either case we are done.
Case 2. There exists t_j such that $t_j = w$, $c(u_jt_j) = 2$, and G contains a $(2,k)_{(u,w)}$-path for each $k \in C\backslash C(u)$, say $t_5 = w$. It follows from (AS#4) that the following two subcases can be obtained.
Case 2.1. $\Delta = 7$, $\{6,7,8,9\} \subseteq B^*$ and G contains a $(2,k)_{(u,w)}$-path for each $k \in \{7,8,9\}$. It follows from (AS#4) that $\{c(u_1y_1), c(u_1y_2), c(u_1y_3), c(u_1y_4)\} = \{6,7,8,9\}$. By Lemma 7, we assume that $c(u_1y_4) = 9$ and $C(y_4) = \{9,1\}$. Then $(uu_1, uv) \to (9,8)$ and recolor u_1y_4 with a color in $\{2,3,4\} \setminus C(u_1)$. Recall that $8 \in B_3 \cap B_4$ and G contains a $(2,8)_{(u,w)}$-path. If G contains no $(6,9)_{(u_1,u_6)}$-path, then we are done. Otherwise, recolor $uu_6 \to 1$, and we are done too.
Case 2.2. $\Delta = 6$, $\{6,7,8\} \subseteq B^*$ and G contains a $(2,k)_{(u,w)}$-path for each $k \in \{6,7,8\}$. Let $c(ww_i) = i+4$, $i \in \{1,2,3,4\}$. Note that $6,7,8 \in C(u_i)$, $i \in \{1,2,3,4\}$ and $1,2,3,4 \in C(w_j)$, $j \in \{2,3,4\}$. One sees that $\{u_1,u_2,u_3,u_4\} \cap \{w_2,w_3,w_4\} = \emptyset$.

If $c(u_1y_4) = \beta \in \{5,6,7,8\}$ and $1 \in C(y_4)$, then re-color u_1y_4 with a color in $\{2,3,4\} \setminus C(u_1)$, $uu_1 \to \beta$, $uv \to \gamma \in \{6,7,8\} \setminus \{\beta\}$, and $uu_5 \to 1$ if $\beta = 5$. Otherwise, $\{c(u_1y_1), c(u_1y_2), c(u_1y_3)\} = \{6,7,8\}$, $5 \notin B_1$ and $5 \notin B_3 \cap B_4$ similarly. Note that swapping the colors of vw and u_5w gives another acyclic edge coloring of H. Hence, in the following discussion, we consider $1,2,3,4$ to be equivalent to each other.

Recall that $1,2,3,4 \in C(w_i)$, $i \in \{2,3,4\}$. If follows that $n_2(w_i) = 2$, $i \in \{2,3,4\}$, and $\{6,7\} \setminus C(w_4) \neq \emptyset$, say $6 \in C(w_4)$. Let r_1,r_2,r_3,r_4,r_5 be the neighbors of w_4 such that $d(r_4) = d(r_5) = 2$, and s_1,s_2,s_3,s_4,s_5 be the neighbors of w_2 such that $d(s_4) = d(s_5) = 2$.

When $\{c(w_4r_4), c(w_4r_5)\} \subseteq \{1,2,3,4\}$, say $\{c(w_4r_4), c(w_4r_5)\} = \{1,\zeta\}$, noting that $8 \in C(r_4) \cap C(r_5)$, swapping the colors of w_4r_4 and w_4r_5, $uv \to 8$, and we are done.

Otherwise, $|\{c(w_4r_4), c(w_4r_5)\} \cap \{1,2,3,4\}| = 1$, and $|\{c(w_2s_4), c(w_2s_5)\} \cap \{1,2,3,4\}| = 1$, say $c(w_4r_4) = 1$, $\{c(w_2s_1), c(w_2s_2), c(w_2s_3), c(w_2s_4)\} = \{1,2,3,4\}$ with $c(w_2s_4) = \xi \in \{1,2,3,4\}$, $6 \in C(s_4)$. If G contains no $(8,6)_{(w_2,w_4)}$-path, then $(w_4r_4, uv) \to (6,8)$ and we are done. Otherwise, $c(w_2s_5) = 8$ and $6 \in C(s_5)$, then swapping the colors of w_2s_4 and w_2s_5, $uv \to 6$, and we are done.

The above entire subsection enumerates all possible cases of partial coloring that can happen to the configuration $(A_{2.3})$, and in each case an acyclic edge $(\Delta+2)$-coloring of G can be achieved.

3.2 Configuration (A_5)

Let us recall the configuration (A_5), and see Fig. 1 for an illustration.

This configuration hasa path $u_1uvv_1w_1$ with $d(u) = 4$, $d(v) = 3$, $d(v_1) = 5$ and $d(w_1) = 3$, and another path $u_1uvv_2w_2$ with $d(v_2) = 5$ and $d(w_2) = 3$, where w_2 could collide into w_1.

Let $c(uu_i) = i$ for $i = 1, 2, 3$, and let $C_1 = C \setminus \{1, 2, 3\}$. By (AS#1), $1 \leq |C(u) \cap C(v)| \leq 2$. We consider the following two cases.

Case 1. $|C(u) \cap C(v)| = 1$. Without loss of generality, assume that $c(vv_1) = 1$ and $c(vv_2) = 4$. Note that $C_1 \setminus \{4\} \subseteq B_1 \subseteq C(u_1) \cap C(v_1)$.

When $4 \notin B_1$, if there exists a $\rho \in C_1 \setminus C(v_2)$ or $\rho \in \{2, 3\} \setminus (C(v_1) \cup C(v_2))$, then $(vv_2, uv) \to (\rho, 4)$ and we are done; otherwise, $C(v_1) = C_1 \cup \{1, \alpha\}$ and $C(v_2) = C_1 \cup C\{4, \beta\}$, where $\{\alpha, \beta\} = \{2, 3\}$, then $(vv_2, uv) \to (\alpha, 4)$ and we are done.

When $4 \in B_1$, it follows that $C(u_1) = C(v_1) = \{1\} \cup C_1$ and $\Delta = 5$. If $C(u_i) = \{i\} \cup C_1$ for some $i \in \{2, 3\}$, say $i = 2$, then $(uu_1, uu_2, uv) \to (2, 1, 5)$ and we are done.

Otherwise, $C_1 \setminus C(u_i) \neq \emptyset$ for each $i \in \{2, 3\}$. Note that we are done if we can re-color vv_1 or vv_2 such that $C(u) \cap C(v) = \{j\}$ and $j \in \{2, 3\}$. And thus we can assume that $2, 3 \in C(v_2)$. Subsequently, if $1 \notin C(v_2)$, then $(vv_1, vv_2) \to (2, 1)$ and uv is colored by some color in $C_1 \setminus C(u_2)$, and we are done.

In the remaining case, $1, 2, 3 \in C(v_2)$ and without loss of generality we assume that $C(v_2) = \{1, 2, 3, 4, 7\}$. If G contains no $(i, j)_{(v_1, v_2)}$-path for some $i \in \{2, 3\}$ and some $j \in \{4, 5, 6\}$, then we are done since $(vv_1, vv_2) \to (i, j)$ gives another acyclic edge coloring of H in which $C(u) \cap C(v) = \{i\}$. Otherwise, G contains an $(i, j)_{(v_1, v_2)}$-path for each $i \in \{2, 3\}$ and each $j \in \{4, 5, 6\}$. Let x_1, x_2, x_3 be the sorted neighbors of v_1 other than v and w_1. Note that $1, 2, 3 \in C(x_i)$, $i \in \{4, 5, 6\}$. It follows that $c(v_1 w_1) = 7$ and assume that $c(v_1 x_i) = i$, $i \in \{4, 5, 6\}$.

If $7 \notin C(x_i)$ for some $i \in \{4, 5, 6\}$, then $(vv_1, vv_2) \to (7, i)$, $v_1 w_1 \to \{1, 2, 3\} \setminus C(w_1)$ and uv is colored by some color in $\{4, 5, 6\} \setminus \{i\}$, and we are done. Otherwise, $7 \in C(x_i)$ and $C(x_i) = \{i\} \cup \{1, 2, 3, 7\}$, $i \in \{4, 5, 6\}$. Then swapping the colors of $v_1 x_4$ and $v_1 x_5$, together with $uv \to 5$ gives an acyclic edge $(\Delta + 2)$-coloring for G.

In the full version of this extended abstract, the other case where $|C(u) \cap C(v)| = 2$ is discussed to acyclically edge color G using at most $\Delta + 2$ colors. They together with the discussion for all the other configurations prove that a triangle-free toroidal graph is acyclically edge $(\Delta + 2)$-colorable.

4 Conclusions

We considered the AECC on toroidal graphs without triangles, and showed that the conjecture holds true, improving the recent result in [6]. Toroidal graphs are a super class of planar graphs. Given that the AECC has been confirmed true for many different cases of planar graphs [16,17,23,26], our result can be the first of many future improvements of the AECC on toroidal graphs.

Acknowledgments. The research is supported by the NSERC Canada, ZJNSF LQ15A010010, NSFC 11601111, and the Ministry of Science and Technology of China Grants G2022040004L and G2023016016L.

References

1. Alon, N., Mcdiarmid, C., Reed, B.: Acyclic coloring of graphs. Random Struct. Algorithms **2**, 277–288 (1991)
2. Alon, N., Sudakov, B., Zaks, A.: Acyclic edge colorings of graphs. J. Graph Theory **37**, 157–167 (2001)
3. Andersen, L.D., Máčajová, E., Mazák, J.: Optimal acyclic edge-coloring of cubic graphs. J. Graph Theory **71**, 353–364 (2012)
4. Basavaraju, M., Chandran, L.S.: Acyclic edge coloring of graphs with maximum degree 4. J. Graph Theory **61**, 192–209 (2009)
5. Basavaraju, M., Chandran, L.S., Cohen, N., Havet, F., Müller, T.: Acyclic edge-coloring of planar graphs. SIAM J. Discret. Math. **25**, 463–478 (2011)
6. Chen, G., Hou, J.: Acyclic edge coloring of triangle-free toroidal graphs. J. Shandong Univ. (Nat. Sci.) **49**, 44–48 (2014)
7. Chen, M., Huang, L., Wang, W.: List vertex-arboricity of toroidal graphs without 4-cycles adjacent to 3-cycles. Discret. Math. **339**, 2526–2535 (2016)
8. Cranston, D.W.: Acyclic edge-coloring of planar graphs: Δ colors suffice when Δ is large. SIAM J. Discret. Math. **33**, 614–628 (2019)
9. Esperet, L., Parreau, A.: Acyclic edge-coloring using entropy compression. Eur. J. Comb. **34**, 1019–1027 (2013)
10. Fiamcik, J.: The acyclic chromatic class of a graph. Math. Slovaca **28**, 139–145 (1978)
11. Giotis, I., Kirousis, L., Psaromiligkos, K.I., Thilikos, D.M.: Acyclic edge coloring through the lovász local lemma. Theor. Comput. Sci. **665**, 40–50 (2017)
12. Hou, J., Liu, G.: Every toroidal graph is acyclically 8-choosable. Acta Math. Sin. (Engl. Ser.) **30**, 343–352 (2014)
13. Molloy, M., Reed, B.: Further algorithmic aspects of the local lemma, pp. 524–529 (1998)
14. Ndreca, S., Procacci, A., Scoppola, B.: Improved bounds on coloring of graphs. Eur. J. Comb. **33**, 592–609 (2012)
15. Shu, Q., Chen, Y., Han, S., Lin, G., Miyano, E., Zhang, A.: Acyclic edge coloring conjecture is true on planar graphs without intersecting triangles. Theor. Comput. Sci. **882**, 77–108 (2021)
16. Shu, Q., Wang, W., Wang, Y.: Acyclic edge coloring of planar graphs without 5-cycles. Discret. Appl. Math. **160**, 1211–1223 (2012)
17. Shu, Q., Wang, W., Wang, Y.: Acyclic chromatic indices of planar graphs with girth at least 4. J. Graph Theory **73**, 386–399 (2013)
18. Shu, Q., Wang, Y., Ma, Y., Wang, W.: Acyclic edge coloring of 4-regular graphs without 3-cycles. Bull. Malays. Math. Sci. Soc. **42**, 285–296 (2019)
19. Tian, F., Yin, Y.: The odd chromatic number of a toroidal graph is at most 9. Inf. Process. Lett. **182**, 106384 (2023)
20. Vizing, V.G.: On an estimate of the chromatic class of a p-graph. Discret. Analiz **3**, 25–30 (1964)
21. Wang, T., Zhang, Y.: Further result on acyclic chromatic index of planar graphs. Discret. Appl. Math. **201**, 228–247 (2016)
22. Wang, W., Ma, Y., Shu, Q., Wang, Y.: Acyclic edge coloring of 4-regular graphs (ii). Bull. Malaysian Math. Sci. Soc. **42**, 2047–2054 (2019)
23. Wang, W., Shu, Q., Wang, Y.: Acyclic edge coloring of planar graphs without 4-cycles. J. Comb. Optim. **25**, 562–586 (2013)

24. Wang, W., Shu, Q., Wang, Y.: A new upper bound on the acyclic chromatic indices of planar graphs. Eur. J. Comb. **34**, 338–354 (2013)
25. Wang, Y., Chen, M., Wang, W.: A note on the list vertex arboricity of toroidal graphs. Discret. Math. **341**, 3344–3347 (2018)
26. Wang, Y., Shu, Q., Wu, J.-L., Zhang, W.: Acyclic edge coloring of planar graphs without a 3-cycle adjacent to a 6-cycle. J. Comb. Optim. **28**(3), 692–715 (2014). https://doi.org/10.1007/s10878-014-9765-6
27. Xu, Y.: On acyclic edge coloring of toroidal graphs. Inf. Process. Lett. **111**, 647–649 (2011)

An Equally-Split Bin Packing Problem

Ding Zou[1,2], Jiayi Lian[1,2,(✉)], Wei Lu[2], Yichao Duan[1], Xingyu Lu[2], Jun Zhou[2], Yuchen Mao[1], and Guochuan Zhang[1]

[1] Zhejiang University, Hangzhou 310058, China
{zoud,jiayilian,yichaod,maoyc,zgc}@zju.edu.cn
[2] Ant Group, Hangzhou 310058, China
{zouding.zd,lianjiayi.ljy,sing.lxy,xiaobo.lw,jun.zhoujun}@antgroup.com

Abstract. We investigate the equally-split bin packing problem (ESBP), which is inspired by a service placement scenario in cloud computing. In this problem, given a set of items, a set of bins already in use, and an infinite number of new bins, one is asked to pack the items using as few bins as possible. Unlike the traditional bin packing problem, ESBP requires that each item must be split equally into at least two pieces, and the pieces of an item must be placed into different bins. We tackle this problem both theoretically and practically. We prove that ESBP is NP-hard and cannot be approximated within a factor better than 4/3 unless P = NP. On the positive side, we propose two algorithms with asymptotic approximation ratios at most $e/(e-1) \approx 1.582$. We also propose a heuristic algorithm, which not only minimizes the number of bins used but also reduces the total number of item pieces. We compared these three algorithms and two other methods on a dataset from Alibaba. The experimental result shows that our heuristic algorithm has the best overall performance.

Keywords: Bin packing · Approximation · Heuristic

1 Introduction

In the equally-split bin packing (ESBP) problem, we have an infinite number of new bins and m old bins already in use, where the capacity of new bins is 1, and the remaining capacity of the jth old bin is $c_j < 1$. Given n items with weight $w_1, \ldots, w_n$, we should pack the items into old and new bins. Unlike the traditional bin packing problem, we require that each item must be split equally into at least two pieces, and that a bin cannot contain more than one piece of the same item. In other words, each item must be packed fractionally and evenly into at least two bins. Our goal is to minimize the total number of bins in use (that is, the m old bins and the new bins that are used).

This work was supported by Ant Group through CCF-Ant Research Fund [Grant No. RF20220212].

D. Zou and J. Lian—Equal contribution.

D. Du et al. (Eds.): COCOA 2024, LNCS 15434, pp. 240–252, 2025.
https://doi.org/10.1007/978-981-96-4445-2_20

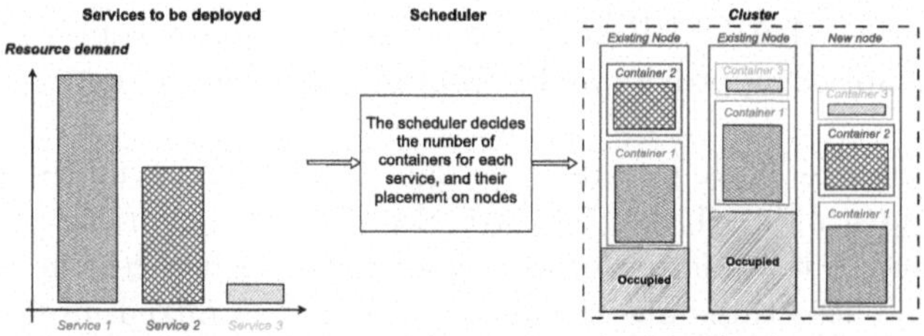

Fig. 1. Illustration of Service Deployment and a Simplified Cloud Architecture

ESBP models a service placement scenario in cloud computing. In cloud computing, one way to organize computational resources is via clusters [4,9, 11]. A cluster consists of hundreds of nodes (also known as machines). When new online services arrive, they will be deployed (via containers) to the existing nodes, and new nodes will be added when necessary. A service may comprise multiple containers, which are distributed across multiple nodes. See Fig. 1 for an illustration. The goal is to use as few nodes as possible. In order to satisfy service level objectives (SLOs), there are constraints that should be satisfied.

1. [Resource Constraint for Services] The total size of the containers (that is, the total amount of resources they provide) for a service must be at least as large as the resource demand of the service.
2. [Resource Constraint for Nodes] The total size of the containers in a node cannot exceed the capacity of the node.
3. [Fault Tolerance] Each service must be deployed (via containers) on at least two nodes. In this way, a service will not be interrupted when some node fails, because the workload of the service can be routed to other nodes.
4. [Load Balancing per Service] For each service, all its containers have the same size (that is, use the same amount of resources) and every node has at most one of its containers. When some service workload is required, a load balancing algorithm, such as Round Robin and IP Hash [3], will be used to distribute the service workload evenly across all containers. This constraint ensures no single container will be overloaded and helps optimize response times for a seamless user experience.

The above is known as the placement of services in cloud computing, and can be modeled as ESBP by viewing computing nodes as bins, and services as items.

1.1 Our Contributions

It is well-known that the classic bin packing problem is NP-hard, while the fractional bin packing problem, in which items are allowed to be split arbitrarily, can be solved trivially. Our problem, ESBP, lies somewhere between the classic bin packing and the fractional bin packing problem: items are allowed to be split, but only into pieces of equal size. It can be trivially solved when there are no old bins. We prove that ESBP becomes NP-hard when it comes with old bins, and that it cannot be approximated within a factor better than 4/3 unless P = NP.

On the positive side, we propose two algorithms with asymptotic approximation ratios at most $e/(e-1) \approx 1.582$. We also present a heuristic algorithm that not only minimizes the number of bins used but also reduces the total number of item pieces. These three algorithms, as well as the mixed-integer linear programming method and other natural heuristics, are compared on a dataset from Alibaba. The experiment result shows that our heuristic algorithm has the best overall performance.

1.2 Related Work

To our knowledge, we are the first to address the ESBP problem. The most relevant problem is bin packing with item fragmentation. Casazza and Ceselli [2] classify this line of research by objectives and fragmentation structures: the objective can be either bin minimization or fragmentation minimization; the fragment can be either size-preserving (the item size equals the total size of its fragments) or size-increasing (more fragments mean a larger total size) [1,2,6]. ESBP differs from the previous work in the sense that every item must be split and split equally and that the pieces of an item must be packed into different bins. Therefore, their methods cannot be directly applied to ESBP.

Another relevant research goes to [7] and [8]. The difference is that in their problem, the splitting scheme of items (that is, the number of pieces of each item) is given as input, while in ESBP, it is part of the decision that the algorithm should make. In experiments, we will feed the splitting scheme produced by our algorithm to the algorithm in [8] and compare the results to that using the default splitting scheme.

1.3 Paper Organization

Section 2 analyzes the complexity of ESBP. Section 3 gives two approximation algorithms and analyzes their approximation ratio. Section 4 presents a heuristic algorithm that has certain advantages over the two algorithms in the previous section. Section 5 presents the result of our experiment.

2 Hardness Results

We prove that ESBP is NP-hard in general, and that it cannot be approximated within a factor better than 4/3 unless P = NP.

Theorem 1. *The Problem ESBP is NP-hard.*

Proof. We give a polynomial-time reduction from the PARTITION problem, which is known to be NP-complete [5]. Given n positive integers $a_1, a_2, \ldots, a_n$,

PARTITION asks if these integers can be partitioned into two subsets with equal sum.

Given a PARTITION instance $\{a_1, \ldots, a_n\}$, we construct an instance of ESBP as follows. Let $M = (\sum_{i=1}^{n} a_i)/2$. For each integer a_i, we create an item i of size $w_i = a_i/M$. There are two old bins, each with a remaining capacity of $1/2$. It is easy to see that the optimal solution for the instance uses at least three bins (two old bins and a new bin). We claim that $\{a_1, \ldots, a_n\}$ can be partitioned into two subsets with equal sum if and only if the optimal solution for the ESBP instance uses exactly three bins. Let B_1 and B_2 be the two old bins.

We first prove the only-if part. Suppose that $\{a_1, \ldots, a_n\}$ can be partitioned into two subsets with equal sum. By our construction, it implies that the items can be partitioned into two subsets I_1 and I_2 such that $\sum_{i \in I_1} w_i = \sum_{i \in I_2} w_i = 1$. Let B_3 be a new bin. We can pack all the items into B_1, B_2, and B_3 as follows. For each item in I_1, we split it into two pieces, putting one piece into B_1 and the other piece into B_3. For each item in I_2, we split it into two pieces, putting one piece into B_2 and the other piece into B_3. Therefore, the optimal packing solution uses exactly three bins.

Next, we prove the if part. Suppose that the optimal packing solution uses exactly three bins: B_1, B_2, and an extra new bin B_3. All three bins must be full as their total remaining capacity equals the total size of items. We claim that no item has its pieces in both B_1 and B_2. Assume that the claim is true. Then each item is split into exactly two pieces, with one in B_1 (or B_2) and the other in B_3. See Fig. 2 for an example. Let I_1 be the set of items whose pieces are in B_1 and B_3, and let I_2 be the set of items whose pieces are in B_2 and B_3. It is easy to see that I_1 and I_2 form a partition of the items, and $\sum_{i \in I_1} w_i = \sum_{i \in I_2} w_i = 1$. This implies that $\{a_1, \ldots, a_n\}$ can be partitioned into two subsets with equal sum.

It remains to prove the claim that no item has its pieces in both B_1 and B_2. Suppose, for the sake of contradiction, that some item i' has its pieces in both B_1 and B_2. The size of the piece of i' in B_3—if there is such a piece—is at most $w_{i'}/3$. For any item other than i', it is split into at least two pieces, and B_3 contains at most one of these pieces. It follows that the total size of the pieces in B_3 is at most

$$w_{i'}/3 + \sum_{i \neq i'} w_i/2 < (\sum_{i=1}^{n} w_i)/2 = 1.$$

This contradicts the fact that B_3 is full. The proof is completed.

The above proof also implies that in ESBP, it is NP-complete to decide if three bins are sufficient to pack a given set of items. The corollary below follows straightforwardly.

Corollary 1. *Unless* $P = NP$, *no polynomial-time algorithm for ESBP can achieve an approximation ratio strictly better than* $4/3$.

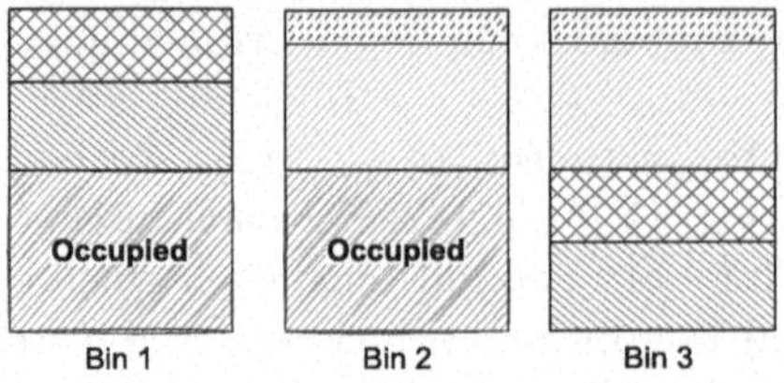

Fig. 2. An optimal packing using three bins.

3 Approximation Algorithms

Since ESBP is NP-hard, it is unlikely to have an efficient exact algorithm. Therefore, we consider approximation algorithms. Our first algorithm is a naive greedy algorithm with an asymptotic approximation of at most $\frac{e}{e-1}$. Then we refine this algorithm and prove that on any problem instance, the refined algorithm is as at least good as the naive algorithm.

Although these two algorithms have the same approximation ratio in the worst case, our experiments show that the refined algorithm uses significantly fewer bins than the naive algorithm.

3.1 A Simple Greedy Algorithm

We start with the base case where there is only a single item with some weight w. Assume that the remaining capacities of the m old bins are $c_1 \leq c_2 \leq \cdots \leq c_m$. It is easy to observe that if the jth bin is the first bin that contains a piece of the item, the item is split into at least $\lceil \frac{w}{c_j} \rceil$ piece. Then we need exactly $j - 1 + \max\{2, \lceil \frac{w}{c_j} \rceil\}$ bins. An optimal solution can be obtained by trying all possible j's and selecting the best solution.

The idea of our first algorithm, *Merge-and-Split*, is very simple: merge (conceptually) all items into one item, and run the above optimal algorithm for a single item. See Algorithm 1 for details.

Although simple, we will show that Merge-and-Split is guaranteed to return a solution that is not too far away from the optimum. Let OPT(I) be the number of bins used by the optimum solution for an instance I. We have a natural lower bound LB(I) on OPT(I) by the observation that the total capacity of the bins used by the optimum must be at least the total weight of items. Therefore, in addition to the m old bins, an optimal solution needs at least $\max\{0, (\sum_{i=1}^{n} w_i - \sum_{j=1}^{m} c_j)\}$ extra new bins. In other words,

$$\mathrm{OPT}(I) \geq \mathrm{LB}(I) = m + \max\left\{0, \sum_{i=1}^{n} w_i - \sum_{j=1}^{m} c_j\right\}. \tag{1}$$

Let MS(I) be the number of bins used by the algorithm Merge-and-Split on instance I.

Algorithm 1. Merge-and-Split

Input: n items with weights $\{w_1, ..., w_n\}$, m old bins with remaining capacities $\{c_1, \ldots, c_m\}$

Sort bins by their capacities $c_1 \le \cdots \le c_m$

$c_{m+1} \leftarrow 1$

$W \leftarrow \sum_{i=1}^{n} w_i$

$j^* \leftarrow 1$

for $j = 2, ..., m+1$ **do**

  ▷ *Start the packing with the jth bin, and if it uses less bins, update* j^* ◁

  if $j - 1 + \lceil \frac{W}{c_j} \rceil < j^* - 1 + \lceil \frac{W}{c_{j^*}} \rceil$ **then**

    $j^* \leftarrow j$

$g \leftarrow j^* - 1 + \max\left\{2, \lceil \frac{W}{c_{j^*}} \rceil\right\}$

If $g > m$, open $g - m$ extra new bins.

Split each item equally into $\max\left\{2, \lceil \frac{W}{c_{j^*}} \rceil\right\}$ pieces, and distribute the pieces into bins $j^*, j^* + 1, \ldots, g$.

Lemma 1. *For any instance I, we have* $\mathrm{MS}(I) \le \frac{e}{e-1}\mathrm{LB}(I) + e - 1 + \frac{e}{e-1}$.

Proof. Let $\{w_1, ..., w_n\}$ be the item weights, and let $\{c_1, \ldots, c_m\}$ be the remaining capacities of the old bins. Define $c_{m+1} = 1$ and $W = \sum_{i=1}^{n} w_i$. Suppose that Merge-and-Split returns a solution that uses g bins (g is defined in line 10 of Algorithm 1). Clearly, $g \le m + \lceil W \rceil$ since we can always pack the items using $\lceil W \rceil$ new bins. To simplify the discussion, we make the following assumptions, which are without loss of generality.

1. $W \ge 2$. If $W < 2$, then $g \le m + 2$. Since $\mathrm{LB}(I) \ge \max\{m, W\}$, it is easy to verify that $g \le \frac{e}{e-1}\mathrm{LB}(I) + e - 1 + \frac{e}{e-1}$. The lemma holds directly.
2. $g > \frac{em}{e-1}$ and $g > \frac{eW}{e-1}$. Otherwise, the lemma already holds since $\mathrm{LB}(I) \ge \max\{m, W\}$.

By the choice of j^* in Algorithm 1, we have that $j - 1 + \lceil W/c_j \rceil \ge g$ for any $j \in \{1, \ldots, m\}$, which implies that $c_j \le W/(g-j)$. Recall that $c = 1$. By the definition of $LB(I)$, we have

$$\begin{aligned}
\mathrm{LB}(I) &= m + \max\{0, (\sum_{i=1}^{n} w_i - \sum_{j=1}^{m} c_j)\} \\
&\ge m + W - \sum_{j=1}^{m} \frac{W}{g-j} \\
&= m + W - W\sum_{j=0}^{m-1} \frac{1}{g-j} + W(\frac{1}{g} - \frac{1}{g-m}) \\
&\ge m + W - W \ln \frac{g}{g-m} - \frac{Wm}{g(g-m)} \qquad (2) \\
&\ge m + \lceil W \rceil - \lceil W \rceil \ln \frac{g}{g-m} - \frac{Wm}{g(g-m)} - 1 \qquad (3)
\end{aligned}$$

The last but one inequality is due to the fact that

$$\sum_{j=0}^{m-1} \frac{1}{g-j} \leq \int_{g-m}^{g} \frac{1}{x} dx = \ln \frac{g}{g-m}.$$

Consider the last term of (3). As $g > \frac{em}{e-1}$ and $g > \frac{eW}{e-1}$,

$$\frac{Wm}{g(g-m)} \leq \frac{W \cdot m}{\frac{e}{e-1} \cdot W(\frac{e}{e-1} \cdot m - m)} \leq \frac{(e-1)^2}{e}. \tag{4}$$

We claim that for $W, m, g > 0$,

$$m + \lceil W \rceil - \lceil W \rceil \ln \frac{g}{g-m} \geq \frac{(e-1)g}{e}. \tag{5}$$

Combining (3), (4), and (5) yields that

$$\mathrm{LB}(I) \geq \frac{(e-1)g}{e} - \frac{(e-1)^2}{e} - 1,$$

which implies that

$$g \leq \frac{e}{e-1}\mathrm{LB}(I) + e - 1 + \frac{e}{e-1}.$$

It remains to prove (5). Define

$$f(\lceil W \rceil, m, g) = \frac{g}{\lceil W \rceil + m - \lceil W \rceil \ln \frac{g}{g-m}}.$$

It suffices to prove that for any $\lceil W \rceil, m, g \in \mathbb{R}^+$ with $\frac{(e-1)g}{e} \geq m \geq g - \lceil W \rceil$, we have $f(\lceil W \rceil, m, g) \leq \frac{e}{e-1}$.

We first calculate the partial derivative of f with respect to m.

$$\frac{\partial f}{\partial m} = \frac{-g(1 - \lceil W \rceil \cdot \frac{g-m}{g} \cdot \frac{g}{(g-m)^2})}{(\lceil W \rceil + m - \lceil W \rceil \ln \frac{g}{g-m})^2} = \frac{g(\frac{\lceil W \rceil}{g-m} - 1)}{(\lceil W \rceil + m - \lceil W \rceil \ln \frac{g}{g-m})^2}.$$

Note that $\frac{\partial f}{\partial m} \geq 0$ for any m with $\frac{(e-1)g}{e} \geq m \geq g - \lceil W \rceil$. Therefore,

$$\begin{aligned} f(\lceil W \rceil, m, g) &\leq f(\lceil W \rceil, \frac{(e-1)g}{e}, g) \\ &= \frac{g}{\lceil W \rceil + \frac{(e-1)g}{e} - \lceil W \rceil \ln \frac{g}{g - \frac{(e-1)g}{e}}} = \frac{e}{e-1}. \end{aligned}$$

Since $\mathrm{LB}(I)$ is a lower bound on $\mathrm{OPT}(I)$. Lemma 1 immediately implies the following theorem.

Theorem 2. *Merge-and-Split is an* $(\frac{e}{e-1}, e-1+\frac{e}{e-1})$*-approximation algorithm. That is, for any instance* I,

$$\mathrm{MS}(I) \leq \frac{e}{e-1}\mathrm{OPT}(I) + e - 1 + \frac{e}{e-1}.$$

We further show that the bound in Lemma 1 is asymptotically tight. Therefore, the asymptotic approximation ratio of $\frac{e}{e-1}$ cannot be improved by any analysis that uses LB(I) as a lower bound on OPT(I).

Lemma 2. *There exists an instance I such that*

$$OPT(I) \geq \frac{e}{e-1} LB(I) - 1.$$

Proof. Consider the following instance I. There is only one item and $m-1$ old bins. The size of the item is $w = m/(e-1)$. The remaining capacity of the j-th old bin is $c_j = w/(w+m-j)$ for $j = 1, \dots, m-1$. It is easy to see that

$$\begin{aligned} LB(I) &= m - 1 + w - \sum_{j=1}^{m-1} c_j = m + w - \sum_{j=1}^{m} \frac{w}{w+m-j} \\ &\leq m + w - w \ln \frac{w+m}{w} = m. \end{aligned}$$

The inequality in the last line follows from

$$\sum_{j=1}^{m} \frac{1}{w+m-j} \geq \int_{w}^{w+m} \frac{1}{x} dx = \ln \frac{w+m}{w}.$$

On the other hand, one can verify that OPT(I) $= m + \lceil w \rceil - 1$. Then,

$$OPT(I) \geq m + w - 1 \geq \frac{e}{e-1} LB(I) - 1.$$

3.2 A Refined Greedy Algorithm

It is easy to observe that Merge-and-Split cannot be a good algorithm in practice because the old bins with small remaining capacity are completely wasted. In this subsection, we propose an algorithm that has the same theoretical guarantee as Merge-and-Split but a much better performance in practice. The algorithm packs the items one by one so that the old bins with a small remaining capacity can also be used. Moreover, as in the traditional bin packing problem, it is always large items that may cause a problem, the items will be packed in decreasing order of their weights.

Let us first consider the following subproblem, namely ESBP with a fixed number of bins: given an ESBP instance I and an integer k, pack all the items using k bins (that is, m old bins and $k-m$ new bins). We propose an algorithm called Max-Min-Capacity for this subproblem. The algorithm packs the items one by one in decreasing order of item weights. When item i is packed, the algorithms choose a way that maximizes the minimum remaining capacity among all the bins containing a piece of item i. The intuition is that we always pack an item using bins with large remaining capacity and that we want the capacities of all the bins to be as close as possible. See Algorithm 2 for details.

We show that if Merge-and-Split uses k bins, then Max-Min-Capacity must succeed with k bins. Then Max-Min-Capacity has the same asymptotic approximation ratio with Merge-and-Split.

Algorithm 2. Max-Min-Capacity

input: n items with weights $\{w_1, \ldots, w_n\}$, m old bins with remaining capacities $\{c_1, \ldots, c_m\}$, and $k - m$ new bins with capacity $c_{m+1} = \cdots = c_k = 1$.
sort the items so that $w_1 \geq \cdots \geq w_n$
for $i = 1 \cdots n$ **do**
    $p^* \leftarrow 0$ and $minCap \leftarrow 0$
    for $p = k, \ldots, 2$ **do**
        ▷ *try to split item i into p pieces, and compute the minimum capacity of the bins containing pieces of item i* ◁
        Let c' be the pth largest remaining capacity of the bins.
        if $c' - w_i/p \geq minCap$ **then**
            ▷ *update if the min capacity becomes larger* ◁
            $minCap \leftarrow c' - w_i/p$.
            $p^* \leftarrow p$
    if $p^* = 0$ **then**
        Cannot pack item i and the algorithm fails.
    else
        Split item i into p^* pieces, distribute the p^* pieces into the p^* bins with the largest remaining capacity, and update the remaining capacity of bins

Lemma 3. *For any ESBP instance I, if Merge-and-Split can pack all the items using k bins, so can Max-Min-Capacity.*

Proof. Let $c_1 \leq c_2 \leq \cdots \leq c_k$ be the original remaining capacities of the k bins. Label the items so that $w_1 \geq \cdots \geq w_n$.

Suppose that Merge-and-Split packs all the items evenly into bin $j^*, j^* + 1, \ldots, k$. Then we have that

$$c_j \geq \frac{\sum_{\ell=1}^{n} w_\ell}{k - j^* + 1} \tag{6}$$

for all $j \in \{j^*, \ldots, k\}$.

We claim that at the moment before Max-Min-Capacity packs item i, the remaining capacity of bin j is at least

$$\frac{\sum_{\ell=i}^{n} w_\ell}{k - j^* + 1}$$

for all $j \in \{j^*, \ldots, k\}$. Suppose that the claim holds. Then when packing item i, Max-Min-Capacity can always split the item into $k - j^* + 1$ pieces and pack the pieces into bins $j^*, \ldots, k$. Therefore, Max-Min-Capacity never fails to pack an item, and must succeed with k bins.

It remains to prove the claims. We prove it by induction on i. When $i = 1$, the claim is true by Eq. (6). Suppose that the claim holds for some i. We show that it must hold for $i+1$. Consider the moment before packing the item i. By the inductive hypothesis, the remaining capacity of bin j is at least

$$\frac{\sum_{\ell=i}^{n} w_\ell}{k - j^* + 1}.$$

for all $j \in \{j^*, \ldots, k\}$. One obvious scheme to pack item i is to place it evenly into bin $j^*, \ldots, k$, and in this way, the minimum remaining capacity of all bins containing item i would become at least

$$\frac{\sum_{\ell=i}^{n} w_\ell}{k - j^* + 1} - \frac{w_i}{k - j^* + 1} = \frac{\sum_{\ell=i+1}^{n} w_\ell}{k - j^* + 1}.$$

Recall that Max-Min-Capacity packs item i in a way that maximizes the remaining capacity of all bins containing item i, so it must pack as at least good as the trivial scheme. We have that after Max-Min-Capacity packs item i, if a bin contains a piece of item i, then its remaining capacity must be at least

$$\frac{\sum_{\ell=i+1}^{n} w_\ell}{k - j^* + 1}.$$

If a bin does not contain a piece of item i, then its remaining capacity is the same as that before packing item i. Therefore, after Max-Min-Capacity packs item i (and before it packs item $i+1$), the remaining capacity of bin j is at least

$$\frac{\sum_{\ell=i+1}^{n} w_\ell}{k - j^* + 1}$$

for all $j \in \{j^*, \ldots, k\}$.

Now we can solve the ESBP problem by finding, in a binary search manner, the smallest integer k that makes Max-Min-Capacity succeed. Note that we can use Eq. (1) as a lower bound and the number used by Merge-and-Split as an upper bound. Lemma 3 guarantees that this approach uses at most the same number of bins as Merge-and-Split.

Corollary 2. *Max-Min-Capacity (with the best k) is as at least good as Merge-and-Split, and therefore, is an* $(\frac{e}{e-1}, e-1+\frac{e}{e-1})$*-approximation algorithm. That is, for any instance* I*,*

$$MC(I) \leq \frac{e}{e-1}\mathrm{OPT}(I) + e - 1 + \frac{e}{e-1}.$$

4 A Heuristic with Fewer Splits

Although Max-Min-Capacity uses much fewer bins, it may split the items into too many pieces. This is not good in practice as each item piece represents a container, and each container brings an overhead cost.

In this section, we propose a heuristic algorithm, called Min-Piece, that not only minimizes the number of bins used but also reduces the total number of item pieces. Although a theoretical analysis is not yet available for Min-Piece, the experiment shows that this algorithm performs well in practice.

Min-Piece is similar to Max-Min-Capacity in the sense that it solves the subproblem which asks to pack all the items into k bins for some given k. Then the best k can be found using binary search. The difference is that, when packing an item, Min-Piece always tries to split items into as few pieces as possible. See Algorithm 3 for details.

Algorithm 3. Min-Piece

input: n items with weights $\{w_1, \dots, w_n\}$, m old bins with remaining capacities $\{c_1, \dots, c_m\}$, and $k - m$ new bins with capacity $c_{m+1} = \cdots = c_k = 1$.

1: sort the items so that $w_1 \geq \cdots \geq w_n$
2: **for** $i = 1 \cdots n$ **do**
3: $p^* \leftarrow 0$
4: **for** $p = 2, \cdots, k$ **do**
5: ▷ *check item i can be packed by splitting it into p pieces* ◁
6: Let c' be the pth largest remaining capacity of the bins.
7: **if** $c' - w_i/p \geq 0$ **then**
8: $p^* = p$
9: **break**
10: **if** $p^* = 0$ **then**
11: Cannot pack item i and the algorithm fails.
12: **else**
13: Split item i into p^* pieces, and pack these p^* pieces into p^* bins using best-fit or worst-fit.

In line 13 of Algorithm 3, one of the following two rules is used to pack the pieces of an item.

- Worst-Fit (WF): Place the pieces into the p^* emptiest bins.
- Best-Fit (BF): Place the pieces into the p^* fullest bins that have enough space for a piece.

5 Experimental Results

We do experiments to evaluate our algorithms: Merge-and-Split, Max-Min-Capacity, Min-Piece(BF), and Min-Piece(WF). We also run two baseline algorithms and compare them with our algorithms. One is a heuristic named SpreadWFD in [8], and the other is solving a MILP formulation of ESBP.

We use the public dataset published by the Alibaba Tianchi Platform [10] and evaluate the above six algorithms on instances with different item numbers and different item sizes. Both the solution quality and the running time of algorithms are measured.

Due to page limits, below we only summarize main result of the experiment. More details are left to a full version. Among our algorithms, Min-Piece(WF) is the best. In most cases, the number of bins used by Min-Piece(WF) matches the lower bound. Merge-and-Split has the worst performance as we expect, but its usage rate is still greater than 70%. This is consistent with Lemma 1. Compared with baseline algorithms, Min-Piece(WF) is very close to MILP on small instances, and runs much faster than MILP on large instances. The experiment also shows that when comparing the default split scheme provided by the data set, the split scheme generated by our algorithm not only has a smaller number of item pieces but also helps to reduce the total number of bins used.

References

1. Bertazzi, L., Golden, B., Wang, X.: The bin packing problem with item fragmentation: a worst-case analysis. Discret. Appl. Math. **261**, 63–77 (2019)
2. Casazza, M., Ceselli, A.: Exactly solving packing problems with fragmentation. Comput. Oper. Res. **75**, 202–213 (2016)
3. Cloud, A.: Load balancing in the age of docker (2024). https://www.alibabacloud.com/tech-news/a/load_balancer/4oc53uwgu1j-load-balancing-in-the-age-of-docker
4. Garefalakis, P., Karanasos, K., Pietzuch, P., Suresh, A., Rao, S.: Medea: scheduling of long running applications in shared production clusters. In: Proceedings of the Thirteenth EuroSys Conference, pp. 1–13 (2018)
5. Karp, R.M.: Reducibility among Combinatorial Problems, pp. 85–103. Springer, Boston (1972)
6. LeCun, B., Mautor, T., Quessette, F., Weisser, M.A.: Bin packing with fragmentable items: presentation and approximations. Theor. Comput. Sci. **602**, 50–59 (2015)
7. Li, S., Wang, L., Wang, W., Yu, Y., Li, B.: George: learning to place long-lived containers in large clusters with operation constraints. In: Proceedings of the ACM Symposium on Cloud Computing, pp. 258–272 (2021)
8. Mommessin, C., et al.: Affinity-aware resource provisioning for long-running applications in shared clusters. J. Parallel Distrib. Comput. **177**, 1–16 (2023)
9. Tech, A.: Sofaserverless system facilitates rapid business development (2022). https://segmentfault.com/a/1190000041833292/en

10. Tianchi: Alibaba data center resource management and scheduling dataset (2018). https://tianchi.aliyun.com/dataset/dataDetail?dataId=6287
11. Wang, L., Weng, Q., Wang, W., Chen, C., Li, B.: Metis: learning to schedule long-running applications in shared container clusters at scale. In: SC20: International Conference for High Performance Computing, Networking, Storage and Analysis, pp. 1–17. IEEE (2020)

The Location-routing Problem in the Pallet Pooling System Considering Carbon Emissions

Xiaoting Shang[1(✉)], Bowen Miao[1], and Qingguo Bai[2(✉)]

[1] Business College, Qingdao University, Qingdao, China
xtshang@qdu.edu.cn

[2] School of Management, Qufu Normal University, Qufu, China
hustbaiqg@alumni.hust.edu.cn

Abstract. This paper studies the location-routing problem in the pallet pooling system. Considering the characteristic of simultaneous delivery and pickup, a mixed integer linear programming model is proposed for the location-routing problem with the limitations on carbon emissions and transportation time in the pallet pooling system. Then, a robust optimization model is proposed to deal with the uncertainty of demand. Considering the computational complexity, a heuristic algorithm combining the genetic algorithm and the ant colony optimization is designed to solve the problem efficiently. Numerical experiments are carried out on classical benchmark instances, and the results validate the feasibility and effectiveness of the proposed model.

Keywords: pallet pooling system · location-routing problem · carbon emissions · demand uncertainty · heuristic algorithm

1 Introduction

1.1 Motivation

The pallet pooling system is a network ensuring the circulation and sharing of pallets at a region, which is designed to achieve high-efficient and low-carbon logistics. The pallet pooling system provides pallets in a shared manner, which is responsible for ensuring the sharing, recycling, and circulation of pallets among manufacturers, wholesalers, retailers, and users. The location of pallet pooling center (PPC) is influenced by various factors, such as distance, demand volume and construction cost. When there is rental or recycling demands in pallet pooling stations (PPSs), the PPC will plan the vehicle routes according to the location of PPSs and demand volumes. The location of PPC and the routing of vehicles are interrelated and inseparable. Hence, this paper explores the integrated optimization of the location-routing problem (LRP) in the pallet pooling system. When the vehicle arrives at the PPS, rental pallets will be unloaded from the vehicle, then recycling pallets will be loaded onto the vehicle, i.e., simultaneous delivery and pickup. For enterprises, transportation time and carbon emissions are important indicators for high-efficiency logistics and low-carbon production, respectively. Therefore, this paper aims to incorporate the limitations on carbon emissions and transportation time for the LRP in the pallet pooling system.

D. Du et al. (Eds.): COCOA 2024, LNCS 15434, pp. 253–266, 2025.
https://doi.org/10.1007/978-981-96-4445-2_21

1.2 Literature Review

The related literature will be reviewed in three paradigms: LRP, vehicle routing problem with simultaneous delivery and pickup (VRPSDP) and related research on pallet pooling system.

LRP aims to integrate facility location problem (FLP) at the strategic level and vehicle routing problem (VRP) at the tactical level to avoid the excessive costs associated with addressing them separately [1]. Yang and Sun [2] studied battery swap station LRP with capacitated electric vehicles under battery driving range limitation. Building on their work, Schiffer and Walther [3] proposed models incorporating additional constraints, such as time windows and charging station capacities. Zhou et al. [4] studied multi-capacity hierarchical LRP using carbon trading policies as a constraint. In various application scenarios, some LRPs also considered time factors. Maghfiroh et al. [5] studied LRPs with time windows, facility capacity, and vehicle capacity. Liu et al. [6] studied the open LRPs with simultaneous pickup and delivery, where vehicles return to a third-party logistics company instead of the starting warehouse. However, research on LRP considering multiple factors such as vehicle/facility capacity, time and carbon emissions is very limited.

VRPSDP is widely used in logistics systems and various industries due to its practical importance and benefits [7]. In 1989, Min [8] was the pioneer in recognizing the possibility of simultaneous delivery and pickup at the same node, proposing a mathematical model to solve a book-distribution problem faced by public libraries. Since then, numerous scholars have studied the mathematical model and solution methods for this issue. Montané and Galvao [9] extended related research by proposing an integer programming model with maximum distance constraints and designing a tabu search algorithm. Building on previous studies, Ai and Kachitvichyanukul [10] proposed a generalized formulation of VRPSDP and a solution method based on PSO. Given its significant practical implications for logistics companies, VRPSDP has been widely studied in various application scenarios in recent years, for example, in the fast fashion retailers [11] and in the last-mile logistics distribution [12]. The above studies indicate that the research on the integration of VRPSDP and FLP remains very limited.

In recent years, pallet pooling has been increasingly promoted in closed-loop systems to improve sustainability and operational efficiency. For example, Roy et al. [13] developed a two-stage integrated framework and concluded that the leased pallet pooling program is superior to the buy/sell scheme in terms of total cost. To address the complexities of pallet demand and supply dynamics, researchers have also focused on mathematical modeling approaches. Ren and Zhang [14] considered the uncertainty of pallet demand and transportation capacity and established a stochastic model for pallet recycling. Wu et al. [15] investigated a decentralized pallet management system and formulated a location-inventory-routing problem for pallet pooling. Furthermore, Yu et al. [16] proposed a new shipbuilding empty pallet pool allocation pattern combining a mathematical planning model. However, research on LRP in the pallet pooling system is very limited.

1.3 Contribution

To our best knowledge, there is no research integrating LRP in the pallet pooling system with the simultaneous delivery and pickup, the limitation on carbon emissions and transportation time in the literature. The main contributions are as follows:

(1) A new mixed-integer linear programming model is established for the LRP in the pallet pooling system, considering the characteristics of simultaneous delivery and pickup, the limitation on carbon emissions and transportation time.
(2) A robust optimization model is proposed to deal with the uncertainty of demand.
(3) Considering the computational complexity, a heuristic algorithm combining the genetic algorithm (GA) and the ant colony optimization (ACO) is proposed.
(4) Numerical experiments are carried out to validate the feasibility of the established model and the efficiency of the proposed algorithm.

2 Problem Formulation

2.1 Problem Description

The pallet pooling system consists of pallet manufacturer, PPCs and PPSs. Pallet manufacturer is responsible for pallet production and the disposal of discarded pallets. PPCs are responsible for the storage, maintenance, and repair of pallets, while PPSs address the temporary rental and recycling of pallets to meet the demands of surrounding areas. The locations of the manufacturer and PPSs are known, and some PPSs will be selected to construct PPCs, with the remaining PPSs being assigned to these constructed PPCs. When requests for rental and recycling pallets are received from PPSs, vehicles departing from the PPC can sequentially visit multiple PPSs. Specifically, a vehicle loaded with rental pallets departs from the PPC and arrives at the first PPS, where it unloads the rental pallets and simultaneously loads the recycling pallets. The vehicle then moves to the next PPS if its capacity allows; otherwise, it returns to the PPC. Additionally, throughout the transportation routes, it is essential for vehicles to achieve simultaneous delivery and pickup while maintaining transportation time efficiency. Furthermore, from an environmental perspective, the carbon emissions generated by the entire location-routing network must be kept within specified limits.

For the convenient of modelling, the following assumptions are considered:

- The transportation cost is proportional to the transportation distance; the carbon emission is proportional to the transportation distance, and related to the loading volume [17].
- Each PPS only can be assigned to one PPC and visited by one vehicle.

Parameters and variables used are explained as follows:

Parameters	
N	Set of PPSs
C_C	The pallet capacity of a PPC
C_T^1/C_T^2	The pallet capacity of a vehicle from pallet manufacturer/PPC to PPC/PPS
FC	The fixed cost to construct a PPC
B	The fixed cost of each route from PPC to PPS
trs_{ij}	The transportation cost between nodes i and j, $\forall i, j \in 0 \cup N$
w_i^+/w_i^-	The demand for rental/recycling pallets at PPS i, $\forall i \in N$
Car_{ij}	The carbon emission of an empty vehicle between nodes i and j, $\forall i, j \in 0 \cup N$
θ	Full-load fuel consumption increment coefficient
L	The limitation of carbon emissions in the pallet transportation
t_{ij}	The transportation time between nodes i and j, $\forall i, j \in N$
s_i	The working time of PPS i, $\forall i \in N$
T_0	The limitation of transportation time
M_0	A large constant
Variables	
r_i^+/r_i^-	Integer, the loading volume of rental/recycling pallets in the vehicle arriving at/leaving from node i, $\forall i \in 0 \cup N$
v_{ij}	Integer, the loading volume of the vehicle during the transportation between nodes i and j, $\forall i, j \in 0 \cup N$
T_i	Integer, the arrival time at node i, $\forall i \in 0 \cup N$
y_{ij}	Binary, equals to 1 if PPS i is assigned to PPC j, equals to 0 otherwise, it is noted that $y_{jj} = 1$ if PPS is selected to construct a PPC, $\forall i, j \in N$
z_{ij}^k	Binary, equals to 1 if PPSs i and j are assigned to PPC k, and PPS i is the preceding node of PPS j, equals to 0 otherwise, $\forall i, j \in N$

2.2 Deterministic Optimization Model

Based on the above problem description, assumptions and notations, the deterministic model for the LRP in the pallet pooling system can be presented below:

$$\begin{aligned} \min \ & \sum_{j \in N} FC \cdot y_{jj} + \sum_{j \in N} trs_{0j} \cdot \left(2 \cdot y_{jj} + \theta \cdot \frac{\sum_{i \in N} w_i^+ \cdot y_{ij}}{C_T^1} \right) \\ & + \sum_{i \in N} \sum_{j \in N} trs_{ij} \cdot \left(\sum_{k \in N} z_{ij}^k + \theta \cdot \frac{v_{ij}}{C_T^2} \right) + B \cdot \sum_{k \in N} \sum_{j \in N} z_{kj}^k \end{aligned} \tag{1}$$

$$\text{s.t.} \ \sum_{j \in N} y_{ij} = 1, \ \forall i \in N \tag{2}$$

$$y_{ij} \leq y_{jj}, \forall i, j \in N \tag{3}$$

$$\sum\nolimits_{j \in N, i \neq j} z_{ij}^{k} = y_{ik}, \forall i, k \in N, i \neq k \tag{4}$$

$$\sum\nolimits_{i \in N, i \neq j} z_{ij}^{k} = y_{jk}, \forall j, k \in N, j \neq k \tag{5}$$

$$\sum\nolimits_{i \in N} \left(w_i^+ + w_i^-\right) \cdot y_{ij} \leq C_C, \forall j \in N \tag{6}$$

$$r_i^+ \geq r_j^+ + w_i^+ - \left(1 - \sum\nolimits_{k \in N, k \neq i} z_{ij}^{k}\right) \cdot M_0, \forall i, j \in N \tag{7}$$

$$r_j^- \geq r_i^- + w_j^- - \left(1 - \sum\nolimits_{k \in N, k \neq j} z_{ij}^{k}\right) \cdot M_0, \forall i, j \in N \tag{8}$$

$$r_j^+ + r_i^- - \left(1 - \sum\nolimits_{k \in N} z_{ij}^{k}\right) \cdot M_0 \leq v_{ij}, \forall i, j \in N \tag{9}$$

$$v_{ij} \leq C_T^2, \forall i, j \in N, i \neq j \tag{10}$$

$$\sum\nolimits_{j \in N} \left(2 \cdot y_{jj} + \theta \cdot \frac{\sum_{i \in N} w_i^+ \cdot y_{ij}}{C_T^1}\right) \cdot Car_{0j} + \sum\nolimits_{i \in N} \sum\nolimits_{j \in N} \left(\sum\nolimits_{k \in N} z_{ij}^{k} + \theta \cdot \frac{v_{ij}}{C_T^2}\right) \cdot Car_{ij} \leq L \tag{11}$$

$$T_i \geq t_{ki} + (z_{ki}^{k} - 1) \cdot M_0, \forall i, k \in N, i \neq k \tag{12}$$

$$T_j \geq T_i + s_i + t_{ij} + \left(\sum\nolimits_{k \in N: k \neq i} z_{ij}^{k} - 1\right) \cdot M_0, \forall i, j \in N \tag{13}$$

$$T_i \leq T_0, \forall i \in N \tag{14}$$

$$r_i^+, r_i^-, v_{ij}, T_i \geq 0, y_{ij}, z_{ij}^{k} \in \{0,1\}, \forall i, j, k \in N \tag{15}$$

Objective function (1) minimizes the total cost: the first term is the construction cost of PPCs; the second term is the transportation cost between pallet manufacturer and PPCs, including round-trip; the third term is the transportation cost between PPCs and PPSs; the fourth term is the fixed route cost. Constraints (2) ensure that each PPS can only be assigned to one PPC. Constraints (3) state that only the constructed PPCs can serve PPSs. Constraints (4)–(5) ensure that the preceding and the subsequent nodes of each PPS are unique, and must be assigned to the same PPC. Constraints (6) indicate that the total demands assigned to the constructed PPC must be less than its capacity, including the rental and recycling demands. Constraints (7) calculate the loading volume of rental pallets when a vehicle arrives at node. Constraints (8) calculate the loading volume of recycling pallets when a vehicle leaves node. Constraints (9) calculate the loading volume of a vehicle during the transportation between two nodes. Constraints (10) guarantee that the loading volume should be less than the capacity of a vehicle. Constraint (11) ensures that the total carbon emissions should less than the maximum upper limit for carbon emissions. Constraints (12) denote that arrival time at the first PPS of a vehicle from the PPC. Constraints (13) represents the time relationship of vehicle arrival between the previous node and the subsequent nodes.Constraints (14) ensure that the final return time of each vehicle to the PPC is less than the maximum transportation time for a single route. Constraints (15) represent the domain of variables.

2.3 Robust Optimization Model

To address the uncertainty of demand, a robust optimization model is proposed. Here, the robust optimization method of Bertsimas and Sim [19] is adopted. We assume that the renting demand $\tilde{w}_i^+$ is an independent random variable, taking a value from the interval $\left[\overline{w}_i^+ - \widehat{w}_i^+, \overline{w}_i^+ + \widehat{w}_i^+\right]$, where $\overline{w}_i^+$ is the nominal value and $\widehat{w}_i^+$ is the maximum deviation from the nominal value. The uncertain representation of recycling demand $\tilde{w}_i^-$ is the same as above. Accordingly, we introduce parameter Γ^0, which represents the number of PPS with uncertain demands served by PPC. The robust optimization model of this study can be proposed as Model [*R*].

$$[R]\ \boldsymbol{min} \sum\nolimits_{j\in N} FC\cdot y_{jj} + \sum\nolimits_{j\in N} 2\cdot trs_{0j}\cdot y_{jj} + \sum\nolimits_{i\in N}\sum\nolimits_{j\in N}\sum\nolimits_{k\in N} z_{ij}^k\cdot trs_{ij} + B\cdot \sum\nolimits_{k\in N}\sum\nolimits_{j\in N} z_{kj}^k \tag{16}$$

$$\boldsymbol{s.t.}\ (2)-(5),\ (12)-(15)$$

$$r_j^+ + r_i^- - \left(1-\sum\nolimits_{k\in N} z_{ij}^k\right)\cdot M_0 \le C_T^2, \forall i,j\in N, i\neq j \tag{17}$$

$$\sum\nolimits_{j\in N} 2\cdot y_{jj}\cdot Car_{0j} + \sum\nolimits_{i\in N}\sum\nolimits_{j\in N}\sum\nolimits_{k\in N} z_{ij}^k\cdot Car_{ij} \le L \tag{18}$$

$$r_i^+ \ge r_j^+ + \overline{w}_i^+ + \widehat{w}_i^+ - \left(1-\sum\nolimits_{k\in N,k\neq i} z_{ij}^k\right)\cdot M_0, \forall i,j\in N \tag{19}$$

$$r_j^- \ge r_i^- + \overline{w}_i^- + \widehat{w}_i^- - \left(1-\sum\nolimits_{k\in N,k\neq j} z_{ij}^k\right)\cdot M_0, \forall i,j\in N \tag{20}$$

$$\sum\nolimits_{i\in N}\left(\overline{w}_i^+ + \overline{w}_i^-\right)\cdot y_{ij} + \max_{\{U|U\subset N,|U|=\Gamma_0\}}\left\{\sum\nolimits_{i\in U}\left(\hat{w}_i^+ + \hat{w}_i^-\right)\cdot y_{ij} + (\Gamma_0 - \lfloor\Gamma_0\rfloor)\cdot \left(\hat{w}_i^+ + \hat{w}_i^-\right)\cdot y_{ij}\right\} \le C_C,\ \forall j\in N \tag{21}$$

Model [*R*] is a nonlinear formulation, which requires advanced computational power. To obtain a tractable formulation, λ and $\beta_i, \forall i\in N$ are introduced as the corresponding dual variables, and strong duality is applied to transform it into the equivalent linear formulation [*R′*].

$$[R']\ \boldsymbol{min}\ (16)$$

$$\boldsymbol{s.t.}\ (2)-(5),\ (12)-(15),\ (17)-(20)$$

$$\sum\nolimits_{i\in N}\left(\overline{w}_i^+ + \overline{w}_i^-\right)\cdot y_{ij} + \sum\nolimits_{i\in N}\beta_{ij} + \Gamma_0\cdot\lambda_j \le C_C, \forall j\in N \tag{22}$$

$$\lambda_j + \beta_{ij} \ge \left(\widehat{w}_i^+ + \widehat{w}_i^-\right)\cdot y_{ij}, \forall i,j\in N \tag{23}$$

$$\lambda_j, \beta_{ij} \ge 0, \forall i\in N \tag{24}$$

2.4 Complexity Analysis

The proposed models are with $O(|N|^3)$ binary variables, $O(|N|^2)$ non-negative variables and $O(|N|^2)$ constraints, in which $|N|$ is the number of PPSs. This model involves a large number of variables and constraints for the greater values of $|N|$, which could potentially make it hard to obtain optimal solutions. On the other hand, LRP is an NP-hard problem [18], and the proposed model is one extension of classical LRP. Hence, efficient algorithms are very necessary.

3 Heuristic Algorithm

Owing to the NP-hard computational complexity of the proposed problem, a heuristic algorithm combining GA and ACO is designed for the problem to obtain high quality solutions efficiently in this section.

3.1 Initialization and Decoding

Initialization. The chromosomes are initialized with greedy strategy. Firstly, the number of PPCs is not less than the minimum of PPCs, which is equal to the sum of demands divided by the capacity of a PPC. Then, the location initialization is based on the transportation costs between the pallet manufacturer and PPSs, with PPCs being constructed first when transportation costs are low. After that, assignment relationship also adopts greedy strategy. The node with higher demand has the priority to assign its nearest PPC. If the nearest PPC does not satisfy the capacity constraints, the PPS is randomly assigned to another PPC that can satisfy the capacity constraints. Finally, the service order (route planning) is randomly generated according to the assignment.

Decoding. To obtain the total cost, the construction cost, the transportation cost and the fixed vehicle cost should be calculated one by one. The construction cost can be calculated according to the location genes. Next, according to the service order, combined with the pallet capacity of a vehicle, route planning can be determined. According to the route planning, transportation cost and the fixed cost of vehicle are calculated. The Big-M method is used to ensure the feasibility of the solution. The probability of a chromosome being inherited by the next generation is determined by its fitness. In this paper, the fitness is defined as the reciprocal of the objective function, multiplied by a suitable constant M_1(where M_1 is a large constant): Fitness $= M_1/f(y, z, r, T)$.

3.2 Genetic Operator

Selection. The roulette selection strategy is used, in which the probability of each chromosome being selected is proportional to its fitness value. Based on the fitness values of chromosomes, the probability of being inherited into the next generation population is calculated.

Crossover. Single-point crossover is performed on the location genes. After crossover, the number of PPCs is checked to see whether it is more than the minimum number of PPCs calculated. If the number of non-zero genes is less than the minimum, a zero gene is randomly selected and changed to 1 until the minimum number is achieved. According to the selected PPCs, PPSs and route planning are rearranged again. The fitness values of parents and offspring are compared, and the chromosomes with higher fitness values proceed to mutation.

Mutation. Two mutation methods are adopted with equal probability. The first is to exchange the genes of assignment: randomly select two PPS with different PPCs, and exchange the PPCs that provide service for them. The second mutation method involves exchanging the location genes: randomly select a PPC and a non-PPC, the location genes, assignment and route planning are exchanged.

3.3 Optimization Strategies

Local Search for Assignment. In order to make rational use of the capacity of PPCs, a local search for assignment is designed, as described in **Algorithm 1**. Each PPS is exchanged with PPSs from different PPCs in turn. After the exchange, the routes are modified and the best chromosome is preferentially retained.

Algorithm 1: Local Search for Assignment

Input: y_{ik}, z_{ij}^{k}
Output: $y_{ik}{}'$, $z_{ij}^{k}{}'$

```
for i ← 0 to |N| do
    for j ← i + 1 to |N| do
        if i and j arc assigned to different PPCs then
            exchange the assignment (y_ik and y_jk)
            modify the route planning (z_ij^k)
            decode and compare the chromosome
        end if
        return the best chromosome
    end for
end for
```

Algorithm 2: Ant Colony Optimization

```
Input: w_i^+, w_i^-, trs_ij, z_ij^k, NcMax
Output: z_ij^k'
for x ← 0 to the number of PPCs do
    if PPC x serves three or more PPSs then
        set τ_0(initial pheromone), the number of ants, Nc ← 0
        do
            for ant k to the number of ants do
                set initial node is PPC x
                while the number of PPSs which PPCx served is not met
                    calculate the probability of ant k selecting node j as the next step
                    use the roulette strategy to choose next node
                end while
            end for
            update τ_ij (pheromone) between node i and j
            Nc++
        while Nc = NcMax
    end if
    return the best route of PPC x
end for
```

Ant Colony Optimization. The ant colony algorithm mimics the intelligent behavior of ants during foraging. Ant colony can find the shortest route to the food source according to the concentration of "pheromone" (τ_{ij}) released by itself in different environments. In order to obtain the shortest service route for a PPC that services three or more PFESs, the ant colony optimization described in **Algorithm 2** is adopted. There are k ants, iterated *NcMax* times. In the initial stage, the pheromone concentration between any two nodes is equal, the starting node of each ant is the currently optimized PPC, and the initial route is provided by the original chromosome. In each iteration, according to *e.q.* (25), a heuristic function (η_{ij}) is calculated, which represents the expected degree of ants moving from node i to node j. According to *e.q.* (26), probability P_{ij}^k of the ant k selecting node j as the next step is calculated. Then the roulette strategy is used to determine the next step. After the entire route is determined, the pheromone concentration (τ_{ij}) between any two nodes is updated according to the ant cycle model (*e.q.* (27),Q is a constant, which represents the total amount of pheromone released by the ant on the route, L_k is the route length that the ant pass through). Through *NcMax* iterations, the optimal routing is finally retained.

$$\eta_{ij} = \frac{1}{trs_{ij}} \tag{25}$$

$$P_{ij}^k = \begin{cases} \frac{[\tau_{ij}(Nc)]^\alpha \cdot [\eta_{ij}(Nc)]^\beta}{\sum_{s \in allow_k} [\tau_{is}(Nc)]^\alpha \cdot [\eta_{is}(Nc)]^\beta}, & s \in allow_k \\ 0, & s \notin allow_k \end{cases} \tag{26}$$

$$\Delta\tau_{ij} = \begin{cases} \frac{Q}{L_k}, & \text{ant } k \text{ from the node } i \text{ visit the node } j \\ 0, & \text{otherwise} \end{cases} \tag{27}$$

4 Numerical Experiments

To verify the proposed model and algorithm, a series of numerical experiments are conducted. The following experiments are carried out on a personal computer (Intel (R) Core (TM) i5-5350U CPU @ 1.80 GHz and RAM 8.0 GB). The proposed algorithm and classical GA were implemented in Visual Studio 2022. The benchmark of Barreto[20] is utilized as the input data.

4.1 Model Analyses and Comparison

To evaluate the rationality and superiority of the proposed model, the spatial distribution of location-routing and transportation time are explored and compared with the two-stage model where the first-stage is to determine the location and the second-stage is to determine the routing. Figure 1(a) shows the network structure of the proposed model, in which two nodes are selected to construct PPCs, with the route planning for both PPCs being radial, extending outward to each PPS. Two routes are planned in PPC1 to serve four PPSs: PPC1 → E → B → PPC1 and PPC1 → D → C → PPC1. Similarly, two routes are planned in PPC2 to serve another four PPSs: PPC2 → G → H → PPC2 and PPC2 → I → A → PPC2. Figure 1(b) illustrates the spatial distribution obtained by the two-stage model, which is more dispersed. In order to satisfy the carbon emissions constraint, the number of PPCs in the two-stage model increases, and the PPSs are assigned to the nearest PPCs. From the viewpoint of transportation efficiency, the space utilization of vehicles has decreased, especially on the PPC2 → H → PPC2 route, with the loading rate on the return trip being only 12.5%, which is much lower than the minimum loading rate of 29.3% in the proposed model. Additionally, the proposed model offers a more compact network structure, reducing carbon emissions by 7.49%, transportation costs by 9.45% and total cost by 25.1%.

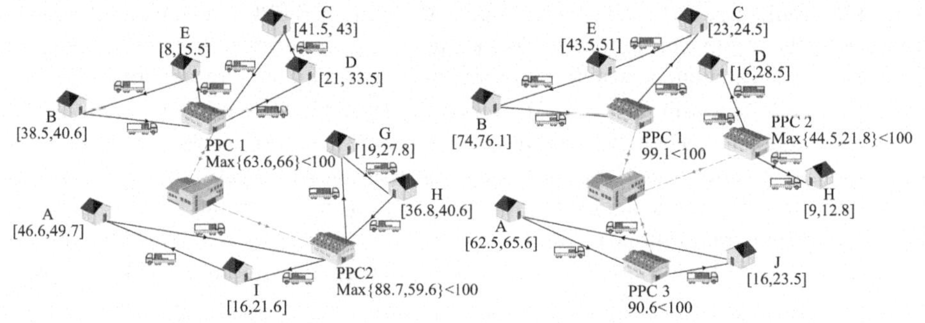

(a) The network structure of the proposed model (b) The network structure of the two-stage model

Fig. 1. Comparison of network results.

4.2 Sensitivity Analyses

This section explores the influence of the main parameters on the total cost and carbon emissions. Here, the pallet capacity of a PPC (C_C), the pallet capacity of a vehicle from PPC to PPSs (C_T^2), the limitation of transportation time (T_0) and the limitation of carbon emissions in the transportation (L) are selected as the main parameters for conducting sensitivity experiments. Note that sensitivity analysis is also conducted on other parameters, but the numerical results are not shown in this paper because their influences are not significant.

The impact of the pallet capacity of a PPC is illustrated in Fig. 2(a), which shows its effects on total cost and carbon emissions. With the increase of C_C, total costs continue to decrease, while the carbon emissions exhibit fluctuations. The location-routing network structure is closely tied to the change of C_C. When C_C increases from 1200 to 1500, although three PPCs are constructed, the network structure between PPCs and PPSs changes, resulting in a slight fluctuation in carbon emissions. Expending C_C from 1500 to 1800, two PPCs can meet all demands, thereby reducing total cost. As C_C continues to increase from 1800 to 2100, the distribution relationship changes, and the corresponding transportation routes are optimized, resulting in a large fluctuation in carbon emissions. However, when C_C reaches 3600, only one PPC is needed, and a significant increase in carbon emissions is in exchange for the reduction of total costs.

Figure 2(b) shows the effect of the capacity of a vehicle from PPC to PPSs. As demonstrated, increasing the pallet capacity of a vehicle leads to a reduction in total cost and carbon emissions, particularly within the capacity range of 300 to 600. This trend occurs because larger pallet capacities allow for more pallets to be transported in a single trip, enabling a vehicle to serve more PPSs with fewer trips, thereby significantly enhancing transportation efficiency and reducing overall carbon emissions and cost. As C_T^2 continues to increase, the rate of decline in total cost and carbon emissions diminishes, eventually stabilizing.

Figure 2(c) illustrates the effect of the limitation of transportation time on total cost and carbon emissions. With T_0 increases, total cost shows a decreasing trend, while carbon emissions fluctuate significantly at several key points. Analyzing the location-routing network at these critical points reveals that when T_0 is limited to below 65, three PPCs are required to meet demand, and the total cost is the highest. However, as T_0 increases from 65 to 70, the reduction in the number of PPCs leads to longer transportation distances and higher in-transit loading volume, resulting in a rise in carbon emissions. Subsequently, the increase of T_0 no longer affects the number of PPCs, but affects the assignment relationship between PPCs and PPSs and the number of routes, resulting in a gradual decrease in carbon emissions and then tends to be stable.

The impact of the limitation of carbon emissions is illustrated in Fig. 2(d), which shows its effects on total costs and actual carbon emissions. When L is less than 44, in order to meet the constraints of the limitation of carbon emissions, three PPCs need to be constructed to reduce the transportation distance and in-transit loading volume. When L is expanded to 46, two PPCs can be satisfied.

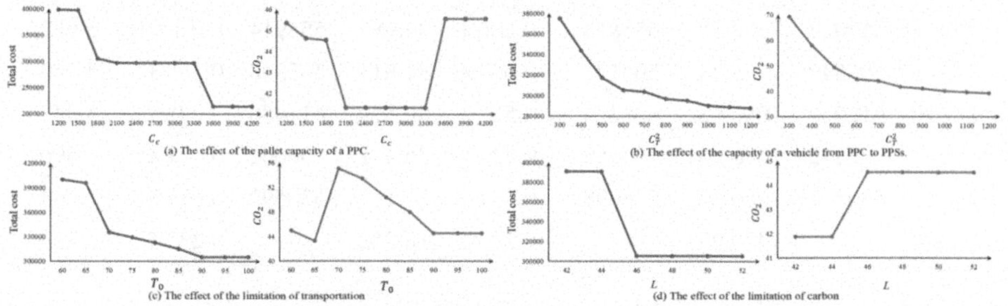

Fig. 2. Sensitive analyses.

4.3 Algorithm Analyses and Comparison

Table 1 shows the computational results of CPLEX, the proposed heuristic algorithm and GA. The comparison is conducted by selecting three cases for each of the five different scales: |N| = 10,|N| = 12, |N| = 15, |N| = 18 and |N| = 20.

For |N| = 10 and|N| = 12, CPLEX achieves optimal solutions from 30.80s to 1375.94s, whereas the proposed algorithm achieves the optimal solution within 10s. Although the GA exhibits a shorter computational time compared to the previous two methods, it struggles to reach the optimal solution, with an average gap of 0.538%. As |N| increases, the computational time for CPLEX grows rapidly, while the computational times for both the proposed algorithm and GA remain relatively stable. For |N| = 15, |N| = 18 and |N| = 20 the values of *UB* represent the results obtained by CPLEX with the limitation of one-hour,two-hour and three-hour, respectively. The computational results indicate that the solution quality of the proposed algorithm is significantly better, with an average gap of −0.455% compared to CPLEX, while the average gap for GA is 0.028%.

Table 1. The computational results of CPLEX, the proposed algorithm and GA

Group	CPLEX		The proposed algorithm			GA		
	UB	CPU(s)	$f_1{}'$	*Gap*	CPU(s)	$f_2{}'$	*Gap*	CPU(s)
|N| = 10	305389	46.84	305389	0.000%	3.300	306802	0.463%	0.436
	271021	33.11	271021	0.000%	3.630	273285	0.836%	0.459
	250573	30.80	250573	0.000%	3.097	252478	0.761%	0.424
|N| = 12	402768	217.45	402768	0.000%	4.390	404754	0.493%	0.900
	370962	527.42	370962	0.000%	4.595	370962	0.000%	1.057
	370434	1375.94	370434	0.000%	9.760	372929	0.673%	3.929
|N| = 15	437245	3602.70	436008	−0.283%	10.738	438381	0.259%	4.306
	439837	3603.61	435976	−0.878%	28.323	438664	−0.267%	2.130
	392993	3603.69	392800	−0.049%	9.798	392919	−0.019%	0.705
|N| = 18	553197	7206.27	551889	−0.236%	36.760	554302	0.199%	3.909
	557022	7202.50	555874	−0.206%	34.987	555914	−0.198%	5.922
	506540	7205.77	503171	−0.665%	31.811	507898	0.268%	5.560
|N| = 20	612050	10803.52	608424	−0.592%	30.559	611456	−0.097%	5.947
	606532	10804.01	603952	−0.425%	36.422	606662	0.021%	4.596
	530131	10801.32	526099	−0.761%	27.108	530592	0.087%	7.164
Average		4470.99		−0.273%	18.352		0.232%	3.163

5 Conclusions

In this paper, the LRP with simultaneous delivery and pickup is studied, which is a new extension of the LRP and the first application of the LRP in the pallet pooling system. A mixed-integer linear programming model is established to minimize the total cost, where the characteristic of simultaneous delivery and pickup, the limitation on carbon emissions and transportation time are considered. Then, considering the uncertainty of demand, the robust optimization model is extended. Further, a heuristic algorithm combining GA and ACO is designed to solve the LRP, which can obtain near-optimal solutions efficiently. Finally, numerical experiments validate the effectiveness of the proposed model and algorithm, providing theoretical support for balancing cost-effectiveness and environmental responsibility. The numerical experiments show that 1) the proposed model can obtain the integrated optimal solution of the location problem and route planning, which is better than the two-stage model; 2) the total cost, carbon emissions, and network structures are highly sensitive to some key parameters; 3) the proposed algorithm shows better performance in terms of solution quality and computational time.

Acknowledgments. This study was funded by the National Natural Science Foundation of China (Nos. 72201144, 72271141) and the Provincial Natural Science Foundation of Shandong (No. ZR2022QG025).

References

1. Prodhon, C., Prins, C.: A survey of recent research on location-routing problems. Eur. J. Oper. Res. **238**(1), 1–17 (2014)
2. Yang, J., Sun, H.: Battery swap station location-routing problem with capacitated electric vehicles. Comput. Oper. Res. **55**, 217–232 (2015)
3. Schiffer, M., Walther, G.: The electric location routing problem with time windows and partial recharging. Eur. J. Oper. Res. **260**(3), 995–1013 (2017)
4. Zhou, Y., Yu, H., Li, Z., Su, J., Liu, C.: Robust optimization of a distribution network location-routing problem under carbon trading policies. IEEE Access **8**, 46288–46306 (2020)
5. Maghfiroh, M.F.N., Yu, V.F., Redi, A.A.N.P., Abdallah, B.N.: A location routing problem with time windows consideration: a metaheuristics approach. Appl. Sci.-Basel **13**(2), 843 (2023)
6. Liu, D., Zhang, H., Liu, Y., Luo, S.: An improved mushroom algorithm for solving the open location-routing problem with simultaneous pickup and delivery. Control Eng. China **30**(10), 1801–1811 (2023). (in Chinese)
7. Zang, X., Zhu, Y., Zhong, Y., Chu, T.: CiteSpace-based bibliometric review of pickup and delivery problem from 1995 to 2021. Appl. Sci.-Basel **12**(9), 4607 (2022)
8. Min, H.: The multiple vehicle routing problem with simultaneous delivery and pick-up points. Transp. Res. Part A: Gen. **23**(5), 377–386 (1989)
9. Montané, F.A.T., Galvao, R.D.: A tabu search algorithm for the vehicle routing problem with simultaneous pick-up and delivery service. Comput. Oper. Res. **33**(3), 595–619 (2006)
10. Ai, T.J., Kachitvichyanukul, V.: A particle swarm optimization for the vehicle routing problem with simultaneous pickup and delivery. Comput. Oper. Res. **36**(5), 1693–1702 (2009)

11. Zhang, Z., Cheang, B., Li, C., Lim, A.: Multi-commodity demand fulfillment via simultaneous pickup and delivery for a fast fashion retailer. Comput. Oper. Res. **103**, 81–96 (2019)
12. Yu, V., Susanto, H., Yeh, Y., Lin, S., Huang, Y.: The vehicle routing problem with simultaneous pickup and delivery and parcel lockers. Mathematics **10**(6), 920 (2022)
13. Roy, D., Carrano, A.L., Pazour, J.A., Gupta, A.: Cost-effective pallet management strategies. Transp. Res. Part E: Logist. Transp. Rev. **93**, 358–371 (2016)
14. Ren, J., Zhang, X.: Pallet recovery stochastic programming model of pallet pool system. Control Decis. **25**(8), 1211–1214 (2010). (in Chinese)
15. Wu, C., Tsang, Y.P., Lee, C.K.M., Ching, W.K.: A blockchain-IoT platform for the smart pallet pooling management. Sensors **21**(18), 6310 (2021)
16. Yu, H., Yang, J., Kang, X., Cong, Z., Yao, S.: Empty pallet allocation optimization in shipbuilding using a pallet pool system. Sustainability **14**(9), 5479 (2022)
17. Zhao, P., Liu, F., Guo, Y., Duan, X., Zhang, Y.: Bi-objective optimization for vehicle routing problems with a mixed fleet of conventional and electric vehicles and soft time windows. J. Adv. Transp. **2021**(1) (2021)
18. Tordecilla, R.D., Montoya-Torres, J.R., Quintero-Araujo, C.L., Panadero, J., Juan, A.A.: The location routing problem with facility sizing decisions. Int. Trans. Oper. Res. **30**(2), 915–945 (2023)
19. Bertsimas, D., Sim, M.: The price of robustness. Oper. Res.. Res. **52**(1), 35–53 (2004)
20. http://prodhonc.free.fr/Instances/instances_us.htm. Accessed 20 July 2024

Scheduling Fully Parallel Jobs with Integer Units

Junyi Zhang, Juan Zou(✉), and Mingyu Ma

School of Mathematical Sciences, Qufu Normal University, Qufu 273165, Shandong, China
zoujuanjn@163.com

Abstract. We consider m identical machines scheduling problems with fully parallel jobs. Each job J_j requires processing time p_j and can be executed on any machine at any time unit. In this paper, four scheduling problems are considered: (1) minimize the maximum cost, (2) minimize the total completion time, (3) minimize the weighted number of tardy jobs, and (4) minimize the total weighted tardiness. For the first two problems, we develop optimal polynomial-time algorithms to solve them, respectively. For the third problem, we design a polynomial-time algorithm if all weights are equal to one; we propose a pseudo-polynomial-time algorithm for the general case. For the last problem, we design polynomial-time algorithms for solving some special cases of this problem, respectively.

Keywords: Identical machines · Fully parallel jobs · Integer parallel units · Polynomial-time algorithm

1 Introduction

In modern manufacturing industry, the parallel machine scheduling problem is an important research area for optimizing resource allocation and improving efficiency. In parallel machine scheduling, splitting the job can significantly improve scheduling flexibility and resource utilization. By splitting the job into multiple parts and assigning them to different machines for parallel processing, the load can be better balanced, waiting time can be reduced, and available machine resources can be maximized. Preemptive (see Baptiste et al. [1], Su et al. [11], Cheng et al. [2], etc.) and splitting job (see Tahar et al. [12], Wang et al. [13], Lee et al. [9], etc.) are two key techniques to achieve this goal. Almost all researchers studied parallel machine scheduling problems under the hypothesis that each job can be processed on at most one machine at a time although preemption is allowed. Recently, Zhang et al. [17] proposed a new splitting model, i.e., scheduling fully parallel jobs. The central idea of the fully parallel jobs model is to split a job into multiple independent processing units, which can be processed in different machines simultaneously, thus significantly improving efficiency. The fully

D. Du et al. (Eds.): COCOA 2024, LNCS 15434, pp. 267–277, 2025.
https://doi.org/10.1007/978-981-96-4445-2_22

parallel jobs model has been widely applied in the context of Flexible Manufacturing System (FMS), MapReduce, and Spark, which are big data processing frameworks. Zhang et al. [17] studied the problem of minimizing the total weighted completion time, they proved that the problem is strongly NP-hard when the number of the machines m is input and gave a 2-approximation algorithm. Wang et al. [14] further considered this problem. When the number of machines m is fixed, they gave a PTAS for the general case and gave a dynamic programming for the special case of $w_j = p_j$. They also improved the approximation ratio to $1 + \frac{m-1}{m+2}$ when m is the input.

However, to the best of our knowledge, the existing literatures are mainly concentrating on minimizing the total weighted completion time and lack research on other objective functions. Therefore, based on different practical situations, we discuss some regular scheduling criteria, including the maximum cost, the total completion time, the total weighted number of tardy jobs, and the total weighted tardiness. For these objective functions, Xing et al. [15] proved that scheduling splitting jobs on parallel machines can be converted to the corresponding single-machine problem. However, for fully parallel jobs, they cannot be directly converted to single-machine problems. Therefore, for each problem under study, we design either a polynomial-time or a pseudo-polynomial-time algorithm to solve it. It aims to design effective scheduling algorithms to optimize resource utilization and improve system performance. This research not only provides new theoretical insights and methods for the fully parallel job scheduling problem but also offers effective solutions for scheduling issues in practical systems.

The remainder of the paper is organized as follows. In Sect. 2, we introduce the notation and formulation. The subsequent sections present our results for maximum cost, total completion time, total weighted number of tardy jobs and total weighted tardiness problems. Finally, in Sect. 7 we summarize the paper and give directions for future research.

2 Notation and Formulation

There are n fully parallel jobs $J = \{J_1, J_2, \ldots, J_n\}$ to be processed on m identical parallel machines $M = \{M_1, M_2, \ldots, M_m\}$. Each job J_j requires processing time p_j and can be executed on any machine at any time unit. Each machine is available at any time $t \geq 0$ and can execute at most one processing unit of one job during one time unit. A schedule S is represented by a tuple $(\varphi_{11}, \varphi_{12}, \ldots, \varphi_{1m}, \varphi_{21}, \varphi_{22}, \ldots, \varphi_{2m}, \ldots)$, where $\varphi_{ti} = j$ means that one processing unit of job J_j is processed on machine M_i at time $[t-1, t)$, $i \in \{1, 2, \ldots, m\}$ and $t \in \{1, 2, \ldots\}$. We say a job is satisfied when it receives the required number of processing units. A schedule is feasible if all jobs are satisfied, i.e., $p_j = |\{\varphi_{ti} | \varphi_{ti} = j\}|, \forall J_j \in J$. For job J_j, its starting time is defined by $S_j = \min\limits_{\varphi_{ti}=j}(t-1)$ and the completion time is defined by $C_j = \max\limits_{\varphi_{ti}=j} t$. Each job J_j has its own due date d_j and weight w_j.

In this paper, we consider the scheduling criteria of the jobs: the maximum cost $f_{\max} = \max_{j=1,\dots,n} \{f_j(C_j)\}$, where each $f_j(\cdot)$ is a nonnegative regular function of the completion time of job J_j; the total completion time $\sum_{j=1}^{n} C_j$; the weighted number of tardy jobs $\sum_{j=1}^{n} w_j U_j$, where $U_j = 0$ if $C_j \le d_j$ and otherwise $U_j = 1$, and the total weighted tardiness $\sum_{j=1}^{n} w_j T_j$, where $T_j = \max\{C_j - d_j, 0\}$. Using the classical three-field notation [4] for scheduling problems, we denote the problems under study by:

(P1) $Pm|fully|f_{\max}$;

(P2) $Pm|fully|\sum_{j=1}^{n} C_j$;

(P3) $Pm|fully|\sum_{j=1}^{n} w_j U_j$;

(P4) $Pm|fully|\sum_{j=1}^{n} w_j T_j$.

From Literature [14], we obtain that the optimal solution of the problem studied in this paper has the following two properties: (i) there is no any idle time on the machine during the whole processing; (ii) each job is processed consecutively one bye one. To find the optimal sequence of the problem, we only consider the schedule in which all the jobs are processed consecutively. We simplify the schedule S to $(J_{[1]}, J_{[2]}, \dots, J_{[k]}, \dots, J_{[n]})$, where $J_{[k]} = J_j$ denotes that the job J_j is in the k-th position in the schedule. Therefore, the completion time of job J_j is $C_j = \left\lceil \frac{\sum_{l=1}^{k} p_{[l]}}{m} \right\rceil$.

Remark 1. *If the single-machine problem $1||f$ is NP-hard in the ordinary sense, then the problem $Pm|fully|f$ is at least NP-hard in the ordinary sense, where f is a non-negative regular cost function.*

3 Minimizing the Maximum Cost

In this section, we consider problem (P1) $Pm|fully|f_{\max}$. We show that problem (P1) can be solved efficiently by modifying the Lawler's Algorithm [6].

Let U denote the set of unscheduled jobs, τ denote the total size of the current unscheduled jobs, and l denote the current position to be scheduled.

Theorem 1. *Algorithm 1 yields an optimal solution for problem $Pm|fully|f_{\max}$ in $O(n^2)$ time.*

Algorithm 1. The Modified Lawler's Algorithm

Input: J, p_j.
Output: a schedule S.
1: $U = J$, $\tau = \sum_{j=1}^{n} p_j$;
2: **for** $i = 1 : n$ **do**
3: $l = n - i + 1$;
4: $t = \lceil \frac{\tau}{m} \rceil$;
5: $f_k(t) = \min\{f_j(t) : J_j \in U\}$;
6: $J_k \rightarrow J_{[l]}$;
7: $U = U \setminus \{J_k\}$;
8: $\tau = \tau - p_k$;
9: **end for**
10: **return**

4 Minimizing the Total Completion Time

In this section, for problem (P2) $Pm|fully|\sum_{j=1}^{n} C_j$, we prove that an optimal solution can be obtained by sequencing the jobs in non-decreasing order of processing time (the SPT rule).

Theorem 2. *The SPT rule solves problem* $Pm|fully|\sum_{j=1}^{n} C_j$ *optimally in* $O(nlogn)$ *time.*

Proof. By contradiction. Consider an optimal sequence S^*. Assume that there are two adjacent jobs in S^*, job J_j followed by job J_k, such that $p_j > p_k$. The start time of job J_j is t. Now swap the two jobs to get a new scheduling $S' = (\ldots, J_k, J_j, \ldots)$. Obviously, it is clear that under such an operation the following facts hold:

(i) the completion time keeps unchanged for all jobs except for jobs J_j and J_k in sequence S'; *vs*.kip0.1cm
(ii) the start time of job J_k is t in sequence S'.

Therefore, we have

$$
\begin{aligned}
&\sum_{j=1}^{n} C_j(S') - \sum_{j=1}^{n} C_j(S^*) \\
&= (C_j(S') + C_k(S')) - (C_k(S^*) + C_j(S^*)) \\
&= (\lceil \frac{t+p_k}{m} \rceil + \lceil \frac{t+p_k+p_j}{m} \rceil) - (\lceil \frac{t+p_j}{m} \rceil + \lceil \frac{t+p_j+p_k}{m} \rceil) \\
&= \lceil \frac{t+p_k}{m} \rceil - \lceil \frac{t+p_j}{m} \rceil \leq 0.
\end{aligned}
$$

Hence, the optimal solution can be obtained by the SPT rule.

5 Minimizing the Weighted Number of Tardy Jobs

In this section, we give a polynomial-time algorithm to obtain the optimal solution for problem $Pm|fully|\sum_{j=1}^{n} U_j$ and design a pseudo-polynomial dynamic programming algorithm for $Pm|fully|\sum_{j=1}^{n} w_jU_j$.

5.1 $Pm|fully|\sum_{j=1}^{n} U_j$

In this subsection, we consider the jobs which all weights are equal to one and prove that the schedule obtained, by the Modified Moore's Algorithm [10], is the optimal one.

Let S^E and S^T denote the sets of early and tardy jobs, respectively, τ denotes the total completion time of the early jobs so far. We define the EDD rule which sequences the jobs in non-decreasing order of due dates.

Algorithm 2. The Modified Moore's Algorithm

Input: d_j, p_j.
Output: S^E, S^T.
1: Sort the jobs in accordance with the EDD rule
2: Sort the jobs with the same due date in accordance with the SPT rule
3: $S^E = \emptyset$, $S^T = \emptyset$, $\tau = 0$;
4: **for** $j = 1 : n$ **do**
5:   **if** $\lceil \frac{\tau + p_j}{m} \rceil \leq d_j$ **then**
6:     $\tau = \tau + p_j$
7:     $S^E = S^E \cup \{J_j\}$
8:   **else**
9:     $j_{\max} = \arg\max\{p_k | J_j \in S^E \cup \{J_j\}\}$
10:     $\tau = \tau + p_j - p_{j_{\max}}$
11:     $S^E = S^E \cup \{J_j\} \setminus \{J_{j_{\max}}\}$
12:     $S^T = S^T \cup \{J_{j_{\max}}\}$
13:   **end if**
14: **end for**
15: **return** $\sum_{j=1}^{n} U_j = |S^T|$

Theorem 3. *Algorithm 2 yields an optimal solution for problem $Pm|fully|\sum_{j=1}^{n} U_j$ in $O(nlogn)$ time.*

Proof. Assume that j jobs have been scheduled so far, where the number of early jobs is k and the total completion time of the early jobs is τ.

We claim that the current solution satisfies the following two properties:

(i) it contains the maximum number of early jobs;
(ii) the total processing time of early jobs is minimal among all the solutions that contain the maximum number of early jobs.

We can show these two properties are true by induction on j. Clearly the two properties hold for $j = 1$. Assume that the two properties hold for j and now consider scheduling job J_{j+1}. The following three cases will be considered:

Case 1: $\lceil \frac{\tau+p_{j+1}}{m} \rceil = \lceil \frac{\tau}{m} \rceil$.

In this case job J_{j+1} must be an early job. Indeed, if we denote by J_i the last early job, noting that at least one of the first j jobs is an early job, then we have $C_{j+1} = C_i \leq d_i \leq d_{j+1}$, i.e., job J_{j+1} must be the early job. Therefore, after scheduling job J_{j+1}, the number of early jobs increases to $k+1$, which is clearly the maximum possible for $j+1$ jobs. Property (ii) still holds, because by assumption, τ is the smallest total processing time of k early jobs, and J_{j+1} has the shortest processing time among unscheduled jobs with the same due date.

Case 2: $\lceil \frac{\tau+p_{j+1}}{m} \rceil > \lceil \frac{\tau}{m} \rceil$ and $\lceil \frac{\tau+p_{j+1}}{m} \rceil \leq d_{j+1}$.

Clearly job J_{j+1} is the early job. Similarly in case 1, the two properties still hold.

Case 3: $\lceil \frac{\tau+p_{j+1}}{m} \rceil > \lceil \frac{\tau}{m} \rceil$ and $\lceil \frac{\tau+p_{j+1}}{m} \rceil > d_{j+1}$.

According to the algorithm, a job with the largest processing time will be deleted from the set $S^E \cup \{J_j\} \setminus \{J_{j_{\max}}\}$. In fact, the number of early jobs remains k. Suppose there are $k+1$ early jobs from the set $\{J_1, \ldots, J_{j+1}\}$. From the algorithm, $k+1$ early jobs are obtained by adding job J_{j+1} to the set of k early jobs from the set $\{J_1, \ldots, J_j\}$ and deleting the largest job among these $k+1$ jobs.

Case 3.1: J_{j+1} is the largest job among these $k+1$ jobs.

In this case the adding job and deleting job are the same one. Thus the set of early jobs from the sets $\{J_1, \ldots, J_j\}$ and $\{J_1, \ldots, J_{j+1}\}$ is the same set, i.e., k early jobs. This is a contradiction.

Case 3.2: J_{j+1} is not the largest job among these $k+1$ jobs.

For convenience of presentation, denote S_j^E as the set of early jobs from the set $\{J_1, \ldots, J_j\}$ and S_{j+1}^E as the set of early jobs from the set $\{J_1, \ldots, J_{j+1}\}$. From the algorithm, $|(S_j^E \cup \{J_{j+1}\}) \setminus \{J_{\max}\}| = |S_{j+1}^E| = k+1$. Therefore, $|S_j^E| = k+1$. This is a contradiction. Therefore, property (i) holds.

Now we consider property (ii). By assumption, the total processing time of k early jobs from the set $\{J_1, \ldots, J_j\}$ is minimal. Therefore, after adding job J_{j+1} to the early jobs and deleting the largest job from the set of early jobs, the new set of early jobs remains minimal in terms of total processing time. We conclude that property (ii) still holds.

Sorting the jobs according to the EDD rule takes time $O(nlogn)$. The loop is executed n times and takes time $O(n)$. Therefore, the time complexity of Algorithm 2 is $O(nlogn)$.

5.2 $Pm|fully|\sum_{j=1}^{n} w_jU_j$

We know that the single-machine scheduling problem $1||\sum w_jU_j$ is NP-hard in a ordinary sense [5]. By Remark 1 in Sect. 2, it is clear that the problem (P3) $Pm|fully|\sum_{j=1}^{n} w_jU_j$ is at least NP-hard in the ordinary sense. We give the following pseudo-polynomial dynamic programming algorithm (denoted DP), which also proves that problem (P3) is NP-hard in the ordinary sense.

It is easy to prove that the early jobs are sorted according to the EDD rule. Therefore, we assume that the jobs are sorted according to the EDD rule. In each phase of DP, one job is considered: it is either scheduled as early as possible (as a early job) or scheduled last (as a tardy job). We define the state variables used in dynamic programming:

j: The number of jobs scheduled so far;
τ: Total processing time of the early jobs so far.

Let $f_j(\tau)$ denote the minimum cost of scheduling job J_j at phase j, assuming that the $j-1$ first jobs are scheduled, and that the completion time of the last early job is $\lceil \frac{\tau}{m} \rceil$.

Algorithm DP

Boundary Conditions

$$f_0(\tau) = 0, \forall \tau \geq 0$$
$$f_j(\tau) = +\infty, j = 1, 2, \ldots, n, \forall \tau < 0$$

Recursive Relations

$$f_j(\tau) = \begin{cases} \min\{f_{j-1}(\tau - p_j), f_{j-1}(\tau) + w_j\}, & \lceil \frac{\tau}{m} \rceil \leq d_j \\ f_{j-1}(\tau) + w_j, & \lceil \frac{\tau}{m} \rceil > d_j \text{ and } \tau \leq \sum_{j=1}^{n} p_j \end{cases}$$

Optimal Solution

$$f^* = \min_{0 \leq \tau \leq \sum_{j=1}^{n} p_j} \{f_n(\tau)\}$$

Theorem 4. *Algorithm DP solves problem* $Pm|fully|\sum_{j=1}^{n} w_jU_j$ *in* $O(n\sum_{j=1}^{n} p_j)$ *time.*

Proof. The sate variable j is bounded by n, and the sate variable τ is bounded by $\sum_{j=1}^{n} p_j$. The running time per iteration of the DP is constant time, It follows that the total running time of the DP is $O(n \sum_{j=1}^{n} p_j)$.

Example 1. We consider an example with 3 machines and 4 jobs:

job	J_1	J_2	J_3	J_4
p_j	8	10	4	7
d_j	5	6	8	9
w_j	10	13	7	5

The optimal solution can be obtained by DP, and the optimal solution is (J_1, J_2, J_3, J_4) where J_1, J_2, J_3 are early jobs and J_4 is a tardy job.

The completion time of early jobs are: $C_1 = \lceil \frac{8}{3} \rceil = 3 \leq 5 = d_1$, $C_2 = \lceil \frac{8+10}{3} \rceil = 6 \leq 6 = d_2$ and $C_3 = \lceil \frac{8+10+4}{3} \rceil = 8 \leq 8 = d_5$. The completion time of tardy job is $C_4 = \lceil \frac{8+10+4+7}{3} \rceil = 10 > 9 = d_4$. Therefore, the total weight number of tardy jobs is $\sum_{j=1}^{n} w_j U_j = w_4 = 5$.

6 Minimizing the Total Weighted Tardiness

In this section, we consider problem $Pm|fully|\sum_{j=1}^{n} w_j T_j$. This problem contains three sets of parameters: processing time, weight and due date. The problem can be refined into eight forms based on whether the parameters of jobs are the same or not. The eight forms are shown in Table 1.

Table 1. The eight forms of the minimizing the total weighted tardiness

	constant weight w		weighs w_j	
	constant processing time p	processing time p_j	constant processing time p	processing time p_j
common due date d	I any sequence	II SPT	III $w_1 \geq w_2 \geq \ldots \geq w_n$	IV NP-hard
due date d_j	V EDD	VI NP-hard	VII assignment	VIII NP-hard

For problem I, it is clear that any schedule is the optimal solution. Problems II, III, and V are "easy" yielding to algorithms which produce optimal sequences in $O(nlogn)$ time. These algorithms sequence the jobs simply by ranking them with respect to one of the parameter sets. For problem II, the processing times are sorted in non-decreasing order; for problem III, the weights are sorted in non-increasing order; and for problem V, the due dates are sorted in non-decreasing order. The optimality can be easily obtained by job-interchange argument. For problems IV, VI, and VIII, the corresponding single-machine problems $1|d_j = d|\sum_{j=1}^{n} w_jT_j$(Yuan [16]), $1||\sum_{j=1}^{n} T_j$(Du [3]) and $1||\sum_{j=1}^{n} w_jT_j$(Lawler [7] and Lenstra et al. [8]) are all NP-hard in the ordinary sense. By Remark 1 in Sect. 2, problems IV, VI, and VIII are also at least NP-hard in the ordinary sense. We formulate problems VII as the assignment problem.

Now, we consider problem $Pm|fully, p_j = p|\sum_{j=1}^{n} w_jT_j$. Since all the jobs have the same processing time, i.e., $p_j = p$, the completion time of the job in the k-th position in the sequence is $\lceil \frac{kp}{m} \rceil$. Define a binary variable x_{ij}, $i = 1, 2, \ldots, n$ and $j = 1, 2, \ldots, n$. If job J_j is scheduled as the i-th job, then $x_{ij} = 1$; otherwise, $x_{ij} = 0$. Denote the cost incurred when job J_j is placed in the i-th position by c_{ij}, where $c_{ij} = w_j \cdot \max\{C_j - d_j, 0\} = w_j \cdot \max\{\lceil \frac{ip}{m} \rceil - d_j, 0\}$. By the above definition, we can obtain an Integer Programming of problem VII, denoted as IP:

$$\min \quad \sum_{i=1}^{n}\sum_{j=1}^{n} c_{ij}x_{ij} \tag{1}$$

$$\text{s.t.} \quad \sum_{i=1}^{n} x_{ij} = 1, \quad j = 1, 2, \ldots, n \tag{2}$$

$$\sum_{j=1}^{n} x_{ij} = 1, \quad i = 1, 2, \ldots, n \tag{3}$$

$$x_{ij} \in \{0, 1\}, \quad i = 1, 2, \cdots, n; j = 1, 2, \ldots, n. \tag{4}$$

Constraints (2) ensures that each job is assigned to one and only one position. Constraints (3) ensures that each position i has at most one job assigned to it.

Theorem 5. *Problem $Pm|fully, p_j = p|\sum_{j=1}^{n} w_jT_j$ can be solved in $O(n^3)$ time.*

7 Conclusions and Future Directions

In this paper, we consider parallel machine environments with fully parallel jobs. Unlike traditional splitting models, the job can only be split in integer form and the split parts can be processed simultaneously on parallel machines. We devised polynomial-time algorithms for problem $Pm|fully|f_{\max}$, $Pm|fully|\sum_{j=1}^{n} C_j$ and $Pm|fully|\sum_{j=1}^{n} U_j$, respectively. We also designed a pseudo-polynomial dynamic programming algorithm for NP-hard problem $Pm|fully|\sum_{j=1}^{n} w_jU_j$. Finally, we studied the problem $Pm|fully|\sum_{j=1}^{n} w_jT_j$ and gave polynomial-time algorithms to find the optimal solution for some special cases, respectively. For future research, we suggest several interesting topics as follows: (i) design approximation algorithms or heuristics for the total weighted tardiness problem; (ii) analyze the scheduling models with job rejection or dynamic job arrivals; (iii) study the more general multi-agent scheduling models, such as competing agents, non-disjoint agents, etc.

Acknowledgements. We thank the editor and the anonymous reviewers for their helpful comments and suggestions on an earlier version of our paper. This research was supported by the Province Natural Science Foundation of Shandong (ZR2023MA051, ZR2022MA019, ZR2022MA038).

References

1. Baptiste, P., Carlier, J., Kononov, A.V., Queyranne, M., Sevastyanov, S., Sviridenko, M.: Integer preemptive scheduling on parallel machines. Oper. Res. Lett. **40**, 440–444 (2012)
2. Cheng, T., Kravchenko, S.A., Lin, B.: Preemptive parallel-machine scheduling with a common server to minimize makespan. Nav. Res. Logist. **64**, 388–398 (2017)
3. Du, J., Leung, J.Y.: Minimizing total tardiness on one machine is NP-hard. Math. Oper. Res. **15**, 483–495 (1990)
4. Graham, R.L., Lawler, E.L., Lenstra, J.K., Rinnooy Kan, A.: Optimization and approximation in deterministic sequencing and scheduling: a survey. Ann. Discret. Math. **5**, 287–326 (1977)
5. Karp, R.M.: Reducibility among combinatorial problems. Complexity Comput. Comput. 85–103 (1972)
6. Lawler, E.L.: Optimal sequencing of a single machine subject to precedence constraints. Manage. Sci. **19**, 544–546 (1973)
7. Lawler, E.L.: A "Pseudopolynomial" Algorithm for Sequencing Jobs to Minimize Total Tardiness. Ann. Discret. Math. **1**, 331–342 (1977)
8. Lenstra, J.K., Rinnooy Kan, A., Brucker, P.: Complexity of machine scheduling problems. Ann. Discret. Math. **1**, 343–362 (1977)
9. Lee, J.H., Kim, H.J.: A heuristic algorithm for identical parallel machine scheduling: splitting jobs, sequence-dependent setup times, and limited setup operators. Flex. Serv. Manuf. J. **33**, 992–1026 (2021)
10. Moore, J.M.: An n Job, one machine sequencing algorithm for minimizing the number of late jobs. Manage. Sci. **15**, 102–109 (1968)
11. Su, L., Cheng, T.C., Chou, F.: A minimum-cost network flow approach to preemptive parallel-machine scheduling. Comput. Sci. Eng. **64**, 453–458 (2013)

12. Tahar, D.N., Yalaoui, F., Chu, C., Amodeo, L.: A linear programming approach for identical parallel machine scheduling with job splitting and sequence-dependent setup times. Int. J. Prod. Econ. **99**, 63–73 (2006)
13. Wang, W.L., Wang, H.Y., Zhao, Y.W., Zhang, L.P., Xu, X.L.: Parallel machine scheduling with splitting jobs by a hybrid differential evolution algorithm. Comput. Oper. Res. **40**, 1196–1206 (2013)
14. Wang, K., Chau, V., Li, M.: Scheduling fully parallel jobs. J. Sched. **21**(6), 619–631 (2018). https://doi.org/10.1007/s10951-018-0563-3
15. Xing, W.X., Zhang, J.W.: Parallel machine scheduling with splitting jobs. Discret. Appl. Math. **103**, 259–269 (2000)
16. Yuan, J.J.: The NP-hardness of the single machine common due date weighted tardiness problem. J. Syst. Sci. Complexity **4**, 328–333 (1992)
17. Zhang, Q., Wu, W.W., Li, M.M.: Minimizing the total weighted completion time of fully parallel jobs with integer parallel units. Theoret. Comput. Sci. **507**, 34–40 (2013)

UAV Target Tracking with Bandit-Based Data Fusion

Yang Lv[1,2], Guochao Fan[1], Mengzhen Li[2], Xiongjun Liu[1], Pengqing Liu[2], and Yapu Zhang[2](✉)

[1] Beijing Jinghang Computation and Communication Research Institute, Beijing 100074, People's Republic of China
fanguochao@yeah.net

[2] Institute of Operations Research and Information Engineering, Beijing University of Technology, Beijing 100124, People's Republic of China
lmz9710@emails.bjut.edu.cn, zhangyapu@bjut.edu.cn

Abstract. Unmanned Aerial Vehicles have emerged as the optimal solution for target tracking due to their low cost and high maneuverability. We study the problem of target tracking in partially observable adversarial environments. Building on previous research, we develop a data fusion algorithm to optimize target tracking. To enable a single UAV to track a hard-to-observe target with limited sensor capabilities, we construct a target environment tracking model along with its corresponding reward function. In the algorithm design, we incorporate the Exp4-IX algorithm from the framework of an adversarial multi-armed bandit with advice, and we prove that the regret bound of this algorithm exhibits sub-linear growth. In numerical experiments, the Exp4-IX algorithm integrates the Previous Position algorithm, Particle Filtering algorithm, and Trajectory Fitting algorithm, and is benchmarked against the Average Fusion algorithm. The results demonstrate its effectiveness in online fusion for smooth trajectory scenarios. This integration allows the UAV to predict the target's position with greater accuracy compared to other algorithms.

Keywords: Unmanned Aerial Vehicles · Target tracking · Multi-armed bandit · Data fusion · Exp4-IX algorithm

1 Introduction

Unmanned Aerial Vehicles (UAVs), characterized by their low cost, small size, and high maneuverability, possess significant application potential. UAVs can be equipped with IoT devices, sensors, and processors, enabling them to perform various functions, including communication, flight, and computation. UAVs are widely employed in various tasks, such as maritime operations [13], disaster relief [14], communication coverage [9], navigation [17], and target tracking [11,12]. Among these, target tracking is one of the most notable applications, as UAVs can cover vast areas and access locations difficult for humans to reach.

D. Du et al. (Eds.): COCOA 2024, LNCS 15434, pp. 278–286, 2025.
https://doi.org/10.1007/978-981-96-4445-2_23

There is extensive research on tracking a single dynamic target with a single UAV, including the Actor-Critic reinforcement learning algorithm [2], YOLO algorithm [5], and DQN algorithm [1]. In the study by [5], the YOLO algorithm is used for real-time target detection when the UAV is stationary, while [1] demonstrates that the DQN algorithm performs better when the line of sight is obstructed by obstacles. When this data is transmitted to the fusion center, it is essential to evaluate each scheme and select the optimal advice to ensure that the target remains within the UAV's surveillance range (Fig. 1).

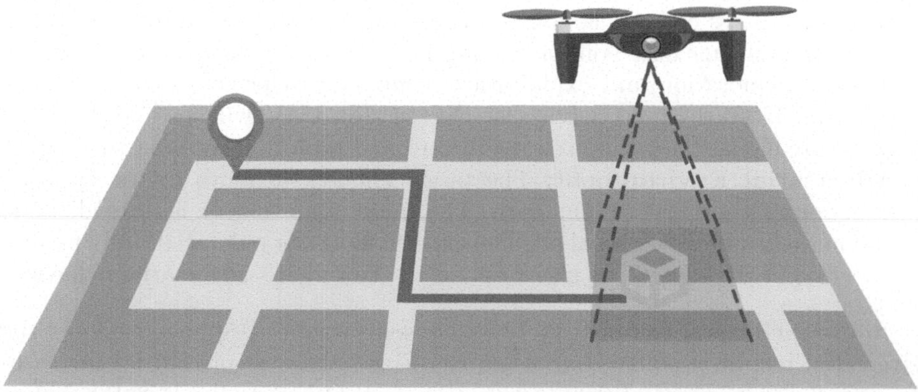

Fig. 1. A single dynamic target with a single UAV in adversarial environments: In target tracking, partial observability refers to the fact that the UAV can only observe the square-shaped region depicted in the image; if the object moves out of this region, it cannot be detected. An adversarial environment refers to a scenario where the target's movement direction may be completely opposite to the UAV's, requiring the UAV to respond swiftly. Under these challenges, we consider the strategy provided by multiple experts, enabling the UAV to anticipate the object's next position in advance.

In scenarios where multiple algorithms or experts provide advice, the modeling approach for tracking a single dynamic target with a single UAV is as follows. The UAV has k possible actions, set as four actions in this paper: forward, backward, left, and right. Let x_t denote the action reward of the UAV at each time step, which will be described in detail later. Our goal is to identify the optimal expert by solving the following optimization problem:

$$\min R_n = \min \mathbb{E}\left[\max_{m\in[M]} \sum_{t=1}^{n} E_m^{(t)} x_t - \sum_{t=1}^{n} X_t\right] \tag{1}$$

where n represents the total number of time steps during the UAV's flight, R_n is the expected difference between the optimal expert's reward and the actual action reward, $E_m^{(t)}$ is the probability provided by the m-th expert for each action at time t, and X_t is the reward after the actual action is taken. If the best expert is found from the beginning, then the regret value is 0.

$\hat{R}_n$ is defined as the regret without expectation, that is, $\mathbb{E}[\hat{R}_n] = R_n$. We demonstrate that for every $\delta \in (0,1)$, there exists an algorithm such that, with probability at least $1-\delta$, the regret $\hat{R}_n$ is upper-bounded.

Tracking a target using a UAV is a dynamic process that involves two key settings: partial observability and an adversarial environment.

a) **Partial Observability:** Although the UAV is equipped with wide-field and long-range radar sensors, its coverage is limited to a rectangular area 60 km ahead in the direction of the aircraft's speed, with a width of 80 km. When the target may not be visible, we use online learning methods to evaluate the effectiveness of UAV actions to assess the accuracy of expert advice and adjust in real-time based on the results. This process requires a proper balance between exploration and exploitation, commonly referred to as the bandit problem. If poor expert advice is selected, the UAV will lose track of the target, reducing the action reward to zero and rapidly increasing the regret.
b) **Adversarial Environment:** The target chooses to move in the opposite direction of the UAV's flight path. The target's movement is unpredictable and may even be adversarial [4]. The experts must not only accurately predict the state of the target but also respond to the target's adversarial behavior.

In the bandit problem of UAV tracking, some studies [15] have explored the adversarial bandit problem, but using the Fixed-Share algorithm [3] increases the overall upper bound of the algorithm's regret. Other studies [6,8,16] assume that the environment follows a randomly independent model; however, in reality, the target's flight trajectory is often adversarial.

Contribution: Our algorithm employs the bandits with expert advice algorithm for single UAV and single target tracking, which integrates multiple algorithms and enhances overall performance. Based on Gergely Neu's Exp4-IX (Implicit eXploration) algorithm [10], we prove that when $\delta < O(1/M)$, our algorithm has a superior high-probability upper bound ([10] Theorem 1). The condition $\delta < O(1/M)$ is relatively easy to satisfy because δ is particularly small.

2 Bandit with Expert Advice

We present the background of the "Bandits with Expert Advice" problem. There are n rounds of interaction between the UAV and the target. In the t-th round, the UAV can choose from k actions, where the directions of movement are restricted to forward, right, downward, and left. In each round, M experts provide flight advice to the UAV based on the previous conditions. Let the advice of the M experts be $E_1^{(t)}, E_2^{(t)}, ..., E_M^{(t)} \in [0,1]^k$, where $\sum_{i=1}^{k} E_{m,i}^{(t)} = 1$ for all $m \in \{1,2,...,M\}$. The reward $x_t \in [0,1]^k$ represents the reward the UAV receives for each action, typically assumed to be between 0 and 1.

Since the UAV's observations are partially observable, suppose the UAV's current position is denoted as (x, y), and it can only detect the area within a range x_s in front of it and a range y_s on both sides. Therefore, the detection range forms a rectangular area. We define the UAV's detection range as:

$$\begin{cases} [x - y_s, x + y_s] \times [y, y + x_s] & \text{if the UAV is moving forward;} \\ [x, x + x_s] \times [y - y_s, y + y_s] & \text{if the UAV is moving right;} \\ [x - y_s, x + y_s] \times [y - x_s, y] & \text{if the UAV is moving downward;} \\ [x - x_s, x] \times [y - y_s, y + y_s] & \text{if the UAV is moving left.} \end{cases} \tag{2}$$

The reward is not simply set based on whether the UAV observes the target (rewarding 0 when observed and 1 when not observed), but rather it is determined by the distance between the UAV and the target. We assume that the closer the distance, the higher the reward, as the target is more likely to appear within the UAV's field of view. Let d_t^i represent the distance after the UAV executes the i-th action, then we define the reward $x_t^{(i)}$ for taking the i-th action in the t-th round as:

$$x_t^{(i)} = \begin{cases} \min\{\frac{1}{d_t^i}, 1\} & \text{if the target is within the field of view;} \\ 0 & \text{if the target is outside the field of view.} \end{cases} \tag{3}$$

This setup ensures that the target is within the UAV's detection range and encourages the UAV to get as close to the target as possible.

In the adversarial environment, our algorithm guarantees a sub-optimal expected upper bound, with the regret value exhibiting sub-linear growth as the number of rounds increases. This indicates that the UAV can not only approach the target but also continuously keep it in sight.

3 Exp4-IX And Regret Upper Bound

Exp4-IX is a standard algorithm for adversarial bandit with advice. It is named Exp4-IX because its full name is the "Exponential-weight algorithm for Exploration and Exploitation with Expert advice - Implicit eXploration". The Exp4-IX algorithm sets the initial weights of the experts as $Q_t = (1/M, 1/M, \ldots, 1/M) \in [0,1]^M$ and executes the following two steps at each time step:

1. Before the UAV executes, obtain the advice from the M experts and use a linear transformation to determine the UAV's flight strategies.
2. After the UAV executes, receive a reward based on the flight state and update the experts' weights.

The detailed algorithm flow is presented in Algorithm 1. To simplify the proof, the reward values x_{ti} are replaced by the loss values $y_{ti} = 1 - x_{ti}$.

It is worth noting that in the algorithm, $\hat{Y}_{ti} = \mathbb{I}\{A_t = i\} y_{ti}/(P_{ti} + \gamma)$, when $\gamma = 0$, the closer P_{ti} gets to 0, the larger the overall variance, leading to very unstable $\tilde{S}_{t,i}$. Therefore, Implicit eXploration is introduced to move Q_t towards a

Algorithm 1: Exp4-IX Algorithm

Input: Number of rounds n, number of actions k, number of experts M, expert advice at different times $E^{(t)} \in [0,1]^{k \times M}$, algorithm parameters η, γ

```
Initialize Q_1 = (1/M, ..., 1/M)^T ∈ [0,1]^{M×1}, S_0 = (0, ..., 0)^T_{M×1} for
t = 1, ..., n do
    // Receive advice from M experts and give flight suggestions to
       the UAV
    Receive advice E^(t)
    Choose the action A_t ~ P_t, where P_t = E^(t) Q_t
    // Receive reward based on UAV flight state and update expert
       weights
    Receive the reward Y_t = 1 - x_{tA_t}
    Estimate the action rewards: Ŷ_ti = I{A_t = i}Y_t / (P_ti + γ)
    Propagate the rewards to the experts: Ỹ_t = E^(t)T Ŷ_t
    Compute experts' cumulative rewards:
        S̃_{t,m} = S̃_{t-1,m} + Ỹ_tm   for all m ∈ [M]
    Update the distribution Q_t using exponential weighting:
        Q_{t+1,m} = exp(-η S̃_tm) / Σ_j exp(-η S̃_tj)   for all m ∈ [M]
```

In mathematical notation:

Initialize $Q_1 = (1/M, \ldots, 1/M)^T \in [0,1]^{M \times 1}$, $S_0 = (0, \ldots, 0)^T_{M \times 1}$ **for** $t = 1, \ldots, n$ **do**

- // Receive advice from M experts and give flight suggestions to the UAV
- Receive advice $E^{(t)}$
- Choose the action $A_t \sim P_t$, where $P_t = E^{(t)} Q_t$
- // Receive reward based on UAV flight state and update expert weights
- Receive the reward $Y_t = 1 - x_{tA_t}$
- Estimate the action rewards: $\hat{Y}_{ti} = \frac{\mathbb{I}\{A_t = i\} Y_t}{P_{ti} + \gamma}$
- Propagate the rewards to the experts: $\tilde{Y}_t = E^{(t)T} \hat{Y}_t$
- Compute experts' cumulative rewards:

$$\tilde{S}_{t,m} = \tilde{S}_{t-1,m} + \tilde{Y}_{tm} \quad \text{for all } m \in [M]$$

- Update the distribution Q_t using exponential weighting:

$$Q_{t+1,m} = \frac{\exp(-\eta \tilde{S}_{tm})}{\sum_j \exp(-\eta \tilde{S}_{tj})} \quad \text{for all } m \in [M]$$

uniform distribution, increasing the randomness of the actions, which can be seen as a form of "forced exploration" [7]. This increases the probability of selecting other actions. After incorporating γ, the upper regret bound of the algorithm becomes at least a high probability bound of $1 - \delta$.

The first term of Theorem 1 is constant. We provide an upper bound for the second term $2\sqrt{nk(\log M + \log(k+1) - \log\delta)}$. Compared to the second term in [10] Theorem 2, $\sqrt{(2nk)/(\log M)}\log(2/\delta)$, our upper bound is smaller when $\delta > O(1/(nk))$. Note that $O(1/(nk))$ is a rather trivial upper bound, so the bound presented in this paper is tighter. All omitted proofs are deferred to the journal version of the paper.

Theorem 1. *For a fixed* $\delta \in (0,1)$, *define* $\eta = \sqrt{(\log M + \log(k+1) - \log\delta)/(nk)}$, $\gamma = \eta/2$. *The following holds with probability at least* $1 - \delta$:

$$\hat{R}_n \leq k \log \frac{k+1}{\delta} + 2\sqrt{nk(\log M + \log(k+1) - \log\delta)} \tag{4}$$

4 Numerical Evaluation for Exp4-IX

In this experiment, we apply the Exp4-IX algorithm to a scenario involving a single dynamic target with a single UAV. We consider different target trajectories and expert strategies for UAV flight. The experiment examines smooth and adversarial trajectories based on the target's motion, while expert advice includes the previous position algorithm, particle filtering algorithm, trajectory fitting algorithm, and the Exp4-IX algorithm. We evaluate the performance of these expert recommendations, along with the Exp4-IX algorithm, under different conditions in UAV target tracking.

The experimental scenario is set within a square area with a side length of 100 km. The time step is 1 s, and the total simulation time is 2000 s seconds. The UAV's movement direction is restricted to four cardinal directions, with a flight speed of 280 m per second. The detection range is a rectangular area extending 10 km forward and 8 km to each side. The target's speed is controlled between 100 and 280 m per second, and the scenario is set with smooth trajectories.

Smooth trajectory generation: Twenty points are selected on a plane, and a smooth curve is fitted through these points. The fitted curve is then further subdivided to ensure that the distance between any two points does not exceed the target's speed. The number of subdivided points does not exceed 2000; any excess points are truncated.

Expert recommendations are categorized as follows:

- **Previous Position Algorithm:** If the UAV detects the target's position in the previous time step, it calculates the probability of moving in each direction based on the spatial relationship between the UAV and the target. If the target was not detected in the previous step, equal probabilities are assigned to all directions to increase the chance of detecting the target.
- **Particle Filtering Algorithm:** A set of particles representing the target's state is generated at the start of the filter. Based on a model, random noise is added to the particles to predict the target's position at the next time step. The UAV's next movement direction is then determined by comparing the predicted position with the UAV's position. The particle weights are updated based on the UAV's observations, and resampling is performed as needed to avoid particle degradation.
- **Trajectory Fitting Algorithm:** The UAV records data based on observed target trajectories at each step. It then predicts the target's trajectory position at each step to determine the next flight direction.
- **Average Fusion Algorithm:** This online algorithm integrates the three aforementioned algorithms by averaging the probabilities provided by the first three algorithms.
- **Exp4-IX Algorithm:** This online algorithm integrates the three aforementioned algorithms, dynamically adjusting their weights to select the optimal strategy (Fig. 2).

The code has been open-sourced on GitHub: https://github.com/lvymath1/UAV_tracking_bandit_algorithm.

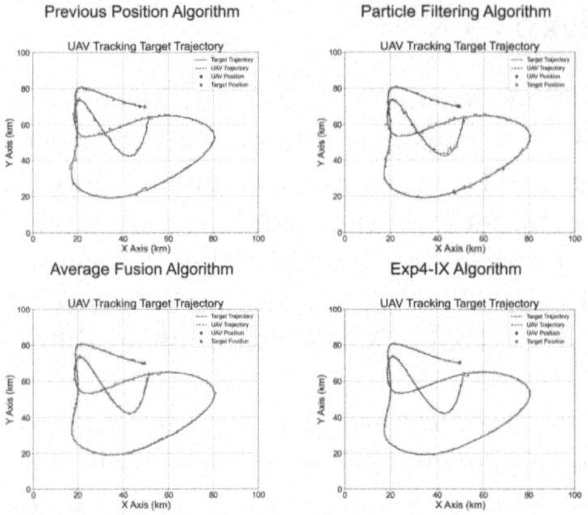

Fig. 2. Effect of the Previous Position Algorithm, Particle Filtering Algorithm, Average Fusion Algorithm, and Exp4-IX Algorithm on UAV target tracking.

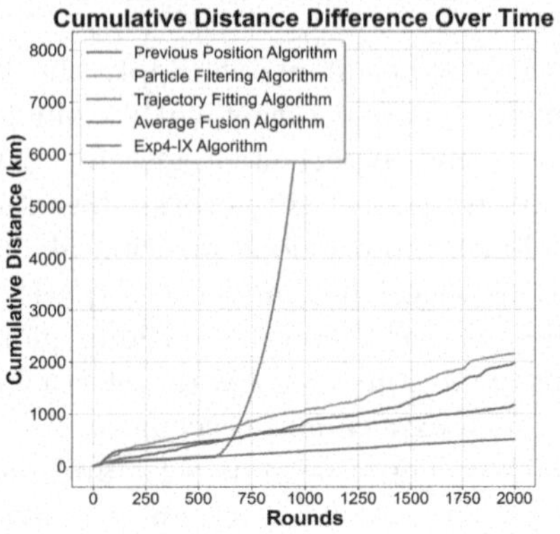

Fig. 3. Image showing the cumulative distance of five algorithms in UAV target tracking.

All simulations were conducted on a Windows laptop equipped with an Intel Core i7-12700H CPU @ 2.30 GHz and 32 GB of RAM, running Python 3.9.

In smooth environments, the target's direction of movement does not change due to the UAV's position. The target's flight path is smooth, with minimal reward fluctuations, making it less adversarial. The Exp4-IX algorithm outperforms other algorithms (see Fig. 3), followed by the Average Fusion algorithm. This indicates that under smooth conditions, the fusion algorithm is superior to individual algorithms, and the Exp4-IX algorithm is a more optimal fusion algorithm. Among the other three algorithms, the Previous Position algorithm performs the best, followed by the Particle Filtering algorithm, but it exhibits greater trajectory volatility compared to Exp4-IX. In the Trajectory Fitting algorithm, the UAV loses track of the target during the tracking process, causing it to predict the trajectory-based only on previously detected target positions, thus gradually deviating from the target.

5 Conclusion

We study the problem of target tracking using UAVs in partially observable adversarial environments and introduce the Exp4-IX algorithm for online data fusion. This algorithm not only has theoretical guarantees but also performs exceptionally well in numerical experiments. Our study demonstrates that Exp4-IX can effectively integrate multiple UAV algorithms, enhancing the UAVs' target-tracking capabilities. However, the algorithm has the following limitations: 1) If the UAV's detection range is not a rectangular area but another shape, the reward function needs to be adjusted. 2) Theoretically, the upper bound of the UAV's regret may be lower than the best expert's error, so if the optimal expert strategy is sufficiently good, Exp4-IX may not perform better than the optimal expert strategy. In future work, we plan to extend the strategy from a single UAV to a collective strategy involving multiple UAVs to track more targets under conditions of partial communication.

References

1. Bhagat, S., Sujit, P.B.: UAV target tracking in urban environments using deep reinforcement learning. In: International conference on unmanned aircraft systems, pp. 694–701. IEEE, Athens (2020)
2. Elhussein, A., Miah, M.S.: A novel model-free actor-critic reinforcement learning approach for dynamic target tracking. In: IEEE Midwest Industry Conference, pp. 1–6. IEEE, Champaign (2020)
3. Herbster, M., Warmuth, M.K.: Tracking the best expert. Mach. Learn. **32**(2), 151–178 (1998)
4. Jin, C., Jin, T., Luo, H., Sra, S., Yu, T.: Learning adversarial Markov decision processes with bandit feedback and unknown transition. In: International Conference on Machine Learning, pp. 4860–4869. PMLR (2020)
5. Lai, Y.C., Huang, Z.Y.: Detection of a moving UAV based on deep learning-based distance estimation. Remote Sens. **12**(18), 1–28 (2020)
6. Landgren, P., Srivastava, V., Leonard, N.E.: Distributed cooperative decision making in multi-agent multi-armed bandits. Automatica **125**(109445), 1–13 (2021)
7. Lattimore, T., Szepesvári, C.: Bandit Algorithms. Cambridge University Press (2020)
8. Lee, K.M.B., et al.: An upper confidence bound for simultaneous exploration and exploitation in heterogeneous multi-robot systems. In: International Conference on Robotics and Automation, pp. 8685–8691. IEEE (2021)
9. Lv, Y., Wu, C., Xu, D., Yang, R.: H-hop independently submodular maximization problem with curvature. High-Confidence Comput. **4**(100208), 1–7 (2024)
10. Neu, G.: Explore no more: improved high-probability regret bounds for non-stochastic bandits. In: Advances in Neural Information Processing Systems, pp. 1–9. NIPS, Montréal (2015)
11. Ojha, S., Sakhare, S.: Image processing techniques for object tracking in video surveillance-a survey. In: International Conference on Pervasive Computing, pp. 1–6. IEEE, St. Louis (2015)
12. Sun, S., Liu, Y., Guo, S., Li, G., Yuan, X.: Observation-driven multiple UAV coordinated standoff target tracking based on model predictive control. Tsinghua Sci. Tech. **27**(6), 948–963 (2022)
13. Wang, G., Wang, F., Wang, J., Li, M., Gai, L., Xu, D.: Collaborative target assignment problem for large-scale UAV swarm based on two-stage greedy auction algorithm. Aerosp. Sci. Tech. **149**(109146), 1–11 (2024)
14. Xu, W., et al.: Approximation algorithms for the generalized team orienteering problem and its applications. IEEE/ACM Trans. Netw. **29**(1), 176–189 (2020)
15. Xu, Z., Lin, X., Tzoumas, V.: Bandit submodular maximization for multi-robot coordination in unpredictable and partially observable environments. arXiv preprint arXiv:2305.12795, pp. 1–14 (2023)

16. Zhang, C., Hoi, S.C.: Partially observable multi-sensor sequential change detection: a combinatorial multi-armed bandit approach. In: Proceedings of the AAAI Conference on Artificial Intelligence, pp. 5733–5740 (2019)
17. Zhu, K., Zhang, T.: Deep reinforcement learning based mobile robot navigation: a review. Tsinghua Sci. Tech. **26**(5), 674–691 (2021)

Fair Maximization of Monotone Submodular Functions in Data Streams

Shuqian Zhu, Longkun Guo(✉), and Jiawei Lin

School of Mathematics and Statistics, Fuzhou University, Fuzhou 350116, China
lkguo@fzu.edu.cn

Abstract. Streaming fair submodular maximization is attracting considerable research interest due to its broad applications in machine learning, particularly for tasks such as feature selection and text summary against large-scale data and fairness considerations. Given a sequence of data points belonging to distinct groups and arriving in a streaming manner, the problem aims to select k data points from the stream to maximize the total revenue of the selected points. In this paper, we first devise an efficient $(\frac{1}{2}-\varepsilon)$-approximation algorithm with $O(\log(\frac{1}{\varepsilon}\log\frac{k}{\varepsilon}))$ passes, an improvement over the previous $O(\frac{1}{\varepsilon}\log\frac{k}{\varepsilon})$ passes. Then, we present a $\left(\frac{1}{3}-\varepsilon\right)$-approximation algorithm that needs only one pass and consumes a buffer of size $O(k+|B|)$ and achieves a ratio strictly greater than $\frac{1}{4}$ while using a buffer of size $O(k\log k)$. Lastly, we conduct extensive experiments using real-world datasets to validate our method, demonstrating that it outperforms all state-of-the-art algorithms in terms of efficiency, effectiveness, and scalability.

Keywords: Streaming · Submodular maximization · Fairness · Approximation algorithm

1 Introduction

Submodularity, a concept encapsulating diminishing marginal utility as referenced in [11], has been extensively explored by researchers such as [15,20,23,25] to identify k nodes with the greatest influence for a given k. Moreover, the studies by [17] in document retrieval aim to efficiently and comprehensively return a concise set of documents that effectively respond to a query. These scenarios highlight the practical applications of submodular maximization, which aims to find k data points from a data set V to maximize a monotone submodular objective function. Despite its NP-hardness [8], the problem is commonly approached using a straightforward greedy algorithm, which repeatedly selects a data point that maximizes the current solution at each decision point and attains an approximation ratio of $1-\frac{1}{e}$.

Emerging machine learning applications over large-scale datasets pose new challenges in solving submodular maximization problems in data streams. Due to the rapid expansion of contemporary datasets derived from various sources and

D. Du et al. (Eds.): COCOA 2024, LNCS 15434, pp. 287–298, 2025.
https://doi.org/10.1007/978-981-96-4445-2_24

formats (including photos, videos, sensor data, and social networks), algorithms requiring complete access to the data stream at every iteration are becoming increasingly impractical because of their high space and time complexity. Consequently, there is a growing need for innovative techniques that can extract valuable insights from massive datasets while maintaining computational feasibility. Streaming algorithms have received significant attention in recent years, as evidenced by contributions from [5,6,12,16,21,24]. These algorithms, designed to keep only a relatively small dataset in memory and traverse the data stream in one or a few passes, provide more efficient solutions compared to traditional greedy algorithms, although they may yield slightly lower-quality results.

On the other hand, it has been revealed that data summarization algorithms commonly used in the online sphere exhibit biases towards sensitive attributes like gender, race, or ethnicity, as noted in works by [7]. For example, in the context of collecting user feedback on product experiences, we recognize that distinct social groups may express diverse opinions and perspectives. Consequently, when summarizing and analyzing data from these diverse groups, it is crucial to ensure that the insights derived accurately represent the viewpoints of each social group. Therefore, investigating solutions under the fairness condition is of paramount importance.

1.1 Our Contribution

This paper proposes streaming approximation algorithms for maximizing a monotone submodular function under a fair constraint in a data stream. The contributions can be summarized as follows.

- Devise an efficient $(\frac{1}{2}-\varepsilon)$-approximation algorithm with only $O(\log(\frac{1}{\varepsilon}\log\frac{k}{\varepsilon}))$ passes over the data stream, improving the previous $O(\frac{1}{\varepsilon}\log\frac{k}{\varepsilon})$ passes by Wang et al. [24].
- For one-pass streaming, we design an algorithm with an approximation ratio $\frac{1}{4}+o(1)$ with a buffer size $O(k\log k)$. The ratio $\frac{1}{4}+o(1)$ is attained by improving our other designated algorithm with a ratio $(\frac{1}{3}-\varepsilon)$ and a buffer size $O(k+|B|)$. In comparison, the algorithm by Wang et al. [24] attains ratio $\frac{1}{2}-\varepsilon$ but consumes a memory of size $O(\frac{k\log k}{\varepsilon}+|B|)$, while its approximation ratio is $\frac{1}{2k}$ if the buffer size is constrained to $O(k\log k)$.
- Extensive experiments were carried out on two practical scenarios to demonstrate the algorithms' efficiency, effectiveness, and scalability.

2 Related Work

To provide clarity and focus, we will specifically concentrate on discussing the most relevant and significant contributions in this field.

Streaming Submodular Maximization. Chakrabarti et al. [4] introduced a pioneering one-pass streaming algorithm with a $\frac{1}{4p}$-approximation for p matroid constraints, which laid the foundation for submodular maximization in streaming settings. Chekuri et al. [6] extended this approach to non-monotone submodular functions. Chan et al. [5] improved the approximation ratio to 0.3178 for partition matroids using a randomized algorithm in the related preemptive online model, while Feldman et al. [9] proposed a randomized algorithm achieving a $\frac{1}{4p}$ approximation for monotone submodular functions under p-matching constraints, requiring $O(k)$ memory per element. Huang et al. [13] further enhanced multi-pass streaming algorithms for both monotone and general submodular functions under p-matroid constraints. Wang et al. [24] introduced a multi-pass algorithm with a $(\frac{1}{2}-\varepsilon)$-approximation and a one-pass version with a reduced $\frac{1}{2k}$-approximation for limited buffer space. Feldman et al. [10] later developed a semi-streaming algorithm for monotone submodular maximization under matroid constraints, achieving a 0.3178 approximation with $\widetilde{O}(k)$ memory, where k represents the matroid's rank.

Swapping-Based Maximum k-Coverage. Fisher et al. [11] introduced a greedy algorithm for the maximum k-coverage problem, achieving an approximation ratio of $1-\frac{1}{e}$. Nemhauser et al. [18] followed with a swapping-based algorithm that achieved a $\frac{1}{2}$ approximation through multiple data stream passes. Saha et al. [19] advanced this by creating a one-pass, swap-based $\frac{1}{4}$-approximation algorithm with $O(kn)$ space complexity. Ausiello et al. [2] developed a single-pass algorithm with similar space requirements and a competitive ratio slightly above $\frac{1}{4}$. Yu et al. [26] later introduced a one-pass algorithm with $\widetilde{O}(n)$ space complexity, achieving an approximation ratio close to 0.3 in simulations.

Optimal Solution Guessing In Maximum k-Coverage. Badanidiyuru et al. [3] introduced an algorithm that achieves a $(\frac{1}{2}-\varepsilon)$-approximation with a memory requirement of $O\left(\frac{k\log k}{\varepsilon}\right)$. The algorithm processes data streams until either k elements are selected or the stream ends, without supporting element removal. Assadi et al. [1] studied space complexity in one-pass algorithms, showing that $\Theta\left(\frac{mn}{\alpha}\right)$ space is needed for an α-approximation of set cover, with $\alpha = o(\sqrt{n})$ and $O(\log n)$ error. McGregor et al. [16] developed a sublinear space algorithm for maximum k-set coverage, aiming for an approximation close to $1-\frac{1}{e}$, and a $\frac{1}{l+1}$-approximation for fairness.

3 Problem Statement

Throughout this work, we consider choosing k items in the stream where the items arrive one by one. When an item s has arrived, we decide whether to take it into the current solution or buffer. Furthermore, once an item is discarded, it will never be picked up again.

Firstly, we consider the following **canonical optimization problem**. Given a data stream V containing numerous items, we wish to find k items in V that maximize the monotone submodular function. We define a non-negative monotone submodular function $f\colon 2^V \to \mathbb{R}$. Then our optimization problem is as follows,

$$\max_{S \subseteq V} \quad f(S)$$
$$\text{s.t.} \quad |S| = k.$$

Now we introduce the **fairness issue** further. Suppose that the data stream V is divided into l (disjoint) groups, each item belongs to a group amongst l groups $V_1, V_2, \cdots, V_l$ and we are allowed to pick k_1 items from group V_1, k_2 items from group V_2, and so on. We want to find such k items so that the monotone submodular function is maximized.

4 Multi-pass Streaming Algorithm

This section focuses on the task of optimizing a monotone submodular function while adhering to fairness constraints in a multi-pass data stream scenario. The efficiency and effectiveness of our proposed solution are evident. It requires $O(\log(\frac{1}{\varepsilon}\log\frac{k}{\varepsilon}))$ passes over the data stream, optimally manages memory by storing only $O(k)$ items, and achieves a $(\frac{1}{2} - \varepsilon)$ approximation ratio.

Algorithmic Techniques. Our technique centers on a well-constructed greedy-based algorithm. It picks not just the item $s_{\max}$ with the highest incremental value $v_{\max}$ in every pass, but also retains items that contribute considerably to strengthening the solution S, i.e., $f(S \cup \{s\}) - f(S) > (1-\alpha)v_{\max}$, all within the confines of fairness restrictions. In post-processing, we leverage reservoir sampling to ensure fairness [22], substantially speeding the attainment of a high-performance solution. The complete approach is explained in Appendix A.

Theoretical Analysis. First, we give the approximation ratio of the algorithm in Theorem 1. And then, the complexity of the algorithm is analyzed in Theorem 2.

Let O represent the optimal solution for the problem of maximizing a monotone submodular function under a fair constraint, i.e., $O = \{o_1, \ldots, o_k\}$. To avoid ambiguity and ensure clear understanding, let $\widetilde{S} = \{s_1, \ldots, s_k\}$ be the final solution returned by the algorithm.

Lemma 1. *For items present in O but not in the final solution $\widetilde{S}$, let $\alpha = o(\varepsilon)$. The following inequality holds:*

$$\sum_{o \in O \setminus \widetilde{S}} [f(\widetilde{S} \cup \{o\}) - f(\widetilde{S})] \leq f(\widetilde{S}) + \varepsilon.$$

Theorem 1. *This algorithm achieves a performance guarantee of a $(\frac{1}{2} - \varepsilon)$ approximation ratio for the problem of maximizing a monotone submodular function, subject to a fairness constraint, within a multi-pass streaming setting.*

Theorem 2. *When $\alpha = o(\varepsilon)$, this algorithm needs $O(\log(\frac{1}{\varepsilon}\log\frac{k}{\varepsilon}))$ passes over the data stream V and stores a maximum of $O(k)$ items. It has a time complexity of $O(n\log(\frac{1}{\varepsilon}\log\frac{k}{\varepsilon}))$.*

5 One-Pass Streaming Algorithm

5.1 Algorithm with Unlimited Buffer

Our one-pass technique, inspired by the methods given by Chekuri et al. [6], is straightforward yet effective. First, we need to present the definition of incremental value.

Incremental Value: Consider a data stream V and a current solution subset $S \subseteq V$. For any item $s \in S$, its contribution $v(s)$ to S is essential. This value is defined as:

$$v(s) = f(S' \cup \{s\}) - f(S'), \text{ where } S' = \{a \in S : a < s\}.$$

Incremental values are important because they capture the total value of the solution.

Fact 1. *The incremental value of an item s in solution S is* ***dynamic*** *and changes as other items are removed from S. This highlights that the incremental values for items in each group are monotonic and non-decreasing.*

Lemma 2. *Given an ordered data stream V, a solution subset $S \subseteq V$, and a monotone submodular function $f : 2^V \to \mathbb{R}$ quantifying the utility of S, it holds that $f(S) = \sum_{s \in S} v(s)$.*

Proof. Enumerate $S = \{s_1, s_2, \cdots, s_l\}$ in the order of their addition to S, and let $S_i = \{s_1, s_2, \cdots, s_i\}$ represent the first i items in S. Therefore,

$$\sum_{s_i \in S} v(s_i) = \sum_{s_i \in S} [f(S_{i-1} \cup \{s_i\}) - f(S_{i-1})] = f(S).$$

Lemma 3. *For two nested subsets $S \subseteq T \subseteq V$ of an ordered data stream V, and an item $s \in V$, the incremental value of s in T is less than or equal to its incremental value in S.*

Proof. Define $S' = \{v \in S : v < s\}$ and $T' = \{t \in T : t < s\}$. By the principle of submodularity,

$$v_T(s) = f(T' \cup \{s\}) - f(T') \leq f(S' \cup \{s\}) - f(S') = v_S(s).$$

Intuition Behind The Algorithm. In the method provided by [6], the assessment of incoming items is based exclusively on a comparison with the minimum incremental value. However, this strategy may not be appropriate if the incremental value constantly increases, particularly in cases where this value becomes substantially large. This implies that arriving items must provide a relatively higher incremental value to prevent exclusion. This criterion can lead to the probable loss of items that, despite delivering large utilities, fail to reach the stringent inflow threshold. To overcome this limitation, we have made several significant modifications to Chekuri's original algorithm, increasing the number

Algorithm 1: Check_Buffer

Function Check_Buffer(S, B, β):
  for *each item* $b \in V_i$ *in* B **do**
    Let C be the item of smallest incremental value in $S \cap V_i$
    if $f(S \cup \{b\} \setminus \{C\}) - f(S) \geq \beta v(C)$ **then**
      $S \leftarrow S \cup \{b\} \setminus \{C\}$
      $B \leftarrow B \setminus \{b\}$
  for *each item* $b \in V_i$ *in* B **do**
    Let C be the item of smallest incremental value in $S \cap V_i$
    if $v(C) < \frac{g(S \cap V_i)}{(1+\beta)k_i}$ *or* $f(S \cup \{b\}) - f(S) < \frac{g(S \cap V_i)}{2k_i}$ **then**
      $B \leftarrow B \setminus \{b\}$
  return S, B

of judgments for items with high utility values. These improvements enable our method to achieve a $(\frac{1}{3} - \varepsilon)$ approximation ratio for the problem, thereby surpassing the performance of Chekuri's algorithm in terms of efficiency.

Algorithmic Techniques. To enhance the theoretical guarantee of solution quality, our algorithm incorporates a mechanism that selects new high-utilities items over existing ones in the streaming context while managing a buffer B to temporarily store items that are neither included in the solution nor safely discarded. Each time there is an adjustment in the solution S, every item in the buffer undergoes reevaluation to evaluate its eligibility. At the end of the streaming process, we apply a local search to find any buffer items that might improve the existing solution. For a complete understanding of this method, please refer to Algorithm 2.

Notation for Analysis:

- S' denotes the solution S at the end of the stream.
- $\widetilde{S}$ denotes the solution S after local search.
- A represents all items added to solution S during the stream.
- For each item s received from V, S_s^- and S_s^+ denote the state of solution S before and after processing s, respectively; A_s^- represents all items selected for S before processing s.
- $g(S \cap V_j)$ represents the sum of incremental values of items from group V_j in solution S; k_j denotes the group size constraint for the j-th group.

Outline of the Analysis: To establish the connection between $\widetilde{S}$ and O, we focus on A and $\widetilde{S}$. Our analysis unfolds in a two-step process. Initially, we aim to ascertain the relationship between $f(S')$ and $f(A)$. Subsequently, we endeavor to determine the upper bound of $f(A \cup \widetilde{S} \cup O)$.

Algorithm 2: One-Pass Streaming Algorithm with Unlimited Buffer

Input: the data stream V, groups $V_1, \cdots, V_l \subseteq V$, total number constraint k, group size constraints $k_1, k_2, \cdots, k_l$, parameters $\beta_1 \geq 1$.
Output: the final solution $\widetilde{S}$.
$S \leftarrow \emptyset$; $B \leftarrow \emptyset$;
for *each item* $s \in V_j$ *received from* V **do**
  if $|S \cap V_j| < k_j$ **then**
    $S \leftarrow S \cup \{s\}$;
  else
    Let C be the item of smallest incremental value in $S \cap V_j$;
    if $f(S \cup \{s\} \setminus \{C\}) - f(S) \geq \beta_1 v(C)$ **then**
      $S \leftarrow S \cup \{s\} \setminus \{C\}$
      Check_Buffer(S, B, β_1);
    else
      if $v(C) \geq \frac{g(S \cap V_j)}{(1+\beta_1)k_j}$ *and* $f(S \cup \{s\}) - f(S) \geq \frac{g(S \cap V_j)}{2k_j}$ **then**
        $B \leftarrow B \cup \{s\}$;
Using local search to find the final solution as $\widetilde{S}$ satisfying the fairness condition and total number constraint.

Lemma 4. *For any item added to solution S before local search, associating solution S' with A results in:*

$$f(A) \leq (1 + \frac{1}{\beta_1}) f(S').$$

Next, we conduct a detailed examination of the items in $O \setminus (A \cup \widetilde{S})$, which forms a subset of O but does not get incorporated into the solution S. For analytical clarity, we categorize this subset into two distinct groups: O_1 and O_2. O_1 is defined as the subset of items that are discarded during the processing of the data stream, O_2 comprises the items that remain in the final buffer B but are not included in the final solution $\widetilde{S}$. Now, we focus on establishing an upper bound for the following expression: $\sum_{o \in O \setminus (A \cup \widetilde{S})} [f(A \cup \{o\}) - f(A)]$.

Lemma 5. *Consider items $o \in O_1$, we propose that:*

$$\sum_{o \in O_1} [f(A \cup \{o\}) - f(A)] \leq \sum_{j \in [l]} \sum_{o \in O_1 \cap V_j} \frac{g(S' \cap V_j)}{k_j}.$$

Lemma 6. *Consider items $o \in O_2$, we propose that:*

$$f(O_2 \cup \widetilde{S}) \leq f(\widetilde{S}) + \sum_{j \in [l]} \sum_{o \in O_2 \cap V_j} \frac{g(\widetilde{S} \cap V_j)}{k_j}.$$

Theorem 3. *Assuming that $\frac{1}{\beta_1} = O(\varepsilon)$, the approximation ratio of Algorithm 2 is $\frac{1}{3} - \varepsilon$.*

In real-world applications, stream processing algorithms must handle vast data streams within constrained time and memory resources (sublinear or independent of n). Algorithm 2 could store an unlimited number of items in the buffer, which isn't ideal. We've addressed this by introducing a strategy to manage the buffer items, ensuring the buffer size remains finite and minimizing costs.

5.2 Algorithm with Limited Buffer

Here, we modify our technique to better handle practical application scenarios involving big data. Recognizing the specific challenges of dealing with large-scale data, we make a modification: setting a restriction on the buffer size. This measure is targeted at optimizing resource management, particularly in high-volume data situations.

Algorithm Techniques. Our strategy includes a buffer management mechanism. Upon reaching its capacity, the buffer employs a method of randomly discarding an item. This approach ensures the buffer size to be finite while retaining items with relatively higher incremental values. Furthermore, we partially adapt Algorithm 2 to align with this new buffer management strategy. In contrast, here we adopt a more cautious approach: checking items in the buffer only when it is full. This adjustment implies that the algorithm will operate more efficiently under normal conditions, resorting to intensive resource management only when necessary. For a detailed procedure, please refer to the steps outlined in Appendix C.1.

Theorem 4. *Assuming that when we choose $\beta_1 = 1$, $\beta_2 = 1 - o(\frac{1}{k})$ and $\varepsilon = (2 - \frac{1}{\beta_2} - \beta_2)n + \frac{1-\beta_2}{k}$, the approximation ratio of the algorithm is $\frac{1}{4} + \varepsilon$.*

Our algorithm represents a significant advancement in the streaming framework established by [6], achieving an approximation ratio strictly better than $\frac{1}{4}$, regardless of whether the space is limited.

6 Experimental Results

In this part, we use numerical analysis to support our theories. Our experiments focus on three main goals. First, we examine how fairness constraints and one-pass data access affect solution utility. Next, we compare the Multi-Pass Fair Streaming Model to other common Multi-Pass Streaming algorithms, showing its improved capabilities. Finally, we demonstrate the effectiveness of the One-Pass Fair Streaming Model, in both its unlimited and limited buffer capacity versions, against existing one-pass streaming algorithms.

The utility achieved by the greedy algorithm, which is an offline, non-streaming algorithm, serves as a logical **benchmark** for optimality. This is due to its theoretical robustness and the practical challenge of outperforming it.

For this purpose, we conducted extensive experiments to evaluate our multi-pass streaming algorithm in the context of maximum graph coverage on large

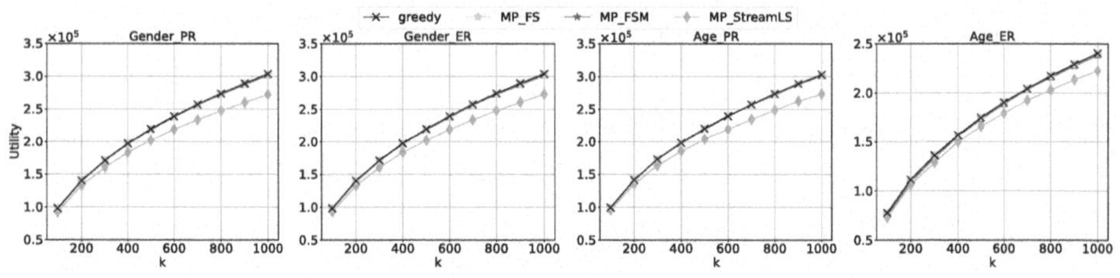

Fig. 1. Solution Utilities of Multi-Pass Streaming Algorithms on POKEC

graphs and personalized recommendations. We compared our algorithm **MP-FS** with the following algorithms: (a) Greedy: the classic $\frac{1}{2}$-approximation offline algorithm proposed by [11]. (b) MP-StreamLS: a $\frac{1}{2+\varepsilon}$-approximation $O(\frac{1}{\varepsilon})$-pass streaming algorithm in [13]. (c) MP-FSM: a $(\frac{1}{2}-\varepsilon)$-approximation $O(\frac{1}{\varepsilon}\log\frac{k}{\varepsilon})$-pass streaming algorithm in [24].

Moreover, we evaluated our algorithms **OPS** (with unlimited buffer size) and **OPS($O(k \log k)$)** (with limited buffer size)—through a comparative analysis against two prominent one-pass streaming algorithms: (a) StreamLS: a $\frac{1}{4}$-approximation streaming algorithm in [6]. (b) SP-FSM: a $(\frac{1}{2}-\varepsilon)$-approximation is achieved for scenarios with an unlimited buffer size, while a $\frac{1}{2k}$-approximation is attained when operating under a bounded buffer size, as presented in [24].

6.1 Maximum Coverage On Large Graphs

The primary objective of this research is to identify a compact subset of nodes in a graph, aiming to encompass a significant portion of the graph's total nodes. Specifically, given a graph $G = (V, E)$ with $n = |V|$ nodes and $m = |E|$ edges, our goal is to select a subset S of V of size k that maximizes the coverage of nodes in the neighborhood of S. This is quantified by the function $f(S) = |\cup_{v \in S} N(v)|$, where $N(v)$ denotes the set of nodes connected to v. This function f is nonnegative, monotone, and submodular.

In this study, we harness the POKEC dataset from SNAP, which serves as a real-world directed graph depicting follower/followee relationships among Pokec users [14]. This comprehensive graph encompasses $1,632,803$ nodes and $30,622,564$ edges, with each node enriched with user demographic data. To better evaluate the performance of our algorithm, we divided these nodes into either $l = 2$ groups based on gender or $l = 7$ groups according to age categories. We systematically varied the total cardinality constraint k from 100 to $1,000$ in our experimental framework. In addition, we determined group-specific cardinality constraints k_i for each group i, employing two distinct methodologies: *proportional representation* **(PR)** and *equal representation* **(ER)**.

Our comprehensive evaluation encompassed both multi-pass and one-pass algorithms. As depicted in Figs. 1 and 2, our findings demonstrate that the **MP-FS** algorithm significantly outperforms other prominent multi-stream algorithms. Notably, MP-FS approaches the quality of the Greedy algorithm while tripling

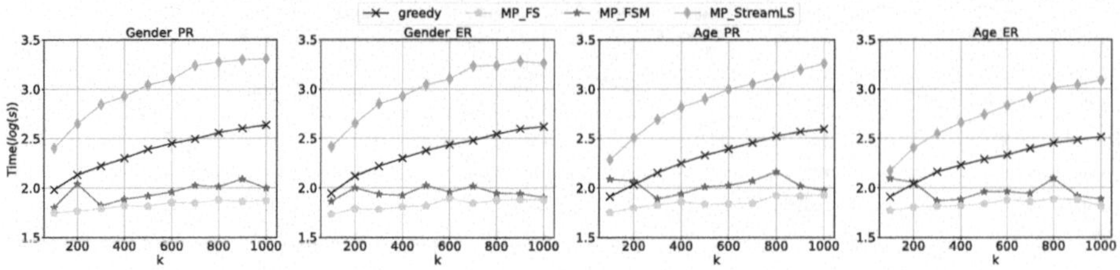

Fig. 2. Running Time of Multi-Pass Algorithms on POKEC

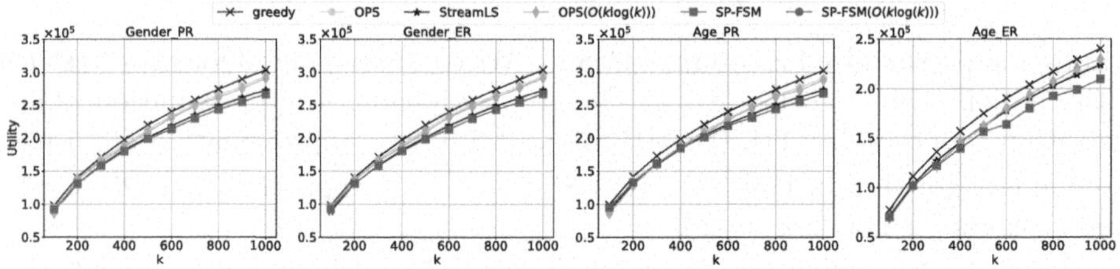

Fig. 3. Solution utilities of One-Pass Streaming Algorithms on POKEC

its processing speed, thus surpassing the efficiency of competing mainstream algorithms. Moreover, for the one-pass algorithm in Fig. 3, we use the Greedy approach as a benchmark for comparison. The first and second graphs specifically analyze gender-based groupings, where **OPS** exhibits a substantial advantage, achieving approximately a 60% improvement over SP-FSM and StreamLS in scenarios without buffer size restrictions, and **OPS($O(k \log k)$)** maintains a 40% lead under buffer size constraints. The third and fourth graphs address age-based groups, showing that OPS achieves a 20% improvement compared to SP-FSM and StreamLS in environments with the unlimited buffer size, and maintains a 10% lead with buffer size limitations. Further experimental details are provided in Appendix D.

In summary, the experimental results demonstrate that our algorithms exhibit strong performance compared to current state-of-the-art algorithms, indicating their potential effectiveness in practical applications.

7 Conclusion

In this work, we addressed the problem of maximizing a monotone submodular function under a fairness constraint in data streams and proposed a suite of streaming algorithms. The first algorithm attains an approximation ratio $(\frac{1}{2}-\varepsilon)$ and requires $O(\log(\frac{1}{\varepsilon}\log\frac{k}{\varepsilon}))$ passes over the data stream. Then, we propose a one-pass streaming algorithm that achieves a noteworthy $(\frac{1}{3}-\varepsilon)$-approximation ratio but with a buffer size up to $O(k+|B|)$ in the worst case. Later, we further enhanced our method by developing an approach that restricts the buffer size to $O(k \log k)$ while maintaining a ratio $\frac{1}{4}+o(1)$. Lastly, we validate the proposed algorithms' efficacy, effectiveness, and scalability through extensive experiments based on two real-world application scenarios.

Acknowledgment. This work is supported by the National Natural Science Foundation of China (Nos. 12271098 and 61772005).

References

1. Assadi, S., Khanna, S., Li, Y.: Tight bounds for single-pass streaming complexity of the set cover problem. In: Proceedings of the forty-eighth annual ACM symposium on Theory of Computing, pp. 698–711 (2016)
2. Ausiello, G., Boria, N., Giannakos, A., Lucarelli, G., Paschos, V.T.: Online maximum k-coverage. Discrete Appl. Math. **160**(13–14), 1901–1913 (2012)
3. Badanidiyuru, A., Mirzasoleiman, B., Karbasi, A., Krause, A.: Streaming submodular maximization: massive data summarization on the fly. In: Proceedings of the 20th ACM SIGKDD International Conference on Knowledge Discovery and Data Mining, pp. 671–680 (2014)
4. Chakrabarti, A., Kale, S.: Submodular maximization meets streaming: matchings, matroids, and more. Math. Program. **154**, 225–247 (2015)
5. Chan, T.H.H., Huang, Z., Jiang, S.H.-C., Kang, N., Tang, Z.G.: Online submodular maximization with free disposal: randomization beats for partition matroids. In: Proceedings of the Twenty-Eighth Annual ACM-SIAM Symposium on Discrete Algorithms, pp. 1204–1223. SIAM (2017)
6. Chekuri, C., Gupta, S., Quanrud, K.: Streaming algorithms for submodular function maximization. In: Halldórsson, M.M., Iwama, K., Kobayashi, N., Speckmann, B. (eds.) ICALP 2015, Part I. LNCS, vol. 9134, pp. 318–330. Springer, Heidelberg (2015). https://doi.org/10.1007/978-3-662-47672-7_26
7. Dash, A., Shandilya, A., Biswas, A., Ghosh, K., Ghosh, S., Chakraborty, A.: Summarizing user-generated textual content: motivation and methods for fairness in algorithmic summaries. Proc. ACM Hum. -Comput. Interact. **3**(CSCW), pp. 1–28 (2019)
8. Feige, U.: A threshold of ln n for approximating set cover. J. ACM (JACM) **45**(4), 634–652 (1998)
9. Feldman, M., Karbasi, A., Kazemi, E.: Do less, get more: streaming submodular maximization with subsampling. Adv. Neural Inf. Process. Syst. **31** (2018)
10. Feldman, M., Liu, P., Norouzi-Fard, A., Svensson, O., Zenklusen, R.: Streaming submodular maximization under matroid constraints. In: Bojańczyk, M., Merelli, E., Woodruff, D.P. (eds.) 49th International Colloquium on Automata, Languages, and Programming (ICALP 2022), Leibniz International Proceedings in Informatics (LIPIcs), Dagstuhl, Germany, vol. 229, pp. 59:1–59:20. Schloss Dagstuhl – Leibniz-Zentrum für Informatik (2022)
11. Fisher, M.L., Nemhauser, G.L., Wolsey, L.A.: An analysis of approximations for maximizing submodular set functions II. Math. Program. Stud. **8**, 73 (1978)
12. Gong, Q., Gao, S., Wang, F., Yang, R.: Multipass streaming algorithms for regularized submodular maximization. Tsinghua Sci. Tech. **29**(1), 76–85 (2024)
13. Huang, C.-C., Thiery, T., Ward, J.: Improved multi-pass streaming algorithms for submodular maximization with matroid constraints. In: Approximation, Randomization, and Combinatorial Optimization. Algorithms and Techniques (APPROX/RANDOM 2020). Schloss Dagstuhl-Leibniz-Zentrum für Informatik (2020)

14. Leskovec, J., Krevl, A.: Snap datasets: Stanford large network dataset collection (2014)
15. Liu, Y., Sun, Y., Li, M.: Approximation algorithms for maximization of k-submodular function under a matroid constraint. Tsinghua Sci. Tech. **29**(6), 1633–1641 (2024)
16. McGregor, A., Vu, H.T.: Better streaming algorithms for the maximum coverage problem. Theor. Comput. Syst. **63**, 1595–1619 (2019)
17. Munro, J.I., Navarro, G., Shah, R., Thankachan, S.V.: Ranked document selection. Theor. Comput. Sci. **812**, 149–159 (2020)
18. Nemhauser, G.L., Wolsey, L.A., Fisher, M.L.: An analysis of approximations for maximizing submodular set functions I. Math. Program. **14**, 265–294 (1978)
19. Saha, B., Getoor, L.: On maximum coverage in the streaming model and application to multi-topic blog-watch. In: Proceedings of the 2009 SIAM International Conference on Data Mining, pp. 697–708. SIAM (2009)
20. Stoica, A.A., Han, J.X., Chaintreau, A.: Seeding network influence in biased networks and the benefits of diversity. In: Proceedings of The Web Conference 2020, pp. 2089–2098 (2020)
21. Tan, J., Sun, Y., Yicheng, X., Zou, J.: Streaming algorithms for non-submodular maximization on the integer lattice. Tsinghua Sci. Tech. **28**(5), 888–895 (2023)
22. Vitter, J.S.: Random sampling with a reservoir. ACM Trans. Math. Softw. (TOMS) **11**(1), 37–57 (1985)
23. Wang, C., Chen, W., Wang, Y.: Scalable influence maximization for independent cascade model in large-scale social networks. Data Min. Knowl. Disc. **25**, 545–576 (2012)
24. Wang, Y., Fabbri, F., Mathioudakis, M.: Fair and representative subset selection from data streams. In: Proceedings of the Web Conference 2021, pp. 1340–1350 (2021)
25. Wang, Y., Yang, X., Zhang, H., Zhang, Y.: Bicriteria algorithms for approximately submodular cover under streaming model. Tsinghua Sci. Tech. **28**(6), 1030–1040 (2023)
26. Yu, H., Yuan, D.: Set coverage problems in a one-pass data stream. In: Proceedings of the 2013 SIAM International Conference on Data Mining, pp. 758–766. SIAM (2013)

Tournament Transitivity of Graphs

Kamal Santra(✉)

Department of Mathematics, Indian Institute of Technology Patna, Bihta 801106, Bihar, India
kamal.7.2013@gmail.com, kamal_1821ma04@iitp.ac.in

Abstract. Let $G = (V, E)$ be a graph where V and E are the vertex and edge sets, respectively. For two disjoint subsets A and B of V, we say A *dominates* B if every vertex of B is adjacent to at least one vertex of A in G. A vertex partition $\pi = \{V_1, V_2, \ldots, V_k\}$ of G is called a *transitive partition* of size k if V_i dominates V_j for all $1 \leq i < j \leq k$. A vertex partition $\pi = \{V_1, V_2, \ldots, V_k\}$ of G is called a *tournament transitive partition* of size k if V_i dominates V_j and V_j does not dominate V_i for all $1 \leq i < j \leq k$. The maximum integer k for which the above partition exists is called *tournament transitivity* of G, and it is denoted by $TTr(G)$. The Maximum Tournament Transitivity Problem is to find a tournament transitive partition of a given graph with the maximum number of parts. In this article, we study this variation of transitive partition from a structural and algorithmic point of view. We show that the decision version of this problem is NP-complete for chordal graphs (connected), perfect elimination bipartite graphs (disconnected), and doubly chordal graphs (disconnected). On the positive side, we prove that this problem can be solved in polynomial time for trees.

Keywords: Tournament transitivity · NP-completeness · Polynomial-time algorithm · Trees · Chordal graphs

1 Introduction

Partitioning a graph is one of the fundamental problems in graph theory. In the partitioning problem, the objective is to partition the vertex set (or edge set) into some parts with desired properties, such as independence, minimal edges across partite sets, etc. In literature, partitioning the vertex set into certain parts so that the partite sets follow particular domination relations among themselves has been studied [1,5–8,11,12,14]. Let G be a graph with $V(G)$ as its vertex set and $E(G)$ as its edge set. When the context is clear, V and E are used instead of $V(G)$ and $E(G)$. The *neighbourhood* of a vertex $v \in V$ in a graph $G = (V, E)$ is the set of all adjacent vertices of v and is denoted by $N_G(v)$. The *degree* of a vertex v in G, denoted as $\deg_G(v)$, is the number of edges incident to v. A vertex v is said to *dominate* itself and all its neighbouring vertices. A *dominating set* of $G = (V, E)$ is a subset of vertices D such that every vertex $x \in V \setminus D$ has a neighbour $y \in D$, that is, x is dominated by some vertex y of D. For two disjoint

D. Du et al. (Eds.): COCOA 2024, LNCS 15434, pp. 299–310, 2025.
https://doi.org/10.1007/978-981-96-4445-2_25

subsets A and B of V, we say A *dominates* B if every vertex of B is adjacent to at least one vertex of A.

There has been a lot of research on graph partitioning problems based on a domination relationship between the different sets. Cockayne and Hedetniemi introduced the concept of *domatic partition* of a graph $G = (V, E)$ in 1977, in which the vertex set is partitioned into k parts, say $\pi = \{V_1, V_2, \ldots, V_k\}$, such that each V_i is a dominating set of G [3]. The number representing the highest possible order of a domatic partition is called the *domatic number* of G, denoted by $d(G)$. Another similar type of partitioning problem is the *Grundy partition*. Christen and Selkow introduced a Grundy partition of a graph $G = (V, E)$ in 1979 [2]. In the Grundy partitioning problem, the vertex set is partitioned into k parts, say $\pi = \{V_1, V_2, \ldots, V_k\}$, such that each V_i is an independent set and for all $1 \leq i < j \leq k$, V_i dominates V_j. The maximum order of such a partition is called the *Grundy number* of G, denoted by $\Gamma(G)$. In 2018, J. T. Hedetniemi and S. T. Hedetniemi [8] introduced a transitive partition as a generalization of the Grundy partition. A *transitive partition* of size k is defined as a partition of the vertex set into k parts, say $\pi = \{V_1, V_2, \ldots, V_k\}$, such that for all $1 \leq i < j \leq k$, V_i dominates V_j. The maximum order of such a transitive partition is called the *transitivity* of G and is denoted by $Tr(G)$. Recently, in 2020, Haynes et al. generalized the idea of domatic partition as well as transitive partition and introduced the concept of *upper domatic partition* of a graph G, where the vertex set is partitioned into k parts, say $\pi = \{V_1, V_2, \ldots, V_k\}$, such that for each i, j, with $1 \leq i < j \leq k$, either V_i dominates V_j or V_j dominates V_i or both [7]. The maximum order of such an upper domatic partition is called the *upper domatic number* of G, denoted by $D(G)$. All these problems, domatic number [1,16,17], Grundy number [4,5,9,14,15], transitivity [6,8,11,12], upper domatic number [7,13] have been extensively studied both from an algorithmic and structural point of view.

The concept of *tournament transitive partition* was introduced by Haynes et al. in 2019 [6] as an open problem. So far, no research has been done on this variation of transitivity. In this article, we study this variation of the transitivity problem from a structural and algorithmic point of view. For two disjoint subsets A and B, we say A *dominates* B if every vertex of B is adjacent to at least one vertex of A. A vertex partition $\pi = \{V_1, V_2, \ldots, V_k\}$ of G is called a *tournament transitive partition* of size k if V_i dominates V_j and V_j does not dominate V_i for all $1 \leq i < j \leq k$. The maximum integer k for which the above partition exists is called *tournament transitivity* of G, and it is denoted by $TTr(G)$. A tournament transitive partition of order $TTr(G)$ is called a *$TTr(G)$-partition.* The Maximum Tournament Transitivity Problem is to find a tournament transitive partition of a given graph with the maximum number of parts. Note that every tournament transitive partition is also a transitive partition. Therefore, for any graph G, $1 \leq TTr(G) \leq Tr(G) \leq n$. The Maximum Tournament Transitivity Problem and its corresponding decision version are defined as follows.

MAXIMUM TOURNAMENT TRANSITIVITY PROBLEM *Instance:* A graph $G = (V, E)$ *Solution:* A tournament transitive partition of G with maximum size

MAXIMUM TOURNAMENT TRANSITIVITY DECISION PROBLEM *Instance:* A graph $G = (V, E)$, integer k *Question:* Does G have a tournament transitive partition of order at least k?

Note that the only tournament transitive partition for the graph K_n is $\{V(K_n)\}$. But on the other hand, according to [8], we know that $Tr(K_n) = n$. This distinction motivates us to investigate this new parameter. The other similarities and differences can be found in the properties of tournament transitivity section. The rest of the paper is organized as follows: Sect. 2 contains basic definitions and notations that are followed throughout the article. This section also discusses the properties of tournament transitivity of graphs. Section 3 shows that the MAXIMUM TOURNAMENT TRANSITIVITY DECISION PROBLEM is NP-complete in chordal graphs (Connected), perfect elimination bipartite graphs (disconnected) and doubly chordal graphs (disconnected). In Sect. 4, we design a polynomial-time algorithm for solving MAXIMUM TOURNAMENT TRANSITIVITY PROBLEM in trees. Finally, Sect. 5 concludes the article.

2 Preliminaries

2.1 Notation and Definition

Let $G = (V, E)$ be a graph with V and E as its vertex and edge sets, respectively. A graph $H = (V', E')$ is said to be a *subgraph* of a graph $G = (V, E)$ if and only if $V' \subseteq V$ and $E' \subseteq E$. For a subset $S \subseteq V$, the *induced subgraph* on S of G is defined as the subgraph of G whose vertex set is S and edge set consists of all of the edges in E that have both endpoints in S, and it is denoted by $G[S]$. The *complement* of a graph $G = (V, E)$ is the graph $\overline{G} = (\overline{V}, \overline{E})$, such that $\overline{V} = V$ and $\overline{E} = \{uv | uv \notin E \text{ and } u \neq v\}$. The *open neighbourhood* of a vertex $x \in V$ is the set of vertices y adjacent to x, denoted by $N_G(x)$. The *closed neighborhood* of a vertex $x \in V$, denoted as $N_G[x]$, is defined by $N_G[x] = N_G(x) \cup \{x\}$. Let $S \subseteq V$, then we define $N_G(S) = \bigcup_{x \in S} N_G(x)$.

A subset of $S \subseteq V$ is said to be an *independent set* of G if every pair of vertices in S are non-adjacent. A subset of $K \subseteq V$ is said to be a *clique* of G if every pair of vertices in K are adjacent. The cardinality of a clique of maximum size is called *clique number* of G, and it is denoted by $\omega(G)$. A *path* of length $k - 1$ in a graph $G = (V, E)$, denoted by P_k, is a sequence of distinct vertices $v_1, \ldots, v_k$ such that $v_{i-1}v_i \in E$ for $2 \leq i \leq k$. Vertices v_1 and v_k are called the *end vertices* of P_k. For any two vertices y, z, we denoted a path starting with y and ending at z by yPz.

Breadth first search (BFS) is a graph searching technique that systematically explores all the vertices of a connected graph. The process starts at a designated vertex and explores all its neighbours before moving on to the next level of neighbours. The ordering in which the vertices are visited is called *BFS ordering.*

An edge between two non-consecutive vertices of a cycle is called a *chord.* If every cycle in G of length at least four has a chord, then G is called a *chordal graph.* A vertex $v \in V$ is called a *simplicial vertex* of G if $N_G[v]$ induces a clique in G.

2.2 Properties of Tournament Transitivity

In this subsection, we present some properties of tournament transitivity that motivate us to study this variation of transitivity. First, we show the following bounds for tournament transitivity. We omit the proof of this proposition due to space constraints.

Proposition 1. *For any graph G, $TTr(G) \leq min\{Tr(G), Tr(\overline{G})\}$.*

Next, we show an upper bound in terms of maximum degree and minimum degree of a graph.

Proposition 2. *For any graph G, $TTr(G) \leq min\{\Delta(G) + 1, n - \delta(G)\}$, where n is the number of vertex and $\Delta(G), \delta(G)$ are maximum and minimum degree of G, respectively.*

Proof. From [6], we know that $Tr(G) \leq \Delta(G)+1$, for any graph G. Also, from the Proposition 1, we have $TTr(G) \leq min\{Tr(G), Tr(\overline{G})\}$. Therefore, $TTr(G) \leq min\{\Delta(G) + 1, \Delta(\overline{G}) + 1\} = min\{\Delta(G) + 1, n - \delta(G)\}$ as $\Delta(\overline{G}) = n - 1 - \delta(G)$.

In the transitive partition $\pi = \{V_1, V_2, \ldots, V_k\}$ of a graph G, if we union any two sets of a transitive partition of order k, we create a transitive partition of order $k - 1$ [8]. However, for the tournament transitive partition, this result is not true for arbitrary values i and j with $i < j$. But if we take into account the first two sets of the partition, the result is always true. Hence, we have the following proposition.

Proposition 3. *If $\pi = \{V_1, V_2, \ldots, V_k\}$ is a tournament transitive partition of a graph G of size k, then the partition $\pi' = \{V_1 \cup V_2, V_3, \ldots, V_k\}$ is a tournament transitive partition of size $k - 1$.*

Given this, we have the following interpolation result.

Proposition 4. *Let G be a graph and k be the order of a tournament transitive partition of G, then for every j, $1 \leq j \leq k$, G has a tournament transitive partition of order j.*

Proposition 5. *Let G be a connected graph and $TTr(G) = k$, $k \geq 3$. Then there exists a tournament transitive partition of G of size k, say $\pi = \{V_1, V_2, \ldots, V_k\}$ of G, such that $|V_k| = 1$, $|V_{k-1}| = 2$. Moreover, $V_{k-1} = \{x, y\}$ and $V_k = \{z\}$ such that $xz \in E(G)$ and $yz \notin E(G)$.*

Next, we find the tournament transitivity of complete graphs, paths, cycles and complete bipartite graphs.

Proposition 6. *For the complete graph K_n of n vertices $TTr(K_n) = 1$.*

Proof. The partition $\{V(K_n)\}$ is a tournament transitive partition of K_n of size 1. Moreover, from the Proposition 2, we have $TTr(G) \leq min\{\Delta(K_n) + 1, n - \delta(K_n)\} = min\{n, 1\} = 1$. Therefore, $TTr(K_n) = 1$.

Proposition 7. *Let P_n be a path of n vertices, and then the tournament transitivity of P_n is given as follows:*

$$TTr(P_n) = \begin{cases} 1 & n = 1, 2 \\ 2 & n = 3, 4 \\ 3 & n \geq 5 \end{cases}$$

Proposition 8. *Let C_n be a path of n vertices, and then the tournament transitivity of P_n is given as follows:*

$$TTr(C_n) = \begin{cases} 1 & n = 3 \\ 2 & n = 4, 5 \\ 3 & n \geq 6 \end{cases}$$

Next, we show that for a complete bipartite graph $K_{m,n}$, $TTr(K_{m,n}) = 1$ if and only if $m = n = 1$. The proof of this proposition is omitted due to space constraints.

Proposition 9. *For a complete bipartite graph $K_{m,n}$, $TTr(K_{m,n}) = 1$ if and only if $m = n = 1$.*

For the transitivity, if H is a subgraph of G, then $Tr(H) \leq Tr(G)$ [8]. But for tournament transitivity, this is not true. For example, let us consider $G = C_5$ and $H = P_5$. Clearly, H is a subgraph of G, and $TTr(C_5) = 2$ (by Proposition 8), $TTr(P_5) = 3$ (by Proposition 7). Another example is $G = K_4$ and $H = C_4$, where $TTr(K_4) = 1$ and $TTr(C_4) = 2$.

In the above examples, neither of the subgraphs is an induced one. However, the subgraph containment relation holds for tournament transitive partition if we consider an induced subgraph instead of a subgraph.

Proposition 10. *If H is an induced subgraph of a graph G, then $TTr(H) \leq TTr(G)$.*

Proof. Let $\pi = \{V_1, V_2, \ldots, V_k\}$ be a TTr-transitive partition of H. Consider $\pi' = \{V'_1, V'_2, \ldots, V'_k\}$ a vertex partition of G such that $V'_1 = (V(G) \setminus V(H)) \cup V_1$ and $V'_p = V_p$ for all $2 \leq p \leq k$. As H is an induced subgraph, V'_i dominates V'_j, but V'_j does not dominate V'_i in π'. So, π' is a tournament transitive partition of size k. Therefore, $k = TTr(H) \leq TTr(G)$.

For the case of transitivity, if G is a disconnected graph with connected component $C_1, C_2, \ldots, C_t$, then $Tr(G) = \max\{Tr(C_i)|1 \leq i \leq k\}$ [6]. But we show that for tournament transitivity, $TTr(G) \geq \max\{TTr(C_i)|1 \leq i \leq k\}$ and the difference $TTr(G) - \max\{TTr(C_i)|1 \leq i \leq k\}$ can be arbitrarily large. First, we prove a theorem that if G is a disjoint union of $t(\leq n)$ number of complete graphs of n vertices, then $TTr(G) = t$. We omit the proof of this theorem due to space constraints.

Theorem 1. *Let G be the disjoint union of $t(\leq n)$ number of complete graphs of n vertices. Then $TTr(G) = t$.*

Proposition 11. *Let G is a disconnected graph with components $C_1, C_2, \ldots, C_t$, then $TTr(G) \geq \max\{Tr(C_i)|1 \leq i \leq k\}$ and $Tr(G) - \max\{Tr(C_i)|1 \leq i \leq k\}$ can be arbitrarily large.*

Remark 1. From the definition of a tournament transitive partition, it is clear that it is also a transitive partition of G. But the difference $Tr(G) - TTr(G)$ can be arbitrarily large. From the Proposition 6, we have $TTr(K_n) = 1$ and from [6], we know that $Tr(K_n) = n$, hence the difference $Tr(K_n) - TTr(K_n) = n - 1$ is arbitrarily large for n. Moreover, we know that $Tr(K_{m,n}) = n + 1$, when $n < m$. But from Proposition 9, we have $TTr(K_{m,n}) = 2$, when either $m \neq 1$ or $n \neq 1$. So, the difference $Tr(K_n) - TTr(K_n) = n - 1$ is arbitrarily large for n.

3 Computational Complexity

In this section, we present NP-completeness results for Maximum Tournament Transitivity Decision Problem. Given a vertex partition $\pi = \{V_1, V_2, \ldots, V_k\}$ of a graph, we can verify in polynomial time whether π is a tournament transitive partition of that graph or not. Hence, the Maximum Tournament Transitivity Decision Problem is in NP. To prove that this problem is NP-hard, we show a polynomial time reduction from the Maximum Transitivity Decision Problem in general graphs, which is known to be NP-complete [10]. The reduction is as follows: given an instance of the Maximum Transitivity Decision Problem, that is, a graph $G = (V, E)$ and an integer k, let us consider $3(\Delta(G) + 1)$ copies of G. Furthermore, we take another three vertices, say x, x' and x''. Now we connect every vertices of first $\Delta(G)+1$ copies of G to x, second $\Delta(G)+1$ copies of G to x' and final $\Delta(G)+1$ copies of G to x''. Finally, add two edges xx' and $x'x''$. Clearly, the resultant graph $G' = (V', E')$ is a chordal graph if G is a chordal graph and having $(3n(\Delta(G)+1)+3)$ vertices and $(3m(\Delta(G) + 1) + 2)$ edges. The construction of G' is illustrated in Fig. 1.

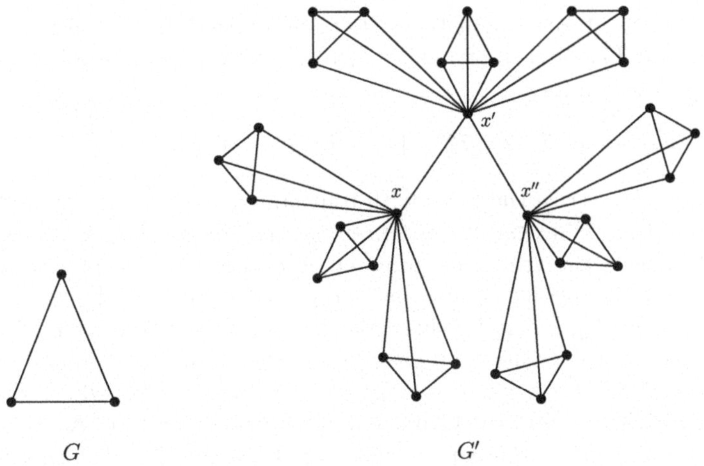

Fig. 1. Construction of G' from G

Let $\mathcal{G}$ be the disjoint union of $\Delta(G)+1$ copies of G, and also we denote the i-th component of $\mathcal{G}$ as G^i. Next, we prove that $Tr(G) = k$ only if $TTr(G) = k$, as shown in the following lemma. The proof of this lemma is omitted due to space constraints.

Lemma 1. *Let $\mathcal{G}$ be the disjoint union of $(\Delta(G)+1)$-copies of G, then $Tr(G) = k$ if and only if $TTr(\mathcal{G}) = k$.*

In the next lemma, we show that G has a transitive partition of size k if and only if G' has a tournament transitive partition of size $k+2$. The proof of this proposition is omitted due to space constraints.

Lemma 2. *The graph G has a transitive partition of size k if and only if G' has a tournament transitive partition of size $k+2$.*

Therefore, we have the following main theorem for this section.

Theorem 2. *The Maximum Tournament Transitivity Decision Problem is NP-complete for chordal graphs (connected).*

Remark 2. The Maximum Transitivity Decision Problem is known to be NP-complete for the perfect elimination bipartite graphs [11] and doubly chordal graphs [12]. Therefore, from Lemma 1, we can say that Maximum Tournament Transitivity Decision Problem is NP-complete for perfect elimination bipartite graphs (disconnected) and doubly chordal graphs (disconnected).

4 Tournament Transitivity in Trees

In this section, we design a polynomial-time algorithm for finding the tournament transitivity of a given tree T. To design the algorithm, in the following lemma we first prove that the tournament transitivity of a tree T can be either $Tr(T)-1$ or $Tr(T)$, where $Tr(T)$ is the transitivity of T. We omit the proof of this lemma due to space constraints.

Lemma 3. *For a tree T, $Tr(T)-1 \leq TTr(T) \leq Tr(T)$.*

Next, we characterize the trees with tournament transitivity equal to $Tr(T)$. We use a function, TRANSITIVENUMBER(x, T), which takes a vertex x and a tree T and returns the transitive number of x in T. The transitive number of a vertex v in T is the maximum integer p such that $v \in V_p$ in a transitive partition $\pi = \{V_1, V_2, \ldots, V_k\}$, where the maximum is taken over all transitive partitions of T. The transitive number of a vertex v in T is denoted by $t(v, T)$. The function TRANSITIVENUMBER(x, T) is presented in [9], where the authors design an algorithm to find the Grundy number of a given tree. As we know that transitivity and Grundy number are the same for a tree [8], the algorithm in [9] also finds the transitivity of a tree. From the description in [9], it follows that TRANSITIVENUMBER(x, T) correctly calculates the transitive number of x in T.

Let us consider two vertices $u, v \in V(T)$ and the path uPv as $(u, v_a, \ldots, v_b, v)$. Additionally, we assume that the transitive number of u is $t(u, T) = t(u)$, and the transitive number of v is $t(v, T) = t(v)$. Let T^c denote the tree T rooted at c and $T^{[c,c']}$ denote the subtree rooted at c, which is obtained by deleting c' from T^c. Let us define the set X as the set of vertices required from the path $\{u, v_a, \ldots, v_b, v\}$ such that $u \in V_{t(u)}$ for some transitive partition π of T. Clearly, $u \in X$ and the vertices of X are consecutive vertices from the path $(u, v_a, \ldots, v_b, v)$. We calculate X iteratively by checking whether the transitive number of u is $t(u)-1$ in the tree $T^{[u,v_\alpha]}$ or not for all $v_\alpha \in \{v_a, v_{a+1}, \ldots, v_b, v\}$. Also, we define Y as the set of vertices required from $\{u, v_a, \ldots, v_b, v\}$ to achieve the transitive number of v as $t(v)$ and calculate Y similarly. For a vertex $w \in X \cap Y$, we say that w agrees for u and v if there exist two transitive partitions of T (not necessarily distinct), say π_p and π_q, such that $u \in V_{t(u)}$ and $w \in V_t$ in π_p and $v \in V_{t(v)}$ and $w \in V_t$ in π_q, respectively. In other words, the index of the sets in which w belongs is the same in both π_p and π_q. In the following lemma, we characterize trees with equal transitivity and tournament transitivity. We omit the proof of this lemma due to space constraints.

Lemma 4. *Let T be a tree and $Tr(T) = k (\geq 3)$. Then $TTr(G) = Tr(T)$ if and only if there exist two vertices y, z of T such that the following conditions hold.*

(a) Transitive number of z in T is k, that is, $t(z, T) = k$.
(b) $y \notin N_T(z)$.
(c) Transitive number of y in T is at least $k-1$, that is, $t(y, T) \geq k-1$.
(d) Either $X \cap Y = \phi$ or $X \cap Y \neq \phi$ and every vertex of $X \cap Y$ agrees for y and z.

[Where X is the set of vertices required from the path yPz such that $y \in V_{k-1}$ for some transitive partition π of T and Y is the set of vertices required form the path yPz such that $z \in V_k$ for some transitive partition π of T]

4.1 Finding the Set X and Y

In this subsection, we calculate the sets X and Y, which are defined as above. Now, we investigate those transitive partitions of T, where $u \in V_{t(u)}$, and show that for those transitive partitions, the vertices of X can be in some specific sets. To show this result, we need to understand how to find the transitive number of x in a rooted tree T^x, and when x achieves that transitive number, then to which sets the children of x belong. The following lemma shows that. To design the algorithm, in the following lemma we first prove that the tournament transitivity of a tree T can be either $Tr(T) - 1$ or $Tr(T)$, where $Tr(T)$ is the transitivity of T. We omit the proof of this lemma due to space constraints.

Lemma 5. *Let $v_1, v_2, \ldots, v_k$ be the children of x in a rooted tree T^x, and for each $1 \leq i \leq k$, l_i denotes the rooted transitive number of v_i in T^x with $l_1 \leq l_2 \leq \ldots \leq l_k$. Let z be the largest integer such that there exists a subsequence of $\{l_i : 1 \leq i \leq k\}$, say $(l_{i_1} \leq l_{i_2} \leq \ldots \leq l_{i_z})$ such that $l_{i_j} \geq j$, for all $1 \leq j \leq z$.*

(a) In this case, the transitive number of x in T^x is $1 + z$.
(b) (i) For all $1 \leq j \leq z$, there exists a transitive partition $\pi = \{V_1, V_2, \ldots, V_{1+z}\}$ such that $x \in V_{1+z}$ and v_{i_j} belongs to any of the sets from $V_j, \ldots, V_{\min\{l_{i_j}, z\}}$.
(ii) If there exists a neighbour of x, say v, other than $\{v_{i_1}, v_{i_2}, \ldots, v_{i_z}\}$, such that the rooted transitive number of v in T^x is at least j, then there exists a transitive partition $\pi = \{V_1, V_2, \ldots, V_{1+z}\}$ such that $x \in V_{1+z}$ and v_{i_j} belongs to any of the sets from $V_1, V_2, \ldots, V_{j-1}$. Otherwise, let r be the minimum index such that $l_{i_t} \geq t+1$ for all $r \leq t \leq j-1$. Then there exists a transitive partition $\pi = \{V_1, V_2, \ldots, V_{1+z}\}$ such that $x \in V_{1+z}$ and v_{i_j} belongs to any of the sets from $V_r, V_{r+1}, \ldots, V_{j-1}$ if and only if such r exists.

Based on the above lemma, we now show that in a transitive partition of T, if $u \in V_{t(u)}$, then the vertices of X can be in some specific sets. A similar result is true for Y as well. We omit the proof of this lemma due to space constraints.

Lemma 6. *Let u and v be two vertices of T and they are connected by the path $(u, v_a, \ldots, v_b, v)$ in T. Let $t(u,T) = t(u)$ and $t(v,T) = t(v)$. Further, let $X = \{x_1, x_2, \ldots x_\alpha\}$ be the set of consecutive vertices, starting with $x_1 = u$, from the path that is required to achieve the transitive number of u is $t(u)$. Similarly, let $Y = \{y_1, y_2, \ldots y_\beta\}$ be the set of consecutive vertices, starting with $y_1 = v$, from the path that is required to achieve the transitive number of v as k. Also, assume that π_p and π_q are two transitive partitions of T such that $u \in V_{t(u)}$ in π_p and $v \in V_{t(v)}$ in π_q, respectively. Then for all $2 \leq j \leq \alpha - 1$, x_j belongs to a unique set V_{p_j} in π_p, where $p_j = t(x_j, T^{[x_j, x_{j-1}]})$ and x_α belongs to a set of sets in π_p as described in Lemma 5(b). Similarly, for all $2 \leq j \leq \beta - 1$, y_j belongs to a unique set V_{q_j} in π_q, where $q_j = t(y_j, T^{[y_j, y_{j-1}]})$ and y_β belongs to a set of sets in π_q as described in Lemma 5(b).*

Based on the above two lemmas, we have Algorithm 1 that describes the process of calculating the sets X and Y for two specific vertices y and z with $t(y,T) \geq k-1$ and $t(z,T) = k$.

Algorithm 1. CALCULATION_X_AND_Y

Input: The tree T and two vertices y and z with $t(y,T) \geq k-1$ and $t(z,T) = k$
Output: Set X and Y corresponding y and z, respectively.

Let the path $P = (v_{a_0} = y, v_a, \ldots, v_b, z = v_{b_0})$;
Initially, $X = \{y\}$ and $Y = \{z\}$;
if $t(y) = t(y,T) = k$ **then**
    $t(y, T^{[y,v_a]}) \geq k-1$;
    $X = \{y\}$;
else
    for all r in $\{v_a, v_{a+1}, \ldots, v_b, z\}$ **do**
        if $t(y, T^{[y,r]}) = k-2$ **then**
            $X = X \cup \{r\}$;
        else
            Stop;
for all s in $\{v_b, v_{b-1}, \ldots, v_a, y\}$ **do**
    if $t(z, T^{[z,s]}) = k-1$ **then**
        $Y = Y \cup \{s\}$;
    else
        Stop;
Calculate the set(s) in π_p to which the vertices of X belong using to the Lemma 6;
Calculate the set(s) in π_q to which the vertices of Y belong using to the Lemma 6;

Based on Lemmas 3 and 4, we can design an algorithm for finding tournament transitivity of a given tree.

Algorithm 2. TOURNAMENT_TRANSITIVITY(T)

Input: A tree T
Output: Tournament transitivity of T, that is, $TTr(T)$

Let $V = \{u_1, u_2, \ldots, u_n\}$ be the set of all vertices of T;
for all $u \in V$ **do**
    TRANSITIVENUMBER(u, T) ;
$S = \{(y,z) | y, z \in V \text{and } y \notin N_T(z) \text{ and } t(y,T) \geq k-1, t(z,T) = k\}$
if $S \neq \phi$ **then**
    for all $(y,z) \in S$ **do**
        Calculate the sets X and Y according to the Algorithm 1;
        if $X \cap Y = \phi$ **then**
            $TTr(T) = Tr(T)$;
        else if $X \cap Y \neq \phi$ and every vertex of $X \cap Y$ agrees for y and z **then**
            $TTr(T) = Tr(T)$;
        else
            $TTr(T) = Tr(T) - 1$;
else
    $TTr(T) = Tr(T) - 1$;
return $(TTr(G))$;

4.2 Running Time of the Algorithm

In this subsection, we analyse the running time of Algorithm 2. In lines 2–4 of the algorithm, we have computed the transitive number of every vertex. According to the running time analysis of [9], it will take $O(n)$ to calculate the transitive number of every vertex. Moreover, for a fixed pair of vertices $\{y, z\}$, we can check whether it is in S or not in a constant time.

Now we analyse the running time of Algorithm 1 for a pair of vertices $\{y, z\}$ such that $y \notin N_T(z)$ and $t(y, T) \geq k - 1, t(z, T) = k$. While computing the set X, the checking in line 8 takes $\sum_{v \in T}(O(deg(v))$ time. This implies that we can compute the set X in time $k(\sum_{v \in T} O(deg(v)))$, where k is the length of the path yPz. In the worst case, X in lines 3–14 can be computed in time $O(n^2)$. Similarly, Y can be computed in time $O(n^2)$. The process of computing X and Y also indicates the set(s) to which the vertices of X and Y belong. Therefore, lines 22 and 23 take constant time. So for all vertices from S, to calculate the sets X and Y, we need at most $n^2 * O(n^2) = O(n^4)$ time. As a result, we get that the running time of Algorithm 2 is $O(n) + O(n^4) = O(n^4)$. Hence, we have the following main theorem.

Theorem 3. *The* MAXIMUM TOURNAMENT TRANSITIVITY PROBLEM *can be solved in polynomial time for trees.*

5 Conclusion

In this paper, we have studied the notion of tournament transitivity in graphs, which is a variation of transitivity. We have shown that the decision version of this problem is NP-complete for chordal graphs (connected), perfect elimination bipartite graphs (disconnected) and doubly chordal (disconnected). On the positive side, we prove that this problem can be solved in polynomial time for trees. It would be interesting to investigate the complexity status of this problem in other graph classes. Designing an approximation algorithm for this problem would be another challenging open problem.

Acknowledgements. Kamal Santra was supported by the Department of Science and Technology (DST) (INSPIRE Fellowship, Ref No: DST/INSPIRE/ 03/2016/000291), Govt. of India.

References

1. Chang, G.J.: The domatic number problem. Discret. Math. **125**(1–3), 115–122 (1994)
2. Christen, C.A., Selkow, S.M.: Some perfect coloring properties of graphs. J. Comb. Theor. Ser. B **27**(1), 49–59 (1979)

3. Cockayne, E.J., Hedetniemi, S.T.: Towards a theory of domination in graphs. Networks **7**(3), 247–261 (1977)
4. Effantin, B.: A note on Grundy colorings of central graphs. Australas. J. Combin. **68**(3), 346–356 (2017)
5. Füredi, Z., Gyárfás, A., Sárközy, G.N., Selkow, S.: Inequalities for the first-fit chromatic number. J. Graph Theory **59**(1), 75–88 (2008)
6. Haynes, T.W., Hedetniemi, J.T., Hedetniemi, S.T., McRae, A., Phillips, N.: The transitivity of special graph classes. J. Comb. Math. Comb. Comput. **110**, 181–204 (2019)
7. Haynes, T.W., Hedetniemi, J.T., Hedetniemi, S.T., McRae, A., Phillips, N.: The upper domatic number of a graph. AKCE Int. J. Graphs Comb. **17**(1), 139–148 (2020)
8. Hedetniemi, J.T., Hedetniemi, S.T.: The transitivity of a graph. J. Comb. Math. Comb. Comput. **104**, 75–91 (2018)
9. Hedetniemi, S.M., Hedetniemi, S.T., Beyer, T.: A linear algorithm for the Grundy (coloring) number of a tree. Congr. Numer. **36**, 351–363 (1982)
10. Hedetniemi, S.M., Hedetniemi, S.T., McRae, A.A., Parks, D., Telle, J.A.: Iterated colorings of graphs. Discret. Math. **278**(1–3), 81–108 (2004)
11. Paul, S., Santra, K.: Transitivity on subclasses of bipartite graphs. J. Comb. Optim. **45**(1), 1–16 (2023)
12. Paul, S., Santra, K.: Transitivity on subclasses of chordal graphs. In: Algorithms and Discrete Applied Mathematics, pp. 391–402. Springer, Cham (2023)
13. Samuel, L., Joseph, M.: New results on upper domatic number of graphs. Commun. Comb. Optim. **5**(2), 125–137 (2020)
14. Zaker, M.: Grundy chromatic number of the complement of bipartite graphs. Australas. J. Combin. **31**, 325–330 (2005)
15. Zaker, M.: Results on the Grundy chromatic number of graphs. Discret. Math. **306**(23), 3166–3173 (2006)
16. Zelinka, B.: Domatically critical graphs. Czechoslov. Math. J. **30**(3), 486–489 (1980)
17. Zelinka, B.: On k-domatic numbers of graphs. Czechoslov. Math. J. **33**(2), 309–313 (1983)

Approximation Algorithm for Min-max Correlation Clustering Problem with Penalties

Yuebo Huang[1], Sai Ji[1(✉)], Xiaoyun Tian[2], and Kun Zhou[3]

[1] Institute of Mathematics, Hebei University of Technology, Tianjin 300401, People's Republic of China
huangyuebo2023@stu.hebut.edu.cn, jisai@hebut.edu.cn

[2] Academy of Mathematics and Systems Science, Chinese Academy of Sciences, Beijing 100190, People's Republic of China
xiaoyuntian@amss.ac.cn

[3] Digital Technology Management, Hoseo University, Chonan 42409, Korea

Abstract. This paper introduces the min-max correlation clustering problem with penalties, which is a generalization of the correlation clustering problem. In this problem, each vertex can be clustered or penalized, and the goal of the problem is to minimize the sum of the number of error edges and the penalty cost at the worst vertex. In this paper, we give an integer programming and linear programming relaxation of the problem, and provide a constant approximation algorithm of the problem based on LP-rounding technique.

Keywords: Correlation clustering · Min-max · Penalties · Approximation algorithm · LP-rounding

1 Introduction

The correlation clustering problem was first described in 2002 by Bansal et al. [6]. It is derived from the document clustering problem and has been demonstrated to be NP-hard. Correlation clustering has a wide range of applications, including protein interaction networks, social computing, communication networks, cross-lingual link identification, and a lot more [3,5,7,8,18,19].

The following is the input for the correlation clustering problem: a complete graph $G = (V, E)$, for each edge $(u, v) \in E$, a label "+" or "−" is assigned depending on the similarity of the two endpoints. If u, v are similar, then the label is positive and otherwise, it is negative. The output is a clustering of every vertex in V which is disjoint, besides, the quantity of clusters is unlimited. Edges with positive labels and endpoints in different clusters, as well as edges with negative labels and endpoints in the same cluster, are referred to as error edges. Edges with negative labels and endpoints in different clusters, as well as edges with positive labels and endpoints in identical cluster, are referred to as correct

D. Du et al. (Eds.): COCOA 2024, LNCS 15434, pp. 311–323, 2025.
https://doi.org/10.1007/978-981-96-4445-2_26

edges. A clustering of set V with the fewest possible error edges is the objective of the problem. Moreover, a clustering of set V with the largest possible number of correct edges is another objective in some variants. This paper concentrates on the minimizing version.

A $(9/\delta^2 + 5)$-approximation algorithm for correlation clustering was provided by Bansal et al. [6] and was designed on the idea of counting "error triangles" (where $\delta = 1/44$). Subsequently, Charikar et al. [10] significantly improved this result and introduced a 4-approximation algorithm by applying the LP-rounding technique. Chawla et al. [11] improved the procedure of rounding based on [4] by introducing a rounding function, and obtained a $(2.06 - \epsilon)$-approximation algorithm. In [13], Cohen-Addad et al. broke through this by employing the Sherali-Adams hierarchy to develop a $(1.994 + \epsilon)$-approximation algorithm. The running time was also developed to be $n^{o\left(1/\varepsilon^2\right)}$. In the last few years, Cohen-Addad et al. [12] have produced the best result for the correlation clustering problem, obtaining a $(1.73+\epsilon)$-approximation through a combinatorial algorithm and its running time is $n^{o(1/\operatorname{poly}(\varepsilon))}$.

Correlation clustering has many generalizations, such as min-max correlation clustering, k-correlation clustering, fair correlation clustering, correlation clustering with penalties, and so on [2,16,20,22,23]. Among them, we concentrate on the min-max correlation clustering problem and the correlation clustering problem with penalties in this paper.

The difference between the min-max correlation clustering problem and its classical problem lies in the optimization perspective. The classical correlation clustering problem aims to maximize the total amount of error edges from a global perspective. However, in min-max correlation clustering problem, the amount of error edges at the worst vertex is optimized. The main technique used in the study of this problem is LP-rounding, and the approximation ratio has been improved from 48 approximations when the problem was initially presented in [21] to the later optimal ratio of 7 [9]. In [17], the ratio has been improved to 5 and the best result up to now is 4 which is attained in [14].

Some variants with penalties are widely studied in combinatorial optimization problems. Different from traditional correlation clustering problem, for any input graph $G = (V, E)$, each vertex v in V has a corresponding cost p_v which will be charged when v is penalized. It can also be interpreted as a penalty cost. Each vertex will either be penalized or be clustered. Finding the optimal solution is the objective, and it will be accomplished by combining decreasing the amount of error edges with the minimization of overall penalty costs. Aboud and Rabani [1] initially introduced correlation clustering problems with penalties and proposed a 9-approximation algorithm via primal-dual schema. It is also the best result so far.

In this paper, we introduce the min-max correlation clustering problem with penalties, representing a combination of the min-max correlation clustering problem and the correlation clustering problem with penalties. In a given complete graph $G = (V, E^+, E^-)$ for this problem, every vertex v in V is provided with a penalty cost p_v which should be paid when v is penalized. We wish to select some vertices from V to penalize and achieve a clustering of the remaining vertices. The objective is to minimize the total amount of error edges and penalty cost at the worst vertex in the graph. We formulize our model by an integer programming which is inspired by Ji et al. in [15]. Moreover, based on the algorithm introduced in [1] with LP-rounding technique, a 21-approximation algorithm is obtained for this problem.

In brief, this paper's main contributions are as follows:

- **From the perspective of model.** For the first time, we focus on the min-max correlation clustering problem with penalties, which can better describe some complex practical problems and has a strong application background. In addition, we model the problem and give the integer programming and linear programming relaxation of the problem.
- **From the perspective of algorithm and analysis.** Following the work of [1,15], we provide a 21-approximation algorithm. Particularly, during the theoretical analysis, a constant approximation ratio analysis is given by combining the structure of the solution and inequality constraints in the programming.

The sections are organized as follows. In Sect. 2, we give the detailed integer programming and the linear programming of the problem. Section 3 contains the constant approximation algorithm and the theoretical analysis. Finally, some discussions are presented in Sect. 4.

2 Preliminaries

At the start of this section, we provide a definition of the min-max correlation clustering problem with penalties. The integer programming of it will be present later.

Definition 1 (Min-max correlation clustering problem with penalties). *Given a complete graph $G = (V, E^+, E^-)$, E^+ represents the set of positive edges while E^- is the set consists of negative edges. Each vertex v in V has a corresponding cost p_v which will be charged when v is penalized. The goal of the problem is to select a penalized vertex set P, and find a clustering $\mathcal{C} = \{C_1, C_2, \ldots, C_k\}$ of $V \backslash P$ to minimize the value*

$$\max_{v\in V} [I(v \in C_i, i \in [k])d_{C_i}(v) + I(v \in P)p_v]$$

where $d_{C_i}(v) = |\{(u,v) \in E^+, u \in V\backslash(C_i \cup P)\}| + |\{(u,v) \in E^-, u \in C_i\}|, [k] = \{1, 2, \ldots, k\}$ and I is an indicator function.

It is first necessary to define a few binary variables as follows, since the variables are used in the later programming for our problem:

For any edge $(u, v) \in E$, let A_{uv} indicates whether it is a cut edge, the value will be 1 if it is a cut, otherwise $A_{uv} = 0$.

For any two vertices $u, v \in E$, let B_{uv} indicates whether these two vertices are both clustered, the value will be 0 if both of them are clustered. If at least one of them is penalized, then $B_{uv} = 1$.

For any vertex $v \in V$, let R_v indicates whether it is clustered, the value will be 0 if it is clustered, otherwise $R_v = 1$.

For any edge $(u, v) \in E$, let Z_{uv} indicates whether it is an error edge, the value will be 1 if it is an error edge, otherwise $Z_{uv} = 0$.

Applying the variables that demonstrated above, we could obtain the integer programming of this problem:

$$
\begin{aligned}
\min \max_{v \in V} & \left\{ \sum_{(u,v) \in E} Z_{uv} + p_v R_v \right\} & \\
\text{s. t. } & A_{uv} + A_{vw} \geq A_{uw}, & \forall u, v, w \in V, \\
& B_{uv} \leq R_u + R_v \leq 2B_{uv}, & \forall u, v \in V, \\
& B_{uv} + Z_{uv} \geq 1 - A_{uv}, & \forall (u, v) \in E^-, \\
& B_{uv} + Z_{uv} \geq A_{uv}, & \forall (u, v) \in E^+, \\
& 1 - A_{uv} \geq Z_{uv}, & \forall (u, v) \in E^-, \\
& A_{uv} \geq Z_{uv}, & \forall (u, v) \in E^+, \\
& A_{uv}, B_{uv}, Z_{uv}, R_v \in \{0, 1\}, & \forall u, v \in V.
\end{aligned}
\tag{1}
$$

The first constraint is a triangle inequality which is able to guarantee that the clustering is justified. The second to the sixth constraints give the conditions under which each edge becomes an error edge. Relaxing each binary variable to between 0 and 1, we obtain the LP relaxation of (1):

$$
\begin{aligned}
\min \max_{v \in V} & \left\{ \sum_{(u,v) \in E} Z_{uv} + p_v R_v \right\} & \\
\text{s. t. } & A_{uv} + A_{vw} \geq A_{uw}, & \forall u, v, w \in V, \\
& B_{uv} \leq R_u + R_v \leq 2B_{uv}, & \forall u, v \in V, \\
& B_{uv} + Z_{uv} \geq 1 - A_{uv}, & \forall (u, v) \in E^-, \\
& B_{uv} + Z_{uv} \geq A_{uv}, & \forall ((u, v) \in E^+, \\
& 1 - A_{uv} \geq Z_{uv}, & \forall (u, v) \in E^-, \\
& A_{uv} \geq Z_{uv}, & \forall (u, v) \in E^+, \\
& A_{uv}, B_{uv}, Z_{uv}, R_v \in [0, 1], & \forall u, v \in V.
\end{aligned}
\tag{2}
$$

3 Approximation Algorithm

This section is the core of our paper, giving the specific algorithm and theoretical analysis. For more details of the algorithm, it is shown in Sect. 3.1. For the theoretical analysis, it is provided in Sect. 3.2.

3.1 Algorithm

Our algorithm is a deterministic greedy clustering algorithm which uses the LP-rounding technique. The algorithm is formed of three main parts. Solving the Programming (2) and obtaining the optimal fractional solution (A^*, B^*, R^*, Z^*) are the first part. The second part is to find out which vertices in V should be penalized. More specifically, the vertex v whether to be penalized or not depends on the value of the variable R^*. The last part is the iterative clustering process. In each iteration, we define the neighbour sets T_v and $T_v^{'}$ of each unclustered vertex v. Then, we pick the vertex v^* with the largest amount of elements in $T_v^{'}$ as the center vertex, and outputs T_{v^*} as a cluster. The process of iteration continues until set S is empty. More details of Algorithm 1 are given below:

Algorithm 1. Greedy Clustering Algorithm

Input: Complete graph $G = (V, E^+, E^-)$, $p_v, \forall v \in V$
Output: A penalized set P and a clustering of $V \setminus P$
1: Solve Programming (2) to obtain (A^*, B^*, R^*, Z^*).
2: Initialize $S := V$, $P := \emptyset$ and $V^* := \emptyset$.
3: Update $P := \{v \in V : R_v^* \geq 1/21\}$, $S := V \setminus P$.
4: **while** $S \neq \emptyset$ **do**
5: Let $T_v := \{u \in S : A_{uv}^* \leq 3/7\}$, $T_v^{'} := \{u \in S : A_{uv}^* \leq 1/7\}$, $\forall\ v \in S$.
6: Pick a vertex $v^* = arg\ max_{v \in S} \left|T_v^{'}\right|$.
7: Let $C_{v^*} := T_{v^*}$.
8: Update $V^* := V^* \cup \{v^*\}$, $S := S \setminus T_{v^*}$.
9: Output $C = \{C_{v^*} : v^* \in V^*\}$.
10: **end while**
11: **return** penalized set P and clustering $C := \{C_{v^*} : v^* \in V^*\}$ of $V \setminus P$

3.2 Theoretical Analysis

In this section, we provide the approximation ratio analysis of Algorithm 1. The penalty costs are analyzed in Lemma 1, and the number of error edges produced by the algorithm are analyzed in Lemmas 2–7.

Penalty Costs. Assuming that Algorithm 1 has k iterations of the process. Let P be the set consists of vertices which are penalized. And let $V^* := \{v_1^*, v_2^*, ..., v_k^*\}$ be the vertices set which are picked at each iteration of the process. In addition, let $C := \{C_{v_1^*}, C_{v_2^*}, ..., C_{v_k^*}\}$ be the clustering of $V \setminus P$ as the output generated by Algorithm 1.

Lemma 1. *The penalty cost of each vertex v is no more than $21 p_v R_v^*$*

Proof. By Step 3 of Algorithm 1, for any vertex $v \in P$, we have $R_v^* \geq 1/21$, which implies that

$$p_v \leq 21 p_v R_v^*$$

Moreover, for any vertex $v \in V \setminus P$, according to Algorithm 1, the vertex v is clustered, so its penalty cost equals zero which is also no more than $21 p_v R^*_v$. The lemma is concluded. □

Error Edges. We classify the generation of error edges into two types, one is positive cut edges, the other is negative edges whose endpoints are in the same cluster. We will discuss separately about the upper bounds of the amount of these two types of error edges.

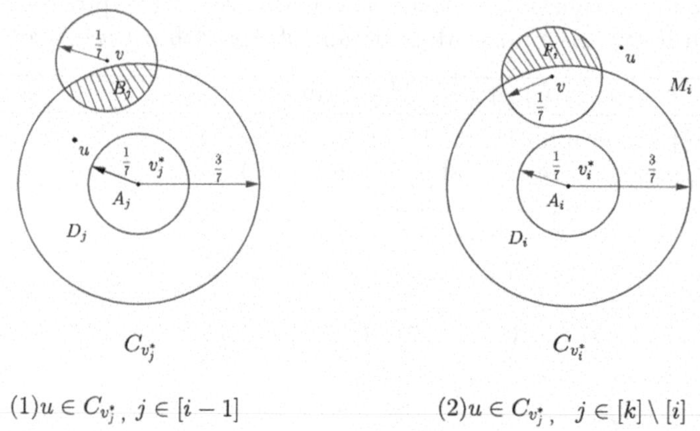

Fig. 1. The relationship between vertex v and cluster $C_{v_j^*}$

Error Edges at Vertex v Derived from Positive Cut Edges. For each vertex $v \in C_{v_i^*}$, $i \in [k]$, the error edges can be divided into two types:

(1) $u \in C_{v_j^*}, j \in [i-1]$;
(2) $u \in C_{v_j^*}, j \in [k] \setminus [i]$.

As illustrated in Fig. 1, for any vertex $v \in C_{v_i^*}$, $i \in [k]$, we define sets as follows:

$$A_j = T'_{v_j^*}$$
$$B_j = \{u \in V : A^*_{uv} \le 1/7\} \cap C_{v_j^*}$$
$$D_j = C_{v_j^*} \setminus (A_j \cup B_j)$$
$$F_i = \left\{u \in V \setminus \cup_{s \in [i]} C_{v_s^*} : A^*_{uv} \le 1/7\right\}$$
$$M_i = \left\{u \in V \setminus \cup_{s \in [i]} C_{v_s^*} : A^*_{uv} > 1/7\right\}.$$

In the following, the number of positive error edges between v and vertices in different sets is analysed separately.

Lemma 2. *Any edge $(u, v) \in V \setminus P$ that participates in the iterative clustering process has B^*_{uv} no more than 2/21.*

Proof. Since the vertices that participate in the iterative process are all unpunished vertices, when vertices u, v are in $V \setminus P$, the value of R^*_u and R^*_v hold that

$$R^*_u \leq \frac{1}{21}, R^*_v \leq \frac{1}{21}.$$

And by the second constraint of Programming (2), the value of B^*_{uv} holds that

$$B^*_{uv} \leq R^*_u + R^*_v \leq \frac{1}{21} + \frac{1}{21} = \frac{2}{21}.$$

The lemma has been proved. □

Lemma 3. *For any vertex $v \in C_{v^*_i}$, $i \in [k]$, the number of error edges generated at vertex v by each positive edge $(u, v) \in E^+$, $u \in D_j$, $j \in [i-1]$ equals one and is no more than $21Z^*_{uv}$.*

Proof. According to the definition of set D_j, when $(u, v) \in E^+$, $u \in D_j$, the value of A^*_{uv} holds that

$$A^*_{uv} > \frac{1}{7}.$$

By Lemma 2 and the fourth constraint of Programming (2), the value of Z^*_{uv} holds that

$$Z^*_{uv} \geq A^*_{uv} - B^*_{uv} \geq \frac{1}{7} - \frac{2}{21} = \frac{1}{21}.$$

Hence, the number of error edges generated by $(u, v) \in E^+$, $u \in D_j$ at vertex v equals one and is not exceeding $21Z^*_{uv}$. □

Lemma 4. *For any vertex $v \in C_{v^*_i}$, $i \in [k]$, the number of error edges generated at vertex v by all positive edges $(u, v) \in E^+$, $u \in A_j \cup B_j$, $j \in [i-1]$ is no greater than*

$$21 \sum_{(u,v) \in E, u \in A_j} Z^*_{uv}.$$

Proof. We consider the next pair of situations.

(1) $u \in A_j$. According to the definition of set A_j, when $(u, v) \in E^+$, $u \in A_j$, $j \in [i-1]$, it holds that

$$A^*_{uv} \geq A^*_{vv^*_j} - A^*_{uv^*_j} \geq \frac{3}{7} - \frac{1}{7} = \frac{2}{7}.$$

By Lemma 2 and the fourth constraint of Programming (2), the value of Z^*_{uv} holds that

$$Z_{uv}^* \geq A_{uv}^* - B_{uv}^* \geq \frac{2}{7} - \frac{2}{21} = \frac{4}{21}.$$

Hence, for all positive edges (u, v), $u \in A_j$ at vertex v, the amount of error edges is no greater than

$$\sum_{(u,v)\in E+, u\in A_j} \frac{21}{4} Z_{uv}^*.$$

(2) $u \in B_j$. For any edge $(u, v) \in E^+$, the lower bound of A_{uv}^* is undetermined, so we cannot analyse the upper bound of the error edge with the help of the fractional optimal solution. Hence, in the following, we analyse the upper bounds of the error edges generated by the vertices in set B_j at v by the help of set A_j. When negative edge (u, v) is in E^-, $u \in A_j$, $z \in B_j$, it holds that

$$1 - A_{uv}^* \geq 1 - (A_{vz}^* + A_{zv_j^*}^* + A_{v_j^* u}^*) \geq 1 - \frac{1}{7} - \frac{3}{7} - \frac{1}{7} = \frac{2}{7}.$$

From Lemma 2 and the third constraint of Programming (2), the value of Z_{uv}^* holds that

$$Z_{uv}^* \geq 1 - A_{uv}^* - B_{uv}^* \geq \frac{2}{7} - \frac{2}{21} = \frac{4}{21}.$$

From Step 3 of Algorithm 1, it is obvious that $|B_j| \leq |A_j|$, hence the amount of error edges generated by the vertices in set B_j at vertex v is

$$\begin{aligned} \left|(u, v) \in E^+ : u \in B_j\right| &\leq |B_j| \leq |A_j| \\ &\leq \sum_{(u,v)\in E+, u\in A_j} \frac{21}{4} A_{uv}^* + \sum_{(u,v)\in E^-, u\in A_j} \frac{21}{4}(1 - A_{uv}^*) \\ &\leq \sum_{(u,v)\in E, u\in A_j} \frac{21}{4} Z_{uv}^*. \end{aligned}$$

Above all, the lemma has been proved. □

Lemma 5. *For any vertex $v \in C_{v_i^*}$, $i \in [k]$, the number of error edges generated at vertex v by each positive edge $(u, v) \in E^+$, $u \in M_i$ equals one and is no more than $21 Z_{uv}^*$.*

Proof. According to the definition of set M_i, when $(u, v) \in E^+$, $u \in M_i$, it holds that

$$A_{uv}^* > \frac{1}{7}.$$

By Lemma 2 and the fourth constraint of Programming (2), the value of Z_{uv}^* holds that

$$Z_{uv}^* \geq A_{uv}^* - B_{uv}^* \geq \frac{1}{7} - \frac{2}{21} = \frac{1}{21}.$$

Hence, for every positive edge $(u, v) \in E^+$, $u \in M_i$, the amount of error edges generated at vertex v equals one and is no more than $21Z^*_{uv}$. □

Lemma 6. *For any vertex $v \in C_{v_i^*}$, $i \in [k]$, the number of error edges generated at vertex v by all positive edges $(u, v) \in E^+$, $u \in F_i$ is no more than*

$$21 \sum_{(u,v)\in E^+,u\in A_i} Z^*_{uv} + 3 \sum_{(u,v)\in E^-,u\in A_i} Z^*_{uv}.$$

Proof. According to the definition of set F_i, the lower bound of any edge $(u, v) \in E^+$, $u \in F_i$, A^*_{uv} is undetermined, so we cannot analyse the upper bound of the error edge through the fractional optimal solution. In the following, we analyse the upper bounds of error edges $(u, v) \in E^+$,$u \in F_i$ generated at vertex v by the help of set A_i.

When $(u, v) \in E^+$, $u \in A_i$, $z \in F_i$, the value of A^*_{uv} holds that

$$A^*_{uv} \geq A^*_{v_i^* z} - A^*_{uv_i^*} - A^*_{vz} \geq \frac{3}{7} - \frac{1}{7} - \frac{1}{7} = \frac{1}{7}.$$

By Lemma 2 and the fourth constraint of Programming (2), the value of Z^*_{uv} holds that

$$Z^*_{uv} \geq A^*_{uv} - B^*_{uv} \geq \frac{1}{7} - \frac{2}{21} = \frac{1}{21}.$$

When $(u, v) \in E^-$,$u \in A_i$, it holds that

$$1 - A^*_{uv} \geq 1 - (A^*_{uv_i^*} + A^*_{v_i^* v}) \geq 1 - \frac{1}{7} - \frac{3}{7} = \frac{3}{7}.$$

From Lemma 2 and the third constraint of Programming (2), the value of Z^*_{uv} holds that

$$Z^*_{uv} \geq 1 - A^*_{uv} - B^*_{uv} \geq \frac{3}{7} - \frac{2}{21} = \frac{7}{21}.$$

From Step 3 of Algorithm 1, we have $|F_i| \leq |A_i|$, hence the amount of error edges generated by the vertices in set F_i at vertex v is

$$\begin{aligned} \left|(u, v) \in E^+ : u \in F_i\right| &\leq |F_i| \leq |A_i| \\ &\leq 21 \sum_{(u,v)\in E^+,u\in A_i} z^*_{uv} + 3 \sum_{(u,v)\in E^-,u\in A_i} z^*_{uv}. \end{aligned}$$

The lemma has been proved. □

Error Edges at Vertex v Generated by Negative Edges. In the following we discuss the upper bound of the error edge at vertex $v \in C_{v_i^*}$, $i \in [k]$ generated by the negative edges $(u, v) \in E^-$, $u \in C_{v_i^*}$.

Lemma 7. *For any vertex $v \in C_{v_i^*}$, $i \in [k]$, the number of error edges generated at vertex v by all negative edges $(u,v) \in E^-$, $u \in C_{v_i^*}$ is no more than*

$$21 \sum_{(u,v)\in E^-, u\in D_i} Z^*_{uv} + 3 \sum_{(u,v)\in E^-, u\in A_i} Z^*_{uv}.$$

Proof. When $(u,v) \in E^-$,$u \in D_i$, it holds that

$$1 - A^*_{uv} \geq 1 - (A^*_{uv_i^*} + A^*_{v_i^* v}) \geq 1 - \frac{3}{7} - \frac{3}{7} = \frac{1}{7}.$$

By Lemma 2 and the third constraint of Programming (2), the value of Z^*_{uv} holds that

$$Z^*_{uv} \geq 1 - A^*_{uv} - B^*_{uv} \geq \frac{1}{7} - \frac{2}{21} = \frac{1}{21}.$$

By the proof of Lemma 6, when $(u,v) \in E^-$, $u \in A_i$, it holds that

$$1 - A^*_{uv} \geq 1 - (A^*_{uv_i^*} + A^*_{v_i^* v}) \geq 1 - \frac{1}{7} - \frac{3}{7} = \frac{3}{7}.$$

By Lemma 2 and the third constraint of Programming (2), the value of Z^*_{uv} holds that

$$Z^*_{uv} \geq 1 - A^*_{uv} - B^*_{uv} \geq \frac{3}{7} - \frac{2}{21} = \frac{7}{21}.$$

In conclusion, for all edges $(u,v) \in E^-$, $u \in C_{v_i^*}$ at vertex v, the amount of error edges is no greater than

$$21 \sum_{(u,v)\in E^-, u\in D_i} Z^*_{uv} + 3 \sum_{(u,v)\in E^-, u\in A_i} Z^*_{uv}.$$

Hence, the lemma has been proved. □

The conclusion that follows can be obtained through connecting Lemma 2–Lemma 7.

Theorem 1. *Algorithm 1 is a 21-approximation algorithm for the min-max correlation clustering problem with penalties on complete graphs.*

Proof. When vertex v is in $C_{v_i^*}$, $i \in [k]$, the number of error edges at vertex v is

$$\left|\{(u,v) \in E^+ : u \in V \setminus C_{v_i^*}\}\right| + \left|\{(u,v) \in E^- : u \in C_{v_i^*}\}\right| \tag{3}$$

From Lemma 2–Lemma 7, here follows

$$
\begin{aligned}
(3) \le\ & 21 \sum_{(u,v)\in E^+, u\in D_j, j\in[i-1]} Z^*_{uv} + 21 \sum_{(u,v)\in E^+, u\in M_i} Z^*_{uv} \\
& + 21 \sum_{(u,v)\in E^+, u\in A_j, j\in[i-1]} Z^*_{uv} + 21 \sum_{(u,v)\in E^-, u\in A_j, j\in[i-1]} Z^*_{uv} \\
& + 21 \sum_{(u,v)\in E^+, u\in A_i} Z^*_{uv} + 3 \sum_{(u,v)\in E^-, u\in A_i} Z^*_{uv} \\
& + 21 \sum_{(u,v)\in E^-, u\in D_i} Z^*_{uv} + 3 \sum_{(u,v)\in E^-, u\in A_i} Z^*_{uv} \\
& \le 21 \left[\sum_{(u,v)\in E^+} Z^*_{uv} + \sum_{(u,v)\in E^-} Z^*_{uv} \right] \\
& \le 21 \sum_{(u,v)\in E} Z^*_{uv}.
\end{aligned}
$$

In addition, according to Lemma 1, the penalty cost of each vertex v is no more than $21 p_v R^*_v$, so the sum is no more than

$$
21 \sum_{(u,v)\in E} Z^*_{uv} + 21 p_v R^*_v.
$$

The theorem is concluded. □

4 Discussions

In this paper, we introduce the min-max correlation clustering problem with penalties and provide a 21-approximation algorithm for it. At present, there are several interesting research directions for the min-max correlation clustering problem with penalties:

- The lower bound of the approximation ratio of the problem.
- Approximation algorithms of problem on general graph.
- Variants of the problem, such as capacitated variant.

Acknowledgements. The first two authors are supported by National Natural Science Foundation of China (No. 12101594) and Science and Technology Project of Hebei Education Department (No. BJK2023076). The third author is supported by National Natural Science Foundation of China (No. 12401426) and the China Postdoctoral Science Foundation under Grant Number (No. 2024M753432).

References

1. Aboud, A., Rabani, Y.: Correlation clustering with penalties and approximating the reordering buffer management problem. Doctoral dissertation, Computer Science Department, Technion (2008)

2. Ahmadian, S., Negahbani, M.: Improved approximation for fair correlation clustering. In: Proceedings of the 16th International Conference on Artificial Intelligence and Statistics, pp. 9499–9516 (2023)
3. Ahmadian, S., Norouzi-Fard, A., Svensson, O., Ward, J.: Better guarantees for k-means and Euclidean k-median by primal-dual algorithms. In: Proceedings of the 58th Annual Symposium on Foundations of Computer Science, pp. 61–72 (2017)
4. Ailon, N., Charikar, M., Newman, A.: Aggregating inconsistent information: ranking and clustering. J. ACM **55**(5), 1–27 (2008)
5. Arthur, D., Vassilvitskii, S.: k-means++: the advantages of careful seeding. In: Proceedings of the 18th Annual ACM-SIAM Symposium on Discrete Algorithms, pp. 1027–1035 (2007)
6. Bansal, N., Blum, A., Chawla, S.: Correlation clustering. Mach. learn. **56**, 89–113 (2004)
7. Braverman, V., Lang, H., Levin, K., Monemizadeh, M.: Clustering problems on sliding windows. In: Proceedings of the 27th Annual ACM-SIAM Symposium on Discrete Algorithms, pp. 1374–1390 (2016)
8. Byrka, J., Fleszar, K., Rybicki, B., Spoerhase, J.: Bi-factor approximation algorithms for hard capacitated k-median problems. In: Proceedings of the 26th Annual ACM-SIAM Symposium Discrete Algorithms, pp. 722–736 (2015)
9. Charikar, M., Gupta, N., Schwartz, R.: Local guarantees in graph cuts and clustering. In: Proceedings of the 19th International Conference on Integer Programming and Combinatorial Optimization, pp. 136–147 (2017)
10. Charikar, M., Guruswami, V., Wirth, A.: Clustering with qualitative information. J. Comput. Syst. Sci. **71**(3), 360–383 (2005)
11. Chawla, S., Makarychev, K., Schramm, T., Yaroslavtsev, G.: Near optimal LP rounding algorithm for correlation clustering on complete and complete k-partite graphs. In: Proceedings of the 47th Annual ACM Symposium on Theory of Computing, pp. 219–228 (2015)
12. Cohen-Addad, V., Lee, E., Li, S., Newman, A.: Handling correlated rounding error via preclustering: a 1.73-approximation for correlation clustering. In: Proceedings of the 64th Annual IEEE Symposium on Foundations of Computer Science, pp. 1082–1104 (2023)
13. Cohen-Addad, V., Lee, E., Newman, A.: Correlation clustering with Sherali-Adams. In: Proceedings of the 63rd Annual IEEE Symposium on Foundations of Computer Science, pp. 651–661 (2022)
14. Heidrich, H., Irmai, J., Andres, B.: A 4-approximation algorithm for min max correlation clustering. In: Proceedings of the 27th International Conference on Artificial Intelligence and Statistics, pp. 1945–1953 (2024)
15. Ji, S., Li, G., Zhang, D., Zhang, X.: Approximation algorithm for the capacitated correlation clustering problem with penalties. In: Proceedings of the 15th Algorithmic Aspects in Information and Management, pp. 15–26 (2021)
16. Ji, S., Xu, D., Du, D., Gai, L., Zhao, Z.: Approximation algorithm for the balanced 2-correlation clustering problem. Tsinghua Sci. Technol. **27**(5), 777–784 (2022)
17. Kalhan, S., Makarychev, K., Zhou, T.: Correlation clustering with local objectives. In: Proceedings of the 33rd Advances in Neural Information Processing Systems, p. 32 (2019)
18. Li, S.: On uniform capacitated k-median beyond the natural LP relaxation. ACM Trans. Algorithms **13**(2), 1–18 (2017)
19. Li, M., Xu, D., Zhang, D., Zhang, T.: A streaming algorithm for k-means with approximate coreset. Asia-Pac. J. Oper. Res. **36**(01), 1950006 (2019)

20. Puleo, G.J., Milenkovic, O.: Correlation clustering with constrained cluster sizes and extended weights bounds. SIAM J. Optim. **25**(3), 1857–1872 (2015)
21. Puleo, G.J., Milenkovic, O.: Correlation clustering and biclustering with locally bounded errors. IEEE Trans. Inf. Theory **64**(6), 4105–4119 (2018)
22. Swamy, C.: Correlation clustering: maximizing agreements via semidefinite programming. In: Proceedings of the 15th Annual ACM-SIAM Symposium on Discrete Algorithms, pp. 526–527 (2004)
23. Tang, Z., Diao, Z.: Approximation algorithms on k-correlation clustering. J. Oper. Res. Soc. China **11**(4), 911–924 (2023)

Improved Approximation Algorithm for Individual Fairness k-Median

Di Wu[1,2], Qilong Feng[1,2](✉), Jinhui Xu[3], and Jianxin Wang[1,2,4]

[1] School of Computer Science and Engineering, Central South University, Changsha, China
csufeng@mail.csu.edu.cn
[2] Xiangjiang Laboratory, Changsha 410205, China
[3] Department of Computer Science and Engineering, State University of New York at Buffalo, Buffalo, NY, USA
[4] The Hunan Provincial Key Lab of Bioinformatics, Central South University, Changsha, China

Abstract. The individual fairness k-median problem is a frequently encountered problem in applications involving center location, which generalizes the standard k-median problem by assigning each point a neighborhood radius, allowing connections only to centers within a constant factor of this radius. In this paper, we present a randomized polynomial-time approximation scheme (PTAS) framework with $(2+O(\epsilon))$-fairness violation for the individual fairness k-median problem, improving upon the previous best approximation ratio of $(7.081+\epsilon)$ and fairness violation of 3. We propose a new dynamic programming approach to deal with the challenges caused by the individual fairness requirements, which is the crucial step in getting the improved ratio.

Keywords: clustering · approximation algorithm · k-median

1 Introduction

Clustering is a fundamental problem in unsupervised learning and has been extensively studied over the past few decades. Among different clustering objectives, the k-median cost function is one of the most commonly studied versions. Given a dataset P of n points in a metric space, the k-median problem aims to open a set of no more than k centers, such that the sum of distance from each point to its nearest opened center is minimized.

The k-median problem is known to be NP-hard [15]. This leads to considerable efforts on the design of heuristic algorithms [8,22,28] and approximation algorithms [17,20,24]. In the metric space setting, the first constant factor approximation algorithm for the k-median problem was given by Charikar et al.

This work was supported by National Natural Science Foundation of China (62332020), Open Project of Xiangjiang Laboratory (22XJ02002), and Central South University Research Program of Advanced Interdisciplinary Studies (2023QYJC023).

D. Du et al. (Eds.): COCOA 2024, LNCS 15434, pp. 324–337, 2025.
https://doi.org/10.1007/978-981-96-4445-2_27

[7], who showed that LP-rounding yields a $6\frac{2}{3}$-approximation. The approximation guarantee was later improved by a series of work [2,5,19,20,24]. The current best approximation ratio of the k-median problem is 2.613, which was proposed by Gowda et al. [14].

To bypass the hardness of approximation results, the researchers have considered low-dimensional inputs like fixed-dimensional Euclidean spaces or more generally metrics of fixed doubling dimension [10,12,13]. The doubling metric space is a metric space with fixed doubling dimension, where the doubling dimension of the metric space is the smallest d such that any ball of radius $2r$ can be covered by 2^d balls of radius r [29]. It is known that identifying a near-optimal solution within $n^{O(1)}g(k)$ time (where g is a positive function) is unlikely for the k-median problem [11]. Cohen-Addad et al. [11] showed that if the Gap-Exponential Time Hypothesis [6,26] holds, then any algorithm for the k-median problem in general metric spaces with an approximation ratio better than $1 + e/2 + \epsilon$ must run in time at least $n^{k^{g(\epsilon)}}$. However, this negative result does not reject the possibility of obtaining a near-optimal solution within bounded time for the problem in doubling metrics [12,13]. Similar results have been achieved for other clustering problems such as capacitated k-median [9], facility location [10], k-median with outliers [10], min-sum k-median [4] using doubling metrics setting.

In many clustering applications, fairness is a significant factor in the assignment of points. One such example is the selection of hospital location [21,25]. We are given a number of k, a population of n people, and a set of locations where hospitals can be constructed. The goal is to construct k hospitals, ensuring that each individual i is served by a hospital no further than the $\lceil \frac{n}{k} \rceil$-th nearest individual from i, such that the sum of the distance from the hospital to each individual is minimized. Motivated by such applications, Jung et al. [21] proposed the notation of individual fairness clustering problem. In this problem, we are given a dataset of n points, where each point x is associated to a *fair radius* r_x, defined as the minimum radius such that the ball of this radius centered at x contains at least n/k points. The goal is to find a set of k centers such that each point has a center within its fair radius, and to minimize the sum of assignment cost. Moreover, Jung et al. [21] proved that the individual fairness clustering problem is NP-hard. They also proved that it is possible to return a feasible solution with 2-fairness violation in any metric space, which implies that there must exist a feasible solution of a set of k centers where each point x is assigned to a center with distance $2r_x$.

Mahabadi and Vakilian [25] generalized this notation to the ℓ_p-norm objective and formalized the individual fairness k-median problem, which can be defined as follows.

Definition 1 (individual fairness k-median [25]). *Let P be a set of n points in a metric space (X, dist), a non-negative integer k, a fair radius function $\gamma : X \to \mathbb{R}$, where each point $x \in X$ is associated with a fair radius $\gamma(x)$ such that the ball of radius $\gamma(x)$ centered at x has at least n/k points from P. The individual fairness k-median problem is to identify a subset $F \subseteq X$ of size at* *most k and an assignment function $\omega : X \rightarrow F$, such that each point $x \in X$ is assigned to its nearest opened center $\omega(x)$ where $\text{dist}(x, \omega(x)) \leq \gamma(x)$, and the clustering cost $\Gamma(X, F)$ is minimized, where $\Gamma(X, F) = \sum_{x \in X} \text{dist}(x, \omega(x))$ (Fig. 1).*

(a) (b)

Fig. 1. (a) An optimal solution for the instance of individual fairness k-median problem ($k = 3, n = 9, \alpha = 1$). The blue points and red points belong to different clusters; (b) An optimal solution for the instance of standard k-median problem ($k = 3, n = 9$). In both (a) and (b), the squared points indicate opened centers. (Color figure online)

For the individual fairness k-median problem, Mahabadi and Vakilian [25] proposed a bi-criteria approximation algorithm that achieves a $p^{O(p)}$-approximation with a 7-fairness violation for the ℓ_p-norm cost in metric space. Particularly, they gave a bi-criteria approximation algorithm with 84-approximation and 7-fairness violation for the individual fairness k-median problem, which was obtained by a local search approach in polynomial time. Chakrabarty and Negahbani [27] later obtained a bi-criteria approximation algorithm based on LP technique, achieving a 8-approximation and a 8-fairness violation for the individual fairness k-median problem in polynomial time. In simultaneous work, Humayun et al. [18] presented a heuristic algorithm based on semi-definite program. Vakilian and Yalçiner [30] proposed an LP-rounding algorithm that provides a $(7.081 + \epsilon)$-approximation solution with 3-fairness violation for the individual fairness k-median problem. This is the best existing approximation ratio and fairness violation for the problem. Most existing algorithms for the individual fairness k-median rely on LP techniques, which inevitably incur a constant approximation loss during the rounding process. Moreover, repeatedly applying the triangle inequality in the rounding process tends to increase the fairness violation.

1.1 Our Results and Techniques

In this paper, we consider the individual fairness k-median problem in metric space with fixed doubling dimension d. We give the first PTAS with $(2 + O(\epsilon))$-fairness violation, which improves the current best approximation ratio and fairness violation by Vakilian et al. [30]. A new dynamic programming approach is proposed to deal with the individual fairness constraints, which is the crucial step in getting the improved ratio.

Theorem 1. *Given an instance* $(X, \text{dist}, k, \gamma)$ *of the individual fairness* k*-median problem in metric space with fixed doubling dimension* d*, with probability at least* $1-\epsilon$*, there exists a randomized* $(1+O(\epsilon))$*-approximation algorithm with at most* $(2+O(\epsilon))$*-fairness violation, where* $\epsilon > 0$ *is an arbitrarily constant. The running time of this algorithm is at most* $O(n^{O(\frac{1}{\epsilon})^{O(d)}})$.

Our algorithm for the individual fairness k-median problem is based on the portal approach and dynamic programming. Portals act as bridges to calculate pairwise distances within the tree embedding. Given an instance $(X, \text{dist}, k, \gamma)$ in doubling metrics, we first construct a split-tree decomposition of the input metrics, partitioning it into blocks. Each block has a subset of points called portals, which allow for distance computations between points across different blocks. We then use dynamic programming on the split-tree to solve the problem.

Several results have applied the portal approach and dynamic programming, such as the results in [1,23,29]. In [1], a quasi-polynomial time approximation scheme (QPTAS) was introduced for the k-median problem in Euclidean space, with the running time of the dynamic programming shown to depend exponentially on the number of portals. Kolliopoulos and Rao [23] improved this result by developing a PTAS for the k-median and facility location problems in d-dimensional Euclidean space. They introduced a structural theorem using an adaptive decomposition scheme that reduced the number of portals to a constant. However, their algorithm is not directly applicable to doubling metrics, as it relies on the specific structure of Euclidean space. Most existing portal approaches and dynamic programming in doubling metrics are applied to solve the standard k-clustering and its related problems [9,10]. In designing dynamic programming to solve the individual fairness k-median problem, a key point is to ensure that each point satisfies the fairness constraint throughout the dynamic programming process. Thus, maintaining the fairness constraint between the subproblem and its children is an important component in designing the dynamic programming process for the individual fairness k-median problem.

To bypass the above obstacles in solving the individual fairness k-median problem, we present a new dynamic programming approach based on our modified portal approach to solve the individual fairness k-median problem, which is based on the relation between the subproblem and the subproblems of its children. By the previous arguments, a key point in designing a PTAS algorithm for the k-median and its related problems is to reduce the number of portals. Our portal approach is based on the badly-cut technique. The basic idea behind this technique is to optimize the solution by estimating the distance between a point and its optimal center, reducing the exponential dependence of the dynamic programming process on $O(\log n)$ factor. When the distance between a point and its optimal center is large, using dynamic programming to determine the center becomes inefficient. In the badly-cut technique, points with such large distances are denoted as badly-cut points to deal with. To bound the distances between badly-cut points and their center in the optimal cluster, the corresponding center should be opened if it is also badly-cut. However, extending this approach to the individual fairness k-median problem is challenging due to fairness constraints. In such settings, it cannot be proven that a near-optimal solution with opened badly-cut centers exists by directly embedding the badly-cut centers into the optimal solution, as the reassignment of points may lead to unbounded fairness violation. To overcome the difficulties caused by the fairness constraints, we construct a modified portal approach based on the reassignment of the points. Moreover, based on the modified portal approach, we propose a new dynamic programming approach to ensure that fairness constraints are maintained between the subproblem and its children.

2 Preliminaries

Definition 2 (doubling metric space). *Given a metric space* (X, dist) *and an arbitrary non-negative real number* r*, for any point* $x \in X$*, let* $B(x,r) = \{y \in X \mid \text{dist}(x,y) \leq r\}$ *denote the set of points in the ball around* x *with radius* r*. The doubling dimension of the metric space* (X, dist) *is the smallest integer* d *such that any ball* $B(x,2r)$ *can be covered by at most* 2^d *balls* $B(y_i, r)$*, where* y_i *is an arbitrary point in the* $B(x,2r)$*. A metric space with bounded doubling dimension is called a doubling metric space.*

For an instance $(X, \text{dist}, k, \gamma)$ of the individual fairness k-median problem, given two points $x, y \in X$, if $\omega(x) = y$, then we say that y *serves* point x. For a positive integer l, let $[l] = \{1, 2, ..., l\}$.

In general, Mahabadi and Vakilian [25] considered the bi-criteria approximation guarantee for the individual fairness k-median problem, called α-fair k-median, which can be defined as follows.

Definition 3 (α-fair k-median [25]). *Given an instance* $(X, \text{dist}, k, \gamma)$ *of the individual fairness* k*-median problem, a solution* (S, ω) *of the given instance is* α*-fair if for any* $x \in X$*,* $\text{dist}(x, S) \leq \alpha \cdot \gamma(x)$ *where* $\text{dist}(x,S)$ *denotes the distance from* x *to its nearest center in* S*.*

Given an instance $(X, \text{dist}, k, \gamma)$ of the individual fairness k-median problem, for any subset $Y \subseteq X$, the *aspect ratio* of Y is defined as $\frac{\max_{y,y' \in Y} \text{dist}(y,y')}{\min_{y \neq y' \in Y} \text{dist}(y,y')}$. Let Δ be the ratio between the maximum and the minimum pairwise distances of P, which is called the *aspect ratio* of P, where $\Delta = \frac{\max_{x,x' \in X} \text{dist}(x,x')}{\min_{x \neq x' \in X} \text{dist}(x,x')}$. We now show that Δ can be polynomially bounded, which induces an arbitrarily small loss in the guarantee.

Lemma 1. *Given an instance* $I = (X, \text{dist}, k, \gamma)$ *of the individual fairness* k*-median problem, and a real number* $\epsilon > 0$*, there exists an instance* $I' = (X, \text{dist}', k, \gamma)$ *with modified metric distance function, such that* $\Delta \leq |X|^3 k \epsilon^{-1} \leq n^4/\epsilon$*. Moreover, any* δ*-approximation solution to* I' *is a* $(1+O(\epsilon))\delta$*-approximation solution to* I *for any constant* $\delta > 1$*. A set of* $|X|^2$ *instances that contains such an instance* I' *can be constructed in polynomial time.*

To simplify the presentation, we will use dist instead of dist$'$ for the metric distance function from this point forward. Due to space limit, all proofs in this paper are provided in the complete version.

3 Decomposition of Metrics

In this section, we introduce the split-tree decomposition for doubling metrics and define the concept of badly-cut points. The concept of badly-cut is based on the split-tree decomposition.

3.1 Split-Tree Decomposition

Let (X, dist) denote a metric space with fixed doubling dimension d. A *σ-net* of X is a subset of points $Y \subseteq X$ such that for any point $x \in x$, there exists a point $y \in Y$ such that $\text{dist}(x, y) \leq \sigma$, and for any $y, y' \in Y$, $\text{dist}(y, y') \geq \sigma$. The following lemma gives the bound on the size of a net in metric space with fixed doubling dimension d.

Lemma 2 ([16]). *Let (X, dist) be a metric space with fixed doubling dimension d and aspect ratio Δ, and let $Y \subseteq X$ be a σ-net. Then, $|Y| \leq (\frac{\Delta}{\sigma})^{O(d)}$.*

A *decomposition* of the metric (X, dist) is a partition of X into subsets, which are called *blocks*. A *hierarchical decomposition* is a sequence of $\ell + 1$ decompositions $\mathcal{D} = \{\mathcal{H}_0, \mathcal{H}_1, ..., \mathcal{H}_\ell\}$ such that every block of $\mathcal{H}_{i+1}$ is the union of blocks in $\mathcal{H}_i$ with $\mathcal{H}_\ell = \{X\}$ and $\mathcal{H}_0 = \{\{x\} \mid x \in X\}$. The blocks of $\mathcal{H}_i$ are called the *level i* blocks. A *split-tree* of the metric space is a complete hierarchical decomposition: the root node is the level ℓ block $\mathcal{H}_\ell = \{X\}$ and the leaves are the singletons such that $\mathcal{H}_0 = \{\{x\} \mid x \in X\}$.

Let (X, dist) be a metric space with fixed doubling dimension d and aspect ratio Δ. Denote a parameter ℓ where $\ell = O(\log \Delta)$. We now introduce the construction of the split-tree. Firstly, we use a greedy strategy to construct a sequence of sets $Y_{\ell-1} \subseteq \ldots \subseteq Y_2 \subseteq Y_1 \subseteq Y_0 = X$ such that Y_i is a 2^{i-2}-net of Y_{i-1} [16]. Then, starting from $\mathcal{H}_\ell = \{X\}$, we recursively partition the blocks until each block only contains one point. We now discuss the relation between the blocks in the $(i+1)$-th level and i-th level, where $0 \leq i \leq \ell - 1$. Let $r_i = 2^i \varrho$, where ϱ is picked uniformly and randomly in $[\frac{1}{2}, 1)$, and let π be a random ordering of the points in X. For each point $x \in X$, let $\pi(x)$ be the order of x in π. For the set Y_i, let $\{y_1, \ldots, y_h\}$ ($h \leq 2^{O(d)}$) be the set of points in Y_i with decreasing order in π. For each block $B_s \in H_{i+1}$ and for each point $y_j \in Y_i$ ($1 \leq j \leq h$), a new block $B_s^{y_j}$ can be obtained at level i by the following way. Let $B_s^{y_j} = \{x \in B_s \mid x \in B(y_j, r_i) \text{ and } (\forall j' > j : \pi(y_{j'}) < \pi(y_j))[x \notin B(y_{j'}, r_i)]\}$. That is, for each point $x \in B_s$, if x is contained in $B_s^{y_j}$, then x is not contained in any block $\{B_s^{y_{j+1}}, \ldots, B_s^{y_h}\}$.

Lemma 3 ([29]). *Given a split-tree of metric space (X, dist) with a sequence of decompositions $\mathcal{H}_0, \mathcal{H}_1, ..., \mathcal{H}_\ell$, the split-tree has the following properties:*

1. *The total number of levels ℓ is $O(\log n)$ (since $\Delta = O(n^4/\epsilon)$).*
2. *Each block of level i has diameter at most 2^{i+1}, namely the maximum distance between any pair of points in a block of level i is at most 2^{i+1}.*
3. *Each level i block is the union of at most $2^{O(d)}$ level $i-1$ blocks.*

4. *For any pair of points $u, v \in X$, the probability that they are in different sets corresponding to the blocks at level i of split-tree is at most $O(d) \cdot \frac{\text{dist}(u,v)}{2^i}$.*

Moreover, we generalize the last property to the following lemma.

Lemma 4. *Let (X, dist) be a metric space with fixed doubling dimension d. Given a randomized split-tree T of (X, dist), for any $x \in X$ and radius parameter $r > 0$, the probability that the points of $B(x, r)$ are in different sets corresponding to the blocks at level i of split-tree is at most $4d \cdot r/2^i$.*

Given an instance $(X, \text{dist}, k, \gamma)$ of the individual fairness k-median problem, let T denote the split-tree obtained by using the method in [29]. For each block B of level i in T, we compute a $\rho 2^{i+1}$-net $\mathcal{P}$ of block B. Each point in $\mathcal{P}$ is called a portal, and $\mathcal{P}$ is also called a portal set. According to Lemma 2, the number of portals at a given block is $\rho^{-O(d)}$. Moreover, the split-tree with the portal sets can be found in time $(1/\rho)^{O(d)} n \log \Delta$ [3,10]. For the portal set $\mathcal{P}$ of a block B of level i in T, there is an important property as follows.

Lemma 5 (**[29]**)**.** *For the portal set $\mathcal{P}$ of a block B of level i in T, assume that the children of B at level $i-1$ are $B_1, B_2, \ldots, B_u$ and each block B_j has a portal set $\mathcal{P}_j \subseteq B_j$. Then, $\mathcal{P}$ is a subset of the portal sets computed for the descendant blocks of B, i.e. $\mathcal{P} \subseteq \mathcal{P}_1 \cup \mathcal{P}_2 \cup \ldots\ \mathcal{P}_u$.*

Given a split-tree T, consider two blocks B_1 and B_2 at level i of T. For points $u \in B_1$ and $v \in B_2$, the distance between u and v on the split-tree, denoted as $\text{dist}_T(u, v)$, is defined as the minimum length of the path composed of the subpath from u to a portal p_i in B_1, the subpath from p_i to a portal p_j in B_2, and the subpath from p_j to v. For a block B at level i of T with a portal set $\mathcal{P} \subseteq B$, if a point x outside of B is assigned to a center c inside B through a portal $p_1 \in \mathcal{P}$, we say that x enters B through p_1. Similarly, if a point x' inside B is assigned to a center c' outside B through a portal $p_2 \in \mathcal{P}$, we say that x' leaves B through p_2.

Lemma 6. *For any metric (X, dist) with fixed doubling dimension d and any $\rho > 0$, given a randomized split-tree T, a pair of blocks B_1 and B_2 of level i, the portal set $\mathcal{P}_1$ of B_1 where $\mathcal{P}_1$ is a $\rho 2^{i+1}$-net of block B_1, the portal set $\mathcal{P}_2$ of B_2 where $\mathcal{P}_2$ is a $\rho 2^{i+1}$-net of block B_2, and any two points $u \in B_1$ and $v \in B_2$, for the distance of u and v on the split-tree T and the distance $\text{dist}(u, v)$ in the metric space, we have: $\text{dist}_T(u, v) = \min_{p_i \in \mathcal{P}_1, p_j \in \mathcal{P}_2} \{\text{dist}(u, p_i) + \text{dist}(p_i, p_j) + \text{dist}(p_j, v)\} \leq \text{dist}(u, v) + 4 \cdot \rho 2^{i+1}$.*

In the following, all the distances considered are the distances on the split-tree T.

3.2 Definition of Badly-Cut Points

For any pair of points $u, v \in X$, if they are in different sets corresponding to the blocks at level i of T, we denote that u, v are cut at level i. For any point

$x \in X$ and radius parameter $r > 0$, if the points of $B(x, r)$ are in different sets corresponding to the blocks at level i of T, we denote that $B(x, r)$ is cut at level i. In the following, we will introduce the concept of "badly-cut" points.

We start by finding an $O(1)$-approximate feasible solution S of the individual fairness k-median problem, where a bi-criteria solution is also acceptable. For each point $x \in X$, let $S(x)$ denote the center in S that x is assigned to, and $\text{dist}^S(x)$ denote the distance between x and $S(x)$. Let opt denote an optimal solution of the individual fairness k-median. For each center c of S, let $\text{opt}(c)$ denote the nearest center to c in opt, and $\text{dist}^o(c)$ denote the distance between c and $\text{opt}(c)$. We give the formal definition of the badly-cut points as follows.

Definition 4 [10]. *Let (X, dist) be a metric space with fixed doubling dimension d. Given a randomized split-tree T and a constant $\epsilon > 0$, a point $x \in X$ is called badly-cut if there exists an integer i such that $B(x, 2^i)$ is cut at level j greater than $i + \log d + 3 + \log\log(1/\epsilon) + \log(1/(\frac{\epsilon^2}{2}))$ (where $2^i \in [\epsilon \text{dist}^S(x), \text{dist}^S(x)/\epsilon]$).*

A center c of S is called badly-cut if there exists an integer i such that $B(c, 2^i)$ is cut at level j greater than $i + \log d + 3 + \log\log(1/\epsilon) + \log(1/(\frac{\epsilon^2}{2}))$ (where $2^i \in [\epsilon \text{dist}^o(c), \text{dist}^o(c)/\epsilon]$).

Let $\lambda(\epsilon, d) = \log d + 3 + \log\log(1/\epsilon) + \log(1/(\frac{\epsilon^2}{2}))$. In the following lemma, we establish the probability bound for each point being badly-cut.

Lemma 7. *Let (X, dist) be a metric space with fixed doubling dimension d. Given a randomized split-tree T, for any point $x \in X$, the probability that x is badly-cut is at most $\frac{\epsilon^2}{2}$.*

4 Algorithm

In this section, we show how to use the portal approach and the dynamic programming process, leveraging the badly-cut technique, to achieve a PTAS for the individual fairness k-median problem. To reduce the time complexity, we first modify the instance by setting the assignment cost of badly-cut points to zero. For any given instance, a necessary condition for applying dynamic programming is the existence of an optimal solution within the instance. Thus, we must prove that a near-optimal solution still exists for the individual fairness k-median problem in the modified instance. We present a modified portal approach that builds on the badly-cut technique outlined in [10]. This approach aims to construct a near-optimal solution with the opened badly-cut centers based on the reassignment of the points.

4.1 Instance Modification

We introduce a preprocessing step of the badly-cut technique. Given an instance $(X, \text{dist}, k, \gamma)$ of the individual fairness k-median problem, we show how to modify the instance by relocating the badly-cut points. We start with identifying badly-cut points of the given instance through a polynomial-time enumeration process based on the split-tree T. In the following, we show that the instance modification induces an arbitrarily small loss on the approximate guarantee with probability at least $1-\epsilon$. Let $\text{cost}(S)$ represent the cost of feasible solution S, and let $\text{cost}(I_T(\mathcal{L}))$ and $\text{cost}(I'(\mathcal{L}))$ denote the cost of any solution $\mathcal{L}$ to instance I_T and I', respectively.

Lemma 8. *Given an instance $I' = (X, \text{dist}, k, \gamma)$ of the individual fairness k-median problem and a real number $\epsilon > 0$, let T be a randomized split-tree and S a feasible solution for I'. We can construct a new instance $I_T = (X, \text{dist}^*, k, \gamma)$ by relocating each badly-cut point $x \in X$ to the position of $S(x)$. With probability at least $1-\epsilon$, the difference in cost between any solution $\mathcal{L}$ for I_T and the corresponding solution for I' is bounded by $O(\epsilon \cdot \text{cost}(S))$. Specifically,*

$$\max_{\mathcal{L}}\{\text{cost}(I'(\mathcal{L})) - \text{cost}(I_T(\mathcal{L})), \text{cost}(I_T(\mathcal{L})) - \text{cost}(I'(\mathcal{L}))\} \le O(\epsilon\text{cost}(S)).$$

4.2 Near-Optimal Solution Construction

In this section, we prove that, based on the modified instance, there exists a near-optimal solution to the individual fairness k-median problem with the opened badly-cut centers in S. We denote $\mu(S)$ as the set of the badly-cut centers in S. Given a modified instance $I_T = (X, \text{dist}^*, k, \gamma)$ derived from I', we must ensure that the centers in $\mu(S)$ remain open in the constructed solution to bound the approximation loss. Moreover, based on the probability of a point being badly-cut, it follows directly that at most ϵk badly-cut centers in S will be opened. To address the situation where the obtained solution includes more than k centers, we propose a modified portal approach. For each badly-cut center c in $\mu(S)$, we open c and close the corresponding center c' from opt, reassigning the points assigned to c' to c within the fairness constraint. This approach guarantees that the resulting solution has a cost of at most $(1 + O(\epsilon))\text{cost}(\text{opt}) + O(\epsilon)\text{cost}(S)$ and a $(2+O(\epsilon))$-fairness violation. The following lemma shows that there exists a near-optimal solution S' containing the opened badly-cut centers of S in I_T.

Lemma 9. *Given a modified instance $I_T = (X, \text{dist}^*, k, \gamma)$ of the individual fairness k-median problem and a real number $\epsilon > 0$, with probability at least $1-\epsilon$, there exists a near-optimal solution S' that includes the opened badly-cut centers from S. The total cost of S' is bounded by $(1 + O(\epsilon))\text{cost}(\text{opt}) + O(\epsilon)\text{cost}(S)$, and S' has at most $(2 + O(\epsilon))$-fairness violation.*

4.3 Dynamic Programming over Split-Tree

In this section, we introduce a dynamic programming approach using the split-tree structure to solve the individual fairness k-median problem. Given a split-tree T, the dynamic programming process proceeds on T from the leaves to the root. For any block B of split-tree T, we define the subtree rooted at block B as T_B, and each subtree is a subproblem in our dynamic programming process.

A table entry in the dynamic programming process of the individual fairness k-median problem is a tuple dp$[B, k_B, \boldsymbol{\psi}, \boldsymbol{\delta}]$, where the parameters are defined as follows: (1) B is the root node of the subtree T_B; (2) k_B $(0 \le k_B \le k)$ is the number of centers opened in B; (3) $\boldsymbol{\psi}$ is a tuple with m elements where m is the size of portal set in B and $m = \rho^{-O(d)}$. For each $i \in [m]$, the element ψ_i in $\boldsymbol{\psi}$ represents the distance between the i-th portal and the nearest open center inside B. ψ_i is power of $(1+\epsilon^2/\log n)$ in the range $[1, \Delta]$; (4) $\boldsymbol{\delta}$ is a tuple with m elements. For each $i \in [m]$, the element δ_i in $\boldsymbol{\delta}$ represents the distance between the i-th portal and the nearest open center outside B. δ_i is power of $(1+\epsilon^2/\log n)$ in the range $[1, \Delta]$.

The base case of the dynamic programming process is at the leaves of the split-tree, where each leaf is a singleton. We need to determine whether this point is opened as a center. For the table entry corresponding to a leaf node B, we consider the following two cases: (1) If the point in B is opened as a center, then the table value of this block is dp$[B, 1, \boldsymbol{\phi}, \boldsymbol{\phi}] = 0$, where $\boldsymbol{\phi}$ is a tuple with m zero elements; (2) If the point in B is not open, then the table value of this block is dp$[B, 0, \boldsymbol{\phi}, \boldsymbol{\delta}] = \delta_1$. Moreover, the first value in $\boldsymbol{\delta}$ is δ_1 and the other values in $\boldsymbol{\delta}$ are 0.

Then, we focus on the subproblem on the subtree T_B, where block B is at level i of split-tree T. Let $B_1, B_2, ..., B_u$ represent the children of block B, where u is at most $2^{O(d)}$. For each child B_j $(j \in [u])$, we define its configuration as $(B_j, k_{B_j}, \boldsymbol{\psi}_j, \boldsymbol{\delta}_j)$, where B_j $(j \in [u])$ denotes the j-th child of B, k_{B_j} is the number of opened centers in B_j, $\boldsymbol{\psi}_j$ is a tuple with m elements, and the i-th element ψ_j^i (ψ_j^i is power of $(1+\epsilon^2/\log n)$ in the range $[1, \Delta]$, $i \in [m]$) in $\boldsymbol{\psi}_j$ denotes the distance between the i-th portal and the nearest open center inside B_j, $\boldsymbol{\delta}_j$ is a tuple with m elements, and the i-th element δ_j^i (δ_j^i is power of $(1+\epsilon^2/\log n)$ in the range $[1, \Delta]$, $i \in [m]$) in $\boldsymbol{\delta}_j$ denotes the distance between the i-th portal and the nearest open center outside B_j.

Since a consistency relationship exists between the configurations of B and its children, the number of centers opened in B can be bounded by the sum of the number of the centers opened in its children. For each ψ_t $(t \in [m])$ in $\boldsymbol{\psi}$, by definition, ψ_t represents the distance between the t-th portal p_t and the nearest open center c within B. This distance must include a portal $p_{t'}$ from a certain child B_j, and ψ_t consists of two parts: (1) the distance between $p_{t'}$ and c; (2) the distance between p_t and $p_{t'}$. For each δ_t $(t \in [m])$ in $\boldsymbol{\delta}$, by definition, δ_t denotes the distance between the t-th portal p_t and the nearest open center c outside B. There must exist a portal $p_{t'}$ in one of the children B_j such that δ_t equals the distance between $p_{t'}$ and c. By Lemma 5, this portal $p_{t'}$ is exactly the same as p_t. The values of k_{B_j}, ψ_j^i and δ_j^i have the following constraints: (1) $k_{B_1} + k_{B_2} + \ldots + k_{B_u} \le k_B$; (2) For each ψ_t in $\boldsymbol{\psi}$, there must be a portal $p_{t'}$ in a certain child B_j such that $\psi_j^{t'} + \text{dist}(p_t, p_{t'}) = (1 \pm (\epsilon^2/\log n))\psi_t$, where $t \in [m], t' \in [m], j \in [u]$; (3) For each δ_t in $\boldsymbol{\delta}$, there must be a portal $p_{t'}$ in a certain child B_j such that $\delta_j^{t'} = \delta_t$, where $t \in [m], t' \in [m], j \in [u]$.

Given a subproblem B at level i of the split-tree T $(i \ne 0)$ and its children $B_1, B_2, \ldots, B_u$, where u is at most $2^{O(d)}$, let $(B, k_B, \boldsymbol{\psi}, \boldsymbol{\delta})$ be the configuration of B and $(B_j, k_{B_j}, \boldsymbol{\psi}_j, \boldsymbol{\delta}_j)$ be the configuration of each children B_j $(1 \leq j \leq u)$, we have

$$\mathrm{dp}[B, k_B, \boldsymbol{\psi}, \boldsymbol{\delta}] = \min_{j\in[u]:k_{B_j},\boldsymbol{\psi}_j,\boldsymbol{\delta}_j} \sum_{j=1}^{u} \mathrm{dp}[B_j, k_{B_j}, \boldsymbol{\psi}_j, \boldsymbol{\delta}_j]$$

$$s.t. \begin{cases} k_{B_1} + k_{B_2} + \ldots + k_{B_u} \leq k_B \\ \forall \psi_t \ in\ \boldsymbol{\psi}, \exists p_{t'} \in B_j, \psi_j^{t'} + \mathrm{dist}(p_t, p_{t'}) = (1 \pm (\epsilon^2/\log n))\psi_t \\ \forall \delta_t \ in\ \boldsymbol{\delta}, \exists p_{t'} \in B_j, \delta_j^{t'} = \delta_t \end{cases}$$

4.4 Analysis

In this section, we analyze the approximation ratio and running time of our algorithm for solving the individual fairness k-median problem in doubling metrics.

We start by analyzing the cost of the solution produced by the dynamic programming process. Let $\mathcal{J}$ denote the solution produced by the dynamic programming process. For each point $x \in X$, let $\mathcal{J}(x)$ represent the center to which x is assigned. We consider the following two cases for a point $x \in X$.

Case 1: the point x is not badly-cut. In this case, the dynamic programming process directly assigns x to its center $\mathcal{J}(x)$. For a point x that is not badly-cut, consider the pair of points x and $\mathrm{opt}^*(x)$, where $\mathrm{opt}^*(x)$ denotes the center serving x in the optimal solution opt^* for the instance $I_T = (X, \mathrm{dist}^*, k, \gamma)$. According to Definition 4, x and $\mathrm{opt}^*(x)$ are separated at level i less than $\zeta = \log \mathrm{dist}(x, \mathrm{opt}^*(x)) + \lambda(\epsilon, d)$. By Lemma 6, if two points u and v are separated at level i, then for each block of the split-tree, the corresponding portal set incurs an additional cost of $4\rho \cdot 2^{i+1}$. Then, the total additional cost for the path between x and $\mathrm{opt}^*(x)$ is at most $\sum_{i=0}^{\zeta} 4\rho \cdot 2^{i+1} \leq 4\rho \cdot 2^{\zeta+2} = 16\rho 2^{\lambda(\epsilon,d)} \cdot \mathrm{dist}(x, \mathrm{opt}^*(x))$. Set $\rho = \frac{\epsilon}{16 \cdot 2^{\lambda(\epsilon,d)}}$. Then, the total additional expected cost of the path between x and $\mathrm{opt}^*(x)$ is at most $O(\epsilon)\mathrm{dist}(x, \mathrm{opt}^*(x))$.

Case 2: the point x is badly-cut. By Lemma 8, for a badly-cut point x, we move x to the position of $S(x)$ where S is a feasible solution of the individual fairness k-median problem. This case splits into two subcases: 1) $S(x)$ is badly-cut. Since our dynamic programming process requires the badly-cut centers of S to be opened, the assignment cost between x and $\mathcal{J}(x)$ is $\mathrm{dist}(x, S(x)) = 0$. 2) $S(x)$ is not badly-cut. For a badly-cut point x at the location of $S(x)$, we consider a pair of points x and $\mathrm{opt}^*(x)$. Since $S(x)$ is not badly-cut, we deal with this case by the same way as **Case 1**. Thus, the total additional cost of the path between x and $\mathrm{opt}^*(x)$ is at most $O(\epsilon)\mathrm{dist}(x, \mathrm{opt}^*(x))$.

Let $\mathrm{cost}'(\mathrm{opt}^*)$ denote the total cost of the solution opt^* based on the dynamic programming process on the split-tree T. By the previous arguments, we have $\mathrm{cost}'(\mathrm{opt}^*) \leq (1 + O(\epsilon))\mathrm{cost}(\mathrm{opt}^*)$. Since our dynamic programming process aims to find the best solution including the opened badly-cut centers of S, $\mathrm{cost}(\mathcal{J}) \leq \mathrm{cost}'(\mathrm{opt}^*) \leq (1 + O(\epsilon))\mathrm{cost}(\mathrm{opt}^*)$. By Lemma 9, with probability at least $1 - \epsilon$, there exists a solution S' containing the badly-cut centers of S with cost at most $(1 + O(\epsilon))\mathrm{cost}(\mathrm{opt}) + O(\epsilon \mathrm{cost}(S))$ in the modified instance $I_T =$

$(X, \text{dist}^*, k, \gamma)$, which implies that $\text{cost}(\text{opt}^*) \leq (1+O(\epsilon))\text{cost}(\text{opt})+O(\epsilon\text{cost}(S))$. Then, the total cost of the solution $\mathcal{J}$ is at most $(1+O(\epsilon))\text{cost}(\text{opt})+O(\epsilon\text{cost}(S))$ for instance I_T. By Lemma 8, with probability at least $1-\epsilon$, the modification of the instance $I' = (X, \text{dist}', k, \gamma)$ to I_T incurs an additional expected cost of $O(\epsilon\cdot\text{cost}(S))$. Thus, the total cost of the solution $\mathcal{J}$ for instance I' is no more than $(1+O(\epsilon))\text{cost}(\text{opt}) + O(\epsilon\cdot\text{cost}(S))$. By Lemma 1, any δ-approximation solution to I' is a $(1+\epsilon)\delta$-approximation to I. Therefore, the total cost of the solution $\mathcal{J}$ for instance I is at most $(1+O(\epsilon))\text{cost}(\text{opt})+O(\epsilon\cdot\text{cost}(S))$. Moreover, through a similar analysis and by applying Lemma 9, the fairness violation of the solution $\mathcal{J}$ for instance I is at most $(2+O(\epsilon))$. Thus, with probability at least $1-\epsilon$, the solution $\mathcal{J}$ for instance I has a cost of at most $(1+O(\epsilon))\text{cost}(\text{opt})+O(\epsilon\cdot\text{cost}(S))$ and a fairness violation of at most $(2+O(\epsilon))$.

Then, we calculate the running time of our algorithm. The running time of our algorithm includes the following components: preprocessing the instance I, constructing the split-tree T, computing the feasible solution S, constructing the instance I_T, and running the dynamic programming process. By the previous arguments, the first four components can be completed in polynomial time.

The running time of the dynamic programming process is related to the number of table entries in the dynamic programming table, and the time to calculate the value of each table entry. We first analyze the number of table entries in the dynamic programming table. For a table entry $\text{dp}[B, k_B, \boldsymbol{\psi}, \boldsymbol{\delta}]$ in the dynamic programming table, by the definitions of the table entry in Sect. 4.3, the total number of entries in our dynamic programming table is at most $O(n^2(\log_{1+\epsilon}\Delta)^{2m}\log n)$. Then, we analyze the time to calculate the value of each table entry by considering all possibilities based on the values of parameters $k_{B_j}, \boldsymbol{\psi}_j, \boldsymbol{\delta}_j$ ($j \in [u]$). The time to calculate the value of each table entry in the dynamic programming process is at most $O(n^u(\log_{1+\epsilon}\Delta)^{2mu})$. We can analyze that the number of portals in a portal set is $m = O(\frac{1}{\epsilon})^{O(d)}$. Thus, the running time of the dynamic programming process can be bounded by $O(n^{O(\frac{1}{\epsilon})^{O(d)}})$. Putting all these together, we can get the following lemma.

Lemma 10. *Given an instance $(X, \text{dist}, k, \gamma)$ of the individual fairness k-median problem in doubling metrics, with probability at least $1-\epsilon$, there exists an algorithm that can find a solution with at most $(2+O(\epsilon))$-fairness violation. This solution has a total cost of at most $(1+O(\epsilon))\text{cost}(\text{opt}) + O(\epsilon\text{cost}(S))$. The running time of this algorithm is at most $O(n^{O(\frac{1}{\epsilon})^{O(d)}})$.*

5 Conclusion

In this paper, we address the individual fairness k-median problem in doubling metrics. We propose a $(1+O(\epsilon))$-approximation algorithm with a $(2+O(\epsilon))$-fairness violation that runs in polynomial time, where $\epsilon > 0$ is an arbitrarily constant. This is the first PTAS for the individual fairness k-median problem in doubling metrics with minimum fairness violation. The main technical contribution is a new dynamic programming approach integrated with a modified portal approach, which is the crucial step in obtaining the better approximation guarantee.

References

1. Arora, S., Raghavan, P., Rao, S.: Approximation schemes for Euclidean k-medians and related problems. In: Proceedings of the 30th Annual ACM Symposium on the Theory of Computing, pp. 106–113 (1998)
2. Arya, V., Garg, N., Khandekar, R., Meyerson, A., Munagala, K., Pandit, V.: Local search heuristics for k-median and facility location problems. SIAM J. Comput. **33**(3), 544–562 (2004)
3. Bartal, Y., Gottlieb, L.A.: A linear time approximation scheme for Euclidean TSP. In: Proceedings of the 54th Annual Symposium on Foundations of Computer Science, pp. 698–706 (2013)
4. Behsaz, B., Friggstad, Z., Salavatipour, M.R., Sivakumar, R.: Approximation algorithms for min-sum k-clustering and balanced k-median. Algorithmica **81**(3), 1006–1030 (2019)
5. Byrka, J., Pensyl, T.W., Rybicki, B., Srinivasan, A., Trinh, K.: An improved approximation for k-median and positive correlation in budgeted optimization. ACM Trans. Algorithms **13**(2), 23:1–23:31 (2017)
6. Chalermsook, P., et al.: From Gap-ETH to FPT-inapproximability: clique, dominating set, and more. In: Proceedings of the 58th IEEE Annual Symposium on Foundations of Computer Science, pp. 743–754 (2017)
7. Charikar, M., Guha, S., Tardos, É., Shmoys, D.B.: A constant-factor approximation algorithm for the k-median problem. J. Comput. Syst. Sci. **65**(1), 129–149 (2002)
8. Church, R.L., Wang, S.: Solving the p-median problem on regular and lattice networks. Comput. Oper. Res. **123**, 105057 (2020)
9. Cohen-Addad, V.: Approximation schemes for capacitated clustering in doubling metrics. In: Proceedings of the 31st ACM-SIAM Symposium on Discrete Algorithms, pp. 2241–2259 (2020)
10. Cohen-Addad, V., Feldmann, A.E., Saulpic, D.: Near-linear time approximation schemes for clustering in doubling metrics. J. ACM **68**(6), 44:1–44:34 (2021)
11. Cohen-Addad, V., Gupta, A., Kumar, A., Lee, E., Li, J.: Tight FPT approximations for k-median and k-means. In: Proceedings of the 46th International Colloquium on Automata, Languages, and Programming, pp. 42:1–42:14 (2019)
12. Cohen-Addad, V., Klein, P.N., Mathieu, C.: Local search yields approximation schemes for k-means and k-median in Euclidean and minor-free metrics. SIAM J. Comput. **48**(2), 644–667 (2019)
13. Friggstad, Z., Rezapour, M., Salavatipour, M.R.: Local search yields a PTAS for k-means in doubling metrics. SIAM J. Comput. **48**(2), 452–480 (2019)
14. Gowda, K.N., Pensyl, T.W., Srinivasan, A., Trinh, K.: Improved bi-point rounding algorithms and a golden barrier for k-median. In: Proceedings of the 34th ACM-SIAM Symposium on Discrete Algorithms, pp. 987–1011 (2023)
15. Guha, S., Khuller, S.: Greedy strikes back: improved facility location algorithms. J. Algorithms **31**(1), 228–248 (1999)
16. Gupta, A., Krauthgamer, R., Lee, J.R.: Bounded geometries, fractals, and low-distortion embeddings. In: Proceedings of the 44th Annual IEEE Symposium on Foundations of Computer Science, pp. 534–543 (2003)

17. Gupta, A., Tangwongsan, K.: Simpler analyses of local search algorithms for facility location. CoRR abs/0809.2554 (2008)
18. Humayun, A.I., Balestriero, R., Kyrillidis, A., Baraniuk, R.G.: No more than 6ft apart: robust k-means via radius upper bounds. In: Proceedings of the 48th IEEE International Conference on Acoustics, Speech and Signal Processing, pp. 4433–4437 (2022)
19. Jain, K., Mahdian, M., Markakis, E., Saberi, A., Vazirani, V.V.: Greedy facility location algorithms analyzed using dual fitting with factor-revealing LP. J. ACM **50**(6), 795–824 (2003)
20. Jain, K., Vazirani, V.V.: Approximation algorithms for metric facility location and k-median problems using the primal-dual schema and lagrangian relaxation. J. ACM **48**(2), 274–296 (2001)
21. Jung, C., Kannan, S., Lutz, N.: Service in your neighborhood: fairness in center location. In: Proceedings of the 1st Symposium on Foundations of Responsible Computing, pp. 5:1–5:15 (2020)
22. Kazakovtsev, L.A., Stupina, A.A.: Fast genetic algorithm with greedy heuristic for p-median and k-means problems. In: Proceedings of the 6th International Congress on Ultra Modern Telecommunications and Control Systems and Workshops, pp. 602–606 (2014)
23. Kolliopoulos, S.G., Rao, S.: A nearly linear-time approximation scheme for the Euclidean k-median problem. SIAM J. Comput. **37**(3), 757–782 (2007)
24. Li, S., Svensson, O.: Approximating k-median via pseudo-approximation. SIAM J. Comput. **45**(2), 530–547 (2016)
25. Mahabadi, S., Vakilian, A.: Individual fairness for k-clustering. In: Proceedings of the 37th International Conference on Machine Learning, pp. 6586–6596 (2020)
26. Manurangsi, P., Raghavendra, P.: A birthday repetition theorem and complexity of approximating dense CSPs. In: Proceedings of the 44th International Colloquium on Automata, Languages, and Programming, pp. 78:1–78:15 (2017)
27. Negahbani, M., Chakrabarty, D.: Better algorithms for individually fair k-clustering. In: Proceedings of the 34th Advances in Neural Information Processing Systems, pp. 13340–13351 (2021)
28. Panteli, A., Boutsinas, B., Giannikos, I.: On solving the multiple p-median problem based on biclustering. Oper. Res. **21**(1), 775–799 (2021)
29. Talwar, K.: Bypassing the embedding: algorithms for low dimensional metrics. In: Proceedings of the 36th Annual ACM Symposium on Theory of Computing, pp. 281–290 (2004)
30. Vakilian, A., Yalçiner, M.: Improved approximation algorithms for individually fair clustering. In: Proceedings of the 25th International Conference on Artificial Intelligence and Statistics, pp. 8758–8779 (2022)

The Price of Fairness for Budget-Feasible EF1 Allocations

Tingwei Hu[1], Lili Mei[2(✉)], Zhen Wang[2,3], and Guochuan Zhang[1]

[1] College of Computer Science, Zhejiang University, Hangzhou 310027, China
{tingweihu,zgc}@zju.edu.cn

[2] School of Cyberspace, Hangzhou Dianzi University, Hangzhou 310018, China
meilili@zju.edu.cn, wangzhen@hdu.edu.cn

[3] Experimental Center of Data Science and Intelligent Decision-Making, Hangzhou Dianzi University, Hangzhou 310018, China

Abstract. Fair allocation of indivisible items among agents with budget constraints has received considerable attention recently. In this model, each agent has a budget, and each item has a specific size and homogeneous or heterogeneous valuations for all the agents. In a fair allocation, each agent receives a bundle of items with total size no more than her budget, meanwhile the EF1 (*envy-freeness up to one item*) fairness holds between any agents. To avoid a trivial and meaningless solution in which all bundles are empty, we invoke the well-studied concept PoF (*price of fairness*), the worst-case ratio between the maximum total valuation without and with the fairness guarantee to capture the efficiency loss induced by fairness.

In this paper, the PoFs are extensively studied in various settings. We first investigate an algorithm outputting an EF1 allocation with the total valuation at least $\frac{1}{2n}$ of the maximum total valuation (without EF1 requirement), where n is the number of agents. It implies that PoF is at most $2n$. Meanwhile, we give an instance to show that PoF is at least $n - \epsilon$. Then a special case that the valuation of an item is identical for all the agents is considered. We establish an algorithm which outputs an EF1 allocation with the total valuation at least $\frac{1}{2}$ of the maximum total valuation. That is to say, an algorithm with PoF of 2 is devised. And we give a lower bound of $2 - \frac{1}{n}$. Finally, the scenario of two agents is discussed. An algorithm with PoF of $\frac{3}{2}$ is constructed when all the agents have identical valuation for an item. For different valuations, we improve the upper bound of PoF to 3.

Keywords: PoF · Budget-Feasible Allocations · EF1

1 Introduction

More recently, fair allocation of indivisible items among agents with budget constraints has received considerable attention [5,6,24] due to many real-world applications. For

This work is partially supported by the National Natural Science Foundation of China (NO. 12201594, 62176080 and 12271477).

D. Du et al. (Eds.): COCOA 2024, LNCS 15434, pp. 338–351, 2025.
https://doi.org/10.1007/978-981-96-4445-2_28

example, when purchasing antiques or artworks, the buyers have limitations on the money they possess. Also, when we assign tasks, the total workload of a person cannot exceed her available time. Motivated by these applications, this paper studies fair resource allocation with budget constraints. In the model, we aim at allocating m indivisible items to n agents in a fair way. Formally, each item j has a size s_j and each agent i has a budget B_i and a valuation v_{ij} to item j. An allocation is budget-feasible if for any agent i, the total size of the items assigned to her does not exceed B_i. The valuation of an agent for her budget-feasible bundle is additive.

In this paper, we focus on the "fairness", which has a long-standing history and can be dated back to the 1940 s [22]. The age-old problem, known as the cake cutting problem, was initialized for the case of divisible items. The researchers have proposed some classical fairness criteria, especially *envy-freeness*, where each agent prefers her own bundle over any other agent's bundle. While in our model due to the indivisible items, even the budget is unbounded, it is easy to see that an envy-free allocation does not always exists, by considering a trivial instance of a single item and two identical agents. The fairness in the context of indivisible items was formally introduced in the seminal work by Budish [10]. He proposed two fairness notions, one of which is the relaxation of envy-freeness, named *envy-freeness up to one item (EF1)*. In an EF1 allocation, each agent prefers her own bundle over that of other agent's, up to the removal of one item from that bundle.

When the budgets are heterogeneous, the bundle of the agent with the larger budget maybe not a budget-feasible solution for the one with the small budget any more. Hence, the traditional EF1 definition is no longer appropriate for our model. The new EF1 concept does not consider the whole bundle of other agent's, but rather any budget-feasible solution for herself, *i.e.*, any agent envies any feasible bundle of the other agent's up to one item. Moreover, due to the budget constraints as well, sometimes we cannot allocate all the items completely. In the context, we use *the charity* to denote the set of all the unallocated items. At this moment, if we still only require the EF1 fairness between the agents as before, then it may emerge a trivial and meaningless solution where all items are in the charity. This paper considers the social welfare, which is the total valuation obtained by all the agents. We adopt the concept of *price of fairness (PoF)*, addressed independently by Bertsimas *et al.* [8] and Caragiannis *et al.* [11], to measure the social welfare loss under the fairness constraints. The PoF is defined as the worst-case ratio between the maximum social welfare and the maximum social welfare with the fairness. In this paper, we establish lower bounds of the PoF and aim to design algorithms to achieve the bounds as much as possible for our model.

1.1 Our Contributions

In this paper, we study PoFs in various setting, which is demonstrated in Table 1. In the table, an upper bound (UB) α means that we can design an algorithm outputting an EF1 allocation with the social welfare at least $\frac{1}{\alpha}$ of the maximum social welfare. The lower bound (LB) β means that there exists an instance in which the ratio between the maximum social welfare and that with EF1 constraint is β.

The upper bounds for n agents also hold for *envy-freeness up to any item* (EFX) fairness criteria, which is stronger than EF1 and will be introduced in related work.

The allocation algorithms for two agents can be revised to polynomial time with PoFs multiplied by the factor ρ, where ρ is the approximation ratio to the corresponding optimization problems.

Table 1. Our results. Heterogeneous valuations denote the setting that each item has heterogeneous valuations for all the agents. Homogeneous valuation denotes the setting each item has identical valuation for all the agents.

Settings	Heterogeneous valuations	Homogeneous valuation
n agents	LB: $n-\epsilon$ (Theorem 2)	LB: $2-\frac{1}{n}-\epsilon$ (Theorem 3)
	UB: $2n$ (Theorem 1)	UB:2 (Theorem 4)
Two agents	UB:3 (Theorem 6)	UB:$\frac{3}{2}$ (Theorem 5)

1.2 Related Work

The research on indivisible items started from the work by Budish [10]. They proposed two fairness concepts: EF1 and maximin share (MMS). In this paper, we focus on the envy-based fairness. To see more on MMS, please refer to the notable progress in [1]. Budish meanwhile showed the existence of EF1 allocations. In 2004, even before the notion of EF1 is proposed, a polynomial time algorithm designed by Lipton *et al.* [19] just outputs an EF1 allocation. After showing the existence, researchers combined the fairness with the social welfare, *i.e.*, maximizing the sum of all agent's valuation under the fairness constraints. They designed approximation algorithms, *e.g.* [2,4,9].

On the other hand, some researchers consider the other objective of Nash social welfare, which is the multiplication of all agents' valuations. Caragiannis *et al.* [12] showed any optimal Nash social welfare solution also satisfies EF1. Meanwhile, they proposed another fairness criterion coined as EFX, saying that each agent prefers her own bundle over any other's by the removal of any item. The existence of EFX allocations is proved up to three agents [13,20]. Chaudhury *et al.* [14] gave an EFX allocation with $n-1$ unallocated items in pseudo-polynomial time.

The model which is the most related to this paper is the budget-feasible allocation problem. As we mentioned previously, this model was first studied by Wu *et al.* [24], where an item has the same size and the identical valuation for each agent. They showed that the optimal Nash social welfare solution is $\frac{1}{4}$−EF1. Last year, Barman *et al.* [5] designed a greedy algorithm outputting an EF2 allocation for the same model, where EF2 means the envy can be eliminated after virtually removing two items. Later, Gan *et al.* [17] established a $\frac{1}{2}$−EF1 polynomial-time algorithm. The existence of the EFX (EF1) allocations is proved in [6], while the algorithm is not polynomial time. They also showed when the items are divisible, an envy-free allocation can be output by a polynomial time algorithm.

In recent years, the price of fairness (PoF) that measures the loss in social welfare due to the fairness constraints has also received an increasing attention. The notion is first introduced independently in [8] and [11] for divisible items. Bei *et al.* firstly considered the PoF of indivisible items [7] without the budget constraint. They discussed the scenario that the total valuation of each agent to all the items is the same (normalized version). For general n, they showed PoF with an upper bound $O(n)$ and a lower bound $\Omega(\sqrt{n})$, respectively. For two agents, the upper bound and the lower bound are $\frac{2}{\sqrt{3}}$ and $\frac{8}{7}$, respectively. Barman *et al.* [3] obtained an upper bound of $O(\sqrt{n})$. For non-normalized version, they proved that PoF is $\Theta(n)$. More recently, Li *et al.* [18] gave a tight upper bound of $\frac{8}{7}$ for two agents. Sun *et al.* [23] and Wu *et al.* [25] studied PoF for indivisible chores.

Finally, our model is also related to the optimization problem of maximizing the sum of all agents' valuations without fairness considerations. This problem has been extensively studied in combinatorial optimization. The generalized model of various sizes and various valuations, referred to as the Generalized Assignment Problem (GAP), was first studied by Shmoys and Tardos [21]. They gave a 2-approximation algorithm. Finally, Fleischer *et al.* [16] obtained a tight approximation ratio of $1+\frac{1}{e-1}$. For the special case of the homogeneous size and valuation, the model is named the multiple knapsack problem. Chekuri and Khanna [15] obtained a $(1+\epsilon)-$ approximation algorithm. To the best of our knowledge, there do not exist approximation algorithms for the special case of the homogeneous size and various valuations, while [15] showed that this problem is APX-hard.

1.3 Organization of Our Paper

In the remainder of this paper, we define the models formally in Sect. 2. The PoFs for n agents are illustrated in Sect. 3. The next section gives improved algorithms for two agents. Section 5 concludes this paper and discusses the future work.

2 Preliminaries

In the budget-feasible fair allocation problem, let $\mathcal{M}$ be a set of m indivisible items and $\mathcal{N}$ be a set of n agents. Each item $g \in \mathcal{M}$ has a size $s(g)$. Each agent $i \in \mathcal{N}$ has a budget B_i and valuations $v_i(g)$ for each item g. The valuation of agent $i \in \mathcal{N}$ to any subset $X \subseteq \mathcal{M}$ is additive, *i.e.*, $v_i(X) = \sum_{g\in X} v_i(g)$. Meanwhile, the total size of the subset X denoted as $s(X)$ is also additive, which is to say $s(X) = \sum_{g\in X} s(g)$.

Let X_i be a bundle allocated to agent $i \in \mathcal{N}$ and $\mathcal{A} = (X_1, X_2, \ldots, X_n)$ be an *allocation profile*. An allocation $\mathcal{A}$ is said to be *budget-feasible* if $s(X_i) \leq B_i$ holds for any agent $i \in \mathcal{N}$. As the previous literature, we consider unallocated items as being donated to a charity. Hence, we use *charity* to represent the set of unallocated items in the context.

In this paper, we focus on the fairness criterion of *envy-free up to one item* (EF1) under budget constraints.

Definition 1 (EF1 under budget constraints). *A budget-feasible allocation profile* $\mathcal{A} = (X_1, X_2, \ldots, X_n)$ *is said to be EF1 if for any two agents* i *and* j *and any subset of items* $T \subseteq X_j$ *with* $s(T) \leq B_i$, *there exists an item* $g \in T$ *satisfying that*

$$v_i(X_i) \geq v_i(T \backslash \{g\}). \tag{1}$$

We will refer to another stronger fairness criterion – *envy-freeness up to any item* (EFX), where Inequality (1) holds for any item $g \in T$.

For the convenience of the statement, if a budget-feasible set T for agent i satisfies that $v_i(T) > v_i(X_i)$, then we say that agent i *envies* the set T. Moreover, if agent i envies the set T and for any item $g \in T$, $v_i(T \backslash g) > v_i(X_i)$, then we say that agent i *strongly envies* T.

We consider the social welfare of maximizing the total valuation of all the agents, *i.e.*, for a budget-feasible allocation profile $\mathcal{A} = (X_1, X_2, \ldots, X_n)$, the social welfare is

$$sw(\mathcal{A}) = \sum_{i=1}^{n} v_i(X_i). \tag{2}$$

For the sake of simplicity, for an instance I of the budget-feasible fair allocation problem, let $msw(I)$ be the maximum social welfare. The allocation is called as *the optimal allocation* in this paper. Let $\mathcal{A}(I)$ be a fair allocation for the instance I. We abuse the notation $sw(\cdot)$. Let $sw(\mathcal{A}(I))$ denote the social welfare of the fair allocation. In this paper, we use the well-studied concept *the price of fairness* (PoF) to measure efficiency loss in the social welfare induced by the fairness.

Definition 2 (PoF under the fairness constraint P**).** *The price of fairness* PoF *under the fairness constraint* P *is the maximum ratio between the maximum social welfare and the maximum social welfare with the fairness requirement* P *over all the instances, i.e., given an instance* I*,*

$$PoF = \sup_{\forall I} \frac{msw(I)}{\max\limits_{\forall \mathcal{A}(I)} sw(\mathcal{A}(I))} \tag{3}$$

In this paper, we aim to design allocation algorithms meeting the EF1 fairness requirement meanwhile the PoF is minimized.

3 PoF for n Agents

In this section, we first discuss the general case that the valuations of an item are different for all the agents. Next, we consider a special case that the valuation of an item is identical to all the agents.

3.1 Heterogeneous Valuations of an Item

As mentioned in the introduction, Barman *et al.* [6] show the existence of EFX allocations, which implies an exponential time algorithm outputting an EFX allocation. Meanwhile, in the allocation output by the algorithm, EFX holds between any two agents and any agent is envyfree to the charity. Our algorithm utilizes Barman's algorithm. To be self-contained, we demonstrate the algorithm first, which mainly employs the concept of a minimal envied set. We give the definition following.

Definition 3 (Minimal envied sets). *For a budget-feasible allocation* $\mathcal{A} = (X_1, X_2, \ldots, X_n)$*, a set of items* T *is said to be a minimal envied set if the following two conditions hold.*

- *The set* T *is envied by some agent* $i \in \mathcal{N}$*, and*
- *For any subset* $T' \subset T$*,* $v_i(T') \le v_i(X_i)$ *for any agent* $i \in \mathcal{N}$*.*

Barman's Algorithm. The algorithm is initialized from an empty allocation. In each iteration, the algorithm identifies a minimal envied set from the charity and then exchange the set with an agent who envies it. It terminates when no agents envy the charity.

It is easy to see that the PoF of Barman's algorithm can be unbounded. For example, there are two agents and one item. The first agent views the item valuable, while the second agent considers the valuation of the item is ϵ. The algorithm can allocate the item to the second agent. Our algorithm overcomes this by giving an initialized allocation. Our algorithm is illustrate in Algorithm 1. Recall that the optimal allocation refers to the allocation achieving the maximum social welfare without considering the fairness constraint.

Algorithm 1: An EF1 allocation for n heterogeneous agents

Input: An instance of the budget-feasible fair allocation problem
Output: An EF1 allocation $\mathcal{A} = (X_1, X_2, \ldots, X_n)$
1 Obtain an optimal allocation $\mathcal{O} = (X_1^*, X_2^*, \ldots, X_n^*)$;
2 Initialize $X_i \leftarrow \arg\max_{g \in X_i^*} v_i(g)$;
3 Initialize the charity $U \leftarrow \mathcal{M} \backslash (\bigcup_{i=1}^n X_i)$;
4 Call Barman's algorithm with the initial allocation;

Theorem 1. *Algorithm 1 outputs an EF1 allocation for* n *heterogeneous agents with PoF at most* $2n$*.*

Proof. It is easy to see that Algorithm 1 outputs an EF1 allocation. Since Barman's algorithm is EFX, we only need to show the initial allocation satisfy EFX. In the initial allocation, each agent has at most one item. Hence, it is EFX naturally EF1.

Next, we demonstrate that for any agent $i \in \mathcal{N}$, the valuation of agent i's bundle X_i is at least $\frac{1}{2n} v_i(X_i^*)$. We compare $v_i(X_i)$ with the valuation of the subset of X_i^* distributing to each agent or the charity in the allocation $\mathcal{A}$. Let T_j be the subset of items both in X_i^* and X_j, *i.e.*, $T_j = X_i^* \cap X_j$, for any agent $j \in \mathcal{N}$. It is easy to see that $v_i(T_i) \le v_i(X_i)$.

Next, we compare $v_i(X_i)$ to $v_i(T_j)$ for any $j \ne i$. Since $\mathcal{A}$ is an EFX allocation, agent i must be EFX to T_j.

If T_j includes at least two items, let $g \in T_j$ be the cheapest item in agent i's valuations. The inequality $v_i(g) \le v_i(T_j \backslash g) \le v_i(X_i)$ must hold. Therefore,

$$v_i(T_j) = v_i(g) + v_i(T_j \backslash g) \le 2v_i(X_i). \tag{4}$$

If T_j contains one item or less, note that X_i is initialized by the most valuable item in X_i^*. Moreover, in the process of Barman's algorithm, the total valuation assigned to agent i is strictly increase. Hence,

$$v_i(T_j) \le v_i(X_i). \tag{5}$$

Finally, consider the items both in charity and X_i^*. let T_0 be the set of items. Note that at the end of Barman's algorithm, each agent is envy-free to the charity. Hence,

$$v_i(T_0) \le v_i(X_i). \tag{6}$$

To sum up, adding up all subsets of X_i^* in allocation $\mathcal{A}$ yields,

$$v_i(X_i^*) = \sum_{j=0}^{n} v_i(T_j) \le 2(n-1) \cdot v_i(X_i) + 2v_i(X_i) \le 2n \cdot v_i(X_i).$$

Here, for agent $j > 0$ where $j \neq i$, we have $v_i(T_j) \le 2v_i(X_i)$, and for $j = 0$ or i, $v_i(T_j) \le v_i(X_i)$. □

The PoF of $2n$ seems large. Barman et al. [3] show that the PoF of EF1 is at least $n - \frac{n}{n+1}$, which implies that Algorithm 1 is asymptotically tight. Actually, by careful construction, we can give a lower bound of $n - \epsilon$.

Theorem 2. *The PoF for n heterogeneous agents with the EF1 fairness is at least $n - \epsilon$, where $\epsilon > 0$.*

Proof. We show the theorem by designing an instance with PoF at least $n - \epsilon$. In the instance, there are n agents with budget $B_i = 1$ for any agent $i \in \mathcal{N}$ and n items. The size of each item is $\frac{1}{n}$. The valuation of agent 1 for each item is $1 - \frac{\epsilon}{n}$. The other agents' valuations are the same. For them, the valuation of each item is $\frac{\epsilon}{n(n-1)}$. The instance is demonstrated in Table 2.

Table 2. An instance with $PoF = n - \epsilon$

items	sizes	valuations for agent 1	valuations for other agents
1	$\frac{1}{n}$	$1 - \frac{\epsilon}{n}$	$\frac{\epsilon}{n(n-1)}$
⋮	⋮	⋮	⋮
n	$\frac{1}{n}$	$1 - \frac{\epsilon}{n}$	$\frac{\epsilon}{n(n-1)}$

The maximum social welfare is $n - \epsilon$, which is obtained by assigning all items to agent 1.

For the fairness allocation, note that the number of agents and items are the same. If one agent is assigned more than one item, then there must exist at least one agent who is empty. Hence, the empty agents must strongly envy the agents who get more than one item. Hence, in any EF1 allocation, each agent gets one item. The social welfare is 1. Therefore, the PoF is $n - \epsilon$. □

3.2 Homogeneous Valuation of an Item

For this special case, we consider the lower bound of PoF with the EF1 fairness constraint. To be simplicity, we use $v(g)$ to denote the valuation of item g.

Theorem 3. *The price of fairness for n homogeneous agents with the EF1 fairness is at least $2 - \frac{1}{n} - \epsilon$, where $\epsilon > 0$. (Due to the space limitation, the proof is in the full version.)*

We design an allocation with PoF at most 2, which is asymptotically tight by Theorem 3. In this algorithm, Barman's algorithm is adopted as well. The algorithm is demonstrated in Algorithm 2.

Algorithm 2: An EF1 allocation for n homogeneous agents

Input: An instance of the budget-feasible fair allocation problem
Output: Allocation $\mathcal{A} = (X_1, X_2, \ldots, X_n)$

Get an optimal allocation $\mathcal{O} = (X_1^*, X_2^*, \ldots, X_n^*)$;
Call Barman's algorithm on all the items in $\mathcal{O}$ and obtain an EF1 allocation $\mathcal{A}_1 = (X_1^1, \ldots, X_n^1)$;
if $sw(\mathcal{A}_1) \geq \frac{1}{2} sw(\mathcal{O})$ **then**
    return $\mathcal{A}_1$.
else
    $U_1 \leftarrow \bigcup_{i=1}^n X_i^1$;
    for *each agent i* **do**
        if $v(X_i^1) \geq v(X_i^* \backslash U_1)$ **then**
            $X_i \leftarrow X_i^1$;
        else
            $X_i \leftarrow X_i^* \backslash U_1$;
    return $\mathcal{A} = (X_1, \ldots, X_n)$.

Theorem 4. *Algorithm 2 outputs an EF1 allocation for n homogeneous agents with PoF of at most* 2.

Proof. We first show that Algorithm 2 satisfies EF1 between any two agents. If the algorithm outputs $\mathcal{A}_1$, then it is EF1 due to the EFX of Barman's algorithm.

Otherwise, note that the bundle allocated to agent i is either the same bundle as in $\mathcal{A}_1$ or a subset of items in the charity of $\mathcal{A}_1$. For the agent i who get the same bundle, according to the EFX between any two agents and the envyfreeness to the charity of Barman's algorithm, it is easy to see that agent i is EF1 to any other agents. For the other agents who get a subset of items in the charity of $\mathcal{A}_1$, note that their bundles are replaced by more valuable bundles. Hence, the EF1 also holds for them.

Now, we show that the PoF of Algorithm 2 is at most 2. It is sufficient to show that the social welfare of the algorithm is no less than $\frac{1}{2}$ the optimal solution.

If $sw(\mathcal{A}_1) \geq \frac{1}{2}sw(\mathcal{O})$, then naturally the PoF of 2 holds. Otherwise, the total valuation of the charity is more than $\frac{1}{2}sw(\mathcal{O})$. Note that $X_i^* \backslash U_1$ for all $i \in \mathcal{N}$ is a n-partition of all the items in the charity. Each agent i gets the more valuable bundle between X_i^1 and $X_i^* \backslash U_1$. Hence, the social welfare is at least the total valuation of the charity. Therefore, $sw(\mathcal{A}) \geq \frac{1}{2}sw(\mathcal{O})$. □

Remark. Both the upper bound and the lower bound in this section hold for PoF with EFX fairness constraint as well. Note that the algorithms in this section are not polynomial time.

4 Improved PoF for Two Agents

In this section, we discuss the setting with only two agents. Without loss of generality, assume that $B_1 \leq B_2$. Moreover, we aim to design polynomial algorithms. To achieve this, a polynomial algorithm to verify EF1 is the primary task we need to overcome.

A Sufficient Condition for EF1. By the definition of EF1, all the budget-feasible solutions need to be checked if we want to verify the EF1 between two agents. Following we give a sufficient condition for EF1. The condition uses the following dense-greedy algorithm. Let $\rho_i(g) = \frac{v_i(g)}{s(g)}$ be the density of agent i to item g. We coin the set of items returned by the following algorithm as dense-fill solution.

Dense-greedy Algorithm. Given an agent i with budget B_i and a set of items S, sort the items in S with a non-increasing order of the agent i's density, arbitrary when tie-breaking happens. Deal with the items one by one. For an item g, if $B_i \geq s(g)$, then put the item in the output set and update B_i to $B_i - s(g)$; otherwise, go to the next item.

Given a budget-feasible allocation profile $\mathcal{A} = (X_1, \ldots, X_n)$, check whether agent i is EF1 to agent j or not. If $B_i \geq B_j$, we only need to check whether existing an item $g \in X_j$ satisfying $v_i(X_i) \geq v_i(X_j \backslash g)$. If $B_i < B_j$, the following lemma gives a sufficient condition.

Lemma 1. *For two agents i and j, let D be any dense-fill solution on item set X_j, where the density is sorted from agent i's view. If there exist a dense-fill solution D satisfying that $v_i(X_i) \geq v_i(D)$, then agent i is EF1 to agent j. (Due to the space limitation, the proof is in the full version.)*

In this section, we fist deal with the easy case that the valuation of each item is identical.

4.1 Homogeneous Valuation of an Item

The PoF with EF1 for two homogeneous agents is discussed first. By Theorem 3, the PoF of this setting is at least $\frac{3}{2} - \epsilon$. In this subsection, we design an algorithm with $PoF = \frac{3}{2}$. In the same way, we also use $v(g)$ to denote the valuation of item g.

In the algorithm, the key point is to reserve at least $\frac{2}{3}$ social welfare of an optimal solution. The algorithm is demonstrated in Algorithm 3.

To show PoF of Algorithm 3 and fairness criterion, we first consider the scenario that in optimal allocation, agent 2 gets at least $\frac{2}{3}$ of the total valuation, which is demonstrated in Algorithm 4.

Algorithm 3: An EF1 allocation for 2 homogeneous agents

Input: An EF1 allocation for 2 homogeneous agents and two budgets B_1 and B_2
Output: Allocation $\mathcal{A} = (X_1, X_2)$

Get an optimal allocation $\mathcal{O} = (X_1^*, X_2^*)$;
$\mathcal{A} \leftarrow \mathcal{O}$;
if $v(X_1) \geq v(X_2)$ **then**
  if $v(X_1) > 2v(X_2)$ **then**
    $X_2 \leftarrow \emptyset$;
    while $\mathcal{A}$ *is not EF1* **do**
      Arbitrarily move one item from X_1 to X_2;
  else
    Pick any item $g \in X_1$;
    while $v(X_1 \backslash g) \geq v(X_2)$ **do**
      $X_1 \leftarrow X_1 \backslash g$;
      Pick any item $g \in X_1$;
else
  if $v(X_2) > 2v(X_1)$ **then**
    Call Algorithm 4 for items in X_2;
  else
    Pick any item $g \in X_2$;
    while $v(X_2 \backslash g) \geq v(X_1)$ **do**
      $X_2 \leftarrow X_2 \backslash g$;
      Pick any item $g \in X_2$;
return $\mathcal{A}$;

Algorithm 4: An EF1 allocation of reallocating items in agent 2

Input: The set of items X_2^* in agent 2 and the budgets B_1 and B_2
Output: An EF1 allocation $\mathcal{A} = (X_1, X_2)$ with $X_1 \cup X_2 = X_2^*$

1. Sort all the items in non-increasing order of the density, arbitrary when tie-breaking happens;
2. Construct two virtual agents with the budget of B_1;
3. The less valuable virtual agent chooses the heaviest density item it can pack until no items fit. Let X_1 and X_2 be two bundles obtained. Assume that X_1 is more valuable than X_2;
4. Allocate X_1 to agent 1 and $X_2^* \backslash X_1$ to agent 2.

Lemma 2. *Algorithm 4 outputs an EF1 allocation between agents* 1 *and* 2. *(Due to the space limitation, the proof is in the full version.)*

Theorem 5. *Algorithm 3 outputs an EF1 allocation for two homogeneous agents with PoF at most* $\frac{3}{2}$.

Proof. We aim to demonstrate, for any instance I and an optimal allocation $\mathcal{O}$, it is always possible to obtain a new allocation $\mathcal{A}$ such that $sw(\mathcal{A}) \geq \frac{2}{3} sw(\mathcal{O})$, and two

agents are EF1 to each other. As mentioned above, the budgets of the two agents holds that $B_1 \leq B_2$. Initially, let the two agents get the optimal solution. We show the theorem by two cases as in the algorithm. The first case is $v(X_1^*) \geq v(X_2^*)$, and the second case is $v(X_1^*) < v(X_2^*)$.
Case 1: $v(X_1^*) \geq v(X_2^*)$. In this case, two sub-cases are discussed. The first one is $v(X_1) > 2v(X_2)$; the other one is $v(X_1) \leq 2v(X_2)$.
Subcase 1.1: $v(X_1^*) > 2v(X_2^*)$. We discard all items in X_2^* and reserve all the items in X_1^*. Since $B_2 \geq B_1$, move items from X_1 to X_2 one by one until EF1 holds. It is easy to see that reserving X_1^* keeps 2/3 total valuation of items in the optimal solution.
Subcase 1.2: $v(X_1^*) \leq 2v(X_2^*)$. In this subcase, $v(X_2^*) \geq \frac{1}{3}msw(I)$. In the algorithm, discard some items in X_1^* and reserve all the items in X_2^*. Note that in the final step, $v(X_1) \geq v(X_2)$. Hence, the total valuation $v(X_1)+v(X_2) \geq \frac{2}{3}msw(I)$. Next, we will show that the two agents are EF1 to each other. It is easy to see that agent 1 does not envy agent 2. For the last item g picked in X_1, we have that $v(X_1 \backslash g) < v(X_2)$. Hence, agent 2 does not envy agent 1 by the removal of item g from X_1.
Case 2: $v(X_1) < v(X_2)$. Similar to Case 1, we divide it to two subcases.
Subcase 2.1: $v(X_2^*) > 2v(X_1^*)$. In this subcase, we put all the items in X_1^* to the charity. By Algorithm 4 and Lemma 2, the two agents are EF1 to each other. Further, it is ensured that all of the items in X_2^* are distributed to the agents. Hence, the social welfare of the allocation $\mathcal{A} = v(X_2^*) > \frac{2}{3}msw(I)$.
Subcase 2.2: $v(X_2^*) \leq 2v(X_1)$. This subcase is symmetric to Subcase 1.2. □

Remark. If we replace the optimal allocation with a PTAS allocation at step 1 of Algorithm 3, the analysis still holds. The PoF now is $\frac{3}{2} + \epsilon$ and Algorithm 3 is in polynomial time.

4.2 Heterogeneous Valuations of an Item

In this subsection, we give an improved algorithm for two heterogeneous agents with PoF of 3 for two agents. Recall that the PoF of Algorithm 1 is 4 for two agents. And the lower bound of PoF is at least 2.

We employ the sufficient condition of EF1 to design polynomial time algorithm. The algorithm is illustrated in Algorithm 5.

Lemma 3. *Given a feasible allocation $\mathcal{A} = (X_1, X_2)$, if agent i ($i = 1, 2$) strongly envies a subset $T \subseteq X_{3-i}$ with $s(T) \leq B_i$, there exists a new allocation $\mathcal{A}'$ satisfying EF1 and $sw(\mathcal{A}') \geq \frac{v_i(X_i)+v_{3-i}(T)}{2}$. (Due to the space limitation, the proof is in the full version.)*

Remark. Lemma 3 implies that we can obtain an algorithm outputting an EF1 allocation with PoF at most 2, when the two agents have the same budget.

Theorem 6. *Algorithm 5 outputs an EF1 allocation for two heterogeneous agents with PoF at most 3. (Due to the space limitation, the proof is in the full version.)*

Remark. Similarly, replace step 1 with an approximation allocation. Then Algorithm 5 runs in polynomial time. And the PoF is at most 3ρ, where ρ is the approximation ratio of GAP. To the best of our knowledge, currently $\rho = \frac{e}{e-1}$.

Algorithm 5: An EF1 allocation for 2 heterogeneous agent

Input: An EF1 allocation for 2 homogeneous agents and two budgets B_1 and B_2
Output: Allocation $\mathcal{A} = (X_1, X_2)$

```
Obtain an optimal solution O = (X1*, X2*);
A ← O;
Obtain a dense-fill solution D ⊆ X2 from the view of agent 1;
if v2(D) ≥ (2/3) v2(X2) then
    X2 ← D;
    if v1(D) > v1(X1) and v2(X1) > v2(D) then
        swap(X1, X2);
    else if agent 1 is not EF1 towards D then
        Apply round-robin for D with agent 2 choosing first;
    else if agent 2 is not EF1 towards X1 then
        Apply round-robin for X1 with agent 1 choosing first;
else if v1(X1) ≥ v1(D) then
    if agent 2 is not EF1 towards X1 then
        Apply round-robin for X1 with agent 1 choosing first;
else
    X2 ← X2\D and X1 ← D;
    if agent 2 is not EF1 towards X1 then
        while agent 2 is not EF1 towards X1 do
            Pick the densest item g ∈ X1 from the view of agent 1;
            X2 ← X2 ∪ g;
        Obtain a dense-fill solution D' ⊆ X2 from the view of agent 1;
        if v1(X1) < v1(D') then
            X2 ← X1 and X1 ← D';
return A = (X1, X2).
```

5 Conclusion and Future Work

In this paper, we give a landscape of PoFs for the budget-feasible allocation problem under the EF1 fairness constraint. PoFs for n agents are considered first, we establish upper bounds of $2n$ and 2 for heterogeneous and homogeneous agents, respectively. While the corresponding lower bounds are n and $2 - \frac{1}{n}$. We then design improved algorithms for two agents. For two homogeneous agents, an algorithm with tight PoF of $\frac{3}{2}$ is designed; For two heterogeneous agents, the PoF is at most 3.

For the future work, whether existing a polynomial time algorithm to compute an EF1 allocation is still an interesting open problem. We believe that it is also interesting to consider PoF for other fairness criteria, *e.g.* EFX, MMS and so on. Finally, it is also an interesting direction to consider approximation algorithms for maximizing the social welfare under the fairness constraints. How to make use of the property of fairness in the optimal solution is a challenge for this line.

References

1. Akrami, H., Garg, J.: Breaking the 3/4 barrier for approximate maximin share. In: Proceedings of the 2024 ACM-SIAM Symposium on Discrete Algorithms (SODA), pp. 74–91 (2024)
2. Aziz, H., Huang, X., Mattei, N., Segal-Halevi, E.: Computing welfare-maximizing fair allocations of indivisible goods. Eur. J. Oper. Res. **307**(2), 773–784 (2023)
3. Barman, S., Bhaskar, U., Shah, N.: Optimal bounds on the price of fairness for indivisible goods. In: Proceedings of the 16th International Conference on Web and Internet Economics (WINE), pp. 356–369 (2020)
4. Barman, S., Ghalme, G., Jain, S., Kulkarni, P., Narang, S.: Fair division of indivisible goods among strategic agents. In: Proceedings of the 2019 International Conference on Autonomous Agents and Multi-Agent Systems (AAMAS), pp. 1811–1813 (2019)
5. Barman, S., Khan, A., Shyam, S., Sreenivas, K.V.N.: Fair allocations under budget constraints. In: Proceedings of the 37th AAAI Conference on Artificial Intelligence (AAAI), pp. 5481–5489 (2023)
6. Barman, S., Khan, A., Shyam, S., Sreenivas, K.V.N.: Guaranteeing envy-freeness under generalized assignment constraints. In: Proceedings of the 24th ACM Conference on Economics and Computation (EC), pp. 242–269 (2023)
7. Bei, X., Xinhang, L., Manurangsi, P., Suksompong, W.: The price of fairness for indivisible goods. Theor. Comput. Syst. **65**(7), 1069–1093 (2021)
8. Bertsimas, D., Farias, V.F., Trichakis, N.: The price of fairness. Oper. Res. **59**(1), 17–31 (2011)
9. Bu, X., Li, Z., Liu, S., Song, J., Tao, B.: On the complexity of maximizing social welfare within fair allocations of indivisible goods https://arxiv.org/abs/2205.14296 (2022)
10. Budish, E.: The combinatorial assignment problem: approximate competitive equilibrium from equal incomes. J. Polit. Econ. **119**(6), 1061–1103 (2011)
11. Caragiannis, I., Kaklamanis, C., Kanellopoulos, P., Kyropoulou, M.: The efficiency of fair division. Theor. Comput. Syst. **50**(4), 589–610 (2012)
12. Caragiannis, I., Kurokawa, D., Moulin, H., Procaccia, A.D., Shah, N., Wang, J.: The unreasonable fairness of maximum Nash welfare. ACM Trans. Econ. Comput. **7**(3), 12:1–12:32 (2019)
13. Chaudhury, B.R., Garg, J., Mehlhorn, K.: EFX exists for three agents. In: Proceedings of the 21st ACM Conference on Economics and Computation (EC), pp. 1–19 (2020)
14. Chaudhury, B.R., Kavitha, T., Mehlhorn, K., Sgouritsa, A.: A little charity guarantees almost envy-freeness. SIAM J. Comput. **50**(4), 1336–1358 (2021)
15. Chekuri, C., Khanna, S.: A PTAS for the multiple knapsack problem. In: Proceedings of the 2000 ACM-SIAM Symposium on Discrete Algorithms (SODA), pp. 213–222 (2000)
16. Fleischer, L., Goemans, M.X., Mirrokni, V.S., Sviridenko, M.: Tight approximation algorithms for maximum general assignment problems. In: Proceedings of the 2006 ACM-SIAM Symposium on Discrete Algorithms (SODA), pp. 611–620 (2006)
17. Gan, J., Li, B., Wu, X.: Approximation algorithm for computing budget-feasible ef1 allocations. In: Proceedings of the 2023 International Conference on Autonomous Agents and Multiagent Systems (AAMAS), pp. 170–178 (2023)
18. Li, Z., Liu, S., Lu, X., Tao, B., Tao, Y.: A complete landscape for the price of envy-freeness. In: Proceedings of the 23rd International Conference on Autonomous Agents and Multiagent Systems (AAMAS), pp. 1183–1191 (2024)
19. Lipton, R.J., Markakis, E., Mossel, E., Saberi, A.: On approximately fair allocations of indivisible goods. In: Proceedings of the 5th ACM Conference on Electronic Commerce (EC), pp. 125–131 (2004)

20. Plaut, B., Roughgarden, T.: Almost envy-freeness with general valuations. SIAM J. Discret. Math. **34**(2), 1039–1068 (2020)
21. Shmoys, D.B., Tardos, É.: An approximation algorithm for the generalized assignment problem. Math. Program. **62**, 461–474 (1993)
22. Steinhaus, H.: The problem of fair division. Econometrica **16**(1), 101–104 (1948)
23. Sun, A., Chen, B., Doan, X.V.: Connections between fairness criteria and efficiency for allocating indivisible chores. In: Proceedings of the 2021 International Conference on Autonomous Agents and Multiagent Systems (AAMAS), pp. 1281–1289 (2021)
24. Wu, X., Li, B., Gan, J.: Budget-feasible maximum Nash social welfare is almost envy-free. In: Proceedings of the 30th International Joint Conference on Artificial Intelligence (IJCAI), pp. 465–471 (2021)
25. Wu, X., Zhang, C., Zhou, S.: Weighted EF1 allocations for indivisible chores. In: Proceedings of the 24th ACM Conference on Economics and Computation (EC), p. 1155 (2023)

On the Min-max Heterogeneous Weighted Delivery Problem

Jianping Li[1(✉)], Ping Yang[1], and Junran Lichen[2]

[1] Department of Mathematics, Yunnan University, East Outer Ring South Road University Town, Chenggong District, Kunming 650504, People's Republic of China
jianping@ynu.edu.cn

[2] School of Mathematics and Physics, Beijing University of Chemical Technology, No. 15, North Third Ring East Road, Chaoyang District, Beijing 100029, People's Republic of China
J.R.Lichen@buct.edu.cn

Abstract. In this paper, we consider the min-max heterogeneous weighted delivery problem (the MMHWD problem). Specifically, given a weighted graph $G = (V, E; w)$ with length function $w : E \rightarrow R^+$ satisfying the triangle inequality, a fixed depot $r \in V$, m items, and k vehicles having nonuniform speeds $\lambda_1, \lambda_2, \ldots, \lambda_k$, each item is initially located at its source vertex s_j and it needs to be delivered to its target vertex t_j, $j = 1, 2, \ldots, m$, each vehicle can move along some edges of G and only deliver one item at a time and each item only can be continuously delivered by one vehicle, it is asked to find a set $\mathcal{C} = \{C_1, C_2, \ldots, C_k\}$ of k tours for these k vehicles, each starting and ending at the same depot r, and collectively delivering all items, the objective is to minimize the maximum completion time of vehicles, where the completion time of a vehicle is its total length divided by its speed. We obtain the two main results. (1) Given any small constant $\delta > 0$, we design an $134.4434(1 + \delta)$-approximation algorithm to solve the MMHWD problem, its time complexity is bounded by a polynomial in the input size and $\frac{1}{\delta}$; (2) We provide an $(\varphi + \frac{9}{5} - \frac{1}{k})$-approximation algorithm to resolve the MMHWD problem, where φ is the ratio of the largest vehicle speed to the smallest one.

Keywords: Combinatorial optimization · Heterogeneous delivery · Nonuniform speeds · Approximation algorithms

1 Introduction

Bärtschi et al. [1] in 2017 proposed an energy-efficient delivery by heterogeneous mobile agents, *i.e.*, the weighted delivery problem. Specifically, given a weighted

This paper is supported by the Natural Science Foundation of China [Nos. 12361066, 12101593]. Junran Lichen is also supported by Fundamental Research Funds for the Central Universities [No. buctrc202219], Ping Yang and Jianping Li are also supported by Project of Yunling Scholars Training of Yunnan Province [No. K264202011820].

D. Du et al. (Eds.): COCOA 2024, LNCS 15434, pp. 352–363, 2025.
https://doi.org/10.1007/978-981-96-4445-2_29

graph $G = (V, E; w)$, where length function $w : E \to R^+$ satisfying the triangle inequality, k mobile agents (*e.g.*, robots, sensors, vehicles and so on), denoted by a_1, a_2, …, a_k, having unitary movement costs c_1, c_2, …, c_k (assuming that all unitary movement costs are greater than or equal to 1), where these agents are initially located on arbitrary vertices p_1, p_2,…, p_k of G, m messages, each of which is initially located at its source vertex s_j and it needs to be delivered to its target vertex t_j, $j = 1, 2, \ldots, m$, each message can be picked up by an agent from any vertex that it visits and it can be carried to any other vertex of G, and dropped there, an agent a_i traverses an edge e and it incurs an energy cost of $c_i \cdot w(e)$ per each time, each agent has its capacity Q of maximum number of messages carried simultaneously. It is asked to find a set $\mathcal{C} = \{C_1, C_2, \ldots, C_k\}$ of k tours to deliver all messages, each starting and ending at its initial location, the objective is to minimize the total energy. When there is only a single message ($m = 1$), Bärtschi et al. [1] designed a polynomial-time algorithm for solving this problem in time $O(n^3)$. When each agent with capacity 1, they [1] designed an $4 \cdot \max \frac{c_i}{c_j}$-approximation algorithm for solving this problem in polynomial time.

Bilò et al. [2] in 2022 considered the heterogeneous weighted delivery problem (the HWD problem). It is asked to find k tours which satisfy the following two conditions: (1) each message is carried by a single agent with exactly one pick-up and one drop-off; (2) each agent returns to its starting location after its last moving. Bilò et al. [2] designed an 8-approximation algorithm for solving the HWD problem in time $\mathcal{O}^*(2^{k(k+1)/2})$, where the $\mathcal{O}^*$ notation hides polynomial factors in the size of the instance, and this 8-approximation algorithm can be transformed into an $\mathcal{O}(k^*)$-approximation algorithm, where k^* is the number of mobile agents used in the optimal solution.

Some classic routing problems related to the delivery problem are the traveling salesman problem (TSP) [8], the Chinese postman problem (CPP) [5]. Fredericksons et al. [3] in 1978 considered the stacker-crane problem (SCP). Specifically, given a mixed graph $G = (V, E \cup A; w)$ with length function $w : E \cup A \to R^+$ satisfying triangle inequality, a fixed depot $r \in V$, for each arc in A, there is a parallel edge of no greater length, it is asked to find a tour to traverse each arc in A exactly once, starting and ending at the same depot r, the objective is to minimize length of that tour. Fredericksons et al. [3] designed an $\frac{9}{5}$-approximation algorithm for solving the SCP, denoted by the algorithm $\mathcal{A}_{SCP}$.

Fredericksons et al. [3] in 1978 considered several k-person routing problems: the k-traveling salesman problem (k-TSP), the k-Chinese postman problem (k-CPP), and the k-stacker-cranes problem (k-SCP), which generalize the TSP, the CPP and the SCP, respectively. Fredericksons et al. [3] designed approximation algorithms for solving the k-TSP, k-CPP and k-SCP, whose approximation ratios are $(\rho + 1 - \frac{1}{k})$, $(2 - \frac{1}{k})$, and $(\sigma + 1 - \frac{1}{k})$, respectively, where ρ is an approximation ratio of any approximation algorithm for solving the TSP, σ is an approximation ratio of any approximation algorithm for solving the SCP.

Gørtz et al. [4] in 2016 considered the heterogeneous traveling salesman problem (HTSP), which is generalized the k-TSP. Specifically, given a complete graph $G = (V, E; w)$ with weight function $w : E \to R^+$ satisfying triangle inequality, a fixed depot r, a set J $(= V\backslash\{r\})$ of customer locations, k vehicles having k nonuniform speeds λ_1, λ_2, ..., λ_k, respectively, it is asked to find a set $\mathcal{C} = \{C_1, C_2, \ldots, C_k\}$ of k cycles for vehicles with nonuniform speeds to cover all vertices in G, each starting and ending at the same depot r, the objective is to minimize the maximum completion time of vehicles, where the completion time of a vehicle is its total travel length divided by its speed. They designed an 90-approximate algorithm for solving the HTSP. Li et al. [7] in 2023 considered the heterogeneous Chinese postman problem (HCPP), which is generalized the k-CPP. They obtained the following two main results. (1) Given an small constant $\delta > 0$, they provided an $20.8765(1+\delta)$-approximate algorithm for solving the HCPP, where its time complexity is bounded by a polynomial in the input size and $\frac{1}{\delta}$; (2) They present an $(1+\varphi-\frac{1}{k})$-approximate algorithm for resolving the HCPP, where φ is the ratio of maximum speed to minimum speed.

Motivated by the aforementioned problems, we consider the min-max heterogeneous weighted delivery problem (the MMHWD problem). Specifically, given a weighted graph $G = (V, E; w)$ with length function $w : E \rightarrow R^+$ satisfying the triangle inequality, a fixed depot $r \in V$, m items, and k vehicles having nonuniform speeds λ_1, λ_2, ..., λ_k, each item is initially located at its source vertex s_j and it needs to be delivered to its target vertex t_j, $j = 1, 2, \ldots, m$, each vehicle can move along some edges of G and only deliver one item at a time and each item can only be continuously delivered by one vehicle, it is asked to find a set $\mathcal{C} = \{C_1, C_2, \ldots, C_k\}$ of k tours for these k vehicles, each starting and ending at the same depot r, and collectively delivering all items, the objective is to minimize the maximum completion time of vehicles, *i.e.*, $\min \max_{1\leq i\leq k} \frac{w(C_i)}{\lambda_i}$. We aim to design some approximation algorithms to solve the above problem in polynomial time.

The remainder of this paper is organized into the following sections. In Sect. 2, we introduce some terminologies and fundamental lemmas; In Sect. 3, given any small constant $\delta > 0$, we design an $134.4434(1+\delta)$-approximation algorithm to solve the MMHWD problem; In Sect. 4, we provide an $(\varphi+\frac{9}{5}-\frac{1}{k})$-approximation algorithm to resolve the MMHWD problem, where φ is the ratio of the largest vehicle speed to the smallest one; In Sect. 5, we summarize our conclusions and put forward some interesting related problems that will be worth of considering in the further research.

2 Terminologies and Key Lemmas

In order to design algorithms to solve the MMHWD problem, we intend to introduce an important definition as follows.

Definition 1. *(Gørtz et al. 2016 [4]) Given a undirected graph* $G = (V, E; w)$ *with two constants* $M > 0$ *and* $\epsilon > 0$*, where length function* $w : E \rightarrow R^+$ *satisfying triangle inequality, let* T_i *be a set of tours in* G*, for each integer* $i \geq 0$*, each starting and ending at the depot* r*. These tours* $\bigcup_{i\geq 0} T_i$ *is said to be* $(\alpha, \beta)_{M,\epsilon}$*-assignable if this set has the following two properties:*

(1) $w(C) \leq \alpha(1+\epsilon)^i M$, *for each integer* $i \geq 0$, $C \in \mathcal{T}_i$, *where* $w(C) = \sum_{e \in C} w(e)$;

(2) $\sum_{j \geq i} w(\mathcal{T}_j) \leq \beta M \cdot \Lambda((1+\epsilon)^{i-1})$, *for each integer* $i \geq 0$, *where* $\Lambda((1+\epsilon)^{i-1})$ *is the total of speeds which are at least* $(1+\epsilon)^{i-1}$, $w(\mathcal{T}_i) = \sum_{C \in \mathcal{T}_j} w(C) = \sum_{C \in \mathcal{T}_j} \sum_{e \in C} w(e)$.

Gørtz et al. [4] designed an algorithm, denoted by the algorithm $\mathcal{A}_{ASSIGN}$, to assign these tours $\bigcup_{i \geq 0} \mathcal{T}_i$ to k vehicles, such that the completion time of the vehicle is assignable. We need a result as follows.

Lemma 1. *(Gørtz et al. 2016 [4]) Given an* $(\alpha, \beta)_{M,\epsilon}$*-assignable collection* $\bigcup_{i \geq 0} \mathcal{T}_i$ *of tours, each starting and ending at the same depot* r*, the algorithm* $\mathcal{A}_{ASSIGN}$ *can assign* $\bigcup_{i \geq 0} \mathcal{T}_i$ *in time* $O(n^6 + \log w(E))$ *to* k *vehicles, such that the completion time of each vehicle is at most* $((1+\epsilon)\alpha + \beta)M$*, where* n *is the number of vertices,* $w(E) = \sum_{e \in E} w(e)$*,* $M > 0$ *and* $\epsilon > 0$ *are defined in Definition 1.*

3 An Approximation Algorithm with Constant Factor for the MMHWD Problem

In this section, we consider the MMHWD problem. We present an instance $I = (G; r; w, \lambda; k, m)$ of the MMHWD problem as follows. Given a weighted graph $G = (V, E; w)$, construct an auxiliary graph $H' = (V', A \cup E''; w')$, satisfying (1) $V' = \{r\} \cup \{s_j, t_j \mid j = 1, 2, \ldots, m\} \cup \cup_{j=1}^{m} V(P(d(s_j, t_j)))$, (2) $A = \{(s_j, t_j) \mid j = 1, 2, \ldots, m\}$, $w'(u, v) = d(u, v)$ for each arc $(u, v) \in A$, where $d(u, v)$ is the length of a shortest path from u to v in G, (3) $E'' = \cup_{j=1}^{m} E(P(d(s_j, t_j)))$, $w'(uv) = w(uv)$ for each edge $uv \in E(P(d(s_j, t_j))$, where $V(P(d(s_j, t_j)))$ (or $E(P(d(s_j, t_j))))$ is denoted as the set of vertices (or edges) on a shortest path from s_j to t_j in G, $j = 1, 2, \ldots, m$. We may first partition the vertex-set V', the edge-set E'' and the arc-set A into levels in the following setting. Assuming the number of layers is $t+1$, *i.e.*, $(1+\epsilon)^{t-1}M < \max\{d(r, v) \mid v \in V'\} \leq (1+\epsilon)^t M$, the partition process is as follows:

(1) Partition V' into subsets V'_0, V'_1, ..., where $V'_0 = \{v \in V' \mid d(r, v) \leq M\}$, and $V'_i = \{v \in V' \mid (1+\epsilon)^{i-1}M < d(r, v) \leq (1+\epsilon)^i M\}$, for each $i \geq 1$;
(2) Partition E'' into subsets E''_0, E''_1, ..., where $E''_0 = \{rv \in E'' \mid d(r, v) \leq M\}$, and $E''_i = \{vv' \in E'' \mid d(r, v) \leq (1+\epsilon)^i M$ *and* $(1+\epsilon)^{i-1}M < d(r, v') \leq (1+\epsilon)^i M\}$, for each $i \geq 1$;
(3) Ignore the direction of all arcs, partition A into subsets A_0, A_1, ..., where $A_0 = \{(r, v) \in A \mid d(r, v) \leq M\}$, and $A_i = \{(v, v') \in A \mid d(r, v) \leq (1+\epsilon)^i M$ and $(1+\epsilon)^{i-1}M < d(r, v') \leq (1+\epsilon)^i M\}$, for each $i \geq 1$;

For convenience, for each $i \geq 1$, we need to define the following symbols: $V'_{\geq i} := \bigcup_{j \geq i} V'_j$, $V'_{\geq i} := \bigcup_{j \geq i} V'_j$, $E''_{\leq i} := \bigcup_{j \leq i} E''_j$, $E''_{\geq i} := \bigcup_{j \geq i} E''_j$, $A_{\leq i} := \bigcup_{j \leq i} A_j$, and $A_{\geq i} := \bigcup_{j \geq i} A_j$.

Given a graph $G = (V, E; w)$, construct another auxiliary complete graph $H = (V', E' \cup A; w')$, where V' is as described above, $E' = \{uv \mid u, v \in V'\}$

and $A = \{(s_j, t_j) \mid j = 1, 2, \ldots, m\}$, and denote $w'(uv) = w(uv)$ for each edge $uv \in E$, $w'(uv) = d(u, v)$ for each edge $uv \in E' \setminus E$, $w'(u, v) = d(u, v)$ for each arc $(u, v) \in A$.

Unlike the HTSP and HCPP, we are asked to find some tours that traverse through the arc in A exactly once to satisfy the delivery of items. The strategy for our algorithm is described in the following ways.

(1) Find an $(\alpha, \beta)_{M,\epsilon}$-assignable collection of subtours that traverse all arcs;
(2) Assign the subtours obtained to k vehicles with nonuniform speeds (see Lemma 1) to minimize the completion time of each vehicle.

We can design an algorithm, denoted by the algorithm $\mathcal{A}_{MMHWD1}$.

Algorithm: $\mathcal{A}_{MMHWD1}$
Input: An instance $I = (G; r; w, \lambda; k, m)$ of the MMHWD problem;
Output: A set of k tours $\mathcal{C} = \{C_i \mid i = 1, 2, \ldots, k\}$.
Begin
Step 1. Set $M = \frac{\max_{(v,v') \in A}(d(r,v) + w'(v,v') + d(v',r))}{\lambda_{\max}}$, where $\lambda_{\max} = \max\{\lambda_i \mid i = 1, 2, \ldots, k\}$;
Step 2. Construct an auxiliary graph $H' = (V', A \cup E''; w')$ as mentioned-above; Using current M and ϵ (the precise value is fixed later), partition V' into subsets V'_0, V'_1, …; partition E'' into subsets E''_0, E''_1, …; Ignore the direction of all arcs, partition A into subsets A_0, A_1, … as mentioned-above;
Step 3. Construct an auxiliary complete graph $H = (V', E' \cup A; w')$ as above;
Step 4. Find a Level-Prim tree (a subgraph) in H [4], denoted by $\mathcal{H}' = \bigcup_{i \geq 0} H'_i$, and connect all vertices in V', where H'_i denotes the edge-set of a minimum spanning tree for $H[V'_{\leq i}]/V'_{<i}$; Set $\mathcal{H}_i = \mathcal{H}'_i \cup A_i$, $i = 0, 1, \ldots, t$;
Step 5. If $(\sum_{j \geq l} w'(H_j) > (7 + \frac{6}{\epsilon}) \cdot M \cdot \Lambda((1+\epsilon)^{l-1})$ holds for $l \in \{1, 2, \ldots, t\})$ then
– Set $M := (1 + \delta)M$, where δ is a small constant; Go to Step 2;
Step 6. Set $\mathcal{S}_0^m (= \mathcal{S}_0) = \{r\} \cup H_0$, and $\gamma = \frac{\epsilon}{(2+\epsilon)(1+\epsilon)}$;
Step 7. For each $i = 1, 2, \ldots, t + 1$, let $\mathcal{S}_i$ to denote the set of all connected components in H_i, Set $\mathcal{S}_i^m = \emptyset$, $\mathcal{S}_i^u = \emptyset$;
Step 8. For (each $i = 1, 2, \ldots, t$, $\eta \in \mathcal{S}_i$) do:
– If $(w'(\eta) \geq \gamma \cdot (1 + \epsilon)^i M)$ then
–– Set $\mathcal{S}_i^m := \mathcal{S}_i^m \cup \{\eta\}$;
– Else
–– Set $\mathcal{S}_i^u := \mathcal{S}_i^u \cup \{\eta\}$;
Step 9. For (each $i = 1, 2, \ldots, t$, $\sigma \in \mathcal{S}_i^u$) do:
– Find a connected component $\pi(\sigma)$ in H_{i-1}, such that $\pi(\sigma) \cap \sigma \neq \emptyset$;
Step 10. For (each $i = 0, 1, 2, \ldots, t$, $\tau \in \mathcal{S}_i^m$) do
– **10.1** Set $Dangle(\tau) = \{\sigma \in \mathcal{S}_{i+1}^u \mid \pi(\sigma) = \tau\}$;
– **10.2** "Double" all edges and arcs twice in $\tau \cup Dangle(\tau)$ to obtain some mixed tours ("double" the arcs twice means adding a parallel and same weighted edge, corresponding to a shortest path in G), partition these mixed tours into some subtours with length of at most $2(1 + \epsilon)^{i+1}M$, denoted by P_1, P_2, …, P_q;

– **10.3** For each path P_j, $j = 1, 2, \ldots, q$, connect the two shortest paths from r to the two endpoints of P_j to obtain some new subtours (apply the Dijkstra algorithm[6] to obtain these shortest paths), denoted by $\mathcal{T}_i(\tau)$;

Step 11. For (each $i = 0, 1, \ldots, t-1$) do:

– Set $\mathcal{T}_i = \bigcup_{\tau \in \mathcal{S}_i^m} \mathcal{T}_i(\tau)$, apply $\mathcal{A}_{ASSIGN}$[4] (see Lemma 1) to assign $\bigcup_{i \geq 0}\{\mathcal{T}_i\}$ to k vehicles to obtain k tours, denoted by $\mathcal{C}' = \{C'_1, C'_2, \ldots, C'_k\}$;

Step 12. Restore the arcs in A and the edges in E' to the shortest path in the set $\mathcal{C}'$ to produce k tours, denoted by C_1, C_2, ..., C_k;

Step 13. Output "k tours $\mathcal{C} = \{C_i \mid i = 1, 2, \ldots, k\}$ related to k vehicles".

End

Using the algorithm $\mathcal{A}_{MMHWD1}$, we have the following lemmas.

Lemma 2. *Given an instance $I = (G; r; w, \lambda; k, m)$ of the MMHWD problem, two constants $M > 0$ and $\epsilon > 0$, if $M \geq OPT$, then for each $l \geq 0$, we have*

$$w'(A_{\geq l}) \leq M \cdot \Lambda((1+\epsilon)^{l-1}),$$

where OPT is the optimal value to this instance I of the MMHWD problem, $A_{\geq l} = \bigcup_{j \geq l} A_j$, $\Lambda((1+\epsilon)^{l-1})$ is the total of speeds that are at least $(1+\epsilon)^{l-1}$.

Proof. Consider an optimal solution for the instance of the MMHWD problem, its traverses each arc in A exactly once. For an arc $a = (v, v') \in A_{\geq l}$, assuming the arc is assigned by a vehicle of speed λ' in the optimal solution. Since $M \geq OPT$, the distance traveled by a vehicle of speed λ' is at most $\lambda' \cdot OPT \leq \lambda' \cdot M$. Note that $a = (v, v') \in A_{\geq l}$, which means that $\lambda' \cdot OPT \geq \max\{d(r, v), d(r, v')\} \geq (1+\epsilon)^{l-1} M$, by the definition of $A_{\geq l}$,it follows that, $(1+\epsilon)^{l-1} M \leq \lambda' \cdot OPT \leq \lambda' \cdot M$, we have, $\lambda' \geq (1+\epsilon)^{l-1}$, this means that the speed of the vehicle traversing through any arc in $A_{\geq l}$ is at least $(1+\epsilon)^{l-1}$.

Since the arc in $A_{\geq l}$ must be traversed by some vehicles with a speed of at least $(1+\epsilon)^{l-1}$, in any feasible solution, which means that, $w'(A_{\geq l}) \leq OPT \cdot \Lambda((1+\epsilon)^{l-1})$. Note that $M \leq OPT$, we have $OPT \cdot \Lambda((1+\epsilon)^{l-1}) \leq M \cdot \Lambda((1+\epsilon)^{l-1})$. Thus, we obtain $w'(A_{\geq l}) \leq OPT \cdot \Lambda((1+\epsilon)^{l-1}) \leq M \cdot \Lambda((1+\epsilon)^{l-1})$.

Lemma 3. *Given an instance $I = (G; r; w, \lambda; k, m)$ of the MMHWD problem, two constants $M > 0$ and $\epsilon > 0$, if $M \geq OPT$, then for all $l \geq 0$, we have*

$$w'(T_{H/V'_{<l}}) \leq M \cdot \Lambda((1+\epsilon)^{l-1}),$$

where $T_{H/V'_{<l}}$ is a minimum spanning tree in graph $H/V'_{<l}$ obtained by contracting vertices $V'_{<l}$ to a single vertex in H.

Proof. Given an instance $I = (G; r; w, \lambda; k, m)$ of the MMHWD problem, where $G = (V, E)$, we define the edge-subset $E_{\geq l}$ is similar to $E'_{\geq l}$, *i.e.*, for each edge $e = vv' \in E_{\geq l}$, $\max\{d(r, v), d(r, v')\} \geq ((1+\epsilon)^{l-1})M$. Considering an edge $e \in E_{\geq l}$, assume that edge e is traversed by a vehicle with a speed of λ'. Note that $M \geq OPT$, where OPT is the optimal value to this instance I of the

MMHWD problem, thus the distance traveled by a vehicle with a speed of λ' is at most $\lambda' \cdot OPT \leq \lambda' \cdot M$. Since $e = vv' \in E_{\geq l}$, then we have $\lambda' \cdot OPT \geq \max\{d(r,v), d(r,v')\} \geq (1+\epsilon)^{l-1}M$, this implies that $(1+\epsilon)^{l-1}M \leq \lambda' \cdot OPT \leq \lambda' \cdot M$, it shows that $\lambda' \geq (1+\epsilon)^{l-1}$. This means that the vehicle of speed traversing through any edge of $E_{\geq l}$ is at least $(1+\epsilon)^{l-1}$.

Since the edges of $E_{\geq l}$ must be traversed by some vehicles of speed at least $(1+\epsilon)^{l-1}$ in any feasible solution, this means that $w(E_{\geq l}) \leq OPT \cdot \Lambda((1+\epsilon)^{l-1})$. Note that $M \leq OPT$, then $OPT \cdot \Lambda((1+\epsilon)^{l-1}) \leq M \cdot \Lambda((1+\epsilon)^{l-1})$, we obtain $w(E_{\geq l}) \leq OPT \cdot \Lambda((1+\epsilon)^{l-1}) \leq M \cdot \Lambda((1+\epsilon)^{l-1})$.

Considering an optimal solution $\mathcal{C}^*$ of the MMHWD problem, all tours start and end at the same depot r and collectively traverse through all vertices in $V'_{\geq l}$. Note that the length function $w(\cdot)$ satisfies the triangle inequality. Therefore, we can construct a big tour C to traverse all vertices in $G/V'_{<l}$ by the optimal solution $\mathcal{C}^*$, such that $w(C) \leq w(\mathcal{C}^* \cap E_{\geq l})$. According to the construction of the auxiliary complete graph H, we can construct a Hamiltonian cycle C' in $H/V'_{<l}$, such that $w'(C') \leq w(C)$, we have $w'(C') \leq w(\mathcal{C}^* \cap E_{\geq l})$.

Due to the subgraph corresponding to the cycle C' contains a spanning tree, such that $w'(T_{H/V'_{<l}}) \leq w'(C')$, it implying that $w'(T_{H/V'_{<l}}) \leq w(\mathcal{C}^* \cap E_{\geq l})$. Note that $w(\mathcal{C}^* \cap E_{\geq l}) \leq w(E_{\geq l}) \leq OPT \cdot \Lambda((1+\epsilon)^{l-1})$, we obtain $w'(T_{H/V'_{<l}}) \leq OPT \cdot \Lambda((1+\epsilon)^{l-1})$. Sine $M \leq OPT$, then $OPT \cdot \Lambda((1+\epsilon)^{l-1}) \leq M \cdot \Lambda((1+\epsilon)^{l-1})$, it shows that $w'(T_{H/V'_{<l}}) \leq OPT \cdot \Lambda((1+\epsilon)^{l-1}) \leq M \cdot \Lambda((1+\epsilon)^{l-1})$.

At Step 4, using an algorithm in [4], find a Level-Prim tree in H, denoted by $\mathcal{H}' = \bigcup_{i\geq 0} H'_i$, where H'_i denotes the edge-set of a minimum spanning tree for $H[V'_{\leq i}]/V'_{<i}$. Then add arc set A into $\mathcal{H}'$, *i.e.*, $\mathcal{H}_i = \mathcal{H}'_i \cup A_i$ for each $i = 0, 1, \ldots, t$. We obtain the following two lemmas.

Lemma 4. *(Gørtz et al. 2016 [4]) Given an instance $I = (G; r; w, \lambda; k, m)$ of the MMHWD problem, an auxiliary complete graph $H = (V', E' \cup A; w')$, two constants $M > 0$ and $\epsilon > 0$, a Level-Prim tree(subgraph) $\mathcal{H}' = \bigcup_{i\geq 0} H'_i$, we have*

(1) The vertex levels along every root-leaf path are nondecreasing;
(2) If $M \geq OPT$, for each $i \geq 0$, $\sum_{j\geq i} w'(H'_j) \leq (6 + \frac{6}{\epsilon}) \cdot w'(T_{H/V'_{<l}})$, where $T_{H/V'_{<l}}$ is a minimum spanning tree in the reduced graph $H/V'_{<l}$ obtained by contracting vertices $V'_{<l}$ to a single vertex in H.

Combing that Lemmas 2-4, we have the following

Lemma 5. *Given an auxiliary complete graph $H = (V', E' \cup A; w')$, two constants $M > 0$ and $\epsilon > 0$, a Level-Prim tree $\mathcal{H} = \bigcup_{i\geq 0} H_i$, we have*

(1) The vertex levels along every root-leaf path are nondecreasing;
(2) If $M \geq OPT$, for each $i \geq 0$, $\sum_{j\geq i} w'(H_j) \leq (7 + \frac{6}{\epsilon}) \cdot M \cdot \Lambda((1+\epsilon)^{i-1})$, where $T_{H/V'_{<l}}$ is a minimum spanning tree in graph $H/V'_{<l}$ obtained by contracting vertices $V'_{<l}$ to a single vertex in H.

Using Steps 6–9 of the algorithm $\mathcal{A}_{MMHWD1}$, we have the following lemma:

Lemma 6. *For each $i \geq 1$, $\sigma \in \mathcal{S}_i^u$, there exists a connected component $\pi(\sigma) \in \mathcal{S}_{i-1}$, and $\pi(\sigma) \in \mathcal{S}_{i-1}^m$.*

Proof. At first, we prove that $\pi(\sigma) \in \mathcal{S}_{i-1}$. At Step 8, for each $i \geq 1$, we have $\sigma \in \mathcal{S}_i^u$, such that $w'(\sigma) \leq \gamma \cdot (1+\epsilon)^i M$. For each $v \in V'(\sigma)$, we have $v \in V_i' \cup V_{i-1}'$. Otherwise, suppose that for any $e = vv' \in E'(\sigma)$, where $v \in V_i'$ and $v' \in V_{<i-1}'$, we have $w'(\sigma) \geq w'(vv') \geq d(r,v) - d(r,v') > (1+\epsilon)^{i-1}M - (1+\epsilon)^{i-2}M > \gamma \cdot (1+\epsilon)^i M$, which is contradicted with $\sigma \in \mathcal{S}_i^u$. Similarly, suppose that for any $a = (v', v) \in A(\sigma)$, where $v \in V_i'$, and $v' \in V_{<i-1}'$, we have $w'(\sigma) > \gamma \cdot (1+\epsilon)^i M$, *i.e.*, for each $v \in V'(\sigma)$, such that $v \in V_i' \cup V_{i-1}'$.

Suppose that the graph H is connected, for any $e = yx \in E'(\sigma)$, $\sigma \in \mathcal{S}_i^u$, where $y \in V_{\leq i-1}'$, $x \in V_i'$. Note that for each $v \in V'(\sigma)$, such that $v \in V_i' \cup V_{i-1}'$, we have $y \in V_{i-1}'$. Therefore, we can find a connected component $\pi(\sigma)$, such that $\pi(\sigma) \cap \sigma \neq \emptyset$, *i.e.*, $\pi(\sigma) \in \mathcal{S}_{i-1}$.

Next, we will prove that $\pi(\sigma) \in \mathcal{S}_{i-1}^m$ in the following setting. If $i = 1$, then $\pi(\sigma) \in \mathcal{S}_0$, *i.e.*, $\pi(\sigma) \in \mathcal{S}_0^m$. If $i \geq 2$, it is clear to see that $\pi(\sigma) \in \mathcal{S}_{i-1}$, there exists a vertex $z \in V'(\pi(\sigma))$, such that $z \in V_{<i-1}'$, we have $d(z,y) + d(y,x) \geq d(x,z)$, by the triangle inequality. And $d(r,z) + d(z,x) \geq d(r,x)$, it follows that $d(z,y) + d(y,x) \geq d(r,x) - d(r,z)$. Since $x \in V_i'$, $z \in V_{<i-1}'$, we obtain $d(r,x) - d(r,z) > (1+\epsilon)^{i-1}M - (1+\epsilon)^{i-2}M = \epsilon(1+\epsilon)^{i-2}M$, which means that $d(z,y) + d(y,x) > \epsilon(1+\epsilon)^{i-2}M$. Note that $\sigma \in \mathcal{S}_i^u$, we have $d(y,x) = w'(yx) \leq w'(\sigma) < \gamma \cdot (1+\epsilon)^i M$. Furthermore, analyzing the length of $\pi(\sigma)$, it can be deduced that, $w'(\pi(\sigma)) \geq d_{\pi(\sigma)}(z,y) \geq d(z,y) > \epsilon(1+\epsilon)^{i-2}M - d(y,x) > \epsilon(1+\epsilon)^{i-2}M - \gamma \cdot (1+\epsilon)^i M = \gamma \cdot (1+\epsilon)^{i-1}M$. *i.e.*, $\pi(\sigma) \in \mathcal{S}_{i-1}^m$, where $\gamma = \frac{\epsilon}{(2+\epsilon)(1+\epsilon)}$.

By using Step 10 of the algorithm $\mathcal{A}_{MMHWD1}$, we have the following lemma:

Lemma 7. *For each $C \in \mathcal{T}_i$, $w(C) \leq (4+4\epsilon) \cdot (1+\epsilon)^i M$.*

Proof. For each $C \in \mathcal{T}_i$, C is composed of two parts, where a path P with a maximum length of $2(1+\epsilon)^{i+1}M$ and two shortest paths from r to the two endpoints of path P, respectively. Combining that $w(P) \leq 2(1+\epsilon)^{i+1}M$, we can derive that $w(C) \leq 2(1+\epsilon)^{i+1}M + 2(1+\epsilon)^{i+1}M = (4+4\epsilon) \cdot (1+\epsilon)^i M$.

Lemma 8. $\Sigma_{C \in \mathcal{T}_i} w(C) \leq \max\{4 + \frac{4}{\epsilon}, 10\} \cdot w'(\tau \cup Dangle(\tau))$.

Proof. Depending on the value of q, there are two cases to be considered.

Case 1: If $q = 1$, *i.e.*, $\mathcal{T}_i(\tau)$ only contains a subtour.

If $i = 0$, it can be obtained $w(C) \leq 2w'(\tau \cup Dangle(\tau))$. If $i \geq 1$, by using Lemma 5, there exists $u \in V'(\tau)$, such that $u \in V_{<i-1}'$. According to the construction of tour C, it consists of two shortest paths from r to u and "double" the edge and arc of $\tau \cup Dangle(\tau)$ twice. Thus, we have $w(C) \leq 2(1+\epsilon)^{i-1}M + 2w'(\tau \cup Dangle(\tau))$. Since $\tau \in \mathcal{S}_i^m$, we have $w'(\tau) \geq \gamma \cdot (1+\epsilon)^i M$ by using Step 8, where $\gamma = \frac{\epsilon}{(2+\epsilon)(1+\epsilon)}$¡£This shows that $(1+\epsilon)^{i-1}M \leq \frac{w'(\tau)}{\gamma \cdot (1+\epsilon)} = \frac{2+\epsilon}{\epsilon} \cdot w'(\tau)$. It follows that $w(C) \leq 2(1+\epsilon)^{i-1}M + 2w'(\tau \cup Dangle(\tau)) \leq 2 \cdot \frac{2+\epsilon}{\epsilon} \cdot w'(\tau) + 2w'(\tau \cup Dangle(\tau)) = (2 \cdot \frac{2+\epsilon}{\epsilon} + 2) \cdot 2w'(\tau \cup Dangle(\tau)) = (4 + \frac{4}{\epsilon}) \cdot w'(\tau \cup Dangle(\tau))$, it implying that $\sum_{C \in \mathcal{T}_i(\tau)} w(C) = w(C) \leq (4 + \frac{4}{\epsilon}) \cdot w'(\tau \cup Dangle(\tau))$.

Case 2: If $q \geq 2$, $\mathcal{T}_i(\tau)$ contains at least two subtours.

We obtain $4w'(\tau \cup Dangle(\tau)) > (q-1) \cdot 2(1+\epsilon)^{i+1}M$ by using Step 10, *i.e.*, $(1+\epsilon)^{i+1}M < \frac{2w'(\tau \cup Dangle(\tau))}{(q-1)}$. Since $E'(\tau \cup Dangle(\tau)) \subseteq E'_{\leq i+1}$ and $A(\tau \cup Dangle(\tau)) \subseteq A_{\leq i+1}$, we have $V'(\tau \cup Dangle(\tau)) \subseteq V'_{\leq i+1}$, it means that the shortest length from r to u is at most $(1+\epsilon)^{i+1}M$, it can be deduced that $\sum_{C \in \mathcal{T}_i(\tau)} w(C) \leq q \cdot 2(1+\epsilon)^{i+1}M + 2w'(\tau \cup Dangle(\tau)) \leq 2q \cdot \frac{2w(\tau \cup Dangle(\tau))}{(q-1)} + 2w'(\tau \cup Dangle(\tau)) = (\frac{4q}{q-1} + 2) \cdot w'(\tau \cup Dangle(\tau))$. Note that $q \geq 2$, we have$\sum_{C \in \mathcal{T}_i(\tau)} w(C) \leq 10 \cdot w'(\tau \cup Dangle(\tau))$. Thus, we have $\sum_{C \in \mathcal{T}_i(\tau)} w(C) \leq \max\{4 + \frac{4}{\epsilon}, 10\} \cdot w'(\tau \cup Dangle(\tau))$.

Lemma 9. *If $\sum_{j \geq i} w'(H_j) \leq (7 + \frac{6}{\epsilon}) \cdot M \cdot \Lambda((1+\epsilon)^{i-1})$ holds for each $i \geq 0$, then the collection $\bigcup_{i \geq 0} \mathcal{T}_i$ obtained at Step 11 is $(\alpha, \beta)_{M,\epsilon}$ - assignable, where $\alpha = 4 + 4\epsilon$, $\beta = (7 + \frac{6}{\epsilon}) \cdot \max\{4 + \frac{4}{\epsilon}, 10\}$.*

Proof. We will prove the $\bigcup_{i \geq 0} \mathcal{T}_i$ satisfies two properties in the Definition 1.

At first, by using Lemma 6, for any $C \in \mathcal{T}_i$, we have $w(C) \leq (4 + 4\epsilon) \cdot (1 + \epsilon)^i M$, which satisfies the first property in the Definition 1, where $\alpha = 4 + 4\epsilon$. Secondly, for each $i \geq 0$, we have $\sum_{j \geq i} w(\mathcal{T}_j) \leq \max\{4 + \frac{4}{\epsilon}, 10\} \cdot \sum_{j \geq i} w'(H_j)$. Since $\{Dangle(\tau) \mid \tau \in \mathcal{S}_j^m\}$ is a partition of $\mathcal{S}_{j+1}^u$, it can be derived that, $\sum_{j \geq i}(w'(\mathcal{S}_j^m)) + \sum_{j \geq i}(w'(\mathcal{S}_{j+1}^u)) \leq \sum_{j \geq i} w'(\mathcal{S}_j)$. Note that $\mathcal{S}_j$ is the set of all connected components in H_j, *i.e.*, $H_{\geq i} = \bigcup_{j \geq i} \mathcal{S}_j$, we obtain $\sum_{j \geq i}(w'(\mathcal{S}_j^m) + w'(\mathcal{S}_{j+1}^u)) \leq \sum_{j \geq i} w'(H_j)$. By using Lemma 6, $\sum_{\tau \in \mathcal{S}_j^m} w'(\tau \cup Dangle(\tau)) = w'(\mathcal{S}_j^m) + w'(\mathcal{S}_{j+1}^u)$. By using Lemma 8, it follows $w(\mathcal{T}_j) = \sum_{\tau \in \mathcal{S}_j^m} w(\mathcal{T}_j(\tau)) = \sum_{\tau \in \mathcal{S}_j^m} \sum_{C \in \mathcal{T}_j(\tau)} w(C) \leq \max\{4 + \frac{4}{\epsilon}, 10\} \cdot \sum_{\tau \in \mathcal{S}_j^m} w'(\tau \cup Dangle(\tau))$. It follows that $w(\mathcal{T}_j) < \max\{4 + \frac{4}{\epsilon}, 10\} \cdot (w'(\mathcal{S}_j^m) + w'(\mathcal{S}_{j+1}^u))$. Thus, it can be concluded that, $\sum_{j \geq i} w(\mathcal{T}_j) \leq \max\{4 + \frac{4}{\epsilon}, 10\} \cdot \sum_{j \geq i}(w'(\mathcal{S}_j^m) + w'(\mathcal{S}_{j+1}^u)) \leq \max\{4 + \frac{4}{\epsilon}, 10\} \cdot \sum_{j \geq i} w'(H_j)$. Combining the Lemma 5, for each $i \geq 0$, we have $\sum_{j \geq i} w'(H_j) \leq (7 + \frac{6}{\epsilon}) \cdot M \cdot \Lambda((1+\epsilon)^{i-1})$, then $\sum_{j \geq i} w(\mathcal{T}_j) \leq (7 + \frac{6}{\epsilon}) \cdot \max\{4 + \frac{4}{\epsilon}, 10\} \cdot M \cdot \Lambda((1+\epsilon)^{i-1})$, which satisfies the second property in the Definition 1, where $\beta = (7 + \frac{6}{\epsilon}) \cdot \max\{4 + \frac{4}{\epsilon}, 10\}$.

Now, using the algorithm $\mathcal{A}_{MMHWD1}$, we obtain the following

Theorem 1. *Given a constant $\delta > 0$, the algorithm $\mathcal{A}_{MMHWD1}$ is an $134.4434(1+\delta)$-approximation algorithm to solve the MMHWD problem, and it runs in time $O(n^6 + \frac{n^4}{\delta} + \log w(E))$, where $w(E) = \sum_{e \in} w(e)$, $n = |V|$.*

Proof. If $M \geq OPT$, for each $i \geq 0$, we have $\sum_{j \geq i} w'(H_j) > (7 + \frac{6}{\epsilon}) \cdot M \cdot \Lambda((1+\epsilon)^{i-1}))$ by the Lemma 5. Set $M := (1+\delta)M$. Otherwise, execute Steps 6–13, which satisfies $M \leq (1+\delta) \cdot OPT$.

When fixing $\epsilon = 1.6746$, by applying Lemma 9, the collection $\bigcup_{i \geq 0} \mathcal{T}_i$ obtained at Step 13 of the algorithm $\mathcal{A}_{MMHWD1}$ is $(\alpha, \beta)_{M,\epsilon}$-assignable, where $\alpha = 10.6984$, $\beta = 105.8295$. By using Lemma 1, the collection $\bigcup_{i \geq 0} \mathcal{T}_i$ obtained at Step 11 is assigned to k vehicles in polynomial time, such that the completion time of each vehicle is at most $134.4434 \cdot M$, which means that $OUT \leq 134.4434 \cdot M \leq 134.4434(1+\delta) \cdot OPT$. Thus, the algorithm $\mathcal{A}_{MMHWD1}$ is an $134.4434(1+\delta)$-approximation algorithm to solve the MMHWD problem.

The time complexity of the algorithm $\mathcal{A}_{MMHWD1}$ can be determined in the following ways. (1)The algorithm will stop before $M > (1+\delta) \cdot OPT$. By assuming that the vehicle of the maximum speed is used to be assigned all arcs in A, we can obtain $(1+\delta) \cdot OPT \leq (1+\delta) \cdot \frac{\sum_{(v,v') \in A}(d(r,v)+w'(v,v')+d(v',r))}{\lambda_{\max}} \leq (1+\delta) \cdot |A| \cdot \frac{\max_{(v,v') \in A}(d(r,v)+w'(v,v')+d(v',r))}{\lambda_{\max}}$. Note that the initial value of M and each iteration increases by a multiple of $(1+\delta)$, thus the maximum number of iterations of Step 2-5 is $O(\frac{1}{\delta} \log |E|)$ $(|A| \leq |E|)$. It implies that Steps 2-5 run in at most $O(\frac{n^4}{\delta})$. Since the Step 6–13 run in time $O(n^6 + \log w(E))$, the algorithm $\mathcal{A}_{MMHWD1}$ can be implemented in time $O(n^4(n^2 + \frac{1}{\delta}) + \log w(E))$.

4 An Approximation Algorithm with Lower Time Complexity for the MMHWD Problem

In this section, we design a new algorithm to solve the MMHWD problem, denoted by the algorithm $\mathcal{A}_{MMHWD2}$. The strategy of the algorithm is to find a tour C_0 to cover a depot r and all arcs, and split C_0 into k subpaths, connect the two endpoints of each path to depot r, to obtain k mixed tours.

Algorithm: $\mathcal{A}_{MMHWD2}$
Input: An instance $I = (G; r; w, \lambda; k, m)$ of the MMHWD problem;
Output: A set of k tours $\mathcal{C} = \{C_j \mid j = 1, 2, \ldots, k\}$.
Begin
Step 1. Construct an auxiliary graph $G' = (V, E' \cup A; w)$, where $E' = E \cup \{s_j t_j \mid j = 1, 2, \ldots, m\}$ and $A = \{(s_j, t_j) \mid j = 1, 2, \ldots, m\}$, we denote $w(uv) = d(u, v)$ for each edge $uv \in E' \backslash E$ and $w(u, v) = d(u, v)$ for each arc $(u, v) \in A$, where $d(u, v)$ is the length of shortest path from u to v in G;
Step 2. Use the Algorithm $\mathcal{A}_{SCP}$ [3] to produce a tour C_0 to cover a depot r and all arcs in A, denoted by $C_0 = (v_1, v_2, \ldots, v_{l(C_0)}, v_1)$, where $v_1 = r$;
Step 3. For $(j = 1, 2, \ldots, k-1)$ do:
– **3.1** Find the last vertex $v_{i_{p'(j)}}$ such that the length of the path from v_1 to $v_{i_{p'(j)}}$ along C_0 such that L_j, where $L_j = \frac{\Sigma_{i=1}^{j} \lambda_i}{\Sigma_{i=1}^{k} \lambda_i}(w(C) - 2w_0) + w_0$, $w_0 = \frac{1}{2} \max\{w(P_{r,u}) + w(u, v) + w(P_{r,v}) \mid (u, v) \in A\}$; Set $r_j = L_j - w(C[v_1, v_{i_{p'(j)}}])$; Let R_j to denote the jth subpath;
– **3.2** If $((v_{i_{p'(j)}}, v_{i_{p'(j)+1}})$ is an edge) then
–– Set $v_{i_{p(j)}} = v_{i_{p'(j)}}$, the terminal vertex of R_j is $v_{i_{p(j)}}$, the initial vertex of R_{j+1} is $v_{i_{p(j)+1}}$;
– Else
–– If $(w(C[v_1, v_{i_{p'(j)}}]) + r_j \leq w(v_{i_{p'(j)}}, v_{i_{p'(j)+1}}) - r_j + w(C[v_{i_{p'(j)+1}}, v_1)))$ then
–– Set $v_{i_{p(j)}} = v_{i_{p'(j)}}$, the initial vertex of R_{j+1} is $v_{i_{p(j)}}$;

-- If $((v_{i_{p(j)-1}}, v_{i_{p(j)}})$ is an arc) then
-- Set the terminal vertex of R_j is $v_{i_{p(j)}}$;
-- Else
-- Set the terminal vertex of R_j is $v_{i_{p(j)-1}}$;
-- Else
-- Set $v_{i_{p(j)}} = v_{i_{p'(j)+1}}$, the terminal vertex of R_j is $v_{i_{p(j)}}$;
-- If $((v_{i_{p(j)}}, v_{i_{p(j)+1}})$ is an arc) then
-- Set the initial vertex of R_{j+1} is $v_{i_{p(j)}}$;
-- Else
-- Set the initial vertex of R_{j+1} is $v_{i_{p(j)+1}}$;

Step 4. Set a set of k tours as $\mathcal{C}' = \{C'_j \mid j = 1, 2, \ldots, k\}$, where $C'_1 = (v_1, v_{i_2}, \ldots, v_{i_{p(1)}}, v_1)$, $C'_2 = (v_1, v_{i_{p(1)}}, \ldots, v_{i_{p(2)}}, v_1)$, $\ldots$, $C'_k = (v_1, v_{i_{p(k-1)}}, \ldots, v_1)$, $v_1 = r$;

Step 5. Restore the arcs in A and the edges in E' to the shortest path in the $\mathcal{C}'$ to obtain k tours, denoted by $\mathcal{C} = \{C_j \mid j = 1, 2, \ldots, k\}$;

Step 6. Output "k cycles $\mathcal{C} = \{C_j \mid j = 1, 2, \ldots, k\}$ related to k vehicles".

End

Now, using the algorithm $\mathcal{A}_{MMHWD2}$, we obtain the following

Theorem 2. *Given an instance $I = (G; r; w, \lambda; k, m)$ of the MMHWD problem, the algorithm $\mathcal{A}_{MMHWD2}$ is an $(\varphi + \frac{9}{5} - \frac{1}{k})$-approximation algorithm to solve the MMHWD problem, and it runs in time $O(n^3)$, where $\varphi = \frac{\lambda_{\max}}{\lambda_{\min}}$, $\lambda_{\max} = \max\{\lambda_i \mid i = 1, 2, \ldots, k\}$, $\lambda_{\min} = \min\{\lambda_i \mid i = 1, 2, \ldots, k\}$, $n = |V|$.*

Proof. Given an instance $I = (G; r; w, \lambda; k, m)$ of the MMHWD problem, let OPT denote the optimal value to the instance I. Suppose that $\mathcal{C}^* = \{C_j^* \mid j = 1, 2, \ldots, k\}$ is an optimal solution, and $\mathcal{C} = \{C_j \mid j = 1, 2, \ldots, k\}$ is the cycles outputted by the algorithm $\mathcal{A}_{MMHWD2}$.

If $k = 1$, its easy to check that $OUT \leq \frac{9}{5}OPT$, implying that the algorithm $\mathcal{A}_{MMHWD2}$ is an $\frac{9}{5}$-approximation algorithm to solve the MMHWD problem.

If $k \geq 2$, consider the algorithm $\mathcal{A}_{MMHWD2}$ the output of the jth tour C_j, $j = 1, 2, \ldots, k$, because the set of tours output by the algorithm $\mathcal{A}_{MMHWD2}$ traversing all arcs in A, we have $\frac{2w_0}{\lambda_{\max}} \leq OPT$, where $w_0 = \frac{1}{2}\max\{w(P_{r,u}) + w(u, v) + w(P_{r,v}) \mid (u, v) \in A\}$.

Note that in an optimal solution $\mathcal{C}^*$ of the MMHWD problem, each tour starting and ending at the same depot r and traverse all arcs in A exactly once, such that m the items are delivered. We can construct a big tour C traversing all arcs in A by using the optimal solution $\mathcal{C}^*$. Therefore, we have $w(C) \leq \frac{9}{5}\sum_{i=1}^{k} w(C_i^*) \leq \frac{9}{5}OPT \cdot \sum_{i=1}^{k}\lambda_i$, by the triangle inequality. Using the definition of w_0 and the triangle inequality, we can deduce that $w(P_{r,v_{i_{p(j)}}}) + w(v_{i_{p(j)}}, v_{i_{p(j)+1}}) + w(P_{v_{i_{p(j)+1}},r}) \leq 2w_0$. It implying that, $\min\{w(P_{r,v_{i_{p(j)}}}) + r_j, w(v_{i_{p(j)}}, v_{i_{p(j)+1}}) - r_j + w(P_{v_{i_{p(j)+1}},r})\} \leq w_0$. Similarly, $\min\{w(P_{r,v_{i_{p(j-1)}}}) + r_{j-1}, w(v_{i_{p(j-1)}}, v_{i_{p(j-1)+1}}) - r_{j-1} + w(P_{v_{i_{p(j-1)+1}},r})\} \leq w_0$.

Considering the worst-case for the length of the tour C_j. At this time, the tour C_j is to start from r to $v_{i_{p(j-1)}}$ along C_0, traversing through $v_{i_{p(j)}}$ to $v_{i_{p(j)+1}}$, and finally returning from $v_{i_{p(j)+1}}$ to r, we easily check that $w(P_{r,v_{i_{p(j-1)}}})+r_{j-1} \leq w_0$, and $w(v_{i_{p(j)}}, v_{i_{p(j)+1}}) - r_j + w(P_{v_{i_{p(j)+1}},r}) \leq w_0$, then we have the following, $\frac{w(C_j)}{\lambda_j} \leq \frac{w_0 + \frac{\lambda_j}{(\sum_{i=1}^k \lambda_i)}(w(C)-2w_0)+w_0}{\lambda_j} = \frac{2w_0}{\lambda_j} + \frac{w(C)-2w_0}{\sum_{i=1}^k \lambda_i} \leq \frac{2w_0}{\lambda_{\min}} + \frac{w(C)}{\sum_{i=1}^k \lambda_i} - \frac{2w_0}{\sum_{i=1}^k \lambda_i} \leq \frac{2w_0 \cdot \lambda_{\max}}{\lambda_{\min} \cdot \lambda_{\max}} + \frac{w(C)}{\sum_{i=1}^k \lambda_i} - \frac{2w_0}{k \cdot \lambda_{\max}} \leq (\frac{\lambda_{\max}}{\lambda_{\min}} - \frac{1}{k}) \cdot \frac{2w_0}{\lambda_{\max}} + \frac{9}{5} \cdot OPT \leq (\frac{\lambda_{\max}}{\lambda_{\min}} - \frac{1}{k} + \frac{9}{5}) \cdot OPT$.

Thus, in this case, we can obtain $\frac{w(C_j)}{\lambda_j} \leq (\frac{\lambda_{\max}}{\lambda_{\min}} - \frac{1}{k} + \frac{9}{5}) \cdot OPT$. When $j = 1, 2, \ldots, k$, the above equation also holds, which means that $OUT \leq (\frac{\lambda_{\max}}{\lambda_{\min}} - \frac{1}{k} + \frac{9}{5}) \cdot OPT$, its implying that the algorithm $\mathcal{A}_{MMHWD}$ is an $(\varphi + \frac{9}{5} - \frac{1}{k})$-approximation algorithm to solve the MMHWD problem, where φ is the ratio of the largest vehicle speed to the smallest one. The whole running time of the algorithm $\mathcal{A}_{MMHWD2}$ is at most $O(n^3)$, where n is the number of vertices.

5 Conclusions and Further Research

In this paper, we obtain the two main results to solve the MMHWD problem. (1) Given any small constant $\delta > 0$, we design an $134.4434(1+\delta)$-approximation algorithm to solve the MMHWD problem, where the running time required is bounded by a polynomial in the input size and $\frac{1}{\delta}$; (2) We provide an $(\varphi + \frac{9}{5} - \frac{1}{k})$-approximation algorithm to solve the MMHWD problem, where φ is the ratio of the largest vehicle speed to the smallest one. The further research is to consider the unrooted min-max heterogeneous weighted delivery problem, we hope to design an approximation algorithm to solve this problem in polynomial time.

References

1. Bärtschi, A., et al.: Energy-efficient delivery by heterogeneous mobile agents. In: 34th Symposium on Theoretical Aspects of Computer Science (STACS 2017), vol. 66, p. 1–14. Schloss Dagstuhl-Leibniz-Zentrum für Informatik (2017)
2. Bilò, D., Gualà, L., Leucci, S., Proietti, G., Rossi, M.: New approximation algorithms for the heterogeneous weighted delivery problem. Theoret. Comput. Sci. **932**, 102–115 (2022)
3. Frederickson, G.N., Hecht, M.S., Kim, C.E.: Approximation algorithms for some routing problems. SIAM J. Comput. **7**, 178–193 (1978)
4. Gørtz, I.L., Molinaro, M., Nagarajan, V., Ravi, R.: Capacitated vehicle routing with nonuniform speeds. Math. Oper. Res. **41**(1), 318–331 (2016)
5. Guan, M.G.: Graphic programming using odd or even points (in Chinese). Acta Math. Sin. 10:263–266. English translation: Chin. Math., vol 1, pp. 263–266 (1962)
6. Korte, B.H., Vygen, J., Korte, B., Vygen, J.: Combinatorial optimization (the Sixth Edition), Springer (2018)
7. Li, J.P., Pan, P.X., Lichen, J.R., Cai, L.J., Wang, W.C., Liu, S.D.: Approximation algorithms for solving the heterogeneous Chinese postman problem. J. Comb. Optim. **45**(1), 17 (2023)
8. Vazirani, V.V.: Approximation algorithms. Springer (2001)

Exact and Approximate Heuristics for the Multi-modal Stable Matching with Applications in Industry Chains

Yang Yang[1,2] and Yicheng Xu[1,2](✉)

[1] Shenzhen Institute of Advanced Technology, Chinese Academy of Sciences, Shenzhen, China
{y.yang3,yc.xu}@siat.ac.cn
[2] University of Chinese Academy of Sciences, Beijing, China

Abstract. In this paper, we tackle the multi-modal globally stable matching problem (MGBSM), a challenge crucial for optimizing industry chain partnerships where agents generate preferences based on various criteria. The goal is to achieve a perfect matching where every agent is paired with a partner, and no unmatched pair prefers each other over their current partners across all criteria. Given the NP-hard nature of this problem, the only exact solution method to date requires exhaustive enumeration of all stable matchings. Our key contribution is the development of efficient algorithms to solve MGBSM, beginning with its transformation into a clique problem in graph theory. By harnessing the properties of both stable matchings and cliques, we design an exact algorithm that efficiently determines the existence of a multi-modal globally stable matching and provides a solution when one exists. Recognizing the limitations of perfect matching, especially in practical industry chain scenarios, we introduce the concept of multi-modal group stable matching (MGSM), which relaxes the perfect matching requirement. We focus on identifying the maximum cardinality MGSM (MaxMGSM) and develop a polynomial-time, approximation-preserving reduction from MaxMGSM to the maximum clique problem. This is followed by a fast polynomial-time approximation algorithm with a provable performance guarantee, offering a practical solution for complex industry chain networks.

Keywords: Industry chain · Stable matching · Multi-modal stable matching · The clique problem

1 Introduction

The matching problem has been extensively studied in fields such as economic algorithms, matching markets, and social choice theory. Given an undirected graph with a vertex set V and an edge set E, the goal of the matching problem is to find a set of edges such that no two edges share the same vertex. If no additional edge can be added to this set without violating the matching condition,

D. Du et al. (Eds.): COCOA 2024, LNCS 15434, pp. 364–375, 2025.
https://doi.org/10.1007/978-981-96-4445-2_30

the set is called a maximal matching. Among all maximal matchings, the one with the largest size is known as the maximum matching, which can be found in polynomial time [12].

In traditional matching problems, all potential choices for an agent are treated equally. For instance, if agent b and c are both connected to agent a, they are considered equally viable partners for a. However, in many real-world scenarios, agents may have preferences over their potential matches. For example, a may prefer b to c. This leads to a variant of the matching problem known as the stable matching problem [7]. In this problem, the agents are divided into two sets, and each agent ranks all agents from the other set in a strict order of preference. A blocking pair is defined as an unmatched pair where both agents prefer each other over their current partners, or they are both unmatched. The stable matching problem in bipartite graphs, also known as the stable marriage problem, aims to pair each agent with a partner such that no blocking pair exists. Gale and Shapley provided an algorithm that can compute a stable matching for any bipartite instance in polynomial time. The concept of stability also extends to general graphs, as in the stable roommates problem, where each agent ranks all other agents in a strict order of preference. The objective is to pair all agents such that no blocking pair occurs and this problem can also be solved in polynomial time [9].

The stable matching problem and its variants have been extensively explored over the years. The stable marriage problem with incomplete lists (SMI) studied by Gusfield and Irving allows agents to rank only a subset of candidates in the other set [8]. The stable marriage problem with ties (SMT) permits agents to rank some options equally when they cannot determine a strict preference [10]. The combination of these two problems, known as the stable marriage problem with ties and incomplete lists (SMTI), has also been investigated [11].

Related Work: While the original stable marriage and stable roommates problems consider only a single criterion for generating preferences, real-world scenarios often involve multiple criteria. Chen, Niedermeier and Skowron introduced the multi-modal stable matching problem, where each agent ranks all available choices in different orders based on various criteria [4]. In their work, three types of stability are defined: global stability, pair stability, and individual stability. In the all-layer multi-modal globally stable matching problem, the goal is to find a perfect matching that remains stable across all criteria. This problem has been proven to be NP-hard when the number of criteria is two or more. The concept of α-layer pair stability requires that no unmatched pair blocks the matching in more than α criteria. Similarly, α-layer individual stability is defined such that, for each unmatched pair, at least one agent prefers their current partner in at least α criteria. While agents in the multi-modal stable matching problem have different but fixed preference lists based on various criteria, the stable matching problem with evolving preferences allows these preferences to change over time. Chen, Skowron, and Sorge explored this problem and introduced two definitions of stability: robustness and near stability [5]. Robustness requires that the matching remains stable regardless of how preferences change over time.

Near stability, on the other hand, allows agents to adjust their preference lists within an acceptable range to maintain stability. Another approach to solving the stable matching problem with evolving preferences, by adjusting the current matching within acceptable bounds, was also investigated in [2].

The application of stable matching extends beyond theoretical models and has significant implications in industry chain management, where matching suppliers with buyers based on multiple criteria is crucial. In industry chains, factors such as cost, quality, reliability, and delivery time play critical roles in forming stable, long-term relationships. The work by Ostrovsky highlights the relevance of stable matching in industry chain networks by showing how stability can be maintained in complex supplier-buyer relationships [13]. Similarly, the research by Yang, Chen and Kumara demonstrates how stable matching can be applied to match customers with additive manufacturing providers in a decentralized market [16]. The study reveals that stable matching outperforms traditional optimization methods, such as Mixed Integer Linear Programming (MILP), particularly in scenarios with varying demand-industry ratios and system complexities. However, traditional stable matching models, which consider only a single criterion, may fall short in these multi-faceted environments. This underscores the importance of multi-modal stable matching, which can account for multiple criteria simultaneously, ensuring stability across various dimensions. Multi-modal stable matching provides a more comprehensive approach to managing industry chains, where the dynamics of supplier-buyer interactions require balancing numerous, often conflicting, factors. Thus, the integration of multi-modal stable matching into industry chain management offers a more robust framework for optimizing complex networks, leading to more resilient and efficient operations. Besides stable matching, other optimization models such as arc routing and non-submodular maximization have shown significant applicability in complex, multi-modal systems [17] [15]. These approaches, along with algorithms for bin packing problems [18] and frequency allocation in cellular networks [3], provide valuable insights for tackling complex, real-world optimization challenges in the scenario of industry chain.

Our Contribution: In this paper, we address the challenging problem of multi-modal globally stable matching (MGBSM) in general graphs. Traditional stable matching algorithms are efficient because they focus on finding a single stable matching with specific properties, allowing them to discard stable matchings that do not meet these requirements. For instance, Gale-Shapley's algorithm generates a stable matching that is most favorable for one side and least favorable for the other. However, when dealing with MGBSM, similar algorithms cannot be applied because a stable matching that satisfies certain properties for one criterion may not even be stable under another criterion. This necessitates considering all possible stable matchings, rather than just those that meet specific properties. Unfortunately, the process of listing all stable matchings for even a single criterion has a time complexity of $\mathcal{O}(3.55^n)$, where n is the number of agents [14]. Therefore, in this paper, we aim to approach the solution to MGBSM in a more efficient manner, both theoretically and practically.

Our first major contribution is the transformation of MGBSM input into an instance of the clique problem. By leveraging the inherent properties of both stable matchings and cliques, we develop an exact algorithm that efficiently determines whether a multi-modal globally stable matching exists in a given input and provides a solution when it does. This algorithm marks a significant advancement over the previously known method, which involved enumerating all possible stable matchings, as it drastically reduces time complexity and demonstrates superior practical performance. Recognizing that a globally stable matching may not always exist, we introduce the novel concept of multi-modal group stable matching (MGSM) and focus on its optimization version, MaxMGSM. Our second key contribution is the construction of an approximation-preserving reduction from MaxMGSM to the maximum clique problem (MCP). Building on this, we present a fast polynomial-time approximation algorithm with a provable performance guarantee for finding a near maximum MGSM. Our approximation algorithm effectively eliminates unnecessary vertices, which not only accelerates the computation process but also leads to a more precise solution space.

Moreover, the practical applications of our work are particularly significant in the context of industry chain management. In complex industry chains, where multiple criteria such as cost, quality, and delivery time must be balanced, traditional single-criterion stable matching models are insufficient. Multi-modal stable matching provides a framework to incorporate these diverse criteria, ensuring that the resulting matching are stable across all relevant factors. By applying our exact and approximation algorithms, industry chains can efficiently identify stable, optimal or near-optimal matchings that satisfy all parties involved, thus enhancing the resilience and efficiency of the entire network. This integration of multi-modal stable matching into industry chain management offers a robust solution for optimizing partnerships in a highly dynamic and competitive environment.

2 An Exact Algorithm for MGBSM

In this section, we introduce the notations and definitions essential for the design and analysis of our algorithms. To highlight the underlying properties of MGBSM, we describe the problem using linear constraints and transform these constraints into an instance of the clique problem. Based on this transformed instance, we then introduce our exact algorithm and provide a detailed analysis of its correctness and time complexity.

We start by introducing the necessary notations for the stable matching problem. Given a set A of n agents, for any agent $a, b, c \in A$ with $a \neq b \neq c$, we use $(a|\ b\ c)$ to indicate that agent a prefers b over c. The preference list and preference table are defined as follows.

Definition 1 (Preference List & Preference Table). *For agent i, a preference list of i is a strict ranking of all other agents based on a specific criterion. A preference table is the collection of preference lists for all agents corresponding to a given criterion.*

We present the formal definition of MGBSM as follows.

Definition 2 (MGBSM). *Given a set* A *of* n *agents and* ω *criteria, each agent generates a preference list based on each criterion. The multi-modal globally stable matching problem (MGBSM) seeks to determine whether there exists a matching in which every agent is paired with a unique partner, and no pair of unmatched agents prefers each other over their current partners under any criterion.*

To better elucidate the underlying properties of MGBSM, we describe the problem using a series of linear constraints. For any $i, j \in A$ with $i \neq j$, we define the variable x_{ij}, which is set to 1 if agent i is paired with agent j, and 0 otherwise. We construct the constraints for MGBSM as follows.

$$\sum_{j \in A \setminus \{i\}} x_{ij} = 1, \forall i \in A, \tag{1}$$

$$\sum_{s \in L_{i,j,k}} x_{is} + \sum_{v \in L_{j,i,k}} x_{jv} \leq 1, \forall i \in A, j \in A \setminus \{i\}, k \in \omega. \tag{2}$$

Type (1) constraints are the matching constraints, which require that each agent i must be matched with exactly one partner from the set $A \setminus \{i\}$. Type (2) constraints capture the stability of MGBSM. In constraint (2), $L_{i,j,k}$ represents the set of agents ranked lower than j in the preference list of i for the k-th criterion, where $k \in [\omega]$.

We apply the following algorithm to transform the above constraints to an instance of a clique problem.

Algorithm 1. Clique Instance Construction

Input: a set of variables x_{ij} in which $i \in A$ and $j \in A \setminus \{i\}$; a set V of vertices, where for any $i, j \in A$ and $i \neq j$, v_{ij} corresponds to x_{ij}; a set of constraints C_1, C_2 ..., C_q;
Output: an instance of the clique problem;
1: construct a complete graph G using vertices in V;
2: **for** $p \in 1, 2, \ldots, q$ **do**
3:     **for** x_{ab} and x_{cd} that are both in C_p, where $a \neq b \neq c \neq d$ **do**
4:         delete edge (v_{ab}, v_{cd}) in G
5:     **end for**
6: **end for**
7: **return** the graph G.

Building on the construction of graph G, we present our exact algorithm for MGBSM as follows.

Algorithm 2. An Exact Algorithm for MGBSM

Input: a graph $G = (V, E)$, where each vertex in V is labeled as v_{ij}, with $i, j \in A$ and $i \neq j$; an empty set M_G of candidate vertices;

Output: a multi-modal globally stable matching or a conclusion that no multi-modal globally stable matching exists;

1: delete vertices with a degree less than $\frac{|A|}{2} - 1$;

2: delete vertices with a degree of $\frac{|A|}{2} - 1$ that have neighbors sharing the same elements in their labels;

3: delete vertices that are connected to all other vertices and add them to the candidate set M_G;

4: repeat steps 1 to 3 until no more vertices are deleted;

5: apply Depth-First Search (DFS) to check the connectivity of the current graph;

6: **if** G is disconnected **then**

7:     decompose G into connected components; create an empty candidate set M_i for each component i;

8:     **for** each connected component G_i **do**

9:         apply steps 1 to 5 to G_i with an updated degree threshold of $\frac{|A|}{2} - 1 - |M_G|$, and update the candidate set M_i;

10:         **if** the number of vertices in G_i is less than $\frac{|A|}{2} - |M_G| - |M_i|$ **then**

11:             delete the component G_i;

12:         **end if**

13:     **end for**

14: **end if**

15: **for** each remaining connected component G_i **do**

16:     execute a clique detection algorithm to find a clique of size $\frac{|A|}{2} - |M_G| - |M_i|$;

17:     **if** such a clique is found **then**

18:         **return** the labels of vertices in the found clique;

19:     **end if**

20: **end for**

21: **return** no multi-modal globally stable matching exists in the given instance.

We present the following theorem and its corresponding proof to establish the correctness as well as time complexity of our algorithm.

Theorem 1. *For a MGBSM instance with n agents and ω criteria, Algorithm 2 can determine the existence of multi-modal globally stable matching in $\mathcal{O}\left(2^{|V|}\right)$ time.*

Proof. In the constructed graph $G = (V, E)$, for $i, j \in A$ with $i \neq j$, each vertex $v_{ij} \in V$ corresponds to a variable x_{ij}, indicating whether agent i and j are selected as part of a multi-modal globally stable matching. Similarly, edge $(v_{ab}, v_{cd}) \in E$ represents whether pair (a, b) and (c, d) can coexist in a stable matching. Therefore, for a given instance with n agents and ω criteria, determining whether there is a perfect matching that is stable for all criteria is equivalent to checking whether there is a clique of size $\frac{|A|}{2}$ in the constructed graph G. Since each vertex in the target clique has a degree of $\frac{|A|}{2} - 1$, we can eliminate vertices with a degree less than $\frac{|A|}{2} - 1$. For any vertex with a degree of exactly $\frac{|A|}{2} - 1$, if it has two neighbors, v_{ab} and v_{bc}, where the labels share the same element b, then the vertex should be deleted. This is because these two neighbors cannot be part of the same clique, as they violate the matching constraint. Since there is no stable matching larger than $\frac{|A|}{2}$, the maximum size of the clique in the constructed graph is also $\frac{|A|}{2}$. Therefore, if a vertex v_{ij} is connected to all other vertices, indicating that pair (i, j) can coexist with all other pairs, then (i, j) must be a part of any multi-modal globally stable matching that exists for the given instance. Since our deletion process may affect the degree of certain vertices, we repeat it until no more vertices are deleted. After this process, the remaining graph may become sparse, potentially resulting in a disconnected graph. To address this problem, we apply the DFS algorithm to decompose the graph into several connected components and then search for cliques within each component separately. As we already have candidate vertices for cliques in the set M_G and M_i, we focus on finding cliques of size $\frac{|A|}{2} - |M_G| - |M_i|$ within these connected components. Our algorithm will eventually return a multi-modal globally stable matching if exists.

In terms of time complexity, since the maximum size of a clique in graph G is no larger than $\frac{n}{2}$, we utilize an algorithm for finding a maximum clique rather than enumerating all maximal cliques. Consequently, the worst-case time complexity of our algorithm is upper bounded by $\mathcal{O}\left(2^{|V|}\right)$, where $|V|$ is the number of vertices left in graph G. Notably, because graph G can be quite sparse, decomposing G into several connected components and applying the maximum clique problem (MCP) algorithm separately allows us to significantly reduce the exponent, which in turn accelerates the processing time of our algorithm in practice. □

Figure 1 illustrates the efficiency of our algorithm in reducing the solution space using randomly generated preference tables. In addition, we conducted experiments on a real-world manufacturer-supplier dataset for cosmetic products. Two preference criteria were generated, based on factors such as lead time, defect rates, order quantities, and manufacturing lead time. After several rounds of vertex deletion using our algorithm, the structure of the multi-modal globally stable matching becomes evident, as shown in Fig. 2.

3 Approximation-Preserving Reduction from MaxMGSM to MCP

In this section, we present the formal definition of the maximum cardinality multi-modal group stable matching (MaxMGSM) and develop an approximation-preserving reduction from MaxMGSM to the maximum clique problem (MCP). Our reduction ensures that the approximation ratio of any algorithm for MCP

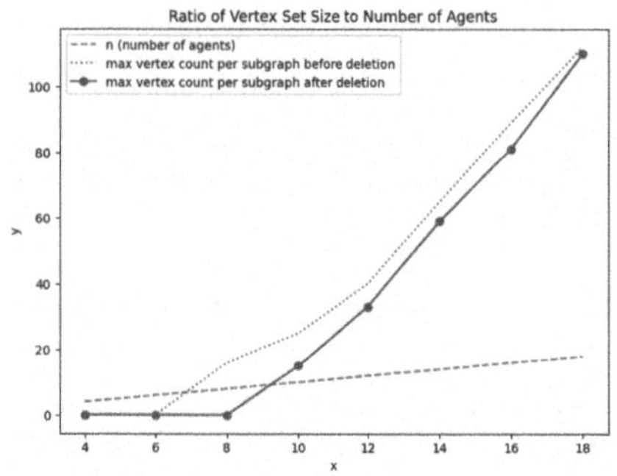

Fig. 1. The performance of vertex deletion using randomly generated preferences.

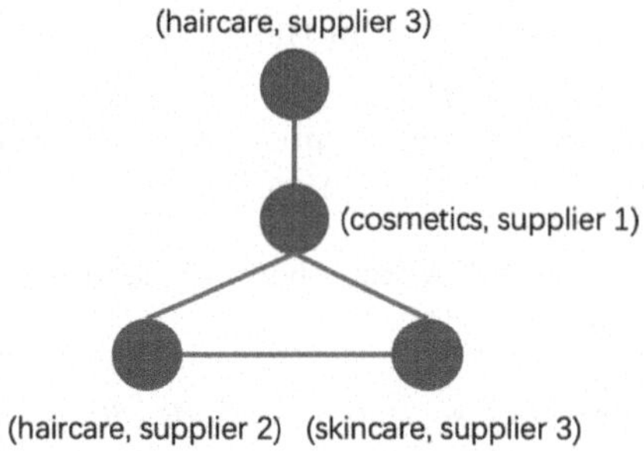

Fig. 2. The constructed graph after vertex deletion on cosmetics data sets with two criteria.

is preserved when applied to MaxMGSM. We begin by defining MaxMGSM as follows.

Definition 3 (MaxMGSM problem). *Given a set A of n agents and ω criteria, each agent generates a preference list based on each criterion. The maximum cardinality multi-modal group stable matching problem (MaxMGSM) aims to find the largest possible matching such that, among all paired agents, no unmatched pair prefers each other over their current partners under any criterion.*

To further capture the properties of MaxMGSM, we formulate the following linear programming model for MaxMGSM.

$$\max \quad \sum_{\substack{i \in A, \\ j \in A \setminus \{i\}}} x_{ij}, \tag{3}$$

$$\text{subject to:} \quad \sum_{j \in A \setminus \{i\}} x_{ij} \leq 1, \quad \forall i \in A, \tag{4}$$

$$\sum_{s \in L_{i,j,k}} x_{is} + \sum_{v \in L_{j,i,k}} x_{jv} \leq 1, \quad \forall i \in A, \quad j \in A \setminus \{i\}, \quad k \in \omega. \tag{5}$$

Unlike MGBSM, where the goal is to find a perfect stable matching, in MaxMGSM we aim to maximize the total number of matched pairs. We relax the stability constraints so that all paired agents are stable within the matching, irrespective of the preferences of agents who are not part of the matching. We proceed by transforming the MaxMGSM instance into an MCP instance using Algorithm 1 from Sect. 2. The following lemma demonstrates that this reduction can be accomplished in polynomial time.

Lemma 1. *A MaxMGSM problem instance can be transformed into a MCP instance in polynomial time.*

Proof. In the constructed graph $G = (V, E)$, each vertex $v_{ij} \in V$ corresponds to a variable x_{ij} in the linear programming formulation. An edge $(v_{ab}, v_{cd}) \in E$ indicates that the pairs (a, b) and (c, d) can coexist in a group stable matching. Given that the number of variables is $\frac{n(n-1)}{2}$ and the number of constraints is polynomial, we can complete our reduction in polynomial time. □

The following theorem establishes that our reduction is approximation-preserving [1].

Theorem 2. *The reduction from MaxMGSM to MCP is approximation preserving.*

Proof. Our reduction from MaxMGSM to MCP is approximation-preserving, as it satisfies the following five properties.

- For any instance p of MaxMGSM, there is a function f that can map p to an instance of MCP;
- for any instance p of MaxMGSM, if a solution exists for p, then a solution must also exist for the corresponding MCP instance;
- for any instance p of MaxMGSM, and for any solution of the corresponding MCP instance, there exists a function g that can transform the solution of the MCP instance into a solution for the MaxMGSM instance;
- mapping f is a function that can be computed by Algorithm 1, while mapping g is a straightforward one-to-one mapping. Both functions can be computed in polynomial time;
- the size of the solution for MCP equals the size of the corresponding MaxMGSM solution transformed by function g.

Based on the aforementioned five properties, we establish that MaxMGSM is AP-reducible to MCP. □

Building on our reduction from MaxMGSM to MCP, rather than directly applying an MCP approximation algorithm to our constructed instance, we present the following approximation algorithm, which significantly accelerates the computation process in practice.

Algorithm 3. An Approximation Algorithm for MaxMGSM

Input: a graph $G = (V, E)$, where each vertex in V is labeled as v_{ij}, with $i, j \in A$ and $i \neq j$; an empty set M_G of candidate vertices; an empty set S of maximal cliques;

Output: a multi-modal group stable matching of almost maximum size;

1: **if** no edges exist in the graph **then**
2:     **return** 0
3: **end if**
4: delete vertices with a degree less than 2;
5: delete vertices that are connected to all other vertices and add them to the candidate set M_G;

6: repeat steps 1 to 5 until no more vertices are deleted;
7: apply Depth-First Search (DFS) to check the connectivity of the current graph;
8: **if** G is disconnected **then**
9: decompose G into connected components; create an empty candidate set M_i for each component G_i;
10: **for** each connected component G_i **do**
11: apply steps 1 to 9 to G_i with an updated degree threshold of $\frac{|A|}{2} - |M_G| - 1$, and update the candidate set M_i;
12: apply the approximation algorithm for finding the maximum clique in G_i;
13: append the found clique S_i to the set of maximal cliques S;
14: **end for**
15: **end if**
16: **return** the labels of pairs that correspond to the clique of size $\max(|S_i| + |M_i|) + |M_G|$, where $S_i \subseteq S$.

Theorem 3. *By combining Algorithm 1 and Algorithm 3, a near-optimal solution for MaxMGSM can be found in polynomial time.*

Proof. In Algorithm 3, since our constructed graph can be quite sparse, we delete isolated vertices and vertices with a degree of one in steps 1 to 4 to reduce the search space of our algorithm. Additionally, any vertex connected to all other vertices is guaranteed to be part of the final clique. Therefore, we remove this vertex from the graph and add it to our candidate set M_G in step 5. After these deletions, the graph is likely to become disconnected, so we apply DFS to decompose the graph into several connected components. This operation significantly accelerates the algorithm, as we can search for maximum cliques within these smaller, more manageable components. After the above steps, we apply Feige's MCP approximation algorithm that approximates maximum clique within a factor of $\mathcal{O}\left(|V| (\log\log |V|)^2 / (\log |V|)^3\right)$, where $|V|$ is the number of vertices in the graph [6]. The algorithm systematically removes subgraphs that are unlikely to contain large cliques and iteratively refines the remaining graph to find increasingly larger cliques. The novel approach focuses on improving the approximation ratio for the maximum clique problem by leveraging the properties of "poor" subgraphs and the structure of the input graph. Given that the maximum size of a clique in our constructed graph is naturally bounded by $\frac{n}{2}$, this algorithm can efficiently find a near-optimal clique in practice. Since constructing the MCP instance takes $\mathcal{O}(n^2)$ time and recovering the MaxMGSM solution requires linear time, Algorithm 3 can return a near-optimal solution for MaxMGSM in polynomial time. □

4 Conclusions

In this paper, we explore the multi-modal stable matching problem through the lens of the clique problem. By transforming instances of multi-modal globally stable matching (MGBSM) into graph instances, we leverage specific clique properties that are not easily discernible using traditional stable matching approaches. Building on these insights, we develop an exact algorithm that effectively finds the final solution by carefully eliminating pairs that can never be part of a stable matching for all criteria. We also generalize the concept of MGBSM to maximum cardinality multi-modal group stable matching (MaxMGSM), which offers a more practical framework since a multi-modal globally stable matching may not always exist. By utilizing stability and matching constraints, we construct an approximation-preserving reduction from MaxMGSM to the maximum clique problem (MCP) in polynomial time. Based on this reduction, we present an approximation algorithm that reduces the search space by removing unnecessary vertices before finding a near-optimal solution. Notably, both of our algorithms are efficient in practice due to the effective reduction of the solution space.

The implications of our work are particularly significant for industry chain management, where matching suppliers with buyers often involves balancing multiple criteria such as cost, quality, delivery time, and reliability. Traditional stable matching models are insufficient for addressing the complexity of these multi-faceted decisions. By applying our multi-modal stable matching algorithms, industry chain networks can optimize resource allocation by ensuring that matchings are stable across all relevant criteria. This approach not only enhances the resilience and efficiency of the industry chain but also enables more strategic partnerships between suppliers and buyers. As global industry chains become increasingly complex, the ability to manage multiple criteria simultaneously through advanced algorithms like ours will be crucial for maintaining competitive advantage and operational stability.

While our study marks the first significant step in designing algorithms for MGBSM and its generalization to MaxMGSM, several avenues for future research remain open. 1) We have observed that our proposed algorithms allow for the deletion of a considerable number of vertices. A more refined analysis of the lower bound on the number of deletable vertices could potentially lead to improved theoretical time complexity. 2) After deleting unnecessary vertices, the constructed graph often becomes disconnected, allowing us to decompose it into several connected components and process them separately in our algorithm. This process presents an opportunity for implementing parallel algorithms, which could significantly enhance computational efficiency. 3) It remains an open question whether MaxMGSM can be approximated within a constant factor. 4) While MGBSM imposes the most stringent requirements for finding a multi-modal stable matching, there are other types of multi-modal stable matching problems with more relaxed constraints that remain practical for real-world applications.

Acknowledgments. This work is supported by National Key R&D Program of China (No. 2022YFB3304402), NSFC 12071460, NSFC 12371321, Guangdong Basic and Applied Basic Research Foundation 2024A1515030197.

References

1. Ausiello, G., Crescenzi, P., Gambosi, G., Kann, V., Marchetti-Spaccamela, A., Protasi, M.: Complexity and approximation: combinatorial optimization problems and their approximability properties. Springer (2012)
2. Bredereck, R., Chen, J., Knop, D., Luo, J., Niedermeier, R.: Adapting stable matchings to evolving preferences. In: Proceedings of the AAAI Conference on Artificial Intelligence, vol. 34, pp. 1830–1837 (2020)
3. Chan, J.W.T., Chin, F.Y., Ye, D., Zhang, Y.: Online frequency allocation in cellular networks. In: Proceedings of the Nineteenth Annual ACM Symposium on Parallel Algorithms and Architectures, pp. 241–249 (2007)
4. Chen, J., Niedermeier, R., Skowron, P.: Stable marriage with multi-modal preferences. In: Proceedings of the 2018 ACM Conference on Economics and Computation, EC 2018, pp. 269–286 (2018)
5. Chen, J., Skowron, P., Sorge, M.: Matchings under preferences: strength of stability and tradeoffs. ACM Trans. Econ. Comput. **9**(4), 1–55 (2021)
6. Feige, U.: Approximating maximum clique by removing subgraphs. SIAM J. Discret. Math. **18**(2), 219–225 (2004)
7. Gale, D., Shapley, L.S.: College admissions and the stability of marriage. Am. Math. Mon. **69**(1), 9–15 (1962)
8. Gusfield, D., Irving, R.W.: The stable marriage problem: structure and algorithms (1989)
9. Irving, R.W.: An efficient algorithm for the "stable roommates" problem. J. Algorithms **6**(4), 577–595 (1985)
10. Irving, R.W., Manlove, D.F.: The stable roommates problem with ties. J. Algorithms **43**(1), 85–105 (2002)
11. Iwama, K., Manlove, D., Miyazaki, S., Morita, Y.: Stable marriage with incomplete lists and ties. In: Automata, Languages and Programming: 26th International Colloquium, ICALP'99 Prague, pp. 443–452 (2002)
12. Monge, A.E., Elkan, C., et al.: The field matching problem: algorithms and applications. In: Kdd, vol. 2, pp. 267–270 (1996)
13. Ostrovsky, M.: Stability in supply chain networks. Am. Econ. Rev. **98**(3), 897–923 (2008)
14. Palmer, C., Pálvölgyi, D.: At most 3.55 n stable matchings. In: 2021 IEEE 62nd Annual Symposium on Foundations of Computer Science (FOCS), pp. 217–227. IEEE (2022)
15. Tan, J., Sun, Y., Xu, Y., Zou, J.: Streaming algorithms for non-submodular maximization on the integer lattice. Tsinghua Sci. Technol. **28**(5), 888–895 (2023)
16. Yang, H., Chen, R., Kumara, S.: Stable matching of customers and manufacturers for sharing economy of additive manufacturing. J. Manuf. Syst. **61**, 288–299 (2021)
17. Yu, W., Liao, Y., Yang, Y.: Exact and approximation algorithms for the multi-depot capacitated arc routing problems. Tsinghua Sci. Technol. **28**(5), 916–928 (2023)
18. Zhang, Y., Chen, J., Chin, F.Y., Han, X., Ting, H.F., Tsin, Y.H.: Improved online algorithms for 1-space bounded 2-dimensional bin packing. In: Algorithms and Computation: 21st International Symposium, ISAAC 2010, Jeju, Korea, 15–17 December 2010, Proceedings, Part II, vol. 21, pp. 242–253. Springer (2010)

Budget Feasible Mechanism for a k-submodular Function in the Clock Auction Model

Hongyang Zhang and Wenchang Luo(✉)

School of Mathematics and Statistics, Ningbo University,
Ningbo 315211, Zhejiang, China
luowenchang@163.com

Abstract. Due to its numerous applications in social marketing and crowdsourcing, the classical topic of designing a budget-feasible mechanism for a submodular valuation function has been well-studied. In this paper, we consider a generalization of this topic: budget-feasible mechanism design for a k-submodular function in the clock auction model. In our problem, each agent has a private cost, and the auctioneer, composed of k departments, attempts to maximize his k-submodular valuation subject to a budget constraint. For the monotone objective, we propose a randomized mechanism with an approximation ratio of $\frac{1}{5+\sqrt{13}}$. Additionally, the randomized mechanism can also achieve an approximation ratio of $\frac{2}{15+3\sqrt{13}}$ for the non-monotone objective. Our mechanism only requires $O(kn)$ value oracle queries, making it more practical.

Keywords: k-Submodular · Budget feasible mechanism design · Clock auction · TripleEagle

1 Introduction

Singer's pioneering work [10] has led to significant interest in pursuing budget-feasible mechanisms on submodular function, due to its theoretical allure and relevance to problems involving budget or resource constraints. In this paper, we consider a generalization of this topic: budget-feasible mechanism design for a k-submodular function in the clock auction model. Unlike budget-feasible mechanisms for a submodular valuation, in this model, the auctioneer is compared to k departments. The auctioneer purchases goods from a group of agents E, assigns them to k departments, and his valuation can be quantified by a k-submodular function $f : (k+1)^E \rightarrow \mathbb{R}_+$. Our goal is to design a truthful mechanism that maximizes the auctioneer's valuation and ensures that the total payment to winners does not exceed the budget B. Every agent $i \in E$ has a private cost c_i for his goods. Hence, there may exist agents who attempt to increase their revenue by inflating their costs.

It is well-known that the auction model encapsulates a broad spectrum of common economic scenarios. From a practical viewpoint, our clock auction model

D. Du et al. (Eds.): COCOA 2024, LNCS 15434, pp. 376–387, 2025.
https://doi.org/10.1007/978-981-96-4445-2_31

for a k-submodular function is applicable to various domains where each agent has a private cost, such as influence maximization with k topics [8], multi-cooperative games [2], sensor placement with k types of sensors [11], and feature selection with k partitions [9]. Clearly our model simplifies to k-submodular maximization with a total knapsack constraint if we only consider the orientation of optimization. Given that it is the generalization of maximum k-cut problem, it is NP-hard for maximizing a k-submodular function.

To solve our model, we draw inspiration from the TripleEagle algorithm [5], which is a clock auction mechanism. This mechanism dynamically sets prices for agents participating in the auction, adhering to the economic principles of supply and demand. We develop a randomized mechanism, which can achieve an $\frac{1}{5+\sqrt{13}}$-approximation for a monotone k-submodular function and a $\frac{2}{15+3\sqrt{13}}$-approximation for a non-monotone k-submodular function.

The paper is organized into five main sections. In Sect. 2, we review the related literature. In Sect. 3, we introduce some definitions and properties related to k-submodular functions, as well as some related concepts in mechanism design. In Sect. 4, we propose a randomized mechanism with offering a detailed analysis of its performance guarantee. Finally, in Sect. 5 we summarize our work and suggest directions for future research.

2 Related Work

For the problem of k-submodular maximization with a knapsack constraint, several notable advancements have been made. Tang et al. [12] achieved an approximation ratio of $\frac{1}{2}(1-\frac{1}{e})$ using partial enumeration and greedy techniques for the monotone objective. Wang and Zhou [13] gave the multilinear extension of the k-submodular function and improved the approximation ratio from $\frac{1}{2}(1-\frac{1}{e})$ to $\frac{1}{2}-\epsilon$ using the continuous greedy technique. In their latest work [6], a continuous greedy algorithm based on a Frank-Wolfe variant was introduced without requiring the monotonicity of the objective. This algorithm provided an approximation guarantee of $\frac{1}{2}-o(1)$ for the monotone case and $\frac{1}{3}-o(1)$ for the non-monotone case. The algorithm proposed in [15] can achieve an $\frac{1}{3}(1-e^{-3})$-approximation using partial enumeration and greedy techniques for a non-monotone objective.

The exploration of budget-feasible mechanisms began with Singer [10], who proposed a randomized $\frac{1}{112}$-approximation mechanism when the objective is a monotone submodular function, which was later enhanced by Chen et al. [4] to $\frac{e-1}{5e}$. Furthermore, Chen et al. [4] presented the randomized and deterministic lower bounds which are $\frac{1}{2}$ and $\frac{1}{1+\sqrt{2}}$ respectively. Jalaly and Tardos [7] developed a randomized mechanism, which is the same as Chen et al. [4] when setting the parameter $r=\frac{1}{2}$. However, new analysis from [7] proves that this mechanism is a $\frac{1}{5}$-approximation. The best result up to now is a randomized budget-feasible mechanism with the approximation ratio of $\frac{2}{5+\sqrt{13}}$ and a deterministic budget-feasible mechanism with the approximation ratio of $\frac{1}{2+\sqrt{6}}$ from [5].

For the non-monotone submodular objective, Amanatidis et al. [1] proposed randomized mechanisms for offline and online settings with approximation ratios of $\frac{1}{505} - o(1)$ and $\frac{1}{1710} - o(1)$, respectively. Balkanski et al. [3] introduced descending clock auctions improving the approximation ratio from $\frac{1}{505} - o(1)$ to $\frac{1}{64}$, which marks the first deterministic budget-feasible mechanism for non-monotone submodular functions. Recently, Han et al. [5] achieved the best-known result of $\frac{1}{12}$ up to now. These methods can be utilized to address the problem with a monotone submodular valuation and achieve better approximation ratios than those of the non-monotone case.

3 Preliminaries

In this section, we introduce some definitions and properties which are related to k-submodular function.

Denote the set of n agents by $E = \{1, 2, \cdots, n\}$. Let $[k] = \{1, 2, \cdots, k\}$ and $(k+1)^E := \{(S_1, S_2, \cdots, S_k) | S_i \subseteq E, \forall i \in [k], S_i \cap S_j = \emptyset, \forall i \neq j \in [k]\}$, which covers all choices that the auctioneer can make. The auctioneer's valuation can be quantified by a k-submodular function $f : (k+1)^E \rightarrow \mathbb{R}_+$ and his total budget is $B (> 0)$. For a fixed $\mathbf{s} \in (k+1)^E$, the support set of $\mathbf{s}$ is denoted as $U(\mathbf{s}) = \bigcup_{i \in [k]} S_i$, where S_i is assigned to the i-th department of the auctioneer for any $i \in [k]$.

Definition 1. *[8] A function $f : (k+1)^E \rightarrow \mathbb{R}$ is called k-submodular if for any $\mathbf{s} = (S_1, S_2, \cdots, S_k)$, $\mathbf{t} = (T_1, T_2, \cdots, T_k) \in (k+1)^E$, one has*

$$f(\mathbf{s}) + f(\mathbf{t}) \geq f(\mathbf{s} \sqcup \mathbf{t}) + f(\mathbf{s} \sqcap \mathbf{t}), \tag{1}$$

where

$$\mathbf{s} \sqcap \mathbf{t} = (S_1 \cap T_1, S_2 \cap T_2, \cdots, S_k \cap T_k),$$

$$\mathbf{s} \sqcup \mathbf{t} = (S_1 \cup T_1 \backslash (\bigcup_{i \neq 1} S_i \cup T_i), S_2 \cup T_2 \backslash (\bigcup_{i \neq 2} S_i \cup T_i), \cdots, S_k \cup T_k \backslash (\bigcup_{i \neq k} S_i \cup T_i)).$$

We say that f is monotone, if $f(\mathbf{s}) \leq f(\mathbf{t})$ for any $\mathbf{s} \preceq \mathbf{t}$, where $\mathbf{s} \preceq \mathbf{t}$ means $S_i \subseteq T_i$ for any $i \in [k]$; f is normalized, if $f(\emptyset) = 0$; and f is non-negative, if $f(\cdot) \geq 0$. Let $\mathbf{s}|_A = (S_1 \cap A, S_2 \cap A, \cdots, S_k \cap A)$ be the truncation of $\mathbf{s}$ with respect to the set $A \subseteq E$. For notational ease, we identify $\mathbf{s}$ with the set $\bigcup_{\tau=1}^{|U(\mathbf{s})|} \{(i_\tau, j_\tau)\}$, where $i_\tau \in S_{j_\tau}$. Denote $\mathbf{s} \cup (i, j) = (S_1, S_2, \cdots, S_j \cup \{i\}, \cdots S_k)$ for any $i \notin U(\mathbf{s})$ and $\mathbf{s} \backslash (i, j) = (S_1, S_2, \cdots, S_j \backslash \{i\}, \cdots S_k)$ for any $i \in S_j$. We adopt the notation $f_{i,j}(\mathbf{s})$ for the marginal value of agent i added into j-th department with respect to $\mathbf{s}$, i.e., $f_{i,j}(\mathbf{s}) = f(S_1, S_2, \cdots, S_j \cup \{i\}, \cdots S_k) - f(S_1, S_2, \cdots, S_k)$ for $i \notin U(\mathbf{s})$. If $(i, j) \in \mathbf{s}$, define $f_{i,j}(\mathbf{s}) = 0$.

In our algorithm, if the pairs in $\mathbf{s}$ are sequentially inserted into $\mathbf{s}$, listed as $(i_l, j_l), (i_{l+1}, j_{l+1}), \cdots, (i_r, j_r)$ in the order that they are added, then we call the set $\mathbf{s}_1 = (i_{\bar{l}}, j_{\bar{l}}) \cup (i_{\bar{l}+1}, j_{\bar{l}+1}) \cup \cdots \cup (i_{\bar{r}}, j_{\bar{r}}) \preceq \mathbf{s}$ as a regular subset of $\mathbf{s}$ for any

$l \leq \bar{l} \leq \bar{r} \leq r$, and also call $\mathbf{s}_1$ as a suffix of $\mathbf{s}$ when $\bar{r} = r$ and a prefix of $\mathbf{s}$ when $\bar{l} = l$, respectively.

In the clock auction model, there exists only one private information, which is the true cost c. Hence, it is a single-parameter environment. In contrast to sealed-bid auctions, each agent will receive numerous offers at declining prices. An agent in the clock auction model stays active if he accepts the final price that is provided; otherwise, he will be removed from the active set. Finally, the mechanism will choose a subset of active agents as the winners and pay them the last prices that were offered. The clock auction can ensure the truthfulness of a deterministic mechanism and the universal truthfulness of a randomized mechanism, which means that the mechanism is a probability distribution over deterministic truthful mechanisms. In our mechanism, let $U(\mathbf{s}) \subseteq E$ be the winner set and $p(i)$ be the payment of agent $i \in U(\mathbf{s})$. We use $c(\mathbf{s})$ and $p(\mathbf{s})$ as shorthand for $\sum_{i \in U(\mathbf{s})} c(i)$ and $\sum_{i \in U(\mathbf{s})} p(i)$, i.e., $c(\mathbf{s}) = \sum_{i \in U(\mathbf{s})} c(i)$ and $p(\mathbf{s}) = \sum_{i \in U(\mathbf{s})} p(i)$. Then, the mathematical model for our problem is

$$\begin{aligned} \max \quad & f(\mathbf{s}) \\ s.t. \quad & p(\mathbf{s}) \leq B. \end{aligned} \tag{2}$$

4 Algorithm and Analysis

Note that Han et al. [5] consider the special case of $k = 1$, which is the budget-feasible mechanism for submodular functions and they present the TripleEagle framework applying to the monotone and non-monotone objectives. Based on the law of supply and demand in economics, this framework determines the last price of each agent. If the agent does not accept the price, he will be removed from the active set. We adapt it to design a budget-feasible mechanism for a k-submodular function ($k \geq 2$). Before offering the last price to each agent, we find the location bringing the maximum marginal gain to the current solution in the TripleEagle framework.

4.1 Randomized Algorithm for k-Submodular Functions

Let $(i^*, j^*) = \arg\max_{i \in E, j \in [k]} f(i, j)$. Given a parameter $\alpha > 1$, RanAlg (i.e. Algorithm 1) calls PriAlg (i.e. Algorithm 2) to process all the agents except i^*. Each of these agents is offered one price by Algorithm 2 and then Algorithm 2 returns a set $\mathbf{t}$ containing the active agent and his corresponding location. After that, if $f(\mathbf{t}) \geq f(i^*, j^*)$, then Algorithm 1 calls Algorithm 2 again to process i^* and finally returns a suffix of $\mathbf{t}$ subject to the budget constraint. Otherwise, Algorithm 1 makes a random decision of returning either (i^*, j^*) or $\mathbf{d}$, where $\mathbf{d}$ is got by trying to add $(i^*, \bar{j})$ to $\mathbf{t}$ when offering i^* a price of $B - p(\mathbf{t})$.

Let's consider the set $\mathbf{t}$ when Algorithm 1 quits, $\mathbf{s}^*$ be the optimal solution of k-submodular maximization under a knapsack constraint and $OPT(E, \infty)$ be the optimal value of k-submodular maximization without any constraints. We

Algorithm 1. RanAlg (α)

Input: $f : (k+1)^E \to \mathbb{R}_+, E = \{1, 2, \cdots, n\}, B, \alpha = \frac{\sqrt{13}+1}{2}$
Output: $\mathbf{s} \in (k+1)^E$

$\mathbf{t} \leftarrow$ *PriAlg* $(\emptyset, E \setminus \{i^*\}, \alpha f(i^*, j^*))$
if $f(\mathbf{t}) \geq f(i^*, j^*)$ **then**
  $\mathbf{t} \leftarrow$ *PriAlg* $(\mathbf{t}, \{i^*\}, \alpha f(i^*, j^*))$
  $\mathbf{s} \leftarrow$ the largest suffix of $\mathbf{t}$ satisfying $p(\mathbf{s}) \leq B$
else
  $Z \sim \mathcal{U}[0, 1]$
  if $Z \leq \frac{\alpha}{2+\alpha}$ **then**
    $p(i^*) = B; \mathbf{s} = (i^*, j^*)$
  else
    $\bar{j} = \arg\max_{j \in [k]} f_{i^*, j}(\mathbf{t})$
    Set $p(i^*) = B - p(\mathbf{t})$ and show i^* the price $p(i^*)$
    if agent i^* accepts $p(i^*)$ **then**
      $\mathbf{d} \leftarrow \mathbf{t} \cup (i^*, \bar{j})$
    else
      $\mathbf{d} \leftarrow \mathbf{t}$
    end if
    $\mathbf{s} \leftarrow \mathbf{d}$
  end if
end if
return $\mathbf{s}$

use notation like $\mathbf{t}_i$ to denote $\mathbf{t}$ right before i is processed. Denote $\Delta = 2$ when f is monotone and $\Delta = 3$ when f is non-monotone.

Lemma 1. *For any regular subset* $\hat{\mathbf{t}} = (i_l, j_l) \cup (i_{l+1}, j_{l+1}) \cup \cdots \cup (i_r, j_r)$ *of* $\mathbf{t}$ *returned by Algorithm 2, we have*

$$f(\hat{\mathbf{t}}) \geq \sum_{\tau=l}^{r} f_{i_\tau, j_\tau}(\mathbf{t}_{i_\tau}) \geq p(\hat{\mathbf{t}}) \cdot \frac{f(\mathbf{t}_{i_l}) + \alpha f(i^*, j^*)}{B}. \tag{3}$$

Proof. Because i_τ accepts the price $p(i_\tau)$, it holds that $0 \leq c(i_\tau) \leq p(i_\tau)$ for any $\tau = \{l, l+1, \cdots, r\}$. For all $(i_\tau, j_\tau) \in \mathbf{t}$, due to $0 \leq c(i_\tau) \leq p(i_\tau) = \frac{B f_{i_\tau, j_\tau}(\mathbf{t}_{i_\tau})}{f(\mathbf{t}_{i_\tau}) + \alpha f(i^*, j^*)}$, we have $f_{i_\tau, j_\tau}(\mathbf{t}_{i_\tau}) \geq 0$, which indicates $f(\mathbf{t}_{i_l}) \leq f(\mathbf{t}_{i_\tau})$. It is clear that $\hat{\mathbf{t}}_{i_\tau} \preceq \mathbf{t}_{i_\tau}$ implies that $f_{i_\tau, j_\tau}(\hat{\mathbf{t}}_{i_\tau}) \geq f_{i_\tau, j_\tau}(\mathbf{t}_{i_\tau})$ from k-submodularity of f. Then we get

$$p(i_\tau) = \frac{B f_{i_\tau, j_\tau}(\mathbf{t}_{i_\tau})}{f(\mathbf{t}_{i_\tau}) + \alpha f(i^*, j^*)} \leq \frac{B f_{i_\tau, j_\tau}(\mathbf{t}_{i_\tau})}{f(\mathbf{t}_{i_l}) + \alpha f(i^*, j^*)}.$$

By k-submodularity and non-negativity of f, we have

$$f(\hat{\mathbf{t}}) \geq \sum_{\tau=l}^{r} f_{i_\tau, j_\tau}(\hat{\mathbf{t}}_{i_\tau}) \geq \sum_{\tau=l}^{r} f_{i_\tau, j_\tau}(\mathbf{t}_{i_\tau}).$$

Algorithm 2. PriAlg $(\mathbf{t}, E_1, \eta)$

Input: $f : (k+1)^E \to \mathbb{R}_+, E_1, B, \eta, \mathbf{t}$
Output: $\mathbf{t} \in (k+1)^E$
for $i \in E_1$ **do**
  $\bar{j} = \arg\max_{j \in [k]} f_{i,j}(\mathbf{t})$
  Set $p(i) = \frac{B f_{i,\bar{j}}(\mathbf{t})}{f(\mathbf{t})+\eta}$ and show i the price $p(i)$
  if agent i accepts $p(i)$ **then**
    $\mathbf{t} = \mathbf{t} \cup (i, \bar{j})$
  end if
end for
return $\mathbf{t}$

Combining the above two inequalities indicates

$$f(\hat{\mathbf{t}}) \geq \sum_{\tau=l}^{r} f_{i_\tau, j_\tau}(\mathbf{t}_{i_\tau}) \geq p(\hat{\mathbf{t}}) \cdot \frac{f(\mathbf{t}_{i_l}) + \alpha f(i^*, j^*)}{B}.$$

Lemma 2. *After executing line 3 of Algorithm 1, we have*

$$f(\mathbf{s}^*) \leq [2f(\mathbf{t}) + \alpha f(i^*, j^*)]\Delta. \tag{4}$$

Proof. After line 3 of Algorithm 1 is executed, all the agents in E must have been processed by Algorithm 2. We construct a solution $\bar{\mathbf{s}}^*$ by the following procedure: initialize $\bar{\mathbf{s}}^* = \mathbf{t}$; for any $i \in U(\mathbf{s}^*) \setminus U(\mathbf{t})$: let $j = \arg\max_{j \in [k]} f_{i,j}(\bar{\mathbf{s}}^*)$ and $\bar{\mathbf{s}}^* = \bar{\mathbf{s}}^* \cup (i, j)$. The initialized step can also be reviewed as a greedy operation, which means we start from the empty set and add the elements in the order that they are retained in $\mathbf{t}$. The procedure to construct $\bar{\mathbf{s}}^*$ is in agreement with the greedy algorithm for k-submodular maximization without any constraint from [14]. This algorithm achieves an $\frac{1}{2}$-approximation for a monotone k-submodular function and an $\frac{1}{3}$-approximation for the non-monotone case. Hence, we have

$$f(\mathbf{s}^*) \leq OPT(U(\mathbf{s}^*) \cup U(\mathbf{t}), \infty) \leq \Delta f(\bar{\mathbf{s}}^*). \tag{5}$$

We use notation like $\bar{\mathbf{s}}_i^*$ to denote $\bar{\mathbf{s}}^*$ right before $i \in U(\bar{\mathbf{s}}^*)$ is added. For any $i \in U(\mathbf{s}^*) \setminus U(\mathbf{t})$, we know $\mathbf{t}_i \preceq \mathbf{t} \preceq \bar{\mathbf{s}}_i^*$ from the construction of $\bar{\mathbf{s}}^*$, which indicates that

$$\begin{aligned} f(\bar{\mathbf{s}}^*) &= f(\mathbf{t}) + \sum_{(i,j) \in \bar{\mathbf{s}}^* \setminus \mathbf{t}} f_{i,j}(\bar{\mathbf{s}}_i^*) \\ &\leq f(\mathbf{t}) + \sum_{(i,j) \in \bar{\mathbf{s}}^* \setminus \mathbf{t}} f_{i,j}(\mathbf{t}_i). \end{aligned} \tag{6}$$

The greedy selection of $(i, \bar{j})$ indicates

$$f(\mathbf{t}) + \sum_{(i,j) \in \bar{\mathbf{s}}^* \setminus \mathbf{t}} f_{i,j}(\mathbf{t}_i) \leq f(\mathbf{t}) + \sum_{i \in U(\mathbf{s}^*) \setminus U(\mathbf{t})} f_{i,\bar{j}}(\mathbf{t}_i), \tag{7}$$

where $\bar{j} = \arg\max_{j \in [k]} f_{i,j}(\mathbf{t}_i)$. Each agent $i \in U(\mathbf{s}^*) \setminus U(\mathbf{t})$ must reject the offered price, i.e.,

$$c(i) > p(i) = \frac{B f_{i,\bar{j}}(\mathbf{t}_i)}{f(\mathbf{t}_i) + \alpha f(i^*, j^*)}. \tag{8}$$

Then we obtain

$$\begin{aligned} f(\mathbf{t}) + \sum_{i \in U(\mathbf{s}^*) \setminus U(\mathbf{t})} f_{i,\bar{j}}(\mathbf{t}_i) &\le f(\mathbf{t}) + \sum_{i \in U(\mathbf{s}^*) \setminus U(\mathbf{t})} c(i) \cdot \frac{f(\mathbf{t}_i) + \alpha f(i^*, j^*)}{B} \\ &\le f(\mathbf{t}) + \sum_{i \in U(\mathbf{s}^*) \setminus U(\mathbf{t})} c(i) \cdot \frac{f(\mathbf{t}) + \alpha f(i^*, j^*)}{B} \\ &\le f(\mathbf{t}) + f(\mathbf{t}) + \alpha f(i^*, j^*) \\ &= 2f(\mathbf{t}) + \alpha f(i^*, j^*). \end{aligned} \tag{9}$$

Since each agent accepts the offered price, it will bring the non-negative marginal gain and we obtain $f(\mathbf{t}_i) \le f(\mathbf{t})$ for any prefix $\mathbf{t}_i$ of $\mathbf{t}$. The second inequality follows the fact that $c(\mathbf{s}^*) \le B$. Therefore, combining inequalities (5)-(9), we have

$$f(\mathbf{s}^*) \le [2f(\mathbf{t}) + \alpha f(i^*, j^*)]\Delta.$$

Lemma 3. *Let* $\mathbf{s}$ *be the solution generated by line 4 of Algorithm 1. For any* $\alpha \in (2, 1+\sqrt{3})$, *we have*

$$f(\mathbf{s}^*) \le \max\{2+\alpha, 3 + \frac{2\alpha+1}{\alpha^2-1}\} \cdot \Delta f(\mathbf{s}). \tag{10}$$

Proof. If $p(\mathbf{t}) \le B$, then we obtain $\mathbf{s} = \mathbf{t}$. When line 4 of Algorithm 1 is executed, we have $f(i^*, j^*) \le f(\mathbf{t})$. Hence, from Lemma 2, we have

$$f(\mathbf{s}^*) \le [2f(\mathbf{t}) + \alpha f(i^*, j^*)]\Delta \le (2+\alpha)\Delta f(\mathbf{s}). \tag{11}$$

On the other hand, when $p(\mathbf{t}) > B$, we can find $\mathbf{h}$, which is the smallest suffix of $\mathbf{t}$ and satisfies $p(\mathbf{h}) > B$. Then $U(\mathbf{h} \setminus \mathbf{s})$ only contains one element and denote it by $(\hat{i}, \hat{j}) = \mathbf{h} \setminus \mathbf{s}$, which is added into $\mathbf{t}$ before the elements in $\mathbf{s}$. From Lemma 2 and k-submodularity, we get

$$\begin{aligned} f(\mathbf{s}^*) &\le [2f(\mathbf{t}) + \alpha f(i^*, j^*)]\Delta \\ &\le [2f(\mathbf{t} \setminus \mathbf{h}) + 2f(\hat{i}, \hat{j}) + 2f(\mathbf{s}) + \alpha f(i^*, j^*)]\Delta \\ &\le [2f(\mathbf{t} \setminus \mathbf{h}) + 2f(\mathbf{s}) + (2+\alpha) f(i^*, j^*)]\Delta. \end{aligned} \tag{12}$$

Suppose that there are $\sigma > 0$ and $\rho \in [0,1]$ such that $f(\mathbf{t} \setminus \mathbf{h}) = \sigma f(i^*, j^*)$ and $p(\hat{i}) = \rho B$, then we get $p(\mathbf{s}) > (1-\rho)B$ from the fact that $p(\mathbf{h}) > B$. Because $\hat{i}$ accepts the offered price, it is valid that

$$f(i^*, j^*) \ge f_{\hat{i},\hat{j}}(\mathbf{t} \setminus \mathbf{h}) = \rho B \cdot \frac{f(\mathbf{t} \setminus \mathbf{h}) + \alpha f(i^*, j^*)}{B} = \rho(\sigma+\alpha) f(i^*, j^*). \tag{13}$$

By Lemma 1, we have

$$f(\mathbf{s}) \geq (1-\rho)B \cdot \frac{f(\mathbf{t}\setminus\mathbf{h}) + f_{\hat{i},\hat{j}}(\mathbf{t}\setminus\mathbf{h}) + \alpha f(i^*,j^*)}{B} \geq (1-\rho^2)(\sigma+\alpha)f(i^*,j^*). \tag{14}$$

Let $x=\sigma+\alpha$, we get $\rho \leq \frac{1}{x}$ and $f(\mathbf{s}) \geq \frac{x^2-1}{x}f(i^*,j^*)$ due to the inequalities (13) and (14). Plugging it into inequality (12) implies

$$\begin{aligned} f(\mathbf{s}^*) &\leq [2f(\mathbf{t}\setminus\mathbf{h}) + 2f(\mathbf{s}) + (2+\alpha)f(i^*,j^*)]\Delta \\ &= [2\sigma f(i^*,j^*) + 2f(\mathbf{s}) + (2+\alpha)f(i^*,j^*)]\Delta \\ &\leq (4+\frac{2-\alpha+2/x}{x-1/x})\Delta f(\mathbf{s}). \end{aligned} \tag{15}$$

Note that $x=\sigma+\alpha>\alpha>0$. When $x\in[\alpha,\frac{2}{\alpha-2})$ and $\alpha<\frac{2}{\alpha-2}$, i.e., $\alpha \in (2,1+\sqrt{3})$, the right side of inequality (15) is maximized when $x=\alpha$, which indicates

$$f(\mathbf{s}^*) \leq (3+\frac{2\alpha+1}{\alpha^2-1})\Delta f(\mathbf{s}), \tag{16}$$

for any $\alpha\in(2,1+\sqrt{3})$. Therefore, from inequalities (11) and (16), the statement holds that

$$f(\mathbf{s}^*) \leq \max\{2+\alpha, 3+\frac{2\alpha+1}{\alpha^2-1}\}\cdot \Delta f(\mathbf{s}).$$

Lemma 4. *Let* $\mathbf{s}$ *be the solution generated by lines 5–19 of Algorithm 1. For any* $\alpha>2$*, we have*

$$f(\mathbf{s}^*) \leq (2+\alpha)\cdot \Delta E[f(\mathbf{s})]. \tag{17}$$

Proof. After line 1 is executed, we must have $f(\mathbf{t})<f(i^*,j^*)$, implying $p(\mathbf{t})<B$. Otherwise, from Lemma 1 and $p(t)\leq B$, we get $f(\mathbf{t})\geq \alpha f(i^*,j^*)$, which contradicts the fact that $f(\mathbf{t})<f(i^*,j^*)$ and $\alpha>2$. Next, we will prove that the set $\mathbf{d}$ generated by Algorithm 1 must satisfy

$$f(\mathbf{s}^*) \leq [2f(\mathbf{d}) + \alpha f(i^*,j^*)]\Delta, \tag{18}$$

according to the following discussion.

Case 1. $i^*\notin U(\mathbf{s}^*)$: For any agent $i\in U(\mathbf{s}^*)\setminus U(\mathbf{d})$, he must reject the offered price in Algorithm 2. Similar to the argument of Lemma 2, we construct a solution $\bar{\mathbf{s}}^*$ by the following procedure: initialize $\bar{\mathbf{s}}^*=\mathbf{t}$; for any $i\in U(\mathbf{s}^*)\setminus U(\mathbf{t})$: let $j=\arg\max_{j\in[k]} f_{i,j}(\bar{\mathbf{s}}^*)$ and $\bar{\mathbf{s}}^*=\bar{\mathbf{s}}^*\cup(i,j)$. Hence, we have

$$f(\mathbf{s}^*) \leq OPT(U(\mathbf{s}^*)\cup U(\mathbf{t}),\infty) \leq \Delta f(\bar{\mathbf{s}}^*),$$

and

$$f(\mathbf{s}^*) \leq [2f(\mathbf{t})+\alpha f(i^*,j^*)]\Delta \leq [2f(\mathbf{d})+\alpha f(i^*,j^*)]\Delta.$$

From [14], we know that a k-submodular function must satisfy pairwise monotonicity, i.e., $f_{i,j_1}(\mathbf{s})+f_{i,j_2}(\mathbf{s})\geq 0$ for any $i\notin U(\mathbf{s}), j_1\neq j_2\in[k]$. Then we must have $f_{i^*,\bar{j}^*}(\mathbf{t}) \geq 0$ since of $\bar{j}^* = \arg\max_{j\in[k]} f_{i^*,j}(\mathbf{t})$. From Lines 10–16 of Algorithm 1, we get $f(\mathbf{t}) \leq f(\mathbf{d})$, which infers the validity of the final inequality.

Case 2. $i^* \in U(\mathbf{s}^*)$ and i^* is a winner: it holds that $\mathbf{d} = \mathbf{t} \cup (i^*, \bar{j}^*)$ and $f(\mathbf{d}) \geq f(\mathbf{t})$, where $\bar{j}^* = \arg\max_{j\in[k]} f_{i^*,j}(\mathbf{t})$. For any agent $i \in U(\mathbf{s}^*) \setminus U(\mathbf{d}^*)$, he must reject the offered price in Algorithm 2. Similar to the argument of Lemma 2, we construct a solution $\bar{\mathbf{s}}^*$ by the following procedure: initialize $\bar{\mathbf{s}}^* = \mathbf{d}$; for any $i \in U(\mathbf{s}^*)\setminus U(\mathbf{d})$: let $j = \arg\max_{j\in[k]} f_{i,j}(\bar{\mathbf{s}}^*)$ and $\bar{\mathbf{s}}^* = \bar{\mathbf{s}}^* \cup (i,j)$. Hence, we have

$$f(\mathbf{s}^*) \leq OPT(U(\mathbf{s}^*) \cup U(\mathbf{d}), \infty) \leq \Delta f(\bar{\mathbf{s}}^*),$$

and

$$f(\mathbf{s}^*) \leq [f(\mathbf{d}) + f(\mathbf{t}) + \alpha f(i^*, j^*)]\Delta \leq [2f(\mathbf{d}) + \alpha f(i^*, j^*)]\Delta,$$

the final inequality comes from the fact that $f(\mathbf{d}) \geq f(\mathbf{t})$.

Case 3. $i^* \in U(\mathbf{s}^*)$ and i^* is a loser: it holds that $\mathbf{d} = \mathbf{t}$ and $c(i^*) > B - p(\mathbf{t})$. Assume that i^* is assigned to $\bar{j}^* \in [k]$ in the optimal solution $\mathbf{s}^*$. Hence, by Lemma 1 and the non-negativity of f, we get

$$c(\mathbf{s}^* \setminus (i^*, \bar{j}^*)) \leq p(\mathbf{t}) \leq \frac{Bf(\mathbf{t})}{f(\emptyset) + \alpha f(i^*, j^*)} \leq \frac{Bf(\mathbf{t})}{\alpha f(i^*, j^*)}. \tag{19}$$

For any agent $i \in U(\mathbf{s}^*) \setminus (U(\mathbf{t}) \cup \{i^*\})$, he must reject the offered price in Algorithm 2. Similar to the argument of Lemma 2, we construct a solution $\bar{\mathbf{s}}^*$ by the following procedure: initialize $\bar{\mathbf{s}}^* = \mathbf{t}$; for any $i \in U(\mathbf{s}^*) \setminus (U(\mathbf{t}) \cup \{i^*\})$: let $j = \arg\max_{j\in[k]} f_{i,j}(\bar{\mathbf{s}}^*)$ and $\bar{\mathbf{s}}^* = \bar{\mathbf{s}}^* \cup (i,j)$. Hence, we have

$$f(\mathbf{s}^* \setminus (i^*, \bar{j}^*)) \leq OPT(U(\mathbf{s}^*) \setminus \{i^*\} \cup U(\mathbf{t}), \infty) \leq \Delta f(\bar{\mathbf{s}}^*),$$

and

$$\begin{aligned} f(\mathbf{s}^* \setminus (i^*, \bar{j}^*)) &< [f(\mathbf{t}) + \sum_{i\in U(\mathbf{s}^*)\setminus(U(\mathbf{t})\cup\{i^*\})} c(i) \cdot \frac{f(\mathbf{t}) + \alpha f(i^*, j^*)}{B}]\Delta \\ &\leq [f(\mathbf{t}) + \frac{f(\mathbf{t})}{\alpha f(i^*, j^*)} \cdot (f(\mathbf{t}) + \alpha f(i^*, j^*))]\Delta \\ &\leq [f(\mathbf{t}) + \frac{1}{\alpha} \cdot (f(\mathbf{t}) + \alpha f(i^*, j^*))]\Delta \\ &= [(1 + \frac{1}{\alpha}) f(\mathbf{t}) + f(i^*, j^*)]\Delta. \end{aligned} \tag{20}$$

The second inequality holds due to inequality (19), and the last inequality is valid because we must have $f(\mathbf{t}) < f(i^*, j^*)$ when Algorithm 1 executes lines

(5)–(19). By k-submodularity, we have

$$\begin{aligned} f(\mathbf{s}^*) &\le f(\mathbf{s}^* \setminus (i^*, \bar{j}^*)) + f(i^*, \bar{j}^*) \\ &\le [(1+\frac{1}{\alpha})f(\mathbf{t}) + f(i^*, j^*)]\Delta + f(i^*, j^*) \\ &\le [2f(\mathbf{t}) + \alpha f(i^*, j^*)]\Delta \\ &= [2f(\mathbf{d}) + \alpha f(i^*, j^*)]\Delta. \end{aligned}$$

The second inequality holds since of inequality (20), $\alpha > 2$ and $\Delta \ge 2$, and the final inequality is derived from $\mathbf{t} = \mathbf{d}$.

Hence, the set $\mathbf{d}$ generated by Algorithm 1 must satisfy

$$f(\mathbf{s}^*) \le [2f(\mathbf{d}) + \alpha f(i^*, j^*)]\Delta.$$

According to the random step in lines 7–18 of Algorithm 1, we get

$$E[f(\mathbf{s})] = \frac{2}{2+\alpha} f(\mathbf{d}) + \frac{\alpha}{2+\alpha} f(i^*, j^*) \ge \frac{f(\mathbf{s}^*)}{(2+\alpha)\Delta}.$$

Theorem 1. *With* $\alpha = \frac{1+\sqrt{13}}{2}$, *Algorithm 1 can return a budget-feasible solution* $\mathbf{s}$ *satisfying* $f(\mathbf{s}^*) \le \frac{(5+\sqrt{13})\Delta}{2} E[f(\mathbf{s})]$ *incurring* $O(kn)$ *oracle queries.*

Proof. If Algorithm 1 returns a solution executing line 4 or 8, This solution is clearly budget-feasible. If Algorithm 1 executes line 9, it is valid that $p(\mathbf{t}) < B$ from Lemma 4. From the description of lines 10–17, we know that the solution is also budget-feasible. Hence, Algorithm 1 can return a budget-feasible solution $\mathbf{s}$. Considering the two cases declared in Lemma 3 and Lemma 4, we obtain

$$f(\mathbf{s}^*) \le \max\{2+\alpha, 3 + \frac{2\alpha+1}{\alpha^2-1}\} \cdot \Delta E[f(\mathbf{s})],$$

for any $\alpha \in (2, 1+\sqrt{3})$. Therefore, with $\alpha = \frac{1+\sqrt{13}}{2}$, the solution $\mathbf{s}$ satisfies $f(\mathbf{s}^*) \le \frac{(5+\sqrt{13})\Delta}{2} E[f(\mathbf{s})]$. Because we expends nk oracle queries to find (i^*, j^*) and k oracle queries to get $\bar{j} = \arg\max_{j\in[k]} f_{i,j}(\mathbf{t})$ for any $i \in E$, it's obvious that Algorithm 1 incurs $O(nk)$ value oracle queries.

Corollary 1. *If* f *is monotone, then it is valid that* $\Delta = 2$ *and the final solution of Algorithm 1 satisfies* $f(\mathbf{s}^*) \le (5+\sqrt{13})E[f(\mathbf{s})]$. *If* f *is non-monotone, then it is valid that* $\Delta = 3$ *and the final solution of Algorithm 1 satisfies* $f(\mathbf{s}^*) \le \frac{15+3\sqrt{13}}{2} E[f(\mathbf{s})]$.

5 Conclusions

In the clock auction model, an auctioneer wants to purchase goods from a subset of agents E and assigns them to k departments. The auctioneer associates a value measured by a k-submodular function and a given budget. Each agent has a private cost of his goods. For the monotone case, we propose a randomized algorithm with an approximation ratio of $\frac{1}{5+\sqrt{13}}$. The randomized algorithm can achieve $\frac{2}{15+3\sqrt{13}}$-approximation for a non-monotone objective. The algorithm only needs $O(kn)$ value oracle queries, which makes it more practical. It is an interesting problem to design a budget-feasible mechanism for a k-submodular function under an individual budget constraint.

Acknowledgments. This work is supported by the National Natural Science Foundation of China (Grant No. 12471302).

Disclosure of Interests. The authors have no competing interests to declare that are relevant to the content of this article.

References

1. Amanatidis, G., Kleer, P., Schäfer, G.: Budget-feasible mechanism design for non-monotone submodular objectives: offline and online. Math. Oper. Res. **47**(3), 2286–2309 (2022). https://doi.org/10.1145/3328526.3329622
2. Bilbao, J.M.: Cooperative Games on Combinatorial Structures. Springer, New York (2000)
3. Balkanski, E., Garimidi, P., Gkatzelis, V., Schoepflin, D., Tan, X.: Deterministic budget-feasible clock auctions. In: Proceedings of the 2022 ACM-SIAM Symposium on Discrete Algorithms(SODA), pp. 2940–2963 (2022). https://doi.org/10.1137/1.9781611977073.114
4. Chen, N., Gravin, N., Lu, P.: On the approximability of budget feasible mechanisms. In: Proceedings of the 2011 ACM-SIAM Symposium on Discrete Algorithms (SODA), pp. 685–699 (2011). https://doi.org/10.1137/1.9781611973082.54
5. Han, K., Wu, Y., Huang, h., Cui, S.: Triple eagle: simple, fast and practical budget-feasible mechanisms. In: Proceedings of the 37th International Conference on Neural Information Processing Systems (NeurIPS) (2023)
6. Huang, L., Wang, B., Zhou, H.: On optimal approximations for k-submodular maximization via multilinear extension. arXiv preprint:arXiv:2107.07103 (2023)
7. Jalaly, P., Tardos, É.: Simple and efficient budget feasible mechanisms for monotone submodular valuations. ACM Trans. Econ. Comput. **9**(1), 4:1–4:20 (2021). https://doi.org/10.1145/3434421
8. Ohsaka, N., Yoshida. Y.: Monotone k-submodular function maximization with size constraints. In: Proceedings of the 28th International Conference on Neural Information Processing Systems (NeurIPS), pp. 694–702(2015)
9. Rafiey, A., Yoshida, Y.: Fast and private submodular and k-submodular functions maximization with matroid constraints. In: Proceedings of Proceedings of the 37th International Conference on Machine Learning (ICML), pp. 7887–7897. PMLR (2020)
10. Singer, Y.: Budget feasible mechanisms. In: Proceedings of 2010 IEEE 51st Annual Symposium on Foundations of Computer Science (FOCS), pp. 765–774 (2010). https://doi.org/10.1109/FOCS.2010.78
11. Singh, A. P., Guillory, A., Bilmes, J. A.: On bisubmodular maximization. In: Proceedings of the Fifteenth International Conference on Artificial Intelligence and Statistics (AISTATS), pp. 1055–1063 (2012)

12. Tang, Z., Wang, C., Chan, H.: On maximizing a monotone k-submodular function under a knapsack constraint. Oper. Res. Lett. 50(1), 28–C31 (2022). https://doi.org/10.1016/j.orl.2021.11.010
13. Wang, B., Zhou, H.: Multilinear extension of k-submodular functions. arXiv preprint: arXiv:2107.07103 (2021)
14. Ward, J., Zivný, S.: Maximizing k-submodular functions and beyond*. ACM Trans. Algorithms **12**(4), 47:1–47:26 (2016). https://doi.org/10.1145/2850419
15. Xiao, H., Liu, Q., Zhou, Y., Li, M.: Approximation algorithms for k-submodular maximization subject to a knapsack constraint. arXiv preprint: arXiv:2306.14520 (2023)

Kernel for Proper Helly Circular-Arc Vertex Deletion: Smaller and Simpler via Graph Isomorphism

Hanchun Yuan[1(✉)] and Zhen Zhang[2]

[1] School of Computer Science and Technology, Zhejiang Normal University, Jinhua 321004, China
hanchunyuan@zjnu.edu.cn

[2] School of Advanced Interdisciplinary Studies, Hunan University of Technology and Business, Changsha 410205, China

Abstract. The modification problems toward subclasses of claw-free graphs have been investigated a lot in recent years. Proper Helly circular-arc graph is an important subclass of claw-free graphs. In this paper, we give an $O(k^8)$ vertex-kernel for proper Helly circular-arc vertex deletion, which is much better than the $O(k^{88})$ vertex-kernel [LATIN 24]. Our kernelization is based on small graph enumeration and isomorphism, which may be helpful for other graph classes with large forbidden induced subgraphs.

Keywords: parameterized complexity · kernelization · vertex deletion · claw-free · proper Helly circular-arc graph

1 Introduction

The existence of a polynomial kernel for claw-free edge deletion is an interesting open problem in the area of parameterized complexity [6]. However, claw-free graphs have a complex structure [5]. Hence, trying to consider the subclasses of claw-free graphs can help understand the structure of claw-free [7]. Actually, the first steps of many *fixed-parameter tractable* (FPT) and kernelization algorithms for modification problems toward subclasses of claw-free are destroying claw and some small forbiddens. There are many subclasses of claw-free graphs, e.g., line graphs, proper circular-arc graphs, anti-prismatic graphs, proper Helly circular-arc graphs, and proper interval graphs. The relationships between some subclasses of claw-free can be seen in Fig. 1.

In recent years, many FPT and kernelization algorithms are proposed for modification problems toward subclasses of claw-free graphs. The modification operation can be vertex deletion, edge deletion, completion (edge addition), and editing (both apply edge deletion and addition). This paper continues this line of investigation. Note that claw-free vertex deletion can be considered as a special version of 3-hitting set, hence has a polynomial kernel (see, e.g., in [1]). The

D. Du et al. (Eds.): COCOA 2024, LNCS 15434, pp. 388–400, 2025.
https://doi.org/10.1007/978-981-96-4445-2_32

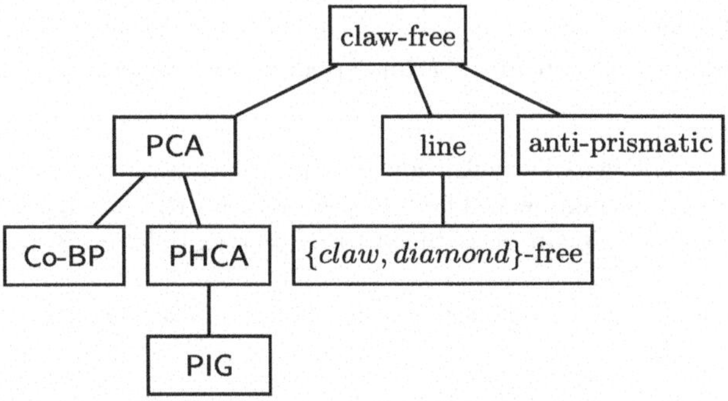

Fig. 1. The relationships between some subclasses of claw-free. Let PCA denote proper circular-arc graphs, PHCA denote proper Helly circular-arc graphs, PIG denote proper interval graphs, and Co-BP denote the complement of bipartite permutation graphs. (When necessary, we will use these abbreviations in the following.)

existence of polynomial kernel for claw-free vertex deletion does not imply that vertex deletion toward subclasses of claw-free graphs also have polynomial kernels. In this paper, we consider proper Helly circular-arc vertex deletion, which is defined as follows.

Proper Helly Circular-arc Vertex Deletion	
Input:	An undirected graph $G = (V, E)$ and $k \in \mathbb{N}$.
Parameter:	k.
Question:	Is there a set $S \subseteq V(G)$ such that $G - S$ is a proper Helly circular-arc graph and $\|S\| \leq k$?

If a graph class has a finite set of forbiddens each of which has finite vertices, then the vertex deletion version admits a polynomial kernel by a reduction to d-hitting set where d is the maximum number of vertices of a forbidden. Then, line vertex deletion and $\{claw, diamond\}$-free vertex deletion admit a polynomial kernel. For example, since d-hitting set admits an $O(k^{d-1})$-vertex kernel [1] and each forbidden of line graph has at most six vertices, we can have an $O(k^5)$-vertex kernel. Note that due to the propogation property, the reduction cannot be applied to edge modification version. However, (some subclasses of) proper circular-arc graphs have infinite forbiddens. For such cases, the kernelization algorithms often use *modulator*. A modulator is a vertex set M such that $G - M$ has no (small) forbiddens and $|M| = poly(k)$. For example, the polynomial kernels for unit interval vertex deletion [10,11] and bipartite permutation vertex deletion [8] find a modulator and then bound the number of vertices in $G - M$. Moreover, Ke et al. [11] proved that using a smaller modulator M, the size of a clique in $G - M$ can be reduced a lot by some observations. Recently, Agrawal et al. [2] gave a polynomial kernel for proper Helly circular-arc vertex deletion. The technique flowchart is similar to the kernel for bipartite permutation vertex deletion [8]. The kernel first construct a modulator M such that $G - M$ is a bipartite permutation graph and M has special properties.

Table 1. The (best known) kernelization results about some subclasses of claw-free. The symbol '?' denotes unknown. The '$O(\cdot)$' represents the number of vertices. Note that Co-BP vertex deletion can be reduced to BP vertex deletion in polynomial time.

graph classes	vertex deletion	edge deletion	completion	editing
PCA	?	?	?	?
PHCA	$O(k^8)$ [Thm. 1]	?	?	?
Co-BP	$O(k^{229})$ [8]	?	?	?
PIG	$O(k^4)$ [11]	?	$O(k^5)$ [3]	?
line	$O(k^5)$ [1]	$O(k^5)$ [9]	?	?
$\{claw, diamond\}$-free	$O(k^3)$ [1]	$O(k^3)$ [12]	?	?

For the edge modification version, Bessy and Perez [3] proved that unit interval completion admits a polynomial kernel. Moreover, the kernel does not use the modulator. The edge modification problems toward proper Helly circular-arc graphs and bipartite permutation graphs are still unsolved. Eiben et al. [9] first proved that line edge deletion admits a polynomial kernel by a modulator and the structure of line graphs. For $\{claw, diamond\}$-free edge deletion problem, Cygan et al. [7] proved that the problem admits a polynomial kernel using modulator. Then, Li et al. [12] improved it by more careful case analysis. In Table 1, one can see the best known results. In this paper, we improve the $O(k^{88})$-vertex kernel [2] for proper Helly circular-arc vertex deletion to $O(k^8)$-vertex kernel.

Theorem 1. *The proper Helly circular-arc vertex deletion problem admits an $O(k^8)$-vertex kernel.*

Overview of the Kernelization Algorithm

Using a 6-ratio approximation algorithm [4] we can find a modulator M such that $|M| \leq 6k$. Then $R = G - M$ is a proper Helly circular-arc graph. By some simple but nontrivial observations, in R, we only consider the components adjacent to M. Let $\mathcal{K}_R$ be a clique partition of R (the definition can be seen in Sect. 2). Note that the kernels in [2,11] also use the same clique partitions. Based on claw-free property, the forbiddens of PCA, and some non-involved reduction rules, we can prove that there are at most $O(k^2)$ *large* (with at least $k+1$ vertices) cliques in R. This part is likely to [2], but the modulator is different.

Let $\mathcal{K}_d$ be a set of d-consecutive cliques such that there is a large clique in $\mathcal{K}_d$. Since the number of large cliques is bounded by $O(k^2)$, then $|\bigcup_{i=1}^{5} \mathcal{K}_i| = O(k^2)$.

In our algorithm, we do not bound the size of each large clique, which is different from known kernels using modulator, e.g., proper Helly circular-arc vertex deletion [2] and unit interval vertex deletion [11]. The key idea is that for each vertex set $T \subseteq M$ with at most five vertices and for each $(6 - |T|)$-consecutive (large) cliques in R, we only preserve $O(k^{6-|T|})$ induced subgraphs with at most five vertices satisfying some properties. Hence, the kernel has $O(k^{|T|}) \cdot O(k^2) \cdot O(k^{6-|T|}) = O(k^8)$ vertices.

2 Preliminaries

All graphs considered in this paper are undirected. For a graph G, let $V(G)$ and $E(G)$ be the set of vertices and edges of G. For an edge $e = uv$, we say that u is *adjacent* to v or v is *adjacent* to u. Let $N(v)$ be the set of vertices adjacent to v and $N[v] = N(v) \cup \{v\}$. A *clique with size* d is a graph with d vertices and any pair of vertices are adjacent. A *path* in a graph G is a sequence of distinct vertices and every consecutive pair of vertices are adjacent. An *induced path* is a path with edges only between two consecutive vertices. A *cycle* is a sequence of distinct vertices with edges between two circular consecutive vertices and a *hole* is a cycle with at least four vertices and with edges only between two circular consecutive vertices. An (A, B)*-cut in* G is a vertex set S such that $(S \cap A) \cup (S \cap B) = \emptyset$ and there is no path from any vertex $a \in A$ and any vertex $b \in B$ in $G - S$. A minimum (A, B)-cut is abbreviated as an (A, B)*-mincut*. Note that an (A, B)-mincut can be found in polynomial time since it can be easily reduced to the (s, t)-mincut problem.

A *Monad*, see Fig. 3-(g), is a graph constructed by a hole H and a vertex nonadjacent to any vertex in H. We say that an induced path *meets* a clique if the clique contains a vertex in the induced path. For a set of d cliques and a path (or a hole) P, if the d cliques contain consecutive vertices of $V(P)$, then call the set of cliques d*-consecutive cliques*. Moreover, if a set of consecutive cliques contain all vertices of a hole, then the set of cliques is called a *clique circle*.

We use M to denote a 6-ratio approximation solution, called *modulator* and let $R = G - M$. Then R is a proper Helly circular-arc graph. Hence there is a P-time obtained *clique partition* (also called "nice clique partition" in [2]) $\mathcal{K}_R$ satisfying that (1) $V(\mathcal{K}_R) = R$, and (2) any two cliques in $\mathcal{K}_R$ are vertex-disjoint, and (3) an edge in R is either in a clique or between two cliques, and (4) any set of consecutive cliques in $\mathcal{K}_R$ form either a clique path, or a clique circle[1].

Proper Helly Circular-arc Graphs. A graph is a *circular-arc graph* if its vertices can be assigned to arcs on a circle such that there is an edge between two vertices if and only if their corresponding arcs intersect. The set of arcs is called an *arc model*, see Fig. 2. An arc model if *proper* if no arc in it properly contains another arc, see Fig. 2-(b). A graph is a *proper circular-arc graph* is it admits a proper arc model, see Fig. 2-(a). In case that there is a point of the circle avoided by all the arcs in an arc model, then one can cut the circle and straighten all the arcs into line segments. Such a graph is an *interval graph*,

[1] This can be proved in [2] by the construction of $\mathcal{K}_R$.

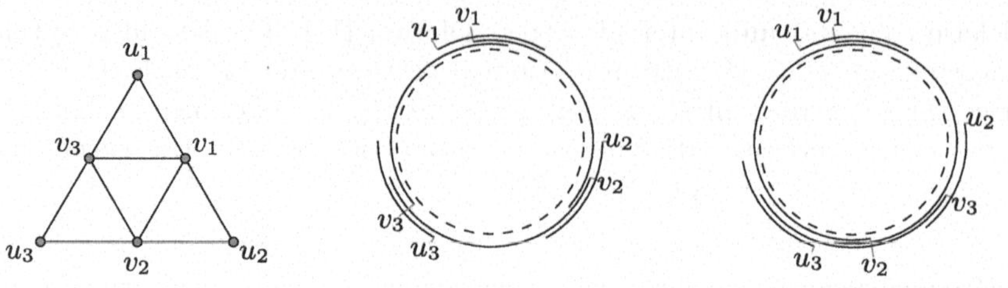

(a) S_3. (b) Proper but not Helly. (c) Helly but not proper.

Fig. 2. S_3 is not a proper Helly circular-arc graph. The arc models in (b) and (c) are both arc models of S_3 and S_3 has no proper Helly arc model.

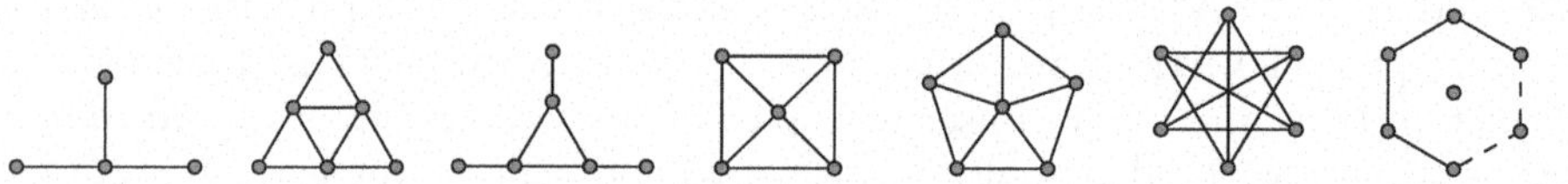

(a) claw. (b) S_3. (c) $\overline{S_3}$. (d) W_4. (e) W_5. (f) $\overline{C_6}$. (g) Monad.

Fig. 3. The forbidden induced graphs of proper Helly circular-arc graphs.

i.e., the intersection graph of a set of closed intervals on the real line. Proper interval graphs are defined analogously. All (proper) interval graphs are (proper) circular-arc graphs. An arc model is *Helly* if every set of pairwise intersecting arcs has a nonempty common intersection, see Fig. 2-(c). A circular-arc graph is a *proper Helly circular-arc graph* if has an arc model that is both proper and Helly. Note that a graph that admits a proper model and a Helly model may not be a proper Helly circular-arc graph, see Fig. 2, S_3 admits a proper model and a Helly model, but it cannot admit a proper Helly model. Proper Helly circular-arc graphs are characterized by forbidden induced subgraphs, see Fig. 3.

Theorem 2 ([13]). *A graph is a proper Helly circular-arc graph if and only if it contains no claw,* S_3, $\overline{S_3}$, W_4, W_5, $\overline{C_6}$, *or Monad.*

In the following, we call the forbidden induced subgraphs with at most six vertices *small forbiddens*; otherwise, we call them *large forbiddens.* If a connected graph has no small forbiddens, then it is a proper Helly circular-arc graph.

Lemma 1 (Corollary 9 in [4]). *Let G be a connected graph. If G does not contain claw,* S_3, $\overline{S_3}$, W_4, W_5, *or* $\overline{C_6}$, *then G is a proper Helly circular-arc graph.*

The following two statements consider the structure relationships between proper interval graphs and proper Helly circular-arc graphs.

Proposition 1 (Folklore). *A disconnected proper Helly circular-arc graph is a proper interval graph.*

Lemma 2 (Corollary 10 in [4]). *Let G be a proper Helly circular-arc graph. The graph G is a proper interval graph if and only if G does not contain a hole.*

3 The Structure of a Clique Partition $\mathcal{K}_R$

For a clique partition $\mathcal{K}_R$, we can relabel the cliques by $K_1, K_2, \ldots$ such that the labels for a set of consecutive cliques are also consecutive. If $\mathcal{K}_R$ has a clique circle, by the forbiddens and the construction of $\mathcal{K}_R$ in [2], then R has only one component and the indices are understood to be modulo $|\mathcal{K}_R|$. Let

$$\mathcal{K}_a^b = \begin{cases} \{K_a, K_{a+1}, \ldots, K_b\}, & \textit{if the set of cliques are consecutive;} \\ \emptyset, & \textit{otherwise.} \end{cases}$$

If a clique $K \in \mathcal{K}_R$ has a neighbor in M, then K is called a *touched clique*; otherwise K is *untouched*. For two touched cliques K_x and K_y, if one of $\mathcal{K}_{x+1}^{y-1}$ and $\mathcal{K}_{y+1}^{x-1}$ is a set of consecutive untouched cliques, then we call K_x and K_y *near touched cliques*. Let K be a touched clique and K' an untouched clique. We call a clique K' *distance-d clique* if there are *exactly* $(d-1)$-consecutive untouched cliques between K' and K, see Fig. 4.

Definition 1 (proximal clique). *Let K be an untouched clique. If K satisfies one of the following conditions, then K is a* proximal clique. *The conditions are:*

- *The distance between K and a touched clique is at most nine.*
- *The number of vertices between K and a distance-9 clique is at most k.*

Moreover, if K only satisfies the second condition, we call K boundary proximal.

If an untouched clique K is not a proximal clique, we call K *a distal clique*. A distal clique that is adjacent to a boundary proximal clique is called a *boundary distal clique*. One can see Fig. 4 to understand the definitions of such cliques. By the definition of boundary proximal cliques, we have the following remark.

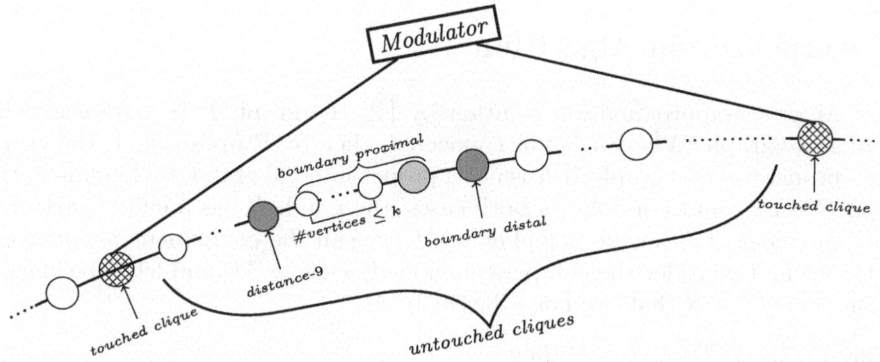

Fig. 4. The types of cliques in a clique partition $\mathcal{K}_R$. The number of vertices between the gray distance-9 clique and the light-gray boundary proximal clique is at most k.

Remark 1. Let $K_{dis\text{-}9}$ be a distance-9 clique, and let K_{bd} be the boundary distal clique such that the cliques between K_{bd} and $K_{dis\text{-}9}$ are all boundary proximal cliques. Then there are at least $k+1$ vertices between $K_{dis\text{-}9}$ and K_{bd}.

For convenience of reference, we restate some features of $\mathcal{K}_R$. For an induced path or a hole P, let $\mathcal{K}_P$ be the set of cliques met by P.

Note 1. Since any edge in R is either in a clique or between two consecutive cliques, for an induced path or a hole P in R, we have that

- each clique in $\mathcal{K}_p$ contains at most two consecutive vertices of P, and
- the cliques in $\mathcal{K}_P$ are consecutive in $\mathcal{K}_R$, and
- if P is a hole, then $\mathcal{K}_P$ is the only component of R.

Note 2 ($\star$[2]). Let[2] H be a hole in G that has a vertex in M and a vertex in R. Then $H \cap R$ is a set of maximal induced paths $\mathcal{P}$ and we have that

- $\mathcal{K}_{P_1} \cap \mathcal{K}_{P_2} = \emptyset$ for any two distinct paths P_1 and P_2 in $\mathcal{P}$, and
- the first and the last cliques (maybe the same) in $\mathcal{K}_P$ are all touched cliques.

Let $C = c_1c_2\cdots c_{|C|}$ be a cycle. We call C a *hole structure at* c_i if (1) the degree of c_i is exactly two, and (2) c_{i-1} and c_{i+1} are nonadjacent.

Lemma 3 ($\star$). *If C is a hole structure at c_i, then C has a hole containing c_i.*

From a hole structure at c_i, an *expansion hole structure at* c_i satisfies the following conditions: (1) c_{i+1} is replaced by a clique K_{i+1}, and (2) c_i is adjacent to all vertices in K_{i+1}, and (3) there is no edge between c_{i-1} and K_{i+1}, and (4) at least one vertex in K_{i+1} is adjacent to c_{i+2}.

Corollary 1 ($\star$). *If C is an expansion hole structure at c_i, then C has a hole.*

4 Kernelization Algorithm

Since M is a 6-approximation solution by [4], the graph R is a proper Helly circular-arc graph. When R is not connected, then by Proposition 1, the graph R is a proper interval graph. If R is not a proper interval graph, by Lemma 2, the graph R must contain a hole. In both cases, the graph R has a clique partition. If a component of R has no neighbor in M, we call the component *nonadjacent to* M. We first consider the components nonadjacent to M, and let $\mathcal{D}$ be the set of components of R that are nonadjacent to M.

Remark 2. If $|V(\mathcal{D})| \geq k+1$, then

- if $|M| \leq k$, then $G - M$ is a proper Helly circular-arc graph and the instance is trivially yes;

[2] Proofs of statements marked with $\star$ are deferred to the full version of the paper.

- if $|M| \geq k+1$, then for any solution S, $G-S$ is disconnected, implying that the resulting graph is a proper interval graph. We can use the $O(k^4)$ kernel for proper interval vertex deletion [11].

In the following, we can assume $|V(\mathcal{D})| \leq k$ and preserve all vertices in $\mathcal{D}$. Now we only consider the components adjacent to M. We considers two cases: proximal cliques and distal cliques. We first give a simple reduction rule that can help us bound the number of touched cliques in $\mathcal{K}_R$.

Reduction Rule 1 *If there is a $(k+3)$-star in G, then remove the center v of the star and decrease k by one.*

The running time of Reduction Rule 1 can be done in polynomial time. Now we prove the safeness of Reduction Rule 1.

Lemma 4 ($\star$). *Let (G', k') be the instance after applying Reduction Rule 1. Then, (G', k') is yes if and only if (G, k) is yes.*

Since each vertex has at most $k+3$ pairwise nonadjacent vertices and the edges in R are only in one clique or 2-consecutive cliques, by Reduction Rule 1, we have that a vertex $v \in M$ touches at most $2(k+3)$ cliques of $\mathcal{K}_R$. Since $|M| \leq 6k$, then there are at most $O(k^2)$ touched cliques. Recall that we call a clique large if it has at least $k+1$ vertices.

Proposition 2 ($\star$). *The number of large non-distal cliques is at most $O(k^2)$.*

4.1 Bounding the Number of Large Cliques

In this section, we first bound the number of distal cliques.

Reduction Rule 2 *Let K_s and K_e be two boundary distal cliques such that all cliques between K_s and K_e are distal cliques. Find a (K_s, K_e)-mincut U and do the following three steps:*
(1) remove all vertices between K_s and K_e, and
(2) add a clique K_x with size $x = \min(k+1, |U|)$, and
(3) add possible edges between K_x and K_s (resp., K_e).

It is clear that Reduction Rule 2 can be done in polynomial time.

Lemma 5 ($\star$). *Let (G', k) be the instance after applying Reduction Rule 2. Then (G', k) is yes if and only if (G, k) is yes.*

Since there are at most three distal cliques between two near touched cliques, by Proposition 2, we have the following corollary.

Corollary 2 *The number of large cliques in R can be bounded by $O(k^2)$.*

Algorithm 1. Mark Procedure.

Input: An instance (G, k).
Output: A graph G with marked vertices.

```
for each set T in M s.t. 1 ≤ |T| ≤ 5 do                                   ▷ O(k^{|T|}).
    for each d consecutive (large) cliques K_d s.t. d ∈ [6 − |T|] do          ▷ O(k^2).
        for each nice feasible graph G_f full of K_d do              ▷ P-time; O(1)-size.
            SUB-PROCEDURE(G^f, T, ∅, ∅).

SUB-PROCEDURE(G^f, T, A, B)
for each set V_B ⊆ B where |V_B| ≤ 6 − |T| − |A| − 1 do
    find maximal vertex-disjoint sets 𝒳 s.t. G[X ∪ A ∪ V_B] ≐ G^f, ∀X ∈ 𝒳;
    if |𝒳| ≥ k + 7 then mark k + 7 of them;
    else mark V(𝒳) and do SUB-PROCEDURE(G^f, T, A ∪ V_B, V(𝒳)).
```

4.2 Bounding the Number of Vertices in R

An induced sugraph G^f in R is called *feasible* if G^f is a C_6 or satisfies:

- G^f has at most five vertices, and
- the cliques in $\mathcal{K}_R$ containing a vertex in $V(G^f)$ are consecutive, and
- each clique has at most three vertices in $V(G^f)$.

In graph theory, two same size graphs G_1 and G_2 are *isomorphic* ($G_1 \cong G_2$) if there exist two vertex orderings σ_1 and σ_2 of $V(G_1)$ and $V(G_2)$ such that σ_1^i is adjacent to σ_1^j if and only if σ_2^i is adjacent to σ_2^j where $i, j \in [|V(G_1|]$. We say two same size feasible graphs G_1^f and G_2^f are *strong isomorphic* if

- $G_1^f \cong G_2^f$, and
- for any set of d-consecutive cliques $\mathcal{K}_d$, $V(G_1^f) \cap K = V(G_2^f) \cap K$, $\forall K \in \mathcal{K}_d$.

Actually, since R is a proper Helly circular-arc graph, we can assume that the vertex ordering is the ordering of corresponding arcs by the left endpoints. A feasible graph G^f is *full* of a set of d-consecutive cliques $\mathcal{K}_d$ if $V(G^f) \cap K \neq \emptyset$ for each $K \in \mathcal{K}_d$ and $V(G^f) \cap K' = \emptyset$ for each $K' \notin \mathcal{K}_d$.

Given a subset T of at most five vertices of M, two feasible graphs G_1^f and G_2^f are *stronger isomorphic by T ($A \doteq B$ by T)* if they satisfy:

- G_1^f and G_2^f are strong isomorphic with orderings σ_1 and σ_2;
- $N(\sigma_1^i) \cap T = N(\sigma_2^i) \cap T$ where $i \in \{1, 2, \ldots, |V(G_1^f)|\}$.

In the following reduction rule, mark all vertices of a graph is abbreviated as *mark a graph.* We apply Algorithm 1 to mark some vertices, see Algorithm 1. We define *nice feasible graph* as a feasible graph not stronger isomorphic (by T) to any marked feasible graphs. *The extra mark procedure when $\mathcal{K}_R$ has exactly three cliques can be omitted, but the proof may become more complex.*

Reduction Rule 3. *Do the mark procedure by Algorithm 1; if $\mathcal{K}_R$ has exactly three cliques, do the* SUB-PROCEDURE*(C_6, $\emptyset$, $\emptyset$, $\emptyset$). Remove unmarked vertices in large cliques in $\mathcal{K}_R$.*

Lemma 6. *There are at most $O(k^8)$ vertices marked by Reduction Rule 3 and the algorithm can be done in polynomial time.*

Proof. The polynomial time is easy to obtain, so we omit it. In Algorithm 1, for each T, there are $O(k^2)$ consecutive $6 - |T|$ cliques containing a large clique by Corollary 2. Note that there are at most $O(1)$ pairwise non-stronger isomorphic feasible graphs since $|T| \leq 5$ and each feasible graph contains at most six vertices[3]. In each level recursion of SUB-PROCEDURE, there is at least one vertex added in A. Then, there are at most $5 - |T|$ levels in recursion. In each level recursion, mark $O(k)$ vertices. For each T, there are at most $O(k^2) \times O(1) \times O(k^{5-|T|}) \times O(k)$ marked vertices.

Then, since there are $O(k^d)$ sets with size d in M, there are $O(k^{|T|}) \times O(k^2) \times O(1) \times O(k^{5-|T|}) \times O(k) = O(k^8)$ marked vertices. Similarly, there are at most $O(k^6)$ marked vertices when $\mathcal{K}_R$ has exactly three cliques. □

In Reduction Rule 3, we can delete the unmarked vertices one by one. If we prove that deleting one unmarked vertex is safe, then Reduction Rule 3 is also safe. We will prove that after deleting some unmarked vertex v_2, the instance $(G - v_2, k)$ is yes if and only if (G, k) is yes.

Let $G' = G - v_2$. Suppose that (G', k) is yes and S' is an optimal solution of (G', k). We want to prove that $G - S'$ is a proper Helly circular-arc graph. Now we first give some useful ingredients.

Lemma 7 ($\star$). *The graph $G - S'$ has no small forbidden induced subgraphs, and if $G - S'$ is connected, then $G - S'$ is a proper Helly circular-arc graph. Moreover, if $G - S'$ is disconnected, then $G' - S'$ cannot have a hole.*

Suppose for contradiction that $G - S'$ is not a proper Helly circular-arc graph. Then, $G - S'$ must contain a hole containing v_2 by Lemma 7. We prove that the existence of a hole H implies $G' - S'$ contains a hole, contradicting Lemma 7.

Lemma 8. *Suppose $G - S'$ is disconnected. If (G', k) is yes, then (G, k) is yes.*

Proof. Suppose for contradiction that $G - S'$ has a hole $H = h_1 h_2 \cdots h_{|H|}$ where $h_1 = v_2$. If h_1 and h_2 are in the same clique, let y_h denote h_1; otherwise, let y_h denote h_2. Simarily, if h_1 and $h_{|H|}$ are in the same clique, let y_t denote h_1; otherwize, let y_t denote $h_{|H|}$. Let $Y_1 = \{y_h\} \cup \{h_1\} \cup \{y_t\}$. Let $Y = Y_1 \setminus M$. We first assume that Y has only unmarked vertices, since other cases are easier.

Claim 1 ($\star$) *The hole H must contain a vertex in M.*[4]

By Claim 1, H has at least one vertex in M. Let $h_j \in H$ be the vertex in M such that j is the smallest index.

Claim 2 *If $h_{j+1} \in M$, then $G - S'$ cannot have a hole.*

[3] Since R is proper Helly circular-arc and each feasible graph (except C_6) has at most 5 vertices, the number of pairwise non-strong isomorphic graphs is small.

[4] This claim uses the marked C_6's when $\mathcal{K}_R$ has exactly three cliques.

Proof. Let $T_1 = \{h_j, h_{j+1}\}$ and $T = T_1 \cup (Y_1 \cap M)$, see Fig. 5. Note that

- either y_h (resp., y_t) is in the same clique as h_2 (resp., $h_{|H|}$), or $y_h \in M$ (resp., $y_t \in M$), and
- we have $|Y \cup T| = |(Y_1 \setminus M) \cup \{h_j, h_{j+1}\} \cup (Y_1 \cap M)| \leq 5$, and
- the cliques containing a vertex in Y are consecutive.

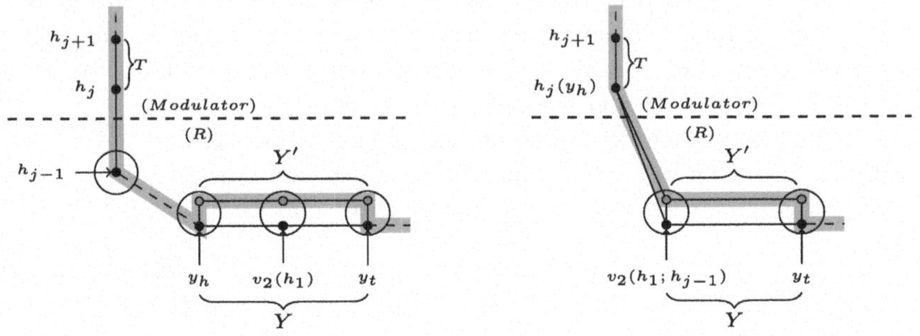

(a) $h_{j-1} \notin Y$. (b) $h_{j-1} \in Y$.

Fig. 5. $h_{j+1} \in M$, and the thick gray lines represent the cycle H' avoiding v_2.

Then, $G[Y]$ is a feasible graph which is an induced path in $G-S'$. Since $|Y \cup T| \leq 5$, and $|S'| \leq k$, there is a marked feasible graph Y' stronger isomorphic (by T) to Y in $G' - S'$. Let $H' = H \setminus \{v_2\} \cup Y'$.

If $y_h \in R$, see Fig. 5-(a), there is a vertex $y_h' \in Y'$ such that y_h' corresponds to y_h. Since y_h' is in the same clique as h_2, y_h' is adjacent to h_2.

If $y_h \in M$, see Fig. 5-(b), there is a vertex $h_1' \in Y'$ such that h_1' corresponds to h_1 and is adjacent to h_2.

Similarly, there is a vertex in Y' adjacent to $h_{|H|}$. Then, H' is a cycle which is not necessary induced (the thick gray lines in Fig. 5).

Since $h_j \in H$ and H is a hole in $G - S'$, we have that h_j is only adjacent to h_{j-1} and h_{j+1} in H. Since Y' is stronger isomorphic (by T) and $h_j \in T$, the vertices (in Y') not corresponding to h_{j-1} cannot be adjacent to h_j. If there is a vertex $h_{j-1}' \in Y'$ corresponding to h_{j-1}, and $v_2 \neq h_{j-1}$, then h_{j-1}' and h_{j-1} are in the same clique in H'; otherwise, h_j has only two neighbors h_{j-1}' and h_{j+1} in H'. Since Y' is stronger isomorphic (by T) and $h_{j+1} \in T$, neither h_{j-1} nor h_{j-1}' is adjacent to h_{j+1}. By Corollary 1, there is a hole containing h_j in $G' - S'$, a contradiction. This concludes the proof of Claim 2. ◁

Claim 3 ($\star$) *If $h_{j+1} \in R$, then $G - S'$ cannot have a hole.*

The case that Y or Z has marked vertices is similar and easier, hence omit it. By Claims 1–3, we have that $G - S'$ has no holes. Since $G - S'$ is disconnected by assumption and has no small forbiddens (by Lemma 7), we have that $G - S'$ is a proper Helly circular-arc graph, hence (G, k) is also yes. □

By Lemmas 7–8, we have the following corollary.

Corollary 3. *Let (G', k) be the instance after applying Reduction Rule 3. Then, (G, k) is yes if and only if (G', k) is yes.*

Proof of Theorem 1. Let M be a 6-ratio solution. Since $|M| \leq 6k$ (otherwise return 'no'), we need to bound $|V(R)|$. Apply Reduction Rules 1–3 in polynomial time.

By Remark 2, in R, the number of vertices in components nonadjacent to M is $O(k)$. For the components adjacent to M, find a clique partition $\mathcal{K}_R$. There are two types of cliques, say small cliques and large cliques in $\mathcal{K}_R$. By Lemma 6, there are at most $O(k^8)$ marked vertices in large cliques. The small cliques are only in boundary proximal cliques except the clique adjacent to a distal clique. Since there are at most $O(k^2)$ touched cliques by Reduction Rule 1 and each touch clique corresponds to at most two sets of boundary proximal cliques, at most $O(k) \times O(k^2)$ vertices are in small cliques and we have an $O(k^8)$ kernel. □

5 Conclusion

In this paper, we give an improved kernel for proper Helly circular-arc vertex deletion by small graph enumeration and isomorphism. This kernel may be improved by careful case analysis, but we are specifically concerned that this method may be useful for other graph classes with large forbiddens. Moreover, although it seems negative, we wonder whether this method can be *extended to edge modification problems.*

Acknowledgement. This research was supported by Zhejiang Provincial Natural Science Foundation of China under Grant No. ZCLQN25F0205, and Zhen Zhang was supported by National Natural Science Foundation of China under Grant No. 62202161.

Disclosure of Interests. We declare that we have no conflicts of interest to this work. We also declare that we do not have any commercial or associative interest that represents a conflict of interest in connection with the work submitted.

References

1. Abu-Khzam, F.N.: A kernelization algorithm for d-hitting set. J. Comput. Syst. Sci. **76**(7), 524–531 (2010)
2. Agrawal, A., Jana, S., Sahu, A.: A polynomial kernel for proper Helly circular-arc vertex deletion. In: 16th Latin American Symposium, LATIN 2024, Puerto Varas, Chile. Lecture Notes in Computer Science, vol. 14579, pp. 208–222. Springer (2024)
3. Bessy, S., Perez, A.: Polynomial kernels for proper interval completion and related problems. Inf. Comput. **231**, 89–108 (2013)
4. Cao, Y., Yuan, H., Wang, J.: Modification problems toward proper (Helly) circular-arc graphs. Inf. Comput. **301**, 105211 (2024)
5. Chudnovsky, M., Seymour, P.D.: Claw-free graphs. IV. Decomposition theorem. J. Comb. Theory B **98**(5), 839–938 (2008)

6. Crespelle, C., Drange, P.G., Fomin, F.V., Golovach, P.A.: A survey of parameterized algorithms and the complexity of edge modification. Comput. Sci. Rev. **48**, 100556 (2023)
7. Cygan, M., Pilipczuk, M., Pilipczuk, M., van Leeuwen, E.J., Wrochna, M.: Polynomial kernelization for removing induced claws and diamonds. Theory Comput. Syst. **60**(4), 615–636 (2017)
8. Derbisz, J., Kanesh, L., Madathil, J., Sahu, A., Saurabh, S., Verma, S.: A polynomial kernel for bipartite permutation vertex deletion. Algorithmica **84**(11), 3246–3275 (2022)
9. Eiben, E., Lochet, W.: A polynomial kernel for line graph deletion. In: 28th Annual European Symposium on Algorithms, ESA 2020, Pisa, Italy. LIPIcs, vol. 173, pp. 42:1–42:15. Schloss Dagstuhl - Leibniz-Zentrum für Informatik (2020)
10. Fomin, F.V., Saurabh, S., Villanger, Y.: A polynomial kernel for proper interval vertex deletion. SIAM J. Discret. Math. **27**(4), 1964–1976 (2013)
11. Ke, Y., Cao, Y., Ouyang, X., Li, W., Wang, J.: Unit interval vertex deletion: fewer vertices are relevant. J. Comput. Syst. Sci. **95**, 109–121 (2018)
12. Li, W., Peng, H., Yang, Y.: Improved kernel and algorithm for claw and diamond free edge deletion based on refined observations. Theor. Comput. Sci. **906**, 83–93 (2022)
13. Lin, M.C., Soulignac, F.J., Szwarcfiter, J.L.: Normal Helly circular-arc graphs and its subclasses. Discret. Appl. Math. **161**(7–8), 1037–1059 (2013)

Approximately Non-k-submodular Maximization Under p-System and ℓ Knapsack Constraints*

Hanlu Ye, Heqing Li, Min Li, Yang Zhou, and Qian Liu(✉)

School of Mathematics and Statistics, Shandong Normal University, Jinan 250014, China
{liminemily,zhouyang}@sdnu.edu.cn, lq_qsh@163.com

Abstract. In recent years, the k-submodular maximization problem has garnered significant attention. However, many practical applications cannot be strictly modeled as k-submodular maximization problems. This highlights the necessity of studying the maximization of approximately non-k-submodular. In this paper, we are committed to solving the problem of maximizing ϵ-approximately α-weakly diminishing returns functions under p-system and ℓ knapsack constraints. We first give a greedy algorithm, achieving a $\frac{1}{(1+\epsilon')(1+\frac{1+\epsilon}{\alpha^2(1-\epsilon)}p+\frac{2(1+\epsilon)^2}{\alpha^2(1-\epsilon)^2}\ell)}$- approximation with the $O(n^2(1+k)\log(2n))$ time complexity, where ϵ' is a very small positive number. Then we further introduce an improved algorithm that not only enhances the approximation ratio to $\frac{\min\{1,\frac{1}{\alpha(1+\epsilon')}\}}{1+\frac{1+\epsilon}{\alpha^2(1-\epsilon)}[(1+\epsilon')(p+\alpha\epsilon')+\frac{7}{4}\ell]}$, but also reduces the time complexity to $O(nk\log\frac{n}{\epsilon'}\log(2n))$.

Keywords: Approximately non-k-submodular · p-System · Knapsack

1 Introduction

Various applications, such as sensor placement involving k types of sensors and influence maximization across k topics, can be effectively modeled as k-submodular maximization (k-SMM) problems. The k-submodular function, a natural extension of the submodular function, was initially proposed by Huber and Kolmogorov [1]. Given a non-empty finite set $\mathcal{N}$, we define $(k+1)^{\mathcal{N}}$ as the family of all k disjoint subsets of $\mathcal{N}$. For $\boldsymbol{g} = (G_1, \ldots, G_k)$ and $\boldsymbol{h} = (H_1, \ldots, H_k)$ in $(k+1)^{\mathcal{N}}$, we denote $(G_1 \cap H_1, \ldots, G_k \cap H_k)$ as $\boldsymbol{g} \sqcap \boldsymbol{h}$, and $((G_1 \cup H_1) \setminus (\bigcup_{j\neq 1} G_j \cup H_j), \ldots, (G_k \cup H_k) \setminus (\bigcup_{j\neq k} G_j \cup H_j))$ as $\boldsymbol{g} \sqcup \boldsymbol{h}$. If $f : (k+1)^{\mathcal{N}} \to \mathbb{R}_+$ satisfies $f(\boldsymbol{g}) + f(\boldsymbol{h}) \geq f(\boldsymbol{g} \sqcap \boldsymbol{h}) + f(\boldsymbol{g} \sqcup \boldsymbol{h})$ for any

This paper is supported by Natural Science Foundation of Shandong Province of China (Nos. ZR2021MA100, ZR2024MA015) and National Science Foundation of China (No. 12371099).

D. Du et al. (Eds.): COCOA 2024, LNCS 15434, pp. 401–412, 2025.
https://doi.org/10.1007/978-981-96-4445-2_33

$\boldsymbol{g}, \boldsymbol{h} \in (k+1)^{\mathcal{N}}$, then f is called a k-submodular function. Additionally, if $f(\boldsymbol{g}) \leq f(\boldsymbol{h})$ for any $\boldsymbol{g}$ and $\boldsymbol{h}$ with $G_j \subseteq H_j, j \in [k]$, where $[k] := \{1, 2, \ldots, k\}$, then f is deemed monotone. Notably, the submodular function represents the case of 1-submodular.

For k-SMM, when the function is non-monotone, Ward and Živný [2] proposed a $\max\{1/3, 1/(1+\max\{1, \sqrt{\frac{k-1}{4}}\})\}$-approximation algorithm for unconstrained settings. And they further introduced a deterministic algorithm that guarantees a 1/3-approximation when the function is monotone. Subsequently, Iwata et al. [3] improved these results, providing approximation ratios of 1/2 and $k/(2k-1)$, respectively.

The total size constraint restricts the number of selected elements to not exceed a specified limit B, where B is a positive integer. And the individual size constraint mandates that each of the k sets contains no more than $B_i, i \in [k]$ elements, respectively. For k-SMM under total size and individual size constraints, if the function is monotone, Ohsaka and Yoshida [4] demonstrated that greedy algorithms achieve approximation ratios of 1/2 and 1/3, respectively. Nie et al. [5] employed a threshold greedy algorithm to reduce the query complexity to $O(\frac{kn}{\epsilon'} \log \frac{B}{\epsilon'})$, maintaining approximation ratios of $(1/2 - \epsilon')$ and $(1/3 - \epsilon')$, respectively. When the function is non-monotone, Nguyen and Thai [6] provided a 1/3-approximation algorithm under total size constraints. Xiao et al. [7] proposed two algorithms under individual size constraints, achieving approximation guarantees of $1/(B_m + 4)$ and 1/4, respectively, where $B_m = \max\{B_1, \ldots, B_k\}$.

In the context of a knapsack problem, each element $e \in \mathcal{N}$ has a corresponding cost c_e, and the objective is to $\max_{S \subseteq \mathcal{N}} f(S)$ subject to the constraint $c(S) = \sum_{e \in S} c_e \leq C$. About ℓ knapsack constraints, the i-th knapsack has an upper bound value C_i, and each element $e \in \mathcal{N}$ has a cost c_{ie}. A set S is deemed feasible if it satisfies $c_i(S) \leq C_i$ for all $i \in [\ell]$. For k-SMM with a knapsack constraint, Tang et al. [11] provided a $\frac{1}{2}(1-e^{-1})$-approximation for monotone objective functions. Later, Xiao et al. [12] ensured approximation results of $\frac{1}{2}(1-e^{-2})$ and $\frac{1}{3}(1-e^{-3})$ when the function is monotone and non-monotone, respectively.

Notably, when the cost of each element $e \in \mathcal{N}$ is uniform, the knapsack constraint simplifies to a cardinality constraint. Furthermore, cardinality constraints can be generalized to matroid constraints.

Definition 1 (matroid). *$\mathcal{N}$ is a finite set, and the set $\mathcal{I}$ composed of independent subsets of $\mathcal{N}$. We call $(\mathcal{N}, \mathcal{I})$ a matroid when the following conditions are met:*

(1) $\emptyset \in \mathcal{I}$;
(2) For any $X \subseteq Y \in \mathcal{I}$, we have $X \in \mathcal{I}$;
(3) For X and Y with $|X| < |Y|$ in $\mathcal{I}$, we can find an $e \in Y \setminus X$ that makes $X \cup \{e\}$ belong to $\mathcal{I}$.

When $(\mathcal{N}, \mathcal{I})$ satisfies the first two conditions, we call it an independence system.

For k-SMM under matroid constraints, Sakaue [8] demonstrated that the greedy algorithm achieves a 1/2-approximation when the function is monotone. Sun et al. [9] proposed a deterministic algorithm that ensures a 1/3-approximation in the non-monotone case. Niu et al. [10] further reduced the time complexity to $O(\frac{n(1+k)}{\epsilon'} \log \frac{r}{\epsilon'})$ through a threshold-decreasing algorithm, where r represents the rank of the matroid, obtaining approximation ratios of $(1/2 - \epsilon')$ and $(1/3 - \epsilon')$ for monotone and non-monotone functions, respectively. Under the constraint of a knapsack and p matroids, Liu et al. [13] achieved a $\frac{1}{2(p+1)}(1 - e^{-(p+1)})$-approximation for monotone objective functions, while Yu et al. [14] provided approximate guarantees of $\frac{1}{p+2}(1 - e^{-(p+2)})$ and $\frac{1}{p+3}(1 - e^{-(p+3)})$ for monotone and non-monotone functions, respectively.

It is important to note that constraints such as cardinality, matroid, and the intersection of p matroids are special cases of p-system constraints. In the context of an independence system $(\mathcal{N}, \mathcal{I})$, with sets X and Y in $\mathcal{N}$ such that $X \subseteq Y$, X is considered a base of Y if $X \in \mathcal{I}$, and adding any $u \in Y \setminus X$ to X would violate independence, i.e., $X \cup \{u\} \notin \mathcal{I}$.

Definition 2 (p-system). *The independence system $(\mathcal{N}, \mathcal{I})$ is called a p-system $(p \geq 1)$ if $|X_1|/|X_2| \leq p$ holds for any two bases X_1 and X_2 of any set $Y \subseteq \mathcal{N}$.*

It is not difficult to find that both cardinality and matroid are special cases where $p = 1$.

Recently, for k-SMM under p-system constraints, Ye et al. [16] provided approximate guarantees of $1/(1 + p)$ and $1/(2 + p)$ for monotone and non-monotone functions, respectively. For k-SMM under the intersection of p-system and ℓ knapsack constraints, Zhang et al. [15] achieved a $\frac{1-\epsilon'}{p+\theta+2\ell}$-approximation, where $\theta = 2$ for monotone functions and $\theta = 3$ for non-monotone functions. Then, they further improved the result with the time complexity of $O(\frac{\log n}{\epsilon'} nk(n + \ell))$.

We know that many practical applications cannot be strictly modeled as k-SMM problems. For instance, in sensor placement involving k types of sensors, environmental interference and inaccurate measurements may lead to noise values, and in influence maximization across k topics, uncertainty about the level of user influence may exist. This underscores the need for the establishment of approximately k-submodular functions. Approximately k-submodular functions are defined by Nguyen and Thai [17], who proposed streaming algorithms under total size constraints. Zheng et al. [18] introduced two types of function definitions: ϵ-approximately k-submodular (ϵ-AS) and ϵ-approximately diminishing returns (ϵ-ADR), where $\epsilon > 0$ and small.

Theorem 1. *[18] If the function is ϵ-ADR, then the function is ϵ-AS. The reverse is not true.*

Under a total size constraint B, they obtained results of $\frac{(1-\epsilon)^2}{2(1-\epsilon+\epsilon B)(1+\epsilon)}$ and $(1 - \epsilon)/2$, respectively. Under individual size constraints $B_i, i \in [k]$ with $B = \sum_{i \in [k]} B_i$, they proved the results of $\frac{(1-\epsilon)^2}{(3-3\epsilon+2\epsilon B)(1+\epsilon)}$ and $(1 - \epsilon)/(3 + \epsilon)$, respectively. For maximizing monotone α-weakly k-submodular($\alpha \in (0, 1]$) functions with matroid and individual size constraints, Matsuoka and Ohsaka [19] achieved a $(1+1/\alpha)^{-2}$-approximation. And for ϵ-AS maximization under matroid constraints, they proved a result of $\frac{1}{2}(\frac{1-\epsilon}{1+\epsilon})\frac{1}{1+\frac{\epsilon r}{1-\epsilon}}$, where r is the rank of the matroid. Wang et al. [20] obtained a $\alpha/(1+\alpha)$-approximation for α-weakly k-submodular maximization with matroid constraints. Jiang et al. [21] proposed ϵ-approximately α-weakly k-submodular (ϵ-A-α-WS) and ϵ-approximately α-weakly diminishing returns (ϵ-A-α-WDR) functions. When $\alpha = 1$, these two functions become ϵ-AS and ϵ-ADR, respectively. When $\epsilon = 0$, the ϵ-A-α-WS function becomes the monotone α-weakly k-submodular function. Under matroid constraints, they provided approximation ratios of $\frac{\alpha(1-\epsilon)^2}{(1+\epsilon)[(\alpha+1)(1-\epsilon)+2r\epsilon]}$ and $\frac{(1-\epsilon)\alpha}{1+\epsilon+\alpha-\epsilon\alpha}$ through the greedy algorithm, respectively, which can supersede the relevant results mentioned above.

Our Contributions. In this paper, we work on maximizing ϵ-A-α-WDR functions under p-system and ℓ knapsack constraints.

- We present a greedy algorithm that achieves a $\frac{1}{(1+\epsilon')(1+\frac{1+\epsilon}{\alpha^2(1-\epsilon)}p+\frac{2(1+\epsilon)^2}{\alpha^2(1-\epsilon)^2}\ell)}$-approximation with the time complexity of $O(n^2(1+k)\log(2n))$.
- Then we propose an improved algorithm that enhances the approximation ratio to $\frac{\min\{1,\frac{1}{\alpha(1+\epsilon')}\}}{1+\frac{1+\epsilon}{\alpha^2(1-\epsilon)}[(1+\epsilon')(p+\alpha\epsilon')+\frac{7}{4}\ell]}$ with the complexity of $O(nk\log\frac{n}{\epsilon'}\log(2n))$. When $\epsilon = 0$ and $\alpha = 1$, the ϵ-A-α-WDR function in this paper can be considered as a k-submodular function. With $p = 1$ and $\ell = 0$, our result $\frac{1}{(1+\epsilon')(1+(1+\epsilon')^2)}$ nearly matches the best known result 1/2 for k-SMM under total size or matroid constraints. When $\ell = 0$, our result $\frac{1}{(1+\epsilon')[1+(1+\epsilon')(p+\epsilon')]}$ almost conforms to the existing $1/(1+p)$-approximation for k-SMM under p-system constraints.

Paper Structure. Section 2 introduces the notations and definitions. Sections 3 and 4 detail our two algorithms. Finally, Section 5 summarizes the work.

2 Preliminaries

Let $\mathbf{0}$ be defined as $(\emptyset, \emptyset, \ldots, \emptyset)$. For $\boldsymbol{g}, \boldsymbol{h} \in (k+1)^{\mathcal{N}}$, the support set $\bigcup_{j=1}^{k} G_j$ of $\boldsymbol{g}$ is denoted by $U(\boldsymbol{g})$ and the size of $U(\boldsymbol{g})$ is denoted by $|U(\boldsymbol{g})|$. We use $\boldsymbol{g}(e)$ to represent the position of element e in $\boldsymbol{g}$, specifically, if $e \in G_j$, then $\boldsymbol{g}(e) = j$. There is a partial order relationship between $\boldsymbol{g}$ and $\boldsymbol{h}$ when $G_j \subseteq H_j$ for all $j \in [k]$, i.e. $\boldsymbol{g} \preceq \boldsymbol{h}$. The notation $\boldsymbol{g} \sqcup (e, j)$ represents the addition of the element $e \in \mathcal{N} \setminus U(\boldsymbol{g})$ to G_j, i.e. $\boldsymbol{g} \sqcup (e,j) = (G_1, \ldots, G_j \cup \{e\}, \ldots, G_k)$. The marginal gain $f(\boldsymbol{g} \sqcup (e,j)) - f(\boldsymbol{g})$ is denoted as $\Delta_{e,j} f(\boldsymbol{g})$. The orthant submodularity corresponds to

$$\Delta_{e,j} f(\boldsymbol{g}) \geq \Delta_{e,j} f(\boldsymbol{h}), \quad \boldsymbol{g} \preceq \boldsymbol{h},\ e \notin U(\boldsymbol{h}), j \in [k].$$

The pairwise monotonicity implies

$$\Delta_{e,i} f(\boldsymbol{g}) + \Delta_{e,j} f(\boldsymbol{g}) \geq 0,\ e \notin U(\boldsymbol{g}),\ i, j \in [k] \ with\ i \neq j.$$

Theorem 2. *[2] A function $f : (k+1)^{\mathcal{N}} \to \mathbb{R}_+$ is termed k-submodular if it is both orthant submodular and pairwise monotone.*

Definition 3 (α-weakly k-submodular). *A non-negative function $F : (k+1)^{\mathcal{N}} \to \mathbb{R}_+$ is α-weakly k-submodular if $\Delta_{e,j}F(\boldsymbol{g}) \geq \alpha\Delta_{e,j}F(\boldsymbol{h})$ for any $\boldsymbol{g} \preceq \boldsymbol{h} \in (k+1)^{\mathcal{N}}$, $e \notin U(\boldsymbol{h}), j \in [k]$ and $\alpha \in (0,1]$.*

This also means that when the function F is α-weakly k-submodular,

$$\sum_{(e,j)\in\boldsymbol{x}} \Delta_{e,j}F(\boldsymbol{g}) \geq \alpha(F(\boldsymbol{g} \sqcup \boldsymbol{x}) - F(\boldsymbol{g}))$$

for $\boldsymbol{g}, \boldsymbol{x} \in (k+1)^{\mathcal{N}}, U(\boldsymbol{g}) \cap U(\boldsymbol{x}) = \emptyset$ is valid.

Definition 4 (ϵ-approximately α-weakly diminishing returns (ϵ-A-α-WDR)). *A function $f : (k+1)^{\mathcal{N}} \to \mathbb{R}_+$ is ϵ-approximately α-weakly diminishing returns for some small $\epsilon \geq 0$ if there exists a monotone α-weakly k-submodular function F that satisfies $(1-\epsilon)\Delta_{e,j}F(\boldsymbol{g}) \leq \Delta_{e,j}f(\boldsymbol{g}) \leq (1+\epsilon)\Delta_{e,j}F(\boldsymbol{g})$ for any $\boldsymbol{g} \in (k+1)^{\mathcal{N}}$, $e \notin U(\boldsymbol{g}), j \in [k]$.*

According to the definition, it is evident that

$$f(\boldsymbol{g} \sqcup \boldsymbol{x}) - f(\boldsymbol{g}) \leq (1+\epsilon)(F(\boldsymbol{g} \sqcup \boldsymbol{x}) - F(\boldsymbol{g}))$$

for any $\boldsymbol{g}, \boldsymbol{x} \in (k+1)^{\mathcal{N}}$ with $U(\boldsymbol{g}) \cap U(\boldsymbol{x}) = \emptyset$.

This paper unites knapsack capacity without losing generality, i.e. $B_i = 1, i \in [\ell]$. Our work can be expressed as $\max_{\boldsymbol{g}\in(k+1)^{\mathcal{N}}} f(\boldsymbol{g})$ subject to $U(\boldsymbol{g}) \in \mathcal{I}$ and $c_i(U(\boldsymbol{g})) \leq 1, i \in [\ell]$.

3 Greedy algorithm

Here, we present Greedy Algorithm.

Density threshold ρ: A certain density threshold ρ, slightly lower than the value/size ratio of the optimal solution, is guessed (i.e. enumerated) by the algorithm.

Under different density thresholds ρ, the algorithm initiates a for-loop to obtain corresponding feasible solutions. Algorithm 1 ultimately selects the best solution from feasible solutions and a single-element optimal solution (e_{max}, j_{max}). For each for-loop, $\boldsymbol{s}_\rho$ starts from $\mathbf{0}$. Elements from $\mathcal{N}$ that meet the following conditions are then filtered: (1) Adding them to $U(\boldsymbol{s}_\rho)$ does not violate independence; (2) After adding them to $\boldsymbol{s}_\rho$, the ratio of the marginal gain to the total cost of the element is not less than ρ, i.e., $\max_j \Delta_{e,j}f(\boldsymbol{s}_\rho)/\sum_{i=1}^{\ell} c_{ie} \geq \rho$. We refer to $\max_j \Delta_{e,j}f(\boldsymbol{s}_\rho)/\sum_{i=1}^{\ell} c_{ie}$ as the density of the element e relative to $\boldsymbol{s}_\rho$; (3) Adding them to $\boldsymbol{s}_\rho$ does not exceed any knapsack capacity. The set of selected elements is recorded as A. An element from A is chosen to add to the optimal location of $\boldsymbol{s}_\rho$ to maximize the marginal gain. The elements are then filtered again to update set A. This loop continues until there are no eligible elements, i.e. $A = \emptyset$.

Mechanism 1. Greedy Algorithm

Input: $f : (k+1)^{\mathcal{N}} \rightarrow \mathbb{R}_+$, p-system $(\mathcal{N}, \mathcal{I})$ and ℓ knapsack-cost functions $c_i : \mathcal{N} \rightarrow [0,1]$.

Output: An $\boldsymbol{s} \in (k+1)^{\mathcal{N}}$ satisfying $U(\boldsymbol{s}) \in \mathcal{I}$ and $c_i(U(\boldsymbol{s})) \leq 1, i \in [\ell]$.

$(e_{max}, j_{max}) \leftarrow \arg\max_{e \in \mathcal{I}, j \in [k]} f((e,j))$

$M \leftarrow f((e_{max}, j_{max}))$

repeat the following for $\rho \in \{\frac{M}{\frac{1}{\alpha}p + \frac{2(1+\epsilon)}{\alpha(1-\epsilon)}\ell}, \frac{(1+\epsilon')M}{\frac{1}{\alpha}p + \frac{2(1+\epsilon)}{\alpha(1-\epsilon)}\ell}, \frac{(1+\epsilon')^2 M}{\frac{1}{\alpha}p + \frac{2(1+\epsilon)}{\alpha(1-\epsilon)}\ell}, \ldots, \frac{2nM}{\frac{1}{\alpha}p + \frac{2(1+\epsilon)}{\alpha(1-\epsilon)}\ell}\}$

$\boldsymbol{s}_\rho \leftarrow \boldsymbol{0}$

for $A = \{e \in \mathcal{N} \setminus U(\boldsymbol{s}_\rho) \mid U(\boldsymbol{s}_\rho) \cup \{e\} \in \mathcal{I}, \frac{\max \Delta_{e,j} f(\boldsymbol{s}_\rho)}{\sum_{i=1}^{\ell} c_{ie}} \geq \rho, and\ c_i(U(\boldsymbol{s}_\rho)) + c_{ie} \leq 1, \forall i\} \neq \emptyset$ **do**

  $(e, j') \leftarrow \arg\max_{e \in A, j \in [k]} \Delta_{e,j} f(\boldsymbol{s}_\rho)$

  $\boldsymbol{s}_\rho \leftarrow \boldsymbol{s}_\rho \sqcup (e, j')$

end for

restart with the next value of ρ

return $\arg\max_{\boldsymbol{s} \in \{\boldsymbol{s}_\rho, (e_{max}, j_{max})\}} f(\boldsymbol{s})$ for all enumerations of ρ

Time Complexity. Algorithm 1 enumerates $\log_{1+\epsilon'}(2n)$ density thresholds ρ. For each ρ, the algorithm tests and evaluates each element in $\mathcal{N} \setminus U(\boldsymbol{s}_\rho)$ to determine A. Then, k position attempts are performed on each element in A to find the (e, j') that can generate the maximum marginal gain when added to $\boldsymbol{s}_\rho$. This operation is repeated $|U(\boldsymbol{s}_\rho)|$ times. Consequently, the complexity is $O((|\mathcal{N} \setminus U(\boldsymbol{s}_\rho)| + |A|k)|U(\boldsymbol{s}_\rho)| \cdot \log_{1+\epsilon'}(2n)) = O(n^2(1+k)\log(2n))$, where n is the size of set $\mathcal{N}$.

Theorem 3. *Algorithm 1 ensures a* $\frac{1}{(1+\epsilon')(1+\frac{1+\epsilon}{\alpha^2(1-\epsilon)}p + \frac{2(1+\epsilon)^2}{\alpha^2(1-\epsilon)^2}\ell)}$*-approximation.*

When $\epsilon = 0$ and $\alpha = 1$, we can derive a conclusion about k-SMM under p-system and ℓ knapsack constraints.

Corollary 1. *We guarantee a* $\frac{1}{(1+\epsilon')(1+p+2\ell)}$*-approximation when f is monotone k-submodular using Algorithm 1.*

This improves the result of Theorem 3 in [15].

For a density threshold ρ, its associated algorithm solution is $\boldsymbol{s}_\rho$. Assuming $\boldsymbol{s}_\rho = \{(e_1, j_1), (e_2, j_2), \ldots, (e_r, j_r)\}$. Let $\boldsymbol{s}_\rho^t = \{(e_1, j_1), (e_2, j_2), \ldots, (e_t, j_t)\}$ represent the solution after t iterations. Note that $\boldsymbol{s}_\rho^0 = \boldsymbol{0}$. Fix an optimal solution $\boldsymbol{o}$, there exists a partition: $\boldsymbol{o}_1, \boldsymbol{o}_2, \ldots, \boldsymbol{o}_r, \boldsymbol{o}_{r+1}$, satisfying

$$\boldsymbol{o} = \boldsymbol{o}_1 \sqcup \boldsymbol{o}_2 \sqcup \cdots \sqcup \boldsymbol{o}_r \sqcup \boldsymbol{o}_{r+1},$$

where $|U(\boldsymbol{o}_t)| \leq p, t \in [r]$, and $U(\boldsymbol{o}_t) \cap U(\boldsymbol{o}_m) = \emptyset,\ t, m \in [r+1], t \neq m$.

Set $U(\boldsymbol{o}_{r+1}) = \{e \in U(\boldsymbol{o}) \setminus U(\boldsymbol{s}_\rho) \mid \frac{\Delta_{e,j'} f(\boldsymbol{s}_\rho)}{\sum_{i=1}^{\ell} c_{ie}} < \rho \ or \ \exists i, c_i(U(\boldsymbol{s}_\rho)) + c_{ie} > 1\}$. Define $T_r = U(\boldsymbol{o}) \setminus U(\boldsymbol{o}_{r+1})$. It is obvious that all elements in T_r can be added to $\boldsymbol{s}_\rho$ to satisfy $\Delta_{e,j'} f(\boldsymbol{s}_\rho) / \sum_{i=1}^{\ell} c_{ie} \geq \rho$ and not exceed any backpack capacity. For $t = r, r-1, \ldots, 2$, perform a for-loop: Let $D_t = \{e \in T_t \setminus U(\boldsymbol{s}_\rho^{t-1}) \mid U(\boldsymbol{s}_\rho^{t-1}) \cup \{e\} \in \mathcal{I}\}$. If $|D_t| \leq p$, set $U(\boldsymbol{o}_t) = D_t$; if $|D_t| > p$, there are two scenarios: If $e_t \notin D_t$, we randomly select a subset $U(\boldsymbol{o}_t) \subseteq D_t$ with $|U(\boldsymbol{o}_t)| = p$; otherwise, we select a subset $U(\boldsymbol{o}_t)$ from D_t that includes e_t and $|U(\boldsymbol{o}_t)| = p$. Set $T_{t-1} = T_t \setminus U(\boldsymbol{o}_t)$. After the for-loop, let $U(\boldsymbol{o}_1) = T_1$. We next prove $|U(\boldsymbol{o}_1)| \leq p$.

We prove the inequality $|T_{r-t}| \leq (r-t)p, t = 0, 1, \ldots, r-1$ by induction. Note that set $X = \{e \in U(\boldsymbol{o}) \setminus U(\boldsymbol{s}_\rho) \mid U(\boldsymbol{s}_\rho) \cup \{e\} \in \mathcal{I}\}$ is included in set $U(\boldsymbol{o}_{r+1})$. Otherwise, set A in Algorithm 1 would not be $\emptyset$, and the algorithm would continue iterating to add elements to $\boldsymbol{s}_\rho$. Therefore, $|T_r| \leq p|U(\boldsymbol{s}_\rho)| = rp$ when $t = 0$. In fact, $U(\boldsymbol{s}_\rho)$ is a base of $Y = T_r \cup U(\boldsymbol{s}_\rho)$ and T_r is independent. Assuming $|T_{r-t+1}| \leq (r-t+1)p$ is valid. There are two scenarios: If $|D_{r-t+1}| > p$, then $|U(\boldsymbol{o}_{r-t+1})| = p$, so $|T_{r-t}| = |T_{r-t+1}| - |U(\boldsymbol{o}_{r-t+1})| \leq (r-t+1)p - p = (r-t)p$; if $|D_{r-t+1}| \leq p$, then $T_{r-t} = T_{r-t+1} \setminus D_{r-t+1}$. According to the definition of D_{r-t+1}, $U(\boldsymbol{s}_\rho^{r-t})$ is a base of $Q = T_{r-t} \cup U(\boldsymbol{s}_\rho^{r-t})$, and since T_{r-t} is independent, $|T_{r-t}| \leq p|U(\boldsymbol{s}_\rho^{r-t})| = (r-t)p$. The proof ends, and when $t = r-1$, $|U(\boldsymbol{o}_1)| = |T_1| \leq p$.

We now construct a sequence $\boldsymbol{o}^0 = \boldsymbol{o}, \boldsymbol{o}^1, \boldsymbol{o}^2, \ldots, \boldsymbol{o}^r$ that satisfies

$$\boldsymbol{s}_\rho^t \preceq \boldsymbol{o}^t, t = 0, 1, \ldots, r.$$

Initially, $\boldsymbol{s}_\rho^0 \preceq \boldsymbol{o}^0$. For $t \in [r]$, define $\boldsymbol{o}^{t-\frac{1}{2}} := \boldsymbol{o}^{t-1} \setminus \boldsymbol{o}_t$, which means removing $\boldsymbol{o}_t$ from $\boldsymbol{o}^{t-1}$. The $\boldsymbol{o}^t$ is obtained by adding element e_t to the j_t-th position in $\boldsymbol{o}^{t-\frac{1}{2}}$, i.e. $\boldsymbol{o}^t = \boldsymbol{o}^{t-\frac{1}{2}} \sqcup (e_t, j_t)$. The following properties hold:

$$(1)\ \boldsymbol{s}_\rho^{t-1} \preceq \boldsymbol{o}^{t-\frac{1}{2}}, t \in [r]; \qquad (2)\ \boldsymbol{s}_\rho^t \preceq \boldsymbol{o}^t, t = 0, 1, \ldots, r.$$

Note that $\boldsymbol{o}^r = \boldsymbol{s}_\rho \sqcup \boldsymbol{o}_{r+1}$.

Lemma 1. *There is* $f(\boldsymbol{o}) - f(\boldsymbol{o}^r) \leq \frac{(1+\epsilon)p}{\alpha^2(1-\epsilon)} f(\boldsymbol{s}_\rho)$.

Then we discuss the elements in $U(\boldsymbol{o}_{r+1})$ as follows:

Case 1. The density of any element in $U(\boldsymbol{o}_{r+1})$ relative to $\boldsymbol{s}_\rho$ does not meet the density threshold ρ, i.e. $\Delta_{e,j'} f(\boldsymbol{s}_\rho) / \sum_{i=1}^{\ell} c_{ie} < \rho,\ e \in U(\boldsymbol{o}_{r+1})$;
Case 2. There exists an element e in $U(\boldsymbol{o}_{r+1})$ whose density relative to $\boldsymbol{s}_\rho$ reaches the density threshold ρ, but adding it to $\boldsymbol{s}_\rho$ will exceed the knapsack capacity, i.e. $\exists e \in U(\boldsymbol{o}_{r+1}), \Delta_{e,j'} f(\boldsymbol{s}_\rho) / \sum_{i=1}^{\ell} c_{ie} \geq \rho$, and $c_i(U(\boldsymbol{s}_\rho)) + c_{ie} > 1, \exists i \in [\ell]$.

Lemma 2. *In Case 1, there is* $f(\boldsymbol{o}^r) - f(\boldsymbol{s}_\rho) < \frac{(1+\epsilon)\rho\ell}{\alpha(1-\epsilon)}$.

Lemma 1 and Lemma 2 can directly derive the following lemma.

Lemma 3. *For Case 1, we have* $f(\boldsymbol{s}_\rho) \geq \frac{f(\boldsymbol{o}) - \frac{(1+\epsilon)\rho\ell}{\alpha(1-\epsilon)}}{1+\frac{(1+\epsilon)p}{\alpha^2(1-\epsilon)}}$.

Lemma 4. *Assuming* $\boldsymbol{s}$ *is the output solution of Algorithm 1. For Case 2, we have* $f(\boldsymbol{s}) > \frac{(1-\epsilon)\alpha\rho}{2(1+\epsilon)}$.

4 Improved Algorithm

To enhance the approximation results and reduce the time complexity described in Section 3, we present the following improved algorithm. The elements in $\mathcal{N}$ are classified into small elements ($c_{ie} \leq 1/2$, $\forall i \in [\ell]$) and large elements ($c_{ie} > 1/2$, $\exists i \in [\ell]$). Let D be the set of large elements that belong to $\mathcal{I}$. The optimal single-element solution $\boldsymbol{s}_D$ is obtained from D, and used as an alternative solution.

The algorithm searches for feasible solutions from $\mathcal{N} \setminus D$. Under different density thresholds ρ, Algorithm 2 loops independently to obtain the corresponding feasible solution $\boldsymbol{s}_\rho$ in a threshold descent manner for the threshold τ. After completing Algorithm 2, it outputs the best of these feasible solutions and $\boldsymbol{s}_D$.

When the algorithm initiates at the density threshold ρ, it first determines a corresponding value M_ρ. The process begins with $\boldsymbol{s}_\rho = \mathbf{0}$ and $\tau = M_\rho$. The algorithm selects elements from $\mathcal{N} \setminus D$ that satisfy the following conditions: (1) Adding them to $U(\boldsymbol{s}_\rho)$ does not violate independence; (2) The maximum marginal gain generated by adding them to $\boldsymbol{s}_\rho$ reaches the current threshold τ; (3) Their density relative to $\boldsymbol{s}_\rho$ can reach the density threshold ρ. These elements are added to $\boldsymbol{s}_\rho$ iteratively. During this addition process, if the knapsack capacity is exceeded, Algorithm 3 (EKAlgorithm($\boldsymbol{s}_\rho$)) is employed to transform the infeasible solution $\boldsymbol{s}_\rho$ into a feasible one, ending the loop. If the knapsack capacity is not exceeded, all these elements are incorporated into $\boldsymbol{s}_\rho$. The threshold is then slightly reduced to $\frac{1}{1+\epsilon'}\tau$. As long as the threshold remains above $\frac{\epsilon'}{n}M_\rho$, the algorithm continues to select elements from $\mathcal{N} \setminus D$ that meet the conditions and add them to $\boldsymbol{s}_\rho$. If the knapsack capacity is not exceeded throughout the process, the loop concludes when the threshold falls below $\frac{\epsilon'}{n}M_\rho$. At this point, $\boldsymbol{s}_\rho$ is deemed feasible.

The processing procedure of EKAlgorithm($\boldsymbol{s}_\rho$) for an infeasible solution $\boldsymbol{s}_\rho$ is as follows: For $t = 1$ to 3, let $\boldsymbol{b}_t = \{(e^1, \boldsymbol{s}_\rho(e^1)), \ldots, (e^{t-1}, \boldsymbol{s}_\rho(e^{t-1}))\}$, noting that $\boldsymbol{b}_1 = \mathbf{0}$. Elements from $U(\boldsymbol{s}_\rho)$ are added to $U(\boldsymbol{b}_t)$ one by one in any order. If the addition of an element to $U(\boldsymbol{b}_t)$ does not exceed the knapsack capacity, it is added to $\boldsymbol{b}_t$ at the same position as in $\boldsymbol{s}_\rho$. Otherwise, this element is marked as e^t and the loop terminates. Finally, the algorithm selects a feasible solution with the maximum cost from $\boldsymbol{b}_t, t \in [3]$, as the output.

Time Complexity. The number of density threshold ρ enumerated in Algorithm 2 is $\log_{1+\epsilon'}(2n)$. There are $\log_{1+\epsilon'} \frac{n}{\epsilon'}$ iterations for the value τ. Each iteration necessitates identifying the optimal location j' for each element in $\mathcal{N} \setminus D$ to be added to $\boldsymbol{s}_\rho$, and verifying if these elements meet the criteria. Therefore, the time complexity is $O(nk \log \frac{n}{\epsilon'} \log(2n))$.

Mechanism 2. Improved Algorithm

Input: $f : (k+1)^{\mathcal{N}} \to \mathbb{R}_+$, p-system $(\mathcal{N}, \mathcal{I})$ and ℓ knapsack-cost functions $c_i : \mathcal{N} \to [0,1]$.
Output: An $\boldsymbol{s} \in (k+1)^{\mathcal{N}}$ satisfying $U(\boldsymbol{s}) \in \mathcal{I}$ and $c_i(U(\boldsymbol{s})) \leq 1, i \in [\ell]$.

$D \leftarrow \{e \in \mathcal{I} \mid \exists i, c_{ie} > \frac{1}{2}\}$ and $\boldsymbol{s}_D \leftarrow \arg\max_{e \in D, j \in [k]} f((e,j))$
$M \leftarrow \max_{e \in \mathcal{I}, j \in [k]} f((e,j))$
repeat the following for $\rho \in \{\frac{M}{\frac{2}{3}\alpha + \frac{1+\epsilon}{\alpha(1-\epsilon)}[\frac{2(1+\epsilon')}{3}(p+\alpha)+\frac{7}{6}\ell]}, \frac{(1+\epsilon')M}{\frac{2}{3}\alpha + \frac{1+\epsilon}{\alpha(1-\epsilon)}[\frac{2(1+\epsilon')}{3}(p+\alpha)+\frac{7}{6}\ell]}, \frac{(1+\epsilon')^2 M}{\frac{2}{3}\alpha + \frac{1+\epsilon}{\alpha(1-\epsilon)}[\frac{2(1+\epsilon')}{3}(p+\alpha)+\frac{7}{6}\ell]}, \ldots, \frac{2nM}{\frac{2}{3}\alpha + \frac{1+\epsilon}{\alpha(1-\epsilon)}[\frac{2(1+\epsilon')}{3}(p+\alpha)+\frac{7}{6}\ell]}\}$
$\tau = M_\rho \leftarrow \max\{f((e,j)) : e \in \mathcal{I} \setminus D, \frac{f((e,j))}{\sum_{i=1}^{\ell} c_{ie}} \geq \rho\}$
$\boldsymbol{s}_\rho \leftarrow \mathbf{0}$
while $\tau \geq \frac{\epsilon'}{n} M_\rho$ **do**
  for $e \in \mathcal{N} \setminus D$ **do**
    if $U(\boldsymbol{s}_\rho) \cup \{e\} \in \mathcal{I}$ **then**
      $j' \leftarrow \arg\max_{j \in [k]} \Delta_{e,j} f(\boldsymbol{s}_\rho)$
      if $\Delta_{e,j'} f(\boldsymbol{s}_\rho) \geq \tau$ and $\frac{\Delta_{e,j'} f(\boldsymbol{s}_\rho)}{\sum_{i=1}^{\ell} c_{ie}} \geq \rho$ **then**
        $\boldsymbol{s}_\rho \leftarrow \boldsymbol{s}_\rho \sqcup (e, j')$
        if $\exists i$; $c_i(U(\boldsymbol{s}_\rho)) > 1$ **then**
          $\boldsymbol{s}_\rho \leftarrow EKAlgorithm(\boldsymbol{s}_\rho)$
          restart with the next value of ρ
        end if
      end if
    end if
  end for
  $\tau \leftarrow \frac{1}{1+\epsilon'} \tau$
end while
restart with the next value of ρ
return $\arg\max_{\boldsymbol{s} \in \{\boldsymbol{s}_\rho, \boldsymbol{s}_D\}} f(\boldsymbol{s})$ for all enumerations of ρ

Mechanism 3. EKAlgorithm($\boldsymbol{s}_\rho$)

for $t = 1$ to 3 **do**
  $\boldsymbol{b}_t \leftarrow \{(e^1, \boldsymbol{s}_\rho(e^1)), \ldots, (e^{t-1}, \boldsymbol{s}_\rho(e^{t-1}))\}$
  for $e \in U(\boldsymbol{s}_\rho)$ **do**
    if $c_i(U(\boldsymbol{b}_t)) + c_{ie} \leq 1, \forall i$ **then**
      $\boldsymbol{b}_t \leftarrow \boldsymbol{b}_t \sqcup (e, \boldsymbol{s}_\rho(e))$
    else
      the element e is denoted as e^t, and exit the loop
    end if
  end for
end for
return $\boldsymbol{b} \leftarrow \arg\max_{\boldsymbol{b} \in \{\boldsymbol{b}_1, \boldsymbol{b}_2, \boldsymbol{b}_3\}} \sum_{i=1}^{\ell} c_i(U(\boldsymbol{b}))$

Theorem 4. *We ensure a* $\min\{1, \frac{1}{\alpha(1+\epsilon')}\}\frac{1}{1+\frac{1+\epsilon}{\alpha^2(1-\epsilon)}[(1+\epsilon')(p+\alpha\epsilon')+\frac{7}{4}\ell]}$*- approximation using Algorithm 2.*

When $\epsilon = 0$ and $\alpha = 1$, we can draw the following inference.

Corollary 2. *Using Algorithm 2, a* $\frac{1}{(1+\epsilon')[1+(1+\epsilon')(p+\epsilon')+\frac{7}{4}\ell]}$*-approximation can be obtained when f is monotone k-submodular.*

This is better than the result in Theorem 4 in [15].

Lemma 5. *The output* $\boldsymbol{b}$ *of Algorithm 3 does not exceed the knapsack capacity and* $\sum_{i=1}^{\ell} c_i(U(\boldsymbol{b})) \geq \frac{2}{3}$.

Fix an optimal solution $\boldsymbol{o}$. Given ρ, assume its corresponding algorithm solution $\boldsymbol{s}_\rho = \{(e_1, j_1), (e_2, j_2), \ldots, (e_r, j_r)\}$, and define $\boldsymbol{s}_\rho^t = \{(e_1, j_1), (e_2, j_2), \ldots, (e_t, j_t)\}$ for $t \in [r]$. Concerning $\boldsymbol{s}_\rho$, two cases are considered:

Case 1. $\boldsymbol{s}_\rho$ is obtained through the processing of Algorithm 3.

Case 2. No situation of exceeding the knapsack capacity occurs in Algorithm 2.

Lemma 6. *For Case 1, there is* $f(\boldsymbol{s}_\rho) \geq \frac{2}{3}\rho$.

Proof. From Step 10 of Algorithm 2 and Lemma 5, it can be directly deduced that

$$f(\boldsymbol{s}_\rho) = \sum_{t=1}^{r} \Delta_{e_t, j_t} f(\boldsymbol{s}_\rho^{t-1}) \geq \rho \sum_{i=1}^{\ell}\sum_{t=1}^{r} c_{ie_t} = \rho \sum_{i=1}^{\ell} c_i(\boldsymbol{s}_\rho) \geq \frac{2}{3}\rho.$$

In Case 2, let $U(\boldsymbol{o}_{r+3})$ represent the set of large elements in $U(\boldsymbol{o})$, i.e. $U(\boldsymbol{o}_{r+3}) = \{e \in U(\boldsymbol{o}) \mid c_{ie} > 1/2, \exists i \in [\ell]\}$. Let $U(\boldsymbol{o}_{r+2})$ represent the set of elements in $U(\boldsymbol{o}) \setminus (U(\boldsymbol{o}_{r+3}) \cup U(\boldsymbol{s}_\rho))$ that were not added to $\boldsymbol{s}_\rho$ due to the density relative to $\boldsymbol{s}_\rho$ not reaching the density threshold ρ, i.e. $U(\boldsymbol{o}_{r+2}) = \{e \in U(\boldsymbol{o}) \setminus (U(\boldsymbol{o}_{r+3}) \cup U(\boldsymbol{s}_\rho)) \mid \Delta_{e,j'} f(\boldsymbol{s}_\rho)/\sum_{i=1}^{\ell} c_{ie} < \rho\}$. Thus, the discussion of the elements in $U(\boldsymbol{o}) \setminus U(\boldsymbol{o}_{r+2} \sqcup \boldsymbol{o}_{r+3})$ is essentially about p-system constraints. Let $U(\boldsymbol{o}_{r+1})$ represent the set of elements in $U(\boldsymbol{o}) \setminus (U(\boldsymbol{o}_{r+2} \sqcup \boldsymbol{o}_{r+3}) \cup U(\boldsymbol{s}_\rho))$ that can be added to $U(\boldsymbol{s}_\rho)$ without violating independence, i.e., $U(\boldsymbol{o}_{r+1}) = \{e \in U(\boldsymbol{o}) \setminus (U(\boldsymbol{o}_{r+2} \sqcup \boldsymbol{o}_{r+3}) \cup U(\boldsymbol{s}_\rho)) \mid U(\boldsymbol{s}_\rho) \cup \{e\} \in \mathcal{I}\}$. Define $T_r = U(\boldsymbol{o}) \setminus U(\boldsymbol{o}_{r+1} \sqcup \boldsymbol{o}_{r+2} \sqcup \boldsymbol{o}_{r+3})$. Using the method of partitioning and constructing a sequence in Section 3, we can obtain

$$\boldsymbol{o} = \boldsymbol{o}_1 \sqcup \boldsymbol{o}_2 \sqcup \cdots \sqcup \boldsymbol{o}_r \sqcup \boldsymbol{o}_{r+1} \sqcup \boldsymbol{o}_{r+2} \sqcup \boldsymbol{o}_{r+3},$$

where $|U(\boldsymbol{o}_t)| \leq p, t \in [r]$, and $U(\boldsymbol{o}_t) \cap U(\boldsymbol{o}_m) = \emptyset,\ t, m \in [r+3], t \neq m$. And a sequence $\boldsymbol{o}^0 = \boldsymbol{o}, \boldsymbol{o}^1, \boldsymbol{o}^2, \ldots, \boldsymbol{o}^r$ that satisfies

$$(1)\ \boldsymbol{s}_\rho^{t-1} \preceq \boldsymbol{o}^{t-\frac{1}{2}}, t \in [r]; \qquad (2)\ \boldsymbol{s}_\rho^t \preceq \boldsymbol{o}^t, t = 0, 1, \ldots, r.$$

Here $\boldsymbol{o}^r = \boldsymbol{s}_\rho \sqcup \boldsymbol{o}_{r+1} \sqcup \boldsymbol{o}_{r+2} \sqcup \boldsymbol{o}_{r+3}$.

The threshold when (e_t, j_t) is added to $\boldsymbol{s}_\rho$ is denoted as τ_t. Algorithm 2 indicates that there are $\tau_1 \geq \tau_2 \geq \cdots \geq \tau_r$ and

$$\Delta_{e_t,j_t} f(\boldsymbol{s}_\rho^{t-1}) \geq \tau_t,\ t \in [r]. \tag{1}$$

While adding each element in $U(\boldsymbol{o}_t)$ to $U(\boldsymbol{s}_\rho^{t-1})$ does not violate independence, due to

$$\Delta_{e,j'} f(\boldsymbol{s}_\rho^{t-1}) < (1+\epsilon')\tau_t,\ e \in U(\boldsymbol{o}_t),\ t \in [r], \tag{2}$$

they were not selected at the previous threshold. Similarly, for elements in $U(\boldsymbol{o}_{r+1})$, we have

$$\Delta_{e,j'} f(\boldsymbol{s}_\rho) < (1+\epsilon')\frac{\epsilon'}{n} M_\rho = \frac{(1+\epsilon')\epsilon'}{n}\tau_1,\ e \in U(\boldsymbol{o}_{r+1}). \tag{3}$$

Lemma 7. *There is* $f(\boldsymbol{o}) - f(\boldsymbol{o}^r) \leq \frac{(1+\epsilon)(1+\epsilon')p}{\alpha^2(1-\epsilon)} f(\boldsymbol{s}_\rho)$.

Lemma 8. *The relationship between* $\boldsymbol{o}^r$ *and* $\boldsymbol{s}_\rho$ *is* $f(\boldsymbol{o}^r) \leq (\frac{(1+\epsilon)(1+\epsilon')\epsilon'}{\alpha(1-\epsilon)} + 1)f(\boldsymbol{s}_\rho) + \frac{1+\epsilon}{\alpha(1-\epsilon)}[\rho(\ell - \frac{|U(\boldsymbol{o}_{r+3})|}{2}) + \frac{|U(\boldsymbol{o}_{r+3})|}{\alpha} f(\boldsymbol{s}_D)]$.

Lemma 9. *For Case 2, there exists* $f(\boldsymbol{s}) \geq \frac{f(\boldsymbol{o}) - \frac{1+\epsilon}{\alpha(1-\epsilon)}\rho(\ell - \frac{|U(\boldsymbol{o}_{r+3})|}{2})}{\frac{(1+\epsilon)(1+\epsilon')(p+\alpha\epsilon')}{\alpha^2(1-\epsilon)} + 1 + \frac{(1+\epsilon)|U(\boldsymbol{o}_{r+3})|}{\alpha^2(1-\epsilon)}}$, *where* $\boldsymbol{s}$ *is the output solution of Algorithm 2.*

Proof. Combining Lemma 7 and Lemma 8, there is $(\frac{(1+\epsilon)(1+\epsilon')(p+\alpha\epsilon')}{\alpha^2(1-\epsilon)} + 1)f(\boldsymbol{s}_\rho) + \frac{(1+\epsilon)|U(\boldsymbol{o}_{r+3})|}{\alpha^2(1-\epsilon)} f(\boldsymbol{s}_D) \geq f(\boldsymbol{o}) - \frac{1+\epsilon}{\alpha(1-\epsilon)}\rho(\ell - \frac{|U(\boldsymbol{o}_{r+3})|}{2})$. According to Algorithm 2, it can be inferred that

$$f(\boldsymbol{s}) \geq \max\{f(\boldsymbol{s}_\rho), f(\boldsymbol{s}_D)\} \geq \frac{(\frac{(1+\epsilon)(1+\epsilon')(p+\alpha\epsilon')}{\alpha^2(1-\epsilon)} + 1)f(\boldsymbol{s}_\rho) + \frac{(1+\epsilon)|U(\boldsymbol{o}_{r+3})|}{\alpha^2(1-\epsilon)} f(\boldsymbol{s}_D)}{\frac{(1+\epsilon)(1+\epsilon')(p+\alpha\epsilon')}{\alpha^2(1-\epsilon)} + 1 + \frac{(1+\epsilon)|U(\boldsymbol{o}_{r+3})|}{\alpha^2(1-\epsilon)}}$$

$$\geq \frac{f(\boldsymbol{o}) - \frac{1+\epsilon}{\alpha(1-\epsilon)}\rho(\ell - \frac{|U(\boldsymbol{o}_{r+3})|}{2})}{\frac{(1+\epsilon)(1+\epsilon')(p+\alpha\epsilon')}{\alpha^2(1-\epsilon)} + 1 + \frac{(1+\epsilon)|U(\boldsymbol{o}_{r+3})|}{\alpha^2(1-\epsilon)}}.$$

5 Discussion

In the context of practical applications, the problem of maximizing approximately non-k-submodular functions is of considerable research interest. This study focuses on ϵ-approximately α-weakly diminishing returns functions in this paper. Under p-system and ℓ knapsack constraints, we propose a greedy algorithm and an improved algorithm, and ultimately guarantee an approximation of $\frac{\min\{1, \frac{1}{\alpha(1+\epsilon')}\}}{1+\frac{1+\epsilon}{\alpha^2(1-\epsilon)}[(1+\epsilon')(p+\alpha\epsilon')+\frac{7}{4}\ell]}$ with the time complexity of $O(nk \log \frac{n}{\epsilon'} \log(2n))$. By assigning specific values to the parameters ϵ, α, p and ℓ, our results align with existing relevant findings and exhibit lower time complexity. Further investigation is necessary to maximize ϵ-approximately α-weakly k-submodular functions under p-system and ℓ knapsack constraints.

References

1. Huber, A., Kolmogorov, V.: Towards minimizing k-submodular functions. In: ISCO, 451–462 (2012)
2. Ward, J., Živný, S.: Maximizing k-submodular functions and beyond. ACM Trans. Algorithms **12**(4), 1–26 (2016)
3. Iwata, S., Tanigawa, S., Yoshida, Y.: Improved approximation algorithms for k-submodular function maximization. In: SODA, pp. 404–413 (2016)
4. Ohsaka, N., Yoshida, Y.: Monotone k-submodular function maximization with size constraints. In: NIPS, pp. 694–702 (2015)
5. Nie, G., Zhu, Y., Nadew, Y., Basu, S., Pavan, A., Quinn, C.: Size-constrained k-submodular maximization in near-linear time. In: UAI, pp. 1545–1554 (2023)
6. Nguyen, L., Thai, M.: Streaming k-submodular maximization under noise subject to size constraint. In: ICML, pp. 7338–7347 (2020)
7. Xiao, H., Liu, Q., Zhou, Y., Li, M.: Non-monotone k-submodular function maximization with individual size constraints. In: CSoNet, pp. 268–279 (2022)
8. Sakaue, S.: On maximizing a monotone k-submodular function subject to a matroid constraint. Discret. Optim. **23**, 105–113 (2017)
9. Sun, Y., Liu, Y., Li, M.: Maximization of k-submodular function with a matroid constraint. In: TAMC, pp. 1–10 (2022)
10. Niu, S., Liu, Q., Zhou, Y., Li, M.: Fast algorithms for k-submodular maximization subject to a matroid constraint (2023). arXiv:2307.13996
11. Tang, Z., Wang, C., Chan, H.: On maximizing a monotone k-submodular function under a knapsack constraint. Oper. Res. Lett. **50**(1), 28–31 (2022)
12. Xiao, H., Liu, Q., Zhou, Y., Li, M.: Approximation algorithms for k-submodular maximization subject to a knapsack constraint (2023). arXiv:2306.14520
13. Liu, Q., Yu, K., Li, M., Zhou, Y.: k-Submodular maximization with a knapsack constraint and p matroid constraints. Tsinghua Sci. Technol. **28**(5), 896–905 (2023)
14. Yu, K., Li, M., Zhou, Y., Liu, Q.: On maximizing monotone or non-monotone k-submodular functions with the intersection of knapsack and matroid constraints. J. Comb. Optim. **45**(3) (2023)
15. Zhang, W., Gong, S., Liu, B.: Efficient algorithms for k-submodular function maximization with p-system and d-knapsack constraint. In: COCOA, pp. 265–276 (2023)
16. Ye, H., Li, M., Zhou, Y., Liu, Q.: Maximizing a k-submodular function with p-system constraints. Math. Found. Comput. (2024)
17. Nguyen, Lan N., Thai, My T.: Streaming k-submodular maximization under noise subject to size constraint. In: ICML, pp. 7338–7347 (2020)
18. Zheng, L., Chan, H., Loukides, G., Li, M.: Maximizing approximately k-submodular functions (2021). arXiv:2101.07157v1
19. Matsuoka, T., Ohsaka, N.: Maximization of monotone k-submodular functions with bounded curvature and non-k-submodular functions. In: ACML, pp. 1707–1722 (2021)
20. Wang, Y., Zhang, D., Zhang, Y., Zhang, Z.: Weakly k-submodular maximization under matroid constraint. In: TAMC, pp. 393–401 (2022)
21. Jiang, Y., Wang, Y., Yang, R., Ye, W.: Maximizing approximately non-k-submodular monotone set function with matroid constraint. In: TAMC, pp. 11–20 (2022)

Alternating Lagrangian Decomposition Combining with Branch and Pricing for Robust and Integrated Airline Aircraft Routing and Crew Pairing

Cong Li[1], Suixiang Gao[1,2(✉)], Wenguo Yang[1], and Zhipeng Jiang[1]

[1] School of Mathematical Sciences, University of Chinese Academy of Sciences, Beijing 100049, China
licong21@mails.ucas.ac.cn, {sxgao,yangwg,jiangzhipeng}@ucas.ac.cn
[2] Zhongguancun Laboratory, Beijing 100089, China

Abstract. Generally, the aircraft routing problem and crew pairing problems are solved in sequence, but the found solution may be suboptimal. In this paper, we integrate the aircraft routing and crew pairing problems as a problem. This problem involves not only short connections that aren't allowed to crews changing aircraft, but also restricted connections that exists the high risk of propagating delay when crews change aircraft. For this problem, We propose a new robust and integrated model with two coupling constraints to improve the robustness of solutions. To solve the robust and integrated model, we propose a heuristic alternating Lagrangian decomposition combining with branch and price algorithm. The solution process iterates between two Lagrangian subproblems that are solved by branch-and-price. The experimental results show the robust-integrated approach is superior to a traditional sequential approach in the aspect of cost savings.

Keywords: Aircraft routing · Crew pairing · Robustness · Alternating Lagrangian decomposition · Branch and price

1 Introduction

For the airline scheduling problem, it is divided into four subproblems: flight scheduling problem, fleet assignment problem, aircraft routing problem, and crew scheduling problem. These problems are sequentially solved due to the complexity of airline scheduling problem. The first problem aims at determining the departure and arrival airports and times of every flight. This process doesn't involve aircraft or crews. The second problem involves assigning one aircraft type to each flight for the maximum profit. The third problem founds the subset of minimum-cost aircraft routs such that each flight is allocated to exactly individual aircraft while considering maintenance requirements. The fourth problem aims at allocating one specific crew to each flight with the goal of minimizing

D. Du et al. (Eds.): COCOA 2024, LNCS 15434, pp. 413–425, 2025.
https://doi.org/10.1007/978-981-96-4445-2_34

costs. In fact, the crew scheduling problem is usually divided into two subproblems: crew pairing problem and crew assigning problem, and these two subproblems are solved in sequence. In crew pairing problem, the subset of the minimum-cost pairings is found such that each flight is allocated by one pairing. For the crew assigning problem, schedules are constructed based on the derived pairings previously, and each pairing is covered by one schedule.

In this paper, we assume that first three subproblems of the airline scheduling problem have been solved, and focus our research on aircraft routing and crew pairing problems. Generally, the aircraft routing problem is solved, and then the crew pairing problem is solved. In the solution process, the interdependent parts of these two problems referring to the variable connection times are often not considered so that the found solutions may not be globally optimal. Many authors have investigated aircraft routing or crew pairing problems. Saddoune et al. [6] solved a crew pairing problem by proposing a set partitioning model. Authors point out shortcomings of traditional three-phase method and then present a rolling-horizon method based on branch-and-price algorithm to solve the proposed model. Clarke et al. [2] conducted research on an aircraft routing problem with the objective of maximizing profit. The authors treated the aircraft routing problem as an asymmetric traveling salesman problem and proposed a corresponding integer programming model based on the constructed Eulerian directed graph.

To find an globally optimal solution about aircraft routing and crew pairing problems, some authors also have conducted research on integrating these two problems as a problem. Cordeau et al. [3] considered short connections that aren't allowed to crews changing aircraft to integrate aircraft routing and crew pairing problems as a problem. They presented an integrated model with a coupling constraint solved by combining a Benders decomposition and branch-and-price algorithms. In 2010, to improve the robustness of the solution, Weide et al. [7] considered not only short connections but also restricted connections that exist the risk of propagating delay when crews change aircraft. The authors presented a robust and integrated model solved by an iterative method. Ruther et al. [5] also considered short and restricted connections simultaneously, and also proposed a novel robust and integrated model solved by a branch-and-price algorithm.

In this study, we integrate aircraft routing and crew pairing problems as a problem, and also consider short and restricted connections simultaneously in the problem. Therefore, we propose a new robust and integrated model with two coupling constraints. In the model, a coupling constraint is based on short connection, another coupling constraint is based on restricted connection. To solve the robust and integrated model, we present a method that combines the heuristic alternating Lagrangian decomposition (ALD) and branch-and-price (BP) algorithms. By the heuristic ALD algorithm, we will divide the robust and integrated model into two Lagrangian subproblems with the penalty cost and variable of restricted connection, and prove that each one of these two Lagrangian subproblems is equivalent to that without the penalty costs and variables of restricted connections, respectively. In addition, we take into account scheduled flights in a several-day period, so the selected pairings are directly allocated to crews. Finally, we conduct our computational experiments on three real-world datasets. And our results of experiments show that, in the aspect of cost savings, our robust-integrated approach is superior to a traditional sequential approach.

2 Problem Description and Formulation

Considered a set F of all flights flown by the same type of aircraft during a several-day period, a set B of crew bases, the number n_b of all crews at crew base $b \in B$, a set A of aircraft bases, the number n_a of all aircraft at aircraft base $a \in A$, we define the robust and integrated aircraft routing and crew pairing problem (RIARCPP) as the problem of seeking least-coat aircraft routs and pairings, as well as allocating aircraft and crews to all flights so that every flight is performed by one aircraft and one crew, and coupling constraints are satisfied. An aircraft rout contains one or multiple flights and satisfies maintenance requirements that include the maximum elapsed time, the maximum flying time, the maximum number of landings (or takeoffs) between two successive maintenance checks. In each aircraft rout, maintenance time is greater than or equal to the minimum maintenance time. A pairing consists of a series of duties and layover whose time is greater than or equal to the minimum layover time, and satisfies the maximum elapsed time and the number of duties contained in the pairing. A duty is composed of a series of flights and subject to the maximum elapsed time, the maximum flying time, the number of landings. Any aircraft rout (pairing) have to begin and end at the same airport (crew base) which is an aircraft base in this paper. When the same aircraft (crew) perform two successive flights, the connection time between these two flights is called turn time (sit time). The turn time (sit time) is required to be no less than the minimum turn time (sit time). The sit time also is required to be not greater than the maximum sit time. Additionally, the minimum turn time isn't allowed to be greater than or equal to the minimum sit time. Finally, we don't take into account deadhead on which crews are treated as passengers in the RIARCPP.

A sit connection or turn connection between two successive flights is a short connection whose time is no less than the minimum turn time and smaller than the minimum sit time. On a short connection, the crew isn't allowed to change aircraft. If a connection time between two successive flights is no less than the minimum sit time and the crew changes aircraft, the high risk of propagating delay may still exist. To decrease the risk and improve the robustness of solutions, we define restricted connections with penalty costs, and the maximum restricted time within the minimum sit time and the maximum sit time. Each restricted connection is a sit connection or turn connection between two successive flights, and its time is between the minimum sit time and the maximum restricted time. On a restricted connection, if the crew changes aircraft, then this can incur a large penalty cost. For the airline, the robustness of solutions is significant, because flights that depart on-time can't have extra recovery cost.

To define flight networks, we represent each flight as a arc from the departure node to the arrival node of the flight, and denote by N_D the set of departure nodes of all flights and N_A the set of arrival nodes of all flights. For each aircraft base $a \in A$ (crew base $b \in B$), we define a flight network $G^a(N^a, E^a)$ ($G^b(N^a, E^b)$) in a similar way used in [3], where $N^a = N_D \cup N_A \cup \{o(a), d(a)\}$ ($N^b = N_D \cup N_A \cup \{o(b), d(b)\}$) and E^a (E^b) is the set of all arcs. The beginning and end of aircraft routs (pairings) are represented by source node $o(a)$ ($o(b)$) and sink node $d(a)$ ($d(b)$), respectively. In the arc set E^a (E^b), the beginning of aircraft rout, the turn connection, the maintenance connection, and the end of aircraft rout (the beginning of pairing, the sit connection, the layover connection, and the end of pairing) can be represented by the aircraft rout beginning arc, the turn arc, the maintenance arc, and the aircraft rout end arc (the pairing beginning arc, the sit arc, the layover arc, and the pairing end arc), respectively; in addition, the empty arc from the source node to sink node is allocated to unused aircraft (crews). In the flight network of each aircraft base (crew base), each source-sink feasible path represents a feasible aircraft rout (pairing), and its cost contains maintenance and turn costs (sit and layover costs described in [6]). We assume that maintenance cost is a fixed cost and don't consider through values [2], so turn cost is 0. We denote by C and D the set of arcs representing short connections and restricted connections in $G^a, a \in A$ (or $G^b, b \in B$), respectively.

Next, we propose a robust and integrated model based above-mentioned flight networks. This model is given as follows:

$$\text{Min} \sum_{a\in A}\sum_{r\in R_a} c_r^a x_r^a + \sum_{b\in B}\sum_{p\in P_b} c_p^b y_p^b + \sum_{(i,j)\in D} \rho_{ij} z_{ij} \tag{1}$$

Aircraft routing constraints

$$\sum_{a\in A}\sum_{r\in R_a} u_{fr} x_r^a = 1 \quad f \in F \tag{2}$$

$$\sum_{r\in R_a\cup\{r_a\}} x_r^a = n_a \quad a \in A \tag{3}$$

Coupling constraints

$$\sum_{b\in B}\sum_{p\in P_b} v_p^{ij} y_p^b - \sum_{a\in A}\sum_{r\in R_a} v_r^{ij} x_r^a \le 0 \quad (i,j) \in C \tag{4}$$

$$\sum_{b\in B}\sum_{p\in P_b} v_p^{ij} y_p^b - \sum_{a\in A}\sum_{r\in R_a} v_r^{ij} x_r^a \le z_{ij} \quad (i,j) \in D \tag{5}$$

Crew pairing constraints

$$\sum_{b\in B}\sum_{p\in P_b} u_{fp} y_p^b = 1 \quad f \in F \tag{6}$$

$$\sum_{p\in P_b\cup\{p_b\}} y_p^b = n_b \quad b \in B \tag{7}$$

$$x_r^a \in \{0,1\} \quad r \in R_a, a \in A \tag{8}$$

$$x_r^a \in \mathbb{Z}^+ \cup \{0\} \quad r \in \cup_{a\in A}\{r_a\} \tag{9}$$

$$y_p^b \in \{0,1\} \quad p \in P_b, b \in B \tag{10}$$

$$y_p^b \in \mathbb{Z}^+ \cup \{0\} \quad p \in \cup_{b\in B}\{p_b\} \tag{11}$$

$$z_{ij} \in \{0,1\} \quad (i,j) \in D \tag{12}$$

where r_a (p_b) represents the empty aircraft rout (pairing) corresponding to the empty arc in G^a (G^b), R_a (P_b) is the set of all aircraft routs (pairings) that don't contain the empty aircraft rout r_a (pairing p_b) in G^a (G^b), c_r^a (c_p^b) is the cost of aircraft rout $r \in R_a$ (pairing $p \in P_b$) equal to the sum of costs of all arcs in r (p), u_{fr} (u_{fp}) is a binary parameter that determines whether flight f is contained in aircraft route r (pairing p), v_r^{ij} (v_p^{ij}) is also a binary parameter that determines whether aircraft route r (pairing p) covers arc (i,j), x_r^a (y_p^b) is a binary variable that determines whether aircraft route r (pairing p) is selected, $x_{r_a}^a$ ($y_{p_b}^b$) is a non-negative integer variable about the empty aircraft route r_a (pairing p_b), z_{ij} is a binary variable that determines whether arc (i,j) is covered, and ρ_{ij} is the penalty cost on arc (i,j).

The objective function (1) implies that the cost of the selected aircraft routs and pairings and the penalty costs when crews changing aircraft on restricted arcs. Constraints (2) and (6) ensure that each flight is assigned to one aircraft rout and one pairing. Constraint (3) (Constraint (7)) ensures that each selected aircraft rout (pairing) is allocated to one aircraft (crew). Coupling constraint (4) signifies that crews aren't allowed to change aircraft on short connections. Coupling constraint (5) means that penalty costs are incurred when crews change aircraft on restricted connections.

3 Solution Algorithm

In this section, we introduce heuristic alternating Lagrangian decomposition (ALD) and branch-and-price (BP) algorithms for solving the RIARCPP. Zeighami et al. [8] have already applied the ALD algorithm to solve an integrated airline crew scheduling problem. Unlike the ALD algorithm used in [8], our heuristic ALD algorithm doesn't approximate an optimal solution by updating the upper and lower bounds of the problem. Instead, it obtains an optimal solution when the value of objective function no longer improves significantly over 2 iterations. Additionally, in heuristic ALD process, the Lagrange multipliers are updated by the absolute value of arc reduced costs, instead of the value of arc reduced costs used in [8]. In BP algorithm, a heuristic branch rule will be presented.

3.1 Alternating Lagrangian Decomposition

By the ALD algorithm, we can divide the robust and integrated model (1)–(12) into two Lagrangian subproblems, where the first Lagrangian subproblem is called ALD-aircraft and the second Lagrangian subproblem is called ALD-crew. This algorithm alternately iterates between the first Lagrangian subproblem and the second Lagrangian subproblem by updating Lagrange multipliers. On the one hand, we derive an optimal solution $x^* = (\bar{x}_r)_{r \in R_a, a \in A}$ by solving the integrated aircraft model containing the first terms of objective function (1) and constraints (2)–(3). Here, coupling constraints (4) and (5) are rewrote as follows:

$$\sum_{b \in B} \sum_{p \in P_b} v_p^{ij} y_p^b - 1 \le 0 \quad (i,j) \in C_A^1 \tag{13}$$

$$\sum_{b \in B} \sum_{p \in P_b} v_p^{ij} y_p^b - 0 \le 0 \quad (i,j) \in C_A^0 \tag{14}$$

$$\sum_{b \in B} \sum_{p \in P_b} v_p^{ij} y_p^b - 1 \le z_{ij} \quad (i,j) \in D_A^1 \tag{15}$$

$$\sum_{b \in B} \sum_{p \in P_b} v_p^{ij} y_p^b - 0 \le z_{ij} \quad (i,j) \in D_A^0 \tag{16}$$

where $C_A^1 = \{(i,j) \in C | \sum_{a \in A} \sum_{r \in R_a} v_r^{ij} \bar{x}_r^a = 1\}$ and $C_A^0 = \{(i,j) \in C | \sum_{a \in A} \sum_{r \in R_a} v_r^{ij} \bar{x}_r^a = 0\}$ are the set of short arcs covered and uncovered by the optimal solution $x^* = (\bar{x}_r)_{r \in R_a, a \in A}$, respectively; similarly, $D_A^1 = \{(i,j) \in D | \sum_{a \in A} \sum_{r \in R_a} v_r^{ij} \bar{x}_r^a = 1\}$ and $D_A^0 = \{(i,j) \in D | \sum_{a \in A} \sum_{r \in R_a} v_r^{ij} \bar{x}_r^a = 0\}$ are the set of restricted arcs covered and uncovered by the optimal solution $x^* = (\bar{x}_r)_{r \in R_a, a \in A}$.

Lemma 1. *For the integrated aircraft model, constraints (13) and (15) are redundant.*

Proof. Because $0 \le \sum_{b \in B} \sum_{p \in P_b} v_p^{ij} y_p^b \le 1$, $-1 \le \sum_{b \in B} \sum_{p \in P_b} v_p^{ij} y_p^b - 1 \le 0$. In addition, $z_{ij} \ge 0$, so constraints (13) and (15) are redundant.

The Lagrangian multipliers of constraints (14) and (16) are denoted as $(\delta)_{aircraft} = ((\delta_{ij})_{aircraft})_{(i,j) \in C_A^0}$ and $(\gamma)_{aircraft} = ((\gamma_{ij})_{aircraft})_{(i,j) \in D_A^0}$, respectively. The ALD-crew is defined as follows:

$$\begin{aligned} \text{Min} \; & \sum_{b \in B} \sum_{p \in P_b} c_p^b y_p^b + \sum_{(i,j) \in D} \rho_{ij} z_{ij} + \sum_{(i,j) \in C_A^0} (\delta_{ij})_{aircraft} \Big(\sum_{b \in B} \sum_{p \in P_b} v_p^{ij} y_p^b - 0\Big) \\ & + \sum_{(i,j) \in D_A^0} (\gamma_{ij})_{aircraft} \Big(\sum_{b \in B} \sum_{p \in P_b} v_p^{ij} y_p^b - 0 - z_{ij}\Big) \\ \text{s.t.} \; & (6)\text{-}(7) \text{ and } (10)\text{-}(12). \end{aligned} \tag{17}$$

On the other hand, we rewrite coupling constraints (4) and (5) by solving the ALD-aircraft and fixing variable y_p^b in the robust and integrated model (1)–(12) as follows:

$$1-\sum_{a\in A}\sum_{r\in R_a} v_r^{ij}x_r^a \le 0 \quad (i,j)\in C_B^1 \tag{18}$$

$$0-\sum_{a\in A}\sum_{r\in R_a} v_r^{ij}x_r^a \le 0 \quad (i,j)\in C_B^0 \tag{19}$$

$$1-\sum_{a\in A}\sum_{r\in R_a} v_r^{ij}x_r^a \le z_{ij} \quad (i,j)\in D_B^1 \tag{20}$$

$$0-\sum_{a\in A}\sum_{r\in R_a} v_r^{ij}x_r^a \le z_{ij} \quad (i,j)\in D_B^0 \tag{21}$$

where $C_B^1 = \{(i,j) \in C|\sum_{b\in B}\sum_{p\in P_b} v_p^{ij}\bar{y}_p^b = 1\}$ and $D_B^1 = \{(i,j) \in D|\sum_{b\in B}\sum_{p\in P_b} v_p^{ij}\bar{y}_p^b = 1\}$ are the set of short and restricted arcs covered by the optimal solution y^* of the ALD-aircraft, respectively; similarly $C_B^0 = \{(i,j) \in C|\sum_{b\in B}\sum_{p\in P_b} v_p^{ij}\bar{y}_p^b = 0\}$ and $D_B^0 = \{(i,j) \in D|\sum_{b\in B}\sum_{p\in P_b} v_p^{ij}\bar{y}_{pb}^b = 0\}$ are the set of short and restricted arcs uncovered by the optimal solution y^* of the ALD-aircraft, respectively; and $y^* = (\bar{y}_p^b)_{p\in P_b,b\in B}$.

Lemma 2. *For the integrated crew model, constraints (19) and (21) are redundant.*

The proof of Lemma 2 is similar to that of Lemma 1, so it isn't given again. Now, the Lagrangian multipliers of constraints (18) and (20) are denoted as $(\delta)_{crew} = ((\delta_{ij})_{crew})_{(i,j)\in C_B^1}$ and $(\gamma)_{crew} = ((\gamma_{ij})_{crew})_{(i,j)\in D_B^1}$, respectively. The ALD-aircraft is defined as follows:

$$\begin{aligned} \text{Min} & \sum_{a\in A}\sum_{r\in R_a} c_r^a x_r^a + \sum_{(i,j)\in D} \rho_{ij}z_{ij} + \sum_{(i,j)\in C_B^1} (\delta_{ij})_{crew}(1-\sum_{a\in A}\sum_{r\in R_a} v_r^{ij}x_r^b) \\ & + \sum_{(i,j)\in D_B^1} (\gamma_{ij})_{crew}(1-\sum_{a\in A}\sum_{r\in R_a} v_r^{ij}x_r^b - z_{ij}) \\ & \text{s.t. } (2)-(3), (8)-(9), \text{ and } (12). \end{aligned} \tag{22}$$

For ALD-aircraft or ALD-crew, the number of their variables rises exponentially if the number of flight expands. Therefore, to avoid enumerating all variables, we use a BP algorithm to solve ALD-aircraft and ALD-crew. The BP algorithm is an algorithm based on branch-and-bound and column generation (CG), and given in Sect. 3.2. In addition, it is very important to know how to calculate and update the Lagrange multipliers, so next we will describe this process.

Firstly, we define the arc reduced cost of arc (i,j) between two successive nodes i and j by the following formula: $\bar{c}_{ij} = \Delta_j - \Delta_i$, where Δ_i is the minimum costs from the source node to node i, and Δ_j is also the minimum costs from the source node to node j [8]. When solving pricing problems of CG using a labeling algorithm (see, Sect. 3.2), Δ_i and Δ_j are computed by the way. The pricing problems of CG not be limited to just one, and arcs $(i,j)\in C$ and $(i,j)\in D$ can occur in each one of these pricing problems, so we take the maximum absolute value of all arc reduced costs on arcs $(i,j)\in C$ and $(i,j)\in D$, respectively.

Secondly, at the start of ALD, the Lagrange multipliers of ALD-aircraft is zero (i.e., $(\delta)_{crew} = 0, (\gamma)_{crew} = 0$). For the ALD-aircraft at iteration t, we use the following formula to calculate and update the Lagrange multipliers of ALD-crew:

$$(\delta_{ij})^t_{aircraft} = |(\bar{c}_{ij})^t_{aircraft}| \quad (i,j) \in C^0_A \tag{23}$$

$$(\gamma_{ij})^t_{aircraft} = \min\{|(\bar{c}_{ij})^t_{aircraft}|, \rho_{ij}\} \quad (i,j) \in D^0_A \tag{24}$$

where $|(\bar{c}_{ij})^t_{aircraft}|$ is the absolute value of the arc reduced cost of arcs $(i,j) \in C^0_A$ or $(i,j) \in D^0_A$ derived by the ALD-aircraft at iteration t. For the ALD-crew at iteration t, its method of calculating and updating the Lagrange multipliers of ALD-aircraft is similar to the aforementioned method, so we will not repeat it here.

In fact, it may be difficult that we directly apply the BP algorithm described in Sect. 3.2 to solve ALD-aircraft and ALD-crew, since the ALD-aircraft and ALD-crew contain the binary variable z_{ij} and penalty cost ρ_{ij}. Hence, we prove that the ALD-aircraft and ALD-crew models are equivalent to those without the binary variable z_{ij} and penalty cost ρ_{ij}, respectively.

Proposition 1. *The ALD-crew model (17) is equivalent to the following model:*

$$Min \sum_{b\in B}\sum_{p\in P_b} c^b_p y^b_p + \sum_{(i,j)\in C^0_A} (\delta_{ij})_{aircraft}(\sum_{b\in B}\sum_{p\in P_b} v^{ij}_p y^b_{pb} - 0)$$

$$+ \sum_{(i,j)\in D^0_A} (\gamma_{ij})_{aircraft}(\sum_{b\in B}\sum_{p\in P_b} v^{ij}_p y^b_p - 0)$$

$s.t.$ (6) – (7) *and* (10) – (11).

Proof. Since $D = D^0_A \cup D^1_A$, we have $\sum_{(i,j)\in D} \rho_{ij} z_{ij} = \sum_{(i,j)\in D^0_A} \rho_{ij} z_{ij} + \sum_{(i,j)\in D^1_A} \rho_{ij} z_{ij}$. We obtain

$$\sum_{(i,j)\in D^0_A} (\gamma_{ij})_{aircraft}(\sum_{b\in B}\sum_{p\in P_b} v^{ij}_p y^b_p - 0 - z_{ij}) =$$

$$\sum_{(i,j)\in D^0_A} (\gamma_{ij})_{aircraft}(\sum_{b\in B}\sum_{p\in P_b} v^{ij}_p y^b_p - 0) - \sum_{(i,j)\in D^0_A} (\gamma_{ij})_{aircraft} z_{ij},$$

so the ALD-crew model (17) is equivalent to the following model:

$$\text{Min} \sum_{b\in B}\sum_{p\in P_b} c^b_p y^b_p + \sum_{(i,j)\in D^1_A} \rho_{ij} z_{ij} + \sum_{(i,j)\in C^0_A} (\delta_{ij})_{aircraft}(\sum_{b\in B}\sum_{p\in P_b} v^{ij}_p y^b_p - 0) +$$

$$\sum_{(i,j)\in D^0_A} (\rho_{ij} - (\gamma_{ij})_{aircraft}) z_{ij} + \sum_{(i,j)\in D^0_A} (\gamma_{ij})_{aircraft}(\sum_{b\in B}\sum_{p\in P_b} v^{ij}_p y^b_p - 0)$$

s.t. (6) – (7) and (10) – (12)

By the formula (24), we know $\rho_{ij} \geq (\gamma_{ij})_{aircraft}$, then $(\rho_{ij} - (\gamma_{ij})_{aircraft}) \geq 0$. Becase the ALD-crew model (17) is a minimization problem and $\rho_{ij} \geq 0$, we have $\sum_{(i,j)\in D^1_A} \rho_{ij} z_{ij} = \sum_{(i,j)\in D^0_A} (\rho_{ij} - (\gamma_{ij})_{aircraft}) z_{ij} = 0$ in the optimal solution.

Proposition 2. *The ALD-aircraft model (22) is equivalent to the following model:*

$$\begin{aligned} Min & \sum_{a\in A}\sum_{r\in R_a} c_r^a x_r^a + \sum_{(i,j)\in C_B^1} (\delta_{ij})_{crew}(1-\sum_{a\in A}\sum_{r\in R_a} v_r^{ij} x_r^b) \\ & + \sum_{(i,j)\in D_B^1} (\gamma_{ij})_{crew}(1-\sum_{a\in A}\sum_{r\in R_a} v_r^{ij} x_r^b) \\ s.t.\ & (2)-(3)\ and\ (8)-(9) \end{aligned}$$

The proof of Proposition 2 is similar to that of Proposition 1, so it isn't also given again. By Proposition 1 and 2, we only need to solve ALD-aircraft and ALD-crew without the binary variable z_{ij} and penalty cost ρ_{ij} by the BP algorithm, respectively. For above all lemmas and propositions, although Zeighami et al. [8] also have related statements and proofs, we have to restate and reprove them due to the differences in our problem.

Let x^* and y^* be optimal solutions of ALD-aircraft and ALD-crew, respectively. We estimate whether (x^*, y^*) satisfies Constraint (4) to update the current optimal solution of the RIARCPP. (x^*, y^*) is a feasible solution of the RIARCPP and the current optimal solution is updated if (x^*, y^*) satisfies Constraint (4); otherwise, (x^*, y^*) is an infeasible solution of the RIARCPP. In this paper, it may be difficult that we find a lower bound of the RIARCPP, so the heuristic ALD algorithm stops if the value of objective function improves by less than 0.01 in the last 2 iterations.

3.2 Branch and Price

The BP algorithm [1] is that CG is used in branch-and-bound algorithm. The CG process depends on a restricted master problem (RMP) and one or several pricing problems, where the RMP is the master problem limited to a subset of variables. By solving the RMP using a solver, an optimal solution and a dual solution are obtained. Then, by solving these pricing problems based on the dual solution, variables with negative reduced costs are found and added to the RMP. When no variables with negative reduced costs are found, the CG stops. In the iterative process of CG, some degenerate solutions may be found such that the objective value of the current RMP doesn't improve significantly and the running time is very long. Therefore, if the objective value of the current RMP improves by less than 0.0001 in the last 10 iterations, then we stop the CG process. Note that we add artificial variables into the RMP to guarantee its feasibilty at the start of CG. Denote ψ_f and ϵ_a (ξ_f and ε_b) as the dual variables of constraints (2) and (3) ((6) and (3)), respectively. The reduced cost of variables x_r^a (y_p^b) is given as follows: $\bar{c}_r^a = c_r^a - \sum_{(i,j)\in C_B^1} v_r^{ij}(\delta_{ij})_{crew} - \sum_{(i,j)\in D_B^1} v_r^{ij}(\gamma_{ij})_{crew} - \sum_{f\in F} u_{fr}\psi_f - \epsilon_a$ ($\bar{c}_p^b = c_p^b + \sum_{(i,j)\in C_A^0} v_r^{ij}(\delta_{ij})_{aircraft} + \sum_{(i,j)\in D_A^0} v_r^{ij}(\gamma_{ij})_{aircraft} - \sum_{f\in F} u_{fp}\xi_f - \varepsilon_b$).

Actually, each pricing problem corresponds to a shortest path with resource constraints and is defined on the flight network of an aircraft base or a crew base. We solve pricing problems by a labeling algorithm described in [4]. Because of each aircraft rout (pairing) corresponding to a source-sink path in G^a (G^b), we use the following three (five) resource constraints to deal with the aircraft rout (pairing) feasibility: the maximum elapsed time, the maximum flying time, the maximum number of landings between two successive maintenance checks (the maximum pairing elapsed time, the number of duties contained in a pairing, the maximum duty elapsed time, the maximum flying time in a duty, the number of landing in a duty). In the labeling algorithm, a label contains one reduced cost component and three or five resource components, and is related with each partial path originating from the source node. The number of all labels found in a large network may be too large to have a short runtime. Therefore, we must use a dominance rule to eliminate labels. We denote by Π the set of resources and c the reduced cost. We denote two different labels as $E_1 = \{V_1^\pi | \pi \in \Pi \cup \{c\}\}$ and $E_2 = \{V_2^\pi | \pi \in \Pi \cup \{c\}\}$, respectively, where V_i^π $(i = 1, 2)$ is the value of reduced cost c or resource π. Label E_1 dominates label E_2 when satisfying $V_1^c \leq V_2^c$ and $V_1^\pi \geq V_2^\pi (\pi \in \Pi)$. Here, all dominated labels can be discarded.

Finally, we present a heuristic branching strategy to find an optimal solution of the RIARCPP. In the heuristic branching strategy, firstly, we fix variables with a fractional value greater than or equal to a predefined threshold (it is 0.65 in this paper) to 1. Secondly, when there are not suitable variables, a variable with fractional value which is maximum among all variables with fractional values is chose and then fixed to 1; in addition, a single branching node will be created at a node contained the branch-and-bound tree. When finding an integer solution, the BP process is stopped.

4 Computational Experiments

In this section, we conduct our experiments on three instances. These three instances were extracted from the first real datasets used in [9] and have four-day periods. The number of crew and crew bases are known, and also considered as the number of aircraft and aircraft bases in each instance, respectively. Furthermore, we choose four airports as maintenance airports in each instance. The information of these three instances is shown in Table 1. We use the GUROBI solver (version 11.0.0) and the equipment that is 13th Gen Intel(R) Core(TM) i9-13900KF central processing unit (CPU) at 3.00 GHz for our experiments. Our implementations are coded in Python.

Table 1. Instances

	Instance 1	Instance 2	Instance 3
NO. Flights	126	135	122
NO. Aircraft/Crews	33	33	33
NO. Aircraft Bases/Crew Bases	3	3	3
NO. Airports	20	16	24

In Table 2, we compare the performance of these three instances solved by the robust-integrated (RINT) and sequential (SEQ) approaches. Using the heuristic ALD and BP algorithms to solve the robust and integrated model (1)–(12) is called the robust-integrated approach. Using BP algorithm to solve the aircraft routing problem and then crew pairing problem is called the sequential approach. In the sequential approach, the interdependent parts of these two problems are neglected, and the sit time must be limited between the minimum and maximum sit time. We provide not only the total iterations of CG (Aircraft CG iter. or Crew CG iter.), the total number of variables found by CG (Aircraft Variables or Crew Variables), the total runtime (in seconds) spent solving the pricing problem (Aircraft PP time or Crew PP time), the total runtime spent on the CG process (Aircraft CG time or Crew CG time), the runtime spent on the BP process (Aircraft BP time or Crew BP time), and the total number of nodes in branch-and-bound tree (Aircraft NO. nodes or Crew NO. nodes) about aircraft routing or crew pairing, but also the total solution cost (Cost), and the total runtime (Total time) for these two approaches. In addition, we also report the iterations of the heuristic ALD (ALD iter.) for the robust-integrated approach.

For one thing, because of the multiple iterations of the heuristic ALD algorithm, we know from Table 2 that total iterations of CG, the total number of variables found by CG, the total runtime spent solving the pricing problem, the total runtime spent on the CG process, the runtime spent on the BP process, and the total number of nodes in branch-and-bound tree of the RINT are larger than those of the SEQ about aircraft routing or crew pairing, respectively, for each instance.

For another, we report the percentage of cost savings yielded by the optimal solutions of RINT, and the ratios of the total runtime of RINT and SEQ in the last two rows of Table 2. For all instances, we observe that cost savings yielded by our RINT approach are very evident, with an average of 35.39%, however, the average total running time of the RINT approach is 7.4 times than that of the SEQ approach. Actually, degenerate solutions are found by solving crew pairing problem using BP such that they contain many artificial variables and have the large costs in our experiments.

Table 2. Computational Results for the Robust-integrated and Sequential Approaches

	Instance 1	Instance 2	Instance 3	Average
SEQ				
Aircraft CG iter.	149	181	88	
Aircraft No. Variables	229	375	144	
Aircraft PP time	71.09	166.21	28.88	
Aircraft CG time	76.21	174.21	31.71	
Aircraft BP time	77.78	176.08	33.09	
Aircraft No. nodes	12	12	10	
Crew CG iter.	106	83	79	
Crew No. Variables	201	171	140	
Crew PP time	27.52	42.88	19.98	
Crew CG time	31.95	47.17	23.49	
Crew BP time	33.03	48.44	24.48	
Crew NO. nodes	1	1	1	
Cost	449367	433401	489800	
Total time	110.82	224.54	57.59	
RINT				
Aircraft CG iter.	702	852	348	
Aircraft No. Variables	1181	1676	572	
Aircraft PP time	311.34	532.47	126.17	
Aircraft CG time	331.42	556.91	137.29	
Aircraft BP time	337.08	561.72	142.33	
Aircraft No. nodes	50	49	39	
Crew CG iter.	663	791	424	
Crew No. Variables	1558	2006	1015	
Crew PP time	502.64	919.04	279.53	
Crew CG time	528.91	947.07	297.91	
Crew BP time	532.88	950.79	302.15	
Crew NO. nodes	8	11	7	
Cost	292300	287139	306296	
Total time	870.03	1512.56	444.52	
ALD iter.	5	5	4	
RINT vs. SEQ				
Cost saving using RINT (%)	34.95	33.75	37.47	35.39
RINT/SEQ (Total time)	7.9	6.7	7.7	7.4

5 Conclusion

We have introduced a new robust and integrated model and a solution algorithm for the RIARCPP in this paper. We have the following contributions: (1) The proposed model involves the short and restricted connections to improve the robustness of solutions. (2) We develop a heuristic ALD combining with BP algorithm and first apply this algorithm to solve the RIARCPP. Furthermore, we prove that each one of ALD-aircraft and ALD-crew is equivalent to that without the penalty costs and variables of restricted connections. (3) In our experiments, the cost savings yielded by the RINT approach are very evident, with an average of 35.39%.

References

1. Barnhart, C., Johnson, E.L., Nemhauser, G.L., Savelsbergh, M.W., Vance, P.H.: Branch-and-price: Column generation for solving huge integer programs. Oper. Res. **46**(3), 316–329 (1998)
2. Clarke, L., Johnson, E., Nemhauser, G., Zhu, Z.: The aircraft rotation problem. Ann. Oper. Res. **69**, 33–46 (1997)
3. Cordeau, J.F., Stojković, G., Soumis, F., Desrosiers, J.: Benders decomposition for simultaneous aircraft routing and crew scheduling. Transp. Sci. **35**(4), 375–388 (2001)
4. Irnich, S., Desaulniers, G.: Shortest path problems with resource constraints. In: Column generation, pp. 33–65. Springer (2005)
5. Ruther, S., Boland, N., Engineer, F.G., Evans, I.: Integrated aircraft routing, crew pairing, and tail assignment: branch-and-price with many pricing problems. Transp. Sci. **51**(1), 177–195 (2017)
6. Saddoune, M., Desaulniers, G., Soumis, F.: Aircrew pairings with possible repetitions of the same flight number. Comput. Oper. Res. **40**(3), 805–814 (2013)
7. Weide, O., Ryan, D., Ehrgott, M.: An iterative approach to robust and integrated aircraft routing and crew scheduling. Comput. Oper. Res. **37**(5), 833–844 (2010)
8. Zeighami, V., Saddoune, M., Soumis, F.: Alternating lagrangian decomposition for integrated airline crew scheduling problem. Eur. J. Oper. Res. **287**(1), 211–224 (2020)
9. Zeighami, V., Soumis, F.: Combining benders' decomposition and column generation for integrated crew pairing and personalized crew assignment problems. Transp. Sci. **53**(5), 1479–1499 (2019)

A Comparative Analysis of NFT-Based Insurance Claim Management Systems Across EVM-Compatible Blockchain Platforms

Minh Nguyen Triet, Le Khanh Bang, Hong Vo Khanh(✉), The Nguyen Anh, Doan Minh Hieu, Pham Thanh Nghiem, Huynh Gia Khiem, and Nam Tran Ba

FPT University, Can Tho, Vietnam
{trietnm3,khanhvh}@fe.edu.vn

Abstract. This study examines the implementation of NFTs and smart contracts for insurance claim management across multiple EVM-compatible blockchain platforms. We analyze the performance and economic efficiency of BNB Chain, Fantom, Polygon, and Celo in handling key operations such as data creation, NFT minting, and NFT transfer. The research aims to address challenges in traditional insurance claim processes, including fraud prevention, operational inefficiencies, and lack of transparency. Through a comparative analysis of transaction fees and computational costs, we evaluate the suitability of each platform for implementing blockchain-based insurance claim management systems. Our findings provide insights into the trade-offs between security, transparency, and cost-effectiveness across different blockchain environments, contributing to the ongoing development of more efficient insurance claim management solutions.

Keywords: Blockchain · Transparency in insurance · Data Privacy and Security · Smart Contracts · RSA-Encrypted NFTs

1 Introduction

The application of blockchain technology in various industries has gained attention for its potential to enhance processes and address existing challenges. In the realm of insurance claim management, blockchain offers promising solutions to longstanding issues related to fraud prevention, operational inefficiencies, and lack of transparency. This study examines the implementation of non-fungible tokens (NFTs) and smart contracts across multiple Ethereum Virtual Machine (EVM))-compatible platforms to improve insurance claim management systems.

Traditional insurance claim management involves a series of steps, including claim submission, verification, processing, and settlement. These processes often require coordination among multiple parties such as policyholders, insurance providers, claim adjusters, and third-party services. The complexity and manual nature of these operations can result in delays, errors, and opportunities for fraudulent activities. Blockchain technology provides a decentralized

D. Du et al. (Eds.): COCOA 2024, LNCS 15434, pp. 426–438, 2025.
https://doi.org/10.1007/978-981-96-4445-2_35

and tamper-resistant platform that can potentially streamline these processes, ensuring that all transactions are transparent and verifiable [5].

Previous research has demonstrated the feasibility of blockchain-based solutions in various insurance applications. For example, systems using machine learning in conjunction with blockchain have shown effectiveness in automating claim processing and securing platforms against fraud and data integrity attacks [4]. Other studies have explored the use of smart contracts to automate policy management and claims processing, aiming to improve efficiency and reduce reliance on intermediaries [15]. The incorporation of NFTs in insurance claim management introduces an additional layer of security and verifiability. These NFTs function as unique digital identifiers for insurance claims, encapsulating relevant information in a tamper-proof format. Smart contracts ensure that all claim-related transactions are automatically recorded on the blockchain, providing a transparent and immutable record of the entire process [2].

An important aspect of blockchain-based insurance claim management systems is their economic efficiency. Transaction fees associated with blockchain operations, such as data creation, NFT minting, and NFT transfer, significantly impact the overall cost-effectiveness of these systems. This study conducts a comparative analysis of transaction fees across four EVM-compatible platforms: BNB Chain, Fantom, Polygon, and Celo. This analysis aims to highlight differences in fee structures and emphasize the importance of selecting a cost-effective platform for implementation [6]. This research provides a comprehensive evaluation of the implementation of NFTs and smart contracts in insurance claim management across multiple blockchain platforms. We perform a detailed performance analysis, focusing on key operations and their associated computational costs. Additionally, we assess the economic efficiency of different blockchain platforms by analyzing transaction fees for these operations. Our objective identify a blockchain solution that balances security, transparency, and cost-effectiveness, thereby enhancing the overall efficiency of insurance claim management systems.

In the rest of the paper, Sect. 2 presents the prior work for the insurance claim management as well as the healthcare system based on blockchain technology. The following sections will discuss the approach and methodology used for our analysis. Through this work, we aim to contribute to the ongoing research on blockchain applications in insurance and provide a framework for future implementations that may further improve the transparency and efficiency of insurance claim management.

2 Related Work

2.1 Blockchain Applications in Insurance Claim Processing

The insurance industry has been exploring blockchain technology to enhance claim management processes. Several studies have demonstrated the potential of blockchain in improving transparency, efficiency, and security in insurance operations. For instance, research has shown how blockchain can be utilized in pay-as-you-go auto insurance models to provide clear and traceable data access for all parties involved [16]. This approach could potentially reduce disputes and streamline the claims process. In the health insurance sector, blockchain-based systems have been proposed to increase transparency between insurers and policyholders. These systems aim to eliminate intermediaries and provide secure access to medical data, potentially speeding up claim processing and reducing administrative costs [13]. Similarly, other researchers have developed blockchain-based frameworks for automating insurance policy processing and claim management through smart contracts [11]. Such systems could offer a tamper-resistant environment for handling claims, thereby increasing trust in the process. The integration of machine learning with blockchain has shown promise in verifying insurance claims and identifying fraudulent activities. Studies have reported significant accuracy in fraud detection using these combined technologies [2]. This approach could help insurers reduce losses from fraudulent claims while expediting the processing of legitimate ones. Researchers have also explored the use of blockchain for streamlining vehicle insurance processes. Platforms utilizing smart contracts for policy management and claim handling have been developed, aiming to enhance user experience and efficiency by reducing reliance on intermediaries [15]. Such systems could potentially lead to faster claim settlements and lower operational costs for insurers.

Moreover, blockchain frameworks have been proposed to automate claims processing and decision-making, potentially leading to cost reductions and improved compliance [1]. These frameworks often leverage smart contracts to execute predefined rules, which could help standardize claim handling procedures across different insurance products. While these blockchain applications show potential, challenges related to scalability and regulatory compliance have been noted [5]. As the technology continues to evolve, addressing these challenges will be crucial for widespread adoption in the insurance industry. Future research could focus on developing solutions that balance the benefits of blockchain with regulatory requirements and scalability needs in insurance claim management.

2.2 Blockchain-Based Solutions for Healthcare Data Management and Privacy

Recent research has explored the application of blockchain technology in healthcare systems, focusing on enhancing data privacy and security. These studies investigate various aspects of healthcare data management, including patient information handling, secure health data exchange, and the integration of artificial intelligence with blockchain systems.

One area of focus has been the development of platforms for managing patient data securely and transparently. For example, researchers have proposed a Hyperledger Fabric-based system for the Internet of Healthcare Things (IoHT) that employs smart contracts to allow patients control over their data access [3]. This approach aims to provide a reliable method for maintaining privacy in healthcare data management. Other studies have examined the integration of blockchain into patient-centric healthcare systems. These works emphasize the use of smart contracts to improve data privacy, foster trust among healthcare providers, and enhance system interoperability [7,8]. The proposed architectures offer insights into secure methods for managing and sharing health records, presenting strategies for handling sensitive patient information.

Blockchain technology has also been applied to specific healthcare domains, such as blood donation systems. Researchers have developed frameworks using private blockchain technology to manage data related to blood quality, availability, and demand [10,12]. These systems aim to improve the efficiency and transparency of blood resource management, potentially enhancing healthcare logistics and supply chain processes. The intersection of blockchain and artificial intelligence in healthcare has been another area of investigation. Studies have explored how blockchain can be used to enhance security and reliability in AI-driven healthcare applications [14]. Proposed frameworks aim to protect against adversarial attacks and maintain data privacy in AI-based decision-making processes. Additionally, researchers have investigated the use of blockchain for sharing personal health and genomic data. Some have proposed models combining blockchain with cloud storage to facilitate health data sharing while prioritizing user control and regulatory compliance [17]. Others have explored decentralized approaches for genomic data sharing, incorporating privacy-preserving technologies to address issues of data accessibility and fair compensation [9].

3 Approach

3.1 Traditional Insurance Claim Process

Figure 1 illustrates the typical workflow of an insurance claim process, depicting the interactions between the insured party, the insurance company, and the claims adjuster. This diagram outlines the key steps involved in filing, assessing, and resolving an insurance claim, providing a clear visual representation of the process that policyholders and insurance professionals navigate. The process begins when the insured party experiences an event that leads to a potential claim, such as property damage or an accident. At this point, the insured individual initiates contact with their insurance company to report the incident and file a claim. This initial step is represented by arrow 1 in the diagram, showing the direct communication between the insured and the insurance company. Upon receiving the claim, the insurance company assigns a claims adjuster to handle the case, as indicated by arrow 2.

The claims adjuster plays a central role in the process, serving as the primary point of contact and investigator for the claim. Their responsibilities include assessing the damage, verifying the policy coverage, and determining the appropriate compensation. To accomplish this, the adjuster often needs to gather additional information from the insured party, which is depicted by arrow 4 in the diagram. This may involve requesting documentation, conducting interviews, or scheduling on-site inspections to evaluate the extent of the damage or loss. Once the claims adjuster has collected all necessary information and completed their assessment, they report their findings back to the insurance company, as shown by arrow 5. Based on this report, the insurance company makes a decision

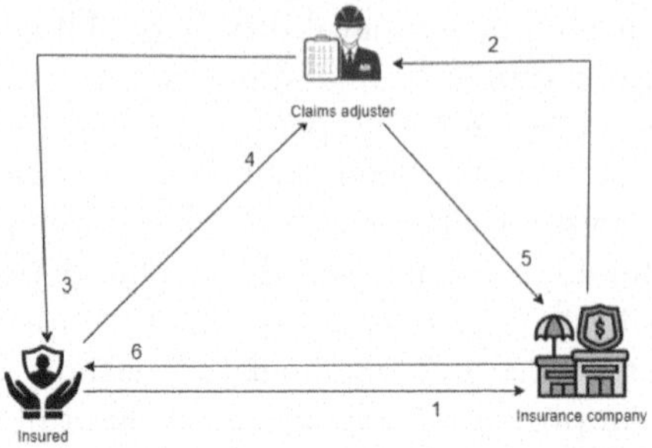

Fig. 1. The traditional process for insurance claim

regarding the claim's validity and the amount of compensation to be provided. If the claim is approved, the insurance company then proceeds to issue payment or arrange for repairs, depending on the nature of the claim and the policy terms. This final step in the process is represented by arrow 3, showing the resolution flowing from the insurance company back to the insured party.

3.2 Blockchain-Based Insurance Claim Process

Figure 2 presents a comprehensive overview of an insurance claim process that incorporates blockchain technology and smart contracts. This diagram illustrates the interactions between various stakeholders, including the insured party, insurance company, claims adjuster, and the technological components that facilitate a more streamlined and transparent claims handling procedure. The process begins with the insured party initiating a claim, as indicated by arrow 1 pointing from the insured to the central system. This central system appears to be a blockchain-enabled platform, as suggested by the presence of the EVM-supported blockchain platform icon in the diagram. Upon receiving the claim, the system interacts with the insurance company, shown by arrows 2 and 3, likely to verify policy details and initiate the claims process. Simultaneously, the system triggers a smart contract, represented by arrow 4, which may contain predefined rules and conditions for claim processing.

The involvement of blockchain technology is further emphasized by the distributed ledger component, connected to the central system via arrow 7. This suggests that claim information is recorded on a decentralized network, potentially enhancing security and transparency throughout the process. The diagram also indicates the creation of an NFT linked to the claim, as shown by arrow 5 leading to the NFT icon. This NFT could represent a unique digital asset associated with the specific claim, possibly used for tracking or verification purposes. The role of the claims adjuster remains crucial in this technology-enhanced process. The adjuster receives information from the central system (arrow 8) and communicates directly with the insured party (arrow 9) to assess the claim. The adjuster's findings are then fed back into the central system (arrow 10), which likely updates the smart contract and distributed ledger accordingly. Throughout

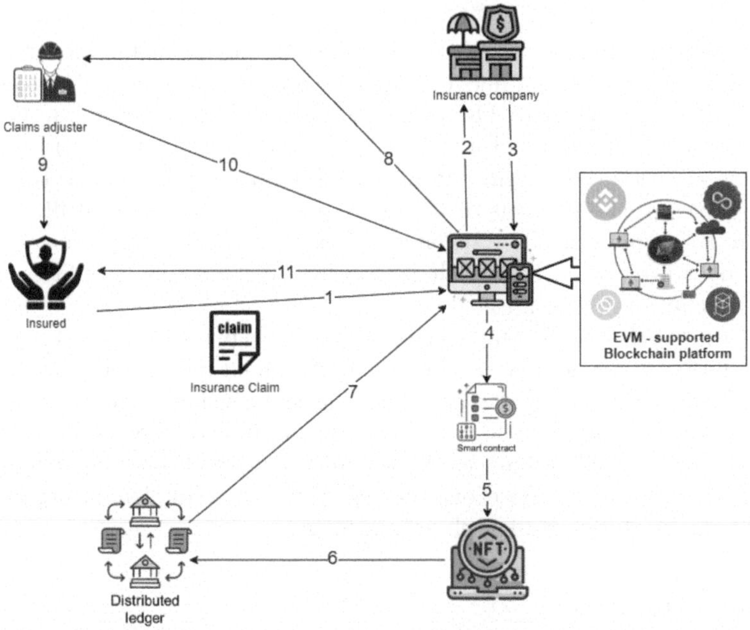

Fig. 2. Blockchain-Integrated Insurance Claim Process

this process, the insured party can interact with the system directly, as shown by arrow 11, potentially to check claim status or provide additional information.

4 Evaluation Scenarios

In the evaluation section of our study, we focus on cross-chain functionality for Insurance Claim Management, specifically examining four EVM-supported platforms, including Binance Smart Chain (BNB Smart Chain)[1]; Polygon[2]; Fantom[3]; and Celo[4].[5]. The decision to consider multiple platforms stems from the need to identify the most suitable blockchain environment for insurance claim processes. Each platform offers unique characteristics and trade-offs, which can significantly impact the efficiency and cost-effectiveness of insurance claim management systems.

By analyzing these diverse platforms, we aim to provide a comprehensive understanding of how different blockchain environments perform in handling insurance claims. This approach allows us to account for variations in transaction speeds, costs, and overall network performance, which are crucial factors

[1] https://github.com/bnb-chain/whitepaper/blob/master/WHITEPAPER.md.
[2] https://polygon.technology/lightpaper-polygon.pdf.
[3] https://whitepaper.io/document/438/fantom-whitepaper.
[4] https://celo.org/papers/whitepaper.
[5] Time of evaluation: 11:00AM - GMT+7 - August 10th 2024.

in determining the viability of blockchain-based insurance solutions. Furthermore, cross-chain analysis helps in assessing the potential for interoperability between different blockchain networks, a key consideration for widespread adoption in the insurance industry. Our evaluation methodology focuses on three main operations that are fundamental to insurance claim management on blockchain: transaction creation, NFT minting, and NFT transfer. These operations represent key stages in the lifecycle of an insurance claim, from initial filing to final resolution. By examining these specific actions across different platforms, we can gain insights into how each blockchain handles the core functionalities required for efficient claim processing.

To conduct a thorough comparative analysis, we examine several key metrics for each platform. These include the gas limit, which determines the maximum amount of computational work that can be done in a single transaction; the gas used by transaction, which reflects the actual computational resources consumed; and the gas price, which influences the overall transaction cost. These metrics are crucial in understanding the economic implications of implementing insurance claim systems on different blockchain platforms.

4.1 Transaction Fee

The analysis of transaction fees across the four EVM-supported platforms is presented in Table 1. BNB Smart Chain stands out with notably higher transaction fees compared to the other platforms. For contract creation, BNB Smart Chain incurs a fee of 0.0273134 BNB, equivalent to approximately $13.91. This cost is substantially higher than its counterparts, with Fantom, Polygon, and Celo all charging less than $0.01 for the same operation. The disparity continues for NFT-related operations, with BNB Smart Chain charging 0.00109162 BNB ($0.56) for NFT creation and 0.00057003 BNB ($0.29) for NFT transfer, while the other platforms maintain fees below $0.01 for these actions. Fantom demonstrates extremely low transaction costs across all operations. The fees for contract creation (0.00957754 FTM), NFT creation (0.000405167 FTM), and NFT transfer (0.0002380105 FTM) all translate to less than $0.01 in USD terms. This cost structure could make Fantom an attractive option for high-frequency or large-scale operations where minimizing transaction costs is a priority.

Table 1. Transaction fee

	Contract Creation	Create NFT	Transfer NFT
BNB Smart Chain	0.0273134 BNB ($13.91)	0.00109162 BNB ($0.56)	0.00057003 BNB ($0.29)
Fantom	0.00957754 FTM ($0.00)	0.000405167 FTM ($0.00)	0.0002380105 FTM ($0.00)
Polygon	0.006840710032835408 MATIC($0.01)	0.000289405001852192 MATIC($0.00)	0.000170007501088048 MATIC($0.00)
Celo	0.007097844 CELO ($0.003)	0.0002840812 CELO ($0.000)	0.0001554878 CELO ($0.000)

Polygon (MATIC) presents a similar cost-effective profile to Fantom, with marginally higher fees. Contract creation on Polygon costs 0.006840710032835408 MATIC ($0.01), while NFT creation and transfer operations are priced at 0.00028 9405001852192 MATIC and 0.000170007501088048 MATIC respectively, both equating to less than $0.01. This pricing model positions Polygon as a competitive option for blockchain-based insurance claim management systems. Celo exhibits the lowest transaction fees among the four platforms when converted to USD. With costs of 0.007097844 CELO ($0.003) for contract creation, 0.0002840812 CELO for NFT creation, and 0.0001554878 CELO for NFT transfer (both less than $0.001), Celo presents an extremely cost-efficient option for blockchain operations.

These findings highlight the importance of considering transaction fees in selecting a blockchain platform for insurance claim management. While BNB Smart Chain offers a well-established ecosystem, its higher transaction costs could be a limiting factor for large-scale or frequent operations. In contrast, Fantom, Polygon, and Celo provide more economical options, potentially allowing for more cost-effective implementation of blockchain-based insurance systems. However, it's crucial to note that transaction fees should be weighed alongside other factors such as network security, scalability, and ecosystem support when making platform decisions for insurance applications.

4.2 Gas Limit

The analysis of gas limits across the four EVM-supported platforms reveals interesting patterns in resource allocation for different blockchain operations. Table 2 presents a detailed comparison of gas limits. For contract creation, we observe that BNB Smart Chain, Fantom, and Polygon have relatively similar gas limits, ranging from 2,731,340 to 2,736,440 units. This similarity suggests a consensus among these platforms regarding the computational resources required for deploying smart contracts. However, Celo stands out with a significantly higher gas limit of 3,548,922 units for contract creation, approximately 30% higher than the other platforms. This higher limit on Celo may indicate a more generous allocation of computational resources, potentially allowing for more complex contract deployments or providing a buffer for unexpected computational needs during contract creation.

Table 2. Gas limit

	Contract Creation	Create NFT	Transfer NFT
BNB Smart Chain	2,731,340	109,162	72,003
Fantom	2,736,440	115,762	72,803
Polygon	2,736,284	115,762	72,803
Celo	3,548,922	142,040	85,673

When it comes to NFT-related operations, we see more variation among the platforms. For NFT creation, BNB Smart Chain sets the lowest gas limit at 109,162 units, while Fantom and Polygon both allocate 115,762 units. Celo again shows a higher allocation with 142,040 units for NFT creation. This pattern continues for NFT transfer operations, where BNB Smart Chain has the lowest gas limit at 72,003 units, Fantom and Polygon both set 72,803 units, and Celo allocates 85,673 units. The consistency between Fantom and Polygon's gas limits for both NFT creation and transfer is noteworthy, suggesting similar approaches to resource allocation for these operations. This consistency could be beneficial for developers working across these platforms, as it allows for more predictable behavior and easier cross-chain compatibility.

Celo's consistently higher gas limits across all operations indicate a platform design that allows for more computational headroom. This approach could provide advantages in terms of flexibility and the ability to handle more complex operations, but it may also have implications for network efficiency and transaction costs. BNB Smart Chain's lower gas limits for NFT operations, compared to the other platforms, could reflect an optimization strategy aimed at reducing resource consumption for these common transactions. This approach might lead to improved scalability for NFT-heavy applications but could potentially limit the complexity of NFT-related smart contract functions.

4.3 Gas Used by Transaction

Table 3 provides a detailed comparison of gas consumption. For contract creation, we observe that BNB Smart Chain, Fantom, and Polygon utilize 100% of their allocated gas limit, consuming 2,731,340, 2,736,440, and 2,736,284 gas units respectively. This full utilization suggests that these platforms have optimized their gas limits to closely match the actual computational requirements of contract deployment. In contrast, Celo uses 2,729,940 gas units, which represents 76.92% of its higher gas limit. This lower percentage utilization on Celo could indicate a more conservative approach to gas limit setting, providing a buffer for potential variations in contract complexity or execution conditions.

Table 3. Gas Used by Transaction

	Contract Creation	Create NFT	Transfer NFT
BNB Smart Chain	2,731,340 (100%)	109,162 (100%)	57,003 (79.17%)
Fantom	2,736,440 (100%)	115,762 (100%)	68,003 (93.41%)
Polygon	2,736,284 (100%)	115,762 (100%)	68,003 (93.41%)
Celo	2,729,940 (76.92%)	109,262 (76.92%)	59,803 (69.8%)

The pattern of gas utilization for NFT creation is similar across all platforms. BNB Smart Chain uses 109,162 gas units, while Fantom and Polygon both consume 115,762 gas units, all representing 100% of their respective gas limits. Celo, maintaining its pattern of lower utilization, uses 109,262 gas units, which is 76.92% of its allocated limit. This consistency in full utilization for BNB Smart Chain, Fantom, and Polygon suggests that these platforms have fine-tuned their gas limits for NFT creation operations. NFT transfer operations show more variation in gas utilization across the platforms. BNB Smart Chain uses 57,003 gas units (79.17% of its limit), while Fantom and Polygon both consume 68,003 gas units (93.41% of their limit). Celo uses 59,803 gas units, which is 69.8% of its gas limit. The lower utilization percentages for NFT transfers compared to contract creation and NFT minting could indicate that these operations have more variability in their computational requirements, prompting platforms to set higher gas limits as a precautionary measure.

The consistent gas usage patterns between Fantom and Polygon across all operations are noteworthy, suggesting similar implementation approaches or shared standards. This consistency could be beneficial for developers working across these platforms, as it allows for more predictable behavior and easier cross-chain compatibility. Celo's lower gas utilization percentages across all operations, compared to its higher gas limits, indicate a platform design that prioritizes flexibility and robustness. This approach provides more headroom for complex operations or unexpected computational needs, potentially reducing the risk of transaction failures due to out-of-gas errors. However, it may also result in less efficient use of network resources if the additional capacity is frequently unused.

4.4 Gas Price

Table 4 presents a detailed comparison of gas prices. In particular, BNB Smart Chain maintains a consistent gas price of 10 Gwei (0.00000001 BNB) across all three operations. This uniformity in gas pricing simplifies cost calculations for users and developers, providing a predictable fee structure regardless of the operation type. However, at 10 Gwei, BNB Smart Chain's gas price is the highest among the four platforms examined, which could lead to higher overall transaction costs, especially for complex or frequent operations.

Table 4. Gas Price

	Contract Creation	Create NFT	Transfer NFT
BNB Smart Chain	0.00000001 BNB (10 Gwei)	0.00000001 BNB (10 Gwei)	0.00000001 BNB (10 Gwei)
Fantom	0.0000000035 FTM (3.5 Gwei)	0.0000000035 FTM (3.5 Gwei)	0.0000000035 FTM (3.5 Gwei)
Polygon	0.000000002500000012 MATIC (2.500000012 Gwei)	0.000000002500000016 MATIC (2.500000016 Gwei)	0.000000002500000016 MATIC (2.500000016 Gwei)
Celo	0.0000000026 CELO (Max Fee per Gas: 2.7 Gwei)	0.0000000026 CELO (Max Fee per Gas: 2.7 Gwei)	0.0000000026 CELO (Max Fee per Gas: 2.7 Gwei)

Fantom offers a lower and equally consistent gas price of 3.5 Gwei (0.0000000035 FTM) for all operations. This pricing strategy positions Fantom as a more cost-effective option compared to BNB Smart Chain, potentially making it attractive for applications that require frequent transactions or have cost-sensitive users. The consistency across operation types, similar to BNB Smart Chain, provides a straightforward fee structure for users and developers. Polygon demonstrates a slightly more complex pricing model. For contract creation, it charges 2.500000012 Gwei (0.000000002500000012 MATIC), while for both NFT creation and transfer, the price is marginally higher at 2.500000016 Gwei (0.000000002500 000016 MATIC). This subtle difference in gas price between contract creation and NFT operations is noteworthy, though the practical impact on overall transaction costs is likely minimal. Polygon's gas prices are lower than both BNB Smart Chain and Fantom, potentially making it the most cost-effective option among these three for high-volume transactions.

Celo presents a unique approach with its gas pricing model. It uses a "Max Fee per Gas" structure set at 2.7 Gwei (0.0000000026 CELO) for all operations. This maximum fee model could provide users with more predictability in terms of maximum transaction costs, while potentially allowing for lower actual fees depending on network conditions. Celo's gas price falls between Fantom and Polygon in terms of cost, offering a balance of affordability and predictability.

These findings underscore the importance of considering gas prices when selecting a blockchain platform for insurance claim management systems. The variations in gas prices can significantly impact the overall cost of operating on these networks, especially for systems that require frequent transactions or complex smart contract interactions. BNB Smart Chain's higher gas price might be justified by its established ecosystem and potentially higher security, but it could be a limiting factor for cost-sensitive applications. Fantom, Polygon, and Celo offer more economical options, with Polygon potentially being the most cost-effective for many use cases.

5 Conclusion

This study has provided a comprehensive analysis of NFT-based insurance claim management systems across four EVM-compatible blockchain platforms: BNB Chain, Fantom, Polygon, and Celo. Our research highlights the potential of blockchain technology to address longstanding challenges in insurance claim processing, including fraud prevention, operational inefficiencies, and transparency issues. The comparative analysis of transaction fees and computational costs across these platforms reveals significant variations in their economic efficiency and performance characteristics. These differences underscore the importance of carefully selecting a blockchain platform that aligns with the specific requirements of an insurance claim management system. Our findings suggest that while all examined platforms offer improvements over traditional claim management processes, each presents unique trade-offs between security, transparency, and cost-effectiveness. The implementation of NFTs and smart contracts across these platforms demonstrates the potential for creating more secure, verifiable, and efficient claim management systems.

References

1. Abramowicz, M.: Blockchain-based insurance. George Washington University Law School (2019)
2. Alnavar, K., Babu, C.: Blockchain-based smart contract with machine learning for insurance claim verification. In: 2021 5th International Conference on Electrical, Electronics, Communication, Computer Technologies and Optimization Techniques (ICEECCOT) (2021)
3. Bang, N., et al.: Blockchain-enhanced ioht: a patient-centric internet of healthcare things platform with smart contract-driven data management. In: International Conference on Advances in Mobile Computing and Multimedia Intelligence, pp. 50–56. Springer (2023)
4. Bhamidipati, N.R., et al.: Claimchain: secure blockchain platform for handling insurance claims processing. IEEE (2021)
5. Bodemer, O.: Transforming the insurance industry with blockchain and smart contracts: enhancing efficiency, transparency, and trust. Engineering: Open Access (2023)
6. Bordekar, J.A., Gupta, S., Tinnaluri, V., Sinha, A.: Insurance claiming process based on blockchain technology using smart contract. Int. J. Adv. Res. Ideas Innov. Technol. (2019)
7. Duong-Trung, N., et al.: On components of a patient-centered healthcare system using smart contract. In: Proceedings of the 2020 4th International Conference on Cryptography, Security and Privacy, pp. 31–35 (2020)
8. Duong-Trung, N., et al.: Smart care: integrating blockchain technology into the design of patient-centered healthcare systems. In: Proceedings of the 2020 4th International Conference on Cryptography, Security and Privacy, pp. 105–109 (2020)
9. Grishin, D., et al.: Accelerating genomic data generation and facilitating genomic data access using decentralization, privacy-preserving technologies and equitable compensation. Blockchain in Healthcare Today (2018)
10. Le, H.T., et al.: Bloodchain: a blood donation network managed by blockchain technologies. Network **2**(1), 21–35 (2022)
11. Loukil, F., Boukadi, K., Hussain, R., Abed, M.: Ciosy: a collaborative blockchain-based insurance system. Electronics (2021)
12. Quynh, N., et al.: Toward a design of blood donation management by blockchain technologies. In: Gervasi, O., et al. (eds.) ICCSA 2021. LNCS, vol. 12956, pp. 78–90. Springer, Cham (2021). https://doi.org/10.1007/978-3-030-87010-2_6
13. Sawalka, S., Lahiri, A., Saveetha, D.: Ethinsurance: a blockchain based alternative approach for health insurance claim. In: 2022 International Conference on Computer Communication and Informatics (ICCCI) (2022)
14. Shinde, R., et al.: Securing AI-based healthcare systems using blockchain technology: a state-of-the-art systematic literature review and future research directions. arXiv preprint arXiv:2206.04793 (2022)

15. Thukral, M.K.: Security and efficiency in vehicle insurance: a blockchain-based solution. IEEE (2023)
16. Vo, H., Mehedy, L., Mohania, M., Abebe, E.: Blockchain-based data management and analytics for micro-insurance applications. In: Proceedings of the 2017 ACM on Conference on Information and Knowledge Management (2017)
17. Zheng, X., et al.: Blockchain-based personal health data sharing system using cloud storage. In: 2018 IEEE 20th International Conference on e-Health Networking, Applications and Services (Healthcom), pp. 1–6. IEEE (2018)

Attaining Equilibria Using Control Sets

Gleb Polevoy[1(✉)] and Jonas Schweichhart[2]

[1] Opole University, Opole, Poland
gleb.polevoy@uni.opole.pl
[2] University of Lübeck, Lübeck, Germany
jonas.schweichhart@uni-luebeck.de
https://repo.uni.opole.pl/info/author/U0ca703dfa7fc54f368e5d27d9d20ce8cd/,
https://www.tcs.uni-luebeck.de/en/mitarbeiter/schweichhart/

Abstract. Many interactions stabilise in a socially suboptimal equilibrium, or are in non-equilibrium states, from which arriving at the desired equilibrium through practical dynamics is either too lengthy or impossible. Thus, we formulate a new theoretically and practically interesting **direct control set** problem for the **planner**: Given a game, any profile s and any desired equilibrium d, find a minimum subset A of the players, such that if they are made to deviate to d, then d becomes a best response for everyone outside A.

We prove that the direct control set optimisation problem is NP-hard, and inapproximable within factor $n^{1-\epsilon}$, for any $\epsilon > 0$. We then study hardness and approximation systematically, considering potential and then harmonic games, which span the finite games. For potential games, finding a direct control set, even with a constant-factor approximation, is still NP-hard, but we solve this problem for singleton congestion games. As for harmonic games, we solve the two-player games, and prove that for more players, the problem is almost as hard as for general games.

Keywords: Equilibrium · Control set · Best response · Potential games · Harmonic games · Zero-sum games

1 Introduction

In a society being stably in a state that we dislike, such as high violence levels or little cooperation, we, an observer who plans what to do, would prefer another stable state. Alternatively, consider being in a temporarily unstable situation, from where reaching the desired stable state through some natural dynamics would be impossible or too long. The fundamental question is how to bring such a society to the preferred stable state. This pertains to many other areas, such as improving institutional and organisational practices, modifying business and international relations, routing, and scheduling.

G. Polevoy and J. Schweichhart—Much of this work was done at the University of Paderborn.

D. Du et al. (Eds.): COCOA 2024, LNCS 15434, pp. 439–452, 2025.
https://doi.org/10.1007/978-981-96-4445-2_36

When the planner knows the players' utilities, she can change the equilibria by subsidies [14], or by subsidies and taxes [18].

All the above methods tamper with the structure of the game, which can be hard or undesirable. To our rescue then come approaches that influence the play without adjusting the rules of the game. For example, Stackelberg strategies [5,6, 19], also called Stackelberg routing [11] for non-atomic routing games, assume the planner controls a certain fraction of the players (such as the compliant drivers, following the recommendations), thereby making the others best respond in a socially beneficial manner close to the social optimum. Similarly, Balcan et al. [2] study influencing dynamics, aiming to converge to a socially efficient NE.

While the abovementioned Stackelberg methods aim to optimise the social welfare by making the already given agents behave in a certain way, we will define a ***direct control set*** $\boldsymbol{A \subset N}$ ***for*** $\boldsymbol{s}$ ***and*** $\boldsymbol{d}$ as the players whose deviation to the desired equilibrium d, while all the others still play s, is **guaranteed** to make each i outside A have d_i is a best response (perhaps one from many) to $(d_A, s_{(N \setminus (A \cup \{i\}))})$. We aim to find a smallest possible direct controlled set, to reach the desired equilibrium in one step, complementing the works by Kempe et al. [10] on infecting a network in contagion games by cascades. We assume we can make some people do what we ask them to do, while they do not have to find the equilibrium behaviour beneficial at first or even after the rest play the desired equilibrium. The equilibrium behavior becomes a best response for everyone, but not necessarily better than the original profile, so there is no optimisation like Stackelberg leader(s) perform(s) in classical Stackelberg games.

We call the control set ***direct***, because the controlled agents are made to play their strategies in the destination profile d. If the planner could also ask those players to play other strategies than d (non-direct control), then, to complete the movement to solution d, the agents in A would have to switch to d themselves, once the rest have best responded by playing their strategies in d. Such a general model of control set might be sometimes more potent, but this would render the analysis even more complicated, while assuming the planner avails over more power, so in this paper, we assume direct control by default.

Applicability. In our model, each player outside the direct control set best responds to the controlled players playing the desired profile d, and the other players outside the control set still playing the original profile s. This limits the applicability of our model, because it is not always the case that all the non-controlled players decide on how to respond to the controlled ones playing d_A simultaneously, i.e. before the other non-controlled players have deviated. The model is still applicable in particular in the following cases:

Synchronous When the responding of the non-controlled players is virtually synchronous, such as in parliamentary voting, legal agreements and other cases with commitments, and some algorithmic trading situations.

Unknown When the response of a non-controlled player remains unknown, before the others have responded as well. That can happen, for example, in secret ballot voting, or in any case with a sufficient observation delay, be it in politics, war, or economy.

Monotonicity When any direct control set remains a direct control set if we extend it. This holds, in particular, where only the magnitude of various behaviours matters, such as in Example 1, or in choosing the most widespread language, behaviour, or protocol.

We further position our work in the literature in Sect. 2, and define the novel ***direct control set*** problem in Sect. 3.[1] We prove in Sect. 4 that the problem is NP-hard, even for approximation within the number of players to any power smaller than 1. After that, we study two main classes that span any game in the Helmholtz-based decomposition by Candogan et al. [3], namely potential and harmonic games; this is up to a nonstrategic game, which does not influence the individual best responses. General potential games are still NP-hard, even for approximation, but we design an exact algorithm for singleton congestion games in Sect. 5. Next, we fully analyse the problem on harmonic games.

The following running example is a stylised version of fighting corruption or another bad behaviour that is stable only when it is widespread.

Example 1 (Fighting corruption). Consider a networks of agents, representing politicians, police agents, and judges, two agents being connected if and only if they can interact. Assume that each node can be either corrupt of not, and being corrupt is profitable if and only if most neighbours in the network are corrupt. Formally, each agent $i \in N$ chooses a strategy $s_i \in \{C(orrupt), H(onest)\}$, and i's utility is

$$u_i(s) \triangleq \begin{cases} 5 & s_i = C \wedge |\{i \in N(i) : s_i = C\}| \geq |N(i)|/2, \\ -10 & s_i = C \wedge |\{i \in N(i) : s_i = C\}| < |N(i)|/2, \\ 1 & \text{otherwise } (s_i = H). \end{cases}$$

This represents the temptation (5 vs 1) of corruption, when one knows enough corrupt actors, and the punishment (-10) otherwise.

Here, everyone being corrupt, namely $s = (C, C, \dots, C)$ is a Nash equilibrium, and assuming no one is completely isolated, nobody engaging in corruption is also an NE, meaning $d = (H, H, \dots, H) \in \text{NE}$. Moving from s to d in the complete graph requires at least $(n-1)/2$ nodes. The symmetry implies the identity of the nodes does not matter.

2 Related Work

Tipping Points. The phenomenon of overtaking the whole network was studied by Schelling in the context of racial segregation [21], and popularly presented by Gladwell [7]. The similar idea of cascading appeared in Heal and Kunreuther [8]. In comparison, we basically aim at engineering a simultaneous tipping in not necessarily symmetric, binary, or super-modular games.

Stackelberg Strategies. While our model minimises the cost of bringing the play from a given profile to the desired equilibrium, Stackelberg routing is dual in that it aims at minimising the price of anarchy by controlling a given fraction of

[1] Our model holds for any solution concept, but we concentrate on Nash equilibria [16].

the flow. Similarly to Stackelberg, consider advertising a strategy recommendation [2] (for routing, for example), which is blindly followed by a given fraction of the flow, and then everyone best-responds, until convergence to an NE.

Focal Points, Defection and Formation Index. Direct control sets both communicate partial information (akin to focal points [20]) and motivate the play. The number of players implying a Nash equilibrium and equilibrium stability was studied by Kalai and Tauman [9,22]. Kalai defined the ***defection index*** of an equilibrium (previously studied by Abraham et al. [1]) as the minimum number of deviators who can motivate some player to deviate. Kalai also defined the ***formation index*** as the minimum number of players such that any such subset forms that Nash equilibrium. Tauman and Kalai [22] bolstered that work by axiomatising best-response equilibria of various viability. These indices are monotonic, because the formation guarantee is required for *any* coalition of a certain size. Unlike us, Kalai's indices refer to any non-compliant coalition of players rather than to a special coalition, thus a profile being ***uniformly m-incentive compatible*** in the sense of Tauman and Kalai [22] implies any coalition of size at least m is a direct control set, but that is unnecessary for that. Algorithmic questions, such as fault-tolerant computation, were studied by Eliaz [4], who proposed ***k-fault-tolerant Nash equilibrium*** which is equivalent to Nash equilibrium with a defection index greater than k, and Abraham et al. [1], who called defection index the ***level of resilience***.

3 Model

Consider a strategic form ***game*** $G = (N, S = S_1 \times S_2 \times \ldots \times S_n, (u_i)_{i=1,\ldots,n})$, where $N = \{1, \ldots, n\}$ is the set of agents, S_i is agent i's strategy set and $u_i \colon S \to \mathbb{R}$ is agent i's utility function. The ***solutions***, forming a ***solution set***, are a set of (strategy) profiles $D \subseteq S$, defined for a game by a rule called ***a solution concept***, such as ***Nash equilibrium***, denoted NE.[2] A Nash equilibrium [16] of G is a strategy profile $s = (s_1, \ldots, s_n) \in S_1 \times \ldots \times S_n = S$ such that (s_{-i} means $(s_1, \ldots, s_{i-1}, s_{i+1}, s_n)$) $\forall i \in N, \forall s'_i \in S_i : u_i(s) \geq u_i(s'_i, s_{-i})$.

We incentivise play by controlling some agents, as illustrated in Fig. 1.

Definition 1. *Consider a profile $s \in S$, a solution concept $D \subseteq S$ and a solution $d \in D$. For example, d can be a Nash equilibrium. A set of agents $A \subseteq N$* ***brings s to d(directly)*** *if these agents can play d while all others play s, so that for each agent outside A, playing d would be a best response (perhaps one from many). In formulas, given the profile $(d_i)_{i \in A}, (s_j)_{j \in N \setminus A}$, for each agent $i \in N \setminus A$, playing d_i is a best response, meaning $u_i(d_i, (d_i)_{i \in A}, (s_j)_{j \in N \setminus (A \cup \{i\})}) \geq u_i(d'_i, (d_i)_{i \in A}, (s_j)_{j \in N \setminus (A \cup \{i\})}), \forall d'_i \in S_i$.*
This set A is called a ***direct control set*** *with respect to G, s and d.*

[2] We use NE to abbreviate the terms "Nash equilibrium\equilibria", when no confusion with the set of equilibria arises.

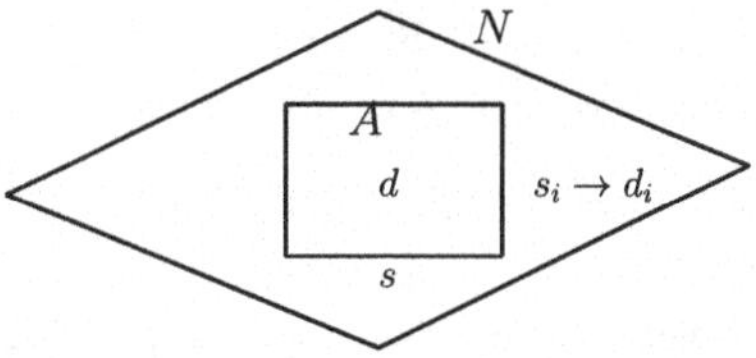

Fig. 1. The diamond denotes all the players N, while rectangle A is a direct control set w.r.t. G, s, d. When all the agents in A play d_A, while the rest play $s_{N \setminus A}$, every i not in A would (weakly) most like to play d_i.

Given weights of agents, representing how hard it is to make that agent play her part in the desired equilibrium, we aim to find a most practical, i.e. lightest, agent set that can bring one solution to another one. We define that next.

Definition 2. *The* ***direct control set problem*** *receives as input a game* $G = (N, S, (u_i)_{i=1,\ldots,n})$*, a profile* s*, a solution profile* d *to* G*, and agents' weights* $w: N \to \mathbb{R}_+$.
A ***solution*** *to the problem is a set of agents* $A \subseteq N$*. A* ***feasible solution*** *is a direct control set with respect to* G, s *and* d*.*
We aim to find a feasible solution A*, minimising its weight* $w(A) \triangleq \sum_{a \in A} w(a)$*.*

The ***decision version*** of this problem is an instance with a natural number k, and the question is whether there exists a direct control set of weight at most k. We denote both the term "direct control set" and this problem as DirConSet.

Remark 1 (Manageable Representation). Naively providing an n-dimensional utility matrix per player, storing all the utilities, as input of the direct control set, would take $\Omega(n \cdot 2^n)$ space, rendering the brute-force algorithm, testing all the possible 2^n subsets of players for being direct control sets, polynomial. Managing such a huge input is, however, impracticable, so we assume a utility oracle instead.

Thus, we assume the game is represented by explicitly storing the strategies of each player, while the utilities are provided by a polynomial oracle, which receives a strategy profile s and a player i as input, and returns $u_i(s)$. This resembles the approach of Papadimitriou and Roughgarden [17]. In some cases, such as for congestion games or polymatrix games, such an oracle is implied by the definition of the game. When reducing another problem to the direct control set, we are required to generate a polynomial utility computation algorithm, implementing this oracle.

Let us exemplify the definitions by looking at Example 1 again.

Example 2 (Fighting corruption - Cont.). Moving from $s = (C, C, \ldots, C)$ to $d = (H, H, \ldots, H) \in \mathrm{NE}$ in the complete graph requires at least $(n-1)/2$ nodes; in other words, a subset A of players if a DirConSet with respect to s and d, i.e. it brings s to d, if and only if $|A| \geq (n-1)/2$. In other words, $A \subseteq N$ is a feasible solution to the problem iff is contains at least $(n-1)/2$ players. In another topology, that direct control sets may differ.

Basic Properties. The minimum possible size of a direct control set is called the ***direct control number***, and it is denoted by Dir1(G, s, d). In Example 1, Dir2$(G, s, d) = \frac{n-1}{2}$. We first prove that for any p, there always exists a game with more than p players, such that its direct control number is exactly p. This proof, introducing a game a bit reminiscent to that in Example 1, is omitted for lack of space, as are some others, too.

Proposition 1. *For any natural numbers $n > p \geq 1$, there exists a game G with n players, possessing Nash equilibria s and d, such that* Dir3$(G, s, d) = p$.

In particular, in some cases, any DirConSet can be of size $n-1$. However, by definition of NE, any $n-1$ players always constitute a DirConSet.

4 Hardness

Exhaustively testing all the subsets for the role of a control set can take exponential time. We now prove that the direct control set problem is intrinsically immensely hard, even for approximation. We cope with that hardness in the coming sections by restricting the problem domain and approximations.

Theorem 1. *The decision version of the* DirConSet *problem is NP-complete.*

Proof. The NP-hardness holds even for approximation, as we show in Theorem 2.
These problems belong to NP, the witness being a direct control set. □

Even approximately solving direct control set is very hard, as we now prove.

Theorem 2. *Finding a minimum direct control set is not approximable within factor $n^{1-\epsilon}$, $\forall \epsilon > 0$, unless $P = NP$. This holds even when we restrict the* DirConSet *problem to bring an NE to an NE in an unweighted game where each player has* 2 *strategies.*

Proof. Consider the problem of deleting a minimum-cardinality vertex set from a connected graph $G = (V, E)$ to obtain a tree, namely a connected acyclic subgraph. This problem is not approximable within factor $|V|^{1-\epsilon}$, for any $\epsilon > 0$, unless P = NP [23], so it suffices to reduce this problem to the DirConSet problem in the following game. Given an instance of the problem of deleting a minimum vertex set from a connected graph $G = (V, E)$ to obtain a tree, we define a game with players V and actions $S_1 = \ldots = S_n = \{a, b\}$ each. Let the utilities of each player $l \in N$ be as follows (this is polynomial to compute, as required in Remark 1):

$$u_l(s) \triangleq \begin{cases} 2 & \text{The subgraph induced by } V \setminus \{i \in V \setminus \{l\} : s_i = b\} \text{ is a tree,} \\ 1 & \text{otherwise, if } s_l = a, \\ 0 & \text{otherwise } (s_l = b). \end{cases}$$

Here, profile $s \triangleq (a, a, \ldots, a)$ is an NE, because the empty set is not a feasible solution, as we can assume w.l.o.g. that the given instance is not a tree, since that can be checked in polynomial time by DFS, and such instances can be reduced to a predefined instance of the DirConSet problem where the empty set is a feasible solution. Profile $d \triangleq (b, b, \ldots, b)$ constitutes an NE, since deleting all the nodes besides one always yields a tree. Finally, a set of vertices is a direct control set w.r.t. $G, (a, \ldots, a), (b, \ldots, b)$ if and only if for any vertex l outside that set, deleting that set (without deleting l) leaves a subgraph inducing a tree. The latter is equivalent to that set solving the given problem of deleting a vertex set to obtain a tree. The above hardness results implies the theorem. □

Having removed all hope for a decent polynomial approximation of the general problem, we will analyse two major subclasses of games according to the Helmholz-like decomposition of games [3], potential (Sect. 5) and harmonic (Sect. 6) ones, which in a way span all the finite games. Indeed, each finite game is a direct sum of a potential, a harmonic and a non-strategic game [3]; the latter component does not influence individual deviations.

5 Singleton Congestion Games

Potential games [13] are defined as the set of strategic form games $G = (N, S = S_1 \times S_2 \times \ldots \times S_n, (u_i)_{i=1,\ldots,n})$ such that there exists a potential. A potential (function) $P : S \to \mathbb{R}$ has to satisfy that for any subprofile $s_{-i} \in \times_{j \in N \setminus \{i\}} S_j$ and for any strategies $s_i \neq t_i \in S_i$, her utility difference when moving from s_i to t_i is equal to that of P, namely $u_i(s_i, s_{-i}) - u_i(t_i, s_{-i}) = P(s_i, s_{-i}) - P(t_i, s_{-i})$. The set of finite potential games is isomorphic to the set of ***congestion games***, which is the set of strategic games $(N, R, (S_i)_{i \in N}, (c_j)_{j \in R})$ on resources R, such that any strategy is a subset of those resources, and c_j are cost functions to be paid by the players for using those resources. Formally, $S_i \subseteq 2^R \setminus \{\emptyset\}$ and $c_j \colon \{1, \ldots, n\} \to \mathbb{R}$. The strategies result in loads $l_j(s) \triangleq |\{i \in N : j \in s_i\}|$ on every resource $j \in R$, which in turn define the eventual costs born by the players, denoted as $C_i(s)$ for every player $i \in N$ and defined as $C_i(s) \triangleq \sum_{j \in s_i} c_j(l_j(s))$.

Potential games are still hard, even for approximation, but we omit the proof for lack of space.

We now deal with ***singleton congestion games***, i.e. congestion games where each strategy contains a single resource, namely $|s_i| = 1$, for every player $i \in N$ and every $s_i \in S_i$. We will tightly bound the necessary size of a direct control set in singleton congestion games, and present a polynomial time algorithm for singleton congestion games G, under additional assumptions.

Our analysis heavily uses the concept of ***the intermediate profile*** and of ***the attraction basin***. The ***intermediate profile*** with respect to instance (G, s, d) and players $A \subseteq N$, denoted by $sd(A)$, is the strategy profile that results from the initial profile s after switching the strategies of all the players in A to their strategy in d, namely $(s_{N \setminus A}, d_A)$. By definition, A is a direct control set w.r.t. (G, s, d) iff for the intermediate profile $sd(A)$ and every player $i \in N \setminus A : d_i \in Br_i(sd(A))$. The concept applies to any game; for singleton congestion games, it is illustrated in Fig. 2.

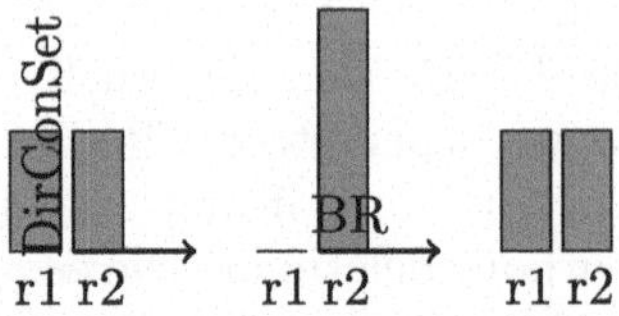

Fig. 2. Consider a game with 2 resources. The leftmost picture describes s, in the centre lies $sd(A)$, and the rightmost picture depicts d.

Furthermore, any pair of players for whom it is a best response to change their strategies in $sd(A)$ have the same possible alternatives. Formally, for any players $i \neq j$ with $Br_i(sd(A)) \setminus sd_i(A) \neq \emptyset$ and $Br_j(sd(A)) \setminus sd_j(A) \neq \emptyset$, it holds that $Br_i(sd(A)) \setminus sd_i(A) = Br_j(sd(A)) \setminus sd_j(A)$. This leads to the following definition, illustrated in Fig. 3:

Definition 3. *Given a strategy profile s of a singleton congestion game $(N, R, (S_i)_{i \in N}, (c_j)_{j \in R})$, we call the set $AB(s) = \bigcup_{i \in N}((\cup_{Q \in Br_i(s)} Q) \setminus s_i)$, i.e. the resources where players want to deviate, the **attraction basin** of s.*

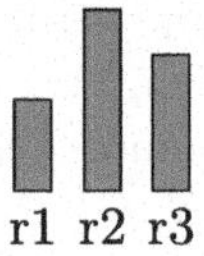

Fig. 3. Consider a game with 3 resources. If all the costs are $c(x) \triangleq x$, then $AB = \{r1\}$.

Denoting, for a resource $r \in R$ and a profile $s \in S$, the players choosing r in s by $P_r(s) \triangleq \{i \in N : s_i = \{r\}\}$ and defining $P_Q(s) \triangleq \cup_{q \in Q} P_q(s)$, we observe that after a deviation, the costs are the same.

Observation 1. $\forall s \in S$ and $\forall q, r \in AB(s)$, $c_q(|P_q(s)| + 1) = c_r(|P_r(s)| + 1)$.

Proof. Otherwise, there exist $q, r \in AB(s)$ such that we have $c_q(|P_q(s)| + 1) < c_r(|P_r(s)| + 1)$. This implies $\{r\} \notin BR_i(s), \forall i \in N$, in contradiction to the definition of attraction basins as a union of the best responses. □

We first provide a tight upper bound on the minimum size of a DirConSet.

Proposition 2. *For any singleton congestion game G, a minimum* DirConSet *with respect to profiles s, d has the size at most $\lfloor n - n/m \rfloor$, and this bound is tight. The positive result holds in the weighted case, while the tightness holds even in the unweighted case.*

We can obtain such a set by taking all the players who play any resource besides one in the destination profile. Tightness is attained by choosing cost functions that never have tied costs, thereby making the attraction basin have size 1 at most, implying that all the players who need to move and who have to end up outside the singleton attraction basin have to be taken to any DirConSet. The same upper bound holds for singleton congestion games with increasing player-specific costs [12] with the same proof.

Proof. We first construct a DirConSet of size at most $\lfloor n - n/m \rfloor$ in the following way. Choose a resource r that maximizes $|P_r(d)|$. Then let $A \triangleq \bigcup_{q \neq r} P_q(d)$, namely all the players who do not choose r in d. Since d is an NE, for every player $i \in P_r(d)$, $\{r\} \in BR_i(d)$, and since the only players not in A are $P_r(d)$, for every player $i \notin A$, $r \in BR_i(sd(A))$. As all the other players belong to A, set A constitutes a DirConSet. Now, $r = \text{argmax}_{q \in R} |P_q(d)|$ implies $|P_r(d)| \geq \lceil n/m \rceil$ $\Rightarrow |A| \leq \lfloor n - n/m \rfloor$.

For tightness, consider a singleton congestion game G where the cost function for each resource j is $c_j(x) \triangleq j + m * (x - 1)$. Now, we take any profile d where the players are almost evenly spread, i.e. $P_j(d) \leq \lceil n/m \rceil$. Any such d is an NE, so we take any (G, s, d) such that for each player i we have $s_i \neq d_i$, meaning that all the players need to move from s to d. Since the cost functions have no equal integer values, at most one resource r is a best response at the same time, which implies that $|AB(sd(B))| \leq 1$ for any DirConSet $B \subset N$. Since every player is supposed to change strategies from s to d, all the players in $N \setminus P_{AB(sd(B))}(d)$ need to be moved by being taken into the DirConSet, and since for every resource $j \in R$, $P_j(d) \leq \lceil n/m \rceil$, $|B| \geq \lfloor n - n/m \rfloor$. □

The possible attraction basins in a singleton congestion game are linked to the possible direct control sets. This will allow finding a minimum DirConSet by computing a DirConSet for all the possible attraction basins and choosing the smallest one. In order to cope with the exponential number of subsets of resources that can form an attraction basin, we notice that in real-world applications, it is unlikely that the cost functions of all resources can be equalized within the parameter interval $\{1, \ldots, n\}$. We assume the following.

Assumption 1 (General Position). *In a singleton congestion game,* $|AB(s)| \leq 1$ *for each profile* $s \in S$.

Then, the number of attraction basins is at most the number of players. Given an attraction basin of the intermediate profile $sd(A)$, we now consider the possible direct control sets A, under condition Eq. (1), which, intuitively, implies monotonicity in the cost functions of the resources, limited to their values in the narrow scope of their loads in d. This condition is checkable in linear time.

Lemma 1. *Consider a singleton congestion game* G*, profile* s *and* $d \in NE$*. Assuming the general position and*

$$C_p(l_p(d) + 1) > C_r(l_r(d)) \Rightarrow C_p(l_p(d)) > C_r(l_r(d) - 1), \forall p \neq r, \tag{1}$$

we claim that for a given intermediate profile $sd(A)$, inducing the attraction basin $AB(sd(A)) = \{r\}$, any minimum direct control set A has to be

$$A = \{P_p(s) \cap P_q(d) : p \neq q, p \neq r, q \neq r\} \tag{2}$$
$$\cup \{P_r(s) \cap P_p(d), : p \neq r\} \tag{3}$$
$$\cup \{i \in P_p(d) \cap P_p(s) : p \neq r, Br_i(sd(A)) = \{r\}\} \tag{4}$$
$$\cup \{\text{Either none, all but one, or all from } P_r(d) \setminus P_r(s)\} \tag{5}$$

In the Lemma, set Eq. (2) describes the deviations outside of the attraction basin, while Eq. (3) refers to moving from playing r to another resource. Next, set Eq. (4) relates to holding still the players who should not deviate, but would like to. Finally, Eq. (5) describes moving to the attraction basic in advance, to prevent someone else who should not deviate from being tempted to move there, and the claim is that instead of considering the exponential number of such sets, it suffices to consider Eq. (5).

Proof. By definition of attraction basin, we need to include those who need to move outside the attraction basin in the DirConSet, thus implying Eq. (2) and Eq. (3). We also need to take to the DirConSet all those whose only best response lies in the attraction basin, while they should not deviate, implying Eq. (4). Those necessary players are also sufficient to make a DirConSet, because the only moves that are not included are those to r, and since $d \in \text{NE} \Rightarrow \forall i \in P_r(d), Br_i(d) = \{r\}$, moving everyone besides those who should move to r implies the best response for all the players in $P_r(d)$ is already $\{r\}$.

The only other optimum DirConSet can be taking there some players from $P_r(d) \setminus P_r(s)$, if that spares the necessity to add to set Eq. (4) some players. It is thus left to prove that the only players adding whom to the DirConSet could (though does not have to) spare more than their cost are the players from Eq. (5). We need to prove that moving to r less than all but one players from $P_r(d) \setminus P_r(s)$ would make $\{r\}$ the unique best response for every player in $P_{N \setminus \{r\}}(sd(A))$, thereby rendering adding any players from $P_r(d) \setminus P_r(s)$ to the DirConSet useless. In other words, $P_r(sd(A))$ lacking two or more players relatively to $P_r(d)$ would imply r is the unique best response for everyone. It thus suffices to show that $C_p(l_p(d)) > C_r(l_r(d) - 1)$ for any resource $p \neq r$.

To this end, we recall the Nash equilibrium condition for $d \in \text{NE}$ implies $C_p(l_p(d) + 1) \geq C_r(l_r(d)), \forall p \neq r$. Provided we show that this inequality is strict (while not assuming at the outset that the equilibrium is strict), the assumption of the lemma will imply that $C_p(l_p(d)) > C_r(l_r(d) - 1), \forall p \neq r$, and we are done. Now, equality in $C_p(l_p(d) + 1) \geq C_r(l_r(d))$ would imply $p \in AB(d)$. Therefore, the general position applied to d as the intermediate profile says that the NE condition, namely $C_p(l_p(d) + 1) \geq C_r(l_r(d))$, can hold with equality for at most one $p \neq r$. So we now handle the case where such a resource p exists. Fix this p.

First, if $P_r(d) \setminus P_r(s) = \emptyset$, then the optimum DirConSet is simply the union of sets Eq. (2), (3), and (4), and we can thus write this case off. Otherwise, take any $i \in P_r(d) \setminus P_r(s)$ and define an auxiliary DirConSet $B \triangleq N \setminus \{i\}$. Now, equality $C_p(l_p(d) + 1) = C_r(l_r(d))$ implies that both p and r are in $AB(sd(B))$, unless $s_i = p$, namely unless i starts at p. Therefore, the general position implies that any $i \in P_r(d) \setminus P_r(s)$ starts at $s_i = p$. So all the moves to r from the intermediate profile $AB(sd(A))$ have to be from p. Now, to prevent $\{r\}$ from being the unique best response in $AB(sd(A))$ for some $j \in P_p(s) \cap P_p(d)$, also all the $P_r(d) \setminus (P_r(s) \cup A)$, who we have proven to play p in profile s, do not have r as the unique best response in $AB(sd(A))$. Therefore, at most one player from $P_r(d) \setminus P_r(s)$ may be on r in the intermediate profile $AB(sd(A))$, because otherwise, moving the second player to r would be strictly suboptimum. So, unless $C_p(l_p(d)+1) \geq C_r(l_r(d))$ holds strictly for p, we have still concluded that moving to r all, or all but one of the $P_r(d) \setminus P_r(s)$ suffices to try all the cases of preventing temptation to move to $\{r\}$. □

We can thus find an optimum DirConSet by iterating over all the possible AB and selecting the cheapest set given by Lemma 1. We can also solve the DirConSet under the following more general assumptions:

1. We can deal with the weighted case by analogy.
2. We can weaken assumption Eq. (1) to this: $c_p(l_p(d) + 1) > c_r(l_r(d)) => c_p(l_p(d)) > c_r(l_r(d) - a)$, for any constant $a \geq 1$. Indeed, then we would just conclude that we need to try either
   (a) bringing no-one to r,
   (b) or bringing everyone to r,
   (c) or bringing everyone besides 1 player to r,
   (d) $\vdots$
   (e) or bringing everyone besides a players to r,
   still amounting to a polynomial number of checks.
3. We can relax the general position to assume that for any $s \in S$, $|AB(s)| \leq b$, for some $b > 0$.

In any such case, we have algorithmically proven the following.

Theorem 3 (Singleton congestion games). *Given a singleton congestion game G, any $s \in S$, and Nash-equilibrium d, there is a polynomial time algorithm that finds a minimum* DirConSet *with respect to (G, s, d) if for all profiles $t \in S$, $|AB(t)|$ is bounded by a predefined constant, and there is a constant $a \in \mathbb{N}$, such that $C_p(l_p(d) + 1) > C_r(l_r(d)) \Rightarrow C_p(l_p(d)) > C_r(l_r(d) - a)$.*

6 Harmonic Games

We are now turning to the other component of the decomposition by Candogan et al. [3], which is harmonic games. Those are finite games which flow representations contain preference cycles. Equivalently, these games are defined next.

Definition 4. *A noncooperative game $G = \{N, S_1, \ldots, S_n, u_1, \ldots, u_n\}$ is* ***harmonic*** *if for any profile $s \in S \triangleq \times_{i=1}^{n} S_i$, there holds $\sum_{i \in N} \left\{\sum_{t_i \in S_i} \{u_i(s) - u_i(t_i, s_{-i})\}\right\} = 0$.*

Since for two-player finite games, harmonic is equivalent to zero-sum [3], while in general, zero-sum games are a strict subclass of harmonic ones, we study DirConSet for harmonic games as follows.

1. Characterising the DirConSet for 2-player zero-sum games, which amounts to characterising the DirConSet for 2-player harmonic games (2 players).
2. Proving that for $n \geq 3$, the DirConSet problem in n-player zero-sum games is hard. Therefore n-player harmonic games are at least as hard.

For 2-player zero-sum games, the following holds. The proof is omitted.

Proposition 3. *Two-player zero-sum game has* $\mathrm{Dir4}(G, s, d) = 0$*, when* $s, d \in$ *NE. Namely, the empty set suffices to bring any Nash equilibrium to another one.*

When in the starting profile s, at least one player does not play a maxmin strategy, then the singleton containing any player is a direct control set, but sometimes so is the empty set.

For any $n \geq 2$, the following inductively proven approximation-maintaining reduction implies the hardness of DirConSet in zero-sum games with n players, given the hardness results from Sect. 4.

Proposition 4. *The* DirConSet *in (general) n-player games is at most as hard for approximation, as in* $n + 1$*-player zero-sum games.*

7 Conclusions and Future Work

After defining the DirConSet problem to optimise the arrival at the desired equilibrium, and proving it is extremely hard even to approximate, we study the spanning classes of the decomposition of Candogan et al. [3], namely harmonic and potential games, analysing salient potential games and characterising the DirConSet problem for harmonic games. The hardness results explain why we rarely observe practical influence by DirConSet, and ensure us from frequent falling under malicious influence of a third party. On the other hand, algorithmic successes in some classes suggest where we can apply such methods, in addition to very small-scale interactions, which allow for an exhaustive search. Various congestion games model scheduling and other common usage of resources. On the other hand, zero-sum games model competitive interactions. One should also test the practical applications of our methods.

We should also model control sets where the desired equilibrium is attained independently of the order of best responding. Alternatively, as a compromise between our problem and the demand to attain the desired equilibrium independently of the best responding order, we can aim at a sequence of best responses leading from the original profile to the desired equilibrium.

Furthermore, characterising the equilibria possessing direct control sets of a certain size may provide structural insight and even contribute to approximation. This issue resembles axiomatic characterisation of Nash equilibria by their viability [22].

For congestion games with increasing player-specific costs [12], the concept of attraction basin makes less sense, since the best responses can differ across players. Therefore, also assuming the general position becomes less motivated. The arising question is thus whether another polynomial algorithm exists for those games.

One can compute direct control sets by adding players until we obtain a direct control set, and selecting a cheapest such set. The exponential complexity of this algorithm can be ameliorated by heuristically trying to select the best order of adding players, till we obtain a cheap direct control set. Alternatively, we can model gradually movement from one profile to another one, through learning. Another important question is incentivising the direct control set players to play the strategies we want them to play. Being a control set member may be non-profitable, even if the social welfare increases when the others best respond. Being in the control set can thus be subsidised, like in mediated equilibrium [15], or connected to another game, played in parallel by those players. The subsidy can be related to the player's loss in the desired equilibrium, and to her contribution to the control set, using cooperative game theory.

References

1. Abraham, I., Dolev, D., Gonen, R., Halpern, J.: Distributed computing meets game theory: robust mechanisms for rational secret sharing and multiparty computation. In: Proceedings of the Twenty-Fifth Annual ACM Symposium on Principles of Distributed Computing, PODC 2006, pp. 53–62. Association for Computing Machinery, New York, NY, USA (2006)
2. Balcan, M.F., Krehbiel, S., Piliouras, G., Shin, J.: Near-optimality in covering games by exposing global information. ACM Trans. Econ. Comput. **2**(4) (2014)
3. Candogan, O., Menache, I., Ozdaglar, A., Parrilo, P.A.: Flows and decompositions of games: harmonic and potential games. Math. Oper. Res. **36**(3), 474–503 (2011)
4. Eliaz, K.: Fault tolerant implementation. Rev. Econ. Stud. **69**(3), 589–610 (2002)
5. Fanelli, A., Flammini, M., Moscardelli, L.: Stackelberg strategies for network design games. Internet Math. **9**, 336–359 (2010)
6. Fotakis, D.: Stackelberg strategies for atomic congestion games. In: Arge, L., Hoffmann, M., Welzl, E. (eds.) ESA 2007. LNCS, vol. 4698, pp. 299–310. Springer, Heidelberg (2007)
7. Gladwell, M.: The tipping point: How little things can make a big difference. Little, Brown (2006)
8. Heal, G., Kunreuther, H.: Social reinforcement: Cascades, entrapment, and tipping. Am. Econ. J. Microeconomics **2**(1), 86–99 (2010)
9. Kalai, E.: Viable nash equilibria: Formation and defection, February 2020,
10. Kempe, D., Kleinberg, J., Tardos, E.: Maximizing the spread of influence through a social network. In: Proceedings of the Ninth ACM SIGKDD International Conference on Knowledge Discovery and Data Mining, KDD 2003, pp. 137–146. Association for Computing Machinery, New York, NY, USA (2003)
11. Krichene, W., Reilly, J.D., Amin, S., Bayen, A.M.: Stackelberg routing on parallel transportation networks, pp. 1–35. Springer International Publishing, Cham (2017)
12. Milchtaich, I.: Congestion games with player-specific payoff functions. Games Econom. Behav. **13**(1), 111–124 (1996)

13. Monderer, D., Shapley, L.S.: Potential games. Games Econom. Behav. **14**(1), 124–143 (1996)
14. Monderer, D., Tennenholtz, M.: K-implementation. J. Artif. Intell. Res. (JAIR) **21**, 37–62 (2004)
15. Monderer, D., Tennenholtz, M.: Strong mediated equilibrium. Artif. Intell. **173**(1), 180–195 (2009)
16. Nash, J.: Non-Cooperative Games. Ann. Math. **54**(2), 286–295 (1951)
17. Papadimitriou, C.H., Roughgarden, T.: Computing correlated equilibria in multi-player games. J. ACM **55**(3) (2008)
18. Polevoy, G., Dziubiski, M.: Fair, individually rational and cheap adjustment. In: Raedt, L.D. (ed.) Proceedings of the Thirty-First International Joint Conference on Artificial Intelligence, IJCAI-22, pp. 447–453. International Joint Conferences on Artificial Intelligence Organization, July 2022, main Track
19. Roughgarden, T.: Stackelberg scheduling strategies. SIAM J. Comput. **33**(2), 332–350 (2004)
20. Schelling, T.: The Strategy of Conflict: With a New Preface by the Author. Harvard University Press (1980)
21. Schelling, T.C.: Dynamic models of segregation. J. Math. Sociol. **1**(2), 143–186 (1971)
22. Tauman, A., Kalai, E.: Beyond dominance and nash: Ranking equilibria by critical mass, July 2023
23. Yannakakis, M.: The effect of a connectivity requirement on the complexity of maximum subgraph problems. J. ACM **26**, 618–630 (1979)

Scheduling on Parallel-batch Machines with Rejection and Resource Matching

Wenhua Li[1], Liang Zhang[1], Ran Lin[1](✉), and Shisheng Li[2]

[1] School of Mathematics and Statistics, Zhengzhou University, Zhengzhou 450001, China
{liwenhua,rlin}@zzu.edu.cn, liangzhang@gs.zzu.edu.cn
[2] School of Mathematics and Information Science, Zhongyuan University of Technology, Zhengzhou 450007, China

Abstract. We consider two scheduling problems with rejection and resource matching on m identical parallel-batch machines. A job is either rejected with a rejection cost or accepted for processing. Accepted jobs are processed on parallel-batch machines with a batch capacity of b. There are multiple kinds of resources, one of which could be matched with an accepted job. Each accepted job must consume exactly one kind of resource and each kind of resource can be consumed by at most one accepted job. Job's processing time may be different when it consumes a different kind of resource. The objective is to minimize the sum of the makespan of the accepted jobs and the total rejection cost of the rejected jobs. When the batch capacity is bounded, we give two approximation algorithms with worst case ratios mb and $2 - \frac{1}{mb}$, respectively. When the batch capacity is unbounded, we give a polynomial time optimal algorithm.

Keywords: Scheduling · Parallel-batch · Resource matching · Rejection cost · Approximation algorithm

1 Introduction

This paper considers two scheduling problems on m identical parallel-batch (p-batch) machines with rejection and resource matching. There are n jobs $\mathcal{J} = \{J_1, \ldots, J_n\}$ and s kinds of resources $\mathcal{R} = \{R_1, \ldots, R_s\}$. Job J_j is either rejected by paying a certain rejection cost δ_j or accepted for processing. Each accepted job must consume exactly one kind of resource and each kind of resource can be consumed by at most one accepted job. That is, each accepted job should be matched with a kind of resource in a feasible schedule. We call this relation between the accepted jobs and the kinds of resources consumed by them a resource matching. The processing time of J_j is $p_{(j,k)}$ if J_j is accepted and consumes resource R_k. The accepted jobs can be grouped into batches and processed on parallel-batch machines where a batch can process at most b jobs simultaneously. The processing time of a batch is the largest processing time among jobs in the batch. We use $b \geq n$ and $b < n$ to denote the unbounded and bounded batch capacity, respectively. The objective is to minimize the sum of the makespan of the accepted jobs and the total rejection cost of the rejected

D. Du et al. (Eds.): COCOA 2024, LNCS 15434, pp. 453–464, 2025.
https://doi.org/10.1007/978-981-96-4445-2_37

jobs. Using the three field notation $\alpha|\beta|\gamma$, the problems in this paper are denoted as follows:

$$(\mathcal{P}1) \quad P|\text{p-batch}, b < n, \mathcal{R}, \text{rej}|C_{\max} + \sum \delta_j;$$

$$(\mathcal{P}2) \quad 1|\text{p-batch}, b \geq n, \mathcal{R}, \text{rej}|C_{\max} + \sum \delta_j.$$

In this paper, we use "1", "Pm" and "P" in the α field to represent single machine, identical machines with fixed number of machines and identical machines with arbitrary number of machines, respectively. We use "rej" and "r_j" in the β field to indicate that jobs can be rejected and jobs have different release times, respectively. We use "$C_{\max}$" and "$\sum \delta_j$" in the γ field to represent the makespan of the accepted jobs and the total rejection cost of the rejected jobs, respectively. The results obtained in this paper are shown in Table 1.

Table 1. Summary of results obtained in this paper

Problem	Algorithm	Ratio	Reference
$(\mathcal{P}1)$	WM-FBLPT	mb	Theorem 1
	MWM-FBLPT	$2 - \frac{1}{mb}$	Theorem 2
$(\mathcal{P}2)$	EM	1	Theorem 3

Here, an application is considered by the following scenario in a service center. The center receives several tasks and a manager needs to assign them to the workers. For a certain task, it will cause different processing time by different worker due to different level of the familiarity. A worker deals with at most one task and a task is assigned to at most one worker. Taking the tasks and workers as the jobs and resources, respectively, this scenario can be modeled as the scheduling problem with resource matching.

Barketau et al. [1] study the scheduling problem with resource matching on m parallel dedicated machines to minimize the maximum lateness of jobs, where jobs are allocated to machines in advance so that each machine is dedicated for processing a fixed set of jobs. They show that the problem is strongly NP-hard even when $m = 1$. Moreover, they show that it is strongly NP-hard to find an approximation algorithm with worst case ratio less than 2 when m is arbitrary. To the best of our knowledge, there is no other result related to the scheduling problem with resource matching.

There is extensive literature for scheduling problem with rejection. Scheduling problems with rejection have been considered on single machine and identical machines by [2,8,12,13,15,16,18,19] and on single parallel-batch machine by [3,9–11]. For an overview of previous results, see Table 2 and Table 3. For more scheduling problems with rejection, we refer the readers to surveys [14] and [17].

According to the above analysis, only the problem studied in [1] is highly relevant. However, the machine environment and objective in [1] are parallel dedicated machines and maximum lateness, while in this paper are parallel batch machines and the sum of the makespan of the accepted jobs and the total rejection cost of the rejected jobs. Comparing to the scheduling problem with rejection

Table 2. Overview of previous results on single machine and identical machines

Problem	Ratio	Time complexity	Reference
$P\|\text{rej}\|C_{\max}+\sum\delta_j$	$2-\frac{1}{m}$	$O(n\log n)$	[2]
	$\frac{3}{2}+\epsilon$	$O(n\log n+\frac{n}{\epsilon})$	[13]
	$\frac{3}{2}-\frac{1}{2m}$	$O(n^3\log n)$	[8]
	$1+\epsilon$	$O((\frac{n^3}{\epsilon})^{\frac{9}{\epsilon^2}})$	[2]
$Pm\|\text{rej}\|C_{\max}+\sum\delta_j$	$1+\epsilon$	$O((\frac{n}{\epsilon})^m)$	[2]
$1\|r_j,\text{rej}\|C_{\max}+\sum\delta_j$	2	$O(n^2)$	[16]
	2	$O(n\log n)$	[12]
	$\frac{4}{3}$	$O(n^2)$	[8]
	$1+\epsilon$	$O(\frac{n^3}{\epsilon})$	[16]
$P2\|r_j,\text{rej}\|C_{\max}+\sum\delta_j$	$\frac{3}{2}+\epsilon$	$O((\frac{n}{\epsilon})^2)$	[19]
$P\|r_j,\text{rej}\|C_{\max}+\sum\delta_j$	2	$O(n^2)$	[15]
	2	$O(n\log n)$	[18]
	$1+\max\{\frac{\sqrt{5}-1}{2},1-\frac{1}{m}\}$	$O(n^2\log n)$	[8]
	$1+\epsilon$	$O(n\log n+m^{O(\frac{1}{\epsilon^2}\log\frac{1}{\epsilon})})$	[18]
$Pm\|r_j,\text{rej}\|C_{\max}+\sum\delta_j$	$1+\epsilon$	$O(\frac{n^{2m+1}}{\epsilon^m})$	[15]
	$1+\epsilon$	$O(n\log n+(\frac{1}{\epsilon})^{3m+6})$	[18]

in the literature, a schedule in our problems needs to give not only a partition which divides the job set into sets of rejected jobs and accepted jobs, but also a resource matching of the accepted jobs. We use a *partition-matching scheme* to represent a partition of job set together with a resource matching for accepted jobs. How to find a suitable partition-matching scheme is the key point of our problems.

The remainder of this paper is organized as follows. Section 2 shows some preliminaries. Section 3 presents two approximation algorithms for problem ($\mathcal{P}1$). Section 4 presents a polynomial time optimal algorithm for problem ($\mathcal{P}2$). Section 5 concludes the paper and discusses some possible future researches.

Table 3. Overview of previous results on single parallel-batch machine

Problem	Ratio	Time complexity	Reference
$1\|\text{p-batch},b<n,\text{rej}\|C_{\max}+\sum\delta_j$	1	$O(n^2)$	[9]
$1\|\text{p-batch},b<n,r_j,\text{rej}\|C_{\max}+\sum\delta_j$	$3-\frac{2}{b}$	$O(n\log n)$	[11]
	2	$O(n^3)$	[9]
	$1+\epsilon$	$O(n(\frac{1}{\epsilon}+1)b^{\frac{1}{\epsilon}+1}(\frac{2n^2}{\epsilon})^{\frac{2}{\epsilon}+2})$	[9]
	$1+\epsilon$	$O((\frac{1}{\epsilon})^{\frac{2}{\epsilon}+2}n^{\frac{3}{\epsilon}+4}b^{\frac{1}{\epsilon}})$	[3]
	$1+\epsilon$	$O(n\log n+\frac{b}{\epsilon^{\frac{3}{\epsilon}+7}})$	[11]
$1\|\text{p-batch},b\geq n,r_j,\text{rej}\|C_{\max}+\sum\delta_j$	2	$O(n^2)$	[10]
	$1+\epsilon$	$O(\frac{n^4}{\epsilon})$	[10]

2 Preliminaries

In this section, we introduce graph theory terminology and a bipartite graph matching model for resource matching. Further, we give some notations and present two main rules to get partition-matching schemes.

2.1 Graph Theory Terminology

For a graph G, let $V = V(G)$ and $E = E(G)$ be the sets of vertices and edges, respectively. For an edge e, we obtain graph $G \setminus \{e\}$ by deleting e from G but leaving the vertices and the remaining edges intact. For a vertex v, we obtain graph $G - \{v\}$ by deleting from G the vertex v together with all the edges incident with v. Graph G is bipartite if the vertex set can be partitioned into two subsets V_1 and V_2 so that every edge in E has one vertex in V_1 and one vertex in V_2. A matching is a set of pairwise disjoint edges. For a matching M, each vertex incident with an edge of M is said to be covered by M. A perfect matching is a matching covering all vertices. For a weighted graph, the weight of a matching M is the total weight of all edges in M. A minimum weight perfect matching is a perfect matching with the minimum weight. The minimum weight perfect matching problem is to find a minimum weight perfect matching or show that there is no perfect matching. The minimum weight perfect matching problem in a weighted bipartite graph $G^{\omega} = (V, E; \omega)$ can be solved in $O(|V||E| + |V|^2 \log |V|)$ time by Hungarian method [6].

For a weighted bipartite graph $G^{\omega} = (V_1 \bigcup V_2, E; \omega)$ with $|V_1| \leq |V_2|$, we call a matching M a *cover matching* if M covers all vertices in V_1. A minimum weight cover matching is a cover matching with the minimum weight. The minimum weight cover matching problem is to find a minimum weight cover matching or show that there is no cover matching in graph G^{ω}. Construct a new graph by adding $|V_2| - |V_1|$ dummy vertices in V_1 and a zero weight edge between each dummy vertex and each vertex in V_2 in graph G^{ω}. Obviously, The minimum weight cover matching problem in G^{ω} is equivalent to the minimum weight perfect matching problem in the new graph. Thus, the minimum weight cover matching problem can be solved in $O(|V_2|(|E| + (|V_2| - |V_1|)|V_2|) + |V_2|^2 \log |V_2|)$ time. Let $M(G^{w})$ be a minimum weight cover matching of graph G^{ω}, if any.

2.2 Problem Transformation

We construct an auxiliary bipartite graph $H = (\mathcal{J} \cup (\mathcal{R} \cup \mathcal{R}'), E \cup E')$ as follows.

- $\mathcal{J}$ and $(\mathcal{R} \cup \mathcal{R}')$ are two components of the vertex set, where $\mathcal{J} = \{J_1, \ldots, J_n\}$ is the job set, $\mathcal{R} = \{R_1, \ldots, R_s\}$ is the resource set, and $\mathcal{R}' = \{R_{s+1}, \ldots, R_{s+n}\}$ is a set of n kinds of virtual resources.
- $E \cup E'$ is the edge set, where $E = \{e_{j,k} | 1 \leq j \leq n, 1 \leq k \leq s\}$ and $E' = \{e_{j,s+j} | 1 \leq j \leq n\}$.

Obviously, there is a one to one mapping between all cover matchings and all partition-matching schemes. We use a cover matching M to represent a partition-matching scheme as below. Job J_j is rejected if $e_{j,s+j} \in M$. Job J_j is accepted and matched with resource R_k if $e_{j,k} \in M$ and $1 \le k \le s$.

Define a weight function ω on edge set $E \cup E'$ in graph H, where $\omega(e_{j,k}) = \frac{p_{(j,k)}}{mb}$ for an edge $e_{j,k} \in E$ and $\omega(e_{j,s+j}) = \delta_j$ for an edge $e_{j,s+j} \in E'$.

Let $H^\omega = (\mathcal{J} \cup (\mathcal{R} \cup \mathcal{R}'), E \cup E'; \omega)$ be the corresponding weighted bipartite graph. Let $\omega(M)$ be the weight of a matching M in graph H^ω, i.e., $\omega(M) = \sum_{e_{j,k} \in M, 1 \le k \le s} \frac{p_{(j,k)}}{mb} + \sum_{e_{j,s+j} \in M} \delta_j$.

For an arbitrary schedule σ, we define the following notations.

- $\mathcal{J}_A(\sigma)$, the set of accepted jobs.
- $\mathcal{J}_R(\sigma)$, the set of rejected jobs.
- $\sigma(j)$, the index of the resource consumed by accepted job J_j.
- $p_{(j,\sigma(j))}$, the processing time of accepted job J_j.
- β_σ, the number of batches.
- $\mathcal{B}_1(\sigma), \ldots, \mathcal{B}_{\beta_\sigma}(\sigma)$, the batches of $\mathcal{J}_A(\sigma)$.
- $p(\mathcal{B}(\sigma)) = \max_{J_j \in \mathcal{B}(\sigma)} p_{(j,\sigma(j))}$, the processing time of batch $\mathcal{B}(\sigma)$.
- $M(\sigma) = \{e_{j,j+s} | J_j \in \mathcal{J}_R(\sigma)\} \bigcup \{e_{j,k} | J_j \in \mathcal{J}_A(\sigma), \sigma(j) = k\}$, the cover matching corresponding to the partition-matching scheme of σ.
- $C_j(\sigma)$, the completion time of accepted job J_j.
- $C_{\max}(\sigma) = \max_{J_j \in \mathcal{J}_A(\sigma)} C_j(\sigma)$, the maximum completion time of accepted jobs.
- $f(\sigma) = C_{\max}(\sigma) + \sum_{J_j \in \mathcal{J}_R(\sigma)} \delta_j$, the objective value of σ.

Let π_1 and π_2 be optimal schedules of problems $(\mathcal{P}1)$ and $(\mathcal{P}2)$, respectively. In this paper, we assume that $\mathcal{J}_A(\pi_1) \neq \emptyset$ and $\mathcal{J}_A(\pi_2) \neq \emptyset$. Otherwise, we can reject all jobs to get an optimal schedule easily. Without loss of generality, we assume that J_λ is a job with the largest processing time $p_{(\lambda,\pi_1(\lambda))}$ in $\mathcal{J}_A(\pi_1)$ and J_η is a job with the largest processing time $p_{(\eta,\pi_2(\eta))}$ in $\mathcal{J}_A(\pi_2)$.

2.3 Two Rules

For an edge $e_{j',k'} \in E$, we denote $H_{j',k'} = (H - \{J_{j'}, R_{k'}\}) \setminus E_{j',k'}$ where $E_{j',k'} = \{e_{j,k} | e_{j,k} \in E, p_{(j,k)} > p_{(j',k')}\}$. In the following, we give two main rules to get the cover matchings.

Rule 1: Minimum weight cover matching (WM)

Find a minimum weight cover matching $M(H^\omega)$ of graph H^ω and let the cover matching be $M(H^\omega)$.

Rule 2: Modified Minimum weight cover matching (MWM)

For a given edge $e_{j',k'} \in E$, construct a cover matching $M_{j',k'}$ as follows. Find a minimum weight cover matching $M(H^\omega_{j',k'})$ of graph $H^\omega_{j',k'}$ and let $M_{j',k'} = M(H^\omega_{j',k'}) \bigcup \{e_{j',k'}\}$.

The time complexities of Rules 1 and 2 are both $O((n+s)(s^2+2ns+n) + (n+s)^2 \log(n+s))$.

Lemma 1. *(i)* $\omega(M(H^\omega)) \leq \omega(M(\pi_1))$. *(ii)* $\omega(M_{\lambda,\pi_1(\lambda)}) \leq \omega(M(\pi_1))$.

Proof. (i) Since $M(\pi_1)$ and $M(H^\omega)$ are both cover matchings of graph H^ω, we have

$$\omega(M(H^\omega)) \leq \omega(M(\pi_1)).$$

(ii) Since $M(\pi_1)$ is a cover matching of graph H^ω and J_λ is a job with the largest processing time in $\mathcal{J}_A(\pi_1)$, we know that $M(\pi_1) \setminus \{e_{\lambda,\pi_1(\lambda)}\}$ is a cover matching of graph $H^\omega_{\lambda,\pi_1(\lambda)}$. Since $M(H^\omega_{\lambda,\pi_1(\lambda)})$ is a minimum weight cover matching of graph $H^\omega_{\lambda,\pi_1(\lambda)}$, we have

$$\omega(M(H^\omega_{\lambda,\pi_1(\lambda)})) \leq \omega(M(\pi_1) \setminus \{e_{\lambda,\pi_1(\lambda)}\}).$$

Thus, we have

$$\begin{aligned}\omega(M_{\lambda,\pi_1(\lambda)}) &= \omega(e_{\lambda,\pi_1(\lambda)}) + \omega(M(H^\omega_{\lambda,\pi_1(\lambda)})) \\ &\leq \omega(e_{\lambda,\pi_1(\lambda)}) + \omega(M(\pi_1) \setminus \{e_{\lambda,\pi_1(\lambda)}\}) = \omega(M(\pi_1)).\end{aligned}$$

□

3 Approximation Algorithms for Problem ($\mathcal{P}1$)

Since problem $P||C_{\max}$ is strongly NP-hard [4], problem ($\mathcal{P}1$) is strongly NP-hard even when $b = 1$ and $p_{(j,k)} = p_j$ for $j = 1, \ldots, n; k = 1, \ldots, s$. In this section, we give two approximation algorithms for problem ($\mathcal{P}1$) with worst case ratios mb and $2 - \frac{1}{mb}$, respectively. The following two algorithms use the classical full batch largest processing time (FBLPT) algorithm [7] for problem $1|\text{p-batch}, b < n|C_{\max}$ and list scheduling (LS) algorithm for problem $P||C_{\max}$ [5]. FBLPT algorithm is a batching rule which sorts the jobs in non-increasing order of their processing times and assigns every b jobs by the sequence into a batch. LS algorithm is a greedy algorithm which assigns the job in an arbitrary list to the machine with the minimum assigned load. The following lemma is a property of FBLPT algorithm provided by Ou et al. [11].

Lemma 2. *Let $\mathcal{J}'$ be a job set. Each job $J_j \in \mathcal{J}'$ has a fixed processing time p_j. If jobs are grouped into batches $\mathcal{B}_1, \ldots, \mathcal{B}_{\lceil \frac{|\mathcal{J}'|}{b} \rceil}$ by FBLPT algorithm, then*

$$\sum_{l=1}^{\lceil \frac{|\mathcal{J}'|}{b} \rceil} p(\mathcal{B}_l) \leq (1 - \frac{1}{b}) \max_{J_j \in \mathcal{J}'} p_j + \frac{1}{b} \sum_{J_j \in \mathcal{J}'} p_j.$$

The description of the two approximation algorithms, called WM-FBLPT and MWM-FBLPT, are as below.

The time complexities of Algorithm WM-FBLPT and MWM-FBLPT are $O((n+s)(s^2+2ns+n)+(n+s)^2\log(n+s))$ and $O(ns((n+s)(s^2+2ns+n)+(n+s)^2\log(n+s)))$, respectively. Let σ_1 and σ_2 be the schedules delivered by Algorithms WM-FBLPT and MWM-FBLPT, respectively. Let $\sigma_2^* = \sigma_2(e_{\lambda,\pi_1(\lambda)})$ be the schedule $\sigma_2(e_{j',k'})$ whose cover matching is $M_{\lambda,\pi_1(\lambda)}$. Obviously, J_λ is a job with the largest processing time $p_{(\lambda,\sigma_2^*(\lambda))}$ in $\mathcal{J}_A(\sigma_2^*)$ and $p_{(\lambda,\sigma_2^*(\lambda))} = p_{(\lambda,\pi_1(\lambda))}$.

Algorithm. WM-FBLPT

1. Run WM to get the cover matching $M(H^{\omega})$.
2. Group the accepted jobs into batches by FBLPT algorithm.
3. Treat each batch as a job, and run LS algorithm.

Algorithm. MWM-FBLPT

1. For each edge $e_{j',k'} \in E$, construct a schedule $\sigma_2(e_{j',k'})$ as follow.
   1.1 Run MWM to get the cover matching $M_{j',k'}$.
   1.2 Group the accepted jobs into batches by FBLPT algorithm.
   1.3 Treat each batch as a job, and run LS algorithm.
2. Choose a schedule with the minimum objective value among all schedules $\sigma_2(e_{j',k'})$.

Lemma 3. *For each feasible schedule σ of problem $(\mathcal{P}1)$, $f(\sigma) \geq \max\{\frac{1}{mb}\sum_{J_j\in\mathcal{J}_A(\sigma)} p_{(j,\sigma(j))} + \sum_{J_j\in\mathcal{J}_R(\sigma)}\delta_j,\ \max_{J_j\in\mathcal{J}_A(\sigma)} p_{(j,\sigma(j))} + \sum_{J_j\in\mathcal{J}_R(\sigma)}\delta_j\}$.*

Proof. Since $C_{\max}(\sigma) \geq \max\{\frac{1}{m}\sum_{l=1}^{\beta_\sigma} p(\mathcal{B}_l(\sigma)),\ \max_{J_j\in\mathcal{J}_A(\sigma)} p_{(j,\sigma(j))}\}$, we have

$$
\begin{aligned}
f(\sigma) &= C_{\max}(\sigma) + \sum_{J_j\in\mathcal{J}_R(\sigma)}\delta_j \\
&\geq \max\{\frac{1}{m}\sum_{l=1}^{\beta_\sigma} p(\mathcal{B}_l(\sigma)) + \sum_{J_j\in\mathcal{J}_R(\sigma)}\delta_j,\ \max_{J_j\in\mathcal{J}_A(\sigma)} p_{(j,\sigma(j))} + \sum_{J_j\in\mathcal{J}_R(\sigma)}\delta_j\}.
\end{aligned} \tag{1}
$$

From the definition of the batch processing time, we have

$$
p(\mathcal{B}_l(\sigma)) \geq \frac{1}{b}\sum_{J_j\in\mathcal{B}_l(\sigma)} p_{(j,\sigma(j))},\ l = 1,\ldots,\beta_\sigma. \tag{2}
$$

Summing up the inequality (2) for $l = 1,\ldots,\beta_\sigma$, we have

$$
\sum_{l=1}^{\beta_\sigma} p(\mathcal{B}_l(\sigma)) \geq \frac{1}{b}\sum_{l=1}^{\beta_\sigma}\sum_{J_j\in\mathcal{B}_l(\sigma)} p_{(j,\sigma(j))} = \frac{1}{b}\sum_{J_j\in\mathcal{J}_A(\sigma)} p_{(j,\sigma(j))}, \tag{3}
$$

as $\sum_{l=1}^{\beta_\sigma}\sum_{J_j\in\mathcal{B}_l(\sigma)} p_{(j,\sigma(j))} = \sum_{J_j\in\mathcal{J}_A(\sigma)} p_{(j,\sigma(j))}$. Combining with inequalities (1) and (3), the lemma holds. □

Since J_λ is the largest job in $\mathcal{J}_A(\pi_1)$, we have $\max_{J_j\in\mathcal{J}_A(\pi_1)} p_{(j,\pi_1(j))} = p_{(\lambda,\pi_1(\lambda))}$. Thus,

$$
f(\pi_1) \geq \max\{\frac{1}{mb}\sum_{J_j\in\mathcal{J}_A(\pi_1)} p_{(j,\pi_1(j))} + \sum_{J_j\in\mathcal{J}_R(\pi_1)}\delta_j,\ p_{(\lambda,\pi_1(\lambda))} + \sum_{J_j\in\mathcal{J}_R(\pi_1)}\delta_j\}. \tag{4}
$$

Table 4. The partial processing times in $\mathcal{I}_1$

Job	Processing time
J_1	$p_{(1,1)} = 1 + \epsilon,\ p_{(1,2)} = mb$
$J_j\ (j = 2, \ldots, mb)$	$p_{(j,j)} = 1,\ p_{(j,j+1)} = 0$

Based on Lemma 1, inequality (4), $p_{(\lambda,\sigma_2^*(\lambda))} = p_{(\lambda,\pi_1(\lambda))}$ and the definition of weight function ω, the following lemma holds.

Lemma 4. *(i)* $f(\pi_1) \geq \frac{1}{mb} \sum_{J_j \in \mathcal{J}_A(\sigma_1)} p_{(j,\sigma_1(j))} + \sum_{J_j \in \mathcal{J}_R(\sigma_1)} \delta_j$.

(ii) $f(\pi_1) \geq \max\{\frac{1}{mb} \sum_{J_j \in \mathcal{J}_A(\sigma_2^*)} p_{(j,\sigma_2^*(j))} + \sum_{J_j \in \mathcal{J}_R(\sigma_2^*)} \delta_j,\ p_{(\lambda,\sigma_2^*(\lambda))} + \sum_{J_j \in \mathcal{J}_R(\pi_1)} \delta_j\}$.

Theorem 1. *For problem ($\mathcal{P}1$), the worst case ratio of Algorithm WM-FBLPT is mb.*

Proof. Due to Lemma 4 (i), we have

$$
\begin{aligned}
f(\sigma_1) &= C_{\max}(\sigma_1) + \sum_{J_j \in \mathcal{J}_R(\sigma_1)} \delta_j \\
&\leq \sum_{l=1}^{\beta_{\sigma_1}} p(\mathcal{B}_l(\sigma_1)) + \sum_{J_j \in \mathcal{J}_R(\sigma_1)} \delta_j \\
&\leq \sum_{J_j \in \mathcal{J}_A(\sigma_1)} p(j, \sigma_1(j)) + \sum_{J_j \in \mathcal{J}_R(\sigma_1)} \delta_j \\
&\leq (mb) f(\pi_1).
\end{aligned}
$$

To show the tightness, we give an instance $\mathcal{I}_1$ with mb jobs and $mb + 1$ kinds of resources. The partial processing times of jobs are shown in Table 4. The rejection costs and the other processing times of jobs are mb. The cover matchings corresponding to σ_1 and π_1 are $M(\sigma_1) = \{e_{1,2}, e_{2,3}, \ldots, e_{mb,mb+1}\}$ and $M(\pi_1) = \{e_{1,1}, e_{2,2}, \ldots, e_{mb,mb}\}$, respectively. So, $f(\sigma_1) = C_{\max}(\sigma_1) = mb$ and $f(\pi_1) = C_{\max}(\pi_1) = 1 + \epsilon$. Therefore, we have $\frac{f(\sigma_1)}{f(\pi_1)} \rightarrow mb$ when ϵ tends to zero. □

The following corollary is a direct result of Theorem 1 when $m = 1$ and $b = 1$.

Corollary 1. *Algorithm WM-FBLPT solves problem $1|\mathcal{R}, rej|C_{\max} + \sum \delta_j$ in $O((n+s)(s^2+2ns+n)+(n+s)^2\log(n+s))$ time.*

Theorem 2. *For problem ($\mathcal{P}1$), the worst case ratio of Algorithm MWM-FBLPT is $2 - \frac{1}{mb}$.*

Proof. Let $\mathcal{B}_x(\sigma_2^*)$ be the batch which starts at time t and completes at time $C_{\max}(\sigma_2^*)$ in schedule σ_2^*. Since all machines must be busy at all times before time t, we have

$$mt + p(\mathcal{B}_x(\sigma_2^*)) \leq \sum_{l=1}^{\beta_{\sigma_2^*}} p(\mathcal{B}_l(\sigma_2^*)). \tag{5}$$

Since J_λ is the largest job in $\mathcal{J}_A(\sigma_2^*)$, we have

$$p_{(j,\sigma_2^*(j))} \leq p_{(\lambda,\sigma_2^*(\lambda))}, \text{ for } J_j \in \mathcal{J}_A(\sigma_2^*). \tag{6}$$

Due to Lemma 2, we have

$$\sum_{l=1}^{\beta_{\sigma_2^*}} p(\mathcal{B}_l(\sigma_2^*)) \leq (1 - \frac{1}{b})p_{(\lambda,\sigma_2^*(\lambda))} + \frac{1}{b} \sum_{J_j \in \mathcal{J}_A(\sigma_2^*)} p_{(j,\sigma_2^*(j))}, \tag{7}$$

as jobs in $\mathcal{J}_A(\sigma_2^*)$ are grouped into batches by FBLPT algorithm. Combining with inequalities (5), (6) and (7), we have

$$\begin{aligned}
C_{\max}(\sigma_2^*) &= t + p(\mathcal{B}_x(\sigma_2^*)) \\
&\leq \frac{1}{m} \sum_{l=1}^{\beta_{\sigma_2^*}} p(\mathcal{B}_l(\sigma_2^*)) + (1 - \frac{1}{m}) p(\mathcal{B}_x(\sigma_2^*)) \\
&\leq \frac{1}{m}((1 - \frac{1}{b})p_{(\lambda,\sigma_2^*(\lambda))} + \frac{1}{b} \sum_{J_j \in \mathcal{J}_A(\sigma_2^*)} p_{(j,\sigma_2^*(j))}) + (1 - \frac{1}{m}) p_{(\lambda,\sigma_2^*(\lambda))} \\
&= \frac{1}{mb} \sum_{J_j \in \mathcal{J}_A(\sigma_2^*)} p_{(j,\sigma_2^*(j))} + (1 - \frac{1}{mb}) p_{(\lambda,\sigma_2^*(\lambda))}.
\end{aligned}$$

That is,

$$C_{\max}(\sigma_2^*) \leq \frac{1}{mb} \sum_{J_j \in \mathcal{J}_A(\sigma_2^*)} p_{(j,\sigma_2^*(j))} + (1 - \frac{1}{mb}) p_{(\lambda,\sigma_2^*(\lambda))}. \tag{8}$$

Due to Lemma 4 (ii), inequality (8) and $f(\sigma_2) \leq f(\sigma_2^*)$, we have

$$\begin{aligned}
f(\sigma_2) \leq f(\sigma_2^*) &= C_{\max}(\sigma_2^*) + \sum_{J_j \in \mathcal{J}_R(\sigma_2^*)} \delta_j \\
&\leq \frac{1}{mb} \sum_{J_j \in \mathcal{J}_A(\sigma_2^*)} p_{(j,\sigma_2^*(j))} + \sum_{J_j \in \mathcal{J}_R(\sigma_2^*)} \delta_j + (1 - \frac{1}{mb}) p_{(\lambda,\sigma_2^*(\lambda))} \\
&\leq (2 - \frac{1}{mb}) f(\pi_1).
\end{aligned}$$

To show the tightness, we give an instance $\mathcal{I}_2$ with mb jobs and mb kinds of resources. The partial processing times and rejection costs of jobs are shown in Table 5. The other processing times of jobs are both $2mb$.

Table 5. The partial processing times and rejection costs in $\mathcal{I}_2$

Job	Processing time	Rejection cost
J_1	$p_{(1,1)} = 1$	$\delta_1 = mb$
$J_j (j = 2, \ldots, mb)$	$p_{(j,j)} = 1$	$\delta_j = \frac{1}{mb} - \epsilon$

The cover matchings corresponding to σ_2 and π_1 are $M(\sigma_2) = \{e_{1,1}, e_{2,mb+2}, e_{3,mb+3}, \ldots, e_{mb,2mb}\}$ and $M(\pi_1) = \{e_{1,1}, e_{2,2}, \ldots, e_{mb,mb}\}$, respectively. So, $f(\sigma_2) = 2 - \frac{1}{mb} - (mb-1)\epsilon$ and $f(\pi_1) = 1$. Therefore, we have $\frac{f(\sigma_2)}{f(\pi_1)} \to 2 - \frac{1}{mb}$ when ϵ tends to zero. □

Remark. For problem $1|\text{p-batch}, b < n, \mathcal{R}, \text{rej}|C_{\max} + \sum \delta_j$, the worst case ratios of Algorithm WM-FBLPT and MWM-FBLPT are b and $2 - \frac{1}{b}$, respectively. But the computational complexity of this problem is unknown.

4 An Optimal Algorithm for Problem ($\mathcal{P}2$)

When the batch capacity is unbounded, there is an optimal schedule such that all accepted jobs are processed in a batch for problem $P|\text{p-batch}, b \geq n, \mathcal{R}, \text{rej}|C_{\max} + \sum \delta_j$. Therefore, problems $P|\text{p-batch}, b \geq n, \mathcal{R}, \text{rej}|C_{\max} + \sum \delta_j$ and ($\mathcal{P}2$) are equivalent. In this section, we give a polynomial time algorithm to solve problem ($\mathcal{P}2$) optimally. The algorithm, called EM, is as below.

Algorithm EM

1. For each edge $e_{j',k'} \in E, j' = 1, \ldots, n; k' = 1, \ldots, s$, construct a schedule $\sigma_3(e_{j',k'})$ as follow.
   - 1.1 Define a new weight function ω' on the edge set $E(H_{j',k'})$ of graph $H_{j',k'}$ where $\omega'(e_{j,k}) = 0$ for an edge $e_{j,k}$ in $E \bigcap E(H_{j',k'})$ and $\omega'(e_{j,s+j}) = \delta_j$ for an edge $e_{j,s+j}$ in $E' \bigcap E(H_{j',k'})$.
   - 1.2 Find a minimum weight cover matching $M(H^{\omega'}_{j',k'})$ of grpah $H^{\omega'}_{j',k'}$.
   - 1.3 Let $M'_{j',k'} = M(H^{\omega'}_{j',k'}) \bigcup \{e_{j',k'}\}$ be the cover matching. Assign all accepted jobs in a batch.
2. Choose a schedule with the minimum objective value among all schedules $\sigma_3(e_{j',k'})$.

Let σ_3 be the schedule delivered by Algorithm EM. Let $\sigma_3^* = \sigma_3(e_{\eta,\pi_2(\eta)})$ be the schedule $\sigma_3(e_{j',k'})$ whose cover matching is $M'_{\eta,\pi_2(\eta)}$. Obviously, J_η is a job with the largest processing time $p_{(\eta,\sigma_3^*(\eta))}$ in $\mathcal{J}_A(\sigma_3^*)$ and $p_{(\eta,\sigma_3^*(\eta))} = p_{(\eta,\pi_2(\eta))}$. Let $\omega'(M)$ be the weight of a matching M in graph $H^{\omega'}_{j',k'}$, i.e., $\omega'(M) = \sum_{e_{j,s+j} \in M} \delta_j$.

Theorem 3. *Algorithm EM solves problem ($\mathcal{P}2$) in $O(ns((n+s)(s^2+2ns+n) + (n+s)^2 \log(n+s)))$ time.*

Proof. Since $M(\pi_2)$ is a cover matching of graph $H^{\omega'}$ and J_η is a job with the largest processing time in $\mathcal{J}_A(\pi_2)$, we know that $M(\pi_2) \setminus \{e_{\eta,\pi_2(\eta)}\}$ is a cover matching of graph $H^{\omega'}_{\eta,\pi_2(\eta)}$. Since $M(H^{\omega'}_{\eta,\pi_2(\eta)})$ is a minimum weight cover matching of graph $H^{\omega'}_{\eta,\pi_2(\eta)}$, we have

$$\omega'(M(H^{\omega'}_{\eta,\pi_2(\eta)})) \leq \omega'(M(\pi_2) \setminus \{e_{\eta,\pi_2(\eta)}\}).$$

From the definition of weight function ω', we have

$$\omega'(M(H^{\omega'}_{\eta,\pi_2(\eta)})) = \sum_{J_j \in \mathcal{J}_R(\sigma_3^*)} \delta_j, \; \omega'(M(\pi_2) \setminus \{e_{\eta,\pi_2(\eta)}\}) = \sum_{J_j \in \mathcal{J}_R(\pi_2)} \delta_j.$$

Thus, we have

$$\begin{aligned} f(\sigma_3) \leq f(\sigma_3^*) &= p_{(\eta,\sigma_3^*(\eta))} + \sum_{J_j \in \mathcal{J}_R(\sigma_3^*)} \delta_j = p_{(\eta,\sigma_3^*(\eta))} + \omega'(M(H^{\omega'}_{\eta,\pi_2(\eta)})) \\ &\leq p_{(\eta,\pi_2(\eta))} + \omega'(M(\pi_2) \setminus \{e_{\eta,\pi_2(\eta)}\}) = p_{(\eta,\pi_2(\eta))} + \sum_{J_j \in \mathcal{J}_R(\pi_2)} \delta_j = f(\pi_2). \end{aligned}$$

That is, $f(\sigma_3) = f(\pi_2)$.

There are at most ns different schedules. Each schedule can be found in $O((n+s)(s^2+2ns+n)+(n+s)^2 \log(n+s))$ time. Therefore, the time complexity of Algorithm EM is $O(ns((n+s)(s^2+2ns+n)+(n+s)^2 \log(n+s)))$. □

5 Conclusions

We consider two scheduling problems with rejection and resource matching on m identical parallel-batch machines to minimize the sum of the makespan of the accepted jobs and the total rejection cost of the rejected jobs. For bounded batch capacity case, two approximation algorithms are provided with the worst case analysis. For unbounded batch capacity, we give a polynomial time optimal algorithm.

However, there are still a lot of problems worthy of further studying. For instance, the computational complexity of problem $1|\text{p-batch}, b < n, \mathcal{R}, \text{rej}|C_{\max} + \sum \delta_j$ needs to be explored. Designing approximation algorithms with better worst case ratios seems to be another attractive work. Furthermore, the generalization of our problems to other machine environments, such as uniform machines and unrelated machines, is also an interesting research topic.

Acknowledgement. The work is supported by the National Natural Science Foundation of China (Grant Nos. 12471305, 12401424 and 12271491) and Key Research Projects of Henan Higher Education Institutions [24A110014].

Declaration. The authors declare that they have no known competing financial interests.

References

1. Barketau, M., Pesch, E., Shafransky, Y.: Scheduling dedicated jobs with variative processing times. J. Comb. Optim. **31**(2), 774–785 (2016)
2. Bartal, Y., Leonardi, S., Marchetti-Spaccamela, A., Sgall, J., Stougie, L.: Multiprocessor scheduling with rejection. SIAM J. Discret. Math. **13**(1), 64–78 (2000)
3. Cao, Z.G., Yang, X.G.: A PTAS for parallel batch scheduling with rejection and dynamic job arrivals. Theoret. Comput. Sci. **410**(27–29), 2732–2745 (2009)
4. Garey, M.R., Johnson, D.S.: Computers and Intractability: A Guide to the Theory of NP-Completeness. W. H. Freeman and Company, San Francisco (1979)
5. Graham, R.L.: Bounds for certain multiprocessing anomalies. Bell Syst. Tech. J. **45**(9), 1563–1581 (1966)
6. Kuhn, H.W.: The hungarian method for the assignment problem. Naval Research Logistics Quarterly **2**(1–2), 83–97 (1955)
7. Lee, C.-Y.: Minimizing makespan on a single batch processing machine with dynamic job arrivals. Int. J. Prod. Res. **37**(1), 219–236 (1999)
8. Liu, P.H., Lu, X.W.: New approximation algorithms for machine scheduling with rejection on single and parallel machine. J. Comb. Optim. **40**(4), 929–952 (2020)
9. Lu, L.F., Cheng, T.C.E., Yuan, J.J., Zhang, L.Q.: Bounded single-machine parallel-batch scheduling with release dates and rejection. Computers & Operations Research **36**(10), 2748–2751 (2009)
10. Lu, L.F., Zhang, L.Q., Yuan, J.J.: The unbounded parallel batch machine scheduling with release dates and rejection to minimize makespan. Theoret. Comput. Sci. **396**(1–3), 283–289 (2008)
11. Ou, J.W., Lu, L.F., Zhong, X.L.: Parallel-batch scheduling with rejection: Structural properties and approximation algorithms. Eur. J. Oper. Res. **310**(3), 1017–1032 (2023)
12. Ou, J.W., Zhong, X.L., Li, C.-L.: Faster algorithms for single machine scheduling with release dates and rejection. Inf. Process. Lett. **116**(8), 503–507 (2016)
13. Ou, J.W., Zhong, X.L., Wang, G.Q.: An improved heuristic for parallel machine scheduling with rejection. Eur. J. Oper. Res. **241**(3), 653–661 (2015)
14. Shabtay, D., Gaspar, N., Kaspi, M.: A survey on offline scheduling with rejection. J. Sched. **16**, 3–28 (2013)
15. Zhang, L.Q., Lu, L.F.: Parallel-machine scheduling with release dates and rejection. 4OR **14**(2), 165–172 (2016)
16. Zhang, L.Q., Lu, L.F., Yuan, J.J.: Single machine scheduling with release dates and rejection. Eur. J. Oper. Res. **198**(3), 975–978 (2009)
17. Zhang, Y.Z.: A survey on job scheduling with rejection. *Operations Research Transactions*, 24(2), 2020
18. Zhong, X.L., Ou, J.W.: Improved approximation algorithms for parallel machine scheduling with release dates and job rejection. *4OR*, 15:387–406, 2017
19. Zhong, X.L., Pan, Z.M., Jiang, D.K.: Scheduling with release times and rejection on two parallel machines. J. Comb. Optim. **33**(3), 934–944 (2017)

Some Combinatorial Algorithms on the Independent Number of k-Regular Connected Hypergraphs

Zhuo Diao(✉) and Haoyang Zou

School of Statistics and Mathematics, Central University of Finance and Economics, Beijing 100081, China
diaozhuo@amss.ac.cn

Abstract. Let $H(V, E)$ be a k-regular connected hypergraph with rank R on n vertices and m edges. A set of vertices $S \subseteq V$ is an independent set if every two vertices in S are not adjacent. The independent number is the maximum cardinality of an independent set, denoted by $\alpha(H)$. In this paper, we prove the following inequality: $\alpha(H) \geq \frac{m-(k-2)n-1}{R}$, and the equality holds if and only if H is a hypertree with R-perfect matching. Based on the proofs, some combinatorial algorithms on the independent number are designed.

Keywords: hypergraphs · independent number · regular · rank

1 Introduction

A hypergraph is a generalization of a graph in which an edge can join any number of vertices. The loop is an edge which contains only one vertex. The multiple edges are edges which contain same vertices. A multiple hypergraph is a hypergraph allowing loops and multiple edges. Let $H(V, E)$ be a multiple hypergraph with vertex set V and edge set E. As for a graph, the order of H, denoted by n, is the number of vertices. The number of edges is denoted by m. The studies of this paper are applicable to multiple hypergraphs.

For each vertex $v \in V$, the degree $d(v)$ is the number of edges in E that contains v. The maximum degree is denoted by Δ. We say v is an isolated vertex of H if $d(v) = 0$. Hypergraph H is k-**regular** if each vertex's degree is k. The **rank** of H is the maximum number of vertices contained in any edge of H, denoted by R.

Let $k \geq 2$ be an integer. A cycle of length k, denoted as k-cycle, is a vertex-edge sequence $C = v_1e_1v_2e_2 \cdots v_ke_kv_1$ with: (1) $\{e_1, e_2, \ldots, e_k\}$ are distinct edges of H. (2) $\{v_1, v_2, \ldots, v_k\}$ are distinct vertices of H. (3) $\{v_i, v_{i+1}\} \subseteq e_i$ for each

This research work is supported by National Natural Science Foundation of China under Grant No. 11901605, the Fundamental Research Funds for the Central Universities, the disciplinary funding of Central University of Finance and Economics.

D. Du et al. (Eds.): COCOA 2024, LNCS 15434, pp. 465–476, 2025.
https://doi.org/10.1007/978-981-96-4445-2_38

$i \in [k]$, here $v_{k+1} = v_1$. For any vertex set $S \subseteq V$, we write $H \setminus S$ for the sub-hypergraph of H obtained from H by deleting all vertices in S and all edges incident with some vertices in S. For any edge set $A \subseteq E$, we write $H \setminus A$ for the sub-hypergraph of H obtained from H by deleting all edges in A and keeping vertices. If S is a singleton set $\{s\}$, we write $H \setminus s$ instead of $H \setminus \{s\}$. A hypertree $T(V, E)$ is a connected hypergraph with no cycles. In a hypertree $T(V, E)$, for any two distinct vertices $\{u, v\} \subseteq V$, there is an unique $u - v$ path P in T.

Given a hypergraph $H(V, E)$, a set of vertices $S \subseteq V$ is an independent set if any two vertices in S are not adjacent. The independent number is the maximum cardinality of an independent set, denoted by $\alpha(H)$. A set of edges $A \subseteq E$ is a matching if every two distinct edges have no common vertex. A perfect matching is a matching in which every vertex of the hypergraph is incident to exactly one edge of the matching.

1.1 Related Works

The problem of determining or estimating the independent number $\alpha(H)$ is fundamental in theoretical computer science, and has numerous applications. The problem of computing $\alpha(H)$ is known to be NP-hard. Chvátal [3] derived an algorithm for determining the upper bound of the independent number of a graph. Boppana and Halldórsson [2] proposed an $O(n/(\log n)^2)$-approximation algorithm for the maximum independent set where n is the vertex number. Fox and Pach [5] gave a subexponential time exact algorithm for computing the independent number of intersection graphs. Hermon, Sly and Zhang [9] proved the mixing time for sampling independent sets on n-vertex k-uniform hypergraphs is $O(n \log n)$ when the maximum degree Δ satisfies $\Delta \leq c2^{k/2}$.

Various bounds on the independent number for graphs have appeared in the literature. For example, Harant and Schiermeyer [8] gave the lower bound for a connected graph G with n vertices and m edges, that $\alpha(G) \geq \frac{1}{2}[(2m + n + 1) - \sqrt{(2m + n + 1)^2 - 4n^2}]$. Hansen and Zheng [7] proved some sharp bounds on the independent number expressed as functions of the vertex number and edge number.

Griggs [6] proved if G is a connected triangle-free graph on $n \geq 3$ vertices and if G is neither an odd cycle nor an odd path, then $\alpha(G) \geq \frac{n}{\Delta(\Delta+1)} + \sum_v \frac{1}{1+d(v)}$, which improved the bound $\sum_v \frac{1}{1+d(v)}$ discovered by Wei in [12] where $d(v)$ is the degree of vertex v. Shearer [11] also gave another form of lower bound for the independent number based on the degree sequence in triangle-free graphs.

For the independent number of hypergraphs, Bollobás, Daykin and Erdös [1] studied the properties of sets of independent edges of a hypergraph. Kostochka, Mubayi and Verstraëte [10] proved the lower bound of the independent number in uniform hypergraphs. The lower bounds for special k-uniform hypergraphs were established by Duke, Lefmann and Rödl [4].

1.2 Our Results

In this paper, for every k-regular hypergraph, we give a sharp lower bound for the independent number. The result is stated in Theorem 1.

Theorem 1. *For every k-regular connected hypergraph $H(V, E)$ with rank R on n vertices and m edges, $\alpha(H) \geq \frac{m-(k-2)n-1}{R}$ holds. Furthermore, the equality $\alpha(H) = \frac{m-(k-2)n-1}{R}$ holds if and only if $H(V, E)$ is a hypertree with R-perfect matching when $R \geq 3$.*

For a hypergraph H with maximum degree Δ, we can construct a Δ-regular hypergraph H' by adding some loops to the vertices and these operations keep the independent number. According to Theorem 1, the next corollary is instant:

Corollary 1. *Let $H(V, E)$ be a rank R connected hypergraph with maximum degree Δ on n vertices and m edges, $\alpha(H) \geq \frac{m-(\Delta-2)n-1}{R}$ holds. Furthermore, $\alpha(H) = \frac{m-(\Delta-2)n-1}{R}$ holds if and only if $H(V, E)$ is a regular hypertree with R-perfect matching when $R \geq 3$.*

The main content of the article is organized as follows:

- In Sect. 2, for regular hypergraphs, we prove the lower bound for the independent number in Theorem 1.
- In Sect. 3, for regular hypergraphs, we characterize the extremal hypergraphs with lower bound attained in Theorem 1.
- In Sect. 4, for a rank R connected hypergraph with maximum degree Δ on n vertices and m edges, a polynomial time algorithm is designed to compute an independent set with cardinality at least $\frac{m-(\Delta-2)n-1}{R}$.

2 The Lower Bound of Independent Number on Regular Hypergraphs

In this section, for every k-regular connected hypergraph, a lower bound of the independent number with rank is proved, as stated in Theorem 2. The content is organized as follows:

- In Definitions 1, 2 and 3, the operations of vertex eliminating and edge eliminating are introduced.
- In Lemma 1, for every k-regular connected hypertree, a lower bound of the independent number with rank is proved.
- In Lemma 2, for every k-regular connected hypergraph, an upper bound of the eliminating vertex feedback number is proved.
- In Theorem 2, for every k-regular connected hypergraph, a lower bound of the independent number with rank is proved.

Theorem 2. *For every k-regular connected hypergraph $H(V, E)$ with rank R, $\alpha(H) \geq \frac{m-(k-2)n-1}{R}$.*

For convenience, we introduce two useful operations: eliminating vertices and eliminating edges.

Definition 1. *In hypergraph $H(V, E)$, given a vertex set $S \subseteq V$, let $H /\!/ S$ be the sub-hypergraph induced by the vertex set $V \setminus S$. We write $H /\!/ S$ for eliminating vertex set S.*

Definition 2. *In hypergraph $H(V, E)$, given an edge set $A \subseteq E$, let $H /\!/ A$ be the sub-hypergraph induced by the vertex set $V \setminus A$. We write $H /\!/ A$ for eliminating edge set A (Fig. 1).*

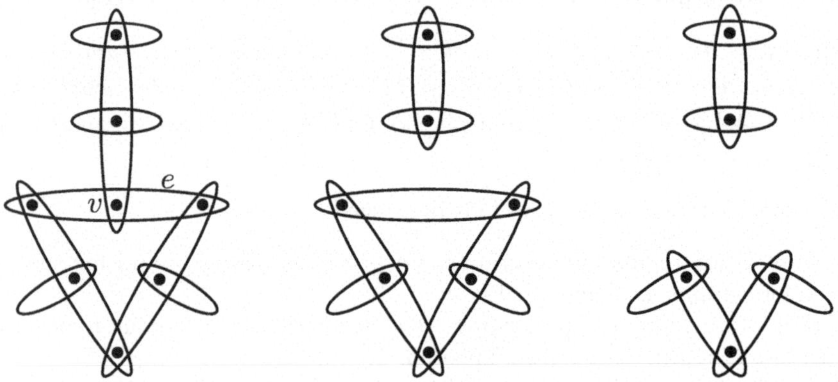

Fig. 1. The operations of vertex eliminating and edge eliminating

Definition 3. *Given a hypergraph $H(V, E)$, a set of vertices $D \subseteq V$ is called an eliminating vertex feedback set if $H /\!/ D$ has no cycle. The minimum cardinality of the eliminating vertex feedback set is called eliminating vertex feedback number, denoted as $D_c(H)$ (Fig. 2).*

Remark 1. The operation of vertex eliminating does not change the degree of other vertices in hypergraphs. Specifically, when a vertex eliminating operation is performed on a k-regular hypergraph, the resulting hypergraph is still k-regular.

Lemma 1. *For every hypertree $T(V, E)$ with rank R on n vertices, $\alpha(T) \geq \frac{n}{R}$.*

Proof. We prove this lemma by induction on the number of vertices.

When the number of vertices is 1, we have $\alpha(T) = \frac{n}{R} = 1$. Assume this result holds when the vertex number is smaller than n. When T contains $n \geq 2$ vertices, take a longest path $P = v_1e_1v_2 \ldots v_te_tv_{t+1}$ from T for $t \geq 1$. Considering the vertices in edge e_t without taking loops into account, v_t is the unique vertex

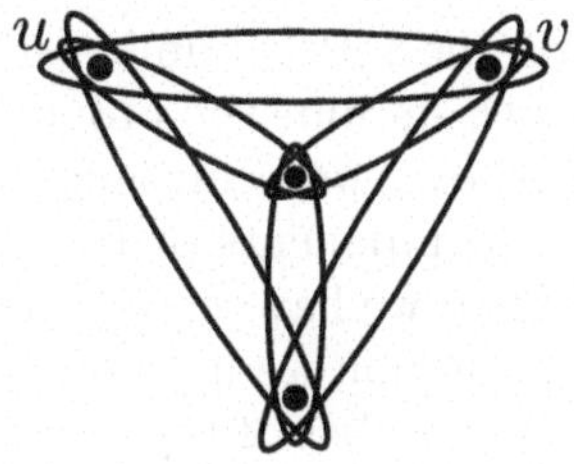

(a) H

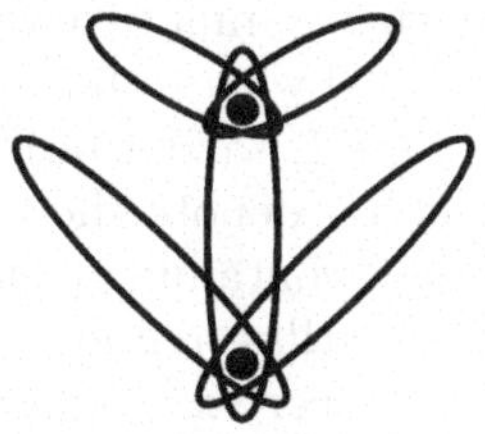

(b) $H /\!/ D$

Fig. 2. The minimum eliminating vertex feedback set $D = \{u, v\}$ of H

with degree more than 1. After eliminating the edge e_t, the components of $T /\!/ e_t$ are $\{T_i \mid 1 \leq i \leq c\}$. Assume that T_i contains m_i edges and n_i vertices. Since every edge of T_i contains at most R vertices, we have $\alpha(T_i) \geq \frac{n_i}{R}$ by induction. Thus,

$$\alpha(T /\!/ e_t) = \sum_{i=1}^{c} \alpha(T_i) \geq \sum_{i=1}^{c} \frac{n_i}{R} \geq \frac{n-R}{R},$$

where the last inequality holds as e_t contains at most R vertices.

On one side, for every independent set I of $T /\!/ e_t$, $I \cup \{v_{t+1}\}$ is an independent set in T, which means $\alpha(T) \geq \alpha(T /\!/ e_t) + 1$. On the other side, the optimal independent set I in T contains exactly one vertex in e_t, assume v_{t+1} without generality. Thus $I \setminus v_{t+1}$ is an independent set in $T /\!/ e_t$, which means $\alpha(T) - 1 \leq \alpha(T /\!/ e_t)$. Then, we have

$$\alpha(T) = \alpha(T /\!/ e_t) + 1 \geq \frac{n-R}{R} + 1 \geq \frac{n}{R}.$$

□

Remark 2. The above Lemma 1 implies a polynomial time algorithm to compute the optimal independent set for a hypertree T, as shown in Sect. 4 Algorithm 1.

Next for every k-regular connected hypergraph, we give an upper bound of the eliminating vertex feedback number.

Lemma 2. *Let $H(V, E)$ be a k-regular connected hypergraph, $D_c(H) \leq (k-1)n - m + 1$.*

Before proving the lemma above, we will prove a series of lemmas which are very useful.

Lemma 3. *For every k-regular connected hypergraph $H(V, E)$ with n vertices and m edges, we have $m \leq (k-1)n + 1$.*

Proof. We prove this lemma by induction on the vertex number n. When $n = 1$, $H(V, E)$ is an isolated vertex and $m \leq k$ holds. Assume the proposition holds for $n \leq t$. When $n = t + 1$, take arbitrarily a vertex v and consider the hypergraph $H' = H /\!\!/ v$. Let $c(H')$ denote the number of components of H' and $l(v)$ denote the number of edges containing exactly v. Then we have $c(H') + l(v) \leq k$. All of the components are still k-regular, and by induction we have $m_i \leq (k-1)n_i + 1$ where n_i and m_i are vertex number and edge number of the i-th component for $i = 1, \ldots, c(H')$. Thus, we have

$$\begin{aligned} m &= \sum_{i=1}^{c(H')} m_i + l(v) \leq \sum_{i=1}^{c(H')} [(k-1)n_i + 1] + l(v) \\ &= (k-1)(n-1) + c(H') + l(v) \leq (k-1)(n-1) + k = (k-1)n + 1. \quad \square \end{aligned}$$

Lemma 4. *For every k-regular connected hypergraph $H(V, E)$ on n vertices and m edges, $m = (k-1)n + 1$ if and only if $H(V, E)$ is a hypertree.*

Proof. **Sufficiency**: If H is a hypertree, we prove $m = (k-1)n+1$ by induction on the vertex number n. When $n = 1$, $H(V, E)$ is an isolated vertex and $m = k$ holds. Assume the lemma holds for $n \leq t$. When $n = t + 1$, take arbitrarily a vertex v and consider the hypergraph $H' = H /\!\!/ v$. Let $c(H')$ denote the number of components of H' and $l(v)$ denote the number of edges containing exactly v. Then we have $c(H') + l(v) = k$. All of the components are k-regular hypertrees, and by induction we have $m_i = (k-1)n_i + 1$ where n_i and m_i are vertex number and edge number of the i-th component for $i = 1, \ldots, c(H')$. Thus, we have

$$\begin{aligned} m &= \sum_{i=1}^{c(H')} m_i + l(v) = \sum_{i=1}^{c(H')} [(k-1)n_i + 1] + l(v) \\ &= (k-1)(n-1) + c(H') + l(v) = (k-1)(n-1) + k = (k-1)n + 1. \end{aligned}$$

Necessity: We prove the necessity part by contradiction. If H is not a hypertree, H contains a cycle C. Take arbitrary one vertex v in C and consider the hypergraph $H' = H /\!\!/ v$. Let $c(H')$ denote the number of components of H' and $l(v)$ denote the number of edges containing exactly v. Then we have $c(H') + l(v) \leq k - 1$. According to the proof of Lemma 3, we derive $m \leq (k-1)n$, which is a contradiction with $m = (k-1)n + 1$. Thus H is a hypertree. $\square$

Next we will prove a generalization of Lemma 2:

Lemma 5. *For every k-regular hypergraph $H(V, E)$ with p components, $D_c(H) \leq (k-1)n - m + p$.*

Proof. Pick arbitrary a minimum eliminating vertex feedback set $D \subseteq V$. Then $H /\!\!/ D$ is acyclic. Suppose that $H /\!\!/ D$ contains exactly t components $H_i = (V_i, E_i)$ with n_i vertices and m_i edges, $i = 1, \ldots, t$. It follows from Lemma 4 that $m_i = (k-1)n_i + 1$ for each $i \in [t]$. Suppose $D = \{v_1, \ldots, v_q\}$. Let $l(v_i)$ denote the number of edges containing exactly v_i in $H/\!\!/\{v_1,\dots,v_{i-1}\}$ for each $i \in [q]$. Thus, we have

$$m - \sum_{v\in D} l(v) = \sum_{i=1}^{t} m_i = \sum_{i=1}^{t}[(k-1)n_i + 1] = (k-1)(n - D_c(H)) + t,$$

which means $(k-1)D_c(H) = (k-1)n - m + \sum_{v\in D} l(v) + t$. To establish the lemma, it suffices to prove $t + \sum_{v\in D} l(v) \leq (k-2)D_c(H) + p$.

In case of $D_c(H) = 0$, we have $D = \varnothing$ and the result holds. In case of $D_c(H) \geq 1$, since D is a minimum eliminating vertex feedback set of H, for each $i \in [q]$, there is a cycle C_i in $H/\!\!/(D \setminus v_i)$ such that $v_i \in C_i$. Let $c(H')$ denote the number of components of H' and $l(v)$ denote the number of edges containing exactly v. Considering $H/\!\!/D$ being obtained from H by eliminating $v_1, v_2, \dots, v_q$ sequentially, for $i = 1, \dots, q$. Since $d(v_i) = k$, the presence of C_i implies that $c(H/\!\!/\{v_1,\dots,v_i\}) - c(H/\!\!/\{v_1,\dots,v_{i-1}\}) + l(v_i) \leq k-2$. Therefore, we have

$$t - p = c(H/\!\!/D) - c(H) = \sum_{i=1}^{q}[c(H/\!\!/\{v_1,\dots,v_i\}) - c(H/\!\!/\{v_1,\dots,v_{i-1}\})]$$
$$\leq \sum_{i=1}^{q}[k - 2 - l(v_i)] = (k-2)D_c(H) - \sum_{v\in D} l(v),$$

which completes the proof. □

Remark 3. The proof and conclusion in the above lemma applies to minimal eliminating vertex feedback sets: for every k-regular hypergraph $H(V,E)$ with p components, $D \subseteq V$ is a minimal vertex feedback set of H, $|\ D\ | \leq (k-1)n - m + p$.

Corollary 2. *For every k-regular connected hypergraph $H(V,E)$ with $D_c(H) = (k-1)n - m + 1$, $D \subseteq V$ is a minimum eliminating vertex feedback set. Suppose $D = \{v_1,\dots,v_q\}$. For each $i \in [q]$, $c(H/\!\!/\{v_1,\dots,v_i\}) - c(H/\!\!/\{v_1,\dots,v_{i-1}\}) + l(v_i) = k-2$ holds, where $c(H')$ is the number of components in H' and $l(v_i)$ is the number of edges containing exactly v_i in $H/\!\!/\{v_1,\dots,v_{i-1}\}$.*

Now we will prove the main theorem in this section, for every k-regular connected hypergraph, a lower bound of the independent number with rank is proved.

Theorem 3. *For every k-regular connected hypergraph $H(V,E)$ with rank R, $\alpha(H) \geq \frac{m-(k-2)n-1}{R}$.*

Proof. According to Lemma 2, we have $D_c(H) \leq (k-1)n - m + 1$. Suppose D is a minimum eliminating vertex feedback set. In hypergraphs, eliminating vertices will not increase the independent number, thus $\alpha(H) \geq \alpha(H/\!\!/D)$. Then we have

$$\alpha(H) \geq \alpha(H/\!\!/D) \geq \frac{n - D_c(H)}{R} \geq \frac{n - [(k-1)n - m + 1]}{R} = \frac{m - (k-2)n - 1}{R}.$$

□

Remark 4. For every k-regular connected hypergraph $H(V, E)$ with rank R on n vertices and m edges, Theorem 2 implies a polynomial-time algorithm for computing an independent set with cardinality at least $\frac{m-(k-2)n-1}{R}$.

3 The Extremal Regular Hypergraphs on Independent Number

In this section, we introduce a class of matching called R-perfect matching where R is the rank of hypergraphs and we prove R-perfect matching hypertrees are exactly the extremal hypergraphs when $R \geq 3$, as stated in Theorem 4.

Theorem 4. *For every k-regular connected hypergraph $H(V, E)$ with rank $R \geq 3$, $\alpha(H) = \frac{m-(k-2)n-1}{R}$ if and only if H is an R-perfect matching hypertree.*

The content is organized as follows:

- In Definitions 4, 5 and 6, a class of matching called R-perfect matching is introduced.
- In Lemma 7, we prove R-perfect matching hypertrees are exactly the extremal hypertrees with $\alpha(T) = \frac{n}{R}$.
- In Theorem 4, we prove R-perfect matching hypertrees are exactly the extremal hypergraphs with $\alpha(H) = \frac{m-(k-2)n-1}{R}$ when $R \geq 3$.

Definition 4. *In a k-regular hypertree $T(V, E)$ with rank R, a matching $M = \{e_1, e_2, \ldots, e_t\}$ is called an R-perfect matching, if $e_i \cap e_j = \varnothing$, $|e_i| = R$ and $t = \frac{n}{R}$. The hypertree is called R-perfect matching hypertree if it has an R-perfect matching.*

Definition 5. *In an R-perfect matching hypertree $T(V, E)$, the edges in R-perfect matching M are called matched edges and other edges are called non-matched edges. The edges connecting different matched edges are called adjacent edges. The non-matched edges are formed by adjacent edges and loops.*

Definition 6. *For an R-perfect matching hypertree $T(V, E)$, the structure tree describes the structure of T as formed by its matched and non-matched edges. Let A denote the set of adjacent edges of T, and B denote the set of matched edges of T. We consider the edges of $A \cup B$ as the vertices, Then we have a natural structure tree $T'(A \cup B, E')$ on $A \cup B$ formed by the edges $e'(a, b)$ with $a \in A$, $b \in B$ and $a \cap b \neq \varnothing$ (Fig. 3).*

Lemma 6. *For every k-regular R-perfect matching hypertree $T(V, E)$, R-perfect matching is unique.*

Proof. We prove this lemma by induction on the matching number t.

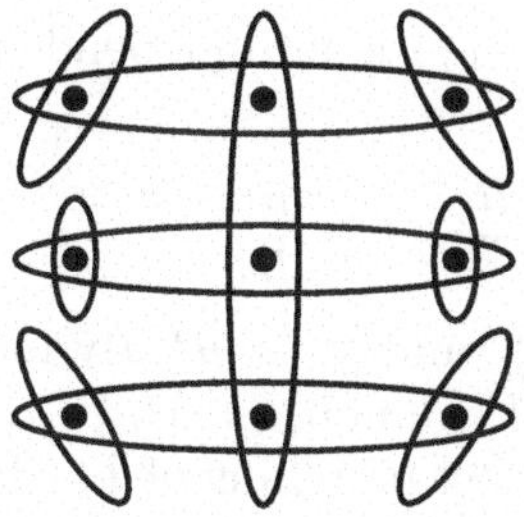

(a)
An R-perfect matching hypertree

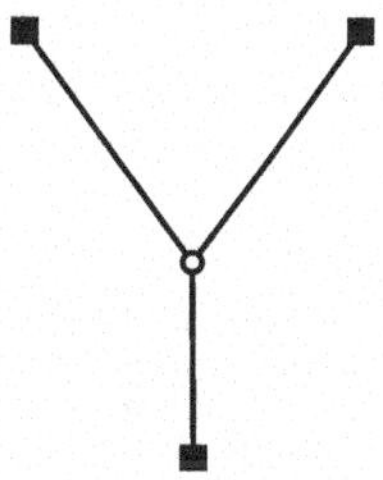

(b) The structure tree

Fig. 3. An R-perfect matching hypertree and its structure tree, where the squares and the hollow dots represent matched edges and adjacent edges, respectively.

- When $t = 1$, the result holds.
- Assume the lemma holds for $t \leq q$. When $t = q + 1$, let us consider the structure tree T' of the R perfect matching hypertree T. v' is a leaf in T' and e' is the corresponding matched edge in T. $T /\!/ e'$ is also an R-perfect matching hypertree and the matching number in $T /\!/ e'$ is exactly q. By our induction, there is an unique R-perfect matching M' in $T /\!/ e'$. Thus $M' \cup \{e'\}$ is the unique R-perfect matching in T. The lemma holds for $t = q + 1$. By induction, the lemma holds for every R-perfect matching hypertree. □

Next we will prove R-perfect matching hypertrees are exactly the extremal hypertrees with $\alpha(T) = \frac{n}{R}$, as stated in Lemma 7.

Lemma 7. *For every k-regular hypertree with rank R, $\alpha(T) = \frac{n}{R}$ if and only if T is an R-perfect matching hypertree.*

Proof. **Sufficiency**: Since T has an R-perfect matching, the number of matched edges is $\frac{n}{R}$. Within any given matched edge, there is at most one vertex included in any independent set of T. Thus, we have $\alpha(T) \leq \frac{n}{R}$. According to Lemma 1, $\alpha(T) \geq \frac{n}{R}$. Therefore, $\alpha(T) = \frac{n}{R}$ holds.

Necessity: We prove the necessity part by induction on the number of edges. When the number of edges is 1, we have $\alpha(T) = \frac{n}{R} = 1$. Assume this result holds when the edge number is smaller than m. When T contains $m \geq 2$ edges, take a longest path $P = v_1 e_1 v_2 \ldots v_t e_t v_{t+1}$ from T for $t \geq 2$. Considering the vertices in edge e_t without taking loops into account, v_t is the unique vertex with degree no less than 2. After eliminating the edge e_t, the components of $T /\!/ e_t$ are $\{T_i \mid 1 \leq i \leq c\}$. Assume that T_i contains m_i edges and n_i. Since every edge of T_i contains at most R vertices, we have $\alpha(T_i) \geq \frac{n_i}{R}$ by Lemma 1. Thus,

$$\alpha(T /\!/ e_t) = \sum_{i=1}^{c} \alpha(T_i) \geq \sum_{i=1}^{c} \frac{n_i}{R} \geq \frac{n-R}{R},$$

where the last inequality holds as e_t contains at most R vertices.

For every independent set I of $T/\!\!/e_t$, $I \cup \{v_{t+1}\}$ is an independent set in T. Then, we have

$$\alpha(T) \geq \alpha(T/\!\!/e_t) + 1 \geq \frac{n-R}{R} + 1 \geq \frac{n}{R}.$$

Combined with $\alpha(T) = \frac{n}{R}$, we have $\alpha(T_i) = \frac{n_i}{R}$ and $|e_t| = R$. By induction, T_i is R-perfect matching hypertree. According to Lemma 6, let M_i be the matched edge set of T_i. Thus, $\bigcup_{i=1}^{c} M_i \cup \{e_t\}$ is the R-perfect matching of T, which implies that T is a R-perfect matching hypertree. □

Remark 5. The above proof also demonstrates a polynomial-time algorithm to decide whether a tree T is a R-perfect matching hypertree. In addition, If T is a R-perfect matching hypertree, the algorithm gives the unique R-perfect matching decomposition.

Next we state our main theorem: R-perfect matching hypertrees are exactly the extremal hypergraphs with $\alpha(H) = \frac{m-(k-2)n-1}{R}$ when $R \geq 3$. Due to the limitation on pages, we omit the proof in this conference paper and postpone them in the full version of the journal paper.

Theorem 5. *For every k-regular connected hypergraph $H(V, E)$ with rank $R \geq 3$, $\alpha(H) = \frac{m-(k-2)n-1}{R}$ if and only if H is R-perfect matching hypertree.*

Remark 6. The condition of Theorem 4 is that $R \geq 3$. For $R = 2$, the result does not hold since an odd cycle is a simple example. It is interesting to characterize the structure of H with $R = 2$ and $\alpha(H) = \frac{m-(k-2)n-1}{2}$.

Corollary 3. *Let $H(V, E)$ be a rank R connected hypergraph with maximum degree Δ on n vertices and m edges, $\alpha(H) \geq \frac{m-(\Delta-2)n-1}{R}$ holds. Furthermore, $\alpha(H) = \frac{m-(\Delta-2)n-1}{R}$ holds if and only if $H(V, E)$ is a regular hypertree with R-perfect matching when $R \geq 3$.*

4 The Combinatorial Algorithms on the Independent Number

In this section, for a rank R connected hypergraph with maximum degree Δ on n vertices and m edges, a polynomial time algorithm is designed to compute an independent set with cardinality at least $\frac{m-(\Delta-2)n-1}{R}$.

Firstly, based on the proof in Lemma 1, for a hypertree T, an optimal independent set can be found in polynomial time as follows.

Secondly, for a connected hypergraph $H(V, E)$ with maximum degree Δ on n vertices and m edges, we can construct a Δ-regular hypergraph H' by adding some loops to the vertices and these operations keep the eliminating vertex feedback set. Assume $D \subseteq V$ is a minimal eliminating vertex feedback set of H and H'. According to Lemma 5 and Remark 3, there is

$$|\, D \,| \leq (\Delta - 1)n - m' + 1 \leq (\Delta - 1)n - m + 1.$$

A simple minimal eliminating vertex feedback set is gotten as follows:

Algorithm 1. Optimal Independent Set of a Hypertree

Input: A hypertree $T(V, E)$.

Output: ALG1(T), an optimal independent set of H.

1: **if** $\mid V \mid = 1$ **then**
2: Let v be the unique vertex in T.
3: **return** $\{v\}$
4: Find a longest path $P = v_1e_1v_2 \ldots v_te_tv_{t+1}$ in T.
5: Let T_1, T_2, ..., T_c be all components of subhypergraph $T /\!/ e_t$.
6: **return** $\{v_{t+1}\} \cup \bigcup_{i=1}^{c} \text{ALG1}(T_i)$

- Take a vertex order $v_1, v_2, ..., v_n$ and put all vertices in V as an eliminating vertex feedback set D.
- Consider the vertices one by one in this order and update the eliminating vertex feedback set D: if $D \setminus v_i$ is an eliminating vertex feedback set, take $D \setminus v_i$ as D; otherwise keep D.
- D is a minimal eliminating vertex feedback set.

The polynomial time algorithm to compute an independent set with cardinality at least $\frac{m-(\Delta-2)n-1}{R}$ is as shown in Algorithm 2:

Algorithm 2. The Independent Set of a Hypergraph

Input: A rank R connected hypergraph $H(V, E)$ with maximum degree Δ on n vertices and m edges.

Output: ALG2(H), an independent set with cardinality at least $\frac{m-(\Delta-2)n-1}{R}$.

1: Take a vertex order $v_1, v_2, ..., v_n$ arbitrarily.
2: Initialize $D = V$
3: **for** $i = 1$ to n **do**
4: **if** $D \setminus v_i$ is an eliminating vertex feedback set **then**
5: Take $D = D \setminus v_i$.
6: Compute an optimal independent set I of the hyperforest $T = H /\!/ D$ by Algorithm 1.
7: Let T_1, T_2, ..., T_c be all components of the hyperforest T.
8: Output $S = \bigcup_{i=1}^{c} \text{ALG1}(T_i)$

Theorem 6. *Let $H(V, E)$ be a connected hypergraph with maximum degree Δ on n vertices and m edges, Algorithm 2 outputs an independent set with cardinality at least* $\frac{m-(\Delta-2)n-1}{R}$.

Proof. D is a minimal eliminating vertex feedback set of H, according to Lemma 5 and Remark 3, we have $\mid D \mid \leq (\Delta - 1)n - m + 1$. $T = H /\!/ D$ is a hyperforest, according to Lemma 1, we have $\alpha(T) \geq \frac{n_T}{R_T}$. Algorithm 1 outputs the optimal independent set of T, thus there is

$$\mid S \mid = \alpha(T) \geq \frac{n_T}{R_T} = \frac{n - \mid D \mid}{R_T} \geq \frac{m - (\Delta - 2)n - 1}{R}.$$

□

References

1. Bollobás, B., Daykin, D.E., Erdös, P.: Sets of independent edges of a hypergraph. Q. J. Math. **27**(1), 25–32 (1976)
2. Boppana, R., Halldórsson, M.M.: Approximating maximum independent sets by excluding subgraphs. BIT Numer. Math. **32**(2), 180–196 (1992)
3. Chvátal, V.: Determining the stability number of a graph. SIAM J. Comput. **6**(4), 643–662 (1977)
4. Duke, R.A., Lefmann, H., Rödl, V.: On uncrowded hypergraphs. Random Struct. Algorithms **6**(2–3), 209–212 (1995)
5. Fox, J., Pach, J.: Computing the independence number of intersection graphs, pp. 1161–1165 (2011)
6. Griggs, J.R.: Lower bounds on the independence number in terms of the degrees. J. Comb. Theory Ser. B **34**(1), 22–39 (1983)
7. Hansen, P., Zheng, M.: Sharp bounds on the order, size, and stability number of graphs. Networks **23**(2), 99–102 (1993)
8. Harant, J., Schiermeyer, I.: On the independence number of a graph in terms of order and size. Discret. Math. **232**(1–3), 131–138 (2001)
9. Hermon, J., Sly, A., Zhang, Y.: Rapid mixing of hypergraph independent sets. Random Struct. Algorithms **54**(4), 730–767 (2019)
10. Kostochka, A., Mubayi, D., Verstraëte, J.: On independent sets in hypergraphs. Random Struct. Algorithms **44**(2), 224–239 (2014)
11. Shearer, J.B.: A note on the independence number of triangle-free graphs, ii. J. Comb. Theory Ser. B **53**(2), 300–307 (1991)
12. Wei, V.K.: A lower bound on the stability number of a simple graph. Technical report, Bell Laboratories Technical Memorandum Murray Hill, NJ, USA (1981)

Approximation Algorithms on k-Correlation Clustering of Uniform Hypergraphs

Zhongzheng Tang[1,2], Yaxuan Li[3], and Zhuo Diao[3(✉)]

[1] Key Laboratory of Mathematics and Information Networks (Beijing University of Posts and Telecommunications), Ministry of Education, Beijing, China
tangzhongzheng@amss.ac.cn

[2] School of Science, Beijing University of Posts and Telecommunications, Beijing 100876, China

[3] School of Statistics and Mathematics, Central University of Finance and Economics, Beijing 100081, China
liyaxuanlee@email.cufe.edu.cn, diaozhuo@amss.ac.cn

Abstract. In this paper, we consider the k-correlation clustering problem on uniform hypergraphs. Given an edge-weighted r-uniform hypergraph $H(V, E)$ where the edges are labeled either positive or negative with non-negative weights, we want to partition the nodes into at most k-clusters to maximize agreements: the total weights of positive edges within clusters and negative edges between clusters. This problem is NP-Hard. We design an approximation algorithm with the approximation ratio $\frac{A_k^r}{k^r - k + A_k^r}$, here $A_k^r = k(k-1)(k-2)\cdots(k-r+1)$ is the number of r permutations from k elements.

Keywords: k-Correlation Clusters · Approximation Algorithms · Choosing the Better of Two Solutions

1 Introduction

Clustering is the action of partitioning individual elements into groups based on their similarity. Clustering problems defined on signed graphs arise in many scientific areas, including portfolio analysis in risk management, biological systems, efficient document classification, detection of embedded matrix structures and analysis of similarity or dissimilarity in relationships. The common element among these applications is the collaborative vs. conflicting environment in which they are defined. The solution of clustering problems defined on signed graphs can be used as a criteria to measure the degree of balance in social networks.

Supported by National Natural Science Foundation of China under Grant No.11901605, No.12101069, the Fundamental Research Funds for the Central Universities, the disciplinary funding of Central University of Finance and Economics.

D. Du et al. (Eds.): COCOA 2024, LNCS 15434, pp. 477–486, 2025.
https://doi.org/10.1007/978-981-96-4445-2_39

Concretely speaking, given an edge-weighted graph where the edges are labeled either positive (similar) or negative (different) with non-negative weights, we want to partition the nodes into clusters to maximize agreements—the total weights of positive edges within clusters and negative edges between clusters. This problem is called Correlation Clustering problem.

When the number of clusters is limited by k, the problem is called k-Correlation Clustering problem. The k-Correlation Clustering problem is a generalization of Max k-Cut problem. If all edges are labelled negative with trivial weight 1, the k-Correlation Clustering problem is equivalent to Max k-Cut problem.

The clustering problem on hypergraphs is a generalization of the clustering problem on graphs. In current research related to hypergraph clustering, the number of theoretical studies is relatively small, while more research focuses on practical applications.

At the theoretical level, several closely related versions of Max k-Cut problem in hypergraph were studied in the literature but very few results have been obtained. Andersson and Engebretsen [2] presented an 0.72-approximation algorithm for the ordinary Max Cut problem in hypergraph. Arora, Karger and Karpinski [3] designed a PTAS for dense instances of this problem under the condition that $|e| \leq d$ for each edge e and some constant d. Ageev and Sviridenko [1] considered the Max k-Cut problem in hypergraph with given sizes of parts, they approximated the problem within a factor of $\min\{\lambda_{|e|} : e \in E\}$ of the optimum where $\lambda_r = 1 - (1 - \frac{1}{r})^r - (\frac{1}{r})^r$ by applying the pipage rounding method. Chandrasekaran, Xu and Yu [5] presented a randomized polynomial time algorithm to solve minimum k-cut problem in hypergraphs.

Zhou [15] et al. proposed a clustering ensemble method via structured hypergraph learning, instead of being constructed directly, which will be more reliable than directly obtaining the final clustering result by partitioning the unreliable hypergraph. Rottenstreich, Kaplan and Hassidim [10] studied the problem of clustering the vertices of a weighted hypergraph such that on average the vertices of each edge can be covered by a small number of clusters, they described and analyzed algorithms for this problem for the case in which the clusters have to be disjoint and for the case where clusters can overlap. Bulo and Pelillo [4] formulated the hypergraph clustering problem in terms of a noncooperative multiplayer "clustering game" and show that a natural notion of a cluster turns out to be equivalent to a classical (evolutionary) game-theoretic equilibrium concept. Xu, Yan, and Zhang [12] studied the partitioning problem of Bollobas and Scott which aims to prove the existence of a bipartition such that each class meets edges of total weightat least $\frac{\omega_1 - \alpha - m}{2} - \frac{2\omega_2}{3}$, where ω_i denote the total weight of edges of size i for $i = 1, 2$, and α be the maximum weight of an edge of size 1. Swan and Zhan generalized [11] the well-known MapEquation, an optimization equation used in the clustering of nonhypergraphs for hypergraphs, and developed an agglomerative algorithm Hypergraph Random Walks (HRW), to find an approximate solution to the generalized MapEquation.

At the practical level, Chang [6] et al. considered the multiclass clustering problem involving a hypergraph model and established a tensor optimization model with an orthogonal constraint. Then, the hypergraph clustering approach was applied to image segmentation and motion segmentation problems. Chen et al. [7] proposed a hypergraph-clustering method dubbed the "high-dimensional data clustering method" based on hypergraph partitioning using an improved Apriori algorithm (HDHPA), which further enhances the effectiveness and accuracy of clustering. Li, Zhao and Tang [9] presented a framework for hybrid image summarization in which social images and corresponding textual information are taken as vertices in a hypergraph and the task of image summarization is formulated as the problem of hypergraph partition. Zhang and Hancock [13] used hypergraph clustering to select the most informative feature subset (mIFS) from a set of objects using high-order similarities, which has both low redundancy and strong discriminating power. The clustering problem on uniform hypergraph is often related to the recovery in the hypergraph stochastic block model (HSBM). For example, Zhao [14] et al. considered the effect of label information on the exact recovery of communities in an m-uniform HSBM and investigated two scenarios of label information, then they derived sharp boundaries for exact recovery under both scenarios from an information-theoretical point of view. Cole and Zhu [8] studied the same problem on a random d-uniform hypergraph H with n vertices partitioned into k clusters of size $s = n/k$, and presented a spectral algorithm which recovers the clusters exactly with high probability, given mild conditions on the parameters of the model.

Our Contributions

- For the edge-weighted k-Correlation Clustering problem on r-uniform hypergraphs, we design a randomized algorithm which has approximation ratio $\frac{(k-A_k^r)a+A_k^r}{k^r}$, here $A_k^r = k(k-1)(k-2)\cdots(k-r+1)$ is the number of r permutations from k elements and a is the weighted proportion of positive edges in all edges. As a varies from 0 to 1, the approximation ratio varies from $\frac{A_k^r}{k^r}$ to $\frac{k}{k^r}$ and the minimum value is $\frac{1}{k^{r-1}}$.
- By the method of conditional expectation, we derandomize the randomized algorithm to obtain a deterministic algorithm whose solution value is as good as the expected value of the randomized algorithm.
- By choosing a better one from the derandomization algorithm and a trivial algorithm, we get a $\max\{a, \frac{(k-A_k^r)a+A_k^r}{k^r}\}$-approximation algorithm. As a varies from 0 to 1, the approximation ratio $\max\{a, \frac{(k-A_k^r)a+A_k^r}{k^r}\}$ varies from $\frac{A_k^r}{k^r}$ to 1 and the minimum value is $\frac{A_k^r}{k^r-k+A_k^r}$ when $a = \frac{A_k^r}{k^r-k+A_k^r}$.

Organization. In Sect. 2, we introduce the formal definitions of the problems. In Sect. 3, we prove the k-Correlation Clustering problem on uniform hypergraphs is a generalization of Max k-Cut problem on graphs. In Sect. 4, we present our algorithms, including a randomized algorithm in the expectation sense, a deterministic algorithm by derandomization with conditional expectations and a choosing better one algorithm.

2 Preliminaries

A hypergraph is a generalization of a graph in which an edge can join any number of vertices. Let $H = (V, E)$ be a hypergraph with vertex set V and edge set E. As for a graph, the order of H, denoted by n, is the number of vertices. The number of edges is denoted by m. Hypergraph H is r-uniform if each edge contains exactly r vertices.

Traditionally, the k-Correlation Clustering problem is defined on graphs: given an edge-weighted graph $G(V, E)$ where the edges are labeled either positive or negative with non-negative weights, we want to partition the nodes into at most k-clusters to maximize agreements—the total weights of positive edges within clusters and negative edges between clusters. Here we generalize the k-Correlation Clustering problem from graphs to hypergraphs:

Definition 1. *The k-Correlation Clustering problem on hypergraphs is defined as follows. Given an edge-weighted hypergraph $H(V, E)$ where the edges are labeled either positive or negative with non-negative weights, we want to partition the nodes into at most k-clusters to maximize agreements-the weights of all agreed edges:*

- *For a positive edge e, e is agreed if all vertices in e are in the same cluster.*
- *For a negative edge e, e is agreed if all vertices in e are in distinct clusters.*

An instance is shown in Fig. 1.

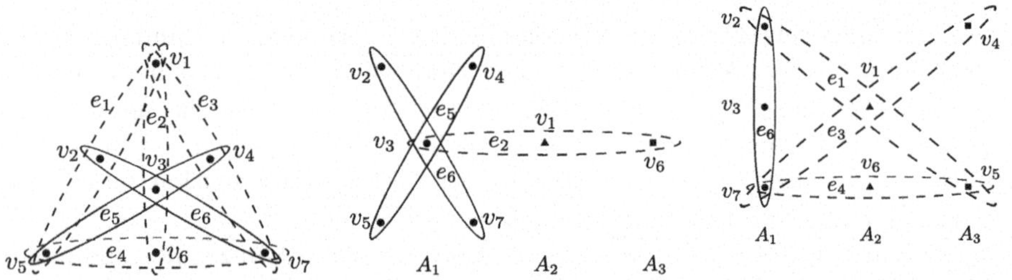

(a) A hypergraph (b) A feasible partition (c) An optimal partition

Fig. 1. For the 3-Correlation Clustering problem on a 3-uniform hypergraph, every edge of the hypergraph in (a) has a unit weight. The solid edges are positive edges while the dashed edges are negative edges. A feasible solution with 3 agreed edges $\{e_2, e_5, e_6\}$ is shown in (b): $A_1 = \{v_2, v_3, v_4, v_5, v_7\}$, $A_2 = \{v_1\}$ and $A_3 = \{v_6\}$. An optimal solution with 4 agreed edges $\{e_1, e_3, e_4, e_6\}$ is shown in (c): $A_1 = \{v_2, v_3, v_7\}$, $A_2 = \{v_1, v_6\}$ and $A_3 = \{v_4, v_5\}$.

In this paper, we consider the k-Correlation Clustering problem on uniform hypergraphs.

3 Computational Complexity

The k-Correlation Clustering problem is a generalization of Max k-Cut problem. If all edges are labelled negative with trivial weight 1, the k-Correlation Clustering problem is equivalent to Max k-Cut problem. Thus the k-Correlation Clustering problem is NP-hard. The next theorem states the k-Correlation Clustering on uniform hypergraphs is also NP-hard.

Theorem 1. *For each $r \geq 2$, the k-Correlation Clustering on r-uniform hypergraphs is NP-hard.*

Proof. We will give a reduction from Max k-Cut problem to k-Correlation Clustering on r-uniform hypergraphs. Given an instance $G(V, E)$ of Max k-Cut problem, we want to partition the vertices at most k-clusters to maximize the number of edges between clusters. The reduction from Max k-Cut problem to k-Correlation Clustering on r-uniform hypergraphs is constructed as following:

Firstly, we can get a r-uniform hypergraph $H'(V', E')$ by adding $r-2$ distinct vertices for each edge in G. Every edge in H' is labelled negative with trivial weight 1. Next, we will show Max k-Cut problem in G is equivalent to k-Correlation Clustering on r-uniform hypergraphs in H':

- For any k-cut of G, there is a vertex partition $V_1, V_2, \ldots, V_k$ of V. For each edge $e \in G$, there is a corresponding edge e' in H. Denote $e = (v_1, v_2)$ and $e' = (v_1, v_2, u_i, 1 \leq i \leq r-2)$, here $\{u_i, 1 \leq i \leq r-2\}$ are $r-2$ added vertices. Putting these vertices $\{u_i, 1 \leq i \leq r-2\}$ into $r-2$ distinct clusters different from the clusters of v_1 and v_2. It is easy to see the edge e is a cut edge of G if and only if e' is an agreed edge in H'. Thus the number of cut edges in G are equal to the number of agreed edges in H'.
- For any k-clusters C of H', there is a vertex partition $V_1', V_2', \ldots, V_k'$ of V'. For each edge $e' \in H'$, there is a corresponding edge e in G. Denote $e = (v_1, v_2)$ and $e' = (v_1, v_2, u_i, 1 \leq i \leq r-2)$, here $\{u_i, 1 \leq i \leq r-2\}$ are $r-2$ added vertices. Here assume $\{u_i, 1 \leq i \leq r-2\}$ are placed into $r-2$ distinct clusters different from the clusters of v_1 and v_2, otherwise we can get a new k-cluster C' by putting $\{u_i, 1 \leq i \leq r-2\}$ into $r-2$ distinct clusters different from the clusters of v_1 and v_2 and the number of agreed edges is non-decreasing. It is easy to see e' is an agreed edge in H' if and only if e is a cut edge of G. Thus the number of agreed edges in H' is equal to the number of cut edges in G.

4 Approximation Algorithms

In this section, we design approximation algorithms to compute the k-Correlation Clustering problem on r-uniform hypergraph. The approximation algorithms are based on random technique, derandomization method with expectation conditions and choosing the better one of two solutions.

The following definition will be used frequently in our discussions. Recalling that E is the edge set, let $W(E)$ denote the total weight of all edges. $W(+)$ is the total weight of all positive edges, and $W(-)$ is the total weight of all negative edges. Let a denote the weighted proportion of positive edges in all edges, i.e. $a = \frac{W(+)}{W(E)}$. Let A be the output of the k-clusters, and $W(A)$ denote the agreements—sthe total weights of positive edges within clusters and negative edges between clusters.

4.1 The Randomized Algorithm

The Randomized Algorithm: Take a vertex order $v_1, \ldots, v_n$ arbitrarily and initialize k vertex sets $A_1 = A_2 = \cdots = A_k = \varnothing$. Place the vertices one by one to the sets $A_1, A_2, \ldots, A_k$ in this order and the rules for placement are as follows: every vertex v_i is put into the sets $A_1, A_2, \ldots, A_k$ with equal probability $\frac{1}{k}$ independently. Finally, output the k-clusters $A_1, A_2, \ldots, A_k$ and the agreement $W(A)$. The randomized algorithm is as shown in Algorithm 1.

Algorithm 1 Randomized Algorithm for $W(A)$

Input: An edge-signed hypergraph $H(V, E)$ and an edge weight function $w : E \to \mathbb{R}^+$.

Output: The k-cluster $A_1, A_2, \ldots, A_k$ and the agreement $W(A)$.

1: Take a vertex order $v_1, \ldots, v_n$ arbitrarily.

2: Initialize $A_1 = A_2 = \cdots = A_k = \varnothing$.

3: **for** $i = 1$ to n **do**

4: Put v_i into one of $A_1, A_2, \ldots, A_k$ with equal probability $\frac{1}{k}$.

5: Output the k-cluster $A_1, A_2, \ldots, A_k$ and the agreement $W(A)$.

Theorem 2. *Given an edge-weighted r-uniform hypergraph $H(V, E)$ where the edges are labeled positive or negative with non-negative weights, a is the weighted proportion of positive edges in all edges, $a = \frac{W(+)}{W(E)}$. The randomized algorithm is a $\frac{(k-A_k^r)a+A_k^r}{k^r}$-approximation algorithm in the expectation sense: $E[W(A)] \geq \frac{(k-A_k^r)a+A_k^r}{k^r} \cdot OPT$, here $A_k^r = k(k-1)(k-2)\cdots(k-r+1)$ is the number of r permutations from k elements.*

Proof. We just need to explain $E[W(A)] = \frac{(k-A_k^r)a+A_k^r}{k^r} \cdot W(E) \geq \frac{(k-A_k^r)a+A_k^r}{k^r} \cdot OPT$. According to the placement rule: every vertex v_i is put into the sets $A_1, A_2, \ldots, A_k$ with equal probability $\frac{1}{k}$ independently. For any edge $e(v_1, v_2, \ldots, v_r)$, define a random variable X_e:

$$X_e = 1, \text{if } e \text{ is in the agreements}$$

$$X_e = 0, \text{if } e \text{ is not in the agreements.}$$

Thus we have

$$W(A) = \sum_{e \in E} W_e X_e = \sum_{e \in E^+} W_e X_e + \sum_{e \in E^-} W_e X_e.$$

Here E^+ is the set of edges labeled positive and E^- is the set of edges labeled negative.

- For each positive edge $e(v_1, v_2, \ldots, v_r)$, e is in the agreement means all the vertices $\{v_1, v_2, \ldots, v_r\}$ are in the same cluster. Thus we have

$$E(X_e) = Pr(X_e = 1) \cdot 1 + Pr(X_e = 0) \cdot 0 = Pr(X_e = 1) = \frac{k}{k^r} = \frac{1}{k^{r-1}}.$$

- For each negative edge $e(v_1, v_2, \ldots, v_r)$, e is in the agreement means all the vertices $\{v_1, v_2, \ldots, v_r\}$ are in different clusters. Thus we have

$$E(X_e) = Pr(X_e = 1) \cdot 1 + Pr(X_e = 0) \cdot 0$$

$$= Pr(X_e = 1) = \frac{k(k-1)(k-2)\cdots(k-r+1)}{k^r} = \frac{A_k^r}{k^r}.$$

Thus we have

$$E[W(A)] = \sum_{e \in E} W_e E[X_e]$$

$$= \sum_{e \in E^+} W_e E[X_e] + \sum_{e \in E^-} W_e E[X_e] = \frac{k}{k^r} \cdot W(+) + \frac{A_k^r}{k^r} \cdot W(-).$$

Combined with $W(+) = a \cdot W(E)$, $W(-) = (1-a) \cdot W(E)$, we have

$$E[W(A)] = \frac{(k - A_k^r)a + A_k^r}{k^r} \cdot W(E) \geq \frac{(k - A_k^r)a + A_k^r}{k^r} \cdot OPT.$$

As a varies from 0 to 1, $\frac{(k-A_k^r)a+A_k^r}{k^r}$ varies from $\frac{A_k^r}{k^r}$ to $\frac{k}{k^r}$ and the minimum value is $\frac{1}{k^{r-1}}$, we have the next corollary instantly:

Corollary 1. *The randomized algorithm is a $\frac{1}{k^{r-1}}$-approximation algorithm in the expectation sense for the maximizing agreements of k-Correlation Clustering problem on r-uniform hypergraph.*

4.2 Derandomization with Conditional Expectation

In this section, by the method of conditional expectation, we derandomize the Algorithm 1 to obtain a deterministic algorithm whose solution value is as good as the expected value of the randomized algorithm.

Initially, set the k-clusters $A_1 = A_2 = \cdots = A_k = \varnothing$ and take an arbitrarily vertex order $v_i, 1 \leq i \leq n$. The cluster of v_i will be made sequentially: v_1 is determined first, then v_2, and so on.

More formally, by the formula of conditional expectations, there is

$$E[W(A)] = \sum_{i=1}^{k} Pr(v_1 \in A_i) E[W(A) | v_1 \in A_i]$$

$$\leq \max\{E[W(A)|v_1 \in A_i], 1 \leq i \leq k\}.$$

We set v_1 to the cluster A_{v_1} so as to maximize the conditional expectation, the choice guarantees an expected value no less than the expected value of the completely randomized algorithm. Similarly, we set v_2 to the cluster A_{v_2} so as to maximize the conditional expectation as follows:

$$E[W(A)|v_1 \in A_{v_1}] = \sum_{i=1}^{k} Pr(v_2 \in A_i)E[W(A)|v_1 \in A_{v_1}, v_2 \in A_i]$$

$$\leq \max\{E[W(A)|v_1 \in A_{v_1}, v_2 \in A_i], 1 \leq i \leq k\}.$$

We continue this process until all n vertices are set and output the k-clusters A^*, so there is

$$E[W(A)] \leq E[W(A)|v_1 \in A_{v_1}] \leq E[W(A)|v_1 \in A_{v_1}, v_2 \in A_{v_2}]$$
$$\leq \cdots \leq E[W(A)|v_i \in A_{v_i}, 1 \leq i \leq n] = W(A^*).$$

According to Theorem 2 and Corollary 1, the next conclusions are instant:

Theorem 3. *Given an edge-weighted r-uniform hypergraph $H(V,E)$ where the edges are labeled positive or negative with non-negative weights, a is the weighted proportion of positive edges in all edges, $a = \frac{W(+)}{W(E)}$. The derandomized algorithm by conditional expectation is a $\frac{(k-A_k^r)a+A_k^r}{k^r}$-approximation algorithm: $W(A^*) \geq \frac{(k-A_k^r)a+A_k^r}{k^r} \cdot OPT$, here $A_k^r = k(k-1)(k-2)\cdots(k-r+1)$ is the number of r permutations from k elements.*

Corollary 2. *The derandomized algorithm by conditional expectation is a $\frac{1}{k^{r-1}}$-approximation algorithm for the maximizing agreements of k-Correlation Clustering problem on r-uniform hypergraph.*

4.3 The One Better of Two Algorithms

The above randomized algorithm has the approximation ratio $\frac{(k-A_k^r)a+A_k^r}{k^r}$ where $a = \frac{W(+)}{W(E)}$ is the weighted proportion of positive edges in all edges. As a varies from 0 to 1, $\frac{(k-A_k^r)a+A_k^r}{k^r}$ varies from $\frac{A_k^r}{k^r}$ to $\frac{k}{k^r}$ and the minimum value is $\frac{1}{k^{r-1}}$. Obviously, the approximate ratio is an inverse proportional linear function of a. This means that the smaller weighted proportion of positive edges, the better approximation ratio. How do we deal with the case of high weighted proportion of positive edges? A trivial algorithm is putting all vertices in one big cluster. The trivial algorithm has the better approximation ratio as the higher weighted proportion of positive edges. Finally, we consider the randomized algorithm and the trivial algorithm together and choose the one better of two algorithms. This algorithm is as shown in Algorithm 2.

Algorithm 2 One Better of Two Algorithms for $W(A)$

Input: An edge-signed hypergraph $H(V, E)$ and an edge weight function $w: E \to \mathbb{R}^+$.

Output: The k-cluster $A_1, A_2, \ldots, A_k$ and the agreement $W(A)$.

1: Derandomize Algorithm 1 by conditional expectations to get the k-cluster $A_1, A_2, \ldots, A_k$ and the agreement $W(A)$.
2: **if** $W(A) > W(+)$ **then**
3: Output the k-cluster $A_1, A_2, \ldots, A_k$ and the agreement $W(A)$.
4: **else**
5: Output the 1-cluster V and the agreement $W(+)$.

Theorem 4. *Given an edge-weighted r-uniform hypergraph $H(V, E)$ where the edges are labeled either positive or negative with non-negative weights, a is the weighted proportion of positive edges in all edges, $a = \frac{W(+)}{W(E)}$. The one better of two algorithms is a* $\max\left\{a, \frac{(k-A_k^r)a+A_k^r}{k^r}\right\}$*-approximation algorithm, which means* $W(A) \geq \max\left\{a, \frac{(k-A_k^r)a+A_k^r}{k^r}\right\} \cdot OPT$*, here $A_k^r = k(k-1)(k-2)\cdots(k-r+1)$ is the number of r permutations from k elements.*

Proof. Let us denote the output of the randomized algorithm as A_1 and denote the output of the trivial algorithm as A_2. According to Theorem 1, that is

$$W(A_1) \geq \frac{(k - A_k^r)a + A_k^r}{k^r} \cdot W(E).$$

For the trivial algorithm, there is only one cluster by putting all vertices together. Thus there is

$$W(A_2) = \sum_{e \in E^+} W(e) = W(+) = a \cdot W(E) \geq a \cdot OPT.$$

The final output is choosing the one better of two algorithms. Thus there is

$$\begin{aligned} &\max\{W(A_1), W(A_2)\} \\ \geq\ &\max\{\frac{(k - A_k^r)a + A_k^r}{k^r}, a\} \cdot W(E) \geq \max\left\{a, \frac{(k - A_k^r)a + A_k^r}{k^r}\right\} \cdot OPT. \end{aligned}$$

As a varies from 0 to 1, the approximation ratio $\max\left\{a, \frac{(k-A_k^r)a+A_k^r}{k^r}\right\}$ varies from $\frac{A_k^r}{k^r}$ to 1 and the minimum value is $\frac{A_k^r}{k^r-k+A_k^r}$ when $a = \frac{A_k^r}{k^r-k+A_k^r}$. Thus we have the next corollary instantly:

Corollary 3. *The one better of two algorithms is a $\frac{A_k^r}{k^r-k+A_k^r}$-approximation algorithm for the maximizing agreements of k-Correlation Clustering problem on r-uniform hypergraphs.*

References

1. Ageev, A.A., Sviridenko, M.I.: An approximation algorithm for hypergraph max k-cut with given sizes of parts. In: European Symposium on Algorithms, pp. 32–41. Springer (2000)
2. Andersson, G., Engebretsen, L.: Better approximation algorithms for set splitting and not-all-equal sat. Inf. Process. Lett. **65**(6), 305–311 (1998)
3. Arora, S., Karger, D., Karpinski, M.: Polynomial time approximation schemes for dense instances of np-hard problems. In: Proceedings of the Twenty-Seventh Annual ACM Symposium on Theory of Computing, pp. 284–293 (1995)
4. Bulo, S.R., Pelillo, M.: A game-theoretic approach to hypergraph clustering. IEEE Trans. Pattern Anal. Mach. Intell. **35**(6), 1312–1327 (2013)
5. Chandrasekaran, K., Xu, C., Yu, X.: Hypergraph k-cut in randomized polynomial time. Math. Program. **186**, 85–113 (2021)
6. Chang, J., Chen, Y., Qi, L., Yan, H.: Hypergraph clustering using a new laplacian tensor with applications in image processing. SIAM J. Imaging Sci. **13**(3), 1157–1178 (2020)
7. Chen, R., Hu, F., Wang, F., Bai, L.: Hypergraph-clustering method based on an improved apriori algorithm. Appl. Sci. **13**(19), 10577 (2023)
8. Cole, S., Zhu, Y.: Exact recovery in the hypergraph stochastic block model: a spectral algorithm. Linear Algebra Appl. **593**, 45–73 (2020)
9. Li, M., Zhao, C., Tang, J.: Hybrid image summarization by hypergraph partition. Neurocomputing **119**, 41–48 (2013)
10. Rottenstreich, O., Kaplan, H., Hassidim, A.: Clustering in hypergraphs to minimize average edge service time. ACM Trans. Algorithms **16**(3), 40 (2020)
11. Swan, M., Zhan, J.: Clustering hypergraphs via the mapequation. IEEE Access **9**, 72377–72386 (2021)
12. Xu, X., Yan, G., Zhang, Y.: Judicious partitions of weighted hypergraphs. Sci. China Math. **59**(3), 609–616 (2016)
13. Zhang, Z., Hancock, E.R.: Hypergraph based information-theoretic feature selection. Pattern Recogn. Lett. **33**(15), 1991–1999 (2012)
14. Zhao, X., Zhao, W., Yuan, M.: Information limits for community detection in hypergraph with label information. Symmetry **13**(11), 2060 (2021)
15. Zhou, P., Wang, X., Du, L., Li, X.: Clustering ensemble via structured hypergraph learning. Information Fusion **78**, 171–179 (2022)

Author Index

D. Du et al. (Eds.): COCOA 2024, LNCS 15434, pp. 487–489, 2025.
https://doi.org/10.1007/978-981-96-4445-2

The manufacturer's authorised representative in the EU is Springer Nature Customer Service Centre GmbH, Europaplatz 3, 69115 Heidelberg, Germany. If you have any concerns regarding our products, please contact ProductSafety@springernature.com

Printed and bound by CPI Group (UK) Ltd, Croydon, CR0 4YY
01/10/2026
02225908-0019